Psychoneuroimmunology
An Interdisciplinary Introduction

Psychoneuroimmunology
An Interdisciplinary Introduction

Edited by

Manfred Schedlowski
Institute of Medical Psychology
University of Essen
Essen, Germany

and

Uwe Tewes
Department of Medical Psychology
Hannover Medical School
Hannover, Germany

Kluwer Academic/Plenum Publishers
New York, Boston, Dordrecht, London, Moscow

Library of Congress Cataloging-in-Publication Data

Psychoneuroimmunologie. English.
    Psychoneuroimmunology : an interdisciplinary introduction / edited
by Manfred Schedlowski and Uwe Tewes.
        p.   cm.
    Includes bibliographical references and index.
    ISBN 0-306-45976-0 (pbk.). -- ISBN 0-306-45975-2 (hardbound)
    1. Psychoneuroimmunology.   I. Schedlowski, Manfred.   II. Tewes,
Uwe.   III. Title.
    [DNLM: 1. Psychoneuroimmunology.   2. Behavioral Medicine.
3. Immune System--physiology.   WL 103.7 P9736p 1999]
    QP356.47.P77713   1999
    616.07'9--dc21
    DNLM/DLC
    for Library of Congress   99-38184
                                                                 CIP

ISBN 0-306-45975-2 (Hardbound)
ISBN 0-306-45976-0 (Paperbound)

©1999 Kluwer Academic/Plenum Publishers
233 Spring Street, New York, N.Y. 10013

Original Title: Psychoneuroimmunologie
Published by: Spektrum Akademischer Verlag
Copyright: ©1996 Spektrum Akademischer Verlag

10 9 8 7 6 5 4 3 2 1

A C.I.P. Record for this book is available from the Library of Congress.

Printed in the United States

Contributors

Michael Antoni • Department of Psychology, University of Miami, Coral Gables, Florida 33124-2040

Katherine L. Applegate • Ohio State University College of Medicine, Department of Psychiatry, Columbus, Ohio 43210

Robert J. Benschop • National Jewish Medical and Research Center, Division of Basic Sciences, Denver, Colorado 80206

Hugo O. Besedovsky • Institute for Normal and Pathological Physiology and Immunophysiology • Marburg University, 35037 Marburg, Germany

Dana H. Bovbjerg • Biobehavioral Medicine Program, Ruttenberg Cancer Center, Mount Sinai School of Medicine, New York, New York 10021

Jan Born • Clinical Research Group–Clinical Neuroendocrinology, Lübeck Medical University, 23538 Lübeck, Germany

Kimberley Brownley • Department of Psychology, University of Miami, Coral Gables, Florida 33124-2040

John T. Cacioppo • Department of Psychology, Ohio State University, Columbus, Ohio 43210-1222

Christopher L. Coe • Department of Psychology, University of Wisconsin, Madison, Wisconsin 53706

Robert Dantzer • INSERM U394, 33077 Bordeaux Cedex, France

Adriana del Rey • Institute for Normal and Pathological Physiology & Immunophysiology, University of Marburg, 35037 Marburg, Germany

Melissa K. Demitrikopoulus • Institute for Biomedical Philosophy, Chamblee, Georgia 30341

Frank Eggert • Institute for Psychology, University of Kiel, 24098 Kiel, Germany

Michael S. Exton • Institute for Medical Psychology, Essen University Clinic, 45122 Essen, Germany

Roman Ferstl • Institute for Psychology, University of Kiel, 24098 Kiel, Germany

Mary Ann Fletcher • Department of Psychology, University of Miami, Coral Gables, Florida 33124-2040

Elliot Friedman • Department of Psychology, Williams College, Williamstown, Massachusetts 01267

Ronald Glaser • Department of Medical Microbiology and Immunology, Ohio State University College of Medicine, Columbus, Ohio 43210

John Hay • Department of Microbiology, State University of New York, Buffalo, New York

Cobi J. Heijnen • Department of Immunology, University Hospital for Children and Youth, "Het Wilhemina Kinderziekenhuis," 3501 CA Utrecht, The Netherlands

Barry Hurwitz • Department of Psychology, University of Miami, Coral Gables, Florida 33124-2040

Alan J. Husband • Faculty of Veterinary Sciences, The University of Sydney, Sydney, NSW 2006 Australia

Gail Ironson • Department of Psychology, University of Miami, Coral Gables, Florida 33124-2040

Michael Irwin • VA Medical Center, Department of Psychiatry (V116A), San Diego, California 92161

Roland Jacobs • Department of Clinical Immunology, Hannover Medical School, 30625 Hannover, Germany

Annemieke Kavelaars • Department of Immunology, University Hospital for Children and Youth, "Het Wilhemina Kinderziekenhuis," 3501 CA Utrecht, The Netherlands

Steven E. Keller • University of Medicine and Dentistry of New Jersey, Departments of Psychiatry and Neuroscience, New Jersey Medical School, Newark, New Jersey 07103

Margaret E. Kemeny • Department of Psychology, UCLA, Los Angeles, California 90095

Janice K. Kiecolt-Glaser • Department of Psychiatry, Ohio State University, College of Medicine, Columbus, Ohio 43210

Maurice G. King • Institute for Behavioural Research in Health, Curtin University, Perth, WA Australia

Nancy Klimas • Department of Psychology, University of Miami, Coral Gables, Florida 33124-2040

Joachim Kugler • Department of Medical Psychology, University Hospital, 52064 Aachen, Germany

Mahendra Kumar • Department of Psychology, University of Miami, Coral Gables, Florida 33124-2040

Arthur LaPerriere • Department of Psychology, University of Miami, Coral Gables, Florida 33124-2040

Mark L. Laudenslager • Behavioral Immunology Laboratory, Department of Psychiatry, University of Colorado Health Sciences Center, Denver, Colorado 80220

Gregory Miller • Western Psychiatric Institute and Clinic, Pittsburgh, Pennsylvania

Bente Klarlund Pedersen • Copenhagen Muscle Research Center, Department of Infectious Diseases, Rigshospitalet, 2200 Copenhagen N, Denmark

Manfred Schedlowski • Institute of Medical Psychology, Essen University Clinic, 45122 Essen, Germany

Steven J. Schleifer • University of Medicine and Dentistry of New Jersey, Department of Psychiatry, New Jersey Medical School, Newark, New Jersey 07103

Reinhold E. Schmidt • Department of Clinical Immunology, Hannover Medical School, 30625 Hannover, Germany

Neil Schneiderman • Department of Psychology, University of Miami, Coral Gables, Florida 33124-2040

Thomas Schürmeyer • Department of Endocrinology and Metabolism, Psychobiological and Psychosomatic Research Center, University of Trier, 54290 Trier, Germany

Uwe Tewes • Department of Medical Psychology, Hannover Medical School, 30625 Hannover, Germany

Heiddis B. Valdimarsdottir • Memorial Sloan-Kettering Cancer Center, New York, New York

Eberhard Weihe • Institute for Anatomy and Cell Biology, Philipps-University Clinic Marburg, 35033 Marburg, Germany

Jürgen Westermann • Department of Functional and Applied Anatomy, Hannover Medical School, 30625 Hannover, Germany

E. Jean Wickings • International Center for Medical Research, CIRMF BP 769, Franceville, Gabon

Julie M. Worlein • Regional Primate Research Center, University of Washington, Seattle, Washington

Robert Zachariae • Aarhus University, Risskov, Denmark

Foreword

I'm sitting on my deck enjoying the tranquility of a spring evening. The air is soft, filled with the mating trills of toads. A pregnant doe is drinking at the edge of the pond and a pair of Canada geese are staking their claim to a nesting site. Spring comes late up here in the Bristol Hills just south of Rochester, but when it arrives, it brings with it the long-anticipated rush of freshness and new growth. It occurs to me that spring is an apt metaphor for the relatively new field of psychoneuroimmunology, as the paradigm shift it espouses has sprouted from multidisciplinary seeds germinating in soil richly fertilized by curiosity and scientific open-mindedness. Today, this spring garden of psychoneuroimmunology is vibrant with a harmoniously integrated collection of buds and multihued blooms that have developed from the initiating seeds of the behavioral sciences, neurobiology, endocrinology, and immunology, and from second- and third-generation seeds bearing the beneficial mutations of new knowledge.

Those who first planted the seeds (Ader, 1996a) and then, in 1980, named this garden *psychoneuroimmunology* (Ader, 1996b), and those investigators who have subsequently explored neural–immune system bidirectional communication, are fluent in several complex scientific languages. However, some of the new students who might be drawn to psychoneuroimmunology by an article in a popular magazine, a personal involvement with some form of immune dysregulation, a trade book on the subject, or one of many "self-help books" for which the immune system or psychoneuroimmunology serves (inappropriately) as a mantra, may find this field daunting. Indeed, without a formal course and/or mentor to provide guidance, these new students may be overwhelmed by hundreds of original data-based publications in peer-reviewed journals, an edited volume of weighty (if not biblical) proportions that reviews nearly every facet of psychoneuroimmunology known in 1990 (Ader, Felten, & Cohen, 1991), and books devoted to relatively narrow subsets of the field and to proceedings of recently held meetings.

Psychoneuroimmunology: An Interdisciplinary Introduction, the very first textbook in this research area, is a sorely needed guide to an initial exploration of this field. This information-rich textbook for upper-level undergraduates and for pre- and postdoctoral students in the biomedical sciences and health professions, recognizes the critical need for the reader to be familiar with the languages of contemporary immunology, neural sciences, endocrinology, and psychology before trying to fully understand ways in which these disciplines have been integrated in psychoneuroimmunology. Indeed, a full seven chapters are devoted to this introductory material. The textbook is comprehensive in that multiple chapters address each of the major topics in the field. The textbook's editors have also recognized the fact that psychoneuroimmunology extends its tendrils into many areas of biology and medicine (e.g., exercise physiology, sleep, AIDS, clinical depression), and have devoted chapters to these subjects of contemporary interest.

As someone who has been exploring behavioral regulation of immunity for more than 20 years, I find *Psychoneuroimmunology: An Interdisciplinary Introduction* of interest from a historical perspective. For example, 15 years ago, most of its chapters could not have been written—there was simply minimal or no information available on most of the subjects covered in this new textbook. For those few chapters that could have been written "way back then," an editor's choice of authors would have been limited to a few individuals from a handful of laboratories (Ader, 1981). By contrast, Drs. Schedlowski and Tewes had the luxury of being able to select authors for each chapter from a large pool of individuals and research groups working in more than two dozen countries in Europe, North and South America, Asia, and Africa. The bottom line is that psychoneuroimmunology has exhibited a remarkable growth worldwide in the past two decades, and there is no sign of a letup.

Should *Psychoneuroimmunology: An Interdisciplinary Introduction* be publicized and widely read outside as well as inside the academic milieu, it could also play a role in contemporary health care. In the United States alone, alternative medicine is now a multi-billion-dollar business. Several therapeutic interventions are now being marketed on the premise that they have beneficial health outcomes. Although these claims have never been proven in clinical trials, the fact that neural–immune system connections exist is often taken as prima facie evidence that the intervention must work, even in a clinical setting in which the immune system is not involved. Anyone who peruses this new psychoneuroimmunology textbook will quickly learn that mechanisms underlying neural–immune system communication do not provide a mantle of scientific respectability covering all forms of therapeutic intervention and all of mind–body medicine. If you question whether such education is necessary, consider a recent advertising supplement to a newspaper (*Democrat and Chronicle* newspaper, 1997) that featured an article on how homes may be therapeutic for mind and body. I don't disagree with that proposition, but I am annoyed by the fact that in that article, psychoneuroimmunology was described "as the art and science of designing interiors to enhance well-being, creativity, and performance." On behalf of those wishing to learn about this remarkable field (psychoneuroimmunology, not interior design), I applaud Professors Schedlowski and Tewes's contribution, and wish their textbook much success.

Nicholas Cohen, Ph.D.
The Center for Psychoneuroimmunology Research
Department of Microbiology and Immunology
The University of Rochester Medical Center
Rochester, New York

References

Ader, R. (Ed.). (1981). *Psychoneuroimmunology.* New York: Academic Press, 661 pp.
Ader, R. (1996a). Historical perspectives of psychoneuroimmunology. In H. Friedman, T. W. Klein, & A. L. Friedman (Eds.), *Psychoneuroimmunology, stress and infection* (pp. 1–24). Boca Raton: CRC Press.
Ader, R. (1996b). Historical perspectives on psychoneuroimmunology. Presidential address: Psychosomatic and psychoimmunologic research. *Psychosomatic Medicine, 42,* 307–322.
Ader, R., Felten, D. L., & Cohen, N. (Eds.). (1991). *Psychoneuroimmunology* (2nd ed.). San Diego: Academic Press, 1218 pp.
Democrat and Chronicle newspaper, Rochester, NY. Supplement of September 27, 1997, entitled HomeWorks.

Preface

One would expect a textbook to provide the reader with an overview about the most important theories, methods, and results of a certain discipline. Textbooks are usually written by those who have studied the discipline and who are eminently qualified in the field. This conventional solution does not exist for a textbook in psychoneuroimmunology. Within the academic domain, there is no teaching field with this name. At present, no student can achieve an academic grade in psychoneuroimmunology. Likewise, there is no specific research institute for psychoneuroimmunology like those that exist for disciplines such as psychology, neuroscience, and anatomy. Rather, scientists working in this field of research come from a variety of disciplines. This integral approach can be demonstrated by examining where the current psychoneuroimmunology research results are being reviewed. Publications appear in a number of highly reputable scientific journals such as the *New England Journal of Medicine, The Lancet, Endocrine Reviews, Physiological Reviews, Immunology Today,* and the *Annual Review of Psychology.* This is just one indication that this field of research is becoming a more authentic discipline within the established scientific assembly.

Psychoneuroimmunology investigates the functional relationships among the nervous system, the neuroendocrine system, and the immune system. Although many of the communication pathways between these systems have yet to be elucidated, it is already well documented that the immune system is influenced and directed by neurochemical signals from both the nervous system and the endocrine system. Conversely, empirical data clearly demonstrate that functions of the latter two systems are affected by products of the activated immune system. The existence of this multidirectional communication pathway provides the experimental basis for the analysis of behavioral effects on immune functions, as well as defining the effects of immunological processes on behavior.

However, the biological consequences of these interactions for the maintenance of health, development, and progression of disease are so far not completely understood. The investigation of this complex network can only be successful by interdisciplinary interaction. This requires the cooperation of such researchers as immunologists, endocrinologists, physiologists, pharmacologists, neuroscientists, psychologists, oncologists, and psychiatrists.

The most elementary of interdisciplinary research is the interaction between two different disciplines. However, communication in research becomes increasingly difficult when more disciplines are involved in the cooperative research. Based on our own experiences with these communication problems, the decision was made to edit a textbook in which immunological, endocrinological, and psychological foundations and methods are summarized in an understandable way, and which provides a comprehensive overview of the different research activities in psychoneuroimmunology.

This book has several goals. It aims to provide an interdisciplinary introduction to pre- and postgraduate students who are particularly interested in the field of psychoneuroimmunology. Furthermore, it aims to provide comprehensive information for those teaching facets of psychoneuroimmunology, and a cross discipline overview for those scientists who are interested in psychoneuroimmunology.

More than 22 distinct research groups have contributed to this volume. Chapters have been written by scientists and researchers from Germany, the United States, Australia, The Netherlands, Denmark, and France. The concepts, methods, and empirical findings are summarized in more than 200 uniformly illustrated figures and 35 tables. Chapters 1–7 describe the basis and methods of the different disciplines and should give the reader an understandable overview about those disciplines with which he/she is not particularly familiar. Chapters 8–12 summarize the different neuroanatomical and neurochemical communication pathways by which the nervous, endocrine, and immune systems interact. Based on this multidirectional interaction, Chapters 13–23 focus on the effects of behavior, such as acute and chronic stress, physical exercise, sleep, and conditioning of immune functions in animals and humans. Psychoneuroimmunological research has dealt primarily with the analysis of basic interactions between these three systems. A more recent focus of psychoneuroimmunology research has been the clinical aspects and perspectives of such domains as oncology, HIV/AIDS, autoimmune diseases, and the reactivation of latent herpesviruses. Chapters 24–27 incorporate some of the few distinctive studies that have been systematically conducted in this applied field.

We are extremely grateful to all of the colleagues who provided such high-quality contributions to this book, and to Mr. Ken Derham from Kluwer Academic/Plenum Publishers for his support. Finally, we are indebted to our Australian colleagues: Ms. Natalie Exton, our editorial assistant, and Drs. Diane Bull and Michael Exton, for their tireless, comprehensive, and exceptional editorial work.

Manfred Schedlowski, Uwe Tewes
Essen and Hannover

Contents

Chapter 17: Does Psychological Depression Cause Immune Suppression in Humans? 327

Michael Irwin and Elliot Friedman

Chapter 18: Exercise and Immune Functions 341

Bente Klarlund Pedersen

Chapter 19: Biobehavioral Influences on Respiratory Immunity 359

Joachim Kugler

Chapter 20: Effects of Psychosocial Interventions on the Immune System 373

Margaret E. Kemeny and Gregory Miller

Chapter 24: Psychoneuroimmunology in Oncology

Dana H. Bovbjerg, Heiddis B. Valdimarsdottir, and Robert Zachariae

Chapter 25: Psychoneuroimmunology and HIV/AIDS

Neil Schneiderman, Michael Antoni, Gail Ironson, Nancy Klimas,
Barry Hurwitz, Mahendra Kumur, Arthur LaPerriere, Kimberley Brownley,
and Mary Ann Fletcher

1 Functional Anatomy of the Immune System

Jürgen Westermann and Michael S. Exton

Jürgen Westermann • Department of Functional and Applied Anatomy, Hannover Medical School, D-30625 Hannover, Germany.   Michael S. Exton • Institute for Medical Psychology, Essen University Clinic, D-45122 Essen, Germany.

1.1. Introduction

Our species is constantly under threat from a large number of pathogens comprising four groups: viruses, bacteria, fungi, and parasites (protozoa and worms). At a homeostatic temperature of about 37°C, and with proteins, fat, and sugar, the body provides an optimal environment for pathogen growth and replication. However, pathogens have difficulty penetrating the barriers that divide the body from the environment. Such a protective surface is comprised of the skin ($\approx$2 m^2), the surface of the respiratory tract ($\approx$150 m^2), the gastrointestinal tract ($\approx$300 m^2), and the genitourinary tract ($<$ 0.5 m^2). One task of the immune system is to guarantee the integrity of the body against all intruders (Janeway & Travers, 1997). Another task within the body is to rapidly recognize cellular degeneration and to prevent the development of cancer. However, some aspects of the immune system are not positive. Allergies and diseases such as rheumatoid arthritis and inflammatory bowel disease are thought to develop as a result of an immune system overreaction. Additionally, transplanted cells and organs can be recognized as being foreign and consequently rejected. Therefore, the different phases of an immune response must be carefully regulated. Recently, it has become clear that other organ systems participate in the regulation of the immune system. However, many open questions regarding these interactions remain. It is possible to answer such questions only when different disciplines such as anatomy, endocrinology, immunology, neurology, and psychology come together to solve these problems that have relevance for the health of humans. Therefore, it is important for those interested in such a multidisciplinary field to have an understanding of the fundamentals of the other disciplines.

The field of immunology has recently witnessed an immense increase in knowledge. For example, Sir Winston Churchill, in a hospital with an infection during World War II, is said to have looked at his hospital chart and asked his physician, "What are these lymphocytes?" He was told, "We do not know, Prime Minister." "Then why do you count them?" Churchill retorted (Janeway, 1988). Even in the 1960s it was believed that lymphocytes served as nutrition for other cells (Ford, 1980; Sprent & Tough, 1994). Today we know that lymphocytes are the main cell type responsible for the specific immune response. Many molecules have been identified that lymphocytes use to communicate with each other. It is possible using molecular biological techniques to trace a signal that is initiated by binding to surface receptors from the cell membrane of a lymphocyte down to its nucleus, and to measure the resulting gene transcription and the subsequent protein production. It is impossible to completely describe the functional anatomy of the immune system within the confines of this chapter. Instead, examples will be used to illustrate which cells are involved in producing an immune response, in which organs the cells are located, and how the cells communicate with each other (Westermann & Pabst, 1992, 1996). Unless otherwise stated, the data presented are related to humans. In addition, we will look at how the morphology of immune cells and organs changes during an immune response, and how one can envision the different phases of an immune response within an organ. Because many investigations of the immune system in humans have to concentrate on the analysis of immunocompetent cells in the blood, a separate section outlining possible pitfalls in interpreting such blood data will follow (Westermann & Pabst, 1990). As the present chapter deals with only a few aspects of the functional anatomy of the immune system, the interested reader is referred to excellent textbooks of immunology (Abbas, Lichtman, & Pober, 1997; Janeway & Travers, 1997) and anatomy (Pabst, 1994; Williams, 1995).

1.2. Cells of the Immune System

Defense reactions are divided into nonspecific (innate, nonadaptive) and specific (acquired, adaptive) immune responses (Abbas *et al.*, 1997; Janeway & Travers, 1997; see Chapter 2). Nonspecific mechanisms are the first line of defense against infection. Macrophages and granulocytes have the task of quickly killing and removing intruding pathogens (antigen). This reaction runs consistently at a steady state and is independent of previous pathogen contact. When this defense is broached, the responsibility is conveyed to the specific defense. Lymphocytes are the cell type exclusively responsible for the specific immune response, and whose function can be simplistically described as dramatically raising the effectiveness of nonspecific defense mechanisms. Initially, the intruding pathogen must be presented to the lymphocyte. This task is undertaken mainly by dendritic cells, which thereby initiate the primary immune response. Such a reaction subsequently involves multiplication of those lymphocytes and the production of those antibodies that are specific for the particular pathogen. Bacteria are then covered by such antibodies, and are more effectively eliminated (phagocytosed) than those without antibody. Furthermore, an immunological "memory" develops: The number of antigen-specific lymphocytes and antibodies increase, and the immune system can respond more quickly and more efficiently to a second encounter with the respective pathogen (secondary immune response).

The cells of the immune system are located in many organs. There are organized lymphoid organs such as the thymus, spleen, lymph nodes, tonsils, and Peyer's patches, which must be differentiated from the diffuse distribution of immunocompetent cells through most other organs of the body. These locations are connected as a common system by blood and lymph vessels.

1.2.1. Lymphocytes

Morphologically, lymphocytes (diameter about 10 μm) are a homogeneous population, and can be characterized by a dense nucleus and small cytoplasm (Fig. 1). Functionally, however, lymphocytes are a very heterogeneous population, which can be classified with the help of monoclonal antibodies. Simplistically, one can divide lymphocytes into T and B lymphocytes.

The precursors of T lymphocytes originate from the bone marrow and migrate into the outer cortex of the thymus (thus the name *T* lymphocytes) (see Chapter 2). There they repeatedly divide, and learn to distinguish between foreign and self proteins (positive and negative selection) through interactions with epithelial cells. T lymphocytes develop into two main groups, the helper and cytotoxic T lymphocytes, which can be distinguished via their membrane molecules CD4 (helper) and CD8 (cytotoxic). Selected T lymphocytes leave the thymus and migrate into peripheral lymphoid organs. T-helper lymphocytes (CD4) initiate an immune response and increase the effectiveness of macrophages in phagocytosing pathogens, whereas the cytotoxic T lymphocytes (CD8) are important for the destruction of virally infected cells and tumor cells. Both cell types contribute to cellular immunity. Memory cells arise after contact with an antigen in peripheral lymphoid organs, and in some cases can be identified via a specific surface molecule (CD45RO; Bell, Sparshott, & Bunce, 1998; Sprent & Tough, 1994; Westermann & Pabst, 1996).

B lymphocytes are selected in the bone marrow (thus the name *B* lymphocytes). The pro- and pre-B cells develop from precursor cells in the bone marrow (Osmond, Rolink,

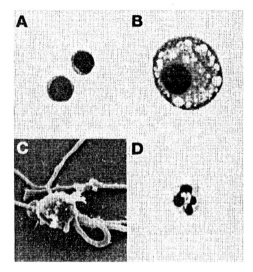

Figure 1. Important immune cells. (A) Lymphocytes; (B) alveolar macrophage; (C) follicular dendritic cell; (D) granulocyte. All are human preparations except B, which is taken from the rat. A, B, D: Light microscopic photograph, same magnification, diameter of a lymphocyte approximately 10 μm, Giemsa stain. C: Scanning electron microscopic photograph. B: Original provided by T. Tschernig, Hannover. C: Original provided by R. Sprenger, Borstel.

& Melchers, 1998). First antibodies are seen in the rough endoplasmic reticulum of the cytoplasm, and later they are expressed on the cell surface. There are different classes of antibodies, which differ in their structure and function: IgM, IgD, IgG, IgA, and IgE (see Chapter 2). IgM and IgG are the most prevalent antibodies, and are produced during primary and secondary immune responses, respectively. IgA is of special relevance for mucosal immunology and IgE for allergic reactions. In the process of an immune response, B lymphocytes develop into plasma cells, which are the main antibody-producing cells (humoral immunity), and into memory B cells, which continuously migrate through the tissues of the body.

Some authors describe the natural killer (NK) cells as a third lymphocyte type (see Chapter 2). These cells are found mainly in the spleen and lungs. They can destroy tumor cells or virally infected cells spontaneously, without previous immunization. In human peripheral blood, approximately 5–10% of leukocytes are NK cells, which are relatively large cells and contain distinct granules in the cytoplasm. They probably originate from the bone marrow and their cytoplasmic granules (perforin, granzymes) make holes in the membranes of target cells and induce death.

Figure 2 shows the estimated total number of T and B lymphocytes and NK cells of humans, and their distribution among different organs. There are approximately twice as many T lymphocytes as B lymphocytes, and the number of NK cells is only about 10% that of the T cells. In addition, the cell types are distributed differently. While almost 40% of all T and B cells can be found in the lymph nodes, nearly 75% of all NK cells are located in the spleen and lungs (Fig. 2).

T and B lymphocytes mediate the specific immune response. In most specific immune responses, the antigen has to be taken up, processed, and presented by an antigen-presenting cell to the rare antigen-specific T lymphocytes. These cells then proliferate in order to increase their numbers. It therefore takes several days for a primary immune response to be effective. Meanwhile, other cells (e.g., macrophages, granulocytes, NK cells) must control the invading pathogens.

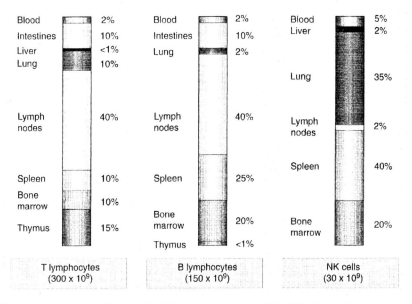

Figure 2. Number and distribution of T and B lymphocytes and natural killer (NK) cells in the body of a young, healthy person. The literature from which these data are compiled is summarized in Westermann and Pabst (1992).

1.2.2. Macrophages

Most macrophages develop in the bone marrow. They are released into the blood and are then termed *monocytes*. Monocytes can then enter the various tissues via the blood. Macrophages are located in all lymphoid and nonlymphoid organs. In addition, they are found beneath the epithelium of all surfaces of the human body. There they are able to recognize invading pathogens. The macrophage ingests pathogens (phagocytosis), which are then killed intracellularly. Another important function of macrophages is the production and secretion of factors (cytokines, chemokines, prostaglandins, and many other substances). These messengers can work on the same cell (autocrine), neighboring cells (paracrine), or via the blood on the whole organism (endocrine). For example, locally cytokines induce the recruitment of leukocytes from the blood, and can systemically induce fever.

In lymphoid organs, macrophages have an irregularly shaped nucleus and an extensive pale basophilic cytoplasm containing numerous azure granules (Fig. 1). Activated macrophages often carry cell remnants, which are usually recognizable with light microscopy. Macrophages can also be characterized with monoclonal antibodies against membrane proteins (e.g., Fc and complement receptors). Different macrophage subpopulations can be found preferentially in distinct compartments of lymphoid organs.

1.2.3. Dendritic Cells

Dendritic cells are important antigen-presenting cells (Hart, 1997). As the name suggests, these cells have many branches (Fig. 1). From precursors in the bone marrow, dendritic cells migrate into the blood where they can be observed in low concentration. From there they migrate both into and beneath the epithelium of most body surfaces. For example, in the skin, dendritic cells are called *Langerhans cells*. After uptake of antigen, they migrate with the afferent lymph to the T lymphocyte region of the draining lymph node (Austyn, 1992). A small part (peptide) of the degraded pathogen (protein) is then presented in combination with the major histocompatibility complex (MHC) class II at the cell surface to T-helper lymphocytes (CD4$^+$), and with the help of other costimulatory molecules, the specific immune response against this pathogen is initiated. It is important to note that most specific immune responses are initiated in the draining lymph node, and organized lymphoid tissues such as the spleen and Peyer's patches, and not at the site of actual pathogen invasion. In contrast to MHC class II molecules, which are mainly expressed on the surface of dendritic cells, macrophages, and B lymphocytes (all having the ability to present antigen to CD4$^+$ T-helper cells), MHC class I molecules can be found on almost all cells of the body. If these cells are infected by viruses, or become a tumor cell, the changed endogenous antigens are presented in association with the MHC class I to the cytotoxic T lymphocytes (CD8$^+$). Their activation initiates the destruction of the infected cell or tumor cell. In this way, the specific immune response against invading pathogens (e.g., bacteria, foreign proteins) is started by only a few cells (MHC class II$^+$), whereas the protection against changed self proteins (e.g. tumor) can be initiated by most body cells (MHC class I$^+$).

Another route runs from the bone marrow and blood to the thymus, spleen, lymph nodes, Peyer's patches, and tonsils, where the dendritic cells can be found in the T lymphocyte regions, and are called *interdigitating dendritic cells* (IDC). The dendritic cells of the germinal centers (B lymphocyte regions) are called *follicular dendritic cells* (FDC). Their precursor cells are still unknown. Antigen–antibody complexes are presented by them for a long time (more than 10 years), which seems to be a requirement for the development of memory B cells in germinal centers.

1.2.4. Granulocytes

Granulocytes are produced in the bone marrow. They have a segmented nucleus and different types of granules in their cytoplasm (Fig. 1). From the bone marrow they are released into the blood. Neutrophil granulocytes, which form the major population of leukocytes in human blood, are recruited to sites of inflammation. Here they support the local macrophages in phagocytosing invading pathogens. After phagocytosis and intracellular killing of the pathogen, neutrophil granulocytes die, thereby forming pus. Eosinophil and basophil granulocytes play a role in allergic responses.

1.3. Organs of the Immune System

This section will demonstrate that the cells of the immune system are spread throughout many organs of the entire body (Fig. 3, Table 1). Additionally, the cells are not randomly distributed within an organ, but are preferentially localized in certain organ compartments.

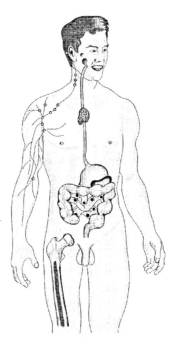

Figure 3. Localization of organs of the immune system in the human body.

An overview demonstrates that B and T lymphocytes are not only found in lymphoid organs, but also in many nonlymphoid organs and body fluids (Table 2).

1.3.1. Thymus

Site and Structure. The thymus consists of two lobes located behind the sternum (Fig. 3). On histological cross sections, further subdivisions into smaller lobes can be seen, with each consisting of the darker-colored *cortex*, reflecting the densely packed T lymphocytes, and the less densely populated, lighter, *medulla* (Fig. 4). Because the main elements of the thymus are lymphocytes and epithelial cells, the thymus is described as a lympho-epithelial organ. While the epithelial cells do not divide in high numbers, many lymphoid cells are in the process of cell division.

Of special interest for the function of the thymus are the different types of epithelial cells in the cortex (Ritter & Boyd, 1993). One example is the so called thymic nurse cell, which can envelop 20–200 lymphocytes with their cytoplasm. The microenvironment in the thymus is also influenced by hormones produced by epithelial cells (e.g., thymosin, thymopoietin, and thymulin). These molecules have a regulating effect on lymphocyte development in the thymus, but can also reach other organs via the blood. T lymphocytes are the predominant cell type located in the medulla of the thymus, although it also contains B lymphocytes, macrophages, NK cells, interdigitating dendritic cells, and mast cells (Westermann, Smith, Peters, Tschernig, Pabst, Steinhoff, Sparshott, & Bell, 1996).

Table 1
Overview of the Important Organs of the Immune System

Organ	Function	Notes
Bone marrow	Production of T-cell precursors Production of B-cell precursors "Education" of B cells (positive and negative selection ensures that B cells reacting against foreign antigens survive and those reacting against self antigens die)	B lymphocytes: • Specific receptors are randomly generated *before* the first contact with the antigen • The receptor recognizes native proteins, lipids, and sugars without the need for processing and presentation via the MHC • The affinity of both the receptor and the subsequently produced antibodies is increased by somatic hypermutation
Thymus	Uptake of T-cell precursors from the bone marrow "Education" of T cells (positive and negative selection ensures that T cells reacting against foreign antigens survive and those reacting against self antigens die)	T lymphocytes: • Specific receptors are randomly generated *before* the first contact with the antigen with no further change in affinity for the antigen • Receptors recognize protein antigens that are presented by the MHC • MHC class I presents peptides (about 10 amino acids long) to $CD8^+$ T cells • MHC class II presents peptides (about 15 amino acids long) to $CD4^+$ T cells
Spleen	Immune response against antigens in the blood delivered via the bloodstream	• Excellent blood supply increases the probability that antigen and specific lymphocyte meet • Loss of the spleen increases the risk of getting severe infections • Production of B and T lymphocytes
Lymph nodes	Immune response against antigens in the lymph delivered via afferent lymph vessels	• Concentration of antigens into the path of migrating lymphocytes increases the probability that antigen and specific lymphocytes meet • Production of B and T lymphocytes
Tonsils	Immune response against antigens in the air delivered via the oral and nasal cavities	• No afferent lymphatic vessels, therefore M cells for antigen uptake present • First lymphatic tissue in the course of the respiratory and gastrointestinal tracts • Production of B and T lymphocytes
Peyer's patches	Immune response against antigens in the food delivered via the gut	• No afferent lymphatic vessels, therefore M cells for antigen uptake present • Production of B and T cells

The thymus reacts especially sensitively to the hormones that are released during stress (e.g., glucocorticoids), whereby the cortex, in particular, becomes less densely populated and the thymus volume thus dramatically decreases. Since every disease produces stress, the size and cellular composition of the thymus are normally only analyzed from thymi taken from humans whose death was sudden (e.g., through accident). If the weight of such thymi is measured across age groups, the organ's weight is seen to be approximately constant through the course of life. The weight of the thymus is about 25 g by 1 year of age, 20 g in young adults, and 20 g by 90 years of age (Pabst, 1994). If, however, the relative size of the different thymic compartments is compared, a distinct decrease of the cortex and an increase of fatty tissue during aging are revealed. Nevertheless, even in old age, residues of functionally capable cortex and medullary tissue remain.

Table 2
Composition of Lymphocyte Subsets in Human Organs and Fluids[a]

Source of cells	Lymphocytes				Ratio	
	B	T	CD4+	CD8+	T/B	CD4+/CD8+
Organs						
Skin	<1	>95	45	45	>95.0	1.0
Thymus	1	95	95	95	95.0	1.0
Liver	10	55	25	40	5.5	0.6
Lymph node	20	70	50	20	3.5	2.5
Spleen	50	40	20	30	0.8	0.7
Tonsils	50	40	30	10	0.8	3.0
Bone marrow	50	50	20	30	1.0	0.7
Greater omentum	70	30	—	—	0.4	—
Lung						
Alveolar space	5	70	40	25	14.0	1.6
Epithelium	<1	>95	25	60	>95.0	0.4
Interstitium	15	85	40	40	5.7	1.0
Small intestine						
Epithelium	1	85	15	85	85.0	0.2
Lamina propria	30	70	50	25	2.3	2.0
Peyer's patches	40	45	30	10	1.1	3.0
Fluids						
Peritoneal fluid	5	95	25	60	19.0	0.4
Cerebrospinal fluid	5	90	60	30	18.0	2.0
Lymph						
Afferent	5	75	40	20	15.0	2.0
Efferent	10	85	65	20	8.5	3.3
Blood	25	70	50	25	2.8	2.0
Breast milk	30	70	40	25	2.3	1.6

[a]The references from which the information is taken are summarized in Westermann and Pabst (1992).

Function. The central task of the thymus is the "education" of the T lymphocytes: T lymphocytes have to learn to distinguish between "self" and "foreign." T-cell precursors arrive from the bone marrow in only small numbers; however, intensive cell proliferation subsequently occurs within the thymus. T lymphocytes can only survive when they fulfill two criteria: T-helper (CD4+) and cytotoxic T lymphocytes (CD8+) must recognize antigens presented by MHC class II+ and class I+ cells, respectively (positive selection); and T lymphocytes must not react too strongly with endogenous structures (negative selection). During this process the T-cell receptor repertoire is generated, comprising about 10^5 different specificities in adult humans. This large number is very surprising as it is known that human DNA (2 m in length in each cell; 10^9 base pairs, but only 1% are used for generating proteins) only contains about 10^5 different genes coding for proteins. It was then found that the huge number of different T-cell receptors was generated by random combination of only a few different DNA segments (T-cell receptor gene rearrangement), enabling the potential generation of about 10^{16} different specificities. A large proportion of the newly produced lymphocytes do not survive these processes (>95%), and only a small number (<5%) exit the thymus via the medulla in order to migrate to other organs (Shortman & Scollay, 1994).

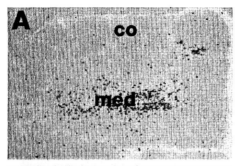

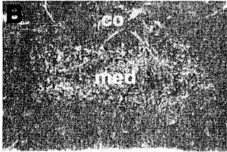

Figure 4. Distribution of B and T lymphocytes in the rat thymus. (A) The thymic compartments cortex (co) and medulla (med) can
clearly be recognized. Although T lymphocytes
predominate (B), B lymphocytes are also present in the thymus (A). They are predominantly
localized in the medulla. (B and T cells are
revealed by appropriate monoclonal antibodies
and the peroxidase technique; the diameter of a
lymphocyte is about 10 μm; the human thymus
has a comparable structure.)

The functions of the thymus explained thus far are primarily concerned with the thymic
cortex and not with the medulla. The permeability of the blood vessels for foreign antigens
is higher in the medulla, and mature B and T lymphocytes, monocytes, NK cells, and
dendritic cells continuously enter the medulla from the blood (Westermann *et al.*, 1996).
This indicates that the medulla and the cortex differ in function, with the medulla operating
as a peripheral lymphoid organ.

In humans, there is a rare, life threatening, genetic disease that results in the lack of a
thymus (di George syndrome). Besides a reduction of the lymphocyte number in the blood
and peripheral lymphoid organs, the response to viruses is drastically diminished.

1.3.2. Spleen

Site and Structure. The spleen is situated in the left upper abdomen (Fig. 3). It has a
beanlike form, and in adults has an average weight of 150 g. In about 20% of humans a
nodulelike accumulation of splenic tissue is present, so-called accessory spleens.

The basic structure of the spleen is formed by reticular fibers and reticular cells, and
therefore the spleen can be recognized as a lymphoreticular organ. On sectioning an unfixed
spleen one can recognize two portions with the naked eye (Steiniger, Barth, Herbst,
Hartnell, & Crocker, 1998). The predominant part (approximately 75%) is red, reflecting its
many red blood cells, and is called the *red pulp*. It contains macrophages, cytotoxic T
lymphocytes (CD8[+]), plasma cells, and the majority of NK cells present in the spleen.
Interspersed in the red pulp are small white nodules. They are 1–3 mm in size, consist
mainly of lymphocyte accumulations, and are called *white pulp*. The white pulp is further
subdivided into the *periarterial lymphatic sheath*, a T-lymphocyte area, and into the

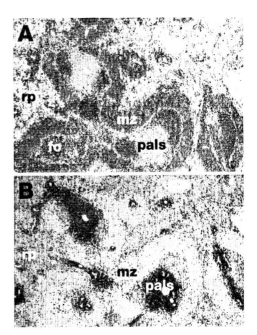

Figure 5. Distribution of B and T lymphocytes in the rat spleen. (A) B lymphocytes are predominantly observed in the marginal zone (mz) and the follicle (fo). (B) T lymphocytes are localized in the periarterial lymphatic sheath (pals). No germinal centers can be observed, reflecting the low pathogen load of the animal. The red pulp (rp) is marked by the weakly colored background staining, which is the result of the endogenous peroxidase activity of the erythrocytes. (B and T cells are revealed by appropriate monoclonal antibodies and the peroxidase technique; the diameter of a lymphocyte is about 10 μm; the human spleen has a comparable structure.)

follicles, a B-lymphocyte area (Fig. 5). Few T-helper lymphocytes (CD4+) can be demonstrated in the follicles. Apart from primary follicles, there are also secondary follicles containing a germinal center. The germinal center is surrounded by a rim of small B lymphocytes, which is called the *corona*, and only originates during a specific immune response involving both B and T lymphocytes. In the germinal centers, intense lymphocyte proliferation occurs, and one can recognize many mitoses. However, the majority of the newly formed B cells die immediately. The surviving B cells become either plasma cells (producing antibodies) or memory B lymphocytes. The white pulp is separated from the red pulp by the *marginal zone*, in which many B lymphocytes are located (Fig. 5).

Follicular dendritic cells exist in the follicles, and interdigitating dendritic cells in the periarterial lymphatic sheath. Macrophages are located in all compartments of the spleen, although their function differs from compartment to compartment. Macrophages preferentially phagocytose remnants of dead lymphocytes in germinal centers, particular antigens from the blood in the marginal zone, and old or nonfunctioning erythrocytes in the red pulp.

Function.   The human spleen has an excellent blood supply. It receives approximately 4% of the entire blood flow of the body despite the fact that it represents only about 0.2% of total body weight. The blood flow is retarded in the marginal zone and the red pulp. Subsequently it is directed through slits between endothelial cells, the size of which can be regulated, in a sievelike system of phagocytosing cells. The spleen can therefore be viewed as a filter installed into the bloodstream. This allows both the removal of old erythrocytes, and the capture of microbiological pathogens and other antigens in the blood. These processes occur predominantly in the red pulp and in the marginal zone, respectively. Furthermore, lymphocytes continually migrate into the spleen, with the migration paths of T and B lymphocytes displaying different characteristics. The lymphocytes leave the circulation and migrate preferentially into the marginal zone. From there the T lymphocytes migrate into the periarterial lymphatic sheath, a T-cell region, and the B lymphocytes into the follicles. The lymphocytes predominantly exit the spleen via the blood, and infrequently via lymph vessels. Furthermore, a continual immigration and emigration of lymphocytes occurs in the red pulp. Since so many more lymphocytes migrate through the spleen than through any other lymphoid organ, there is a great probability that an antigen-specific lymphocyte meets "its" antigen in the spleen and thus an effective immune response can be triggered.

The following functions of the red pulp can be deduced from the splenic structure:

1. The red pulp is important for the final maturation of young erythrocytes (reticulocytes). Chromatin remnants are leached during the passage through the endothelial slits. If a functional red pulp is absent, one can then observe these chromatin remnants in peripheral blood erythrocytes as "Howell–Jolly bodies."
2. Similarly, erythrocytes infected with pathogens causing malaria are eliminated from the blood during passage through the spleen. Therefore, humans without a spleen are especially endangered by malaria.
3. Additionally, old erythrocytes are eliminated in the red pulp. In diseases that lead to a decreased flexibility of the erythrocyte membrane, erythrocytes cannot pass through the slits, are trapped, and phagocytosed, resulting in reduced numbers of red blood cells (anemia).
4. If erythrocytes or thrombocytes are loaded with antibodies during disease, these cells are very effectively eliminated by the spleen, resulting in reduced numbers of red blood cells and platelets (anemia and thrombocytopenia, respectively).
5. Cellular elements, microorganisms, and particular antigens that enter the blood directly (not traversing filtering lymph nodes) are also phagocytosed by spleen macrophages. Thus, immune responses against these foreign products are subsequently initiated.

Therefore, in the red pulp one can differentiate an unspecific filter function (elimination of abnormal corpuscular components; 1, 2, 3) from a specific filter function (phagocytosis of antibody-coated cells and initiation of a specific immune response to antigens in the blood; 4, 5).

The human spleen does not act as a reservoir for erythrocytes, unlike the spleen of other species (e.g., dog). However, the red pulp contains approximately 30% of all blood platelets that can be mobilized through hormones such as adrenaline. During disease, various blood cells can be produced in the spleen instead of the bone marrow, a function the spleen also fulfills in the first month of human development.

1.3.3. Lymph Nodes

Site and Structure. The lymph nodes are filter stations integrated into the lymph. The size of dormant lymph nodes range from millimeters to over 1 cm in length. There is no exact calculation of the number of lymph nodes in humans. The recognized figure ranges between 300 and 700, with the majority of nodes found in the abdomen and the pelvic region (Fig. 3). In old age, only the size and not the number of lymph nodes is reduced. Lymphatic vessels are found in most tissues, with afferent lymphatic vessels reaching lymph nodes. When a lymph node is the first filter station for a particular tissue, it is called the *draining (regional) lymph node*. Subsequent stations that filter lymph from various draining lymph nodes are known as *collecting lymph nodes*. Because certain tumor cells first develop metastasis in the draining lymph nodes, knowledge of the position of these lymph nodes is important for the diagnosis and therapy of tumors.

Connective tissue radiates from the capsule to the inner lymph node, thus subdividing the lymph node into segments. Reticular fibers and reticular cells build a three-dimensional network within these segments. The lymph nodes are therefore considered a lymphoreticular organ. The capsule is penetrated by several afferent (influx) lymph vessels, which drain the lymph into the marginal sinus that rests underneath the entire capsule. From there, the lymph flows through the tissue into the efferent (exiting) lymph vessel leading out of the lymph node, finally reaching the blood via the thoracic duct.

The lymph node can be divided into the *cortex, paracortex,* and *medulla* (Fig. 6). In the cortex the lymphocytes are arranged into primary and secondary follicles, which consist of B lymphocytes and few T-helper lymphocytes (CD4$^+$). The paracortex is comprised mainly of T lymphocytes. A special feature of the paracortex is the high endothelial venules (HEV), whose endothelium allows lymphocytes from the blood to enter the lymph node tissue at a high rate.

Following a stay of several hours, the lymphocytes leave the lymph node via the

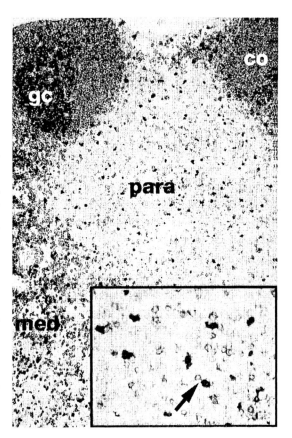

Figure 6. Distribution of B lymphocytes, migrating T lymphocytes, and proliferating cells in the rat lymph node. Most B lymphocytes (blue) are in the cortex (co), where they form follicles; a germinal center (gc) is seen containing many proliferating cells (red); few B lymphocytes are found in the medulla (med) and in the paracortex (para). The inset shows a magnification of the paracortex. B lymphocytes (blue), proliferating cells (red), and migrating T cells (yellow, injected 48 h previously) are present. The arrow indicates an interaction between a proliferating cell (red) and a migrating T lymphocyte. (The different cell types are revealed by appropriate monoclonal antibodies and either the peroxidase technique or the alkaline phosphatase anti-alkaline phosphatase technique; the diameter of a lymphocyte is about 10 μm; human lymph nodes have a comparable structure.)

efferent lymph vessel. The dendritic cells in the paracortex are interdigitating dendritic cells. Numerous plasma cells are resident in the medulla of the lymph node, particularly those that produce IgG, some IgM-producing cells, and very few that produce IgA and IgE.

Function. Like the spleen, the lymph nodes have the task of presenting as much antigen to as many lymphocytes as possible, thus increasing the probability that an antigen is recognized by "its" specific lymphocyte. However, in contrast to the spleen, which collects antigens from the blood, in the lymph node antigens from the lymph are presented. Lymph is generated in the tissues, and afferent lymph vessels carry the lymph into the draining lymph node. Thus, antigens (e.g., bacteria, tumor cells) from a widespread region are concentrated into a small lymph node through which many lymphocytes migrate. These antigens are then presented to the lymphocytes, which can reach the lymph nodes by two routes: via afferent lymph vessels coming from the drained tissue, and via the HEV coming from the blood (Fig. 7). Many more lymphocytes enter the lymph nodes via the HEV than through the afferent lymph (afferent lymph is thus rich in antigen, poor in cells). An immune response commences when the antigen is recognized by "its" lymphocyte. Thus, this system can be viewed as a spider (lymph node) sitting in the middle of a huge web (afferent lymphatics). The interior of the lymph node can be likened to a library: Many books (antigens) are presented to a large number of visitors (lymphocytes) in order to get them interested (immune response). During an immune response the blood supply of the lymph node increases quickly and large secondary follicles with active germinal centers develop. Thus, within a few days the lymph node swells by a multiple of its initial volume, which leads to a painful tension of the lymph node capsule.

Lymph nodes that are stimulated relatively often by antigens from the afferent lymph, such as mesenteric lymph nodes, have more and larger germinal centers. Plasma cells arise

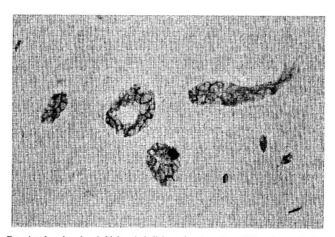

Figure 7. Entry into lymph nodes via high endothelial venules. An activated T lymphocyte (red) enters a lymph node via a high endothelial venule (blue). The high endothelial venules are identified by their localization in the paracortex, their high endothelium, and their staining for an endothelial cell marker. Naive and memory B and T cells enter lymph nodes in much higher numbers. (The different cell types are revealed by appropriate monoclonal antibodies and the alkaline phosphatase anti-alkaline phosphatase technique; the diameter of a lymphocyte is about 10 μm.)

that can be found in the medulla and produce predominantly IgG. Additionally, lymphocytes that are produced in response to a certain antigen not only rest in this single lymph
node, but leave it via the efferent lymph and reach the blood (efferent lymph is thus rich in
cells, poor in antigen). There they unite with lymphocytes that are continually migrating
through the body. The magnitude and regulation of lymphocyte traffic are described in more
detail in Section 1.4.

1.3.4. Tonsils

Site and Structure.   The tonsils (lymphoepithelial organs) are situated at the transition of the oral and nasal cavity in the throat (Fig. 3). Four tonsils are distinguished: the
unpaired pharyngeal tonsil, the two palatine tonsils, the unpaired lingual tonsil, and the tubal
tonsil. Additionally, lymphatic nodules 1–3 mm wide are seen at other sites of the oral cavity
such as the palate, the floor of the mouth, and the larynx.

All tonsils have a comparable structure, thus only the palatine tonsils shall be described
as an example in more detail (Brandtzaeg & Halstensen, 1992; Gebert, 1997). The covering
epithelium of the palatine tonsils is unsuitable for antigen uptake. The actual contact surface
between antigens and immune cells, therefore, is in the crypts. This contact is facilitated by a
spongelike loosened epithelium containing many lymphocytes. The 15–20 crypts of a palatine tonsil branch out, resulting in a substantial surface enlargement that amounts to
approximately 300 cm^2 per tonsil. Specialized cells for antigen uptake are found in the
loosened epithelium of the crypts, which are similar to the M cells of the Peyer's patches.
The tonsils do not have afferent lymph vessels like the lymph nodes. One can recognize
follicles which are adjusted in the direction of the crypts, often being secondary follicles
consisting predominantly of B lymphocytes and containing many germinal centers. The
T-cell region, *interfollicular region*, is characterized by HEV and is situated between the
follicles. The entry of lymphocytes derived from the circulation is located here. The
distribution of T-lymphocyte subpopulations is comparable to that of the paracortex in
lymph nodes.

When the class of the antibodies produced in the palatine tonsil is examined, a typical
ratio of IgG : IgA : IgM : IgD is 64 : 30 : 4 : 2. Thus, as in lymph nodes, IgG dominates. In
contrast to lymph nodes, however, IgA follows in second position instead of IgM. Since
IgA is the highly dominant immunoglobulin in the intestine, these data show that the tonsils
hold a functionally intermediate position between the intestinal immune system and the
lymph nodes. Efferent lymph vessels connect the tonsils with the lymph nodes of the neck.

Among the secondary lymphoid organs, the tonsils show the clearest age dependency
in their growth. The tonsil weight reaches a maximum during early childhood, the pharyngeal tonsil is largest usually around kindergarten age, and the palatine tonsil is of
maximum weight at elementary school age. Children obviously have contact with many
viral and bacterial pathogens, which stimulate the activity and growth of the tonsils.
Subsequently, tonsil weight, as well as lymphocyte density, diminishes rapidly in all parts of
the tonsil. Data regarding absolute tonsil weights differ considerably, because they mostly
originate from pathological samples.

Function.   The large contact surface of the specialized epithelium at the beginning of
the respiratory and gastrointestinal tracts allows antigen uptake and elicitation of specific
immune responses. The tonsils are integrated into the entire immune system by the continual influx of lymphocytes through the HEV, and the exit via the efferent lymphatics.

Complaints during the enlargement of tonsils can be explained by their position. For example, an enlarged pharyngeal tonsil hangs from the roof of the pharynx, in front of the opening of the nasal cavity, into the pharynx, resulting in handicapped nasal breathing in such children. If it also obstructs the opening of the connection between the middle ear and the pharynx (tuba auditiva, Eustachian tube), the hearing ability is drastically reduced and even otitis media may result. Thus, these children breathe through the open mouth and have a deficiency in their hearing ability. Both their appearance (open mouth) and their inability to understand simple questions (hearing loss) often falsely classify them as unintelligent. Surgical removal of the pharyngeal tonsils solves this problem and prevents possible damage to hearing, as well as the child's future school career.

1.3.5. Peyer's Patches

Site and Structure.   Organized lymphoid tissue can be found in the intestinal wall as single follicles, or as accumulations of follicles with a specialized epithelium, the Peyer's patches (Fig. 3). These lymphoid tissues in the gut wall are known as *gut-associated lymphoid tissue* (Prindull & Ahmad, 1993). Peyer's patches are present within the entire small intestine. Solitary follicles can also be found in the small and large intestine, and rectum. In young adults the total number of Peyer's patches with more than 25 follicles is approximately 100, and even in humans over 90 years old they number about 50 (Pabst, 1994). Like the tonsils, the Peyer's patches do not have afferent lymph vessels, and are therefore covered with a specialized epithelium situated against the intestinal lumen, which enables antigen contact.

The Peyer's patches are subdivided into distinct compartments (Fig. 8). The *follicles* are always secondary follicles with an active germinal center (Griebel & Hein, 1996). In other peripheral lymphoid organs, a mixture of active developing and regressing secondary follicles can usually be seen, whereas in the Peyer's patches, all follicles are active. Between the follicles, the *interfollicular region* consisting of tightly packed small T lymphocytes and HEV can be recognized. A special region of the Peyer's patches is the *dome*, a caplike accumulation of T and B lymphocytes against the intestinal lumen.

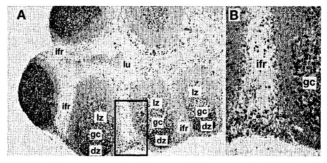

Figure 8.   Distribution of B lymphocytes and proliferating cells in Peyer's patches of the rat. (A) Many follicles are present. They contain B lymphocytes (blue) and germinal centers (gc). In the germinal centers a dark zone (dz) and a light zone (lz) can be seen which contain many proliferating cells (red); only few B lymphocytes are present in the interfollicular region (ifr); because the diameter of the follicle is much larger than the base of the dome, the dome is not seen in this section. (B) The boxed area indicated in A is shown at higher magnification. (lu, lumen of the gut; the different cell types are revealed by appropriate monoclonal antibodies and the alkaline phosphatase anti-alkaline phosphatase technique; the diameter of a lymphocyte is about 10 μm; human Peyer's patches have a comparable structure.)

Function.   The epithelium covering the dome shows four specific features:

1.   No crypts and villi can be found.
2.   The mucus-producing goblet cells are almost entirely missing.
3.   The epithelial cells do not produce the secretory component necessary for IgA to be transported by transcytosis into the intestinal lumen.
4.   M cells can only be found in the dome epithelium.

The lack of mucus and IgA facilitates the adhesion of antigens from the intestinal lumen to the dome epithelium. Through the M cells, large molecules can then come in contact with cells of the immune system despite the epithelial barrier (Gebert, Rothkötter, & Pabst, 1996). The M cells have deep invaginations at their basolateral surface in which lymphocytes can be found. Additionally, the basal membrane in the dome has large pores through which macrophages stretch their cell processes. Thus, all cell types needed for the initiation of an immune response are situated in local proximity. B lymphocytes are extensively produced in the follicles (germinal centers) of the Peyer's patches. Both B and T lymphocytes leave the Peyer's patches and reach the mesenteric lymph nodes via afferent lymphatic vessels. From there they gain access to the blood and are able to migrate to other organs.

1.3.6. Nonlymphoid Organs

Lymphocytes can be found in many other organs, and body fluids, apart from those described above (Table 2). Different organs and their compartments vary in their composition of B and T lymphocytes and their subpopulations. This shall be demonstrated with the distribution of lymphocytes within the intestinal mucosa.

The intestine comprises a surface of approximately 300 m^2, which is necessary for the efficient absorption of nutrients. The abundance of food antigens and the huge quantity of microorganisms of the intestinal lumen are separated from the inside of the body only by a monolayer of intestinal epithelial cells. Lymphocytes are situated between the epithelial cells and are called *intraepithelial lymphocytes*. In healthy human intestine, the number of intraepithelial lymphocytes is 20, 13, and 5 (per 100 enterocytes) in the jejunum, ileum, and colon, respectively (Pabst, 1994). The entire number of intraepithelial lymphocytes is therefore very large. Intraepithelial lymphocytes are present before birth, and are exclusively cytotoxic T lymphocytes (CD8$^+$) (Fig. 9). In contrast, in the subepithelial connective tissue of the lamina propria, T-helper lymphocytes (CD4$^+$) predominate (Fig. 9). About 75% of all human plasma cells are located in the intestinal wall, and most of the remaining 25% in the spleen, lymph nodes, and bone marrow. IgA is the predominant immunoglobulin of the mucosa. It is contained in the secretions of the gastrointestinal and respiratory tracts (Krug, Tschernig, Holgate, & Pabst, 1998; Pabst & Tschernig, 1997). IgA is also found in salivary glands, tear ducts, and mammary glands. Nonetheless, the immunological function of the intestinal wall can only be understood if it is recognized that macrophages, many mast cells, and granulocytes are present in the lamina propria, a site which is densely innervated.

1.3.7. Development

In the 10th week of human development, the thymus is the first lymphoid organ that can be detected. By the 14th week, the thymus shows all of the compartments of a fully developed organ. The tonsils (16th week), the lymph nodes (20th week), and the spleen (24th week) complete their development much later. In these organs, the B-cell region (containing primary follicles) is the compartment that is formed at the end of development.

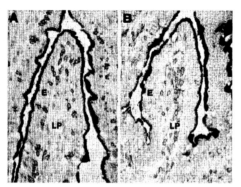

Figure 9. Distribution of cytotoxic T lymphocytes (CD8[+]) and helper T lymphocytes (CD4[+]) in the intestinal mucosa of the pig. (A) Cytotoxic T lymphocytes (CD8[+]) predominate within the epithelium (E) of the intestine (intraepithelial lymphocytes). (B) Helper T lymphocytes (CD4[+]) predominate in the lamina propria (LP) of the intestine. (The T-cell populations are revealed by appropriate monoclonal antibodies and the alkaline phosphatase technique; unspecific coloring of the brush border is produced by endogenous alkaline phosphatase; the diameter of a lymphocyte is approximately 10 μm; the human intestine has a comparable structure; original provided by H. J. Rothkötter, Hannover.)

1.4. Migration, Proliferation, and Apoptosis

When considering normal histological sections, it is not apparent that many lymphocytes continually immigrate and emigrate. In addition, it is also often overlooked that lymphocytes proliferate and die in lymphoid organs. The static impression that is easily obtained while observing a histological preparation is therefore misleading. Migration, proliferation, and cell death have to be thoroughly regulated, as for every lymphocyte that has immigrated or is newly produced, another one must emigrate or die in order to have constant lymphocyte numbers. In the following sections, some aspects of the migration, proliferation and apoptosis of lymphocytes are discussed.

1.4.1. Migration

The total number of lymphocytes in the body of a healthy young adult has been estimated at approximately 500×10^9 (Westermann & Pabst, 1992). Lymphocytes are concentrated in lymphoid organs, but are also diffusely spread in many other organs such as the bone marrow, skin, lungs, intestinal wall, and liver (Table 2). In contrast to all other blood cells, lymphocytes have the ability to leave the circulation and subsequently return to the blood (Abernethy & Hay, 1992; Pabst & Binns, 1989). This process allows connections between the different lymphoid and nonlymphoid organs, and enables the detection of antigens at every possible site of the body (Fig. 10). Therefore, the blood and lymph

→

Figure 10. Each day about 5×10^{11} lymphocytes leave the blood to migrate into various organs and return into the blood. (A) Most of the blood lymphocytes enter organs such as the lung, spleen, liver, and bone marrow via flat venules; the vast majority return into the blood via normal venules; only few lymphocytes leave these organs via the afferent lymph. (B) In humans about 0.3×10^{11} lymphocytes per day migrate from the blood into organs such as the lymph nodes, tonsils, and Peyer's patches; they use specialized venules, the so-called high endothelial venules (HEV); the lymphocytes return into the blood via the efferent lymph (classic pathway of lymphocyte recirculation); thus, only a small fraction use the HEV route relative to the number of the lymphocytes leaving and returning into the blood each day ($\sim 5 \times 10^{11}$). (C) Only very few lymphocytes migrate under normal circumstances into organs such as the skin, synovia, muscle, and brain; they enter these tissues via normal venules and return into the blood either via normal venules or via afferent and efferent lymph; however, during acute and chronic inflammation HEV-like structures may develop, leading to a large increase in lymphocyte influx (Westermann & Pabst, 1996).

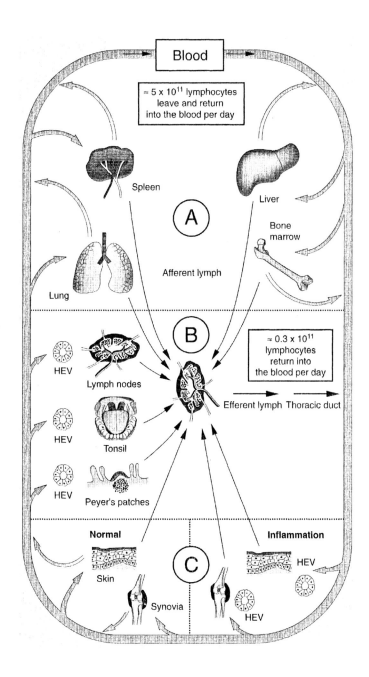

represent the connecting pathways for migrating lymphocytes. An impression of the extent of migration can be obtained if one understands that the 500×10^9 lymphocytes that pass through the blood per day approximates the number of all lymphocytes existing in the body (in magnitude).

Three dominant pathways in which lymphocytes leave the blood can be distinguished (Fig. 10; Westermann & Pabst, 1996):

1. Approximately 90% of all lymphocytes immigrate from the blood into the spleen, lung, liver, and bone marrow. There they mainly interact with the endothelium of the vasculature and thereby form the marginal pool. The spleen receives more than 50% of these lymphocytes. The latter leave the circulation into the marginal zone, immigrate into the white pulp, and return into the blood after some hours via the red pulp.

2. Approximately 10% of lymphocytes use the HEV to reach the lymph nodes, tonsils, and Peyer's patches. Lymphocytes stick to the specific surface molecules of these endothelial cells and migrate through the endothelium into the lymphoid organs.

3. Under normal circumstances, only a few lymphocytes immigrate into organs such as the skin, the synovia of joints, or muscles. During the course of acute and chronic inflammations, HEV also arise in places where they are not normally found. In this way the number of immigrating lymphocytes is increased.

The described migration pattern is comparable for B and T lymphocytes, for cytotoxic T lymphocytes (CD8$^+$) and T-helper lymphocytes (CD4$^+$). In addition, naive, activated, and memory T lymphocytes also follow these routes. However, their migration routes differ once within a lymphoid organ. For example, B and T lymphocytes enter lymph nodes via HEV in comparable numbers. Then most T cells migrate through the paracortex and the medulla into the efferent lymph to join the bloodstream. In contrast, B lymphocytes migrate from the paracortex into the cortex, then back into the paracortex and via the medulla into the efferent lymph. The different routes also explain the longer time required by B cells to pass through a lymph node. The time needed to migrate from the blood into the lymph nodes, from there into the efferent lymph, and then into the blood again, is 30 h for B cells and 15 h for T cells. Naive and memory T cells also enter lymph nodes in comparable numbers (Fig. 11). In the tissue, however, memory T cells predominate among the few T cells that migrate into the cortex and germinal centers, and memory T cells are able to migrate faster through the paracortex (Fig. 11). This may help to explain why memory T cells are more efficient in providing help for the B-cell response (migration into cortex and germinal centers), and why they respond quicker to antigens (faster migration through the paracortex). In contrast to the current view, our experiments do not support a selective immigration of a certain lymphocyte subpopulation over another population at the level of the HEV (Westermann, Geismar, Sponholz, Bode, Sparshott, & Bell, 1997), indicating that security is the main purpose of the multiple adhesion pathways that regulate lymphocyte immigration at different sites. This ensures that immigration of lymphocytes occurs under all circumstances (Fig. 12). In contrast, the interaction within the tissue of many molecules and tissue structures facilitates preferential migration routes (Fig. 12). Particular regions of lymphoid organs are predominantly integrated into the migration routes of lymphocytes, such as the paracortex of the lymph nodes, the marginal zone, and the periarteriolar lymphatic sheath of the spleen, while other organs and regions, such as the thymus and the secondary follicles, are only minimally included in the migration pathways (Fig. 13). Very little is presently known about the mechanisms regulating these processes. Adhesion

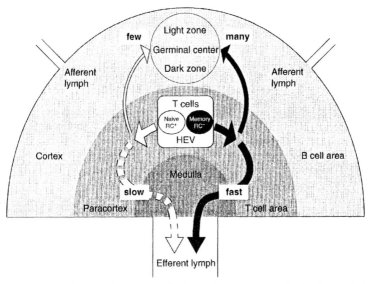

Figure 11. Migration of naive and memory T cells through the compartments of a lymph node. Both naive (CD45RC⁺) and memory (CD45RC⁻) T cells enter lymph nodes from the blood across high endothelial venules (HEV). This optimizes the chance of specific T cells finding the remnants of a previous antigenic insult, because lymph nodes concentrate antigen (reaching it via the afferent lymph) directly in the path of circulating lympho- cytes. Memory T cells preferentially migrate into the B cell area (including germinal centers) and survey the T-cell area faster than naive T cells (wider arrows). "Memory" may thus reside not only on an increased frequency of antigen-specific cells and a decreased threshold to stimulation, but also on the ability to specifically enter the B-cell area and to monitor large quantities of lymphoid tissues in a short time.

molecules on lymphocytes play an important role, as well as endothelium, tissue cells, and extracellular matrix (Pabst & Westermann, 1994).

Thus, the migration of lymphocytes leads to a continuous change of their location, and enables an effective immunosurveillance of the body. In addition, after vaccination with tetanus toxoid, antigen-specific lymphocytes are not restricted to the lymph node where the initial immune response started, but can spread to many lymphoid and nonlymphoid organs to provide protection against tetanus toxin.

1.4.2. Proliferation and Apoptosis

When comparing lymphocytes with erythrocytes or granulocytes, specific characteristics become obvious. Under normal circumstances, erythrocytes and granulocytes are only produced in the bone marrow. In contrast, lymphocytes are produced not only in the bone marrow but also in the thymus, spleen, lymph nodes, Peyer's patches, and most other organs. The magnitude of proliferation depends on the organ and the particular compartment within the organ (Fig. 13). In the thymic cortex, numerous lymphocytes proliferate, while only few can be seen in the medulla (Fig. 13). The number of proliferating cells in the spleen corresponds to that in the medulla of the thymus. However, even in the spleen there are organ compartments that contain many proliferating lymphocytes: the germinal centers (Fig. 13).

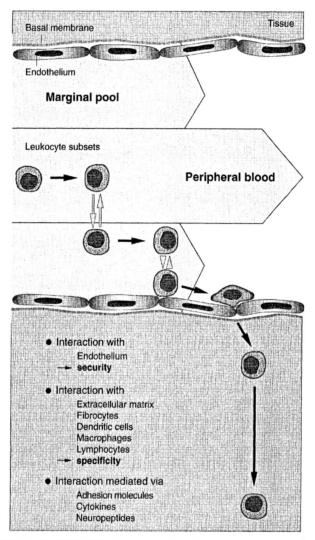

Figure 12. Delivery, entry, and migration of lymphocytes through lymphoid organs. Lymphocytes are delivered to the tissues mainly by the peripheral blood. They interact with the marginal pool, which brings them close to the vascular endothelium. Here different adhesion molecules mediate the adherence to the endothelium and the subsequent entry into the tissue. Our data indicate that most lymphocyte populations are randomly recruited at that step. Within the tissue, however, different lymphocyte subpopulations take different migration routes.

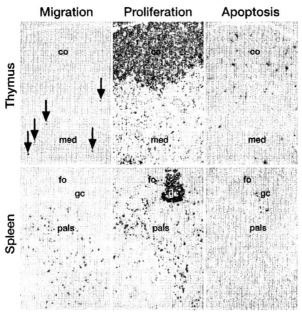

Figure 13. Migration, proliferation, and death of lymphocytes in thymus and spleen. Labeled T lymphocytes are injected into a rat, and 24 h later the thymus and spleen are removed. Immigrated T cells (blue), cells that were proliferating (red), and cells that were dying (blue) at the time of organ removal are identified with monoclonal antibodies on consecutive sections of the same organ. (Migration) Few labeled T cells (arrows) are found in the medulla (med) and none in the cortex (co) of the thymus. However, numerous T lymphocytes are found in the periarterial lymphatic sheath (pals) of the spleen 24 h after injection. (Proliferation) Many proliferating cells (red) can be seen in the cortex (co) of the thymus, and also in the medulla (med). All compartments of the spleen contain proliferating cells; an accumulation of proliferating cells in the spleen is seen within germinal centers (gc). (Apoptosis) In dying lymphocytes the DNA of the nucleus is modified (Tunel[+]), which can be revealed immunohistochemically; as with proliferating cells, dying cells are preferentially found in the cortex (co) of the thymus, and in germinal centers (gc) of the spleen, although at a much lower frequency than that of proliferating cells. (fo, follicle; alkaline phosphatase technique; the diameter of a lymphocyte is approximately 10 μm).

Cell death is also frequently observed among lymphocytes. Self-destruction of the cell, called *apoptosis*, is often induced via surface molecules (Ueda & Shah, 1994). The number of apoptotic cells is very high in the thymic cortex (Fig. 13). Few apoptotic cells are seen in the medulla of the thymus and the spleen, the exception being germinal centers, which contain many apoptotic cells (Fig. 13). Thus, in the compartments in which the repertoire of B- and T-cell receptors is generated (germinal centers and thymic cortex, respectively), cell proliferation and cell death occur at high rates.

Fig. 13 shows that at each moment in each tissue, lymphocytes migrate in and out, proliferate, and die. Thus, a histological section taken from a lymph node will not reveal the same lymphocytes as a section taken a few minutes earlier. It is important to be aware of these dynamic changes when looking at histological sections, for only then is a correct interpretation of the observed phenomena possible.

1.5. Morphological Changes during an Immune Response

In the course of immune responses, alterations in the morphology of the lymphocytes, the architecture of the lymphoid organs, the proliferation rate and the migration paths of the lymphocytes occur along with many other parameters (e.g., expression of adhesion molecules, production of cytokines) (see Chapter 2).

Lymphocytes are not terminal cells like the granulocytes of the blood. Stimulation by an antigen initiates a transformation. A small lymphocyte with a dense nucleus and a small cytoplasm then becomes a large cell with a loose nucleus, distinct nucleolus, and extensive basophil cytoplasm (Fig. 14). This lymphoblast divides and thus initiates cell multiplication. A naive T lymphocyte is usually stimulated by an antigen presenting cell in a lymphoid organ such as the lymph node or the spleen. This cell presents the processed antigen together with a costimulatory signal, inducing cell growth and subsequent cell proliferation (Fig. 15). Two types of T cells are generated during this process: One type are effector T cells that either kill infected cells (CD8[+]), activate macrophages (CD4[+]), or initiate a B cell response leading to the production of antibodies (CD4[+]; Abbas, Murphy, & Sher, 1996). The other cell type is the memory T lymphocyte. So far it is difficult to distinguish these two T cell subtypes by, for example, analyzing the expression of surface molecules and intracellular cytokines (Ahmed & Gray, 1996; Bell *et al.*, 1998). For B lymphocytes the situation is different. When naive B cells recognize the native unprocessed antigen and receive help from antigen-specific T lymphocytes, they proliferate in the T-cell region of lymphoid organs first and then migrate into B-cell regions such as the cortex of lymph nodes and the follicles of the spleen. There they form a germinal center (Fig. 16; Kosco-Vilbois, Zentgraf, Gerdes, & Bonnefoy, 1997; Toellner, Gulbranson-Jugde, Taylor, Man-Yuen Sze, & Mac-Lennan, 1996). In the germinal center, B cells proliferate, change the isotype of the B-cell receptor (and the isotype of the antibodies produced later; isotype switching), and increase the ability of the receptor to bind the antigen (affinity) by somatic hypermutation. Thus, many antigen-specific B cells are generated that can produce high affinity antibodies of the IgG, IgA and IgE isotypes. Two types of B lymphocytes finally leave the germinal centers: effector and memory B cells. The effector B cells become plasma cells: Their nucleus is usually situated noncentrally and they often have a cartwheel structure because of the

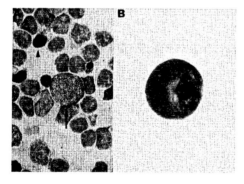

Figure 14. After activation lymphocytes change size and function. (A) When a small resting lymphocyte (arrowhead) is activated, its size increases (arrow), and the lymphocyte starts to proliferate. (Cytopreparation from a human spleen; the diameter of a small lymphocyte is about 10 μm; the black points over the nucleus of the lymphocyte indicate that a radioactive marker has been incorporated into DNA during cell division; Giemsa staining; original provided by R. Pabst, Hannover.) (B) Cytopreparation from human bone marrow showing an antibody-producing plasma cell. (In the right corner the shadow of an erythrocyte can be identified; Giemsa staining; original provided by R. Leo, Hannover.)

Figure 15. Cytospot preparation of lymph node T cells before and after stimulation. T lymphocytes are shown in blue. The stimulation via the T-cell receptor and the costimulatory molecule CD28 leads to large T lymphocytes that have incorporated DNA precursors (brown) during cell division. (Alkaline phosphatase anti-alkaline phosphatase technique and peroxidase anti-peroxidase technique; original provided by U. Bode, Hannover.)

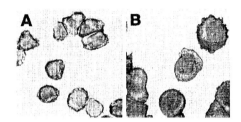

arrangement of the chromatin (Fig. 14). The cytoplasm appears strongly basophilic, because of the distinct rough endoplasmic reticulum and the widespread Golgi apparatus, which leads to a lighter coloring beneath the nucleus (Fig. 14). These are morphological signs of an intensive protein synthesis necessary to produce sufficient amounts of antibody. Mature plasma cells are found mainly in the gut, the spleen, and the bone marrow. They are terminal cells and are unable to divide. Thus, in contrast to T lymphocytes, among B lymphocytes the three main functional subtypes can be distinguished: naive B cells bear low-affinity receptors of the IgM and IgD isotype, whereas memory B cells have high-affinity receptors of the IgG, IgA, or IgE isotype. The effector cell stage, the antibody-producing plasma cells, can be recognized, because of their characteristic morphology. The changes in the shape of lymphocytes and the generation of new structures such as germinal centers, strongly influence the architecture of lymphoid organs. For example, an immunologically inactive spleen looks very different from an immunologically active spleen (Fig. 17). The presence of germinal centers indicates that an immune response is ongoing. The size and the number of germinal centers demonstrate the strength of the immune reaction. Since germinal centers are only generated when B and T lymphocytes collaborate, the presence of germinal centers also clearly shows that the molecules necessary for this interaction are present and functioning. Thus, looking at a histological section of a lymphoid organ already provides clues as to which processes are operating within the immune system.

Activation of lymphocytes also alters their migration pattern. This will be described using B lymphoblasts as an example, which become IgA-producing plasma cells. Whereas

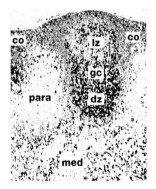

Figure 16. Structure of a germinal center within a lymph node. After activation in the T cell region, B cells migrate into the cortex (co) of the lymph node and start to proliferate. Thereby they generate a germinal center (gc) comprised of two main compartments, the dark zone (dz) and the light zone (lz). In the germinal center many B cells are produced, the isotype of the B cell receptor is switched, and its affinity increased. Two types of B cells leave the germinal center: B cells that become plasma cells and produce antibodies, and memory B cells that continuously migrate through the body. (B cells are revealed in blue and proliferating cells in red; para, paracortex; med, medulla; alkaline phosphatase anti-alkaline phosphatase technique.)

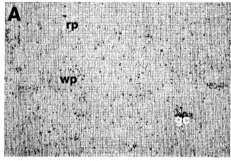

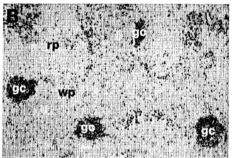

Figure 17. Splenic architecture before and after stimulation. (A) Before stimulation, only few proliferating cells (red) and only one small germinal center (gc) are seen on a cryostat section of a spleen. (B) Three days after stimulation, the number of proliferating cells in the white pulp (wp) and the red pulp (rp) has increased considerably, and many germinal centers (gc) can be recognized. (Alkaline phosphatase anti-alkaline phosphatase technique.)

small resting B cells randomly and continuously migrate to lymphoid and nonlymphoid organs, IgA-producing B lymphoblasts are preferentially found in the lamina propria of the gut. After antigen from the gut has reached the Peyer's patches and the mesenteric lymph nodes, an immune response is initiated and activated B lymphoblasts are generated (Fig. 14). These cells reach the blood via the intestinal lymph vessels and the thoracic duct (Fig. 10). Later they are preferentially found in the lamina propria of the entire intestine, where they produce IgA antibodies. However, the distribution of the IgA blasts is not limited to the intestinal mucosa; they are also found in the mucosa of the respiratory tract, of the salivary glands, of the tear ducts, and of the mammary glands. Therefore, IgA is the most prevalent antibody class in the mucosa and the secretions of the body. Recently, it was found that the preferential accumulation of activated lymphocytes at certain sites reflects preferential survival of these cells within the tissue, rather than specific migration into these sites (Bode, Wonigeit, Pabst, & Westermann, 1997). Such information may help to improve vaccination strategies that are already therapeutically used. A patient is vaccinated via the intestine through a swallowed capsule containing bacterial fragments. Because IgA producing B lymphoblasts also migrate into the mucosa of the bronchial tract, they can prevent infections caused by these bacteria at a site distant from the vaccination site. These systems are also called *mucosa-associated lymphoid tissue*, because of the similarity of immune responses in the mucosa of the respiratory, gastrointestinal, and genitourinary tracts, and their integration via the blast migration. This shows that despite the distribution of lymphoid cells all over the body, the mobility of lymphocytes facilitates the integration into a common defense system.

1.6. Anatomy of an Immune Response

The different phases of nonspecific and specific immune responses and their inter-action *in vivo* are complicated (see Chapter 2), and in many parts not well understood. They depend, among many other variables, on the site of penetration and on the type of invading pathogen. In addition, it is beyond the scope of this chapter to describe the cell surface molecules and cytokines that are involved in the regulation of an immune response. However, it should be noted that a high mobility of the cell membrane and its surface molecules is necessary before the T-cell receptor can bind to the antigen presented by MHC molecules (Fig. 18). Thus, it is difficult to provide a general outline of an immune reaction. Bearing this in mind, Figures 19–21 nevertheless try to describe schematically the different phases of a type IV immune response, concentrating on topographic aspects: where it takes place, and which cell types and molecules are involved. First, the pathogen has to overcome the epithelial barrier (Fig. 19). Then it is faced with local macrophages and the complement system (Fig. 19). If the pathogen is still alive, both macrophages and complement system recruit leukocytes from the blood, leading to the typical signs of inflammation (Fig. 19). At the same time, locally produced cytokines are released into the circulation and induce fever (Fig. 20). In addition, a specific immune response is initiated in the draining lymph node (Fig. 20). Antigen-specific lymphocytes and antibodies greatly enhance the efficiency of macrophages and granulocytes, finally leading to the elimination of the pathogen (Fig. 21).

In contrast to the majority of immune responses, which are T cell dependent, Fig. 22 outlines an immune response that is not, or only to a small extent, T cell dependent. The combination of different morphological and functional techniques made it possible to attribute the different phases of an immune reaction against a distinct bacterium (pneumo-coccus; thymus independent type 2 antigens) to single compartments of the spleen (van

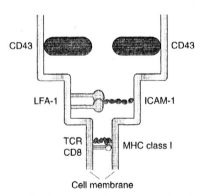

Figure 18. Compared with other cell surface molecules the T-cell receptor is small. Before the specific receptor of a T cell can recognize the antigen presented by the MHC (green), first the negative charge of the cell surface between the two interacting cells has to be overcome (red). Then the T cell uses adhesion molecules (yellow, e.g., LFA-1) to interact with the vascular endothelium for tissue recruitment (1), to establish the firm adhesion that is necessary for the specific interaction of the T cell receptor with the MHC (2), and to provide costimulatory signals that are needed for the final activation (3). Because the T-cell receptor and the MHC are relatively small molecules, the cell membrane must be flexible and the cell surface molecules mobile in order to allow their interaction. (The size of the different molecules are proportional to each other; the width of the cell membrane is about 4 nm; adapted from Springer, 1990.)

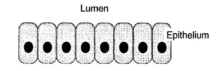

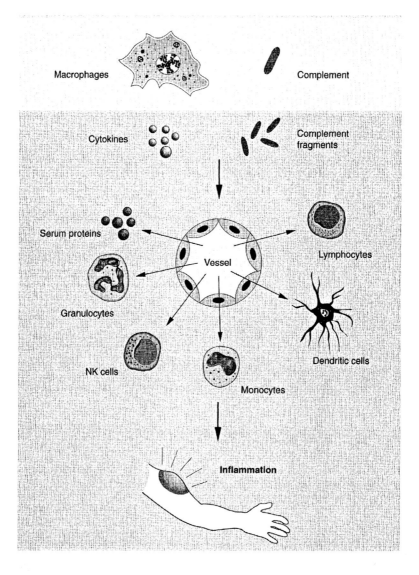

Rooijen, 1990). The pathogens reach the marginal zone where they are presented to the B lymphocytes by macrophages (Fig. 22). B lymphocytes with a corresponding specific receptor on their surface are activated, leave the marginal zone, and immigrate into the PALS. There they develop into plasma cells and finally reach the red pulp, where they then produce the respective antibodies. Thus, the marginal zone is especially important for the immune reaction against pneumococcus. Since the marginal zone of children is not fully developed in the first years of life (Pabst, 1994), children are particularly threatened by an often lethal pneumococcal infection. In addition, after removal of the spleen (splenectomy), even adults are at risk of getting a pneumococcal infection. Therefore, the spleen is no longer seen as a dispensable organ, and surgeons try to preserve the spleen following injury.

1.7. The Blood as a Diagnostic Window of the Immune System

Information on lymphocyte distribution in organs and organ compartments, and on their migration routes and proliferation rates is also of clinical relevance. This will be explained below using the determination of blood lymphocyte numbers as an example.

In order to assess the functional state of the immune system, blood samples are often

←————————————————————————————————

Figure 19. Local infection: Schematic overview of nonspecific defense mechanisms. One example of the different phases of nonspecific defense mechanisms is schematically outlined. (Top) **Epithelial barrier**, mechanisms of protection: *Microbiological*: normal flora competes for nutrients and attachment to epithelium, and can produce antibacterial substances. Antibiotic treatments that kill the normal flora make an individual susceptible to infections. *Mechanical*: epithelial cells joined by tight junctions, and longitudinal flow of fluid across the epithelium, e.g., transport of particles from the airways to the oral cavity. Skin burns lead to severe infections. *Chemical*: antibacterial effects of lysozyme in tears, saliva and sweat. Low pH in the stomach destroys pathogens as does the enzyme pepsin. **Effect**: Most pathogens are not able to penetrate the intact epithelium of the gastrointestinal (300 m^2), respiratory (150 m^2), genitourinary (<0.5 m^2) tracts, and the skin (2 m^2). (Middle) **Local macrophages and complement** destroy pathogens: *Phagocytosis* of invading pathogens by local macrophages. *Lysis* and enhancement of phagocytosis of pathogens by the complement system (alternative pathway). **Effect**: Many invading pathogens are immediately removed at this step. (Bottom) **Enhancement of the local nonspecific defense mechanisms** by cytokines (e.g., TNFα, IL-1, IL-6; produced by activated macrophages) and complement fragments (e.g., C3a, C4a, C5a; produced via the alternative pathway): *Upregulation of adhesion molecules* on blood vessel endothelium leads to: 1. Extravasation of leukocytes: see below. 2. Blood clotting: cuts off the blood flow and prevents pathogens from entering the blood and from spreading all over the body. *Leukocyte recruitment into the tissue* 1. granulocytes → phagocytosis → removal of extracellular pathogens: accumulation of dead granulocytes is responsible for the formation of pus; absence or inability to migrate into tissues results in life threatening infections; even the uptake of pathogens without destruction (due to a genetic defect) buys time and is life saving. 2. NK cells → lysis (removal of intracellular pathogens). 3. Monocytes → become macrophages → phagocytosis. 4. Dendritic cells → uptake of local antigen and transport into the draining lymph node → induction of specific immunity. 5. Lymphocytes → search for "their" specific pathogen → production of cytokines. *Increased production of cytokines* e.g., TNFα retains infection at the local level. *Clinical signs of inflammation*: 1. Redness due to dilatation of blood vessels. 2. Heat due to dilatation of blood vessels. 3. Swelling due to edema, serum protein leakage, and leukocyte recruitment. 4. Pain due to swelling and certain mediators, e.g., prostaglandins and leukotrienes. Aspirin inhibits the production of prostaglandins and reduces pain. *Time course*: Considerable numbers of all cell types are present in the tissue within a couple of hours after pathogen arrival. **Effect**: Removal of pathogen; if not possible: Localization of pathogen and induction of a specific immune response. *In vivo* the situation is more complicated. For example, pathogens may gain direct entry into the blood. Then the spleen would play a more prominent role. In addition, the role of interferon (IFN)-α and -β has not been mentioned. These interferons are produced by virally infected cells and inhibit virus replication, increase MHC class I expression, and activate NK cells to kill virally infected cells.

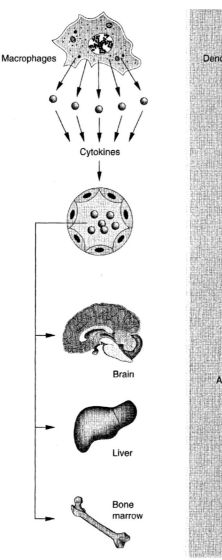

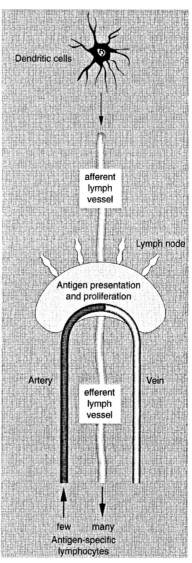

taken, with the number of B and T cells measured by flow cytometry (see Chapter 7). However, alterations in the number of lymphocytes in the blood are much harder to assess than alterations in the number of erythrocytes. For example, a decrease in the number of erythrocytes in the blood can either reflect an insufficient production in the bone marrow or a shortened life span in the blood. These conclusions can be drawn because blood cells such as erythrocytes, granulocytes, or monocytes are normally only produced in the bone marrow, and after a period of maturation they leave for the blood (Table 3). There these cells either spend their whole life (erythrocytes), or immigrate into the different tissues where they remain for shorter (granulocytes) or longer (monocytes) periods of time (Table 3). In contrast, lymphocytes are not only produced in the bone marrow but also in other organs, and pass into the blood very quickly in order to immigrate into the various tissues (Table 3). They do not die in the tissue, but can emigrate into the blood. Thus, a decrease in the number of lymphocytes in the blood can be explained by multiple reasons: The production of these cells in one or more organs could be decreased; the cells possibly leave the blood rapidly; or they are retained in an organ. Additionally, a combination of these factors is

←——

Figure 20. Systemic effects of an infection on nonlymphoid and lymphoid organs. The systemic effects of cytokines and the induction of the specific immune response is schematically outlined. (Left) **Cytokines reach the following organs via the blood,** when local infection cannot be controlled. *Brain*: Induction of fever by TNF-α, IL-1 and IL-6. Fever inhibits the growth of pathogens and enhances the strength of immune responses. *Liver*: Production of acute phase proteins within two days of infection. C-reactive protein activates complement to increase the phagocytosis of pathogens. Manose-binding protein increases phagocytosis and activates the complement system (classical pathway). *Bone marrow*: Increased release of granulocytes into the blood. More granulocytes are available for extravasation and subsequent phagocytosis. **Effect**: Within one or two days the local immune system is supported by these effects. (Right) **Induction of the specific immune response.** *Uptake of local antigen* by dendritic cells → migration into the draining lymph node via afferent lymphatics. Destruction of the afferent lymphatics prevents the induction of specific immunity showing that the induction of a specific immune response almost always is initiated within lymphoid organs. *Presentation of processed antigen* by dendritic cells to naive T lymphocytes which continuously and randomly enter the lymph node in large numbers via specialized blood vessels (high endothelial venules). Within two days all specific T cells are trapped in the draining lymph node. *Proliferation of specific T lymphocytes* in the draining lymph node and generation of: 1. Cytotoxic T cells → lysis of cells. 2. T-helper 1 cells → activation of macrophages. 3. T-helper 2 cells → activation of B lymphocytes. The cytokines of the local microenvironment influence the type of cell produced: IL-12 and IFN-γ secreted by macrophages and NK cells, respectively, induce the generation of T-helper 1 cells, removal of intracellular pathogens; IL-4 secreted by mast cells and certain T lymphocytes induce the generation of T-helper 2 cells, removal of extracellular pathogens; the regulation of this process is very complex: T-helper 1 cells suppress the generation of T-helper 2 cells and vice versa (feedback). The route of pathogen entry, the amount and the strength of interaction are also important, e.g., large amounts of pathogens and strong interaction with the T cell receptor favors the generation of T-helper 1 cells. *Release of antigen-specific T cells* into the blood searching the body tissues for specific antigen. 1. Five days after antigen arrival specific T cells start to leave the draining lymph node via the efferent lymphatics. This leads to an increase in the number of specific T cells available for extravasation into the tissues (factor 10–100). Development of memory T cells which continuously migrate through lymphoid and nonlymphoid organs. *Activation of naive B cells* within the lymph node by native antigen and antigen-specific T-helper 2 cells. Proliferation and subsequent migration into lymph node follicle. *Generation of germinal centers* in which antigen-specific B lymphocytes: 1. Proliferate → increased numbers. 2. Hypermutate → increased affinity of the B cell receptor and the subsequently produced antibodies. 3. Switch the isotype → change from IgM to IgG, IgA, and IgE antibodies. 4. Become plasma cells → migration into bone marrow and spleen, production of specific antibodies, and release into the blood. 5. Become memory cells → continuous migration through lymphoid and nonlymphoid organs. **Effect**: Increased numbers of antigen-specific T cells and high amounts of antibodies are available in the blood for subsequent distribution to the tissues where they increase the efficiency of the nonspecific defense mechanisms. *In vivo* the situation is more complicated. Antigens may also be presented by macrophages and B lymphocytes. B cells do not always need the help of T-helper 2 cells. Antigen presentation may also lead to anergy.

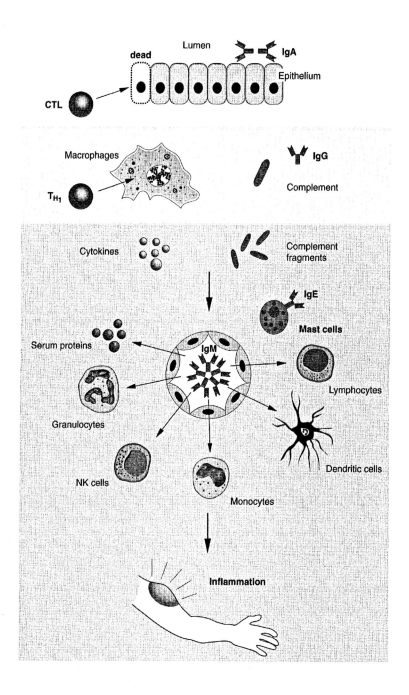

possible. Furthermore, only about 2% of all lymphocytes in the body are located in human blood at any given time point. An alteration in the number of lymphocytes in the blood is thus much harder to assess than alterations in the number of erythrocytes, granulocytes, or monocytes, particularly ordinary situations already substantially influence the number of lymphocytes in the blood (Table 4). For example, the number of T lymphocytes in the blood can vary up to 100% according to whether the blood sample is taken in the morning or evening.

It is therefore unlikely that blood lymphocytes can be regarded as a simple mirror automatically representing changes in the number of the other 98% of lymphocytes distributed throughout the many tissues of the body. A more precise understanding of how the migration, proliferation, and apoptosis of single lymphocyte subpopulations are regulated, is necessary for a better interpretation of changes in the composition of blood lymphocytes occurring in many diseases (Table 4). This will enable clinicians to use the

←——

Figure 21. The specific immune response enhances the effectivity of nonspecific defense mechanisms. The contributions of the specific immune response (bold cells and antibodies) to nonspecific defense mechanisms (light outline, same as in Fig. 19) are schematically outlined. **Primary immune response**. The cells of the specific immune system use the abilities of the cells mediating the nonspecific defense mechanisms by increasing their efficiency. The specific immune system comes into play when the pathogen is not recognized by macrophages and granulocytes, and when too many pathogens are present. *Extravasation of effector T lymphocytes into the tissue (cellular immune response)*. 1. Cytotoxic T cells: lysis of infected cells → removal of intracellular pathogens, e.g., viruses. 2. T-helper 1 cells: activation of macrophages → removal of intracellular pathogens, e.g., bacteria. 3. T-helper 2 cells: activation of B lymphocytes → production of antibodies → removal of extracellular pathogens, e.g., parasites. *Distribution of antibodies (humoral immune response)*. 1. **IgA** on the surface of the epithelial barrier prevents adhesion of toxins, viruses and bacteria. 2. **IgG** within the tissue: opsonizes pathogens (labeling for and enhancement of phagocytosis by macrophages); neutralizes toxins before they are pathologic to the body tissue; activates the complement system leading to opsonization, leukocyte recruitment and target cell lysis (classical pathway); facilitates lysis of cells by NK cells (antibody-dependent cell-mediated cytotoxicity, ADCC); mainly produced by B lymphocytes in the bone marrow and spleen; reaches the tissues via the blood. 3. **IgE** binds to mast cells which are often positioned in the vicinity of blood vessels, and sensitizes it for release of inflammatory mediators: increase of blood vessel permeability; extravasation of serum proteins (edema); recruitment of leukocytes. 4. **IgM** in the blood: activates the complement system (classical pathway); neutralization of toxins. **Effect:** Antigen-specific T cells and antibodies increase the efficiency of the nonspecific defense mechanisms in the tissues. The peak of a primary immune reaction occurs after about two weeks of pathogen invasion. **Secondary immune response**. When the same pathogen is met a second or third time, the specific immune response is more rapid and more effective. This is called immunological memory, and is mediated by both the cellular and humoral effector arm of the immune system. *Changes in the T cell population compared to a primary immune response*. 1. Frequency of antigen-specific T cells increases 10 to 100 times → increased probability of meeting the "right" antigen. 2. Surface molecules are up- or downregulated: increased extravasation into tissues → more efficient effector function; no further need for costimulation → activation in nonlymphoid tissues possible. *Changes in the B cell population compared to a primary immune response*. 1. Frequency of antigen-specific B cells increases about 10 times → increased probability of meeting the "right" antigen. 2. Alterations in the antibodies produced: antibody concentration in the blood about 100 times higher; switch from IgM to IgG → due to smaller size better penetration into the tissues; antibody affinity is about 100 times higher → better binding to very low amounts of antigen. **Effect:** In contrast to a primary immune response which peaks 2 weeks after pathogen invasion, a secondary immune response already peaks after 1 week providing 100 times more antibodies with a 100 times higher affinity. 1. *Clinical implications*. Artificial induction of memory by vaccines is one of the most outstanding contributions of immunology to medicine. 2. Immunological memory is shaped by the pathogens in the environment → movement from one place to another (e.g., Europe to Africa) is often associated with disease until the immune system has developed a "new" memory. 3. Many questions regarding immunological memory have to be answered: is persisting antigen necessary for the maintenance of memory?; are memory lymphocytes short or long lived?; are memory T cells different from naive T cells, as is the case for memory and naive B cells? However, *in vivo* the situation is even more complicated. For example, only T cells with the αβ receptor are described and not T cells with the γδ receptor. In addition, CD5⁺ B cells are not mentioned. Furthermore, the mechanisms leading to immunological memory *in vivo* are unclear.

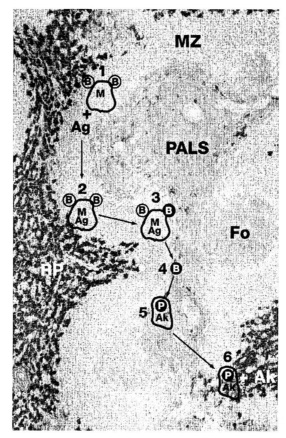

Figure 22. Anatomy of an immune response against pneumococcus in the spleen. (1) The antigen (thymus-independent type 2 antigen) reaches the spleen via the marginal zone (MZ). (2) There it is phagocytosed by macrophages, prepared and presented to B lymphocytes (B, open circle). (3) B lymphocytes that bear specific receptors for the antigen on their cell surface are activated (B, black circle). (4) Activated B lymphocytes migrate in the periarterial lymphatic sheath (PALS). (5) There they differentiate into plasma cells (P) containing specific antibodies in their cytoplasm (Ak). (6) The plasma cells eventually migrate into the red pulp (RP), where they secrete their produced antibody. (Cryostat section of a rat spleen, with the macrophages of the red pulp revealed in brown, so that the single splenic compartments can be better identified; Fo, follicle.)

blood more efficiently as a diagnostic window for the immune system (Westermann & Pabst, 1990).

1.8. Interactions between the Immune System and the Nervous System

The immune system and the nervous system have many features in common. Both detect and respond to alterations in the environment in a highly specific manner, both show

Table 3
Life History of Erythrocytes and Leukocytes[a]

Cell type	Site of production	Numbers in the blood per μl (mean transit time)	Site of function (life span)
Erythrocytes	Bone marrow	$\approx 5 \times 10^6$ (about 120 days)	Blood (about 120 days)
Granulocytes	Bone marrow	$\approx 3.5 \times 10^3$ (about 1 day)	Migration into inflamed tissues (hours to days)
Monocytes	Bone marrow	$\approx 0.5 \times 10^3$ (about 2 days)	Migration into normal and inflamed tissue (as macrophages up to many years)
Lymphocytes[b]	Bone marrow Thymus Spleen Lymph nodes Tonsils Many other sites	$\approx 2.5 \times 10^3$ (about 30 min)	Migration into normal and inflamed tissues and emigration from these tissues (many years)

[a]The references from which the present information is taken are summarized in Westermann and Pabst (1990).
[b]In contrast to the other blood cells, only lymphocytes can proliferate outside the bone marrow and alternate between proliferating and resting phases. Whereas the large majority of other blood cells are kept in the circulation, only 2% of all lymphocytes are located in the blood. Only few data on the life history of NK cells are available.

Table 4
Examples of Conditions Affecting the Composition
of Lymphocyte Subsets in the Blood[a]

	Main finding[b]	
Daily life		
Age (increases)	CD4+CD45R+	↓
Gender (women)	CD4+CD45R+	↑
Pregnancy	CD4+	↓
Sexual practice	CD8+	↑
Smoking	CD4+	↑
Sport	CD8+	↑
Stress		
Acute	CD8+	↑
Chronic	CD8+	↓
Drugs and hormones		
Cimetidine	CD4+CD8+	↑
Cyclophosphamide	B	↓
Cyclosporin A	CD8+	↑
Endotoxin	CD4+	↓
Epinephrine	CD8+	↑
Factor VIII/IX concentrate	CD8+	↑
Glucocorticosteroids	CD4+	↓
Insulin	B	↑
Interleukin 2	CD25+	↑

(continued)

Table 4   (*Continued*)

	Main finding[b]	
Medical procedures		
Anesthesia and surgery	CD4+	↓
Splenectomy	B	↑
Technique of subset determination		
Antigen polymorphism	e.g., CD4+	∅
Time of blood sampling		>50% variation
Site of blood sampling		>100% variation
Storage of blood sample	e.g., CD4+	↑
Counting device	e.g., CD8+	↓
Data presentation	e.g., CD4+	↓
Diseases		
Skin		
Candidiasis	CD4+	↓
Kala-azar	CD4+	↓
Nervous system		
Down's syndrome	CD4+CD45R+	↑
Multiple sclerosis	CD4+CD45R+	↓
Schizophrenia	B CD5+	↑
Respiratory system		
Pneumonia	CD4+	↓
Pulmonary sarcoidosis	CD8+	↑
Alimentary system		
Celiac disease	CD4+	↓
Crohn's disease	CD8+	↓
Urogenital system		
Pyelonephritis	CD4+	↓
Immune system		
AIDS	CD4+	↓
Rush hyposensitization	CD4+CD45R+	↑
Endocrine system		
Graves' disease	CD8+DR+	↑
Hashimoto's disease	CD4+DR+	↑
Hematopoietic system		
Idiopathic thrombocytopenic purpura	CD4+	↓
Joints and connective tissues		
Systemic lupus erythematosus	CD4+	↓

[a]The references from which the information is taken are summarized in Westermann and Pabst (1990).
[b]↑, increase; ↓, decrease in numbers.

memory, and both consist of complex organizations of cells having similar appearances but different functions. A chief distinction between these two systems is their anatomical organization. Compared with the nervous system, the immune system, and the lymphocytes in particular, are noted for their mobility (Janeway, 1988). For a long time both systems were considered to act independently. Meanwhile, there is good evidence that the immune system and the nervous system influence each other (Blalock, 1994). For example, interleukin-2, an important cytokine of the immune system, is capable of stimulating the release of adreno-corticotropic hormone from the pituitary gland. In the reverse direction, thyroid-stimulating hormone, a pituitary hormone, is able to increase the strength of the humoral immune response. The immune system and nervous system are also connected morphologically. For most lymphoid organs it has been demonstrated that nerve fibers from the sympathicus and vagus not only reach blood vessels, and are thus able to influence the circulation, but also stay in tight contact with lymphocytes, macrophages, dendritic cells, and epithelial cells (Straub, Westermann, Schölmerich, & Falk, 1998). Therefore, interactions between these organs and the nervous system are also possible via nerve fibers. A more precise description of the innervation of the lymphoid organs is provided in Chapter 8.

An attractive vision suggesting an even closer association between these two systems is that of Blalock (1994). He claims that the immune system should be our sixth sense, and establishes this position with the following example: A virus cannot be seen with the naked eye, is too small to be felt, causes no taste or smell, and makes no sound, but nonetheless we sense its presence when we have influenza.

ACKNOWLEDGMENTS. We thank all of our colleagues in the Department of Functional and Applied Anatomy for their continuous support, especially K. Bankes-John, I. Dressendörfer, S. Lopez-Kostka, A. Reuße, and F. Weidner. The critical comments of Dr. U. Bode, Dr. R. Pabst, Dr. T. Tschernig (Department of Anatomy, Medical School of Hannover, Germany), and Dr. K. Resch (Department of Pharmacology, Medical School of Hannover, Germany), the help in preparing the figures by D. Stelte and H. Reyland, and the correction of the English by Ms. S. Fryk are gratefully acknowledged. The authors' studies were supported by the Deutsche Forschungsgemeinschaft (SFB 244 A7 and We 1175/4-2). M.S.E. was supported by a Postdoctoral Fellowship from the Alexander von Humboldt Foundation.

2   Foundations in Immunology

Roland Jacobs and Reinhold E. Schmidt

Roland Jacobs and Reinhold E. Schmidt • Department of Clinical Immunology, Hannover Medical School, D-30625 Hannover, Germany.

2.1. Introduction

The real importance of the immune system is evident when it does not function well. Nowadays perhaps the most well-known immune defect is the acquired immunodeficiency syndrome (AIDS). This disease made it clear to the public that we are permanently surrounded by pathogens like bacteria, protozoa, fungi, and viruses. All of these infectious agents are challenging the body and when they are not stopped, these particles will invade and make us ill.

How does a healthy body prevent infectious organisms from invading? The first line of defense is the physical barrier of an intact skin. In addition, a special milieu on the skin is maintained by apathogenic microorganisms, preventing pathogens from settling there. Special types of outer skins are the mucous membranes that line the lung, intestine, and oral cavity. Mucous makes it more difficult for bacteria and viruses to penetrate these outer membranes. In addition, there are antimicrobial components in the mucus and in the saliva. The latter fluid is supplemented with proteins designed for binding pathogens. These proteins are the immunoglobulins, or antibodies, and will be discussed below. However, when this protective layer is disrupted or the skin is wounded, the second line of defense becomes the focus of attention: the immune system.

The immune system is comprised of billions of so-called white blood (leukocytes) and tissue cells. The first contact between intruder and the immune system is established unspecifically by phagocytes. These cells are relatively large cells that are able to move around and ingest pathogens. There are different types of phagocytes: The skin is infiltrated by Langerhans cells, in the bloodstream there are monocytes and granulocytes. The liver is populated by Kupffer cells, the brain by glia cells, and the lung by astrocytes. These phagocytes do not simply eat up any kind of unknown substances, but also offer parts of these substances to other more specialized cells of the immune system on their surfaces. This feature of phagocytes is called *processing*. In parallel, a lot of mediators are produced and released during this process, tempting other cells to initiate an inflammatory reaction.

We now turn to introducing the specific arm of the immune system. One should keep in mind, however, that terms such as *innate* or *unspecific immune system* do not relate to the capabilities of the involved cells. Moreover, an effective innate defense is indispensable for the development of a specific response. Unlike the phagocytes attacking all foreign particles, each cell of the specific immune response is optimized to recognize unique molecules (*antigens*) of the intruders. Therefore, a cell that can recognize a particular antigen of a pathogen will bind to it or react to it, but will leave another one alone.

The main source of such specialized cells are the lymphocytes. This cell type comprises about 30% of all white cells in peripheral blood. Lymphocytes can be divided roughly into three subpopulations. Besides the more popular B and T cells, there are also the natural killer (NK) cells. To a certain extent, NK cells take their position between the innate and specific immune defense.

2.2. Cells of the Immune System

In a brief outline the most important cells of the immune system, and their features, will be described here. For a more detailed description and images of the different cell types, refer to Chapter 1.

The majority of the phagocytosing cells are comprised of monocytes/macrophages and granulocytes. Monocytes represent about 5 to 14% of the white blood cells. In tissues the same cells can differentiate into macrophages. Granulocytes make up about 45 to 80% of all leukocytes. According to histological staining, granulocytes are further divided into basophil, eosinophil, and neutrophil granulocytes. Most of the peripheral blood granulocytes belong to the latter group (~90%).

Basophils and eosinophils are rather specialized in that they release bioactive substances from their granules. Polycellular pathogens, like worms and other parasites, which cannot be ingested by single leukocytes are drenched in enzymes and amines, thus initiating an extracellular digestion of the intruders. All phagocytes are able to engulf particles unspecifically. Phagocytosis, however, is more efficient if the particles are *opsonized*. Opsonization describes a coating of the pathogens with molecules that makes it easier for the phagocytes to bind and pick up foreign substances. Opsonins can be antibodies or fragments of the complement system such as C3b. On the one hand, these molecules can bind to the pathogens via specific antigen-binding sites, and on the other hand, to phagocyte receptors, promoting adhesion between each other.

The cells engulf foreign substances by invagination of the outer membrane, including the particles in the so-called phagosome. In a second step, these phagosomes fuse with enzyme-filled lysosomes. Within the lysosomes the pathogens are digested and killed. Components of the digested pathogens can then be transported in the opposite direction to the cell surface. In this way antigens are processed and presented to the lymphocytes, thus initiating the specific response.

During the whole process of ingestion and killing of antigens, the phagocytes release cytokines. These substances draw further leukocytes (phagocytes and lymphocytes) to the site of defense, promoting a state of inflammation.

2.3. Lymphocytes

2.3.1. B Lymphocytes

B lymphocytes (B cells), like all other blood cells, originate from pluripotent hematopoietic stem cells of the bone marrow. In chickens the site of B-cell maturation is the bursa of Fabricius, a gland near the cloaca. B cells are so named for the bursa. In humans, the equivalent of the avian bursa is the bone marrow, where the B cells mature, so the term *B cell* is apt for humans as well.

The main task of B cells is the production and secretion of antibodies (= immunoglobulins). Antibodies are glycoproteins with characteristic structure (Fig. 1). According to their structural backbone, immunoglobulins (Ig) are divided into classes and subclasses. In humans, one can distinguish five classes: IgA, IgD, IgE, IgG, and IgM. Subclasses are denoted by appending numbers and letters, e.g., IgG1, IgG2a, or IgA1. Normally, a single antibody consists of two heavy (H) and two light (L) chains. The H chains determine to which class or subclass the Ig molecule belongs. The IgG1 immunoglobulin contains two $\gamma 1$ chains and the IgM molecule two, or as explained later, ten μ chains. Each H chain is assembled of four (α, γ, δ, μ) or five (ε) subunits, which are called *domains*. One of these domains is variable (V region), the others are constant (C region) in each subclass (Table 1).

There are only two kinds of light chains, namely, the kappa (κ) and the lambda (λ) chain. Both L chains of a single antibody molecule are always of the same type, either κ or λ. The two L chains do not exhibit any functional difference. However, κ-type L chains are found twice as often as λ chains among all immunoglobulins. Similar to H chains, the L

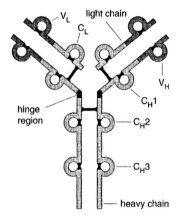

Figure 1. Molecular structure of immunoglobulins. The typical Ig molecule consists of two heavy (H) and light (L) chains. The light chains are connected to a heavy chain by disulfide bonds. The hypervariable regions of the variable domains of one H and one L chain form a single joint antigen-binding site. The whole molecule is stabilized in a Y-shape by additional disulfide bridges between the two H chains. The H chain comprises three constant domains (C_H1–C_H3) and one variable region (V_H). Besides the variable domain (V_L) the L chain contains only one constant region (C_L). Within the heavy chain there is a proline-rich region enabling a certain flexibility of the "arms" as a result of the molecular characteristics of proline.

chains consist of different domains, one variable (V) and one constant (C) region. Two disulfide bonds connect one heavy and one light chain in such a way that the variable domains of both chains are located close to each other. This functional unit represents the antigen-binding site. The combination of both variable regions determines the specificity of the antibody. Two heterodimers each consisting of one L and one H chain are also connected by disulfide bonds between the constant domains of the H chains. These bonds give the whole Ig molecule a Y-like structure.

In vitro, the Ig molecule can be degraded into different fragments by proteolytic enzymes (Fig. 2). Degradation with pepsin yields two fragments: a divalent (two antigen-binding sites) F(ab)$_2$ fragment (= *f*ragment *a*ntigen *b*inding) and the Fc (*f*ragment *c*rystallizable) fragment. Splitting the molecule with papain reveals three fragments: two identical monovalent (one binding site) Fab, and one Fc fragment.

After secretion by plasma cells—the term for terminally differentiated B cells—the complete antibody molecule is capable of binding to a particular structure. This structure is called an *epitope* and represents a part of an antigen.

Permanently challenged by a multitude of different antigens, the immune system is forced to provide specific antibodies to every conceivable antigen. The genes for the antibodies specific for millions of antigens must be encoded by DNA. However, if there really was a particular gene for each specificity, then the DNA molecule would have to be much longer than it actually is. Therefore, another pathway is used to guarantee a genomic code for every possible specificity. The DNA that is initially identical in each cell of an

Table 1
Numbers of Genes Encoding
the Different Regions of an Immunoglobulin

V region		J region		C region		D region	
V_H	$\gg 100$	J_H	6	C_H	9	D_H	> 4
V_κ	>70	J_κ	5	C_κ	1		
V_λ	20–30	J_λ	8	C_λ	4 or 5		

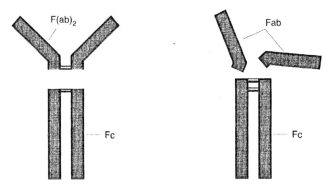

Figure 2.   Antibody fragments. Like any other protein, immunoglobulins can be enzymatically digested. (Left) Splitting with pepsin reveals two main fragments, one F(ab)$_2$ and one Fc fragment. (Right) Two identical Fab portions and one Fc are the result of proteolysis by papain.

organism is changed during maturation of a single B cell. In other words, the genome of B cells undergoes DNA recombination.

The genes for different protein elements that form the whole Ig molecule are located on different chromosomes. In humans, the gene for the H chain is part of chromosome 14, the gene for the κL chain is located on chromosome 2, and the gene for the λL chain is found on chromosome 22. Each Ig gene complex is composed of three (L chain) or four (H chain) regions. These gene segments are denoted V (*v*ariable), D (*d*iversity), J (*j*oining), and C (*c*onstant) (Table 1).

The principle of somatic recombination is to bring two distant genomic regions into close proximity by cutting and subsequently joining the DNA molecule. In the cells there are at least three mechanisms to perform recombination:

1.   The different gene segments are marked by special sequences acting as recombination signals. In part, these signal sequences on the single DNA strand correspond complementarily. When two recombination signals bind to each other, the DNA between forms a loop. At the base the loop is excised by nucleases, and thereafter the free ends of the DNA molecule are joined together again by ligases (Fig. 3a).

2.   Another way to translocate DNA segments is by the exchange of chromatid sections (crossing over). When sister chromatids are close to each other but unequally oriented, an exchange of chromatid fragments of different length, after breaking and reassembling, reveals a deletion on one chromatid and an extension of the other. Moreover, again the sequence of gene segments is changed by this crossing over (Fig. 3b).

3.   A third mechanism of translocation is by inversion. One chromatid forms a loop as mentioned above, and after breaking at the bottom, the loop is then inserted again but in the opposite direction (Fig. 3c).

As a result of these cleaving and joining procedures, two gene segments that are normally located far away from each other can be moved into close proximity, thus enabling a huge variety of encoding by a limited kit of genes.

In the cell, all of these mechanisms of translocations of chromatid segments are

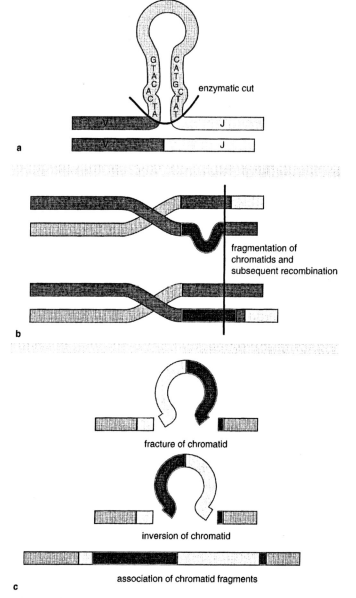

Figure 3. Somatic recombination of Ig genes. The excision of DNA segments translocates different gene sections into close proximity. These recombined genes are transcribed into mRNA and thereafter translated into proteins. (a) At complementary DNA sequences a single strand binds to itself and the arising loop is excised at the bottom. (b) A crossover of two sister chromatids that are not yet in line correctly can cause two gene segments to come together. (c) The cut-off loop of a chromatid can subsequently be reintegrated the wrong way, revealing a newly oriented gene sequence.

working in parallel. However, deletion after loop formation seems to be the most frequent kind of recombination.

The whole process of recombination follows a precise sequence. First an IgH gene complex for the heavy chain is established by linking the D_H to the J_H complex. After successful recombination, the V_H complex is added. All steps are performed simultaneously on both chromosomes (diploid set of chromosomes). Only when a recombination is finished correctly can the next one follow. Otherwise another recombination trial will be started.

Theoretically, in each cell two Ig genes could be processed, leading to the expression of antibodies with two specificities. Practically, a successful recombination of one functional Ig gene product inhibits any further recombination trial (allelic exclusion). After completing the gene complex that encodes the heavy chain, the gene for the κL chain is formed by adding V_κ to a C_κ segment. A successful recombination terminates this process. If all κL segments have been used, a recombination of λL sections will be tried. In the case of failing to rearrange a functional gene for the complete Ig molecule, the cell will die. In another case, the recombined Ig gene can be transcribed into an RNA and finally translated into an Ig protein.

By free choice of assembling gene segments, an extreme variability of combinations is guaranteed. In addition, this variability is enhanced by somatic mutation. In contrast to other gene sequences, recombined Ig gene complexes undergo mutations quite frequently, thus creating antibodies with altered specificity. Obviously, repair mechanisms are less effective in regard to recombinant Ig DNA compared with other genes.

Another feature of antibody production is the Ig class switch. As mentioned above, there are different classes and subclasses of antibodies. Terminally matured B cells are called *plasma cells*. They have rearranged Ig genes and secrete immunoglobulins with only one specificity. However, these plasma cells still have the opportunity to change the class of "their" antibodies. The nature of the Ig switch is a connection of the same V_H-D_H-J_H complex, with different C_H sequences revealing antibodies with the same specificity but a different constant domain or Ig class.

The first H chain to be expressed by B lymphocytes is the μ chain. The RNA transcript of the μ chain still contains the $C_H\delta$ gene. This $C_H\delta$ sequence, however, is a silent gene (= not translated) so the μ chain can be translated. Later on, the primary RNA is alternatively spliced, thus deleting the μ transcript and enabling the transcription of the δ chain. At this step the B cell can express both IgM as well as IgD at the surface. If this B cell now fails to meet its specific antigen, it will die. However, if the B cell does contact the antigen, development of the cell will continue. At this time point, gene segments encoding all Ig classes are together on a single RNA molecule: C_μ-C_δ-$C_{\gamma 1}$-$C_{\gamma 2}$-$C_{\gamma 3}$-$C_{\gamma 4}$-C_ε-$C_{\alpha 1}$-$C_{\alpha 2}$. By alternative splicing the different classes can be translated. In case of deleting the C_μ-C_δ complex, for example, a C_γ can be fixed to the V_H-D_H-J_H segment, revealing the production of an IgG molecule.

All B cells produce IgM and IgD antibodies initially. After antigen contact the B cells switch to translate IgG, IgA, or IgE immunoglobulins. It is absolutely possible that the descendants of a single cell clone undergo different class switches. So, one sister cell may switch to IgA production whereas another one may produce IgG first and then IgA. After switching to IgA production, however, no other classes can be made by this cell because all other C_H sequences have been deleted before.

2.3.2. Structure and Function of Different Antibody Classes

The structure and function of a single antibody molecule have already been described briefly. The molecule consists of a covalently bound tetrameric aggregation of two identical light and two identical heavy protein chains. Such a molecule is called *divalent* (two antigen-binding sites) and *monomeric* (despite the tetrameric chemical composition).

Most of the secreted antibodies correspond to this structure. On the contrary, the IgM molecule in blood appears as a pentameric version formed by five covalently bound monomeric IgM units. The single units are connected at the Fc portion, revealing a starlike formation with ten binding sites. Eighty percent of all IgA antibodies are monomeric and the remaining 20% form a dimeric complex with four antigen-binding sites.

The different types of antibodies (IgG, IgM, and so on) are also called *isotypical variants* or, for short, *isotypes*. The biochemical features of each class characterize the different isotypes for particular immunological strategies (Table 2).

2.3.3. Antibody Classes

IgM (Fig. 4a) is the predominant immunoglobulin of the primary immune response after challenge by an unknown antigen for the very first time. The pentameric backbone, with ten antigen-binding sites, is very effective in agglutination (clotting) of antigenic particles. Moreover, IgM is the most potent isotype in activating the complement cascade.

Complement is a system of serum proteins supporting many immunological functions, and will be explained in more detail later. The monomeric IgM variant is anchored on the surface (sIgM) of B cells and serves as the specific B-cell receptor (BCR). Together with the sIgM molecule, IgD is coexpressed on the outer membrane of B cells. It is likely that IgD plays a controlling role in Ig switching, although nothing definitive is as yet known.

IgG is the predominant antibody of the secondary immune response. It is produced rapidly in high concentrations after meeting an already known antigen. There must have already been a primary response to that exact specific antigen. During the first contact, B cells specific for the appropriate antigen expand and produce IgM antibodies. Moreover, a few of these specific B cells do not proliferate but rather develop into memory cells. The latter are long living (up to 10 years) cells with the same specific BCR. These primed cells are ready to react at once when the appropriate antigen turns up again. They start to divide into further memory cells and many effector B cells, producing IgG antibodies without having to perform all of those recombination steps. This memory effect of lymphocytes (the same is true for T lymphocytes) is the basis for successful vaccination.

Incidentally, IgG is the only class of immunoglobulins that are capable of passing through the placenta, and thereby providing specific antibody-mediated protection for newborns, who cannot as yet produce their own IgG antibodies (Table 3).

IgA is the main immunoglobulin in saliva, colostrum, and mucous membranes (Fig. 4b). Because of the associated secretory component, dimeric IgA is protected from many

Table 2
Biochemical Characteristics of Immunoglobulin Classes

	Molecular weight	Sedimentation coefficient	Valence	Structure	Percentage of total immunoglobulin
IgG	150,000	7 S	2	Monomeric	80%
IgM	900,000	19 S	10	Pentameric	6%
IgA	160,000	7, 9, 11 S	2, 4	Monomeric	13%
	385,000			Dimeric	
IgD	185,000	7 S	2	Monomeric	0–1%
IgE	200,000	8 S	2	Monomeric	0.002%

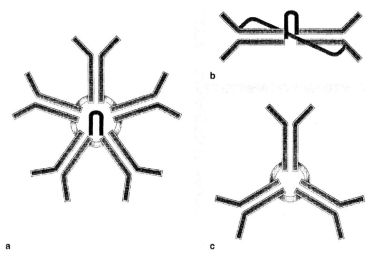

Figure 4. Shapes of immunoglobulins. (a) Monomeric IgM represents the antigen receptor of B cells; released into blood, IgM is composed of five subunits that are connected by disulfide bridges and equipped with a joining piece. (b) IgG, IgD, and IgE are almost exclusively present in a monomeric shape, whereas most of the IgA is aggregated by a joining peptide to a dimeric variant. (c) Trimeric, tetrameric, and pentameric IgA are rather rare entities. An additional secretory component protects IgA on mucous membranes from enzymatic digestion.

proteolytic enzymes and is able to remain undigested on the mucous membrane in this unsuitable environment. Via breast milk, a mother transfers IgA to her infant, serving to provide protection for the mucous membranes of the newborn.

The Fc portion of the IgE molecule has a high affinity to the IgE receptor (FcεR) on mast cells and basophils. After binding to the FcεR of those cells, IgE waits for its specific antigen. If there is contact with that antigen, the IgE molecule changes its conformation, thus providing a signal to the cells to degranulate. In other words, the cell releases enzymes and other substances from intracellular granules. One of these substances is histamine, the substance that gives rise to the typical features of allergy (hay fever, asthma).

The physiological relevance of this IgE-mediated mechanism seems to lie in the defense of worms and other metazoic parasites. After recognition of the antigens on the outside of the parasite, by the surface-bound IgE, IgE then degranulates and various enzymes and other chemical substances are poured out over the intruder, thus damaging the parasite and attracting further reactive cells.

Table 3
Distribution and Functional Features
of Human IgG Subclasses

IgG1	IgG2	IgG3	IgG4	Ig subclass
65%	23%	8%	4%	Percentage of total IgG
+++	++	++++	±	Fixation of complement
+++	+	+++	±	Binding to Fc receptors
++	±	++	++	Cross placenta

2.3.4. T Lymphocytes

T lymphocytes comprise the second biggest cell population of the specific immune system. Like B cells, they originate from pluripotent stem cells in the bone marrow. Chemotactic signals lead them to the thymus (T cells are so named for this organ) via the bloodstream. Here, the T cells mature in the absence of antigen but most of these immature T cells die (for details see Chapter 1). Only a few cells can leave the thymus as mature T cells. When these T cells meet their specific antigen during circulation in the blood or lymphatic tissue, they start to proliferate and differentiate. One portion of these cells develops into effector T cells to provide a quick cellular immune response after a second challenge by the antigen.

Like B cells, T lymphocytes bear a highly specific receptor on their surface, enabling the binding of a certain antigen. The main task of B cells is to release antibodies, a soluble form of their BCR, into the serum. In contrast, T cells do not release any receptorlike proteins. The T-cell receptor (TCR) is a heterodimeric protein tightly anchored in the membrane.

There are two kinds of TCR: the phylogenetically older γ/δ T-cell receptor (TCR1) and the earlier described α/β T-cell receptor (TCR2). About 80 to 90% of all T cells express TCR2. The α chain is encoded on chromosome 14, and the genetic code for the δ chain is encrypted within the α segment. Both genes for the β and the γ chain are part of chromosome 7, each on a different arm. Analogous to the genetic organization of the Ig gene complexes, the code for the TCR is spread over different gene regions.

For the α chain there have been described $\gg 50$ V_α, $\gg 50$ J_α, and one C_α region; an additional D_α gene is not expressed. The complete information for the β chain is composed of one each out of $\gg 70$ $V_{\beta1}$, 1 $D_{\beta1}$, 1 $J_{\beta1}$, 1 $D_{\beta2}$, 6 $J_{\beta2}$, and 1 $C_{\beta2}$ region. The genes encoding the γ/δ TCR are also assembled from different gene segments.

The mechanisms that manage the rearrangement of immunoglobulin genes are identical to those of the TCR recombination. The time course of the synthesis of the TCR starts with the initiation of a γ/δ recombination. When there is a successful δ-region composition, no further rearrangement of this chromosome will occur. If the following γ-recombination is also successful, this particular T cell will express a γ/δ TCR. However, if no δ-rearrangement of one or both chromosomes succeeded, the cell tries to assemble a gene for the β chain. If it fails to arrange a functional gene for the β chain, the cell will die.

As mentioned above, activated T cells do not secrete soluble antibodies like B cells do. Rather, they communicate with other cells or kill them. The direct contact between effector and target during defense is essential for the so-called cellular immune function. T lymphocytes recognize their specific antigen nearly always in context with the major histocompatibility complex (MHC) proteins of the counterpart (target cell).

In humans, the MHC is identical to the HLA (human leukocyte antigen) system and is identical to the classical transplantation antigens. Two of these MHC products are relevant for T cells: class I and class II molecules. MHC class I proteins are present on all nucleated cells whereas class II molecules are expressed mainly by antigen presenting cells (APC). Cytotoxic T cells, usually CD8[+], can attack only cells expressing the same MHC class I proteins.

The MHC is an extremely heterogeneous gene system providing almost every individual with their own personal pattern of MHC molecules. So, usually, T cells can only communicate with cells from the same individual—a feature termed *MHC restriction*. Of course, the body's own cells are only attacked by CD8[+] cells if they become infected or are otherwise altered, e.g., by cancer. Potentially, each cell of the body can be infected or altered. So it is useful that CD8[+] cells are class I restricted (note that class I proteins are present on all nucleated cells). To make binding possible the antigen has to be part of the membrane of the target cell.

In contrast, T-helper cells (CD4$^+$) recognize an antigen together with MHC class II proteins, and they usually do not kill other cells. When an antigen is detected on the surface of a target cell, the helper cell can bind to it and interfere in immunoregulation by secreting cytokines (chemical substances that have an effect on other cells).

All types of T cells have to recognize two different molecules at the same time: the endogenous MHC molecule (class I or class II) and the antigen. How does the cell manage the corecognition of two molecules? During a class I-restricted immune response, foreign proteins, such as virus proteins or tumor-associated proteins, are degraded into peptides within the cytosol. The digested peptides consist of 9–11 amino acids. They are translocated by transport proteins (TAP transporter) into the endoplasmic reticulum (ER). At this site a trimeric complex consisting of one β_2-microglobulin, one class I heavy chain (both comprise the complete class I molecule), and the antigen is formed, thereby arranging the antigen into a groove of the class I heavy chain. This complex is then transported from the ER through the cytoplasm, and at last inserted into the outer membrane. The antigen lying in the groove of the class I protein enables an effector T cell to bind simultaneously the antigen and the MHC product and kill that target.

Lysis of a cell makes sense only when an altered cell must be destroyed or a penetrated virus needs to be withdrawn from the essential protein synthesis apparatus of the cell. When particles that are not capable of division are incorporated, e.g., by healthy phagocytes, it is not necessary to destroy these cells. This causes another strategy of antigen processing. The antigen is taken up by phagocytes via endocytosis and degraded in endosomes. Simultaneously, MHC class II molecules are produced in the ER by associating an α and a β chain with an invariant chain. The latter one is important for the correct folding of the growing class II chains, and inhibits an early peptide binding already in the ER.

Moreover, the aggregation of the invariant chain with the class II α/β dimer mediates a suppressive signal. This makes transport of the complex (which is inside a vesicle) into the cytoplasm, via the Golgi complex, possible. At last the invariant chain is responsible for the melting of the vesicle with the endosome containing the degraded antigen. After fusion of the endosome and the vesicle, the invariant chain is removed enzymatically from the groove of the class II molecule, where the degraded antigen fragment can now be bound. The complex of antigen and class II protein is then transported to the surface and presented to the TCR of helper cells.

Within the CD4$^+$ helper cell compartment, Th0, Th1, and Th2 cells can be distinguished. They can be identified by their pattern of cytokines released during an immune response. Th0 cells secrete preferentially anti-inflammatory IL-4 and proinflammatory IL-2 and IFNγ. Th1 cells are proinflammatory cells providing IL-2, IFNγ, and TNFα/β. Th0 and Th1 lymphocytes, along with the cytokines, mostly stimulate other T cells.

Th2 cells are able to produce IL-4, IL-5, IL-6, and IL-10. All of the latter cytokines are essential for B-cell differentiation and proliferation. It is still a matter of discussion whether these different T-helper cell types are really differentially determined cells, or whether they are morphologically identical cells at varying activation states.

2.4. NK Cells

The exact ontogeny of NK cells has not been entirely elucidated. They originate from the bone marrow and bear surface proteins belonging to both the T-cell lineage (CD2) and

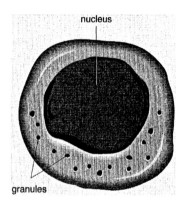

Figure 5. NK cells. NK cells are also described as large granular lymphocytes (LGL). They are larger lymphocytes with a high cytoplasm-to-nucleus ratio. The cytoplasm contains azurophilic granules.

the myeloid cells (CD11b, CD16). Morphologically, they are members of the large granular lymphocytes (LGL) containing azurophilic granules in the cytoplasm (Fig. 5).

NK cells produce several cytokines, like IFNγ and TNFα, and have a cytotoxic effect on infected or neoplastic cells. Cytolysis can be managed directly or mediated by antibodies specific to the target cells. For the antibody-dependent cellular cytotoxicity (ADCC), the CD16 (FcγRIII) molecule is the decisive structure on NK cells.

Cross-linking of target antigens by specific antibodies and binding of the Fc portion of the antibodies by the Fc receptor (CD16) of NK cells initiate this kind of killing. In contrast to T cells, NK cells are not MHC restricted. However, they interact in a different way with MHC class I molecules. If NK cells recognize their own MHC proteins on the surface of another cell, cytotoxic attack is inhibited. Based on the observation that the less class I molecules are expressed, the more vulnerable are target cells of different MHC to allogeneic (from another individual) NK cells, two hypotheses were introduced, as follows.

One model assumes that NK cells can bind to MHC molecules of a cell and if the class I proteins are not altered, an inhibitory signal is provided to the NK cells (altered self model). The second model postulates stimulatory molecules covering the MHC molecules in cases such as infection making the target sensitive to NK cell lysis (masking model). Both models clearly imply that the MHC plays an important role in NK cell function and ensures the protection of normal cells from lysis by NK cells. Latest investigations favor the altered self model, by showing that NK cells can actively recognize certain MHC proteins. However, a unique NK cell receptor has not been described.

In mice, the Ly49 molecule is able to bind distinct MHC proteins. Blockade of Ly49 on the NK cell or the appropriate class I molecule on the target cell inhibits a potential lysis. In humans, analogous molecules have been defined as all mediating a certain NK cell receptor function. For example, there is a correlation between the expression of the p58 molecule on NK cells and the recognition of HLA-C products.

The same is true for the binding of distinct HLA-B proteins by CD94 (Kp43) of the effector cell. In addition, NKB1 molecules can contact HLA-B51 and -B58. In all of these cases, an inhibition of lysis is possible by blocking either the recognizing molecules of NK cells or the corresponding MHC class I proteins of the target cells. Several MHC interacting molecules are coexpressed by single NK cells, which means that these NK cells own a repertoire of receptors to distinguish a pattern of MHC molecules.

It is assumed that NK cells—analogous to T cells—recognize antigen-induced alteration of the class I protein (groove), thus turning on cytotoxic functions. Even point mutations within the groove can be noticed by some NK cells, at least *in vitro*, emphasizing the high specificity of the receptors. In contrast to T cells, NK cells also exert innate immune function, despite the specific recognition mentioned above. NK cells do not need antigen sensitization prior to activation. Therefore, an effective immune response against bacteria, viruses, and tumors is immediately possible. On the other hand, there is no NK cell memory like that of T and B cells.

2.5. Cell Surface Molecules

Each cell presents a multitude of different molecules (also called *antigens* or *markers*) on its outer cell membrane. Most of them are essential for various functions like adhesion or signal transduction. Because each distinct cell population has particular tasks, they are equipped accordingly with a special molecule pattern. Vice versa, from this specific protein pattern the different cell populations can be identified with purified antisera or monoclonal antibodies raised against representative proteins of the appropriate cell type.

Antisera are produced by animals after challenging with the specific xenogeneic (= from another species) cell population. Monoclonal antibodies are immunoglobulins produced by a hybridoma, a fusion product from an antibody-secreting B cell and a B-cell tumor cell (myeloma cell), thereby getting the ability to produce antibodies from the B cell and immortality from the myeloma cell. Hybridomas can permanently secrete antibodies of one specificity *in vitro*. In contrast to polyclonal antisera, monoclonal antibodies have exactly one specificity and are therefore more useful in distinguishing different molecules on a single cell type.

Many of the monoclonal antibodies developed in laboratories all over the world are periodically sent to differentiation workshops where the antibodies are accurately tested and classified as to their specificity. Based on the results the antibodies are assigned a CD (cluster of differentiation) number. In this way the antibodies and their specificity, respectively, become standardized. The appropriate CD number is synonymous for the recognized molecule as well as for the recognizing antibody. For example, the CD3 antigen on T cells can be bound by CD3 antibodies.

Each cell expresses simultaneously variants of about 170 CD molecules defined so far. Often it is necessary to define several surface molecules of a cell for an exact identification. For example, CD20 is exclusively expressed on B cells, CD2 is present on NK and T cells, and CD3 is present on T-helper and T-cytotoxic cells. An incomplete list of the most important CD proteins is shown in Table 4.

2.6. Binding of Antigens

T-cell receptors bind their antigen specifically with the variable regions in the same fashion as antibodies. There are at least four different mechanisms for binding an antigen (Fig. 6). All of these mechanisms have in common bonds that are weak, not noncovalent, and only effective at a short distance. The nature of these bonds is described for the formation of antigen–antibody (Ag–Ab) complexes. However, the forces responsible for the binding of antigens by T-cell receptors are exactly the same.

Table 4
Antigens on Lymphocytes

	Cell type	Name/function
CD1	T-cell progenitors	Associated with β_2-microglobulin
CD2	T and NK cells	Receptors for LFA-3 and sheep red blood cells
CD3	T cells	TCR-associated signal-transducing complex of 5 chains
CD4	T-helper cells	MHC class II and HIV receptor
CD8	T-cell/NK-cell subpopulations	MHC class I receptor
CD11a	Leukocytes (broad)	α chain of LFA-1
CD14	Monocytes	LPS receptor
CD16	NK cells, monocytes, granulocytes	FcγRIII (Fc receptor III for IgG)
CD18	Leukocytes (broad)	β chain for CD11a, b, c
CD19	B and precursor B cells	Modulates B-cell responsiveness
CD20	B cells	Ca^{2+} channel, involved in B-cell activation and proliferation
CD21	B cell	EBV receptor, involved in B-cell activation and proliferation
CD25	Activated T, NK, and B cells	α chain (p55) of IL-2 receptor
CD28	T cells	B receptor (costimulator in T-cell activation)
CD32	Monocytes, granulocytes	FcγRII (Fc receptor II for IgG)
CD45	Leukocytes (broad)	LCS (leukocyte common antigen)
CD45RA	T-cell subpopulation, B cells, monocytes, granulocytes	Restricted LCA, resting T cells
CD45R0	T-cell subpopulation, B cells, monocytes, granulocytes	Activated (memory) T cells
CD56	NK and activated T cells	NCAM (neural cell adhesion molecule)
CD64	Monocytes, macrophages	FcγRI (high-affinity Fc receptor I for IgG)
CD69	Activated lymphocytes, monocytes	AIM (activation inducer molecule) involved in lymphocyte signal transduction
CD94	NK cells, T-cell subpopulation	Kp43: one of NK cell receptor equivalents with specificity for certain MHC molecules (HLA-B7)
CD122	NK cells, activated T and B cells	β chain (p75) of IL-2 receptor

Hydrogen bonds are caused by the different polarity of two molecules coming close together. The close distance allows the positively charged hydrogen (H^+) to form a bridge preferentially with molecules like COO^-, NH^-, and O^-.

Electrostatic bonds arise from the attraction of positively charged side groups of two molecules. An ionized amino group (NH_3^+) of one protein can be attracted by the carboxyl group (COO^-) of another one.

van der Waals' forces result from the fact that even in apolar uncharged molecules, single electrons are permanently moving, thus incidentally concentrating electrons in different parts of the molecular electron clouds which gives rise to dipoles for a short time.

Hydrophobic forces arise from the higher affinity of apolar molecules for each other than for polar compounds. For example, these forces join apolar oil drops together into a coherent layer in water (polar).

Normally, all of these chemical and physical bonds are involved simultaneously in antigen binding. Because of the weak nature of these bonds, a stable Ag–Ab complex is only achieved by an optimal complementarity between both participants. The strength of a single Ag–Ab bond—the sum of all attracting and repelling forces—is termed *affinity*. As already mentioned, a single immunoglobulin has at least two antigen-binding sites, and an antigen usually has several identical accessible epitopes. This means that there is more than one

hydrogen bonds

electrostatic
bonds

van der Waals'
forces

Figure 6. Antigen binding. The
bonds between antigens and antibodies or T-cell receptors, respectively, are comprised of several weak
noncovalent forces. The main binding
mechanisms are hydrogen bonds,
electrostatic bonds, van der Waals'
forces, and hydrophobic forces.

hydrophobic
forces

water exclusion

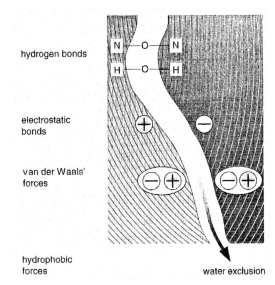

bond possible between an antibody and an antigen, with each bond possessing its own
affinity. The totality of all affinities between an antigen and the antibody is called *avidity*.
Basically, avidity is higher than the single affinities because it is very unlikely that all Ag–
Ab bonds will come loose at the same time. Thus, an antigen and antibody can remain
connected despite dissociation of single bonds (Figure 7).

2.7. Signal Transduction

Binding of the antibody is not only a simple marking of the antigen, but rather induces
conformational changes of the immunoglobulin. Subsequently, this event initiates all
further steps essential for an effective immune response. Immunocompetent cells bearing
Fc receptors are then enabled to bind the antibodies complexed to the antigen via this
receptor. The Fc receptor activated in this way then passes this signal onto associated
proteins. In the same manner, the TCR provides signals into the cells after recognizing the
specific antigen.

Usually, a T cell needs at least the simultaneous triggering of two independent
receptors (TCR + CD8, TCR + CD4) to be activated. The receptors (up to thousands per
cell) are not firmly localized in the cell membrane, but rather can be moved in the membrane
by components of the cytoskeleton after an event such as binding to the antigen. Subsequently, further receptors are concentrated at the site of interaction with the antigen, a
process called *capping*.

Signals, provided by receptor–antigen interaction, are transmitted to other intracellular
proteins that can activate or inactivate further molecules. Some of these molecules are
regulating proteins, turning genes on or off, thus for instance arranging the cell for
production of mediators. The IL-2 production of an activated T-helper cell is briefly outlined
in Fig. 8.

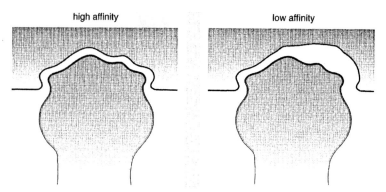

high affinity   low affinity

Figure 7. Binding affinity. An optimal affinity is guaranteed only by exact complementarity between epitope and paratope when the weak binding forces are supported by closest distance.

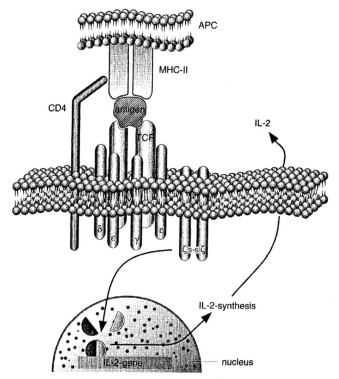

Figure 8. Signal transduction. This example shows an antigen presented to a T-helper cell by an APC via MHC class II molecule. The T-cell receptor (TCR) recognizes the antigen and delivers the signal to single chains of the TCR associated CD3 complex. CD4 functions as a coreceptor binding to the MHC class II molecule. In the end, transcription factors in the nucleus enhance the mRNA synthesis leading to an increased IL-2 production.

The antigen within the MHC class II molecule of the antigen-presenting cell is bound by the TCR. The subsequent conformational alteration of the receptor induces the activation of a protein tyrosine kinase (PTK) of the helper cell. Kinases can transfer phosphate residues to other molecules. The TCR itself does not have an intrinsic enzymatic activity but is closely surrounded by proteins of the CD3 complex which are receptive to signals from the TCR. Simultaneously, an associated PTK of the CD4 molecule that binds to MHC class II of the APC phosphorylates further cell metabolites. One of these phosphorylated intracellular molecules is phospholipase C, which converts phosphatidylinositol diphosphate from the cell membrane into diacylglycerol (DAG) and inositol triphosphate (IP_3). The DAG activates the next enzyme (protein kinase C), whereas the IP_3 causes intracellular stores to release calcium (Ca^{2+}). Free Ca^{2+} ions support synergetically DAG in activating protein kinase C.

These are the main events leading to a translocation of transcription factors into the nucleus where they bind directly to the DNA, triggering activation of the target gene (e.g., IL-2 gene), and thus instructing the cell to produce the appropriate cytokine.

2.8. Cytokines

As already mentioned, leukocytes can exert immunological functions by direct contact, but also by secreting cytokines. These soluble mediators are low-molecular-weight substances that play a regulative role in immune response. *Cytokine* is the generic term for all mediators used by immunocompetent cells for communication with each other. The purpose of this hormone-like communication is a well-balanced regulation of the immune function.

Cytokines are synthesized as required, or stored in intracellular granules to be released directly after reception of the relevant signal. Because of their very short half-lives, cytokines are only effective on cells in close proximity, ensuring a local limitation. All cytokines act via specific receptors that pass signals into the cells. The distribution of these receptors is responsible for the restriction of cytokine effects among particular cell types. Many receptors increase in number after stimulation, enabling determination as to the state of activation of cells. For this reason, such receptors are termed *activation markers* (e.g., IL-2 receptor).

According to the source of production, cytokines are also subdivided into *lymphokines* (produced by lymphocytes) and *monokines* (produced by monocytic cells). Another popular term is *interleukin* (IL) plus an appended number, independent from either source or mediated effects (IL-1, IL-2, …). In addition, many cytokines have trivial names (Table 5).

2.9. Complement

Complement is a group of serum proteins most of which are constitutively present as inactive preenzymes. This system consists of at least nine different proteins (C1 to C9) and is started by enzyme precursors after binding to membranes (e.g., of bacteria) or antibodies. During the next course, any following preenzyme is converted by limited proteolysis into its active form by preceding components, which are usually specific proteases. Products of proteolysis are a smaller peptide and a bigger fragment that binds to a membrane nearby, thereby becoming the next active complement component itself. Because each activated

Table 5
Cytokines

	Source	Function
IL-1	Monocytes, macrophages, fibroblasts	Proliferation of T and B cells, chemotaxis of neutrophils, macrophages, lymphocytes
IL-2	T cells	Growth factor for T, B, and NK cells, enhancement of NK activity
IL-3	T cells	Growth and differentiation of hematopoietic stem cells, growth of mast cells
IL-4	CD4$^+$ (Th2) T cells	Proliferation and differentiation of B and T cells
IL-5	CD4$^+$ (Th2) T cells	Proliferation of B cells and eosinophils
IL-6	CD4$^+$ T cells, monocytes, fibroblasts	Growth and differentiation of B cells and hematopoietic precursors
IL-7	Bone marrow stroma cells	Proliferation of B and pre-B cells, growth and early T cells, maturation of megakaryocytes
IL-8	Monocytes, fibroblasts, endothelial cells	Chemotaxis and activation of neutrophils
IL-9	CD4$^+$ T cells	Proliferation of preactivated T cells
IL-10	CD4$^+$ (Th2) T cells, B cells	Inhibition of IFNγ and IL-2 secretion, growth factor for T and B cells
IL-11	Bone marrow stroma cells	Shortens the G0 phase of hematopoietic stem cells
IL-12	T cells	Growth factor for T and NK cells, enhancement of NK activity
IL-13	T cells	Induces B-cell growth and differentiation, suppresses secretion of proinflammatory cytokines by monocytes/macrophages
IL-14	T cells	Proliferation of activated B cells, inhibition of Ig production by B cells *in vitro*
IL-15	Activated macrophages, monocytes	Effects are synergistic with IL-2
IL-16	CD8$^+$ T cells	LCF, lymphocyte chemoattractant factor for CD4$^+$ cells
IL-17	CD4$^+$ T cells	Enhances production of IL-6 and IL-8
IL-18	Monocytes/macrophages	Costimulant for Th1-like cells, IL-18 induces production of IFNγ and IL-2 and enhances NK activity
IFNα, IFNβ	Leukocytes, fibroblasts	Antiviral, activation of T and NK cells, up-regulation of MHC class I molecules
IFNγ	T cells, NK cells	Stimulation of IL-1 and IL-2 synthesis, induction of MHC class I and II proteins
TNFα	Monocytes, T cells, NK cells	Activation of phagocytes, induction of IFNγ, IL-1, IL-6
TNFβ	T cells	Activation of phagocytes, induction of IFNγ, IL-1, IL-6

complement component can activate several enzymes, the whole course exhibits the dynamic of a cascade. This cascade can be started directly, for example by bacterial membrane proteins (alternative pathway), or mediated by antibodies bound to an antigen (classical pathway). The main effects of the complement system are cell activation, cytolysis, and opsonization (making substances easier to bind to). In both classical and alternative pathways, the splitting of C3 is the central event of the complement cascade.

2.9.1. Alternative Pathway

The alternative pathway can be classified as a part of the innate immune response. Complement is permanently active at a low level. An intervening antigen only enhances the intensity of complement by disturbing passively controlled mechanisms. Pathogens like bacteria, fungi, and viruses, and other substances able to increase complement activity, are called *complement activators*.

The central complement protein, C3, is spontaneously hydrolyzed to form a $C3(H_2O)$ complex. This complex associates reversibly with factor B (an inactive serine protease). In this constellation, factor B is activated by factor D, thereby splitting factor B into two fragments, Ba and Bb. The smaller Ba fragment is released whereas the generated $C3(H_2O)Bb$ (initiating C3 convertase) converts C3 into C3a and C3b. The smaller C3a is set free and C3b associates with factor B. The latter is cut again by factor D to form C3bBb (amplifying C3 convertase) that is further stabilized by factor P (properdin). This C3bBbP complex can bind to additional C3b fragments bound to surfaces of activating substances. This aggregate acts as the C5 convertase of the alternative pathway (Fig. 9) mediating the splitting of a C5 molecule. This event provides the signal for the lytic reaction of complement.

2.9.2. Classical Pathway

The classical pathway of complement activation is part of the specific immune system because the binding of antibodies is essential for the initiation of the cascade (Fig. 10). The initiating molecule is C1, which consists of the C1q protein and the C1r2C1s2 tetramer. In blood, the free C1 is coupled to C1 inhibitor, suppressing the activation of the C1 complex. C1q can bind to the Fc portion of complement fixing antibodies. The particular structure of C1q enables this molecule to bind to the Fc fragments of two immunoglobulins in close proximity.

Because of the pentameric structure of the IgM molecule, there are always two Fc fragments near each other. However, the binding site for C1q is covered, thus inhibiting any contact. When the star-shaped IgM binds to an antigen, the molecule is forced into a crablike conformation. C1q is now not inhibited from binding, and one IgM antibody is able to initiate the cascade.

Statistically, there are about 1000 IgG antibodies bound to one cell before two IgG

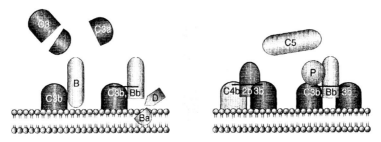

Figure 9. Alternative pathway of the complement cascade. Some substances (e.g., bacteria, viruses) cause C3 to spontaneously hydrolyze to $C3(H_2O)$, to which the inactive factor B can reversibly bind. On binding, factor B can be split by factor D into Ba and Bb. This $C3(H_2O)Bb$ complex represents the initiating C3 convertase. This enzyme splits C3 and the arising C3b associates with factor B, which is then divided itself by factor D. After stabilization by properdin (P), the amplifying C3 convertase C3bBbP is formed. By adding further C3b molecules, this complex serves as C5 convertase of the alternative pathway. Adapted from I. M. Roitt.

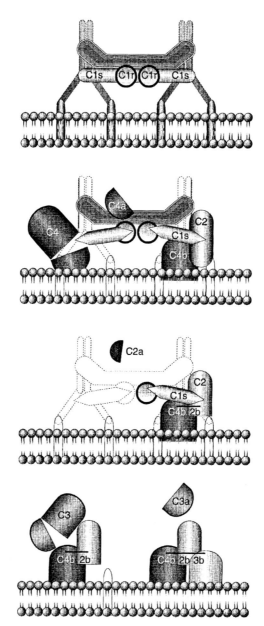

Figure 10. Classical pathway of the complement cascade. C1q fixes to antibodies that had bound their specific antigen. As a consequence, the activated C1s splits a C4 molecule. The larger C4b fragment complexes a C2 molecule, which thus becomes accessible to proteolysis by C1s. This produces a C4b2b complex (= C3 convertase of the classical pathway) that can divide C3 into C3a and C3b. The latter fragment associates with the C3 convertase to form a C4b2b3b complex (C5 convertase).

molecules come so close together that complement activation is possible. Subsequently, C1r and C1s are activated and the latter splits a C4 molecule into C4a and C4b, which forms a complex with C2. Thereby, C2 itself becomes sensitive to proteolysis by C1s. The resulting C4b2b fusion (= C3 convertase of the classical pathway) splits C3 into C3a and C3b. C3b then becomes part of the C4b2b3b complex (C5 convertase). From this point on, the reaction sequence of the alternative and classical pathways are identical to each other and lead to the lysis of target cells.

2.9.3. The Cytotoxic Reaction and the Membrane Attack Complex

The C5 convertase of the classical pathway (C4b2b3b) and the C5 convertase of the classical pathway (C3bBb3b), respectively, splits C5 into two pieces. The smaller C5a diffuses into serum and the C5b fragment attaches to the membrane. To this membrane-bound C5b the proteins C6, C7, and C8 associate in succession, thereby inserting C8 into the membrane. Afterwards, the generated C5b678 complex mediates the polymerization of several C9 proteins to form a tubelike aggregate penetrating the membrane (Fig. 11). The unhindered exchange of substances between the intra- and extracellular spaces finally causes lysis of the cell. The association of C5b678 with several C9 molecules is therefore termed the *membrane attack complex.*

Besides this killing component, some of the complement proteins have other functions. C3b and C4b can enhance the phagocytotic potential of monocytes and granulocytes, cells that have receptors (C3bR, C4bR) for these fragments. When C3b or C4b is bound to a particle or target cells, the effector cells can ingest protein-marked substances by the more effective receptor-mediated phagocytosis. Such coating with substances that make phagocytosis easier is called *opsonization.*

As already mentioned, during the second part of the cascade, the smaller soluble fragments C3a, C4a, and C5a are released. These peptides can mediate contraction of smooth muscle cells, permeability of blood vessels, and the release of histamine by mast cells, thus supporting immune functions. The single reactions are also found during anaphylaxis or hypersensitivity, a pathological overreaction of the immune system. Therefore, the fragments C3a, C4a, and C5a are also termed *anaphylatoxins.*

Figure 11. Lytic pathway. The C5 convertase divides C5 into C5a and C5b, which then associates with a nearby membrane. In succession, C6, C7, and C8 aggregate to the membrane-fixed C5b protein. This complex mediates the polymerization of several C9 molecules, which are then inserted tubelike into the membrane to form pores. The integrity thus disturbed by these pores, the cell dies.

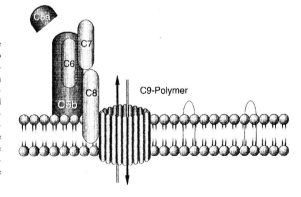

2.10. Immunological Defense

When an antigen gains entry into the body despite all barriers (skin, mucus), it is confronted by phagocytes. In the case of skin, these are the Langerhans cells, which take up the particles and digest them. Fragments of the antigen are brought to the cell surface of the phagocyte—the antigen is processed. The cells present the processed antigen together with MHC class II molecules to B and T cells with the corresponding specificity. After antigen contact, appropriate B cells react with proliferation and differentiation into antibody-producing plasma cells. T-helper cells, on the other hand, proliferate and produce cytokines that are needed by both cell types for an effective immune response. Released antibodies mark bacteria, viruses, or other foreign particles making them accessible for further immune mechanisms, such as complement activation or receptor-mediated phagocytosis.

Other mechanisms are performed in the case of viral infection or tumor control. Viruses consist nearly exclusively of genetic code and do not have their own metabolism. Therefore, they must rely on the metabolic pathways of the host cell. Usually viruses enter the cell via receptors. For the moment, they are safe there from recognition by antibodies. The viruses downregulate the metabolism of the cell and condition the cell to produce viral proteins, such as virus-specific enzymes, and envelopes instead.

For the infection of further cells, the viral genome is packed into envelope proteins and released from the host cell by exocytosis. Virus-encoded proteins are also integrated into the cell membrane where they can be noticed by antibodies. In addition, viral antigens can be recognized by T or NK cells when presented together with MHC class I proteins.

By these mechanisms the metabolic pathways for further viral replication are withdrawn and virus particles are set free, making them accessible to antibodies. Moreover, cytokines released by the involved cells during these processes regulate the whole immune response to enhance their efficiency. Tumor cells also alter the pattern of expressed membrane proteins so that they can be attacked by NK and T cells in the same manner as virus-infected cells.

There are different mechanisms to kill a target cell. The cells can be drenched with cytolytic enzymes, or the membrane can be destroyed by pore-forming proteins (perforin) that are released by cytotoxic cells. Last, but not least, the targets can be induced to dissolve their own nucleus—an event that is called *apoptosis*.

In contrast to NK cells, T and B cells develop a so-called memory after each antigenic stimulus. This means that a small part of the activated cell does not differentiate terminally, but rather remains dormant although specifically preactivated and in circulation for years. After a subsequent contact with the antigen, they can proliferate and provide a sufficient immune response much quicker than after the first challenge.

All of these defense mechanisms are almost continuously occurring in our bodies as we are constantly threatened by microorganisms and other pathogens. Ergo, an organism remains healthy only when the immune system is able to withstand the constant attacks.

2.11. Allergies

Allergies or hypersensitivities are basically quite normal immune responses, except they are dysregulated or misdirected. Antigens that are able to provoke hypersensitivity are called *allergens*. Usually these reactions are classified into four types of hypersensitivities according to their pathomechanisms.

2.11.1. Anaphylactic Hypersensitivity: Type I Reaction

When an antigen finds its way into the body, it is taken up by APC. These cells present the processed antigen on their surface to T cells, which thereby become activated to stimulate specific B cells for antibody production. In contrast to the normal immune reaction, which is identical up to this step, a Type I reaction is characterized by the secretion of mostly IgE isotypes. These antibodies bind to the Fcε receptor of basophils and mast cells. With this activation, these particular cells release vasoactive mediators such as histamine, serotonin, or leukotrienes that lead to characteristic features like swelling, flush, pain, hay fever, asthma, and anaphylactic shock.

2.11.2. Antibody-Dependent Cytotoxic Hypersensitivity: Type II Reaction

A Type II reaction is a response that is conditional on antibodies (IgM or IgG) specific for endogenous structures. The binding of the body's own antigens causes different kinds of cell damage. Granulocytes, monocytes, and NK cells recognize antigen-coupled antibodies via their Fc receptors. Triggered this way, the effectors attack the target cells and kill them. On the other hand, the antigen-bound immunoglobulin can initiate the classical complement pathway that leads to cell lysis by the cytotoxic membrane attack complex.

Actually, it should not be possible for the body's own components to be recognized by the leukocytes, because all potentially autoreactive cells are eliminated during education of the individual's immune system. However, autoreactivity is a quite frequent phenomenon, and cross reactivities are often responsible for autoimmunity. Cross-reactivity occurs when the body produces antibodies and promotes specific T cells to defend against an intruder antigen. Unfortunately, if this antigen has a certain similarity to endogenous structures, the specific receptors can bind them, too, even if it is with a lower affinity.

2.11.3. Immune Complex-Mediated Hypersensitivity: Type III Reaction

A Type III reaction is caused by immune complexes. These complexes consist of several antibodies cross-linked to each other by antigens. They emerge constantly from normal immune reactions and are normally taken away to the spleen or liver by cellular Fc receptors. There the complexes are metabolized and cleared. In the case of some autoimmune diseases, like chronic polyarthritis, a part of the complexes are deposited because of overproduction. Comparable to the events in autoimmunity, the uncleared immune complexes of Type III reactions can activate complement, granulocytes, monocytes, or NK cells leading to tissue damage.

2.11.4. Cell-Mediated Hypersensitivity: Type IV Reaction

In contrast to the above three types of hypersensitivity, a Type IV reaction becomes apparent after a delay of about 24 h. It is thus termed *delayed-type hypersensitivity* (DTH). The symptoms are mediated by the cellular compartment of the immune system and the reaction is antibody independent. After a second antigen contact, sensitized T cells proliferate and secrete mediators attracting macrophages to the site of antigen presentation where they release substances that cause tissue damage. Many hyperreactions against metals (e.g., nickel) are of the DTH type. Various such triggering substances are called *haptens* (incomplete antigens) because they couple with the body's own proteins and can only be recognized in those combinations by sensitized T lymphocytes.

Although allergic reactions underlie perfectly normal immune mechanisms, the reaction is disturbed causing pathological consequences. Moreover, there is usually not just one mechanism responsible for allergies but rather a combination of different types I to IV.

2.12. Glossary

ADCC   antibody-dependent cellular cytotoxicity. Target cells are lysed when coated with antibodies and effector cells bind to the Fc portion of the immunoglobulins.

Apoptosis   inducing a cell to commit "suicide" by fragmenting its nucleus.

B cell   lymphocyte that produces antibodies when differentiated into a plasma cell.

Clone   identical descendants of a common mother cell.

Colostrum   first milk that is secreted by the mammary gland in the first few days after birth.

Degranulation   release of cellular granules into the extracellular space.

Deletion   loss of a DNA segment by mutation or recombination

Epitope   part of an antigen to which an antibody or T cell receptor can bind with the appropriate paratope.

Fab fragment   fragment antigen binding, part of the antibody molecule that can bind the antigen.

Fc fragment   fragment crystallizable, fragment composed of the constant segments of the antibody's heavy chains. The Fc portion of an antibody can be bound by Fc receptors present on certain cells.

Hapten   substance that is not immunogenic on its own but is rather dependent on coupling to a carrier to provoke an immune response. Carriers can be exogenous or endogenous proteins.

Paratope   antigen-binding site of an adequate receptor (TCR or antibody) that binds to the epitope of an antigen.

3 Principles of Endocrinology

Thomas H. Schürmeyer and E. Jean Wickings

3.1. Introduction

Cells in multicellular organisms must adopt specialized functions. By forming organs, cells can take on specialist functions, which would not be possible in individual cells; functions such as the conversion of food sources to energy (digestion), locomotion (running, jumping, flying), and the processing of more complex information signals (sight, hearing, touch, thought), as well as their storage (learning, memory). The coordination of all of these specialized functions within an organism is essential, as is the exchange of information between all of the different organs and their regulation by the brain. The brain transmits messages by two efferent pathways, through the nervous and the endocrine systems. These

Thomas H. Schürmeyer • Department of Endocrinology and Metabolism, Psychobiological and Psychosomatic Research Center, University of Trier, D-54290 Trier, Germany.   E. Jean Wickings • International Center for Medical Research, CIRMF-BP 769, Franceville, Gabon.

two are closely linked, and some "glands" (the hypothalamus, posterior pituitary, and adrenal medulla) histologically are derived from the nervous system. The transmission of impulses through the nervous system proceeds by way of "neurotransmitters" that are in themselves classical hormones (e.g., noradrenaline). For further reading, see, for example, Bentley (1980), Brown (1994), and Williams (1996).

3.2. General Principles

3.2.1. Hormones

Historical Aspects. Hormones are chemically defined substances that are secreted by endocrine glands or glandlike cells and exert their effect at a distance from their point of origin. A substance is said to have an endocrine effect when removal of a gland or its destruction results in a loss of hormone effect, but which can be restored by replacement with the hormone itself or an extract of the gland.

M. Bayliss and E. H. Starling first employed the term *hormone* in 1902 for the as yet undefined secretin. The first therapeutic use of gland extracts had occurred some years earlier (Table 1).

The chemical structures of hormones began to be elucidated in this century (Table 2), and required heroic efforts; Butenandt extracted 1500 liters of urine in order to isolate 15 mg of the testosterone metabolite androsterone. R. Guillemin prepared just 1 mg of thyrotropin-releasing hormone from 5 million sheep hypothalami, and E. Rinderknecht and R. E. Humbel extracted plasma proteins from nearly 1 million liters of human blood in order to characterize the structure of insulinlike growth factor (otherwise known as somatomedin C).

Steroid and Thyroid Hormones. Hormones can be divided into several main groups according to their chemical structure. *Steroid hormones* are synthesized from cholesterol in the ovaries, testes, adrenal glands, and placenta. They are made up of four carbohydrate rings and can be categorized as estrogens, gestagens, androgens, gluco- and mineralocorticoids. Their chemical synthesis requires enzymes located in the endothelial reticulum and the mitochondria, which occur in differing activities according to the individual gland, so directing the synthesis of the various end products. Steroids are not stored after production, but are secreted directly after synthesis (Table 3). The active vitamin

Table 1
First Therapeutic Uses of Endocrine Gland Extracts

Year	Authors	Extract	Hormone	For treatment of
1889	C. E. Brown-Sequard	Testis	Testosterone	Sexual activity[a]
1891	G. R. Murray	Thyroid	Thyroxine	Hypothyroidism
1909	W. B. Bell	Posterior pituitary	Oxytocin	Postpartum bleeding
1911	J. Hofbauer	Posterior pituitary	Oxytocin	Inadequate labor
1913	R. van der Welden & F. Farini	Posterior pituitary	Adiuretin (vasopressin)	Diabetes insipidus
1921	F. G. Bantin & C. H. Best	Pancreas	Insulin	Diabetes mellitus

[a]This result could not be confirmed, as almost no testosterone is stored in the testis.

Table 2
First Discovery of the Chemical Structurse of Some Important Hormones or Their Metabolites

Year	Authors	Hormone	Action
1926	C. R. Harrington	Thyroxine	Main thyroid hormone
1937	T. Reichstein	Deoxycorticosterone	Adrenal steroid precursor
1943	C. H. Li, G. Sayers	ACTH[a]	Pituitary hormone directing adrenal function
1951	V. du Vigneaud	Oxytocin	Labor-inducing hormone of the posterior pituitary
1954	S. A. Simpson	Aldosterone	Main adrenal mineralocorticoid
1955	F. Sanger	Insulin	Main pancreatic hormone
1969	R. Guillemin	LHRH or LH?	Central control of gonadal function
1971	C. H. Li	GH	Pituitary growth hormone
1977	C. H. Li	β-LPH	Precursor of β-endorphin
1983	S. Shibahara	hCRH	Human ACTH-releasing hormone

[a]Independent discovery by two authors in the same year.

D metabolite, $1,25(OH)_2D_3$, a renal hormone, has a chemical structure similar to that of a steroid.

Thyroid hormones are characterized by two iodized carbohydrate rings. Their mode of transport, receptor activation, and structure resemble those of the steroids.

Peptide Hormones and Catecholamines. *Peptide hormones* are amino acid chains the length of which may be only a few residues or up to several hundred residues. Their stability and function depend on their three-dimensional structure and on the number and place of attached sugar residues. In contrast to steroids, peptide hormones usually have a short half-life of a few minutes. They are secreted in episodic fashion and bind to receptors on the target cell surface, transmitting their signal through activation of "second messenger" systems rather than by influencing gene transcription directly. These second messengers may be intracellular enzymes or electrolyte channels in the cell membrane. Peptide hormones are catabolized by proteolytic enzymes.

Table 3
Plasma Concentration, Production Rate, and Metabolic Clearance Rate of Various Steroid Hormones

Hormone[a]		Plasma concentration	Production rate (mg/24 h)	Metabolic clearance rate (liters/m^2/24 h)
Estradiol	M	<40 gm/ml	<0.1	1000–1500
	F	>40 pg/ml	0.06–0.2	1000–1500
Progesterone	M	<0.5 ng/ml	<0.6	2000–3000
	F(I)	>1.0 ng/ml	<2.0	2000–3000
	F(II)	>6.0 ng/ml	15–40	2000–3000
Testosterone	M	>3.0 ng/ml	5–10	400–800
	F	<1.0 ng/ml	<0.3	200–400
Cortisol[b]		20–250 ng/ml	8–25	200–300
Aldosterone		40–180 pg/ml	0.08–2.0	1000–2000
Dehydroepiandrosterone[b]		2–25 ng/ml	2–15	1000–1500
Dehydroepiandrosterone sulfate		<4000 ng/ml	10–20	8–22

[a]M, male; F, female; I, follicular phase, and II, luteal phase of the menstrual cycle.
[b]Diurnal phase.

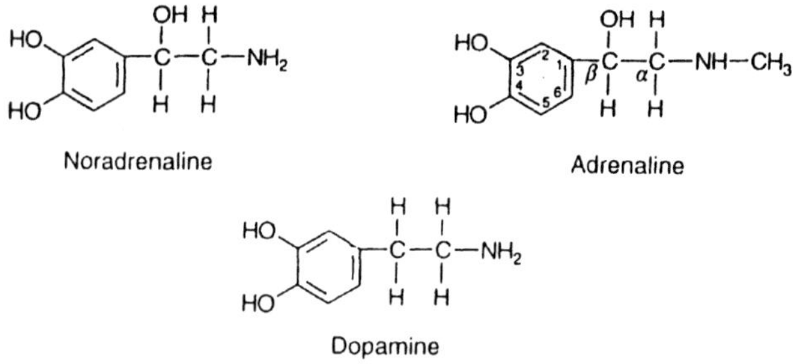

Figure 1. Chemical structures of the catecholamines noradrenaline, adrenaline, and dopamine.

The *catecholamines* adrenaline, noradrenaline, and dopamine are derivatives of the amino acid phenylalanine and synthesized via tyrosine and dopa (Fig. 1), and they have a dual function as hormone and as neurotransmitter, as do many peptides. They have a short half-life and exert their action via membrane-bound receptors rather than by any direct effect on the genome, in a similar manner to that seen for peptide hormones.

Transport Proteins.   The main steroid and thyroid hormones, and also vitamin D, circulate in blood bound to transport proteins (Table 4). The exact function of these proteins, which are synthesized in the liver, is unknown but may facilitate a more economical mode of hormone production, an easier diffusion of the hormone from the gland of origin, or a more efficient targeting of the hormone toward specific tissues. Drugs (e.g., hormonal contraceptives) and disease can affect the binding protein production, and thereby alter the levels of free, biologically active hormone in the circulation. It is only this free fraction of a hormone that is responsible for the hormonal effect at a target cell, and the levels of free hormone in circulation can be deduced by measuring the level of binding protein.

Catecholamines and peptide hormones are released from the secretory cells by exocytosis. They have no transport protein, for they require an episodic mode of action, and any protein binding would neutralize the pulsatile mode of transmission of information. Occasionally the protein binding of peptide hormones may have a specific function, as is probably the case for corticotropin-releasing hormone (CRH)-binding protein, which protects the maternal organism during pregnancy from the effects of the large quantities of CRH secreted by the placenta.

3.2.2. Hormone Receptors and Hormone Action

Intracellular Receptors.   Receptors for steroid hormones, vitamin D, some insect hormones such as ecdysone, and plant growth regulators occur in the cell cytoplasm. After binding the hormone, the hormone–receptor complex is transported into the cell nucleus where it binds to DNA, thereby influencing the transcription of specific genes via the genomic hormone regulatory elements (HREs), which have a common nucleotide sequence. The function of these HREs resembles closely that of the viral "enhancer" elements, which regulate viral replication. These steroid receptors are very long proteins, which can be divided into six domains, starting at the N-terminus. Sections A, B, D, and F are variable regions, depending on the different classes of receptor, whereas the C and E domains are very similar. The C domain is comprised of 65 to 70 amino acids, and incorporates two zinc

Table 4
Steroid and Thyroid Hormone Binding Proteins

Binding protein	Hormone bound
Thyroxine-binding globulin (TBG)	Thyroxine, triiodothyronine
Thyroxine-binding prealbumine	Thyroxine, triiodothyronine
Sex hormone-binding protein	Estradiol, testosterone
Cortisol-binding hormone (CBG)	Cortisol, progesterone

atoms to form the so-called "zinc finger," which binds the receptor to DNA. There is a considerable homology in this region between hormone receptors and the tumor-enhancing, translation products of some oncogenes. There is less similarity between the various E domains of the different steroid receptors. Whereas the steroid hormone receptors are located in the cell cytoplasm, thyroid hormone receptors are always to be found in the cell nucleus. The translation product of the v-*erb-A* viral oncogene is almost identical to the human thyroid hormone receptor, which forms one class of hormone receptors with those for steroids.

Membrane-Bound Receptors. In contrast to the intracellular steroid-receptor families, receptors for peptide hormones and many neurotransmitter proteins are located in the cell membrane, with the C-terminal portion, which forms the specific hormone-binding site, protruding from the cell surface. Adjacent to the extracellular hormone-binding site is an approximately 20 amino acid long section, which is anchored in the cell membrane, and finally the intracellular N-terminal section, which contains an enzyme cleavage site and is responsible for transmitting the hormone signal to the interior of the cell. The extracellular section of the receptor also has a cystine-rich portion, which causes polymerization of hormone–receptor complexes, a so-called "patching," resulting in the internalization of these complexes where they are deactivated. The intracellular cytoplasmic portion of the receptor activates cellular messenger systems, such as adenylate-cyclase or phospholipase C, via a G protein. The amino acid sequence of this part of the receptor is common to the various peptide hormone receptors.

The different receptor families are characterized by their midportion, which may traverse the cell membrane once or several times. Receptors for insulin, growth hormone, and various lymphokines and growth factors (e.g., epidermal growth factor) belong to the first family with a single passage through the cell membrane. This class of receptor acts via activation of a G protein, but in addition is also internalized into the cell by endocytosis and can exert an effect directly on the cell nucleus.

Receptors for the neurotransmitters GABA, glycine, stimulatory amino acids, and the nicotinic acetylcholine receptor traverse the cell membrane four times. They have no intracellular second messenger system, but associate to act as ion channels for potassium and chloride, for example.

Receptors for the muscarinic acetylcholine receptor, β-adrenoreceptor, also various neuropeptides as well as the light receptors in the retina have a sevenfold passage across the membrane. These receptor types are linked to the G protein and activate the classical second messenger systems, adenylate-cyclase and phospholipase C, as well as acting as ion channels. Three G protein subunits form the G protein complex: the γ subunit, which anchors the complex in the cell membrane; the α subunit, which binds the receptor, also GTP and acts as a GTPase; and the β subunit, which activates the individual second messenger system.

Receptors with 24 passages of the membrane act as ion channels (e.g., for sodium), and their activating ligands are presumed to be endothelins.

The different intracellular receptor transduction systems are characterized by their second messengers. Cyclic AMP (cAMP) is the product of the adenyl cyclase enzyme system, localized on the internal surface of the cell plasma membrane and which is activated by the G protein. cAMP activates a host of cellular enzymes as well as gene expression. For example, after the hormone glucagon binds to its receptor on the liver cell, cAMP is produced, which in turn stimulates those enzymes implicated in the breakdown of glycogen so that glucose is released and blood sugar levels rise. If the phospholipase C system is activated rather than adenyl cyclase, two second messengers are produced: diazylglycine and inositol triphosphate (IP-3). Diazylglycine activates calcium- and phospholipid-dependent cellular enzymes and also certain genes. IP-3 mobilizes intracellular calcium, which then activates other enzymes, including cAMP phosphodiesterase, which deactivates cAMP. Such is the close interrelationship between the cAMP and IP-3 pathways, which mutually influence each other.

Receptor Kinetics.   Once the membrane-bound receptor has been activated by binding its hormone, it is internalized by the cell and taken into the cytoplasm where it is either degraded or taken up by the Golgi apparatus to be recycled as a new cell receptor. The new production of receptors or their regeneration is a time-dependent process with only a limited capacity. If a high concentration of hormone is continuously presented to the cell receptors, they rapidly become saturated and internalized, so that no free receptors are available on the cell surface for activation in response to hormone binding, resulting in a "desensitization" of the cell. To avoid this situation, the physiological stimulation of most membrane-bound receptors occurs not in a continual fashion but episodically, in response to pulsatile hormone secretion, thus giving the receptors a chance to regenerate themselves. Some therapeutic applications of hormones secreted episodically can override this natural regeneration process, either by being given as a depot injection or by continual release. For example, if LHRH is given in a long-acting form, a short-term stimulation of the pituitary–gonadal axis occurs in response to the initial stimulation of the pituitary receptors, followed by a long-term suppression or desensitization of the pituitary. In this state, neither the LHRH analogue nor any endogenously produced LHRH can activate its receptor, leading to a "downregulation" of the system. This suppression can also have repercussions at the level of the central nervous system (CNS).

Hormone Action.   Hormones have four principal actions. They may influence the hormone secretion of other endocrine organs, as is the case of the hypothalamic releasing hormones, which stimulate the release of pituitary hormones, and the pituitary hormones themselves. The neurotransmitter catecholamines and steroid hormones also exert a feed-back effect on the production of their respective tropic hormones. A second hormone action is the influence on energetic balance, such as the effect of thyroid hormones and catecholamines. Third, hormones affect the tonus of smooth muscles (e.g., oxytocin). Fourth, hormones act on the permeability of biological membranes, affecting ion, water and metabolite exchange (e.g., adiuretin, vasopressin, aldosterone, vitamin D).

3.3. Glands of the Endocrine System

Only a review of the most important endocrine organs and their function will be given here (Fig. 2).

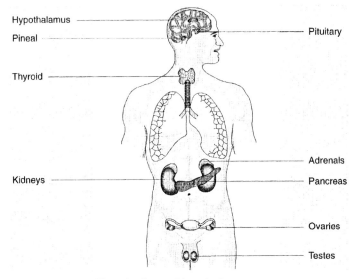

Figure 2.   Organs of the endocrine system.

3.3.1. Pineal

The exact function of the human pineal gland is not fully understood, although there are a number of unproven hypotheses concerning its role in hypothalamic and pituitary function, in the sleep–wake cycle, and in certain psychological functions. In some animal species, the pineal gland is known to act as a "third eye," which transmits information to other centers on the circadian light–dark cycle and stimulates changes in seasonal pelage coloration and reproductive function.

The main secretory product of the pineal is melatonin (Fig. 3), although it also secretes other protein hormones whose function in humans is unknown. Melatonin secretion is suppressed in humans by light impulses, which pass from the eye by means of a neural connection to the pineal. In some animals there are light receptors in the pineal itself. There is a marked daily secretion pattern of melatonin in early childhood, with high levels of plasma melatonin being secreted at night and low levels during the day. This period of childhood between 1 and 3 years corresponds to a time when other circadian rhythms are being established. There is no change in the size of the pineal gland during childhood, so that the size in relation to body size is continually decreasing during development, and the same is true for melatonin secretion, so that circulating levels of melatonin are 80% lower after puberty than in young children. Every third person shows signs of calcification of the 5 mm

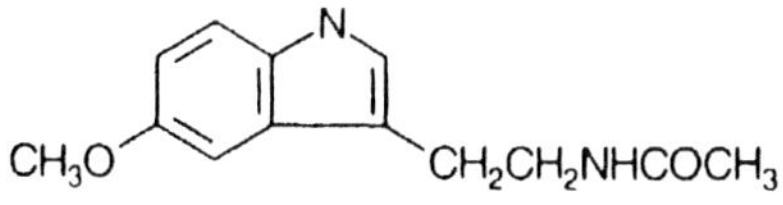

Figure 3.   Chemical structure of melatonin.

long pineal by the age of 18 years. Light causes a 60% reduction in melatonin secretion in normal subjects and also in those blind persons in whom the neural connection between the retina and the optic cortex has been severed but not the connection between the retina and the pineal.

3.3.2. Hypothalamus

Situated below the third ventricle, the hypothalamus is a small brain area that contains several hundred nerve cell nuclei (Fig. 4). These nerve cell nuclei secrete releasing hormones into the portal venous system, which stimulate pituitary function. The secretory function of these hypothalamic cells is regulated through a neuronal network, which connects the hypothalamus with other brain areas assimilating external or internal signals. For example, information reaching the brain from the eye indicating a threatening or dangerous situation influences the release of hypothalamic hormones, which in turn regulate the production of pituitary and adrenal "stress" hormones. The main hypothalamic releasing hormones are listed in Table 5. With the exception of dopamine, all releasing hormones are small peptides, which stimulate pituitary cells. Dopamine is a catecholamine, which suppresses pituitary prolactin secretion. Disturbances of hypothalamic function through injury or tumor in this region of the brain result in the loss of anterior and posterior pituitary function. The exception to this is prolactin secretion, which is controlled by the suppressive effects of dopamine, and hence any disruption in dopamine action will result in an increase in prolactin release.

3.3.3. Pituitary

The human pituitary weighs about 0.7 g and lies in the sella turnica, a bony depression in the sphenoid at the base of the skull, in front of the optic chiasma above the pituitary fossa behind the nose. Its existence has been known for about 2000 years although its function and

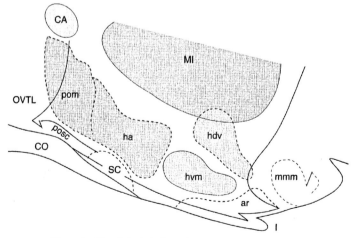

Figure 4. Hypothalamic brain nuclei in the rat (paramedial view).

Table 5
Hypothalamic Releasing Hormones
and Their Corresponding Pituitary Hormones and Target Organs

Releasing hormone	Pituitary hormone	Target organ
Anterior pituitary		
Corticotropin-RH (CRH)	ACTH	Adrenal cortex (cortisol)
	β-LPH, β-endorphin	?
	MSH	Skin melanocytes
Thyrotropin-RH (TRH)	TSH	Thyroid (thyroxine)
Growth hormone-RH (GHRH)	GH	Liver (somatomedin C)
Luteotropin-RH (LHRH)	LH	Gonads (estradiol, testosterone)
	FSH	Ovaries, testes
Dopamine	Prolactin	Mammary glands
Posterior pituitary		
	ADH (vasospressin)	Kidneys
	Oxytocin	Uterus, mammary glands

the consequences of its malfunction were only recognized in this century. The two lobes of the pituitary have different embryonic origins; the posterior lobe is an extension of the paraventricular and supraoptic brain nucleus of the hypothalamus, and therefore a part of the brain. The anterior pituitary originates from the pharynx, separating from it during embryonic development and migrating to the posterior pituitary structure.

The hypothalamus and pituitary are linked via the pituitary portal blood system (Fig. 5). The superior pituitary arteries feed a capillary network at the base of the hypothalamus in the region of the median eminence, from where blood flows through small veins of the pituitary stalk in a second, consecutive capillary network. From this point, blood flows back through the larger veins to the heart. It is, therefore, possible that small amounts of the hypothalamic releasing hormones can be released into this complex portal system and can reach the pituitary undiluted by the main blood circulation, so exerting their stimulatory effect on the pituitary. In contrast, the pituitary hormones are released in high enough concentrations that dilution in the main venous drainage system and passage through the lungs does not affect their capacity to reach and stimulate their target organs and tissues.

The anterior pituitary amplifies the action of the hypothalamus, secreting at least nine hormones, which stimulate and regulate the function of other glands (Table 5). The most important hormones of the anterior pituitary are thyroid-stimulating hormone (TSH), luteinizing hormone (LH), follicle-stimulating hormone (FSH), growth hormone (GH), prolactin, and adrenocorticotropic hormone (ACTH).

Signals are passed down the "hypothalamus–pituitary–end organ" axis in a discrete digital fashion similar to that employed by nerve cells, i.e., to depolarize or not. The alternative analogue form of stimulation results in a continual signal, which may vary in intensity between fixed limits, reaching the target organ (Fig. 6).

The pituitary hormones LH, TSH, and ACTH regulate other glands; GH acts directly on the metabolism of certain tissues and stimulates the formation of IGF-1 (insulin growth factor-1, or somatomedin C) in the liver. Factors influencing the secretion of GHRH-dependent GH include sleep, physical activity, stress, and a low blood sugar level. Sleep and stress-related situations also stimulate the release of prolactin, a peptide hormone, which stimulates mammary secretions from tissues already sensitized by estrogen action. The stimulation of lactation occurs via neural pathways, which are activated by the infant

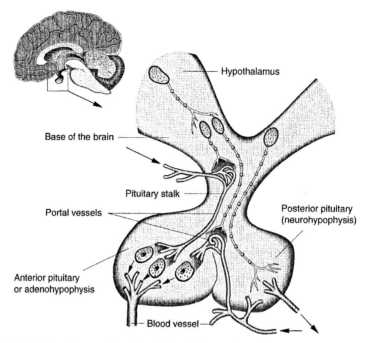

Figure 5. The pituitary forms a prolongation of the hypothalamus at the base of the brain. The neurohypophysis or posterior pituitary functions only as support tissue for the attachment of the hypothalamic nerve endings, through which hormones are secreted into the bloodstream. In contrast, the adenohypophysis or anterior pituitary is a functioning gland, which secretes hormones in response to specific stimulation of the pituitary cells by the nerve cells of the hypothalamus. The hypothalamic hormones or releasing factors are released into the portal blood system connecting the hypothalamus with the pituitary.

suckling at the nipple and which transmit this signal via the hypothalamus to the pituitary. This function of prolactin is found only in mammals; prolactin is also found in amphibians, reptiles, and birds where it regulates electrolyte balance, growth, reproduction, and lipid metabolism.

The secretion of ACTH involves the production of a long precursor molecule, proopio-melanocortin (POMC), which is then broken down into several peptide fragments, including ACTH. β-Lipoprotein (β-LPH), a peptide with a relatively long half-life, is released simultaneously, and is converted to β-endorphin, a peptide with morphinelike properties (see Chapter 10). Hence, ACTH, β-LPH, and β-endorphin are always secreted together, although the exact function of the latter two peptides outside of the CNS is not known. Psychological, neurological, and immunological actions have been variously suggested for these molecules. A differentiation must also be made between the pituitary action of these peptides and the effect of POMC-derived signals within the CNS. POMC-derived fragments, including ACTH(4–10), various endorphins and melanocyte stimulating hormone (MSH), enkephalins and CLIP (corticotropinlike intermediate lobe peptide) are produced in the brain and by animals, including humans, that lack the intermediate lobe of the pituitary.

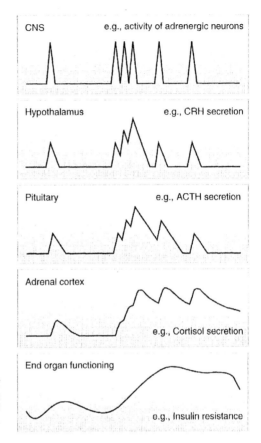

Figure 6. An example of the conversion of digital signals into analogue form following transmission of signals to subsequent organs of the hypothalamus–pituitary–end organ axis.

All of these products may have specific actions within the CNS influencing mood, memory, and behavior.

The hormones of the posterior pituitary, adiuretin (ADH) and oxytocin, are synthesized in the hypothalamus and transported via nerve cells to the posterior pituitary where they are released into the bloodstream. ADH acts on the renal glomeruli causing a back absorption of fluid and is essential in the regulation of fluid balance; its secretion is influenced by loss of fluids (hypovolemia) or an increase in osmolarity, e.g., in sodium ion concentration. Oxytocin stimulates contractions of the uterus, thus accelerating birth and delivery of the afterbirth, and of the mammary gland, thus inducing lactation. In animals within the CNS oxytocin may influence social and mating behavior.

If the anterior pituitary is damaged by injury or by tumor growth, then not only is the function of the pituitary lost, but also that of the pituitary-dependent target organs. Decreased LH and FSH secretion is not life threatening, causing a secondary hypogonadism. Loss of GH secretion in adults influences lipid metabolism, muscle strength, and general well-being. If GH loss occurs in childhood, GH replacement therapy is mandatory until puberty to avoid dwarfism. In contrast, reduced TSH secretion results in a serious secondary reduction in thyroid function within a few weeks, which can be reversed by appropriate replacement treatment. The most serious consequence of the loss of pituitary function is the effect on ACTH secretion and the resulting loss of adrenal function, which becomes life threatening within several days if not treated. As adequate treatment with ACTH itself is very difficult, because of its pulsatile release pattern, the patient is usually given a dose of cortisol (20–30 mg/day orally) sufficient to cover the normal daily production rate. This compensates for the loss of endogenous cortisol production without producing any undesirable side effects. The general well-being of a patient receiving only cortisol replacement is not compromised by the loss of the other POMC-derived peptides, which are lost along with ACTH, and points against a basic requirement of the organism for β-LPH and β-endorphin outside the CNS.

Should the posterior pituitary be damaged, the direct secretion of ADH from the hypothalamus in part compensates for its function. In the case of injury, resulting from a tumor or bleeding into this area, to the area of hypothalamic secretory activity a potentially fatal situation arises because of ADH loss. There is a rapid loss of large quantities of fluid, resulting in diabetes insipidus, as there is no antidiuretic action of ADH on the kidneys. Giving large volumes of fluid (up to 40 liters/day) can effect treatment in the acute phase, and long-term treatment with an ADH analogue, desmopressin, can then be instituted in the form of nose drops.

Tumors may occur in the anterior pituitary region, but rarely in the posterior region, which are nearly always benign. They can, however, cause pressure on the optic nerve, so causing a reduction in the visual field or even blindness. Such tumors occasionally produce large amounts of hormones, resulting in specific clinical disease. The most common of these is the prolactin-secreting tumor (prolactinoma), which causes galactorrhea and menstrual cycle disturbances in women and potency problems in men. GH-producing tumors in adulthood cause acromegaly, whereby the facial features become distorted, and the hands, feet, nose, ears, tongue, and internal organs are all enlarged. Glucose metabolism is also affected. ACTH-secreting tumors are usually small, reaching only a few millimeters in diameter, but lead to Cushing's disease. LH-, FSH-, and TSH-secreting tumors are very rare.

3.3.4. Thyroid

The thyroid is the largest endocrine gland in humans, reaching nearly 20 to 30 g, and is the major regulator and coordinator of metabolism in various tissues. It has its embryonic origin on the floor of the pharynx, from where it migrates during development to the throat to form two lobes one on each side of the windpipe, which are joined by a small isthmus. The gland itself is formed of thousands of small (<1 mm) follicles, which contain a colloid secreted by the single layer of epithelial cells into the central antrum of the follicle.

In contrast to some animal species like hamsters, frogs, and fish, humans can only increase or decrease their basal metabolic rate within a small, fixed range (e.g., during infections causing fever episodes, cold, or fasting). The role of the thyroid is, therefore, to maintain and coordinate the metabolic activities of the different organs. In those species that briefly increase their metabolic rate severalfold (e.g., during the development of tadpoles), or that adapt to a change in their environment (e.g., hibernation in certain mammals), the thyroid plays a different role. In humans it also has an important function during fetal development and during childhood for growth and maturation, especially of the CNS. Congenital hypothyroidism, or cretinism, shows the classical clinical status of dementia, short stature, muscular hypotrophy, and delayed pubertal development.

In adults the thyroid controls and coordinates basal energy turnover through a direct action on metabolically active organs such as the liver, and has a permissive role on the function of other hormones. In other words, the action of catecholamines or glucocorticoids, for example, is only complete in the presence of adequate thyroid hormone secretion. The most important permissive role is to sensitize the organism toward the action of the stress and circulatory hormones, adrenaline and noradrenaline. Any increase in energy consumption or heat production requires a concomitant increase in oxygen intake through dilation of the bronchus and increase in respiration rate. It also causes an increase in the mechanism for heat and energy distribution throughout the organism by increasing the heart rate and cardiac capacity. Lastly, there is an increase in the rate of heat loss through dilation of the skin capillaries with accompanying increase in skin temperature, and an increased energy intake, which is registered as increased appetite and higher intestinal motility. Any test of neural reflexes in this situation will demonstrate a more rapid response, and the renal action of aldosterone and ADH is enhanced, so that sodium and water are retained, leading to an increase in blood pressure. In response to all of these altered functions, the energy depots of the organism will be depleted as a result of catabolism, i.e., protein from bone and muscle, as well as lipids from fatty tissue are lost.

Regulation. The thyroid is regulated by the action of the pituitary proteohormone, TSH. This hormone is composed of two approximately 100-amino-acid-long chains. The α chain varies across species, and in humans, the α chain is identical to that of LH. This α chain contains a disulfide bridge, also sugar residues (mannose, glucosamine, galactosamine), which are linked to the protein chain through the amino acid asparagine. These structures are responsible for the three-dimensional character of the hormone molecule and for its biological function. The β chain of TSH is responsible for the regulation of the thyroid, but exerts its biological activity only in the presence of the α chain. TSH has a biological half-life of 30 to 90 min and is secreted in an episodic fashion. The production of TSH is stimulated by the hypothalamic releasing factor TRH, a tripeptide. The hypothalamic hormones somatostatin and dopamine exert a suppressive effect on TSH release, but the mechanism of action is as yet unknown. A feedback loop exists whereby the thyroid products of TSH stimulation regulate the release of TRH and TSH (Fig. 7). At the local level of the thyroid, a large excess of iodine results within a few days in the suppression of thyroid hormone production. This so-called Wolff–Chaikoff effect is of pharmacological interest only, as it has no physiological basis.

Hormone Synthesis. The thyroid hormone thyroxine is synthesized from the amino acid tyrosine and iodide ions (Fig. 8). Thyroid cells act as iodine traps, in that they absorb

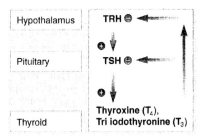

Figure 7. The hypothalamic–pituitary–thyroid feedback loop.

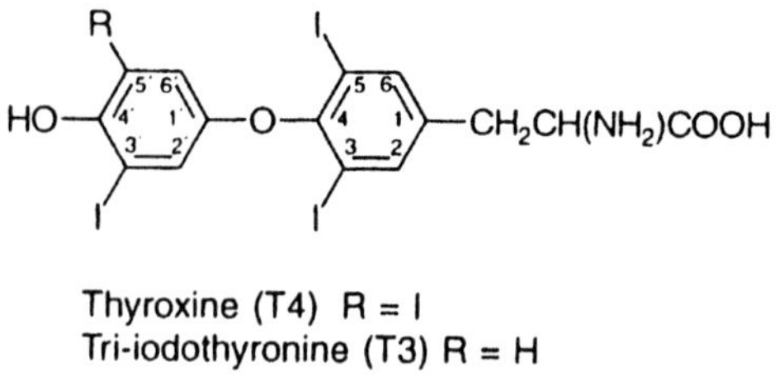

Thyroxine (T4) R = I
Tri-iodothyronine (T3) R = H

Figure 8. Chemical structures of thyroxine and triiodothyronine.

iodide from the blood very effectively, enriching the intracellular concentration up to 300- to 400-fold. Other tissues, such as the pancreas, the cerebral plexus choroidea, the stomach epithelium, and the mammary glands, can also take up iodine, but with a lower efficiency. The thyroid epithelial cells secrete a large protein, thyroglobulin, into the follicle. The tyrosyl residues are iodinated by the thyroid peroxidase on the surface of the intrafollicular microvilli, and two diiodo- or monoiodo-tyrosine molecules are coupled to T_4 or T_3, in which form they are stored. As required, the iodinated and transformed thyroglobulin is packaged in vesicles or colloid droplets and absorbed by pinocytosis by the microvilli of the basal membrane into the thyroid cells from the follicular lumen. The internalized droplets are fused with lysozymes, which break down the thyroglobulin. The T_3 and T_4 so released then diffuse into the bloodstream. All steps of thyroid hormone synthesis are stimulated by TSH, in particular the pinocytosis of thyroglobulin from the follicular lumen.

Hormone Transport.   The absolute ratio of T_4 to T_3 in circulation is 60:1, but because 99.95% of T_4, as opposed to 99.5% of T_3, is bound to transport proteins, the effective ratio of free T_4 to T_3 is 6:1. Of the protein-bound T_3 and T_4 fractions, approximately 60% binds to thyroxine binding globulin, 30% to thyroxine-binding prealbumin, and the remainder to albumin. The synthesis of these transport proteins occurs in the liver and is regulated by the suppressive influences of androgens and glucocorticoids and the stimulatory effect of estrogens. Hence, thyroid hormone levels increase during pregnancy. Women taking estrogen-containing oral contraceptives also experience an increase in the concentrations of protein-bound T_4. The free fraction is, however, not altered and hence the high levels of total thyroid hormones measured in such women are not associated with any physiological effect on the organism. Certain drugs do affect the binding of thyroid hormones to the transport proteins, such as those used in the treatment of epilepsy (diphenylhydantoin) and of infections (e.g., phenylbutazone). The overall levels of T_3 and T_4 remain unchanged, but the active, free fraction is elevated, resulting in the patient showing symptoms of hyperthyroidism.

Because of the high degree to which the thyroid hormones are bound to proteins, they have a low turnover rate, with only about 10% of T_4 and 50% of T_3 being broken down each day. They have correspondingly long half-lives; T_4 about 7 days and T_3 about 1.3 days. The breakdown of thyroid hormones occurs through dehalogenation in the liver and kidneys. A considerable proportion of thyroid hormones is sequestered in an enterohepatic circulatory loop, similar to that for estrogens. The hormone is esterified with glucuronate or sulfate in the liver and excreted in bile. Intestinal bacteria hydrolyze these bile salts, so that approximately 50% of the hormone is reabsorbed. Many drugs can influence this enterohepatic circulation, as can intestinal diseases and eating habits.

Hormone Action. T_4 must be dehalogenated to T_3 in order to exert its hormonal effect, as T_3 is bound to specific intranuclear receptors with a tenfold higher efficiency than T_4. No activation of the hormone–receptor complex is required, in contrast to steroid hormones. Receptor concentrations per milligram DNA are markedly high in such metabolically active tissues as the liver, kidneys, heart, and brain, also in the hypothalamus, but lower in the spleen and testes. Binding of the hormone–receptor complex to DNA causes an increase in RNA synthesis after 4 to 24 h, followed by the synthesis of specific proteins. Thyroid hormones do not act rapidly, but their effect is prolonged reaching a maximum after several days.

3.3.5. Parathyroid Glands and Calcium Metabolism

Adult organisms contain about 1.2 kg calcium, of which 99% is in the form of bones. Daily calcium turnover is approximately 700 mg, with absorption through the intestines and excretion via the intestines, kidneys, and sweat. The regulation of calcium turnover and the maintenance of serum calcium levels are effected by the parathyroid hormone (PTH), calcitonin produced by the C cells of the thyroid, and 1,25-dihydroxycholecalciferol, the active form of vitamin D produced in the kidneys. Calcitonin acts on bone mineralization, stabilization of membranes, the activity of various enzymes, and the establishment of electrical gradients between cellular compartments. At birth, an infant has about 30 g of calcium, and the element must therefore be accumulated during infancy and childhood through the small intestine, from calcium-containing foodstuffs, such as milk, which is an especially rich source of this mineral (1.5 to 2 g/liter). More than 99% of calcium filtered through the renal glomeruli can be reabsorbed in the tubules.

Milk is a doubly rich source of nutrients, as it also contains, as do egg yolks and fish oils, vitamin D_3, which was first discovered in 1936. This is the precursor for the renal hormone 1,25-dihydroxycholecalciferol [1,25(OH)$_2$D$_3$], which was first characterized chemically in 1971 (Fig. 9). Plants contain mainly vitamin D_2, which must first be converted to vitamin D_3. Some plants on the American continent do, however, synthesize 1,25(OH)$_2$D$_3$, which can cause a fatal hypercalcemia in animals ingesting their leaves. The prohormone vitamin D_3 is stored in the fat tissue after absorption through the intestine, and is converted to 1,25(OH)$_2$D$_3$ by the actions of a hepatic microsomal 25-hydroxylase and a proximal renal

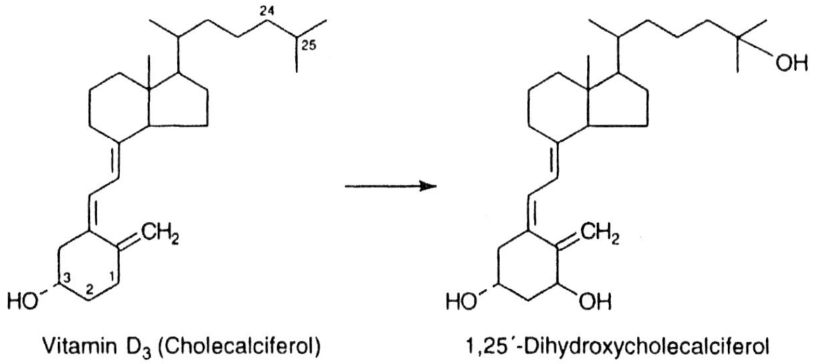

Vitamin D₃ (Cholecalciferol)   1,25´-Dihydroxycholecalciferol

Figure 9. Dihydroxycholecalciferol, the hormonally active form of vitamin D_3.

tubular 1α-hydroxylase. The activity of the renal enzyme is regulated by serum concentrations of calcium, phosphate, and PTH. The chemical structure of $1,25(OH)_2D_3$ is related to that of the steroids, and before exerting its physiological effect binds to a specific receptor, first characterized in 1987. $1,25(OH)_2$-D_3 stimulates intestinal calcium intake, the reabsorption of calcium through the kidneys, and bone mineralization.

It was first realized at the beginning of this century that the surgical removal of the parathyroid glands, four small glands situated behind the thyroid weighing only 40 mg and measuring just a few millimeters, resulted in a disruption of calcium metabolism. The parathyroid glands synthesize the 84-amino acid-long peptide hormone PTH, isolated and characterized in 1959, of which the 34 residues at the N-terminus of the molecule are responsible for its hormonal activity. PTH stimulates calcium release from bones and calcium reabsorption from renal tubules, also the activation of vitamin D_3. Calcitonin, a 32-amino-acid peptide hormone, was discovered in 1960, and is a PTH antagonist, suppressing calcium release and stimulating renal excretion. The C cells secreting calcitonin lie between the thyroid follicles, and in lower vertebrates form a specific gland, the ultimobranchial body. Calcitonin release is stimulated by high serum calcium levels and by the gastrointestinal hormones gastrin, glucagon, secretin, and cholecystokinin; hypocalcemia and somatostatin suppress its secretion.

3.3.6. Pancreas and Gastrointestinal Hormones

Insulin, Glucagon, and Somatostatin. The endocrine pancreas contains about 2 million secretory cells, which synthesize the three most important pancreatic hormones, insulin, glucagon, and somatostatin. The pancreatic islet cells, first described in 1869 by Langerhans, are about 150 μm in diameter and are comprised of the insulin-secreting B cells, surrounded by the glucagon-producing A cells and the more diffusely distributed somatostatin-producing D cells. About 2% of all pancreatic cells are endocrine cells, and 80% of the islet cells are B cells. J. de Meyer described the role of a hormone produced by the pancreatic islets in diabetes mellitus in 1909, to which he gave the name *insulin*. The identification of insulin followed in 1921 by F. G. Banting and C. H. Best, and the amino acid sequence was elucidated in 1955 by F. Sanger. Insulin was introduced shortly after its discovery into the treatment of diabetics, who before that time had had only a short life expectancy. Today, insulin treatment allows most diabetics to lead an almost normal life. In the United States and Europe 5–10 % of the adult population suffers from diabetes mellitus.

The primary role of insulin is in the transformation of foodstuffs such as carbohydrates, fats, and proteins into forms that can be utilized by cells. Most important is the regulation of the highly metabolically active cells of the liver, muscle, and fat tissue by insulin and its antagonists, glucagon, adrenaline, and cortisol. Insulin stimulates the cellular uptake of the energy source, glucose, from blood inducing a fall in blood sugar levels. This action of insulin is counteracted by GH in muscle and adipose tissue. When glucose is available to cells, other energetic resources can be conserved, so that little intracellular catabolism of proteins and fats occurs. Some glucose is also stored in the form of glycogen. Hence, insulin acts as an anabolic hormone. In addition to its effect on glucose, insulin also stimulates the uptake of amino acids and the synthesis of proteins by muscle and liver cells, as does GH. Glucocorticoids such as cortisol cause breakdown of proteins to amino acids, which are then used in gluconeogenesis, i.e., they act as diabetic agents to stimulate the turnover of glucose. The two other insulin antagonists glucagon and adrenaline stimulate the catabolism of glycogen (glycogenolysis) by liver and muscle tissue, causing a rise in blood glucose levels.

Insulin is built up from a 21-amino-acid A chain and a 30-amino-acid B chain, linked by two disulfide bridges. The hormone is stored in granules in the B cell in the form of a dimer of two insulin molecules incorporating two zinc atoms. The A and B chains are formed from one prohormone molecule, proinsulin, which is cut into three sections in the Golgi apparatus of the cell. The middle section, a 35-amino acid sequence known as the connecting or C-peptide, is always secreted in parallel with insulin, although it has no apparent biological function. The pharmaceutical preparations of insulin used to treat diabetics do not contain the C-peptide. Insulin is released in response to a postprandial increase in blood glucose levels. The entry of glucose into the B cells causes a rise in intracellular ATP levels, which closes calcium channels on the cell membrane. The resulting depolarization of the membrane causes the release of insulin-containing granules. This principle is also used in the treatment of some diabetics. In those cases where a sufficient number of functional islet cells are present, the ingestion of sulfonylurea drugs causes a release of insulin in response to the glucose-induced activation of cell membrane calcium channels.

The discovery by C. P. Kimball and J. R. Murlin of glucagon, another islet hormone, which also causes a rise in blood sugar levels, followed that of insulin in 1923. Glucagon was chemically characterized in 1957 by W. W. Bomer and co-workers. The main action of this 29-amino-acid-long peptide hormone is at the level of the liver, where it is transported from the pancreas by the hepatoportal vein, which allows glucagon to reach the liver in considerably higher concentrations than those measured in circulation. It stimulates hepatic glucogenolysis and gluconeogenesis from amino acids and suppresses glycogen production. In this manner glucagon reduces the drop in blood glucose levels between meals and, in conjunction with adrenaline, ensures that there is sufficient energy available for acute physical activity. Glucagon causes a breakdown of fats, i.e., it has a lipolytic action. In a clinical situation, glucagon is used in the treatment of hypoglycemia, such as that arising after an overdose of insulin.

The third pancreatic islet hormone, somatostatin, is a 14-amino-acid-long peptide and is produced elsewhere in the body, for example, in the intestine and in the brain. Its main function is to regulate the release of other hormones, such as insulin and glucagon from the pancreas, and also GH from the pituitary. Because of this action, oktreotide, a long-acting, synthetic analogue of somatostatin, is used in the treatment of diseases that result in the increased production of several hormones. The increased GH secretion seen in acromegaly can be successfully treated with this analogue, as can hormone-producing gastrointestinal tumors that cannot be surgically removed.

Diabetes Mellitus. The underlying cause of diabetes mellitus is an insufficient action of insulin, reducing cellular uptake of glucose and increasing blood glucose, although this is not available to the organism as an energy source. To compensate for the apparent lack of glucose, cells catabolize proteins and fats, which in the latter case results in the production of ketone bodies causing ketoacidotic coma. The high blood sugar levels can also cause a hyperosmotic coma. The two most frequently encountered types of diabetes are Type I and Type II diabetes mellitus. Type I diabetes seems to be the result of a childhood or even prenatal viral infection, which in an organism already genetically predisposed for abnormal immune response, can cause the formation of autoantibodies against the pancreatic islet tissue. Hence, a continual and increasingly severe destruction of the islet cells finally results, usually at puberty, in an insulin-dependent diabetic state. The first clinical symptoms are weakness and fatigue, linked to an unusual thirst. In this state, the tendency toward a ketoacidotic coma can be dangerous. Because the islet cells themselves are compromised, a treatment with sulfonylurea compounds to stimulate endogenous insulin production is useless, and insulin treatment must be instituted. Approximately 0.3% of the U.S. and European population suffer from Type I diabetes mellitus.

Type II diabetes mellitus is more prevalent, with an incidence of 5 to 10%. The precise origin of diabetes in these patients is difficult to define, although there appears to be a genetically inherited tendency toward developing insulin intolerance. Sufficient quantities of insulin are produced in the pancreas, but there appears to be a peripheral resistance to its action, e.g., in muscle tissue. Hence, insulin production increases, in an attempt to maintain a normal level of energy metabolism. These patients also have a tendency toward a deficient fat metabolism, with increased cholesterol and triglyceride levels and arterial hypertension. These "metabolic syndrome" symptoms may precede the development of diabetes by many years, or can also develop after the onset of diabetes and there is an increased risk for severe cardiac and circulatory complications. The insulin resistance can exist for many years before overt symptoms of the disease are noticed. Precipitating factors that increase insulin requirement, such as obesity, pregnancy and infection, exhaust the capacity of the already damaged islet cells to secrete insulin, resulting in diabetes mellitus.

Treatment in the first instance should be directed toward such negative factors as obesity and lack of exercise, and sulfonylurea compounds can in a certain proportion of cases increase endogenous insulin production. A more efficient approach to treatment is to give drugs that intensify the action of insulin in target tissues, such as biguanides (e.g., metformin), which cannot, however, be used in all instances because of their potential side effects. Finally, there is always the possibility of treatment with insulin.

Gastrin, Cholecystokinin, and Secretin.　Digestive function is coordinated by a host of hormonelike regulatory peptides, of which there may be up to 100, produced by the gastrointestinal tract. These same peptides are often encountered in the CNS, and have here as yet unknown functions. According to the hotly disputed theory of Pearse (1975), these regulatory peptides or hormones are produced by APUD cells, which are found in all parts of the body and have a neuroectodermal origin. The thyroidal C cells and the somatostatin-producing cells of the pancreatic islets are examples of these APUD cells. Other gastrointestinal hormones produced by APUD cells are gastric inhibitory peptide (GIP), vasoactive intestinal peptide (VIP), motilin, and substance P.

The clinically most important of the gastrointestinal hormones are gastrin, secretin and cholecystokinin. One of the functions of gastrin is to stimulate the secretion of gastric acids. The release of gastrin from the G cells of the stomach antrum and duodenum occurs as an automatic reflex in response to food intake, especially proteins, and also after stimulation of the vagus nerve and intravenous application of adrenaline or calcium. The hormone has a short half-life of several minutes, and is removed by the kidneys, liver, and small intestine. Cholecystokinin (pancreozymin), a peptide resembling gastrin, stimulates exocrine pancreatic secretion of digestive enzymes, secretion of pepsin and gastric acids, and enhances the contraction of the smooth muscles of the upper small intestine and the gallbladder. It is produced in the I cells of the upper small intestine on the arrival of proteins and fats in this part of the intestine. Secretin stimulates the secretion of electrolytes and water by the pancreas, bile secretion by the liver, and pepsin secretion by the stomach. It also accelerates stomach emptying, the secretion of gastric acids, and transport through the duodenum. Secretin is synthesized in the S cells of the upper duodenum and in the upper jejunum, when stomach secretions arrive in the duodenum. It is a 24-amino-acid-long peptide, chemically related to glucagon, GIP, and VIP, has a short half-life, and is inactivated after a few minutes in the kidneys.

3.3.7. Adrenal Cortex

In adults the adrenals weigh only about 4 g, with the adrenal cortex accounting for only 20%. The adrenal cortex and adrenal medulla have different embryonic origins and are found associated in the same organ only in mammals. The adrenal cortex has three cellular regions, each producing one main secretion product, aldosterone, cortisol or dehydroepiandrosterone (DHEA) (Fig. 10).

Cortisol. Cortisol, the most important hormone of the adrenal cortex, is secreted by the middle cellular region of the cortex, the zona fasciculata, which also produces other steroids resembling cortisol, e.g., corticosterone. Only approximately 15% of the zona fasciculata is required for survival of the individual. However, in the absence of any cortisol production, severe disturbances of electrolyte balance arise, causing sodium loss, potassium retention, and metabolic acidosis, which can result in death within a few days. The association of the loss of adrenal function with these symptoms was first recognized by Thomas Addison in 1855, and primary adrenal cortex insufficiency is still known as Addison's disease.

Cortisol binds to two different receptors, so that its dual action as a mineralocorticoid and a glucocorticoid can be differentiated. The action of cortisol on electrolyte balance is relatively weak compared with that of the pure mineralocorticoid, aldosterone. As a glucocorticoid cortisol acts on energy metabolism, causing an increase in blood glucose levels by stimulating gluconeogenesis from amino acids released by the breakdown of proteins, i.e., cortisol has an anabolic action. Chronic protein catabolism related to long-term cortisol stimulation leads to the overall reduction of muscle mass, the breakdown of structural proteins in skin and bone (i.e., to osteoporosis and skin atrophy), growth arrest in children,

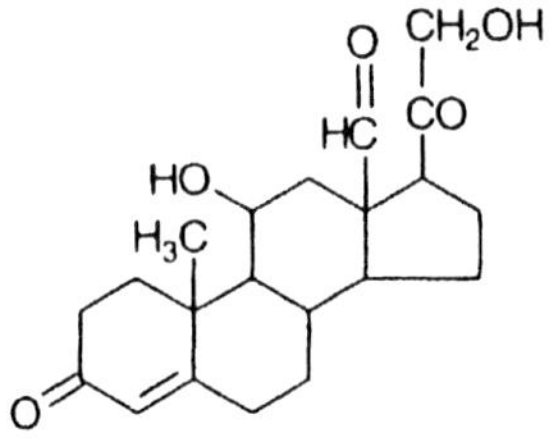

Aldosterone 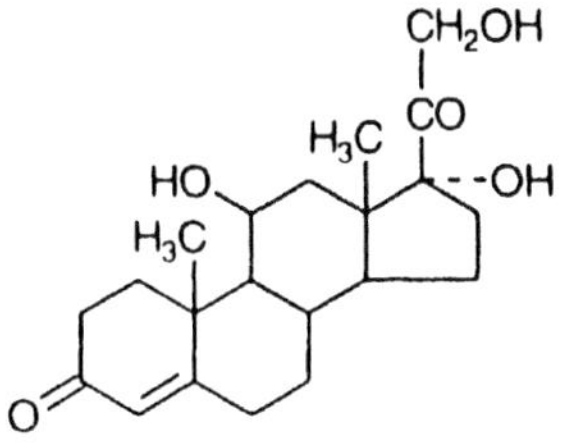   Cortisol

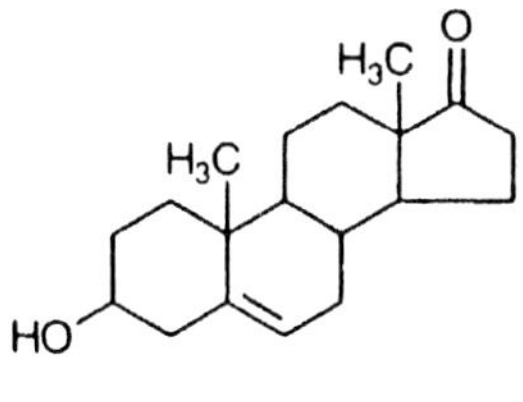

Dehydroepiandrosterone

Figure 10.   The main hormones of the adrenal cortex: aldosterone, cortisol, and DHEA.

and compromised wound healing following injury. Because the adipose tissue of the trunk
has a higher sensitivity to insulin than does that of the limbs, the cortisol-induced increase in
blood glucose and insulin levels causes a centripetal fat distribution. High levels of cortisol
have a suppressive action on inflammatory and immunological responses, as well as less
well characterized effects on the subject's psychological status. Treatment with high doses
of cortisol can produce euphoria, increasing appetite and the feeling of well-being. There is
also a potential to become addicted to cortisol, so that when cortisol levels return to normal,
the patient may experience withdrawal symptoms of reduced pain threshold and psychological disturbances.

Diurnal cortisol secretion occurs in irregular episodic form, with increased secretory
action seen in the early hours of the morning, from 2 AM onwards, and associated with food
intake during the day. Cortisol is also secreted in response to stress. Each secretory episode
of cortisol is preceded by a pulse of ACTH release from the pituitary, except in those rare
cases of cortisol-secreting tumors where steroid production is independent of ACTH
stimulation. The action of ACTH on the adrenal cortex is influenced by several other factors.
However, the most important regulatory feedback on cortisol release involves the stimulation of ACTH release by the hypothalamic releasing hormone CRH and the feedback
suppression on pituitary activity by cortisol itself. The presence and action of CRH was only
discovered 10 years ago (Fig. 11).

The action of CRH on pituitary function is enhanced by arginine-vasopressin, which is
identical to ADH produced by the hypothalamus and posterior pituitary and which in some
species has a more important effect on secretion of ACTH and cortisol than in humans.
Hypothalamic CRH release is stimulated by the cerebral adrenergic and serotonergic neural
transmitters and is suppressed by the feedback action of cortisol.

The activation of the hypothalamus–pituitary–adrenal axis is directed by an endogenous Zeitgeber, which is influenced by physical and mental activity, food intake, and stress
stimuli. In response to stress, the adrenergic system, in particular the locus coeruleus of the
brain, is stimulated within minutes. The synapses of this neural pathway within the
hypothalamus activate CRH release, then ACTH release, and after 10 min cortisol production is also stimulated. The effect of cortisol action, in contrast to that of the short-lived
catecholamines, lasts over several hours. There are several hypotheses that suggest that the
release of cortisol in response to stress represents not a primary response to the stressor, in
contrast to that of the catecholamines, but that cortisol is produced in order to limit stress-
induced changes, such as aggressive immunological reactions, through its immunosuppressive and anti-inflammatory actions. Functional disturbances of this interaction of immune

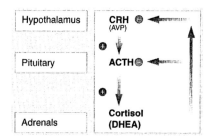

Figure 11.   The hypothalamic–pituitary–adrenal feedback loop.

activating and suppressive actions of the endocrine regulatory mechanisms caused by a relative hypocortisolism may, therefore, play an important role in the onset or progression of autoimmune diseases and disturbances of psychic adaptation to stress (Benschop, Jacobs, Sommer, Schurmeyer, Raab, Schmidt, & Schedlowski, 1996a; Schmid-Ott, Jacobs, Jäger, Klages, Wolf, Werfel, Kapp, Schürmeyer, Lamprecht, Schmidt, & Schedlowski, 1998).

Renin, Angiotensin, and Aldosterone. R. Tigerstedt and P. G. Bergman demonstrated for the first time in 1898 the potential of a kidney extract to cause a rise in blood pressure in rabbits, and named this hypothetical renal factor *renin*. Further experiments by H. Goldblatt in 1934 showed that this factor could be released by constricting the renal arteries. Two more substances also implicated in the action of renin on blood pressure—angiotensin and hypertensin—were described in 1940. In fact, these two factors were found to be identical, and the name *angiotensin* was retained. Today we know that renin is an enzyme that is produced mainly in the juxtaglomerular cells of the renal glomeruli, but also in the uterus, in blood vessel walls, and in the brain.

Angiotensin I is a peptide of 12 amino acids, which is produced by cleavage of the longer precursor, angiotensinogen, under the action of renin. Angiotensinogen is released from the liver into the circulation, where it is cleaved by the proteolytic action of renin. In turn, angiotensin I is converted into the shorter angiotensin II through cleavage of 4 amino acids by the angiotensin-converting enzyme (ACE) found in the vascular endothelium of the lungs and kidneys. This octapeptide angiotensin II has a pronounced effect on blood pressure through the mechanisms noted in Table 6. Angiotensin has a short half-life in plasma of only a few minutes. Drugs based on the venom of South American vipers, which act as ACE inhibitors and hence prevent the formation of angiotensin II, and different agents acting as angiotensin II receptor antagonists, have recently been developed and represent the most efficient and tolerated medication for the treatment of hypertension.

Aldosterone, a steroid resembling cortisol in structure, is produced in the outermost zone of the adrenal cortex, the zona glomerulosa, under the influence of ACTH or angiotensin II. The stimulatory effect of ACTH on aldosterone secretion lasts only for several hours, whereas the effect of angiotensin II can be of longer duration. An increase in renin secretion, for example during disturbances of renal blood flow, causes a rise in plasma angiotensin II levels, which can give rise to a secondary hyperaldosteronism. As a consequence, aldosterone stimulates the reabsorption of sodium through the distal portion of the renal tubules over several days, until an escape mechanism, probably involving renin and angiotensin, reestablishes normal function. The mineralocorticoid-dependent excretion of potassium and hydrogen ions is not affected by this escape mechanism, so that the patient develops hypokalemia and alkalosis as a result of hyperaldosteronism.

Table 6
Mechanisms of Action of Angiotensin II

Blood pressure increases through	Constriction of arterioles
	Stimulation of aldosterone secretion
	Stimulation of the adrenal medulla
	Stimulation of vasopressin secretion
	Increase of sodium reabsorption
	Increase in fluid intake (thirst)
Contraction	Of smooth muscles cells in the stomach, intestine, and uterus
Suppression	Of learning processes (during animal experimentation)

Adrenal Androgens.   The androgens produced by the adrenals are synthesized in the zona reticularis, the zone of the adrenals whose function is still not fully understood. The adrenal androgens have an important role in the maintenance of estrogen secretion from the fetoplacental unit during embryonic development. There is little production during early childhood, but about 2 years before puberty androgen production starts up again, reaching a maximum in young adults and then falling off after age 30. The function of these androgens and the mechanism by which their production is controlled are unknown. Short-term fluctuations in the secretion of DHEA occur in parallel to those of cortisol and are controlled by ACTH.

Hypercortisolism and Adrenal Cortex Insufficiency.   The most common cause of hypercortisolism, or Cushing's syndrome, is the high-dose corticosteroid therapy used for severe rheumatism, and allergic or immunological problems. An endogenous hypercortisolism, in contrast, is rare, but usually caused by a small (a few millimeters long) ACTH-secreting tumor of the pituitary, which can be surgically removed successfully in most instances. Even more rare are benign or malignant, cortisol-secreting adrenal tumors, which must be removed, or malignant tumors of the bronchi, pancreas, thymus, or C cells of the thyroid, which secrete ACTH in an autonomous manner. The consequences of hypercortisolism are muscle and skin atrophy, osteoporosis, genetically related obesity, Type II diabetes mellitus, hypertension, and hypokalemia.

An adrenal cortex insufficiency with reduced cortisol secretion is often the result of a destruction of the adrenal cortex through autoantibodies, occasionally of a tuberculosis bacterium or tumor. The patients suffer from lethargy, nausea, low sodium levels and high potassium levels, low blood pressure, and a reduced blood glucose level. This syndrome is potentially lethal, requires infusions of electrolytes and glucose, and treatment with cortisol. Most patients require lifelong cortisol replacement therapy at a dose equivalent to normal adrenal secretion levels (20–30 mg/day). This type of treatment has no adverse side effects. Not all patients require a mineralocorticoid replacement in addition to cortisol, but when this is the case the longer-acting fludrocortisone preparation is the treatment of choice instead of the natural steroid, aldosterone.

3.3.8. Adrenal Medulla and the Sympathetic Nervous System

Adrenaline, the first hormone ever to be synthesized in the laboratory (1904), is produced in the light epithelial cells that make up 80% of the adrenal medulla. Noradrenaline comes from the adrenergic synapses of the sympathetic nervous system, and the autofluorescing, "dark" cells of the adrenal medulla, which take up a silver stain. The hormonal roles of adrenaline and noradrenaline were recognized well in advance of their action as neurotransmitters in the central and peripheral nervous system. Cholinergic fibers of the splanchnic nerve innervate the adrenal medulla, and its blood supply derives from the vessels of the adrenal cortex. Cortisol arriving from the adrenal cortex plays an important role in controlling medullary secretory activity. There is no anatomical evidence to support the common hypothesis that cortisol secretion is regulated by the adrenal medulla; it is possible, however, that adrenal androgen production may be under the influence of the adrenal medulla because of the proximity of the medulla and the zona reticularis.

Catecholamines are synthesized from the amino acids phenylalanine or tyrosine, which are converted first into dopa and then into dopamine, which is finally converted into noradrenaline (Fig. 1). The enzyme PNMT reductase, required for the production of adrenaline from noradrenaline, is only found in the cells of the adrenal medulla and its activity can be stimulated by cortisol, which is produced in the neighboring cortex and present in high concentrations in the medulla. Adrenaline, noradrenaline, and dopamine are stored in granules, are released as required, and can be taken up again by the cell. Catecholamine release occurs following stimulation of the adrenal medulla via the cholinergic afferent synapses, which are activated by physical or emotional stress, cold, or reduction in blood pressure or blood glucose levels.

In contrast to adrenaline, noradrenaline is taken up almost exclusively by the sympathetic nervous system. The plasma half-lives of the catecholamines are less than 10 min, and only about 5% is excreted in urine without being metabolized. A large proportion is converted in the liver and kidneys by the mitochondrial enzyme monoamine oxidase (MAO) and the cytoplasmic enzyme catechol-O-methyltransferase into metadrenaline and vanillylmandelic acid (otherwise known as 4-hydroxy-3-methoxymandelic acid).

The cell membrane receptors for the catecholamines belong to the seven-finger peptide receptor class. They can be divided into groups of α-, β-, and dopamine receptors, which have different clinical roles; noradrenaline binds more avidly to α-receptors, whereas adrenaline binds to both α- and β-receptors. Some of the more important α- and β-receptor responses are listed in Table 7.

The adrenal medulla is not an essential endocrine organ, as the loss of adrenaline production can be compensated for by the organism. However, a serious, treatment-resistant hypotension (Shy–Drager syndrome) can arise if noradrenaline secretion is disrupted through a general neuropathy. Potentially fatal tumors of the adrenal medulla, such as pheochromocytomas, cause cardiac arrhythmia, headaches, and hypertension.

Catecholamines have stimulated acute interest from a pharmacological viewpoint, and many valuable drugs have been developed following the elucidation of their function. Dopa production can be suppressed by methyl-p-tyrosine, likewise the conversion of dopamine to noradrenaline by disulfiram. The uptake of the catecholamines into storage granules can be blocked by reserpine, and the class of compounds known as the MAO inhibitors can reduce their breakdown. Adrenaline is used to treat localized bleeding, where its vasoconstrictive properties are called into play, and it is also effective in treating severe allergic reactions, specifically as an inhalant to relieve severe bronchial constriction. Under certain conditions,

Table 7
Examples of Adreneric Receptors and Their Actions

α-Receptors	β-Receptors
Contraction of the blood vessels in skin, muscle, GI tract, and heart	Dilation of muscle and coronary vessels
Increase in systolic blood pressure	Increase in systolic blood pressure through increased cardiac volume
Increase of diastolic blood pressure through increased peripheral resistance	Increase in cardiac output
Decrease in cardiac frequency	Increase in cardiac frequency and involuntary reflexes
Some increase in basal metabolic rate	Increased basal metabolic rate
	Glycolysis (increase in blood sugar levels)
	Dilation of the bronchi
	Suppression of labor pains
Contraction of the bladder sphincter	
Mydriasis (dilation of the pupil)	
Secretion of saliva	

noradrenaline can be used to treat critical loss of blood pressure, although this treatment may cause dangerous cardiac arrhythmia.

Synthetic analogues of the catecholamine molecules have been developed, which bind to one specific type of receptor, blocking its action. Phenylepinephrine binds exclusively to the α-receptor, isoproterenol to the β-, and salbutamol only to the β_2-receptors. The latter compound is very effective in dilating the bronchi during asthma attacks. Fenoterol, another β_2-bronchodilator, can also be used to block premature labor. β-Blockers such as propranolol are prescribed for cardiac arrhythmia and hypertension, but this treatment may give rise to secondary effects such as obstructive lung disease, an erectile impotency, or a rise in blood glucose levels. α-Blockers, such as phenoxybenzamine, also cause a drop in blood pressure, and they are used preoperatively in patients undergoing excision of catecholamine-secreting tumors of the adrenal medulla. They are also prescribed for neurological disturbances of bladder function. Yohimbin, an α_2-blocker, is useful in the treatment of certain types of impotency.

3.3.9. Gonads

The reproductive hormones are in no way essential for the survival of an individual, but do serve a useful function in making life more interesting; they are, however, responsible for ensuring the survival of the species and for determining at least in part the psychosocial framework of our existence.

Embryonic gonocytes migrate into the mesodermally derived genital sinus after only 4 weeks of development, and start to differentiate into testes or ovaries at week 7 postconception. At 20 weeks, the seminal epithelium, alternatively about 600,000 oocytes, which form the lifetime stock of eggs in the human female, are already present in the embryonic gonads. The fetal testis is able to produce testosterone at 3 months resulting in the development of a male phenotype (Fig. 12). The penis, urethra and prostate develop out of the genital tubercle

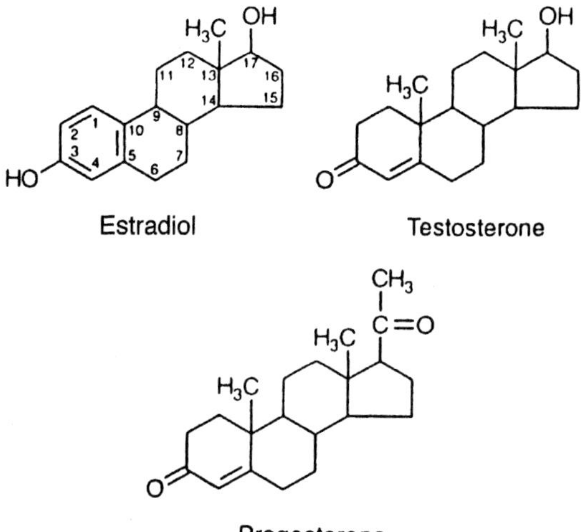

Estradiol

Testosterone

Progesterone

Figure 12. Chemical structures of estradiol, testosterone, and progesterone.

and urogenital sinus, while the epididymis, seminal vesicles and vas deferens develop from the Wolffian ducts. The Müllerian ducts degenerate in the male embryo. In the absence of fetal testosterone production, the internal and external genitalia develop as a female phenotype. The clitoris forms from the genital tubercle and the uterus, Fallopian tubes, and part of the vagina develop from the Müllerian ducts, and the Wolffian ducts atrophy. During this critical period of anatomical development, the sexual differentiation process is very sensitive to the influence of exogenous androgens, such as anabolic steroids, or anti-androgens, which block the action of testosterone, whereby the female embryo can be masculinized or conversely the male embryo feminized. In male fetuses there is a second secretory phase of testosterone at the time of birth, and high levels of testosterone can be measured in the circulation at this time. Extrapolating from results of various animal experiments, it is believed that the testosterone secreted at this time causes the central imprinting of the male sexual behavior patterns seen in adulthood.

Human sex steroids can be classified in three groups, the androgens, estrogens, and progestogens, which each have specific receptors in the cytoplasm and nuclei of the target cells. The three most important and most active steroids of each class are testosterone, estradiol, and progesterone, respectively (Fig. 12). The steroids are chemically and structurally closely related to each other and to the mineralocorticoids and glucocorticoids. Testosterone is synthesized in the interstitial Leydig cells of the testis, which lie between the seminiferous tubules. Estrogens are synthesized in the granulosa cells of the ovarian preovulatory Graafian follicle and postovulation by the cells of the corpus luteum. Progesterone is synthesized by the corpus luteum, which forms after the egg has been released from the follicle. The testis secretes small amounts of estrogens, and the ovary produces small amounts of testosterone. The adrenals produce low levels of all three sex steroid classes. In the target tissues of the sex hormones, the steroids bind to their specific cytoplasmic receptor. Depending on the type of cell, there can be between 5000 and 100,000 receptors expressed in the cell. After the binding of their ligand, the activated receptors are transported into the cell nucleus, where the steroid–receptor complex influences the transcription of certain genes.

Gonadal function is regulated by the pituitary gonadotropins, LH and FSH. In men LH stimulates testosterone production by the interstitial Leydig cells and was originally named *interstitial cell stimulating hormone*. In women LH is released as a single preovulatory secretory peak to induce follicle rupture and oocyte release, and controls steroid production by the thecal cells of the preovulatory follicle and progesterone production in the corpus luteum. FSH is required for normal ovarian follicular development and oocyte maturation, and regulates estradiol production by the granulosa cell compartment of the follicle. In men FSH initiates and maintains spermatogenesis, in association with testosterone produced by the Leydig cells. Both gonadotropins are glycosylated (12–17%) proteohormones, with molecular masses of approximately 26,000 to 30,000 kDa. The carbohydrate residues are essential for hormonal activity and stability. The gonadotropin hormones are made up of two subunits each of about 100 amino acids, as is the case for TSH. These subunits taken singly have no biological activity; the β unit determines the specific hormonal activity whereas the α unit is almost identical for LH, FSH, and TSH in a given mammalian species.

During pregnancy in primates the placenta secretes chorionic gonadotropin (CG), a hormone similar in structure and action to LH, which maintains the life span of the corpus luteum (luteotropic action). The β unit has a 30-amino-acid extension at the C-terminus, and hence both LH and CG bind to the same receptors. The diagnostic tests for pregnancy are based on the very sensitive methods of detection of βhCG in urine.

LH and FSH receptors are found on the cell membranes of the testicular Leydig and

Sertoli cells and on the ovarian granulosa, thecal and luteal cells, but in few other tissues. The hormone–receptor complex stimulates the intracellular production of cAMP within minutes of the hormones binding to the receptor. After 15 to 60 min, nuclear RNA synthesis increases and the enzymes necessary for steroid synthesis are produced 1–2 days later. An increase in the mitotic processes of spermatogenesis is seen 6–9 h after FSH action. The pituitary secretion of LH and FSH is stimulated by the hypothalamic peptide LHRH. As for the secretion of thyroid and adrenal hormones, there is a feedback mechanism active at the level of gonadotropin release, which is controlled by the sex steroids. Inhibin, the recently discovered testicular protein, exerts a negative feedback on the secretion of FSH from the pituitary, but as yet no clinical role has been found for this hormone (Fig. 13).

Andrological Endocrinology. The male steroid hormone testosterone is synthesized in the Leydig cells, which make up approximately 12% of the testicular mass. Following a brief period of activity at birth, the Leydig cells remain dormant until puberty, when the testis is reactivated by hypothalamic–pituitary stimulation. Some of the effects of testosterone are listed in Table 8. Testosterone is not stored in the Leydig cells but released immediately after LH-induced synthesis. The half-life of testosterone is short (<60 min) and only about 2% of the steroid circulates in the free form, the remainder being bound to various transport proteins, albumin, transcortin (CBG) and specifically to sex hormone-binding globulin (SHBG). SHBG is synthesized in the liver under the control of estrogens and binds not only testosterone but also its metabolite dihydrotestosterone with a threefold higher affinity. Estradiol is weakly bound, but when considering the ratio of free estradiol to free testosterone the levels of SHBG can exert a profound effect on the availability of estradiol compared with that of testosterone. Elevations in SHBG occur as a result of hormonal contraception and during pregnancy when estradiol production is very high. Female fetuses are protected *in utero* from the virilizing effects of androgens in maternal circulation by SHBG, which effectively neutralizes them.

Gynecological Endocrinology.

Menarche, Menstrual Cycle, and the Menopause. A regular menstrual cycle, such as that seen in humans and many primate species, occurs rarely in nonprimate species, where breeding or ovulation is governed by seasonal or other factors. In girls at puberty there are approximately 200,000 oocytes in the ovary, each surrounded by a layer of granulosa cells, forming primordial follicles. Only 400–500 of these will actually mature and be released as oocytes during the reproductive lifetime of a woman. The menarche, which marks the onset of sexual maturation, occurs from 9 years onwards, and regular ovulatory menstrual cycles

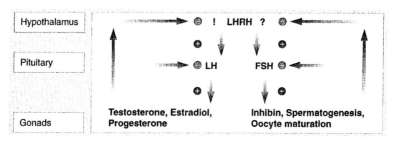

Figure 13. Hypothalamic–pituitary–gonadal feedback mechanism.

Table 8
Examples of Action of Androgens in Different Compartments

Brain	LHRH secretion ↓, libido ↑
Musculature	Protein synthesis ↑, muscle division in man
Liver	Enzyme induction
Kidney	Erythropoietin secretion ↑
Bone marrow	RNA synthesis ↑, erythrocytopoiesis ↑
Bone	Completion of development (end of growth)
Larynx	Growth (change of voice)
Thymic filaments	Involution
Skin	Sebum production ↑, division of hair (body hair growth, deglabration)
Prostate and seminal vesicle	Growth, secretion ↑
Penis and scrotum	Growth, pigmentation
Seminiferous epithelium	Initiation and maintenance of sperm maturation

commence shortly thereafter. During each 28-day cycle, one follicle is programmed to develop into a mature Graafian follicle under gonadotropin action during the follicular phase of the cycle (days 1 to 12). The 2-cm-diameter preovulatory follicle with its layer of granulosa cells has a lumen containing the oocyte–cumulus complex which is filled with a fluid rich in steroids and nutrients for the developing oocyte. Surrounding the granulosa cell layer is a layer of stromal cells forming the theca interna and the theca externa. Under the influence of the midcycle LH peak, the mature follicle ruptures to release the egg and the ruptured follicle collapses to form the corpus luteum, excluding the theca externa layer. The corpus luteum, so called because of the yellow pigmentation contained in the lutein (ex-granulosa) cells, secretes progesterone during 10 days, if pregnancy does not intervene, after which it degenerates.

The menstrual cycle can be regarded from an endocrinological point of view as an initial 14-day follicular phase during which time the oocyte matures, a short periovulatory phase when gonadotropin levels are high, and a luteal phase when progesterone is produced (Fig. 14). After ovulation the oocyte, which has a life span of about 2 days if not fertilized, is taken up by the isthmus region of the Fallopian tubes, through which it is transported until encountering sperm and being fertilized. During the follicular phase, the endometrium of the uterus proliferates and then matures under the influence of progesterone until nidation or implantation of the fertilized embryo occurs. If fertilization and implantation do not occur, the corpus luteum degenerates after 10 days and the uterine endometrium atrophies and is sloughed off during a period of bleeding for 3 to 5 days (menses). A new cycle follows immediately. At about 50 years of age, the ovary's stock of primordial follicles, which can be programmed to mature, is exhausted and hence follicular development with its ensuing steroid production ceases. At this time, symptoms of hormone deficiency develop, and the menopause sets in.

Pregnancy and Parturition. An oocyte fertilized in the Fallopian tube is transported to the uterus, where it arrives 3 days later. After fertilization, the activity of the uterus is slowed by the action of progesterone secreted from the corpus luteum under the influence of blastocyst-derived hCG and implantation occurs 8 to 13 days postfertilization. During this time, the fertilized oocyte divides several times and folds to form the blastocyst. Because fertilization occurs more frequently than implantation and pregnancy, these first few days appear to form a critical period in the development of the human embryo. After implantation, the blastocyst differentiates into the embryo and placenta, which then takes over hCG

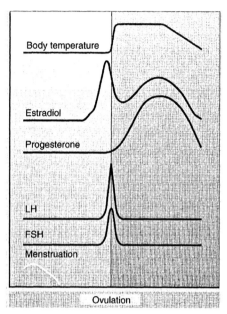

Figure 14. Patterns of hormone secretion during a normal ovulatory menstrual cycle.

production from the blastocyst. The continuing activity of the corpus luteum at this time, stimulated by the LH-like CG, reduces uterine motility, so that the implanting blastocyst is not rejected. After about 3 months, the corpus luteum degenerates as hCG production falls, but progesterone levels continue to rise as production of both progesterone and estradiol is taken over by the syncytiotrophoblast cells of the placenta. The fetus contributes significantly to steroid production at this time, as the fetal adrenals supply the androgen precursors necessary for estradiol formation. Thus, the fetoplacental unit ensures the production of those hormones essential for the maintenance of pregnancy (Fig. 15).

The placenta synthesizes a number of hormones, e.g., placental lactogen, which exerts an effect similar to that of prolactin and prepares the mammary glands for lactation. It also has a weak GH-like action and coordinates the growth of the fetus and of the placenta. At the same time, it also increases the maternal requirement for insulin, so that any genetically determined predisposition toward diabetes mellitus will be unmasked during pregnancy: The daily insulin treatment of diabetic mothers-to-be must also be increased. Two hormones of placental origin are particularly important at parturition. One is relaxin, a peptide hormone related to insulin, which induces/produces a softening of the symphysis cartilage, the uterus, and the vagina and suppresses uterine contractions. The posterior pituitary hormone oxytocin counteracts the latter effect by stimulating contractions, and is frequently used to accelerate birth. Oxytocin is also released in response to nipple stimulation during suckling and it is this action that historically midwives have used to stimulate the delivery of the afterbirth. Laying the newborn baby on the nipple stimulates the uterine contractions necessary to expel the placenta.

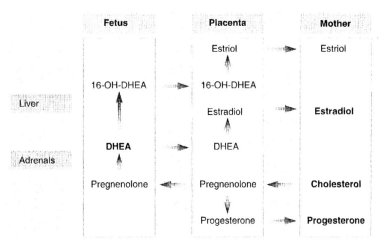

Figure 15. Hormone synthesis by the fetoplacental unit during pregnancy.

Hormonal Contraception, Antiestrogens, and Antiprogestogens. Gregory Pincus developed the first hormonal contraceptive pill in the late 1950s based on the observations that pregnant women do not ovulate, and that animals with persisting luteal steroid production were infertile. The first clinical tests were then carried out in Puerto Rico. The efficacy of hormonal contraception is based on several endocrine principles. Estrogens and progestogens suppress LH and FSH secretion, blocking ovulation. At the same time, the time-dependent development and maturation of the uterine endometrium is interrupted. The high estrogen content of the "morning-after" pill causes increased uterine and tubal motility so that no implantation of a fertilized oocyte can occur. The hormone composition of the "minipill" increases the viscosity of the cervical mucus and prevents sperm passage into the uterine compartment. The side effects of hormonal contraception are caused by the high estrogen content and by the androgenlike action of the progestogen preparations used. The most severe side effects are the increased risk of thromboses, especially in women who smoke, the potential stimulation of any hormone-sensitive tumor and the psychological effects on mood, depression, libido, and sexual activity. Estradiol itself is not suitable for use in oral form, and the estrogen preparation most commonly used is ethinyl estradiol; the progestogens are usually progesterone derivatives, 17α-hydroxyprogesterone, 19-nortestosterone, or the antiandrogen, cyproterone acetate.

Antiestrogens, for example tamoxifen, clomiphene, or cyclophenyl, bind to estrogen receptors, thus blocking them. They are used in the treatment of hormone-dependent tumors, i.e. endometrial and mammary tumors, and also for the treatment of infertility. Antiprogestogens block the progesterone-dependent maintenance of pregnancy, inducing abortion. These compounds have not been licensed for use in some countries and some antiprogestogens can also act as cortisol antagonists, and their misuse may cause adrenal insufficiency.

4   Concepts in Psychology

Uwe Tewes

4.1. Introduction

Research in psychoneuroimmunology frequently refers to psychological constructs that are proposed to be important for the regulation of immunological and endocrinological processes. These constructs are often poorly defined, with different experts sometimes using varying definitions of the same concept. This is especially true for the theoretical concepts of stress and emotion. Some of these concepts are conceptually and theoretically blurred—the same notion may be used to explain different concepts and constructs. Furthermore, the concepts may have something in common with each other that is not readily obvious to everyone.

    The construct of anxiety is a typical example of this terminological inexactness. In psychoanalysis, anxiety may be a symptom or a signal that the control of the dynamics of

Uwe Tewes • Department of Medical Psychology, Hannover Medical School, D-30625 Hannover, Germany.

drives is threatened. Most analytic models of personality define anxiety as the readiness of individuals to act with fear to dangerous situations. This readiness is graded differently between individuals and may be genetically determined.

Learning theorists have defined anxiety as a conditioned flight reaction or avoidance response. One could argue that definitions and theoretical constructs have only secondary importance, insofar as it is possible to operationalize anxiety into a measurable form, which can then be correlated with endocrinological and immunological parameters. However, this merely descriptive point of view is not suitable for the development of causal explanations in psychoneuroimmunology.

When examining whether anxiety may influence immune functions, and which physiological pathways are utilized for such modulation, the biological phenomenology of anxiety, and how these changes may alter immune functions, must be considered. The attempt to find an answer to such a complex question requires an adequate theory of anxiety. In animal models one may, for example, assume that the social rank position of an individual correlates with immunological status. This correlation may have the effect that low immune functioning in animals with low ranks may be of importance for selectional processes. Animals with low immune functioning would have lower chances of reproduction and survival. Low immune functioning in animals with low ranks may result from the permanent threat of those animals that have a higher rank. Such constant threat may lead to more frequent alarm reactions that may threaten homeostasis of the endocrine system as well as immunocompetence. If one wants to test the hypothesis that the biological mechanisms of adaptation and maladaptation are also preserved in humans, then it would be unreasonable to base research purely on psychoanalytical concepts.

A further problem results from the fact that many different psychological constructs are somehow related, and that it is not easy to discriminate between primary and secondary causes. In an examination of the influences of psychological factors on the endocrine and immune systems, the effects of stressors such as fear-inducing situations, depression, or feelings of helplessness on these systems are investigated. Meanwhile, there is a wealth of evidence demonstrating that depression, fear, and emotions that are associated with real or supposed helplessness can modulate immune functions. If an individual is placed in threatening situations in which it has no chance to escape, or if it is confronted by a conflicting decision between two alternatives that are similarly threatening, then reactions like fear or anxiety are probable.

If an individual is increasingly exposed to similar situations, feelings of helplessness may generalize and the individual may conclude that he has no influence on what will happen to him. This is called *learned helplessness*, and it may proceed to depression if it is of long duration. If we reveal that a suppression of immune functions is associated with anxiety, depression, or the feeling of helplessness, then we can develop several research hypotheses that might be the actual causes for these associations. For example, it is possible that depression causes immunomodulation, and that feelings of helplessness and panic are only correlated to immune alterations because they are symptoms of depression, or because they may evoke a depressive state.

Alternatively, it might be possible that feelings of helplessness and anxiety produce an alteration of the immune functions while simultaneously inducing depression. In this case the correlation between depression and immunomodulation would not reflect a causal relationship. Even if one could show that emotional states of depression are usually followed by an alteration of immune functions, it would still be unclear whether it was the emotional state that caused the changes, or whether depression resulted in other alterations that may have an unfavorable influence on the immune system. For instance, depression could result in alterations in health behavior, such as increased alcohol consumption, increased smoking, and/or an increase in sedentary lifestyle.

A further explanation may be that changes in the immune system cause a depressive mood, and that the feelings of anxiety and helplessness are only corresponding symptoms of the depression.

These reflections show that psychoneuroimmunological research requires psychological concepts that can be transformed into experimental designs, thus providing a means to elucidate causal relationships. It is important that empirical psychologists develop uniform and convincing theoretical concepts for the explanation of psychoneuroimmunological processes.

Psychologists in recent years have focused psychoneuroimmunological research energies on a number of different paradigms that have been proven to be successful. These are primarily:

1. Classical conditioning of immunological processes
2. Effects of acute and chronic stress on the immune system
3. Correlations between psychological disturbances and immune function
4. Correlations between personality traits, personality types, and immune function

These paradigms will be exemplarily explained here, and are also illustrated in other chapters of this book.

4.2. Classical Conditioning and the Immune System

The construct of classical conditioning describes an associative learning process, and is based on a very simple experimental design. An individual is exposed to a stimulus that elicits an innate, unconditioned reflex. For a hungry dog, this stimulus may be, for example, the sighting of food (Fig. 1). The automatic reaction, the unconditioned reflex, in such an instance is an increased salivary response. The stimulus (food) that elicits this reflex is called the *unconditioned stimulus*. In humans, the unconditioned stimulus may be, for example, a cold current of air blown into the eye, producing the unconditioned reaction of blinking. Presentation of the unconditioned stimulus can be paired with a neutral stimulus such as an acoustic stimulus, which ordinarily does not elicit the unconditioned reflex. After repeated pairings of the neutral stimulus with the unconditioned stimulus, the exclusive presentation of the former neutral stimulus produces the same reflex ascribed to the unconditioned stimulus. The former neutral stimulus is now a conditioned stimulus. The reaction that is released by the conditioned stimulus is called the *conditioned response* or *conditioned reflex*. After repeated presentation of the conditioned stimulus, the production of the response is lost as a result of habituation. In Chapter 23 this type of learning and its application to psychoneuroimmunology are described in more detail.

In the late 1920s a pupil of Ivan Pavlov, the discoverer of classical conditioning, tried to apply the paradigm to immunological research. These investigations demonstrated for the first time that immunological functions could be reproduced as a conditioned response to a neutral stimulus (Metal'nikov and Chorine, 1926). Modern conditioning experiments are based on the research of Ader and Cohen (1975). They demonstrated that the immunosuppressive effect of cyclophosphamide in rats could be reproduced following the tasting of a saccharin solution, if this neutral stimulus was previously paired with cyclophosphamide.

There is a wealth of animal experiments, especially with rats or mice, showing that immune functions can be altered by means of classical conditioning. In such experiments,

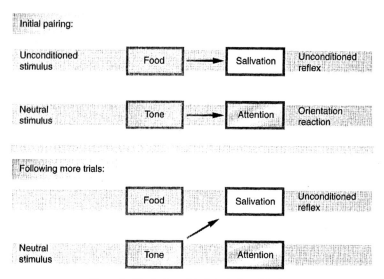

Figure 1. The basic model of classical conditioning.

conditioned alterations in the number and function of immunocompetent cells were generally investigated. However, the effect of conditioned alterations of immune functions on the progress of autoimmune disease and other clinically relevant models have also been investigated (for a systematic overview, see Chapter 23).

Although there is a lot known about the neurobiological basis of classical conditioning of the nervous system, very little is known about the biological basis of conditioning of the immune system. Experiments that condition immune functions are predominantly pharmacological in nature. In these experiments, the animals are exposed to substances (unconditioned stimulus) that either stimulate or suppress immune function. These substances are paired with neutral substances (conditioned stimulus) that by themselves have no effect on immune functions. Presently little is known regarding the biological establishment of these associations.

The immunological alterations observed in such classical conditioning experiments may be the side effects of other changes in the organism. This would be true in cases where the process of conditioning would lead to changes in endocrine regulation, where hormone secretion would subsequently affect the immune system. Indeed, it appears that endocrine factors may constitute one pathway by which conditioning produces changes in immune functions (see Chapter 23). It is interesting that only in recent investigations has it been possible to reliably reproduce conditioned alterations of immune functions in humans. However, the fact that these results are not easily reproducible in humans demonstrates that they cannot be caused by placebo effects, as placebo effects are not elicitable in animals.

4.3. Stress

4.3.1. Definition

An organism is continuously supplied with energy, through the utilization of water, oxygen, and nutrients. Most of the body's functions are controlled in a dynamic steady state, by means of homeostatic processes. Many physiological processes have a role in maintaining the organism's internal equilibrium. Such processes have setpoints. Some processes function on high frequency rhythms, for instance heart contraction, respiration, and neural impulses. Other physiological processes function on low-frequency rhythms, like the sleep–wake cycle, the menstrual cycle, and changes in emotions following seasonal rhythms. This complex system of homeostatic regulation of bodily processes shows partial tolerance for discrepancies between the setpoint and the actual level of system functioning. There is a biological necessity for this tolerance, as the organism needs time to return to a level near the set point, through counterregulation or changes in behavior.

The dynamic equilibrium of the body can be disturbed by internal and/or external influences. Typical causes may be:

1. Mechanical influences (e.g., injury)
2. Physical influences (e.g., heat, noise, and light)
3. Chemical influences (e.g., poison)
4. Biological influences (e.g., infection)
5. Lack of food, water, or sleep
6. Disturbances in the exchange of information with the environment
7. Disturbances in social interaction
8. Severe threat in dangerous situations

All of these influences may lead to a disturbance in an organism's homeostatic regulation. Whenever homeostasis cannot be reestablished by involuntary mobilizations of biological resources, or by spontaneous changes in behavior, an alarm reaction is elicited within the organism. This reaction and its experience is often defined within the literature as a *stress response*. The cause of this reaction is generally called a *stressor*.

The concept of stress was first described by Cannon (1929), who regarded it as an emergency reaction that helps the organism to mobilize energy for fight-or-flight responses in dangerous situations. The theory of Cannon describes what we today call an *acute stress reaction*, and it was Selye (1936) who first used the term *stress* for the description of various threatening conditions that if they last long enough, produce chronic changes in the homeostasis of the organism. Selye (1936) named the homeostatic changes that were caused by chronic stress the *general adaptation syndrome*. Such chronic changes produced dramatic alterations in the organism, including smaller thymus glands, enlarged adrenal glands, weight loss, and the formation of gastric ulcers.

Selye's theory of stress is based on experiments with animals. The transfer of his concept of stress to human behavior led to an inflation in the generalization of this concept.

In a real struggle for survival, the emergency reaction is biologically advantageous for the organism. Both animals and humans may experience life-endangering situations that require an acute mobilization of all biological resources. However, humans often react in such a manner to situations/stimuli that are not life-endangering, such as when completing exams, when a large sum of money is lost, when unpleasant or burdening information is received, when relationship or occupational problems arise, or in any number of other situations that may be perceived as threatening. In these cases it is not the biological organism that is existentially endangered, but only the psychological well-being of the individual. The mobilization of bodily resources finds its expression in increased heart rate, blood pressure, and respiration rate, increasing levels of adrenaline, noradrenaline, cortisol, and other hormones in the blood, restriction of gastric processes and restriction of blood flow into smooth muscle, and increased blood flow to skeletal muscle and brain tissue. Such biological changes are paradoxical for an organism in situations that are not truly threatening.

The generalization of the stress model to human behavior has some disadvantageous implications if we try to verify this concept. In animal experiments, different stressors can be exactly defined and described, and the reactions in animals that all stem from the same stressor are similar. Typical stressors that are used in these experiments are exposure to cold temperatures, noise, electric shock, forced running, destruction of social relationships, and crowding (see Chapters 5 and 13). In experiments with human subjects, the stressors can only be applied in a very mild form, for ethical reasons. In contrast to laboratory animals such as inbred mice or rats, the interindividual difference between human subjects in reaction to the same stressor is considerable. The same point holds true for field research, where one tries to observe and explore individuals in acute critical situations in everyday life. The large variability between individuals experiencing similar situations necessitated an increasing concentration of research on individual differences in the coping mechanisms of humans. In this regard, it soon became apparent that the cognitive appraisal of stressful situations is extremely important in modulating the severity of the stress reaction (Cohen, Doyle, Skoner, Rabin, & Gwaltney, 1997). The generalization of the concept of stress, however, restricts its experimental validation. In this case we have to consider three types of variables (see Fig. 2).

The dependent variables are the biological and psychological reactions to the stressor. The stressor is the independent variable that can be manipulated systematically. There are many differences in stressor type and intensity, and in humans the observable reaction can show us whether a critical incident is a threatening stimulus. Thus, the observed reaction serves as a criterion for the dependent variable as well as for the intensity of the independent variable. In this model, the concepts of coping and appraisal are defined as intervening variables. These are mechanisms of suppression and sensitization, as well as processes of primary and secondary appraisal (Lazarus, 1991).

4.3.2. Acute versus Chronic Stress

A further problem results when we try to differentiate between acute and chronic stress (see Chapters 15 and 16). Stressors may be differentiated not only by quality and intensity but even by duration and frequency. There are no clear criteria for the differentiation between acute and chronic stress in the literature. A stressor may be defined on the one hand as the application of a very short unique stimulus, such as momentary electric shock; or a chronic burden over years, such as caring for an Alzheimer's disease patient, occupational strain, or an antagonistic relationship (e.g., Kiecolt-Glaser, Marucha, Malarkey, Mercado, & Glaser, 1995). A very short quarrel could be an acute stressor, but if this quarrel were to be observed as a symptom of a chronic conflict, then it would only be a symptom of chronic stress. It depends on the context or background, and the cognitive appraisal, whether we

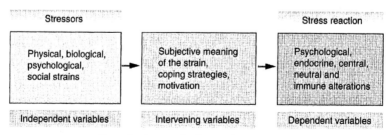

Figure 2. An example of stress-relevant variables.

ascribe such a situation as an acute or a chronic stressor. The differentiation between acute and chronic stress by criteria that are specific for the stimulus does not make sense. One should try to differentiate these stimuli by mathematical criteria that consider aspects of the biological stress reaction. That is, if a stimulus leads to a short discrepancy between the setpoint and the actual level of arousal in a biological system that is for a short duration outside the boundaries of tolerance, and that is very soon regulated to a normal level, then we could call it an *acute stressor*. However, if the reaction leads to an enduring increase of the setpoint for the biological system, for example the prolonged increased secretion of cortisol, then we could refer to it as a *chronic stressor*. Whether a stimulus is a stressor and whether such a stressor is acute or chronic, can only be observed by the reaction of the organism.

4.3.3. Characteristic Features of Stressors

The question of whether a stimulus functions as a stressor, that is, whether it evokes a reaction that is specific for stress, normally depends on the conditions under which the individual is exposed to the stimulus (e.g., running for an animal such as a rat that likes moving would not be a stressor). However, if this animal is *forced* to run, it would constitute an enormous stressor for the organism. For this reason, experimental stress research attempts to reveal what specificities of a stimulus are required for it to function as a stressor. As previously mentioned, a stimulus can be defined as a stressor when this stimulus leads to a severe discrepancy between the homeostatic setpoint and actual functioning, with this inconsistency experienced as aversive or unpleasant by the individual.

Typical stressors may be:

1. Intensive sensory stimuli like strong light, noise, or flooding of stimuli in several modalities; as well as the opposite phenomena, such as sensory deprivation or restriction of all stimuli in the environment (e.g., Monjan & Collector, 1977).
2. Lack of fulfillment of needs that are regulated by homeostasis, such as sleep or food and fluid ingestion (e.g., Irwin, McClintick, Costlow, Fortner, White, & Gillin, 1996). The restorative function of sleep can be observed under experimental conditions, but for those who have a continual disruption of the sleep–wake rhythm (e.g., shiftworkers), a severe burden to homeostatic functioning is evident.
3. Low demands for achievement under basal conditions, as is common in some workplaces, serve as stressors. Conversely, extreme demands of achievement under time pressure produce stress responses (Frimerman, Miller, Laniado, & Keren, 1997).
4. Social stressors are particularly disturbing when involving individual contacts within socially dwelling individuals, such as social isolation, separation, and loss of close relatives (Bartrop, Luckhurst, Lazarus, Kiloh, & Penny, 1977).

Ursin and Olff (1993) point out that there is neither any single event nor any class of events that can be defined as a stressor per se. However, there might be characteristic features of stimuli that increase the probability that these stimuli evoke a stress reaction. These characteristic features are particularly the novelty, uncontrollability, and the unforeseeability of stimuli. Each new stimulus and each new situation raises the level of activation in the organism, and requires an adaptation reaction that is monitored by neurophysiological and endocrine processes and is therefore accompanied by a stress response. Aversive stimuli are thus experienced by the individual as extremely burdening when they appear to be uncontrollable, and when they initiate feelings of helplessness as a result of enduring potential danger, because the consequences cannot be foreseen by the individual.

4.3.4. The Stress Response

The first systematic research on stress reactions was done by Selye (1936). He revealed that animals in laboratory experiments reacted to diverse burdens with a stereotyped syndrome of illness, the "general adaptation syndrome." Typical stressors in his experiments were injuries of the body such as infections or physical trauma, cold shocks, poisoning, or restriction of movement. An organism responds to very brief stimuli that lead to feelings of pain or anxiety with a quick reaction that was described by Cannon (1932) as the "emergency reaction." This presenting response is accompanied by an activation of the sympathetic nervous system, and the mobilization of bodily resources, which can be monitored by the secretion of adrenaline and noradrenaline. The response allows the organism to react very quickly with a flight-or-fight reaction. Typical symptoms of upregulated sympathetic output are an increased heart rate, dilation of the blood vessels in the muscles and brain, contraction of vessels in the skin and viscera, dilation of the pupils, and a desynchronized EEG—all corresponding to a high level of activation which can lead to a defensive reaction. Long-lasting aversive stimulation leads to an enduring response, which can be considered as separate phases that merge together. The first phase of this syndrome corresponds to the emergency reaction as it was described by Cannon (1932). Selye (1936) described three different phases:

1. Alarm reaction, with shock and countershock
2. Resistance phase
3. Exhaustion phase

Within the countershock phase of the alarm reaction, after initial releasing of adrenocorticotropic hormone from the pituitary, glucocorticoids are already released from the adrenal cortex. Even though the mobilization of sugar is supported by glucocorticoids, the mucous membrane is threatened because of the increases in stomach secretion, gastric acid production, and gut motility. The release of glucocorticoids is enhanced in the second phase, and the adrenal cortex may even become hypotrophied. The phase of exhaustion describes the state of compensation of the whole system, with resistance against further stressors reduced, thus resulting in possible total breakdown/collapse of functioning should a stressor remain.

While Selye (1936) based his description on disturbances of body functions following physical stress, stress is today recognized as a complex physiological phenomenon that also incorporates psychological stressors. Under stress we may experience unpleasant emotions such as anger or anxiety that may initiate different responses, according to the type of stressor and its relevance for the organism. Behavior may be incorporated into these emotions (e.g., expressed hostility or anger), although this may be inhibited by prior experience of the individual and the situation itself. In phylogenesis, flight and fight were originally biologically suitable, but under the conditions of today's civilization such a reaction may have deleterious effects on the individual. A substantial disadvantage of the stress mechanism is that inadequate processing is a new stressor in itself, creating a vicious circle where stress intensity may increase via a positive feedback mechanism. On the physiological level, the initial phase of the stress reaction is accompanied by symptoms of sympathetic arousal and general activation. Subsequently, symptoms of parasympathetic counterregulation may be appended. The physiological indicators of the stress reaction are therefore similar to those apparent during strong activation of the organism.

For Ursin and Olff (1993), the stress reaction is considered nothing more than an alarm to potential endangering or real disturbances of the homeostatic equilibrium. The reaction itself contains several sequences of processes of homeostatic counterregulation, all with different time latencies. Within milliseconds, changes in the EEG potential of the central nervous system are recorded. Changes in the sympathetic nervous system output are recorded almost simultaneously, which lead to the typical peripheral physiological changes such as an increase in muscle tone and heart rate. Catecholaminergic activation occurs within seconds, activated by secretion of adrenaline from the adrenal medulla and the release of noradrenaline from sympathetic nerve terminals, leading to an increased blood supply to skeletal muscles and the heart, an increased heart rate, and cessation of gastro-intestinal motilities. Simultaneously, more energy is provided by the increase of lipolysis in fatty tissue and proteolysis in the liver. The secretion of cortisol has a latency of 10–20 min. It is synthesized in the adrenal cortex and leads to an increase in amino acid concentration in the blood, thus supporting the acceleration of tissue healing processes. Additionally, the absorption of blood glucose is terminated, while more glucose is provided to the blood. The different time latencies of the catecholaminergic and glucocorticoid systems are postulated by Ursin and Olff (1993) (see Fig. 3).

Schedlowski, Jacobs, Stratmann, Richter, Hädicke, Tewes, Wagner, and Schmidt (1993b) empirically confirmed the model of Ursin and Olff (1993) in a study examining parachutists who made their first tandem parachute jump. Figure 4 shows the adrenaline and cortisol concentrations of 45 first-time parachute jumpers, with measurements made at 10-min intervals. The results showed more acute peaks in adrenaline and cortisol than those postulated by Ursin and Olff (1993), although the response latencies are similar.

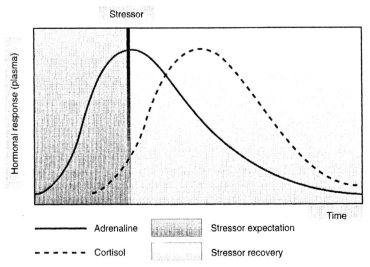

Figure 3.   Contrasting time lags of catecholamine and cortisol response to a stressor. Modified from Ursin and Olff (1993).

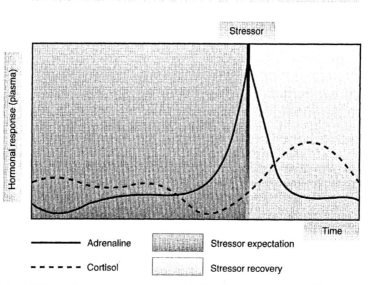

Figure 4. Adrenaline and cortisol plasma concentration in 45 tandem parachute jumpers. Blood was continuously sampled in 10-min intervals from a period 2 h before to 1 h after after the jump, and hormone concentrations measured. Modified from Schedlowski *et al.* (1993b).

4.3.5. Stress Theories

The basic postulate of Selye (1936), that the stress reaction is an unspecific answer of the organism to any type of burden or threat, has stimulated stress research over decades to search for the mechanisms of such a global mobilization of forces by the organism. More recently, interest shifted from this question and researchers have increasingly attempted to investigate the different components of stress reactions that may be independent of each other. Figure 5 simplistically represents the stress model of Pinel (1993).

The importance of the sympathetic nervous system in the stress response was neglected by Selye for a long time. Pinel (1993), like other authors such as Dunn (1989) and Sachar (1980), proposes a dual axis model of a stress reaction which is shown in Fig. 6.

During central nervous system recognition and interpretation of a stressor, the output is not only an activation of the pituitary gland but also an arousal of the sympathetic nervous system via the limbic system and hypothalamus, leading to noradrenaline secretion from nerve terminals of the sympathetic nervous system and a release of adrenaline by the adrenal cortex. These catecholamines enhance the resistance and the efficacy of the organism during the fight-or-flight reaction.

Other authors such as Ursin and Olff (1993) propose that there might be a third system that can be activated by stress. This is known as the testosterone system. During stress the hypophysis (pituitary) is stimulated to produce luteinizing hormone, which stimulates the testis for testosterone production. The increase of testosterone may provide males with an edge in life-threatening situations and give them social advantages as testosterone increases

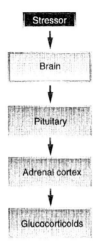

Figure 5. Selye's single-axis stress model. Modified from Pinel (1993).

glucose uptake in the muscles (Sapolsky, 1991). Ursin and Olff (1993) posit that these three systems, the cortisol system, the catecholaminergic system, and the testosterone system, can be independently activated. The type of response that is incurred by the individual depends on his cognitive appraisals and his individual learning history of stimulus–response outcomes. Nevertheless, as will be described later in this book (Chapter 9), there are a number of other hormones that are activated during a stress response, most of which can be proposed to offer some benefit to a flight-or-fight response, as well as being able to modulate immune functions.

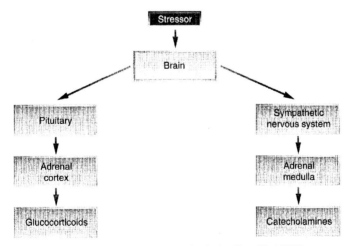

Figure 6. The dual-axis stress model of Sachar. From Pinel (1993).

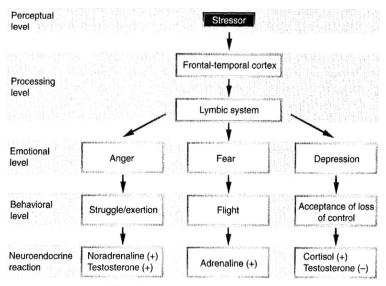

Figure 7. Henry's triple-axis stress model.

Henry (1986) has attempted to define these three types of physiological reactions according to different emotional experiences and to thus explain how they control behavior under duress (Fig. 7). It appears that chronic fear or enduring depression is exceptionally harmful for the organism. During constant strain the enhancement of adrenaline synthesis induces impairment of the cardiovascular system and muscles. Furthermore, immunological functions are weakened. In the long term the enhanced secretion of cortisol can cause high blood pressure, muscular atrophy, or reduction in immunological resistance.

4.3.6. Expectancy of Control and Possibility of Control

An aversive stimulus is perceived as stressful if the individual does not believe that he may have an influence on the frequency and intensity of the stimulus. Laboratory animals that were exposed by chance to strong aversive stimuli, such as electric shock, demonstrated short, intense stress reactions that were much stronger than the reactions of those animals that were exposed to the same number of intensive aversive stimuli but perceived those stimuli as a consequence of their own behavior (punishment). This model can be transferred to the behavior of humans. Figure 8 shows the electrodermal skin response as a reaction to stress in a learning experiment with humans. Weaker electrodermal reactions point to a more intensive stress reaction, because in this case the electric conductivity is raised by enhancement of the activity of the set glands. In a preexperiment in which two groups of test subjects were exposed to mild electric shocks by chance, both groups showed similar levels of the skin response. In the second phase of this experiment, one group was persuaded that they could reduce the frequency of the aversive stimuli by a better learning output, while the other group thought that they were not able to control the aversive stimuli by their learning behavior. The first group had a clearly higher skin resistance, indicating that these subjects

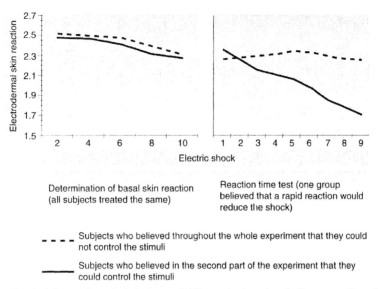

Figure 8. The influence of perceived shock controllability on the electrodermal skin response. From Geer, Davison, and Gatchel (1970).

underwent lower stress responses, even though both groups received the same frequency and intensity of electric shock.

In psychological experiments, the expectancies of control may be manipulated in two ways. As in the example mentioned above, the test persons may be convinced to believe that a control of the stimuli is possible, though it really is not the case. Another possibility is to manipulate the actual possibility of stimulus control. These combinations of believed and actual control can be simplistically illustrated by a 2 by 2 crossover design (see Fig. 9).

Lack of control, or lack of belief of control, leads to an extreme feeling of helplessness. The belief of control, or the rare possibility of control, reduces the subjective strain, even under strong aversive conditions. The other two areas of this scheme, the subject's belief of control in connection with lack of control, or the lack of belief of control in association with the possibility of control, are particularly appropriate for research into the cognitive elements of the stress response. In experiments with humans on a learning or achievement test, the belief of control can be manipulated by giving wrong feedback of the results independent of the real success in these tests. This can be alternatively effected by conveying to the test subjects that they can reduce the frequency and intensity of aversive stimuli by correct behavior. For experiments with animals, numerous experimental conditions are known to explore these effects. Animals can be stressed by reducing the possibility of moving, or by forcing them to move. Appetitive behavior, such as the search for food, can be disturbed by aversive stimuli such as electric shocks to the foot, or animals can be socially isolated or placed in a crowded social environment. The belief of control in these situations can be influenced by the reduction of aversive stimuli after correct behavior such as lever pressing.

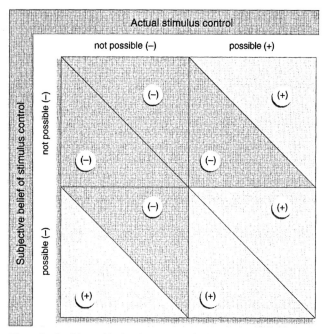

Figure 9. Possible combinations between perceived and actual stimulus control.

The belief of control in humans is influenced by the individual experiences and learning of these situations. If somebody has endured particular experiences over a long time period, and the person interprets that the consequences of his own behavior changing the outcome are totally unforeseeable, then the belief of control may diminish totally. The subjective feeling of helplessness may generalize under these conditions though there may objectively be a high possibility of controlling the situation. This discrepancy between belief of control and real possibility of control is called *learned helplessness* (Seligman, 1975). By this learning mechanism the processes whereby an individual perceives his own behavior and the possibilities of failure are changed. For example, a student who has a very specific explanation for his failure on an exam, may think that he failed the exam because it was given on Friday the 13th. This student would repeat the exam with a much more positive attitude than another student whose expectations of failure are so far generalized that he is convinced of being too stupid to learn the information required for the exam. Figure 10 describes the components of such an attribution schema developed by Seligman.

4.4. Depression

4.4.1. Symptoms

Depression is as a rule differentiated within the literature as the experience of depression that is regarded either as an affective disorder or as a personality trait. Depression

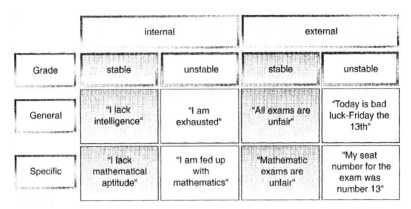

Figure 10. Attribution schema developed by Seligman for explaining depressive reactions following exam failure.

primarily encompasses an increased readiness of the individual to react to strain with unpleasant feelings, feelings of unease and helplessness. The risk of developing major depressive episodes is increased in first-degree relatives, being highest in monozygotic twins. Major depressive episodes frequently follow psychological stress and related life events. However, numerous factors such as environmental stimuli, personality, psycho-dynamics, and drugs may influence disease pathogenesis. These factors may produce neurobiological changes that induce or influence depression.

In most of the systems for the classification of psychological and psychiatric disorders, such as ICD 10 or DSM-IV, there is consensus regarding the symptoms that are ascribed to the group of depressive disorders. Depressive states have their effects on all levels of experience and behavior. On the cognitive level it presents as a change of personal self-image so that the individual develops a negative image of himself, caused by self-doubt, self-reproach, and brooding. The individual is no longer able to cope with his cognitive thoughts in the same way as prior to depression. Concentration, attention, and the ability to cope with pressure by cognitive demands, are reduced. On an emotional level, the general mood shows a negative change. The individual feels sad, unhappy, helpless, and apathetic. On the motoric level, the activity changes toward lethargy or agitation. On the motivational level, it is expressed as a loss of interest and urge. The individual has problems making decisions or completing tasks. On the physiological level, a reduction of all essential energetic functions can normally be observed. The individual shows a lack of needs for sleep, food, and the sex drive is suppressed. On the level of social behavior, the individual avoids contact with others or tries to be ignored when in the company of others.

In most of the cases, the individual observes these changes by himself, and all of these feelings of helplessness, loss of perspective for the future, feelings of deficiency, shortcoming, failing, and generally the feeling that everything is too much for him, may induce suicidal inclinations.

In addition to such heterogeneous symptoms, the phenomenon of disturbances may have different causes. Depressive reactions may be the consequence of straining experiences like hatred of an occupation, loss of relationships, or the breakdown of personal goals. Furthermore, the causes and effects cannot always be easily distinguished. Failure in sexuality may perhaps be a symptom of a depressive disturbance, but it can also be the cause of a depressive episode. Depressive disturbances may be the consequence of the abuse of legal or illegal drugs, or they may be symptoms of neurological or endocrine dysfunction. In women it may be induced during pregnancy or following childbirth. Depressions may be symptoms of a neurotic personality change or a psychiatric disturbance. The construct of depression comprises heterogeneous feelings and symptoms of other disorders may also be present. Therefore, the disorder of depression itself is considered to be a syndrome. In general, there is agreement about the existence and nature of this syndrome, but the lists of relevant symptoms vary within the literature. There is no single symptom that is necessary or sufficient for the diagnosis of depression. A distinction is commonly made between endogenous and exogenous depression, the latter also known as *reactive depression*. The endogenous depressions are divided into unipolar versus bipolar depression. In bipolar depression, episodes of mania alternating with depressive episodes are observed, while in unipolar depression, only episodes of depressive feelings are present. There is presently considerable disagreement regarding the physiological or behavioral basis of depression. Previously, behavioral and cognitive approaches dominated the investigation of depression, whereas the biological perspective has recently received greater focus. A more thorough description of depression and its relationship to immune function is provided in Chapter 17.

4.4.2. Behavioral and Cognitive Approaches

Among cognitive approaches, the theory of Beck (1967) is well known. Beck thought that one cause of depressive mood is the fact that these patients tend to self-doubt, self-reproach, and are anxious when confronted with events by which they feel threatened. Each negative experience of the individual is confirmation of his pessimistic view of himself, and the hopelessness of all of his life. Under these conditions the individual increasingly tends toward illogical fallacies in his thoughts, reinforcing this pessimistic view and forcing him deeper into depression. The patient's thought processes become increasingly distraught and irrational. Such thoughts typically include the postulation of relationships between causes and effects that are incorrect. Such errors include thoughts that a negative event happened more often to him than to others, that he is ignored in a restaurant by a waiter, or that it is raining because he is organizing a party in his garden.

A further error of thought of the depressive individual is overgeneralization. From a sudden but rare trouble with his partner, he draws the conclusion that he is not able to develop partnerships in harmony. If anything turns out to be a failure for him, he concludes that this is a confirmation of his general worthlessness and insignificance.

Selective abstraction is a further error in which the individual tends to attribute the guilt or the responsibility for a conflict exclusively to himself. This occurs even when he was not primarily responsible, so that he may attribute the blame for the failure of his working team exclusively to himself.

Another thought of depressive individuals is the "overstatement," whereby the individual tends to accept minimal adversity from typical examples of his total incompetence. The opposite train of thought is the "understatement," which shows similar differences between reality and self-assessment. However, in this case the individual is never satisfied, let alone proud with good success or achievement, but judges whatever he has achieved as worthless.

This theoretical approach leads to the therapeutic consequence that it must usually be sufficient to reprogram the thought structure of a depressive patient toward a better consciousness of reality, and to eliminate the illogical cognitive mistakes so as to improve the mood of the individual. This approach is an antithesis to the classical theory, which explains the characteristic thoughts of depressive people as a symptom, and not as a cause of the depression. Beck's (1967) theory drew the attention of behavioral psychologists because this approach can be verified by psychological experiments. Illogical mistakes of thought are readily verifiable, and therefore it is possible to investigate whether the individual's mood is improved when the therapists are successful in correcting the thoughts of the individual. However, because depressive moods are normally limited in time, it is not always easy to differentiate whether positive changes in mood are therapeutic effects or spontaneous remissions.

The theory of learned helplessness (Seligman, 1975) is a more behavioral approach that is based on the same observations that were discussed in the section on stress. This theory postulates that it is not being under stress itself that leads to a depressive mood, but more the feeling of helplessness that results from the fact that the individual believes he has no control over threatening stimuli or the situation. This hypothesis can be examined by experimental methods. Seligman (1975) showed that the experience of uncontrollable threatening situations produces similar symptoms in animals and humans. The animal becomes more passive, and less capable of learning. Furthermore, it incurs weight loss. Similar symptoms are seen in depressive humans. It can be experimentally shown that when feelings of helplessness were created in subjects, their performance was similar to that demonstrated by depressive individuals as opposed to healthy subjects. The term *learned helplessness* indicates that depressive individuals generalize negative experiences. When feelings of helplessness are created within a specific situation in either animals or human subjects, they will subsequently show more passive reactions in similar situations.

4.4.3. Biochemical Approach

The hypothesis that depression may be mediated via biogenic amine metabolism disturbance was inferred from pharmacological studies approximately 30 years ago. During this time, the effect of controlling the catecholaminergic and serotonergic systems was investigated, in the belief that antidepressant drugs were effective in alleviating deficiencies in these systems. However, more recent evidence shows that there are causes other than catecholamine and indolamine deficiencies. Furthermore, the temporal lag between antidepressant administration and therapeutic effect was not accounted for by the original hypothesis. Thus, a common pathophysiology of depression has not as yet been identified. The catecholaminergic, dopaminergic, and serotonergic systems are intricately organized, each with different neuronal morphology and related neuropeptides. This complexity combined with the knowledge of heterogeneous mechanisms of antidepressant medication has led to depression being conceptualized by pathophysiological heterogeneity.

Despite an increased knowledge of the complex regulation of neurotransmitter systems, receptors, second messenger induction, and gene expression, a comprehensive understanding of the etiology of depression remains to be achieved.

The biochemical approach may be most suitable for the development of hypotheses regarding the correlation between depression and immune function. However, as is revealed in Chapter 17 of this book, the results are inconsistent, which might be caused by the heterogeneity and instability of the depressive syndrome over time.

It must be noted that behavioral and biochemical approaches are in no way mutually exclusive, as intensive and enduring learning experiences may cause changes in homeostatic regulation, thus provoking depressive reactions by biochemical regulative processes.

4.5. Differences in Personality

Personality is generally defined as the manner in which the individual differs in the extent and structure of his characteristic traits over time. The question of whether particular personality types are more prone to specific illnesses has been posed for centuries, but it is only in the last 20–30 years that attempts have been made to provide a scientific answer. The most researched personality type proposed to be predictive of illness development is that associated with cardiovascular heart disease (Type A personality; Rosenman, 1986). Nevertheless, as a result of the knowledge that psychological and neural processes influence immunity, personality researchers have investigated the role that personality factors play in modulating immune function and disease. In this respect, the greatest focus has been in the examination of a "cancer-prone personality."

Prospective, longitudinal, and retrospective investigations have been used to examine the relationship between personality and cancer, and thus a clear picture does not easily emerge from the literature. However, work by Greer and colleagues has led the way in attempting to forge a link between personality and cancer. Specifically, this group conducted psychodiagnostic investigations of women 3 months following breast cancer surgery (Greer, Morris, & Pettingale, 1979; Greer, Pettingale, Morris, & Haybittle, 1985). At 5- and 10-year follow-ups it was revealed that those women who had reacted to the surgery with feelings of helplessness and stoic calm showed a significantly higher death rate than those women who showed a strong will to survive and fought against the illness.

Prospective studies have revealed similar factors as predictors of cancer development (Grossarth-Maticek, Kanazir, Schmidt, and Vetter, 1982a). However, these characteristics may prove problematic in investigating a general "cancer personality," as they do not fit directly into well established scales that are currently used for general personality assessment. Nevertheless, the factors most closely relate to depressive mood, a trait that is incorporated into major personality inventories.

Using a prospective study, Linkins and Comstock (1990) investigated depressed mood among a large (>2000) rural U.S. population. These subjects were investigated 12 years later, and the development of cancers related to depression scores. Results demonstrated that depressed mood significantly increased the risk of cancer development among smokers, with the increase relating to both smoking-associated and non-smoking-associated cancers.

Although results such as these are striking, they must be interpreted with some deal of caution. An inherent problem with studies examining the association between personality and cancer/immune function is that they are either correlational, or in the case of prospective designs, predictive. Thus, no direct cause-and-effect relationship can be defined, as is the case in purely experimental research design. For example, the personality differences revealed by Linkins and Comstock (1990) may have themselves been caused by health status and possible disposition to cancer. Alternatively, the level of depression in the subject population may have changed between the time this characteristic was measured and the time of cancer induction.

Thus, the multitude of research designs, coupled with the difficulty in establishing cause and effect, makes this area of psychoneuroimmunological research a complex and perhaps problematic field. Nevertheless, the development of models relating depressive personality to cancer development offers hope for establishing models of cause and effect. Such models may be developed by introducing well-established therapeutic strategies to alter depressive personality into research of this type. Studies showing that psycho-behavioral interventions not only can benefit mood but also prevent cancer development, will take a large leap forward in establishing the credibility of these models.

4.6. Summary

This chapter has demonstrated that there is ample scientific evidence that psychological factors influence immune functions, and may also influence the course of disease. The knowledge base ranges from an understanding of the mechanisms and functional meaning of immune alterations (stress), to unclear comprehension of the relationship between psychological factors and immune function (personality).

These differences are related to the ability to functionally manipulate the variable in question. As has been demonstrated, psychology deals with constructs that can sometimes prove difficult to measure and define. Nevertheless, careful experimental design and extrapolation of results has allowed psychological processes to be scientifically investigated. Similar scientific rigor will allow scientists to continue examining how psychological processes impact on immune function and disease.

5   Psychological Methods

Uwe Tewes and Manfred Schedlowski

5.1. Introduction

This chapter provides an overview of those psychological methods that are relevant for interdisciplinary research in psychoneuroimmunology. Within this scope, the field of psychology is confronted with entirely different problems to disciplines such as endocrinology, immunology, anatomy, or neurophysiology. The dependent variables that concern researchers from such disciplines are normally readily and reliably countable or measurable. The functions of variables such as molecules, cells, or tissue can be described and analyzed. However, psychological variables are not so readily definable and measurable. Nevertheless, psychological processes and modes of behavior that in former times were exclusively examined and described by psychological methods, can today be found in modern textbooks of physiology, endocrinology, or neurology. Thus, psychologists con-

Uwe Tewes   •   Department of Medical Psychology, Hannover Medical School, D-30625 Hannover, Germany.
Manfred Schedlowski   •   Institute of Medical Psychology, Essen University Clinic, 45122 Essen, Germany.

ducting psychoneuroimmunological research have developed robust research methods, not only to examine psychological influences on the immune system, but also to reveal how immunological and endocrine changes may alter the behavior, experience, and cognitive processes of organisms.

5.2. Test Theory

Similar to biological scientists, experimental psychologists work with parameters that are quantifiable. However, in contrast to the natural sciences, psychological measurements are normally estimates of traits. Psychological traits and processes cannot normally be observed directly. Psychological measurements are predominantly a matter of hypothetical constructs. A specific construct such as anxiety, depression, or intelligence offers the theoretical background for the development of methods for measurement, such as psychological tests. The results of these measurements should allow us to draw conclusions about the underlying constructs, in accordance with the theoretical assumptions on which the construct is based. Following such conclusions, hypotheses can be generated for the construction of further empirical research, so that the construct and theory can be further evaluated and modified by empirical methods. Thus, a continuous process of feedback between theory and empirical research develops. If it is not possible to confirm the hypothesis by empirical research, then it is conceivable that the construct for which the hypothesis was generated was incorrect. On the other hand, it may be that the tests chosen for the experiment were not valid tools to measure the construct. In such an instance, one has to modify the test and not the construct. Thus, the defining of a psychological construct and its measurement are complex tasks that take a long time to develop. Nevertheless, this process is crucial to psychological research, as hypothetical constructs such as emotions, personality traits, or cognitive abilities can never be directly observed. They can only be realized by the results of psychological tests.

When utilizing psychological tests in interdisciplinary research, it must be considered that in psychology, in contrast to the natural sciences, new methods of measures are not necessarily better than older ones. The continuous feedback between theory and empirical research is an ongoing process, so that reliable methods can only be established after a long period of development. Thus, most established methods have been utilized for many years, whereas for newly developed measurement techniques it is equivocal whether they are really measuring what they propose to measure.

The selection of psychometric tests for psychoneuroimmunological research should proceed in a pragmatic way. Most of the psychodiagnostic textbooks include detailed and objective discussions regarding the disadvantages and advantages of various tests and methods of measurement. In addition, each test manual contains information about the statistical characteristics of the tests. Tests have to fit several criteria of quality, such as objectivity, reliability, and validity.

The criterion of *validity* informs the user of the test how it is suitable for the measurement of the theoretical construct that was the basis of its development, or in what way it may differentiate between criteria groups that would be predicted by the theoretical construct, for example, between persons with low anxiety and high anxiety. Simplistically, validity refers to the criterion that the test actually measures the construct that it proposes to measure.

The criterion of test *reliability* reflects how accurately the test differentiates between subjects, in whatever it claims to measure. In other words, reliability refers to that portion of the variation in test scores that results from the error of measurement. Because the reliability is reduced by an increased error of measurement, reliability is the highest border of validity. A test is not able to reflect a specific criterion or construct more exactly than is allowed by its reliability. If a particular construct is stable, then a reliable test would be able to measure the construct on numerous occasions and achieve the same result. Thus, there is no variation in test scores resulting from error of measurement.

The criterion of *objectivity* is a measure of the independence of the test results from the investigator. The results of the test should not be dependent on the particular investigator who administers the test instructions and evaluates the results. Multiple choice tests or computer aided tests allow the greatest objectivity.

The sample of the population on which the test was standardized should also be considered in test selection. The norms of the test provide information as to whether the results of a specific subject lie within normal scores, or whether the particular individual significantly deviates from the "normal population." A test that is objective, reliable, and valid, but that was only administered to a small sample not representative of the whole population, allows an experiment to differentiate between the subjects of the sample, but does not allow the results to be generalized to the intended population.

5.3. Psychological Scaling and Measurement

5.3.1. Methodological Considerations

When examining the association between the concentration of a hormone in the blood, such as cortisol, and the degree of a psychological characteristic, such as depression, the results may demonstrate that the parameters show a common variance, meaning that there is a relationship or correlation between both parameters. Such a result would appear to indicate that depressive subjects have a higher cortisol level in the plasma than subjects with lower depression. However, scientists who are familiar with empirical methods would not be satisfied with such a general interpretation. For them it would be of great importance to know how close this relationship is and what implications result. It may be easy for an endocrinologist to estimate the reliability of his measurements by examining inter- and intra-assay variance. Even if different laboratories are using different methods that may influence the measurement of the hormone, this would have no influence on the correlation between cortisol and depression within the same working group, as constant differences (such as samples measured using a particular technique in the same laboratory) have no effects on the relative differences within the sample. This would be a type of error that is controllable. If the cortisol measurement was careful, then one could assume that two subjects with identical measurements really have the same levels of cortisol. However, conclusions like this are not feasible for psychological parameters. If two subjects have exactly the same measurement on a depression scale, this does not necessarily mean that the degree of the trait is identical. A psychological test only provides an estimation of the true score of the trait that it is intended to measure. This error of measurement reduces the relationship that can be ascertained by empirical methods.

The endocrinologist may decide for himself whether he is interested in measurements that are constant over time, or whether he prefers to take several measurements over a longer period, from which he can calculate the mean, or whether he thinks that rhythms of very short phases have to be taken into account, thus affording a high density of measurement over time. The instruments of measurement are always the same, and the interpretation of these quantitative results is not changed by these alternatives. The psychologist who wants to measure anxiety or a depressive mood first has to decide whether he wants to measure a construct like anxiety or depression as a personality characteristic, or as an emotional reaction that varies from one situation to another. In both cases he has to use different instruments whose results are not absolutely comparable. The results of different psychological tests are not as comparable with respect to validity as the results of endocrine measurements.

Personality traits or emotional states are normally measured by questionnaires. The subject has to answer questions or statements about himself, which have different item (question) formats.

> *Example*: "I am afraid of closed spaces"
>   True or False

If a self-rating scale contains several items of this type, then the reliability of the scale can be increased using a category scale. The subjects are requested to rate the items on a four or more point scale (commonly called a Likert scale).

> *Example*: "I am afraid of closed spaces"
>   Never   Sometimes   Quite often   Very often

These item formats are unipolar. The same items may be given as a bipolar type.

> *Example*: "When I am in closed spaces I feel ..."
>   Very anxious   A little anxious   A little relaxed   Very relaxed

In most cases one omits a neutral category in the middle so as to motivate the subject to commit himself to a particular direction. The subject receives a score for each item based upon the number of categories (e.g., "never" receives a score of 1, through to "very often," which receives a score of 4).

The raw score of the test/questionnaire represents the total of the scores for each item, for the total number of items. This remark may seem very trivial, but it is often overlooked that the very same total score may be obtained by different items.

Most of the tests are based on the principles of so-called classical test theory. This model discriminates between the true score and the observed score, such that it is assumed that the observed score differs more or less from the true score. The true score can never be observed directly, but it can be estimated by a number of independent measurements over several items on the same subject. The reliability of a test is equal to the proportion of error variance within the total variance of the observed scores. Information regarding the reliability of the test is included in the test manual, and from this information one can draw conclusions about the extent of error within a single measurement. If a subject receives a total score of 10 points on a test, one can make an estimation of the probability of the true score for a specific range around the observed score. Thus, one can make a statement like the following:

> The subject has a score of 10 points on an anxiety test. Considering the error of measurement, one can assume with a probability of 95% that the true score can be expected in the area of 8.5–11.5.

These remarks illustrate that every empirical correlation between a hormone, such as cortisol, and a psychological trait, such as depression, results in an underestimation of the true relationship. But it is possible to make a projection as to the correlation that could have been expected if the psychological test had contained no measurement error. This is called a correction for attenuation. Empirical correlation is divided by the square root of the coefficient of the reliability of the psychological test. If there is a correlation, of say 0.5, between the psychological test and the endocrine parameter, and the reliability of the test amounts to 0.8, it may be concluded that the true correlation is 0.78. This is a substantial deviation from the observed correlation. The squared correlation tells us what percentage of the variance of one parameter is determined by the other. The empirical correlation of 0.5 tells us, therefore, that 25% of the variance of the cortisol measurement is determined by depression, or vice versa. If the proportion of the error of measurement is eliminated by the correction for attenuation, and if the true correlation is squared so the proportion of the determined variance amounts to 61%, this is more than twice as much as the observed correlation.

Computations that may be done on a trial basis have led to substantial discrepancies between empirically computed correlations and supposed true correlations between criteria. In such a case, for further research it may be advisable to utilize psychological tests that are more reliable. In most test manuals, two types of coefficients of reliability are mentioned, the retest reliability and the internal consistency (alpha coefficient). The retest reliability is estimated by two separate measurements of the same subjects with the same test. The coefficient alpha is a measurement of homogeneity of a scale, and is estimated by the covariance of the test items. In connection with constant traits like intelligence, the error of measurement should be computed by the retest method. If the parameters that are to be measured fluctuate greatly over time, for example when measuring emotions, one should compute the coefficient alpha.

Furthermore, it is particularly advisable to supplement classical types of scales when acute changes in mood or emotional state are to be measured. This is especially the case when one wants to measure changes in anxiety, arousal, or pain. In that case, analogue scales are commonly applied. A continuous line like the one below is used:

0 ——————————————————————————————— 100

The subject is then requested to mark on the line the intensity of his acute state of arousal. In this case only the ends of the scale are defined, with arousal intensity subsequently measured quantitatively by determining the distance between the zero point and the subject's mark on the scale.

5.3.2. Psychological Constructs and Scales Commonly Used in Psychoneuroimmunological Research

Numerous standardized questionnaires have been developed to measure distinct emotions. This section briefly summarizes a selection of psychological states and traits that are often employed in psychoneuroimmunological studies, and that have been demonstrated to be associated with immune measures.

Within an experimental setting a researcher might be interested in examining whether feelings that an individual experiences during a stress-inducing task, such as anger, anxiety, or tension, are associated with changes in immune functions. One instrument to measure these *mood* states is the Profile of Mood States (Lorr & McNair, 1982). This questionnaire consists of 65 adjectives, which measure levels of tension/anxiety, depression/dejection, anger/hostility, vigor, fatigue, and confusion. If one is more specifically interested in the level of acute *anxiety*, the State-Trait Anxiety Inventory can be employed (Spielberger, Vagg, Barker, Donham, & Westberry, 1980). This scale can be used to measure both acute state anxiety and habitual trait anxiety. The questionnaire is thus divided into two different scales, each containing 20 items. In the state scale, the subject is requested to describe his present emotional feelings, whereas in the trait scale, the subject is asked to describe his habitual mood. Thus, this questionnaire not only can measure specific negative feelings (anxiety) occurring in a specific situation, but via the trait scale can also measure the stable individual tendency to generally react to situations that are recognized as dangerous, with an increase in state anxiety.

The degree of *depressive* mood in healthy subjects or patients with depressive disorders has been shown to be correlated with immune functions (see Chapters 4 and 17). A self-reporting questionnaire that is often used as a single scale for assessing the degree of depression is the Beck Depression Inventory (Beck, Ward, Medelson, Mock, & Erbaugh, 1961). This scale contains 21 items, with each item consisting of four ordered statements regarding symptoms of depression. The ordered statements describe different grades of the seriousness of the depressive encroachment, and the subject must choose the statement that best describes his actual situation.

The concept of *social support* is based on the assumption that the support received from family members, friends, and colleagues, particularly during significant life events, positively affects not only psychological, but also somatic well-being (Cohen, Doyle, Skoner, Rabin, & Gwaltney, 1997; Sarason, Sarason, Potter, & Antoni, 1985). There are a number of questionnaires that measure the level of social support an individual receives. An example is the Social Support Questionnaire, which contains 27 items that measure the extent to which individuals believe that others love and value them, even if they may perform poorly or behave in a less than desirable fashion.

Instead of measuring the amount of social support, one can assess the amount of *loneliness* a subject experiences. Loneliness is defined as an unpleasant, straining state, emerging when a social network is perceived as deficient. Loneliness can be assessed with a scale such as the University of California at Los Angeles Loneliness Scale (Russell, Peplau, & Cutrona, 1980). However, the validity of loneliness scales is still difficult to assess because there is an absence of external validity criteria for loneliness, as for example, loneliness is not simply synonymous with isolation.

People differ in their belief as to how well they can control both specific life situations and life in general. The construct of *locus of control* of reinforcement was first described by Rotter (1975). Locus of control is commonly measured on three subscales: internal locus of control, external locus of control, and fatalism. Subjects scoring high on the internal locus of control subscale are described as internally oriented, with a high degree of self-determination in life situations. High scores on the external locus of control subscale are interpreted as a subjective feeling of powerlessness and a "powerful others external control orientation." The fatalism subscale describes an external locus of control based on fatalism. High scores on this scale reflect the subjective feeling that life is based predominantly on destiny, coincidence, and luck.

The relationship of everyday *stress* to health and well-being is discussed not only in psychoneuroimmunology research. The extent of stress load can be analyzed by a large number of different scales and questionnaires, which have been developed and standardized to measure stress. The decision concerning which measure is appropriate in a specific study depends on a number of considerations, such as the population under investigation, the study design, and logistic issues. For a comprehensive review of methodological considerations when measuring stress, see Cohen, Kessler, and Gordon (1995). An example of such a stress scale is the Daily Hassles Scale (De Longis, Coyne, Dakof, Folkman, & Lazarus, 1982). This questionnaire consists of a list of 117 hassles in relation to work, health, family, friends, the environment, and practical considerations. In responding to this scale, subjects first indicate the hassles they have had in the past month (hassles frequency), and the rate the severity of each hassle (hassle intensity).

5.4. Stress Experiments

Measuring the effects of stress on immune functions is one main approach in psycho-neuroimmunological research investigating the functional relationships between the brain, the neuroendocrine system, and the immune system in animals and humans. Most of the experimental settings applied in psychoneuroimmunology to operationalize the stress concept have been transferred from different research areas such as psychology, physiology, neuroendocrinology, cardiovascular research, or pain research, in which these stress models have been proven to elicit a stress response with clearly defined cardiovascular and neuroendocrine alterations. As the different stress models employed in animal and human experiments will be described in more detail in subsequent chapters, only a brief summary of different stress models used in psychoneuroimmunology will be given here.

Stressors can be defined according to three characteristics: duration, quantity, and quality of the stressor (see also Chapter 4). The duration of a stressor can be most simply defined as acute or chronic. For example, an acute stressor in animal experiments would be a short single electric foot shock, and in human studies a few minutes of public speaking in front of an audience, or a parachute jump (Table 1). In contrast, chronic stressors affect the well-being of the organism over longer periods (days to years), such as a long-lasting exposure to noise in animal experiments, or the care of a disabled family member in human studies. However, stressors in real-life situations are not always exclusively acute or chronic, and can be a series of events that occur over an extended period of time (stressor

Table 1
**Models Commonly Employed in Animal and Human Experiments
to Investigate the Impact of Stress on the Immune System**

			Duration	
	Quality	Stress model	Acute	Chronic
Animal	Physical	Electric	✓	✓
		Chemical	✓	✓
		Thermal	✓	✓
		Exercise	✓	✓
	Psychological	Conditioned fear	✓	
		Escapable vs. inescapable shock	✓	
	Social	Crowding	✓	✓
		Intruder	✓	✓
		Maternal deprivation	✓	✓
Human	Physical	Exercise	✓	✓
	Psychological	Mental arithmetic	✓	
		Public speaking	✓	
		Parachute jump	✓	
		Exam	✓	✓
		Loss of a spouse		✓
		Caregiving		✓
		Divorce		✓

sequences), or chronic intermittent stressors that may occur once a day, once a week, or once a month. These time dimensions of stress situations are not only of phenomenological value, as acute or chronic stress can have different, sometimes opposite, effects on immune functions (Chapters 15 and 16).

The second important characteristic is the quantity of a stressor applied in an experimental setting. For example, in animal experiments the amount of noxious stimulus such as electric foot or tail shock can be precisely operationalized when investigating the impact of acute stress. In human studies the mental or emotional load ranges from solving mental tasks in the laboratory (e.g., mental arithmetic) to a first-time parachute jump (Table 1). Differences in the quantity of a stressor can induce different neuroendocrine and immunological responses.

Finally, stressors can be categorized as a function of their quality in mainly physical, social, and psychological demands. In animal experiments, physical stressors include exposure of the animals to thermal, chemical, and electric stimuli, or forced exercise (Table 1). For the induction of psychological stress in animals one needs a more sophisticated approach. Psychological stressors can be operationalized with different paradigms, such as conditioned fear reactions or the approach–avoidance conflict. In the approach–avoidance conflict model, a hungry rat is presented food on a separate platform in the cage. Each time the animal attempts to obtain the food, it is exposed to an aversive stimulus, e.g., an electric shock. The animal quickly learns the association between food presentation and the aversive stimulus, thus producing the approach (to get the food)–aversion (to avoid the electric shock) conflict. This conflict induces psychological stress reactions, which can be measured in the neuroendocrine and immune responses.

Social stress in socially living animals such as monkeys or rats is normally induced by changing the social environment of the animals (see Chapters 13 and 14). There are different models of how social stress can be induced. In the "maternal deprivation" model, the pups or the dam is removed from the cage. In socially living animals there is a limit to the number of individuals that can comfortably live together. "Crowding stress" occurs when the number of animals in one cage exceeds this limit, even when there is abundant access to water and food. Social stress can also be induced with the so-called "intruder model," in which a new animal is placed into an established group of animals. All of these models induce a quite complex dynamic in the social environment of the animals, producing pronounced changes in endocrine and immunological parameters.

In humans, the effects of physical stress on the immune system are predominately analyzed using the "exercise stress" model (see chapter 18). Exercise stress allows a precise quantification of the applied stressor (acute versus chronic exercise, low versus high intensity) in highly controlled experimental settings in the laboratory (e.g., bicycle ergometer) or field studies (e.g., marathon run). In addition, cross-sectional as well as longitudinal studies can be designed for untrained and trained individuals.

Psychological stress models in humans can be distinguished according to whether they are chronic or acute in nature. In studies examining the effects of chronic stress, one is interested in assessing whether, and to what extent, longer-lasting psychological load such as exam stress, loss of a spouse, caregiving, divorce, or unemployment affects functions of the immune system. Experiments investigating acute stress commonly test immune reactivity before and after a single short-lasting psychological stressor. These acute stress situations can be further characterized as emotional stressors such as public speaking or parachute jumping, which predominantly induce feelings like fear, anxiety, or anger. In contrast, mental stressors such as arithmetic activate cognitive processes (attention, concentration) without a strong emotional involvement (see Chapters 15 and 16).

5.5. Conditioning Experiments

The discovery that changes in immune functions could be behaviorally conditioned provided the field of psychoneuroimmunology with an initial surge (Ader & Cohen, 1975) (see Chapter 23). Conditioning of immune functions is based on an associative learning paradigm initially developed by Ivan Pavlov (Pavlov, 1927). In a series of classic experiments, Pavlov showed that a dog presented with food will salivate (unconditioned response). However, Pavlov paired the food presentation with the sounding of a tone. After a number of trials, Pavlov re-presented the tone *without* exhibiting the food, and found that the dog salivated to the tone itself (conditioned response). That is, an association was formed between a neutral stimulus (tone) and a stimulus producing a physiological response. Such learning resulted in the organism producing the physiological response to the neutral stimulus *without cognitive control*. As such, Pavlov named the response a *conditioned reflex*, and this behavioral paradigm became known as *classical* or *Pavlovian conditioning*.

Although Soviet laboratories attempted to apply such a paradigm to inducing changes in the immune system, it was not until the 1970s that U.S. researchers showed that immune functions could be classically conditioned (Ader & Cohen, 1975). Three paradigms have emerged as the principal methods used to condition alterations in immune function: *conditioned taste aversion* learning, *odor conditioning*, and *fear conditioning*. Similar to Pavlov's original technique, these methods induce an associative learning between a benign *conditioned stimulus* (CS) and an immunomodulatory *unconditioned stimulus* (UCS) (Table 2). After single or multiple presentations of these stimuli in close temporal proximity, re-presentation of the CS alone produces changes in immune function that mimic those produced by the UCS itself.

5.5.1. Conditioned Taste Aversion

The use of a conditioned taste aversion paradigm to alter immune functions has been methodologically based on the work of Ader and Cohen (1975). This paradigm allows animals to consume a novel ingestive CS that has some hedonic properties. Shortly after consumption the organism is injected with a drug that alters immune functions. At a subsequent time, animals that have received such a pairing are reexposed to the CS alone. These animals avoid consumption of the CS (=conditioned taste aversion), and demonstrate

Table 2
**Comparison of Stimuli and Responses in the Paradigms Used by Pavlov
and Those in Conditioned Taste Aversion, Odor Conditioning, and Fear Conditioning**

	Pavlov	Conditioned taste aversion	Odor conditioning	Fear conditioning
CS	Tone	Sweet/sour taste (e.g., saccharin)	Distinct odor (e.g., camphor, peppermint)	Environmental stimuli (e.g., light, color, tone)
UCS	Food	Drug that alters immune function	Drug that alters immune function	Electric shock
UCR	Salivation	Changes in immune function	Changes in immune function	Changes in immune function
CS	Tone	Taste	Odor	Environmental stimuli
CR	Salivation	Aversion of CS (= CTA); changes in immune function	Changes in immune function	Changes in immune function

concomitant changes in immune function that would normally be ascribed to the immuno-modulatory drug. This method of conditioning has been commonly used, primarily because of the rapid acquisition of the CS–UCS association that is possible with this technique.

5.5.2. Odor Conditioning

A variation of the conditioned taste aversion paradigm is odor conditioning. This is conducted using a similar methodology to that employed in conditioned taste aversion, but with the novel taste replaced by a unique odor. Camphor and peppermint are typical odors that have been employed in this paradigm, although other distinct odors such as dimet-hylsulfide (sulfur smelling) and triethylamine (fishy smelling) have been utilized (Ghanta, Rogers, Hsueh, Demissie, Lorden, Hiramoto, & Hiramoto, 1994; Russell, Dark, Cummins, Ellman, Callaway, & Peeke, 1984). In contrast to conditioned taste aversion, animals exposed to odor conditioning do not modulate the ingestion of the CS. The odor is either placed under the nose of the animal using a cotton bud, or environmentally exposed within an enclosed chamber. Thus, where conditioned taste aversion may provide a behavioral indicator of the effectiveness of conditioning, no such marker is available in this paradigm.

5.5.3. Fear Conditioning

Fear conditioning commonly implements a specialized *conditioning chamber* (Lysle, Cunnick, & Maslonek, 1991). Such a chamber allows the provision of the CS and UCS. That is, the chamber contains novel environmental cues such as color, light, and noise that act as a CS. The chamber is additionally able to impart the UCS, small electric shocks that are conducted via the chamber floor. Stressors of this type produce distinct changes in immune function. Subsequently, on reintroduction to the chamber, the learned expectation of shock produces changes in immune function that mimic the actual immune alterations produced by the shock itself. Although a regularly used paradigm, this method is less regularly implemented than conditioned taste aversion techniques, as the results may be influenced by CS reexposure acting as a stressor.

5.6. Intervention Studies

Psychosocial interventions such as relaxation or stress management training have been shown to positively affect well-being. Regular relaxation training, such as progressive muscle relaxation, biofeedback, hypnosis, or meditation, leads to a reduction in physical tension, which can be measured as decreased heart rate, blood pressure, muscle tension, and skin conductance. In parallel, negative feelings such as anxiety and depression are reduced. Therefore, these techniques have been employed in psychoneuroimmunological research to investigate whether psychosocial intervention positively influences parameters of the immune system, and the course of a disease (see Chapters 20 and 25).

One of the most common relaxation techniques in psychoneuroimmunological re-search is progressive muscle relaxation. In this technique, the subject is asked to tense, and subsequently relax, individual muscle systems. After regular training the subject learns to relax both muscularly and psychologically, to differentiate different grades of muscular ten-sion, and to associate cognitive and bodily tension. The relaxation training is often combined with so-called guided imagery, where the subject creates mental images that improve the relaxation effects and increase feelings of hope and well-being (see Chapter 20).

In clinical populations such as cancer or HIV patients, relaxation training is often combined with a cognitive–behavioral stress management training (see Chapters 20 and 25). These structured group intervention programs commonly consist of three components: health education, stress management, and coping with the disease. In the health education section, patients are given information concerning the etiology, diagnosis, and therapy of the disease, and information regarding general health-related behavior (e.g., nutrition, exercise, sleep). In the stress management component, patients learn to identify their individual stress situations and stress reactions, and subsequently learn to avoid stress-inducing events and/or improve their resources to cope with stress (e.g., relaxation training). In order to improve the technique of coping with the disease, the training commonly focuses on an active coping method, on both a behavioral (exercise, improved nutrition, relaxation training, social support) and a cognitive level (to set new priorities in life, positive thinking, problem solving), instead of avoidance strategies such as suppressing emotion, avoiding social contacts, or drug and alcohol consumption. These intervention techniques have been employed in a number of studies investigating the effects of psychosocial interventions on immune functions and disease outcome in cancer and HIV patients (Fawzy, Kemeny, Fawzy, Elashoff, Morton, Cousins, & Fahey, 1990; Fawzy, Fawzy, Hyun, Elashoff, Guthrie, Fahey, & Morton, 1993a; Schedlowski, Jungk, Schimanski, Tewes, & Schmoll, 1994; Spiegel, Bloom, Kraemer, & Gottheil 1989; see Chapter 20).

5.7. Correlational Studies

A wealth of research in psychoneuroimmunology has been conducted by computing the correlations between psychological, endocrine, and immunological parameters. For example, one may compute the correlation between scores of personality traits (such as depressive moods, anxiety, or helplessness) and several endocrine or immunological parameters. However, the validity of data such as these is very restricted, as correlations provide no information of causal relationships. If one found, for example, a correlation between depression and suppression of immune functions, one could give three explanations for this relationship:

1. A disturbance of a mood such as depression may have a negative effect on immune functions.
2. Suppression of immune functions may have psychological side effects such as depressive moods.
3. Physical illness may suppress immune function while simultaneously increasing the depression of the patient.

The results of correlational studies can therefore only lead to the development of hypotheses which should subsequently be examined by experimental methods. However, if one has a theoretical model regarding causal relationships between psychological, immunological, and endocrine parameters, correlation coefficient matrices may be analyzed to reveal whether these theoretical concepts can be confirmed by empirical data. In these cases, the hypothesis of causal relationships is proved by structural equation models in which a system of causal relationships is described by several structural equations, in order to test the correspondence of the theoretical model with empirical data. However, the validity of correlational studies is reduced because biological parameters normally show modulation of amplitude and frequency. Furthermore, they are modified by seasonal, circadian, and ultradian rhythms, and they show different latencies. For example, acute stress may lead to an immediate increase of adrenaline, while an increase in cortisol might be observed with a latency of 20 min. If all measurements for correlational computations are made at this same time, these differential effects influence the measurements in the same way as errors. Instead of cross-sectional correlational studies, it is preferable to conduct repeated measurements over time which can be evaluated by analysis of variance with repeated measures. Another appropriate method may be the analysis of time series.

5.8. Conclusions

When planning research in psychoneuroimmunology, the selection of psychological parameters must be chosen by criteria other than those used for the selection of biological measurements. Numerous parameters may be recorded for the description of biological states, so far as this is justified by theoretical and conceptual reasoning. In this case, more data and more information lead to a higher quality of the description of the biological system by quantitative measurements. However, this approach cannot be simply transferred to the measurement of psychological states and traits. Psychological constructs are not immediately assessable. Results of psychological tests only allow indirect conclusions. The data normally have a proportion of measurement error. At first glance it may be effective to integrate numerous psychological scales that we find in multivariate personality questionnaires into psychoneuroimmunological studies. This decision may be associated with the hope of finding interesting results. However, a wiser approach is to reduce the psychological parameters to a smaller selection that can be justified with good reasons. If one wants to test the hypothesis that adequate coping with stress has an influence on immunological parameters, then it would make more sense to utilize a one-dimensional scale with low error of measurement on which individual differences can be reflected precisely. An increase of the number of psychological parameters may soon lead to a confusing and incalculable number of possible relationships and interactions, which do not allow unambiguous and reasonable scientific interpretations of the results.

However, it should be noted that when measuring psychological parameters, one may fall back on statistical methods for the reduction of data that may lead to interpretable results. In such cases it may be possible to identify personality types by means of statistical methods (e.g., nonhierarchical cluster analysis), and it is possible to examine whether these different personality types contrast in endocrine and immunological processes. Nevertheless, causal interpretations in this case are not feasible, as it is impossible to say whether the effects are related to an influence of the personality differences on biological processes, or whether psychobiological changes modulate personality.

Behavioral psychologists are used to developing conceptual and theoretical models to examine how biological and psychological systems interrelate, and to examine the functional significance of this relationship. This attempt is illustrated by several chapters in this volume, like those investigating the impact of sleep, stress, depression, and conditioning paradigms on immune functions. However, it must be noted that studies in which psychological variables are only included on a trial basis, combined with the expectation that these might lead to some interesting results, normally do not lead to any valid scientific findings. Nevertheless, behavioral scientists are accustomed to operationalizing their concepts into experimental designs that lead to unambiguous, clear results, and that allow statements about the fundamental relationships between the different components of these systems.

6   Endocrinological Methods

Thomas H. Schürmeyer and E. Jean Wickings

Thomas H. Schürmeyer • Department of Endocrinology and Metabolism, Psychobiological and Psycho-somatic Research Center, University of Trier, D-54290 Trier, Germany.   E. Jean Wickings • International Center for Medical Research, CIRMF-BP 769, Franceville, Gabon.

6.1. Introduction

The concentrations of hormones found in body fluids are so low that normal chemical methods are not sufficiently sensitive to detect them. The first successful methods used in this century were bioassays. For example, growth hormone activity was determined by the ability of a tissue extract to influence the growth of tadpoles. Similar bioassays exist for nearly all hormones, and they can be very sensitive, but are in general extremely laborious and tedious to perform, and unsuitable for many clinical applications.

The development of the radioimmunoassay (RIA) by Yalow and Berson in 1959 signaled the breakthrough in methodology. The first hormone to be measured by RIA was insulin; subsequently, methods were rapidly developed by other endocrinologists for the measurement of practically all known hormones and many of their metabolites, and also drugs in all of the important body fluids (e.g., blood, urine, amniotic fluid). With the advent of RIAs and later of immunoassays using other labels, an enormous expansion occurred in the field of endocrinology and related disciplines, which adapted this methodology to suit their needs. More recently, immunoradiometric methods have been developed from the basic RIA technique (see Section 6.3). At this stage, it became possible to label either the hormone (antigen) or the antibody with chemical (fluorescent, luminescent) as well as radioactive tags. One such luminescent tag is the molecule luminol, which, as the substrate for the luciferase enzyme, is responsible for the characteristic light of glowworms.

6.2. Immunoassay and Applied Methods

6.2.1. General Principles and Practical Aspects

A sample containing the hormone to be measured is first diluted with a known amount of labeled hormone (radioactive, luminescent, fluorescent, enzyme-labeled). Next, a specific antibody raised against the hormone is added (Fig. 1, Step I), which binds both labeled ("tracer") and unlabeled hormone. An incubation period follows, during which time equilibrium is reached between the quantity of bound and unbound hormone. If the concentration of unlabeled hormone is small, then the amount of tracer hormone bound is high, and conversely, if the amount of unlabeled hormone in the sample is high, then only a small amount of tracer will be bound. The antibody-bound hormone is separated from the unbound or free hormone fraction by addition of precipitating or absorbing reagents (Step III). At this stage, it is possible to measure the amount of tracer associated with either fraction, depending on the separating reagent used. In the case of steroids, activated charcoal suspension is frequently used to absorb and precipitate the free fraction (Step IIIa), which allows, after a short centrifugation, the supernatant to be removed for quantification of the amount of tracer present (Step IVa). Alternatively, the hormone-specific antibody can be precipitated, either by classic protein precipitation methods or by addition of a second antibody specifically directed against the first, which on combining with that antibody forms an insoluble complex (Step IIIb). This latter method is preferred for protein hormone assays. Following centrifugation the supernatant containing the unbound hormone is removed and the amount of tracer associated with the bound fraction can be quantified (Step IVb). By incorporating samples with known quantities of hormone in an assay setup, a standard curve can be constructed against which unknown samples can be compared (Fig. 2) (see Section 6.6.2).

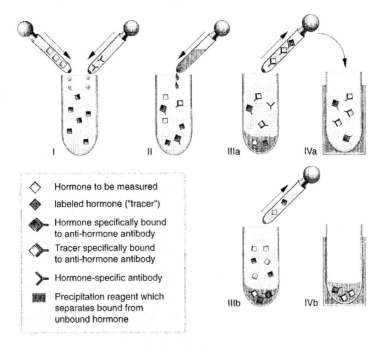

Figure 1. Principle of immunoassays.

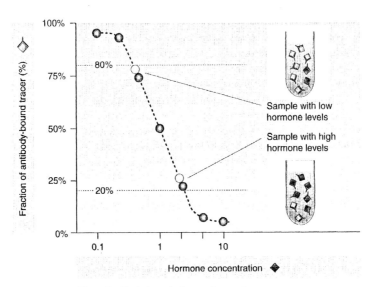

Figure 2. Typical standard curve from an immunoassay.

6.2.2. Extraction

It may be necessary to include an extraction step prior to the immunoassay in order to remove any interfering substances from the sample or to concentrate the sample. When chromatographic methods of purification, as described below, are required, the inclusion of an extraction step becomes essential. Figure 3 shows the type of extraction frequently used with samples for the determination of steroid hormones. The sample is extracted (Step I) with an organic reagent (ether is the most frequently used) by manual shaking (Step II). The steroid is more soluble in the lighter organic phase, and the interfering factors remain in the heavier aqueous layer. The aqueous phase is separated either by freezing or by centrifugation (Step III), and the organic layer can be quantitatively removed (Step IV). The organic phase is dried off under a gentle stream of warm air to leave the extract residue at the bottom of the tube (Step V). This is reconstituted in an aqueous solution, usually the assay buffer (Step VI), leaving a clean extract containing the hormone, but no water-soluble interfering factors. According to the volume of sample initially extracted and of the buffer used to reconstitute the residue, the hormone may also be in a more concentrated solution.

6.2.3. Chromatography

Not all interfering factors can be removed by a simple extraction step, as some are also soluble in organic reagents, and some may also be closely related structurally to the hormone of interest and hence not differentiated by the hormone-specific antibody (see Section 6.6.2). This problem arises, for example, if cortisol and cortisone have to be separated in order to measure their individual concentrations in a sample. In such instances, various chromatographic separation methods are available, but only thin layer chromatography will be described here. Alternatively, high-pressure liquid chromatography (HPLC; see Section 6.5.1), column chromatography, immunoaffinity chromatography, and electrophoresis and electrofocusing can be used.

To apply thin-layer chromatography techniques, the sample to be separated must be

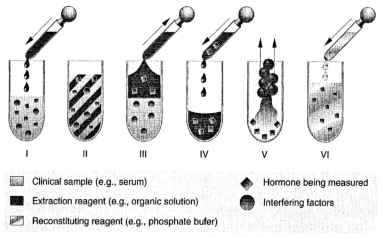

| I | II | III | IV | V | VI |

⬚ Clinical sample (e.g., serum)   ◆ Hormone being measured

■ Extraction reagent (e.g., organic solution)   ● Interfering factors

▨ Reconstituting reagent (e.g., phosphate bufer)

Figure 3. Principle of hormone extraction using organic reagents.

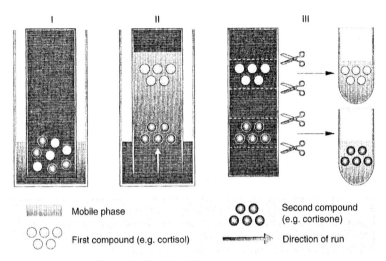

Figure 4.   Principle of thin-layer chromatography.

taken up in the smallest possible volume following extraction, in order to be loaded onto the lower edge of the plate. The solid support used in thin-layer chromatography may be filter paper, or glass plates coated very thinly with silica gel. After loading the sample and drying off the solvent used for loading, the support is placed or hung in an airtight tank with a predetermined mixture of liquid "mobile" phase (Fig. 4, I). This solvent mixture rises through the solid support by capillary action, and the hormones in the loaded sample are separated according to their solubility in the components of the mobile phase (e.g., cortisol has the solubility of an alcohol, whereas cortisone has that of an aldehyde) (Fig. 4, II). When the solvent front reaches the top of the solid support, the plate or sheet is removed and rapidly dried, and the regions corresponding to, for example, cortisol and cortisone can be cut out individually (Fig. 4, III). The hormone can be recovered from the solid support into an appropriate solvent, which is then dried as described above, and the sample can be measured in an immunoassay.

The inclusion of chromatographic purification or extraction procedures necessarily results in a loss of the specific hormone, which must be corrected for in the final analysis. To correct for procedural losses, a small amount of labeled hormone (usually radiolabeled in this instance) is added to the clinical sample at the outset, and it is assumed to behave identically to the hormone in all extraction and chromatographic steps. The absolute amount of hormone associated with this radioactive "recovery" label is so small as to be negligible when calculating the final concentration of the hormone measured in the immunoassay. The overall losses incurred during extraction are on the order of 10 to 30%, as measured by a recovery of 70 to 90% of the radioactive "recovery" tracer, but as high as 40 to 60% if a chromatographic step is included. The level of hormone measured in the immunoassay must be multiplied by the recovery factor in order to estimate correctly the concentration of hormone in the clinical sample (Baxter, 1980; Péron & Caldwell, 1972; Wickings & Nieschlag, 1975).

6.3. Immunoradiometric Assay and Related Methods

6.3.1. General Principles and Practical Aspects

Antibodies recognize only a small section (epitope) of the target antigenic molecule. If the antigen is a large molecule (e.g., peptide hormones ACTH and PTH or glycoprotein hormones such as LH, FSH, TSH, or STH), the antibody will not be capable of differentiating between the whole molecule and fragments of the hormone incorporating the antibody-specific epitope, nor between different hormones containing the same epitope (e.g., glycoprotein hormones containing identical species-specific subunits). These types of hormones are preferentially measured using radioimmunometric assays whereby two specific antibodies are employed which may be directed toward the two ends of the hormone molecule. One of these hormones can be labeled as described in Section 6.2.1, whereas the second can be used as the target for the second precipitating reaction. Figure 5 shows such a system in the form of a "coated-tube" assay. The coated-tube assay works on the principle that the first antibody is bound in excess on the walls of the reaction tube. The sample and the second labeled antibody are added to this tube (Step I). During the ensuing incubation, the first "coated" antibody binds the hormone from the sample, which in turn binds the second labeled hormone (Step II). Any unbound second antibody is then aspirated or removed by washing and the fraction bound is counted (Step III). The amount of radiolabeled antibody bound is proportional to the hormone concentration. If both antibodies are present in excess, then hormone fragments, which bind to only one antibody, do not interfere in the assay. In addition to this, immunoradiometric assays are more sensitive than RIAs, but require relatively large amounts of radioactivity and antibodies, which increase the cost of each test.

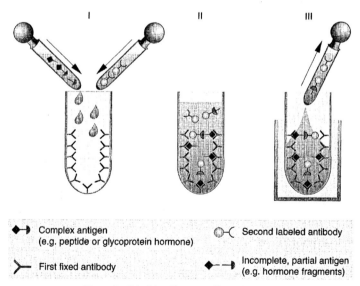

Figure 5.   Principle of immunoradiometric assays.

6.3.2. Antibody Production

Polyclonal antibodies are raised in experimental animals by immunizing with the substance subsequently to be measured in the assay. Some antigens (e.g., glycoprotein hormones such as FSH or TSH) are sufficiently large to be antigenic when injected in conjunction with a suitable adjuvant (killed tuberculosis or whooping cough bacteria) without further modification. Very small molecules, such as steroids, catecholamines, thyroid hormones, or hypothalamic releasing factors, are not recognized as antigen without first being coupled to a much larger carrier molecule. Carriers such as thyroglobin, tetanus toxoid, or albumin have all been used successfully. The way in which the "hapten" or hormone molecule is coupled to its carrier can influence the properties of the antibodies raised, depending on which epitopes of the hormone are left exposed to the immune system. If a peptide is coupled through its N-terminus, then only the C-terminus of the hormone is left exposed to the antibody recognition site. Extrapolating to the other families of hormone haptens, the antibodies raised will be either hormone-specific or reactive against a group of structurally related hormones. Antibodies produced by this mode of immunization are polyvalent in nature, because many epitopes of the hormone-carrier molecule will be recognized, including those to the carrier only and to the chemical bridge between the hapten and the carrier. Depending on the reaction to the first immunization, it is frequently necessary to give subsequent "booster" injections of the antigen over a period of several months at intervals of several weeks, in order to stimulate the animal's immune response to produce sufficient quantities of the desired antibody. The amount of antibody produced is measured in antiserum dilutions or "titers" (useful in the range of 1/100,000), and the properties essential in an assay system (sensitivity, specificity, lack of cross-reactivity) are also tested in samples collected regularly during the immunization process. Once characterized, aliquots of the antiserum are stored frozen or lyophilized, in order to protect them against degradation. The above procedure describes the production of polyclonal antibodies, as the immunization procedure activates many different elements of the animal's immune system, resulting in a mixture of different antibody types. These antibodies come from different "clones" of cells, and may, therefore, differ in their individual properties.

The production of "monoclonal" antibodies contrasts that of polyclonal antibodies. In this method, mice are immunized against the prescribed hormone and once the immune system has been stimulated the spleen is removed and homogenized. Mouse spleen cells are then fused with tumor cells from immortalized cell lines, and the successfully fused cells are separated and placed in culture. Very few cells survive these drastic measures, but those that do are diluted out on culture so that single cells can be identified as the origin of plaques in culture. The antibodies produced in culture by these cell clones are described as monoclonal, having all originated from a single cell and having identical properties. The production of monoclonal antibodies is technically laborious and fraught with difficulties. Single cells grown in dilute culture do not reproduce well, and culture conditions such as the addition of special cell growth factors and optimization of temperature, pH, CO_2 concentration, and cell density, as well as strict observation of sterile technique, are essential for success. After a successful fusion and growth of cells, the tedious business of testing every single clone for antibody production begins. With luck, one clone culture will produce a suitable antihormone antibody in high concentrations, which can then be "milked" over a long period (Fig. 6). The great advantage of monoclonal antibodies is that their characteristics do not change during the immunization process. Hence, all antibodies collected from that clone will be identical, in contrast to polyclonal antibodies, which change their properties with time and after each booster injection.

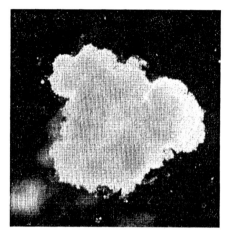

Figure 6. Antibody-producing clone of cells on the edge of the culture dish.

However, the increased selectivity of the monoclonal approach, and an enhanced specificity of an antibody clone may be a disadvantage when measuring hormones of the type that may have subtle interindividual structural differences not affecting their biological activity (e.g., glycoprotein hormones). The considerable investment required to produce monoclonal antibodies must be one of the reasons why polyclonal antibodies are still predominantly used in clinical and research assays.

6.4. Radioreceptor Assays and Applied Methods

Radioreceptor assays are in principle very similar to RIAs, except that the antibody used in immunoassays is replaced by a hormone-specific receptor. In an immunoassay, the ability of labeled and unlabeled hormone to compete for the antibody binding site, in relation to the concentration of the unlabeled hormone fraction, forms the basis of the technique. In the case of a radioreceptor assay, the ability of a hormone to bind a ligand to its specific receptor is tested. Using this approach, it is possible to determine the receptor concentration present in the reaction tube, for example, from tissue extracts, and the binding characteristics of the receptor can be examined. The most important property is the binding affinity of the receptor for its primary, biological ligand, which can be determined in the presence and absence of competing ligands which the primary ligand may or may not be able to displace from the receptor binding site. Other important characteristics are the number of binding sites per receptor molecule and the proportion of specific to nonspecific binding of the ligand in the receptor preparation. The results of such binding studies are presented in the form of Scatchard plots, whereby the proportion of bound to free hormone (B/F) at equilibrium is plotted on the y axis and the concentration of bound hormone (B) is plotted on the x axis. This latter bound fraction is the same as the concentration of occupied receptor sites. Using the following formula,

$$B/F = k(R - B)$$

the total number of receptors in the sample (R) and the association constant (k) can be

assay can then be used in the same way as an immunoassay to determine the concentration of a ligand like 1,25-dihydroxycholecalciferol, cAMP, or catecholamines in an unknown sample. Some bioassays are a natural extension of radioreceptor assays and measure the effect of hormones on specific cells.

6.5. Chromatographic Methods

6.5.1. High-Pressure Liquid Chromatography

The basic principle of thin layer chromatography previously described (see Section 6.2.3) also applies in HPLC, where the separation of molecular forms is achieved by elution of a sample from a solid inorganic support using a mixture of organic solvents. The solid support in HPLC is in the form of capillary columns packed with silica or cross-linked dextran. The efficiency of separation of samples depends on the effective size of the interface between the stationary phase and the mobile phase in which the sample is progressively dissolved. Hence, the physical as well as the chemical interactions between the packing material of the support phase and the compounds to be separated play an important role in determining the efficiency of separation. This is achieved either by a nonspecific absorption of one or more components onto the solid phase, or by the distribution of the components along the support phase, and the inert column may act as a sieve, sorting the components retained on the column according to size. As indicated by the name, the organic solvents are forced through the column at high pressure. This allows certain solid support phases to be used at high pressure but which would otherwise be impermeable to the organic phase. The run time is also appreciably shortened by the use of high pressures, and, because of the high reproducibility in running conditions, methods can be automated (Fig. 7; Loeber, 1984).

The high degree of purification of a sample following HPLC means that direct detection methods (e.g., electrochemical and spectrophotometric detectors) can often be applied at the end of the run for certain classes of hormones. The measurement of catecholamines

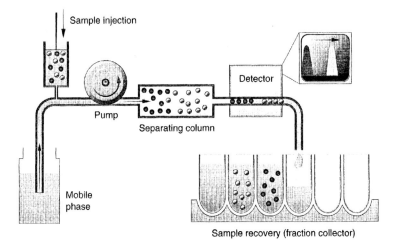

Figure 7. A schematic diagram demonstrating the principle of HPLC.

in brain nuclei of experimental animals in neuropsychiatric studies, and in certain instances, the measurement of urinary steroids in clinical cases can be effected by direct detection following HPLC. Until recently it was only possible to measure blood concentrations of catecholamines by HPLC and electrochemical detection. This method, however, is more susceptible to interference than, for example, the separation of individual catecholamines by HPLC coupled with a robust radioreceptor assay or immunoassay of each column fraction. Figure 7 shows a schematic representation of an HPLC separation coupled with fraction collection at the end of the column to preserve the individual components eluted from the column. Figure 8 shows a typical spectrum of eluted steroids following spectrophotometric quantification.

6.5.2. Gas Chromatography

Gas chromatography (GC) is closely related to the more recent HPLC methods, in that the mobile phase used is a gas instead of a fluid. Substances suitable for separation on GC must be stable at high temperatures and in their vapor phase. Stationary supports, through which the mobile gas phase carries the substances to be separated, may be solid or liquid (e.g., silica gel or silicone oil). On emerging from the end of the column, the different fractions containing the pure substances can be collected, and for some molecules it is possible to incorporate a detection system directly at the end of the column. Flame spectrophotome-

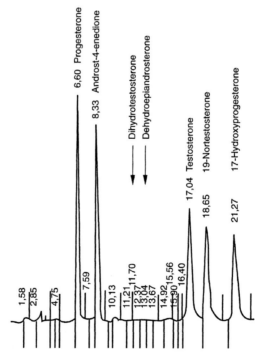

Figure 8. The separation of individual steroids by HPLC and quantification by spectrophotometry.

try and mass spectrometry (MS) are the most commonly used methods of detection and quantification, and GC-MS has become the "gold standard" for the detection, identification, and measurement of many steroids. Quality control samples used in many of the national and international immunoassay control programs are subjected to MS-GC in order to provide reference values of the hormone concentrations present in the samples. This method is too laborious to find applications in routine clinical practice and in large series of clinical research samples.

6.6. Quality Control

All methods undergo a mandatory control of reliability before being introduced into clinical or routine use. The criteria rigorously checked include specificity, sensitivity, accuracy, and precision. Any method whether published or available commercially is to be regarded with caution if it is not accompanied by an analysis of these four criteria. Any assay that does not satisfy these strict criteria will not provide useful clinical or biological information.

6.6.1. Specificity

No method—whether RIA, immunoradiometric assay, or any method employing antibodies, molecular biological techniques using RNA or DNA probes, or methods for the determination of receptor binding using isolated hormone receptors—is totally specific for the particular hormone, peptide, or oligonucleotide in question. All of these methods are based on chemical interactions between the substance to be measured and the chemical agent employed in the assay method. If a biological sample contains compounds that are structurally closely related to the one substance to be measured, a nonspecific assay reagent will not be capable of distinguishing between them, and the estimate of the specific substance will be overestimated because of the contribution of these "cross-reacting" compounds in the assay. The effect that one of these compounds may have on the assay method must be checked, and the concentration of the cross-reacting compound in the sample must be taken into account. Thus, cortisol concentrations measured using an RIA may under certain circumstances be rendered invalid if the antibody reagent employed shows a high cross-reactivity to compounds and drugs structurally related to cortisol. If, for example, the subject has not been treated with the synthetic preparation prednisone, the cortisol values measured will be correct. If the patient has taken this drug, then the high degree of cross-reactivity of prednisone with the anticortisol antibody will interfere with the measurement of cortisol. Even antibodies that show only a low cross-reactivity ($<1\%$) will bind so much prednisone that patients injected with this drug, for example for an asthma attack, will show dramatically elevated levels of "cortisol" as a result of the high concentration of the drug in the bloodstream. In such cases it is mandatory to separate the interfering components in serum from cortisol in order to obtain a correct determination of cortisol levels. Such preliminary extraction and separation (chromatography) steps increase the cost and time involved of a single hormone test considerably.

Hormone receptors generally have a lower specificity for their ligands than do antibodies. Naturally occurring hormones are often poor therapeutic compounds because of their unsuitable pharmacokinetics, although they may be rendered more effective by chemical modification, or structurally related molecules may be synthesized. These compounds show similar binding and activation properties as the natural hormone to the receptor. Most hormonal contraceptive preparations contain ethinyl estradiol, rather than estradiol. Such structurally related or altered molecules are often of practical value as ligands in radioreceptor assays if a difference in affinities between ligands at the receptor-binding site is under examination.

When measuring peptide hormones in immunoassays, it should always be remembered that the antibody recognizes only a short sequence of amino acids (epitope) and not the whole peptide molecule. Thus, smaller peptide fragments or precursors will also be recognized in the assay, yielding falsely high values. β-Lipoprotein is a precursor of β-endorphin that occurs in high concentrations in the blood, and almost all assays for β-endorphin cross-react with β-lipoprotein, as the latter molecule contains the β-endorphin epitope. It should be mentioned that the majority of psychoneuroendocrinological publications purporting to measure β-endorphin used methods that cross-reacted with β-lipoprotein. One way of avoiding this problem is to use immunoradiometric assays for determining blood levels of these hormones.

6.6.2. Sensitivity

The sensitivity of a method is defined as the smallest amount of hormone that can be distinguished from zero with 95% confidence limits. This depends on several factors, such as the quality of antibodies used, the binding affinity of a receptor preparation, the proportion of antibody to antigen, and the proportion of labeled to unlabeled hormone in the reaction tube. In many cases, the reaction conditions (e.g., incubation duration and temperature, pH, volume of sample) also play an important role. The correct choice of these parameters depends very much on experience and "instinct" of the technician setting up the method. The sensitivity can be readily determined by carrying out replicate measurements of the blank value of the assay (no added unlabeled hormone) and calculating the spread of the resulting concentrations. The lowest detection limit is the average value of measurements made with no hormone added, plus two standard deviations of the mean. The optimal working region of the standard curve lies between 20 and 80% binding of the added tracer, whereby the sensitivity is usually lower than this (around 10% binding). Hence, results that lie between the sensitivity of the assay and 20% binding, also those between 80 and 90% binding, should be regarded with caution. The high degree of variation in the extreme ranges of the standard curve can be shown in a precision profile over the whole region of the curve (see below).

6.6.3. Accuracy

The accuracy of a method is determined by additivity experiments. A known concentration of unlabeled hormone is added to a sample with no measurable hormone concentration; in the absence of interfering factors, the final value measured should be that of the hormone added. If a sample of known concentration is diluted, then the final values measured should be in proportion to the dilution factor. After the addition of known concentrations of unlabeled hormone to a sample of known value, the final hormone concentrations should be the sum of the inherent level and the added amounts. The recovery of added amounts of unlabeled hormone should lie ideally between 80 and 120% of the expected concentration. By performing this type of analysis over the range of the standard curve, the accuracy of a method can be determined, in order to assess any problems leading to an over- or underestimation of the hormone concentration.

6.6.4. Precision

The precision of each method must be determined to estimate the reproducibility of a given value on repeated measurement in the same assay (intra-assay variation) and in consecutive assays (interassay variation). Interassay variation is generally higher than intra-assay variability, and even greater is the variability between different methods for measuring the same hormone. This intermethod variability is often a component of quality control (QC) programs, where the reference method for measuring the absolute concentrations of, for example, steroid hormones is gas chromatography and mass spectrophotometry (GC-MS), whereas a QC scheme will most likely employ immunoassay or immunoradiometric methods.

Every reputable scientific journal insists that methods used in publications be defined in terms of their precision (intra- and interassay coefficients of variation). The intra-assay variability is easily determined by calculating the difference between duplicate measurements for each sample, and evaluating them over a whole assay. The calculated value should be below 10%. Interassay variation is assessed by including the same samples in each assay and calculating the spread of concentrations measured over a period of time. Samples, usually serum samples, containing low, medium, and high levels of the hormone should be assessed each time, as the precision of the assay can vary with the level of hormone measured. An average coefficient of interassay variation, usually calculated from optimal values measured in the middle of the standard curve, is then quoted in publications. For steroid and thyroid hormones this variability should be below 15% for assays not including a chromatography step after extraction, but can be as high as 20% in assays incorporating a chromatographic separation. The calculation of a precision profile for each assay shows the degree of variation in the reproducibility of hormone determinations over the range of the standard curve (Fig. 9). Variation is highest at the two ends of the standard range, and

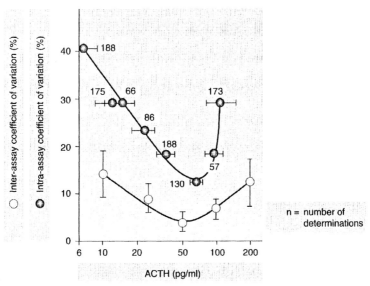

Figure 9. A typical example of precision profiles for an ACTH radioimmunoassay.

hence samples containing high or low levels of a hormone will give the least reproducible results on analysis. To reduce this lack of precision, samples with expectedly low or high levels of a hormone should be concentrated or diluted correspondingly, so as to utilize the portion of the standard curve showing maximum precision. Users of commercially available reagents (or kits) should be aware that the coefficients of variation quoted by the manufacturers cannot be expected to meet the rigorous standard required of clinical and especially of research applications. Each method should be characterized within the laboratory framework where it is being applied, and these in-house values may then be quoted in publications.

6.7. Dynamic Tests of Hormone Secretion

There are a number of functional tests that measure the ability of an endocrine gland to respond to stimulatory or suppressive substances (either drugs or synthetic releasing factors), with the appropriate increase or decrease in the circulating levels of the target hormone. Such dynamic tests do not permit any conclusions to be drawn regarding the basic functioning of a hormonal system, nor in attempting to quantify the response within a clinical framework, but belong in the domain of pharmacology, because of the nature and amount of the stimulus/suppressor used. They are not capable of elucidating the very sensitive changes involved in the physiological equilibrium of the endocrine system. Since the introduction of sensitive methods of determination of hormone levels, these dynamic functional tests have largely been abandoned for diagnostic purposes, being called on only in certain instances of borderline basal values or in the event of a difficult differential diagnosis. In the field of diagnostic psychoneuroendocrinology and in the differential diagnosis and long-term follow-up of psychiatric patients, such insensitive tests are now only of historical interest. The frequent use of functional tests in their heyday led to a plethora of results that had little value in elucidating the physiological and pathophysiological events underlying a clinical situation. A very brief description of these tests will be given here, as they are still occasionally encountered in the field of endocrinology.

6.7.1. Stimulation Tests

A classical and useful hormone stimulation test is the hCG test, which is used to test for the presence of functionally intact testicular tissue in boys with nondescended testes, where the testes may be in the abdominal cavity or nonexistent (anorchism). The boy is given a single injection of human chorionic gonadotropin (hCG), which functionally resembles the pituitary hormone LH. An increase in circulating levels of testis-derived androgen, testosterone, should be seen in response to hCG in cases where functional Leydig cells are present. Equally useful is the ACTH test for patients with only residual adrenal tissue, or who are incapable of producing cortisol because of a congenital metabolic disturbance. If the plasma cortisol level does not reach at least 20 μg/dl 60 min after an injection of 250 μg ACTH(1–34), then the patient has in all probability a primary adrenal insufficiency (morbus Addison). The adrenals of patients with adrenogenital syndrome secrete androgens rather than cortisol because of a congenital enzyme defect, resulting in the production of large amounts of the precursor 17α-hydroxyprogesterone in heterozygotes for the defect, as well as in the homozygotes. The ACTH test is not suitable for differentiating individual endocrine responses to stress.

The secretion of pituitary hormones can be stimulated by the injection of specific

Table 1
Hypothalamic Stimulating and Suppressing Factors,
with Their Target Pituitary Hormones

Hypothalamus	Pituitary
Releasing factors	
LHRH (luteotropin, gonadorelin)	LH (luteinizing hormone)
	FSH (follicle stimulating hormone)
THR (thyrotropin-releasing hormone)	TSH (thyrotropin-stimulating hormone)
GHRH (somatotropin-releasing hormone)	STH (stomatotropin, growth hormone)
CRH (corticotropin-releasing hormone)	ACTH (adrenocorticotropic hormone)
Inhibiting factor	
Dopamine	Prolactin
Somatostatin	STH (somatotropin, growth hormone)

releasing hormones secreted by the hypothalamus, and which can be synthesized in pure form in sufficient quantities for test purposes (Table 1). Intravenous injection of the hypothalamic hormones causes the almost immediate release of the corresponding pituitary hormones. No information can be deduced from these tests on the functioning of the hypothalamus, nor of the central nervous system. In certain psychiatric cases, a reduced or increased response in such tests, such as that seen in primarily nonpsychiatric illnesses, can be observed, but can only be regarded as one of the symptoms, and not as a cause of the underlying illness.

The hypothalamic production of these releasing factors can be readily affected by pharmacological means, but also by the framework in which the test is carried out. However, the tests elucidate only a part of the overall response to the stimulus applied. For example, the secretion of the releasing factor CRH, which is particularly interesting from a psycho-logical point of view, can be altered by drugs that affect serotonin, noradrenaline, and acetylcholine secretion within the brain. In turn, such drug-induced changes affect the resulting ACTH and cortisol secretion. In a stress situation it is impossible to exclude the possibility that only one of these interdependent factors may be affected. Hence, it is impossible to interpret these pharmacological tests within the framework of the complex underlying physiology.

One standardized endocrinological "stress" test consists of measuring either ACTH or cortisol following insulin-induced hypoglycemia. The hypoglycemia seen 30 to 45 min after an intravenous injection of 0.15 IU insulin/kg body weight causes the unpleasant, and for children and aged subjects often dangerous side effects of profuse sweating, increased heart rate and blood pressure, and disturbances of concentration and vision. It also induces the release of catecholamines from the sympathetic nervous system, stimulating ACTH and cortisol secretion. This test determines the ability of the hypothalamic–pituitary–adrenal axis to respond adequately to a given metabolic stress. No direct information about the response to a physical, psychological, or pain-induced stress can be deduced from the insulin-induced "stress" test, although certain useful conclusions may be drawn, given that no standard tests of the reaction to pain exist.

6.7.2. Suppression Tests

One of the most commonly known endocrinological suppression tests is the dexa-methasone suppression test, which has been employed for many years in psychiatric practice. Dexamethasone is a synthetic glucocorticoid with 35 times the potency of the adrenal steroid cortisol. When this suppression test is carried out for endocrinological reasons, the patient receives 2 mg dexamethasone in tablet form at 10 PM, a dose that corresponds to double the normal daily production of endogenous glucocorticoids. In normal subjects, this dose of synthetic glucocorticoid suppresses all endogenous production of cortisol, and hence circulating concentrations of cortisol measured the next morning are normally below 5 μg/dl, that is to say, dramatically suppressed. Dexamethasone does not pose a technical problem in the cortisol assay, as the synthetic steroid does not cross-react with the natural hormone. An intact endocrinological feedback mechanism ensures that any unnecessary endogenous production of cortisol is shut down. Patients with an ACTH-secreting tumor in the pituitary, lung, or pancreas, or with a cortisol-secreting tumor in the adrenals will not show any suppression of cortisol production following dexamethasone treatment, and hence, the test differentiates well between these clinical situations.

In the field of psychiatry, dexamethasone suppression tests used to be carried out with a lower dose of 1 mg of the synthetic steroid in order to identify those patients with an endocrine depression. At this dose, such patients frequently fail to show any suppression of endogenous cortisol secretion. However, some normal subjects with no signs of depression also fail to demonstrate any decrease in cortisol production with this dose of dexamethasone. Patients with any severe acute illness or with chronic inflammation are also resistant to the suppressive effects of this lower dose of dexamethasone. It is also known that the suppressive effect of dexamethasone depends on the time of day when the treatment is given, and that disturbances in the circadian rhythm of cortisol secretion are frequently found in depressive patients. Eating disturbances can also influence the outcome of a dexamethasone suppression test, and because many depressed patients also show eating disturbances, we may explain the results of the dexamethasone test seen in depressed patients solely in terms of their eating problems.

Other groups of patients who frequently show no response to a low dose of dexamethasone are the elderly, patients with chronic renal insufficiency requiring dialysis, with anorexia mentalis, and alcoholics. The only factor that all of these groups have in common with depressive patients is the disturbance in the circadian pattern of sleep–wake, rest–activity, and eating–fasting cycles, which is often encountered in psychiatric patients. Hence, we can only describe the result of the dexamethasone suppression test as a symptom of the disturbance, without reference to the underlying pathology. This endocrine suppression test has proved to be unreliable and insensitive when used in the differential diagnosis of complex psychoimmunoendocrinological problems.

6.8. Investigation of Secretory Function

Nearly all hormones are secreted in episodic bursts, rather than in a continual fashion. There are several hypotheses as to why this mode of secretion is preferred, using findings from the field of information processing theory linked with endocrinological data describing the interaction between a hormone, as the information carrier, and its receptor, as the information receiver. Thus, the intermittent nature of hormone secretion appears thrifty in nature, as the organism is spared the cost of continual, uninterrupted mode of hormone secretion, which would flood the system and occupy and exhaust the supply of cellular receptors. This latter exhaustion of receptor capacity is known as *desensitization* and can have a therapeutic application.

This pulsed transfer of information increases the stability of the system and is less sensitive to any interference. In addition, changes in the intensity of the signal are more effective in modulating the message transmitted. These are just some of the advantages that oscillating systems have exploited not only in modern information technology but also in nature for information transfer.

A further advantage of the episodic mode of hormone secretion is that a change in the amplitude or the frequency, and under certain conditions, the sequence of these secretory episodes can result in the activation of different pathways simultaneously. In most endocrine systems, changes in the amplitude code for a change in the intensity of a signal. The physiological significance of a change in the signal frequency, such as to direct the signal toward a specific organ or system, is not fully understood and more work is required in this field. This phenomenon of episodic hormone secretion is of special interest in neuropsychological and psychiatric problems as there is no doubt that the episodic secretion of pituitary hormones is directed by the hypothalamus. This provides a "partial window into the brain," namely, into the functioning of the hypothalamus, a brain center closely linked to other nuclei.

Any attempt to analyze the episodic nature of hormone secretion will be extremely laborious because of the experimental test conditions and the number of hormone determinations ensuing. Figure 10 shows hormonal profiles of ACTH, cortisol, and corticosterone measured over 24 h, at intervals of 10 min, in a healthy volunteer; i.e., three parameters measured in 144 blood samples collected over 24 h, or 432 test results. Any comparison of

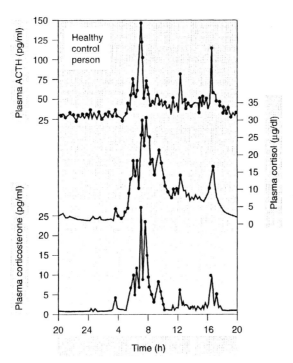

Figure 10. Secretion profiles of ACTH, cortisol, and corticosterone over a 24-h period in a normal volunteer.

healthy subjects and patients for either diagnostic or therapeutic purposes would require the analysis of thousands of samples. For this reason, investigations of the episodic mode of hormone secretion have been restricted to small groups of patients or volunteers only.

Another reason against applying such diagnostic tests of secretory function in routine clinical practice is the difficulty associated with analyzing the resulting hormone profiles. What are the criteria to distinguish between normal and pathological modes of hormone secretion? Some changes are apparent, but that which the brain is capable of doing in an instant has eluded the rigors of most of the more common mathematical analytical methods. The simpler heuristic methods (e.g., Clifton & Steiner, 1981; Santen & Barden, 1973), the PULSAR method (Merriam & Wachter, 1984), as well as the CLUSTER method (Veldhuis, Evans, Rogol, Drake, Thorner, Merriam, & Johnson, 1984) are all based on arbitrarily assigned, descriptive criteria, such as amplitude height, episode duration, rate of change of hormone values during secretion, or the number of apparent secretory episodes. More modern model-based analyses, such as DETECT (Oerter, Guardabasso, & Rodbard, 1986) or DESADE (Ranft, Prank, & Brabant, 1988), use known metabolic half-lives of hormones as more objective criteria, and are capable of calculating secretion rates at specific points of the profile, as well as total daily output from data collected during the experimental period. In addition, they are capable of determining the proportion of hormone secreted episodically to that secreted continually. However, none of these programs completely mimic brain function in detecting a pattern of episodic hormone secretion, and considerable research investment will be required before we are as discriminating as the brain, and we can use such analyses in the investigation of psychoneuroendocrinological disorders (Schürmeyer, 1989, 1992).

6.9. Conclusion

New development possibilities have arisen from the techniques described above, which may enable us not only to understand the endocrine changes occurring in isolated organs but also to investigate in an interdisciplinary fashion the human organism in its entirety, using the new concepts of information transmission. Developments in molecular biology now give us the tools to investigate at a genetic level such phenomena as the predisposition for disease, which may determine the way in which an organism responds. At this time it is impossible to say if, and in what measure, our system-oriented view of psycho-immunoneuroendocrinological mechanisms will prevail, or whether a more molecular-oriented approach will be more productive, and finally, whether we can resolve the findings from these two frequently opposing viewpoints.

7 Methods in Immunology

Roland Jacobs

7.1. Introduction

The human immune system, which is a network of interactive components, aims to recognize and eliminate common pathogenic agents or organisms. This cooperation can be separated theoretically into different subunits. For example, the specific arm of the immune response can be distinguished from the unspecific. Likewise, the distinction between cellular and humoral defense mechanisms can also be useful. In the living organism, all single components depend on each other to enable a coordinated and interactive immune response. However, for the investigation of the underlying mechanisms, examining the components of this complex system makes our task easier.

Presented in this chapter are the most important test systems for the analysis of immunological functions that have psychoneuroimmunological relevance. Most of these assays are based on the enormous specificity of proteins or polysaccharides that serve as receptors or ligands, respectively. Although the methods are described for the human system, each assay is feasible for other species if the appropriate reagents (e.g., antibodies) are available. For example, antibodies against human cell proteins are mainly raised from

Roland Jacobs   •   Department of Clinical Immunology, Hannover Medical School, D-30625 Hannover, Germany.

mice that have previously been immunized with human cells. However, for analyses of mice, antibodies from a different species (e.g., rabbit) are used. The cell lines used for several bioassays are often species specific: for the determination of NK activity in humans, the human cell line K562 is established. In mice, the murine cell line YAC is used for the same purpose.

In psychoneuroimmunology, human research is predominantly focused on the peripheral blood of volunteers or patients (e.g., Schedlowski, Falk, Rohne, Wagner, Jacobs, Tewes, & Schmidt, 1993a), which is treated differently according to the particular inquiry. For determination of hormones or cytokines from plasma, blood has to be mixed with anticoagulating substances (e.g., heparin, citrate, or EDTA), whereas supplements are not necessary to gain serum from blood. The solid (cells) and liquid (serum, plasma) components of blood can be separated by centrifugation. The resultant liquid supernatant can be used, for instance, for the determination of immunoglobulin concentration. This is usually done by nephelometry, a technique that is based on the formation of immune complexes after adding an appropriate antiserum. Laser light is scattered by these complexes but not by unbound immunoglobulins, thus enabling quantitative (light scatter) and qualitative (specificity of complexing antiserum) determinations.

Another source for investigation of immunologically relevant humoral substances (e.g., IgA) is saliva. Sufficient amounts of saliva can be obtained by applying cotton swabs in the mouth. After soaking itself full, the swab is transferred into a salivette tube and centrifuged. If the swab remains above a septum, the saliva is collected in the bottom of the tube and is ready for analysis (Fig. 1).

For analyzing the cellular components, blood has to be supplemented with anticoagulants (e.g., heparin). For an exact quantification of the subsequent analyses, a white blood cell count has to be prepared. This can be performed with a microscope by counting all of the different cell types (granulocytes, monocytes, and lymphocytes) or can be done using an automatic cell analyzer. After determination of the total leukocyte counts, and the percentages of all cell subsets, it is a simple mathematical routine to calculate the absolute numbers of the different cell types.

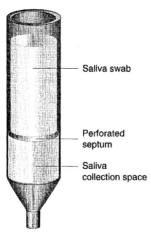

Saliva swab

Perforated septum

Saliva collection space

Figure 1. Saliva collection tube (salivette). Saliva is extracted from a soaked cotton swab by centrifugation in a particular collection tube. The lower reservoir is separated from the upper rest of the tube by a perforated septum holding back the swab during centrifugation. Saliva can then easily be aspirated from the reservoir with a pipette after removal of the swab.

7.2. Isolation of Lymphocytes

For many functional analyses of blood cellular components, the blood must first be separated. The blood sample is mixed with an equal volume of a physiological solution such as phosphate-buffered saline (PBS) or medium. If the cells are to be subsequently cultured, it is more suitable to use the appropriate medium for cultivation at the stage of separation. For density gradient centrifugation, this suspension is layered onto a separating solution with a specific density of 1.077 g/ml (like Ficoll). Alternatively, the separating solution can be placed below the diluted blood; both methods are feasible as long as the solutions do not mix. The actual separation is then done by centrifugation for about 20 min at 1000g. Thereafter, several layers can be distinguished: At the bottom the red blood cells are sedimented, and directly above them are the granulocytes, which can be recognized as a small slightly dull coating. Above this sediment the separating solution is arranged with a cloudy layer above it: the interphase containing mononucleated cells (monocytes and lymphocytes), and the final uppermost layer is diluted plasma containing thrombocytes and other subcellular particles (Fig. 2). The interphase is carefully collected with a pipette and washed twice, first at 1000g and then at 700g, each for 10 min to remove the platelets and remaining separating solution. Finally, the cells are resuspended in medium or PBS.

For most purposes the cells must be adjusted to a particular concentration requiring an exact cell count. Therefore, an aliquot of the cell suspension is diluted with an exact volume of low strength concentrated acetic acid (3%) or trypan blue solution. A Neubauer chamber is filled with this dilution and all of the cells in one square corner (consisting of 16 smaller ones; Fig. 3) are counted through a microscope. The accuracy of the determination is enhanced by counting two diagonally opposed squares and calculating the arithmetic mean.

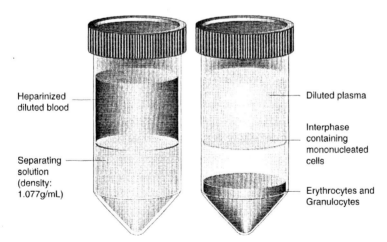

Figure 2. Isolation of lymphocytes. Diluted blood (1:2) is layered onto a separating solution (e.g., Ficoll, left side). By centrifugation at about 1000g characteristic layers are established (right side). Red blood cells accumulate at the bottom of the tube, and granulocytes lie directly on the erythrocytes. Above that and on the separating solution is a small cloudy layer comprised of mononucleated cells (lymphocytes and monocytes). The uppermost layer consists of diluted plasma supplemented with small particles like platelets.

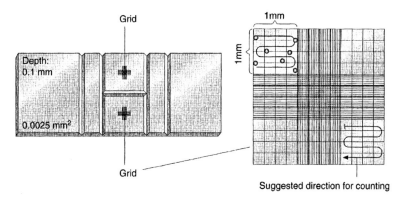

Figure 3. Cell counting. Two exactly dimensioned grids are shaped into the glass of a Neubauer chamber. On the right side a magnified grid is displayed. For lymphocyte counting the bigger corner squares are used. Sixteen subsquares comprise one corner square with 1 mm side length and 0.1 mm depth revealing 0.1 mm^3 (= 0.1 μl) volume when covered with a glass slide. All cells counted in 16 subsquares (7 cells in this example) are dispensed in 0.1 μl. For calculating the precise number the cells have to be multiplied by the factor of dilution (e.g., 20). Further multiplication by 10,000 yields the number of cells suspended in 1 ml. In this example a total number of 1.4 × 10^6 cells/ml can be calculated (7 × 20 × 10,000 = 1.4 × 10^6).

Dilution with acetic acid causes the erythrocytes to burst but leaves the white blood cells unaffected, thus making the cell counts much easier. On the other hand, trypan blue is only taken up by dead cells, which enables a distinction of viable cells from dead ones. When fresh blood is isolated, the viability of the white blood cells should be 100%. From the number of counted cells, the volume of the aliquot, and the dilution factor, the exact number of separated cells can be calculated.

7.3. Isolation of Granulocytes

Granulocytes can be recovered from the sediment that is obtained after density gradient separation. After dilution of this cell fraction with plasma expander or hydroxyethyl-starch solution (1:2 to 1:5), the mixture is left to stand for about 45 min. During this time, the red cells sink to the bottom of the tube and the granulocytes accumulate in the supernatant. The few contaminating red blood cells can be easily removed by lysis with ammonium chloride or by osmotic shock with water. After appropriate washing, the granulocytes are ready for use. Alternatively, granulocytes can be isolated with solutions that have different properties (e.g., density or ingredients) than Ficoll (e.g., Percoll). The separating procedure is very similar to the isolation of lymphocytes. However, both kinds of isolation are not very gentle, and the rather sensitive granulocytes can be activated this way, thus making an interpretation of certain results more difficult. Some of the main functions of neutrophil granulocytes are phagocytosis and bactericidal defense. The phagocytotic potential can be measured after offering labeled bacteria. According to the kind of label, the readout is performed in a counter (radioactive bacteria) or a cytofluorometer (fluorescent bacteria).

7.4. Storage of Cells

When it is not possible to perform the desired assays on the day of blood collection, the unseparated blood sample can be stored at room temperature overnight. In the case of longer delays, the blood must be separated, and the isolated mononucleated cells have to be frozen. Before freezing, the mononucleated cells have to be resuspended in a small volume of medium containing serum or plasma (about 50%) and 10% dimethyl sulfoxide as cryogenic protectant. After interim storage in a $-80°C$ freezer overnight, the cryo tubes are transferred to liquid nitrogen for long-term storage (for years). Thawing of the cells should be carried out very quickly because of the toxicity of the cryo protectant at normal temperatures. After rapid thawing at 37°C, the cells are stepwise diluted with a physiological solution, washed three times, and the viability is estimated with trypan blue. Cells conserved this way are suitable for functional investigation, but only of limited feasibility for phenotypic analyses, which should be preferably performed with freshly isolated cells.

7.5. Phenotypic Analyses

By using labeled monoclonal antibodies in a fluorescence-activated cell sorter (FACS) analysis, it is possible to determine lymphocyte subsets of any blood sample within a short time. Normally, the monoclonal antibodies are derived from immunized mice and exert a single specificity for a certain cell surface protein on human lymphocytes. Well-known molecules of this kind are, for example, CD3 (T cells) and CD4 (T-helper cells).

Aliquots of separated mononucleated cells are incubated together with antibodies of a single specificity (CD2, CD3, and so on) each for about 30 min. Thereafter, all unbound antibodies are removed by washing three times (Fig. 4). When labeled antibodies are used, the cells are then ready for FACS analysis. The first step of an indirect immunofluorescence is identical to the direct procedure, except that unlabeled antibodies are used. The final staining is performed in a second step by incubating all of the samples with a labeled antiserum that binds specifically to all of the antibodies of the first step. The application of an antiserum instead of monoclonal antibodies is used because an antiserum has several

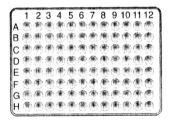
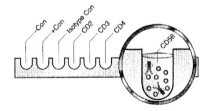

Figure 4. Phenotyping. The incubation of cell aliquots with appropriate monoclonal antibodies in a microtiter plate is well established in practice (left side). Alternatively, single tubes for each test can be used, making the whole procedure more fussy. Cells are dropped into the appropriate number of wells of the plate aliquots. Thereafter, antibodies with different single specificities are added to each of the wells. After incubation for about 30 min at 4°C, the cells are sedimented to the bottom of the wells by centrifugation. After removal of the supernatants containing the unbound antibodies, the cells can be resuspended and analyzed.

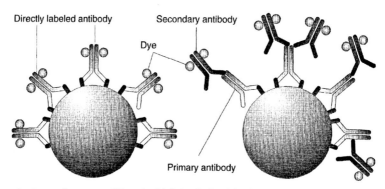

Figure 5. Immunofluorescence. When using labeled antibodies (left side), the tests can be analyzed directly after appropriate incubation and three washes. When utilizing unlabeled antibodies, the final dye is performed in a second step with labeled antibodies that bind specifically to the first antibody representing the antigen for the secondary antibody. Thereafter, the fluorescences can easily be analyzed by cytofluorometry.

specificities enabling, for example, the equal recognition of all isotypes (e.g., IgG3, IgM) at the first step. After the final wash, the cells can be analyzed by cytofluorometry (Fig. 5). To estimate whether the monoclonal antibodies have been bound unspecifically to the cells, controls have to be performed simultaneously in each assay. Therefore, an antibody that does not recognize an epitope on the cells, which is of the same isotype, is analyzed in parallel to the antibody with the desired specificity. For indirect phenotypic analyses, an additional test lacking the first antibody has to be done to evaluate the unspecific binding of the second antibody. Moreover, incubation with the appropriate antibodies should be performed in the presence of human IgG to prevent an unspecific binding of murine antibodies via Fc receptors on human cells. If this is not done, then an incorrect positive staining could occur because of the not negligible proportion of lymphocytes that express Fc receptors (e.g., NK cells). In this case, the use of human IgG can solve the problem. Because of the higher affinity of human immunoglobulins to human Fc receptors, murine antibodies are less likely to bind (Fig. 6).

Phenotypic analyses can also be performed without prior isolation of mononucleated cells. However, interfering red blood cells have to be removed by lysis (e.g., by ammonium chloride) before the cells can be analyzed. This procedure is faster but less gentle and more antibody consuming.

Care should be taken on the selection of molecules that are analyzed. The different lymphocyte subsets are mostly not determined by one single molecule, but rather by a characteristic antigen pattern. An example can illustrate this: Cytotoxic lymphocytes are either T cells or NK cells. Typical markers for NK cells are CD16 and CD56 (Jacobs, Stoll, Stratmann, Leo, Link, & Schmidt, 1992). However, some T cells can also express one of these markers (Uciechowski, Gessner, Schindler, & Schmidt, 1992). On the other hand, NK cells never express CD3 on their surface (Hercend & Schmidt, 1988), but like T cells they coexpress CD2 and a part of both subsets express CD8. This means that there is little use analyzing only one molecule as representative for one cell subset. The selected antibodies should be able to recognize the most important lymphocytes like T, B, and NK cells. An antibody panel including CD2, CD3, CD8, CD16, CD20, and CD56 fulfills this claim (Fig. 7).

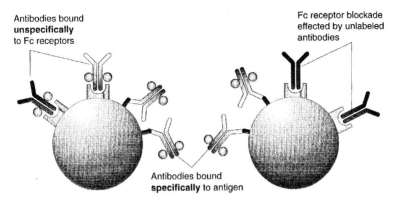

Figure 6. Fc receptor blockade. Several leukocytes (e.g., NK cells, monocytes) coexpress Fc receptors in addition to other surface antigens. These receptors can bind the Fc portion of numerous antibodies regardless of their antigen specificity, producing wrong positive results because binding via Fc receptors is indistinguishable from specific antigen binding. For this reason the Fc receptors should be saturated with species-specific immunoglobulins during phenotypic analysis. Because of the higher affinity of immunoglobulins from the same species, xenogeneic antibodies are prevented from this kind of unspecific binding.

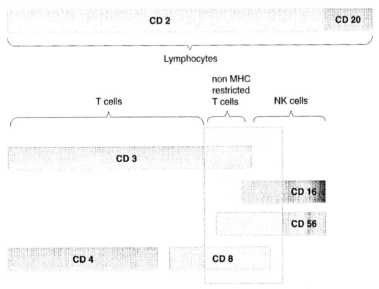

Figure 7. Antigen distribution on human lymphocytes. The surfaces of different lymphocytes are marked by a characteristic antigen distribution. Nearly all lymphocytes express CD2 (about 90%) or CD20 (5–10%). However, a single cell does not express exclusively one single surface protein. Each cell exerts a characteristic antigen pattern. For example, some NK cells (CD16$^+$, CD56$^+$) coexpress CD8 and another NK cell subset does not. A meaningful definition of a single cell type is only possible by determining a specific antigen pattern. This sketch illustrates that for instance CD8 and CD56 are markers appearing on different cell types. By definition, T cells are CD3$^+$CD56$^-$, non-MHC-restricted T cells are CD3$^+$CD56$^+$, and NK cells are CD3$^-$CD56$^+$.

For a more exact determination of lymphocyte subsets, two color fluorescences can be performed. For example, one part of the antibodies is labeled with fluorescein isothiocyanate (FITC) and the other part is linked to phycoerythrin (PE). FITC emits green fluorescence and PE shines red after excitation. Simultaneous incubation of one cell sample with fluorescent antibodies of varying specificities allows the investigation of the coexpression of different molecules on a single cell. The staining procedure is identical to the direct immunofluorescence, with the exception of using two differently labeled antibodies instead of one for each sample. In addition to negative controls for each color that is used, positive controls (cells that are clearly positive for each color) can be valuable for the fine adjustment (compensation) of the cytofluorometer device. By applying antibodies linked to dyes with emission spectrum differing from PE or FITC, it is possible to analyze simultaneously three different molecules on single cells.

7.6. Fluorescence Analysis

The cells tagged by fluorescent antibodies are aspirated into the tube system of the cytofluorometer (FACS) and injected into a sheath jet. By passing a nozzle before entering the measuring cuvette, the cells from the suspension are accelerated and lined up. One cell after another can now be hit by a laser beam. Using appropriate optical devices and electronic detectors, the light scattered forward (FSC) and sideward (SSC) can be measured. The laser beam is scattered differently by different cells, according to the cells' properties. The scatter signal of the FSC is mainly influenced by the size of the cell and the SSC signal depends on the granularity (cell surface and inclusions) of the cells (Fig. 8).

On the computer screen connected to the analyzer, both scatter signals are displayed graphically (Fig. 9). In this dot plot the x axis represents the size and the y axis the granularity of the analyzed cells, revealing accumulation of dots depicting distinct cell populations with specific characteristics (size and granularity). An example of the analysis of peripheral blood is shown in Fig. 9. In the lower left corner, small and unstructured particles are displayed. The majority of these particles are cell debris, platelets, and red blood cells. To the right, lymphocytes are accumulated, and monocytes farther to the right. The granulocyte subset is located above the lymphocytes and monocytes. From these populations the relevant cells can be selected by software gating. Only these gated cells are then taken into consideration for the following analyses. As mentioned above, the single cells are scanned by a laser light when passing the measuring cuvette. In addition to detecting granularity and size, corresponding photodetectors can also register fluorescent light emitted from labeled antibodies that are bound to the cells. After software discrimination of autofluorescence, signals that are emitted by each cell (even if the cell is untouched from any dye) can be recorded as the realistic fluorescence intensity. The density of the different molecules is proportional to the fluorescence intensity: The more molecules there are on the cell, the farther right the mean fluorescence signal is displayed (Ormerod, 1990).

Single-color immunofluorescences are usually displayed in histogram style (Fig. 10). For two-color fluorescences, the dot plot or contour style is more suitable. Contour graphs are very similar to dot plots, except that the different rings in a contour graph describe cells with similar properties. Both the dot plot and the contour layout can be interpreted as histograms that are viewed from above. For statistics the whole area is split into four rectangles (Fig. 11). The lower left region contains all cells that do not express either of the antigens. The lower right area shows all cells that express only the antigen that is detected by the green fluorescent (fluorescence 1 = FL1) antibody. The upper left rectangle encloses the

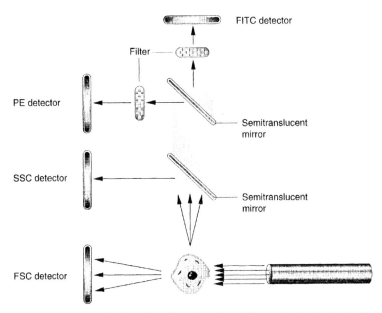

Figure 8. Fluorescence analysis. When a focused laser beam hits a cell in a cuvette, it scatters in two directions: forward (forward scatter, FSC) and sideward (side scatter, SSC). The degree of these scatterings depends on the properties (cell size and granularity) of the particular cell type. When the cells are labeled with antibodies, the fluorochrome is stimulated by the laser and emits subsequently light signals of a specific wavelength. Using semitranslucent mirrors and filters, the spectrum of emitted light is split and directed to photodetectors that transform the light signals into electric impulses. Adapted from N. P. Carter and E. W. Meyer.

Figure 9. Dot plot. The screen of a computer connected to the cytofluorometer displays the single cells as dots in a system of coordinates. In this example the *x* axis represents the cell size and the *y* axis corresponds to the granularity of the cells. Cells with a certain size and granularity are displayed at the same place in the system of coordinates, forming a distinct mass on the screen. By gating with appropriate software, the different cell types can be analyzed individually.

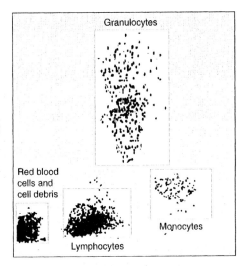

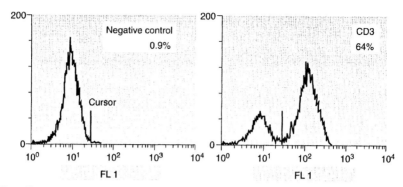

Figure 10. Histogram display. Histograms are suitable especially for displaying single-color fluorescences. In the left panel the graph of a negative control is depicted, stressing that even unlabeled cells exhibit a low but detectable autofluorescence. This autofluorescence can be discriminated by software excluding these data from analysis and regarding only the signals to the right of the cursor as positive. The histogram in the right panel depicts the graph of $CD3^+$ lymphocytes.

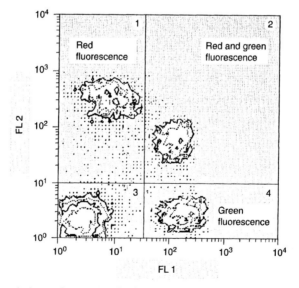

Figure 11. Two-color immunofluorescence. All cells that are stained solely by green fluorescent antibodies are displayed in the lower right quadrant (4), and all exclusively red fluorescent cells are depicted in the upper left quadrant (1). The cells that express both antigens shine red and green simultaneously as seen in the upper right quadrant (2), whereas cells lacking both antigens are displayed as double negative cells in the lower left quadrant (3).

cells bound by the red labeled (FL2) antibody. Finally, the upper right region contains those cells that coexpress both antigens (double positive). Additional information can be obtained from the contour and dot plot graphs.

As an example, four cell populations are shown in Fig. 11: double-positive, double-negative, and two single-positive cell subsets. The population that is stained red consists of two subsets. One subset expresses the antigen strongly (many molecules per cell, bright expression) and the second population at a lower level (dim expression, note the logarithmic scale). The latter cells coexpress the green-stained antigen that is expressed to the same degree as in the single green positive cells.

Not only surface antigens can be detected with labeled antibodies, but also proteins that are hidden in the cells. Because of their size, the antibodies cannot pass the cell wall, and therefore the membrane has to be prepared for permeability prior to the staining, being careful not to destroy the cell completely. This technique is widely used to detect intracellular cytokines that are produced by activated cells. For example, the intracellular staining of a cytokine is briefly described. After density gradient separation, the mononucleated cells are stimulated with phorbol myristate acetate (a strong polyclonal activator) and ionomycin for about 4 h at 37°C, in the presence of an agent like monensin that prevents the release of the produced cytokine by blocking intracellular transport processes. Thereafter, the cells are fixed with 4% paraformaldehyde for 10 min to preserve cell properties, and finally washed. The following resuspension in a mild detergent (e.g., saponin) causes perforation of the membranes. The extra- and intracellular proteins of the cells can then be stained and analyzed as in normal phenotyping.

7.7. Separation of Leukocyte Subpopulations

A simple density gradient centrifugation (e.g., Ficoll) is suitable for separating cells according to their physical properties, but is insufficient to divide cell subsets with identical physical characteristics. For some experimental questions, the cells obtained from the interphase have to be further processed to fulfill desired criteria (Klein, 1991). The easiest way to deplete monocytes from the mononucleated cell fraction is an incubation of the cells in plastic dishes at 37°C. Monocytes have the intrinsic capability to adhere to inert surfaces, and after appropriate incubation the nonadhering lymphocytes can be harvested with a pipette and the adhering monocytes remain in the dish (Fig. 12).

More effectively, monocytes can also be depleted from the mononucleated cell fraction by applying a nylon wool column. The column consists of a syringe that is stuffed with nylon wool and filled with medium. The medium is then exchanged with the cell suspension and incubated for about 30 min at 37°C, in an upright position. The cells slowly seep through the wool and all of the monocytes, and a lot of B cells, remain attached to the nylon. When the column is finally opened, the nonadhering cells are allowed to drain into a collection tube (Fig. 13).

In contrast, the separation of different lymphocyte populations is a little more complicated. The possibility to distinguish different cell populations with monoclonal antibodies makes them an effective tool for this task. Basically, there are two ways to approach a subset isolation via antibodies: positive enrichment or depletion. For positive selection the desired cells are marked with antibodies and isolated. The converse procedure is termed *depletion*; all cells that are superfluous are targeted by the antibodies and after isolation only the cells of interest are left. Both approaches have advantages and drawbacks. Positive selection gives a higher purity with a minimum of antibodies employed. However, the

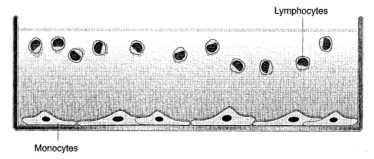

Figure 12. Separation by adherence. Monocytes, for example, adhere spontaneously to inert surfaces at 37°C. After incubation of mononucleated cells for about 1 h, the nonadhering lymphocytes can be simply separated from the adhering monocytes by harvesting the supernatant.

antibodies remain on the surface and may interfere with a functional investigation, as cross-linking of several molecules (e.g., CD3 or CD16) induces activation (proliferation, cytokine production) of positive cells (Schmidt, Michon, Woronicz, Schlossman, Reinherz, & Ritz, 1987). Therefore, where possible, the desired cells should be left untouched during separation by performing negative selection (depletion of all undesired cells). However, depletion gives a lower purity and a poor cell yield, despite higher antibody consumption.

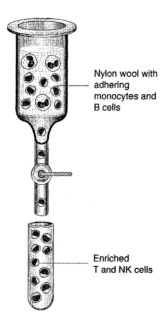

Nylon wool with adhering monocytes and B cells

Enriched T and NK cells

Figure 13. Separation with nylon wool. A syringe stuffed with nylon wool and closed at the bottom is filled air bubble free with medium and the cell solution is dropped into the syringe. After incubation at 37°C for about 30 min, the stopper is opened and the syringe is carefully flushed with medium. The monocytes and a lot of B cells remain in the nylon wool and the lymphocytes (predominantly T and NK cells) are flushed with the medium.

7.8. Separation of Lymphocyte Subsets

Usually, prior to selection of particular lymphocyte subsets, an adherence step is performed to remove monocytes (see above). Before describing different kinds of antibody-mediated cell selection, a method that uses sheep red blood cells instead of antibodies should be mentioned. This method is relatively old but nevertheless very effective in specifically selecting CD2⁺ cells (T and NK cells). Sheep red blood cells express a surface molecule that can incidentally bind the human CD2 antigen. When coincubating human lymphocytes with sheep erythrocytes, rosettes—a lot of red blood cells surrounding CD2⁺ T and NK cells—will form. Rosettes can easily be separated from nonrosetting cells by a Ficoll centrifugation. All cells that were not bound by red blood cells (mainly B cells) are found in the interphase fraction, and by virtue of the altered physical properties of the erythrocytes, the rosetted CD2⁺ lymphocytes pass through the separating solution and accumulate at the bottom of the tube. Later on, the sheep red blood cells can be lysed and the pure CD2⁺ cell fraction is ready for use. Again it is important to remember that cross-linking of CD2 by sheep red blood cells during the separation phase can induce activating signals in the lymphocytes.

The diversity of available monoclonal antibodies allows purification of every imaginable cell population. An antibody comprises two antigen binding sites, Fab and an Fc portion; both components have different biological functions. Fab sites are responsible for the binding of an antigen, or in this case for the binding of a particular lymphocyte structure. The Fc part is isotype specific, which means that every (murine) IgG1 antibody has the same Fc structure. In the living organism, the antigen-bearing antibodies can be bound by Fc receptors from granulocytes, monocytes, or NK cells. The antigen–antibody complexes are cleared this way by the Fc receptor expressing cells (Roitt, 1994).

In the laboratory, the Fc structure is predestined for labeling or tagging. Through the linking of dyes, enzymes, radioactive substances, or paramagnetic particles, antibodies make excellent immunological tools. The application of dye-labeled antibodies enables cell sorting by FACS, and the selection of cell subsets via magnetic beads is less complicated. The procedure is very similar to the antibody staining for immunofluorescences; however, higher cell concentrations are employed and all steps are done under sterile conditions if necessary.

After incubation with the antibodies that are labeled with paramagnetic particles, the test tube is applied to a strong magnet. This serves to attach all paramagnetic particles, including all positive cells, to the tube wall. The negative cells can then be aspirated with a pipette and the positive cell fraction can be recovered after removing the tube from the magnetic field (see Fig. 14).

A further development of this technique is the magnetic activated cell sorter (MACS). The beads linked to the antibodies are extremely small, thus enabling a reanalysis of separation success with a FACS, without interference from the magnetic beads. The magnet induces an effective magnetic field in a column that is stuffed with paramagnetic wool or microspheres (Fig. 15). Prepared cells are dropped into the column and the cells that are bound by antibodies are held back in the metal matrix by magnetic forces. Unbound cells can pass the matrix and can be collected as the negative fraction. After removing the column from the device, the magnetic field collapses. The cells are no longer attached to the matrix and can then be recovered by rinsing.

Another antibody-mediated separation technique is panning. Antibodies are fixed with the Fc part to an inert surface (plastic dish) so that the antigen binding sites can project freely

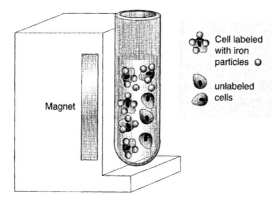

Figure 14. Magnetic cell separation. Cells are bound by specific antibodies that are covalently linked to ironlike particles. The cells prepared this way will be fixed to the wall of a tube when the tube is brought in contact with a magnet. All negative cells are not affected by the magnet and can easily be separated by aspirating with a pipette. Afterwards the tube can be removed from the magnet and the positive cells can be resuspended.

into the surrounding medium (Fig. 16). Cells are then added into the dish, and after incubation for about 30 min the negative cells can be aspirated with a pipette. The positive lymphocytes remain fixed by the antibodies and might be detached by rigorous shaking.

When there is the opportunity to use an electronic cell sorter (FACS), it is even possible to separate lymphocytes that express the same antigen but in varying densities. The sorter works similarly to the cytofluorometer, and the preparation for cell sorting is exactly the same as that used for phenotypic analyses. However, before passing the laser beam, each single cell is packed into one droplet of a physiological fluid (Fig. 17). Each cell within a

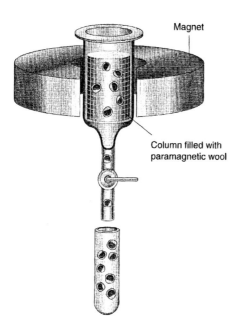

Figure 15. Cell separation by MACS. Plastic columns are filled with paramagnetic wool or microspheres serving as a matrix for a strong magnet that magnetizes the wool or microspheres, only for as long as the columns are in the magnetic field. Antibodies linked to extremely small metal beads are utilized with this equipment to separate cell subsets efficiently. Whereas positively stained cells are held back by the magnetic environment, the untouched cell can pass the matrix. After removal from the magnetic field, the positively stained cells can be gained by flushing the column. Because of the subminiature size of the beads, they do not negatively affect cytofluorometric analyses, enabling exact evaluation of the success of a separation.

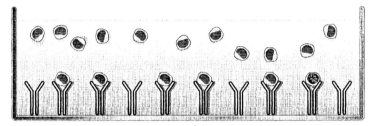

Figure 16. Panning. Antibodies fixed to the bottom of a petri dish or culture flask can bind cells that express the appropriate surface antigen. After about 30 min of incubation the supernatant with enriched negative cells can be recovered.

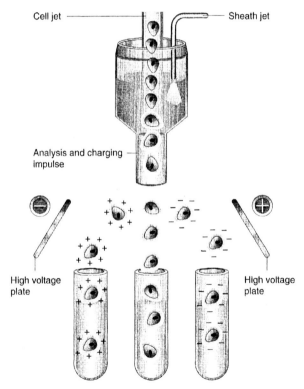

Figure 17. Cell separation by FACS. The device aspirates the cells stained with labeled antibodies and injects them into a sheath stream thereby aligning the cells. In a next step each single cell is packed into a droplet. The properties of cells in the droplets are then analyzed. When a cell corresponds to the properties that were chosen, the droplet is charged positively or negatively. By passing an electrostatic field the cells can now be attracted to the opposite pole. Finally, the differently directed cells are collected in separate tubes.

droplet is then analyzed with regard to the different parameters (size, granularity, and fluorescence). If a cell corresponds to criteria that can be adjusted by the operator via software, the droplet containing this particular cell is electrically charged. Further on, the cell passes high-voltage plates that attract the cell according to the charge of the droplet. The direction of the cell is thereby changed and the droplet is collected in a different tube. In contrast to the techniques described before, the very efficient sorting of cells by FACS is very time consuming (several hours) and requires an experienced operator.

Another way to enrich a particular subset is to lyse all of the unwanted cells. The unwanted cells must be coated with specific antibodies and complement in order to perforate the membrane. Preferably, antibodies of the IgM type are used for complement lysis as this isotype is more efficient than others in complement fixation. The reason for the better complement activation is the pentameric structure of the IgM molecule. For successful complement activation, at least two monomeric antibody molecules have to bind the antigen tightly to each other. The pentameric IgM structure enables the initiation of the complement cascade by the binding of only one IgM to the antigen, as the molecule is equivalent to five monomeric molecules binding the antigen next to each other. It is presumed that at least 1000 IgG molecules are necessary for it to be likely that two IgG bind the antigen close enough to each other to initiate IgG-induced complement lysis.

For this kind of depletion, lymphocytes are incubated with the desired antibodies for 30 min and subsequently the unbound antibodies are removed by washing. After addition of the complement, in the presence of DNase, the respective cells die as a result of membrane perforation. The enzyme is essential to prevent clotting of the cells by cutting the long DNA molecules that are set free from already destroyed cells, for otherwise the DNA-associated proteins would clump the intact cells together. This technique is very efficient but has the disadvantage that the lysed cells are no longer available (Figure 18).

There is no absolute favorable method for all cell separations, and the most suitable technique depends on the prevailing question. Important criteria for the right type of selection are desired purity of the cells, costs, time, and a possible interference with subsequent assays by separation-induced activation of the cells.

7.9. Lymphocyte Functions

Lymphocytes comprise a group of differently specialized cells whose cooperation guarantees a well-regulated immune response. The varying properties can be assessed in

membrane attack complex

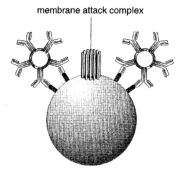

Figure 18. Lysis by complement. After incubation with preferably IgM antibodies of the desired specificity, the cells are washed. Addition of complement initiates the cascade by the bound antibodies, thereby forming the membrane attack complex (MAC) that perforates the membrane of the target cell, killing the cell.

respective *in vitro* assays. One of the standard test systems is the cytotoxicity assay that can be performed in several variations, and the NK assay (Tagasuki, Mickey, & Terasaki, 1973) is described here as an example.

NK cells are capable of spontaneously killing certain tumor or virus-infected cells. In contrast to specific T-cell-mediated mechanisms, there is no prior sensitization required and NK cells are not MHC restricted, although MHC molecules play a role in target recognition (see Chapter 2).

The cell line K562 has been established as the standard target cells for NK assays. The cell line is raised from a patient suffering from erythromyeloid leukemia. K562 cells lack the expression of MHC class I molecules and can easily be maintained in culture. Prior to the assay, the target cells are suspended in a radioactive Na_2CrO_4 solution. After about 1 h, the radioactive chromium is incorporated by the target cells, which thereafter are washed and adjusted to a particular cell concentration (e.g., 5×10^5/ml). The isolated lymphocytes (effectors) are adjusted to the desired cell concentration (e.g., 3×10^6/ml) and several dilutions (1:2, 1:4, and 1:8) are prepared. These dilution levels are quite usual but they can be chosen to one's liking. Triplicates of wells of a V-bottom microtiter plate are filled with 50 μl of the target suspension and 100 μl of the respective effector cell dilution. For determination of the spontaneous (minimal value) and maximal chromium release, three wells are each filled with target cells and medium or a cell lysing detergent, respectively. The cells are then gently centrifuged to guarantee a tight contact between targets and effectors. During the next 4 h of incubation, the effectors attack the targets and the chromium of each lysed cell is set free into the medium (Fig. 19). The plates are then centrifuged again and the supernatant is transferred into test tubes, and the amount of radioactive chromium can be measured in a γ-counter (Schmidt, MacDermott, Bartley, Bertrovich, Amato, Austen, Schlossmann, Stevens, & Ritz, 1985).

In a similar test system, the so-called antibody-dependent cell-mediated cytotoxicity (ADCC) is assessed. However, the latter requires NK-resistant targets that have been loaded with a target-cell-specific antiserum prior to the assay (Fig. 20). In this assay the target cells are not bound directly, but via antibodies that bridge the antigen of the target and the Fc receptor of the NK cell. The mathematical calculation of both assays is identical. The means of the measured triplicates are used in the following formula:

$$\frac{\text{specific release} - \text{spontaneous release}}{\text{maximal release} - \text{spontaneous release}} \times 100 = \text{specific lysis} \ (\%)$$

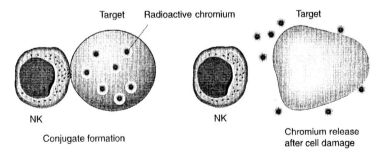

Target   Radioactive chromium   Target

NK   NK

Conjugate formation   Chromium release after cell damage

Figure 19. NK cytotoxicity. NK cells bind to target cells that were previously pulsed with chromium-51. The NK cells organize their large granules and pour out those substances onto the target cells, damaging the cells fatally.

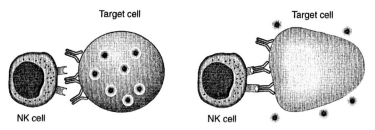

Figure 20. Antibody-dependent cell-mediated cytotoxicity (ADCC). A lot of cells that are resistant to NK cytotoxicity can be killed by NK cells via ADCC. Lysis of the particular cell is mediated by antibodies that bind specifically to the antigen on the target cell surface and also to the Fc receptor (CD16) on NK cells. The assay is performed like a normal NK assay, but the target cells have to be preincubated with the antiserum prior to the test.

To compare results from different assays, the percentages can be transformed into lytic units (LU). By mathematical transformation with Von Krogh's equation, the sigmoid scopes of cytotoxicity graphs that are obtained when plotting percentage of lysis versus E:T ratio are linearized. From this transformation a direct relation between the extent of cell lysis and the number of targets is achieved (Pross, Baines, Rubin, Shragge, & Patterson, 1981). The calculation of Von Krogh's equation is very complicated and the assays cannot be transformed in every case. Therefore, a simpler method for the calculation of lytic units has been designed (Bryant, Day, Whiteside, & Herberman, 1992). For both methods some variables are laboratory specific so that useful comparisons can only be drawn between assays of the same laboratory. With this reservation the calculation of lytic units is a suitable method for the analysis of cytotoxicity assays.

Targets killed by cytotoxic effector cells die from necrosis. Another totally different kind of cell destruction is programmed suicide (apoptosis). This event is physiological and occurs naturally under the influence or withdrawal of hormones, and is responsible for the elimination of cells that are no longer needed in the body. The suicide of the cells can also be induced experimentally by triggering the CD95 (Fas, APO-1) antigen on target cells. Several medicaments designed for treatment of cancer induce apoptosis in tumor cells. Apoptosis is associated with a characteristic pattern of physiological changes in the affected cells including increasing size, nuclear chromatin condensation, and an extensive DNA cleavage into fragments. The latter point can be used to measure the extent of apoptosis. Usually, the TdT-mediated dUTP nick end labeling (TUNEL) technique is employed to determine suicidal cell death. In principle, the cells are gently made permeable by a mild detergent. The resulting membrane pores allow passage of a mixture of FITC-labeled nucleotides and the enzyme terminal deoxynucleotidyl transferase (TdT). This enzyme catalyzes the binding of FITC-tagged nucleotides to the free 3'-OH-DNA ends of broken DNA strands. The DNA cleavage is dyed this way and the extent of apoptosis can be analyzed by fluorescence microscopy or flow cytometry.

Besides the cytotoxic potential, other lymphocyte capabilities can be measured. For example, the proliferative capacity can be assessed. *In vivo*, immunocompetent cells proliferate after activation by antigen, other cells, and/or cytokines. *In vitro*, the lymphocytes are usually stimulated unspecifically by mitogens like the lectins phytohemagglutinin (PHA), pokeweed mitogen (PWM) or concanavalin A (Con A). Lectins are proteins (usually of plant origin) that bind several highly specific sugars on glycoproteins or glycolipids. After stimulation of the lymphocytes or mononucleated cells for 48 h in an incubator (37°C), [³H]thymidine is added to the culture. Proliferating cells incorporate all components from the medium, thereby integrating the radioactive nucleotide into freshly synthesized DNA. Twenty-four hours later, the test is terminated with a harvester that aspirates the contents of the culture well and lyses all of the cells. The cellular components are sucked through a filter, holding back all bigger molecules like DNA fragments (Fig. 21). The filters are transferred into test tubes and scintillation fluid is possibly added to amplify the weak signals in a β-counter. The counted decay is proportional to the amount of incorporated [³H]thymidine and therefore a measure of the intensity of proliferation. Usually, the assay is performed in 96-well U-bottom plates in triplicates, and spontaneous proliferation (minimum value) is estimated from a parallel culture minus the mitogen supplement.

A special variation of this kind of proliferation assay is the mixed lymphocyte reaction (MLR). In contrast to the mitogen-driven cultures described earlier, the effector cells in the MLR are offered allogeneic cells as a stimulus. The assay is used to evaluate the risk of transplant rejection [host-versus-graft reaction (HvG)] or the probability of a graft-versus-host reaction (GvH) in bone marrow transplant experiments. The assay is performed by coculture of the cells from donor and recipient. After 4 to 5 days of culture, radioactive [³H]thymidine is added for 24 h. The cultures are then harvested and the incorporated [³H]thymidine is determined in a β-counter. When the cells from both parties are allowed to proliferate, it is impossible to distinguish whether cells of the donor or the recipient are more active (two-way MLR). This can be overcome by deactivating the cells of one party by irradiation (one-way MLR), which thus serve as stimulator cells but cannot proliferate themselves. This enables an exact interpretation of the activation level of both directions (HvG and GvH). Background proliferation is assessed by coculture of the appropriate cells with irradiated autologous stimulators and positive controls performed by using irradiated third-party cells for stimulation.

Proliferation is only one feature of all cellular events after lymphocyte activation. Like during the specific immune response *in vivo*, the cells begin to produce a diverse pattern

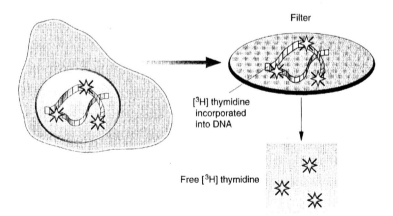

Figure 21. Proliferation assay. Proliferation can be estimated by measuring [³H]thymidine. This nucleotide is added to the assay and is therefore incorporated preferentially by the cell during DNA synthesis. To terminate the assay the cells are lysed and the cell fragments are sucked through a filter. Free [³H]thymidine can pass the filter, but [³H]thymidine that was inserted into DNA is held back by the filter. The amount of incorporated radioactivity can then be measured from the filters in a β-counter.

of cytokines. These substances (e.g., IFN, IL-2) can be obtained from the supernatants of stimulated cells. As described for the proliferation assay, the cells are stimulated with mitogens; however, no radioactive substances are added. Instead, the supernatants are collected and frozen or analyzed immediately.

The analyses of the cytokines can be performed with an immunoassay, which is usually commercially available, or in a bioassay with particular cell lines that can, for instance, only proliferate in the presence of the respective cytokine. Several dilution levels of the cytokines are added to a defined number of indicator cells. The cultures are then treated as in a proliferation assay. When identical cultures are performed in parallel with cytokines of known activity, the assay can also be analyzed quantitatively.

Apart from the cytokines, antibodies that are produced by activated B cells can be extracted from culture supernatants. For polyclonal stimulation of B cells, bacterial endo-toxins like lipopolysaccharide or the protein *Staphylococcus aureus* type Cowan (SAC) are established. The amount of immunoglobulins produced can be analyzed in immunoassays. Because of the mitogens used, the test methods described above are polyclonal, activating various cell subsets simultaneously. To provoke a specific response *in vitro*, the lympho-cytes have to be stimulated with their specific antigen. However, the antigen-specific cells are very rare in a normal lymphocyte sample, thus requiring prolonged culture periods or modification of other parameters, for otherwise the quantity of products (e.g., cytokines, antibodies) may be too small or proliferation may be undetectable.

7.10. Immunoassays

Immunoassays are all based on the highly specific antibody–antigen (Ab–Ag) bind-ing. A lot of systems are available that utilize radioactive (radioimmunoassay) or enzyme (enzymeimmunoassay) labeling. The assays can be performed in a liquid phase (antigen and antibody are in solution) and for determination of the specific reactions the Ab–Ag complexes have to be precipitated with appropriate reagents (e.g., ammonium sulfate). Next, the tubes are centrifuged and the supernatants removed (Fig. 22), and the specific complexes in the sediment can then be analyzed. In another approach the immunoassay works with a solid phase that is usually the plastic wall of a test tube, cuvette, or 96-well plate. The solid phase is either coated with specific antibodies for binding the respective antigen, or vice versa, the antigen is fixed to the wall if specific antibodies should be tested. In a competitive setting, labeled molecules of known concentration are mixed with the desired molecules of unknown quantity (Fig. 23).

The forces that cause the formation of Ab–Ag complexes are not covalent but a combination of different binding types that are only effective at short distances: hydrogen bonds, electrostatic bonds, van der Waals' forces, and hydrophobic forces. The complexes that are formed in this way are not stable and they are perpetuated by a balance of free and bound antigens or antibodies, respectively. Therefore, the labeled reagents can compete with the unlabeled substances and can be isolated according to the concentration of the substances.

After an adequate time of reaction and subsequent washes, the amount of remaining labeled substance is assessed. In the case of a radioactive system, the activity of the isotopes can be determined in a counter. When enzyme-labeled substances are analyzed, a substrate has to be offered to the enzymes in a second step. Usually, a change of the substrate color is catalyzed by the enzyme in a dose-dependent fashion that can be quantified photometrically.

Figure 22. Immunoassay. After incubation, the antigen–antibody complexes have to be separated from free antigens and antibodies, usually by precipitation of protein (e.g., with ammonium sulfate). Centrifugation will separate the complexes to the bottom whereas the uncomplexed reagents remain soluble.

Labeled antibody

Antigen antibody complex

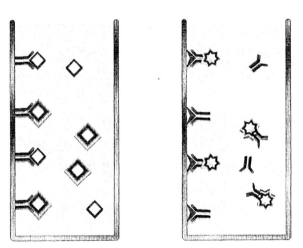

Figure 23. Competitive immunoassay. Labeled and unlabeled antigen compete for the limited binding sites of the antibodies fixed to the wall of the test tube (left panel). In another approach (right panel), labeled and unlabeled antibodies compete for the limited antigen attached to the tube wall. The optical or radioactive signals that can be measured are inversely proportional to concentration of the sample.

The registered signals are inversely proportional to the amount of substance to be determined in these assays.

The sandwich technique allows the fixation of free antigen to a solid phase. The wall of a tube is coated with the specific antibodies that bind the antigen. After an adequate time of incubation, and several thorough washes, labeled antibodies with the same specificity as the ones coated to the wall are added. Excess antibodies are removed by washing and the radioactivity or the change of color in the test tube is measured analogously to the competitive assays. The scores in this system are proportional to the amount of the determining substance (Figure 24).

The sensitivity of this assay can be enhanced by applying a second-step system. The first part of the assay is identical to the sandwich technique, except that the soluble antibody is not labeled. The second step comprises the addition of labeled antibodies that bind specifically to the first one, and thereafter the tube can be measured. The sensitivity enhancement results from the fact that one antigen is bound by several antibodies, each representing an antigen for further second step antibodies. Theoretically, by using several anti-antibodies the sensitivity of the test could be improved infinitely, but each Ag–Ab complex formation is associated with usually negligible unspecific bonds between all involved molecules. These unspecific complexes would also be potentiated with each newly added anti-antibody, making the results unreliable.

In practice, the enzyme based assays are preferable to the radioactive tests despite the higher sensitivity of the latter ones. Radioactive reagents are expensive and require particular equipment (counter) and a decontaminating disposal.

The assessment of the results of cytokine production in a bioassay, or the proliferative features in response to mitogen stimulation from blood samples that were drawn at different time points should be done very carefully. Usually, the composition of lymphocyte subsets

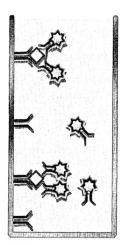
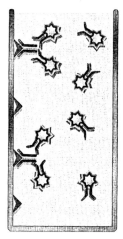

Figure 24. Sandwich technique and indirect immunoassay. In the sandwich technique the antigen is caught by fixed antibodies and in a second step the attached antigen is bound by labeled antibodies with the same specificity (left panel). The right panel shows an indirect system. An initial antibody that binds the antigen that is fixed to the tube wall is detected by a second antibody with specificity for the first one. At the readout the optical or radioactive signals are proportional to the concentration of the desired substance.

changes extremely quickly under physical or psychological stress. The varying lymphocytes differ widely from each other with regard to their proliferative potential, the capability to produce cytokines, and so on. For example, NK cell numbers in peripheral blood can be more than doubled within minutes of application of catecholamines. Stimulated NK cells exhibit a strong proliferation and produce considerable amounts of TNFα but no IL-2. When determining cell proliferation and cytokine production, it is possible that wide changes can be registered at different time points and that this may reflect cell trafficking rather than cell activation. In this example, measured alteration of the function may only confirm the phenotypic analysis, and is, therefore, redundant. To assess the activity of single cell populations one would have to analyze sorted cells. The situation is clearly different from determining cytotoxicity which is only exerted by particular cells. Because of the known effector numbers (blood cell count and phenotyping), the activity on a single cell level can be calculated.

7.11. Glossary

CD   cluster of differentiation; nomenclature that clusters antigens according to their biochemical and functional properties in order to standardize the antigens across the species borders, and these antigens are recognized by a given group of monoclonal antibodies

Con A   concanavalin A; a lectin that is mitogenic for T cells

cpm   counts per minute; a measure of radioactive decay

Cryo tube   small plastic tube for freezing cells; can withstand the temperature of liquid nitrogen ($-196°C$)

FACS   fluorescence-activated cell sorter; a laser-driven device for analyzing properties of cells

FITC   fluorescein isothiocyanate; green fluorescent dye

g   unit for the force of gravity

Humoral   pertaining to plasma and lymph; *humoral immune response* is used alternatively for *antibody-mediated defense*

Medium   physiological nutrient solution for culturing cells

PE   phycoerythrin; red fluorescent dye

PHA   phytohemagglutinin; lectin that is mitogenic for T cells

PWM   pokeweed mitogen; lectin that can activate T and B cells

8 Molecular Anatomical Basis of Interactions between Nervous and Immune Systems in Health and Disease

Eberhard Weihe, Michael Bette, Thorsten Fink,
Horacio E. Romeo, and Martin K.-H. Schäfer

8.1. Introduction

Traditionally, the nervous and immune systems have been regarded to function largely independently of each other. Interdisciplinary investigations of the recent past, however,

Eberhard Weihe, Michael Bette, Horacio E. Romeo, and Martin K.-H. Schäfer • Institute for Anatomy and Cell Biology, Department of Molecular Neuroimmunology, Philipps-University Clinic Marburg, D-35033 Marburg, Germany.   Thorsten Fink • Institute of Forensic Medicine, Johannes Gutenberg University Mainz, AM Pulverturm 3, D-55131 Mainz, Germany.

have demonstrated that the nervous and immune systems have close functional interrelationships at several nodal intersections (Ader & Cohen, 1993; Besedovsky & del Rey,
1996b; Felten & Felten, 1994; Ottaway & Husband, 1994; Weihe, Nohr, Michel, Müller,
Zentel, Fink, & Krekel, 1991a). Transmitters of the nervous system act on immune cells and
messengers of the immune system influence the nervous system. Typical neuronal messengers appear to be synthesized in immune cells, and typical immune cell messengers appear
to be produced in neurons. This is a further dimension of complex neuroimmune interactions. Functional interactions between the nervous and immune systems are fundamen-
tally important not only for physiological regulations, but also for diverse diseases such as
chronic painful inflammations (arthritis, pancreatitis), degenerative diseases, autoimmune
diseases (Multiple sclerosis), and viral infections affecting the nervous system (e.g., neuro-
AIDS) (Akaike, Weihe, Schäfer, Fu, Zheng, Vogel, Schmidt, Koprowski, & Dietzschold,
1995; Anisman, Baines, Berczi, Bernstein, Blennerhassett, Gorczynski, Greenberg, Kisil,
Mathison, Nagy, Nance, Perdue, Pomerantz, Sabbadini, Stanisz, & Warrington, 1996a,b;
Eiden, Rausch, Da Cunha, Murray, Heyes, Sharer, Nohr, & Weihe, 1993; McGeer, Rogers,
& McGeer, 1994; Nottet & Gendelman, 1995; Weihe, Nohr, Müller, Büchler, Friess, &
Zentel, 1991b; Wekerle, 1993). The following presentation provides fundamental insights
into the complex functional and molecular anatomy of the dialogue between the nervous and
immune systems in health and disease, and outlines perspectives for novel therapeutic
strategies targeted to restore the disturbed neuroimmune interactions.

8.2. Methodological Aspects

Detailed molecular anatomical knowledge of the principles of neuroimmune interactions is achieved by applying modern neurobiological, neuroanatomical, immunobiologi-
cal, and molecular biological methods. Transmitter substances in nerves supplying the
immune organs and tissues can be unequivocally identified by immunohistochemistry using
transmitter-specific antibodies on both the light and electron microscopic level (Felten &
Felten, 1994; Fink & Weihe, 1988; Weihe, Müller, Fink, & Zentel, 1989a; Weihe et al.,
1991a). The specific spatial relationships of transmitter-identified nerve endings with subtypes of immune cells can be characterized by costaining for neuronal messengers and
immune cell markers on identical tissue sections (Müller & Weihe, 1991; Nohr & Weihe,
1991). The regional and cellular localization of neurotransmitter receptors in cells of the
immune system is determined by receptor autoradiography. The immune cell-specific
synthesis of receptor proteins is evaluated and measured on the mRNA level by radioactive
and nonradioactive in situ hybridization histochemistry, and on the protein level by
immunocytochemical localization of the receptor (see Weihe, Schäfer, Nohr, & Persson,
1994a). However, the low abundance of the expression of neurotransmitter receptors in
immune cells limits their detectability and localization employing immunocytochemical or
in situ hybridization techniques. According to our experience, immune cells tend to exhibit
nonspecific immunostaining and nonspecific in situ hybridization signals. In particular,
eosinophilic cells hybridize with many cRNA probes nonspecifically. Therefore, in our
opinion, reports of abundant positive immunostaining for neurotransmitter molecules, or
hybridization for neurotransmitter mRNAs in immune cells, should be interpreted with
caution. Data on messenger and receptor expression from in vitro experiments and measurements in cell lines must be reevaluated under physiological and pathophysiological in vivo
conditions. To investigate the pathway-specific origins of transmitter-identified nerve supply to the various immune organs and lymphoid tissues, anterograde and retrograde tracing

techniques are combined with immunocytochemistry or *in situ* hybridization (Romeo, Fink, Yanaihara, & Weihe, 1994). Using this approach, different categories of innervation of the immune organs have been recognized: (1) afferent peptidergic sensory nerve fibers projecting to the CNS via dorsal root or cranial (e.g., vagal) sensory ganglia and (2) efferent sympathetic (noradrenergic) and (3) parasympathetic (cholinergic) nerve fibers projecting to the peripheral lymphoid tissues (see Table 1). The analysis of molecular reactions in immune cells and neurons to experimental neuromanipulation (chemical or surgical denervation) or to immune manipulation (immune stimulation with LPS, bacterial superantigens like SEB or SEA, experimental inflammation, viral infection, autoimmunity, or immune suppression with cyclosporine) provides evidence for the functional significance of neuroimmune interactions in health and disease, and reveals underlying cellular and molecular mechanisms (Behar, Ovadia, Polakiewicz, & Rosen, 1994; Bette, Schäfer, van Rooijen, Weihe, & Fleischer, 1993; Bode, Zimmermann, Ferszt, Steinbach, & Ludwig, 1995; Dietzschold, Schwäble, Schäfer, Hooper, Zheng, Petry, Sheng, Fink, Loos, Koprowski, & Weihe, 1995; Eiden *et al.*, 1993; Weihe *et al.*, 1991b; Wekerle, 1993).

8.3. Molecular Anatomical Prerequisites for Neuroimmune Interactions

In the periphery, the "hardware" for neuroimmune interactions consists of the wiring of the immune organs, and lymphoid tissues and cells, with the different categories of peripheral nerves (Weihe *et al.*, 1991a). Different transmitters of the nervous system (noradrenaline, acetylcholine, neuropeptides) and their specific receptors on target immune cells represent the "software" by which nerves influence immune cells (Madden & Felten, 1995). Immune cells synthesize cytokines acting on cytokine receptors located on the membranes of neurons or on support (glial) cells, to influence neurons directly or indirectly

Table 1
Messengers in the Neuroimmune Dialogue

Messengers of the nervous system
   Classical neurotransmitters
      Noradrenaline (postganglionic sympathetic neurons)
      Acteylcholine (postganglionic parasympathetic neurons)
      Excitatory amino acids (e.g., aspartate, glutamate; primary afferents)
   Neuropeptides
      Substance P (sensory neurons, vagal and primary spinal afferents)
      Calcitonin gene-related peptide (sensory, vagal, primary spinal)
      Neuropeptide Y (postganglionic sympathetic)
      Vasoactive intestinal polypeptide (postganglionic parasympathetic)
      Opioids (postganglionic sympathetic, postganglionic parasympathetic and sensory)
   Growth factors and neurotrophins
   Cytokines
Messengers of the immune system
   Cytokines
      Interleukins
      Interferons
      Tumor necrosis factor
   Neuropeptides
   Growth factors and neurotrophins

Table 2
Receptors Involved in the Neuroimmune Dialogue

Receptors in immune cells	Receptors in neurons	Receptors in glia cells
Cholinergic receptors	Cytokine receptors (?)	Cytokine receptors
Noradrenergic/adrenergic	Growth factor/neurotrophin receptors	Growth factor/neurotrophin receptors
receptors	Histamine receptors	Classical neurotransmitter receptors
Neuropeptide receptors		Neuropeptide receptors

by stimulation of glial-derived messengers (Hopkins & Rothwell, 1995; Rothwell & Hopkins, 1995; see Tables 1 and 2). This is the "software" for immune–neural influences. In addition, neurons themselves may produce cytokines that may act in an autocrine or paracrine fashion. Neuropeptides and endocrine messengers may be produced in immune cells to regulate immune cell functions in an autocrine, paracrine, or even intracrine manner (see Weihe *et al.*, 1994a). Furthermore, diverse molecules of the complex families of growth factors and neurotrophins and their receptors, which are differentially expressed in immune cells, mast cells, neurons, glial cells, or other cell types (Acheson & Lindsay, 1994), are likely to participate directly or indirectly in the interactive neuroimmune "concert."

8.4. Basic Principles of the Neuroimmune Dialogue in the Periphery

The possible peripheral neuroimmune interactions depend on close spatial relationships of nerve endings with different cells of the immune system (lymphocytes, macrophages, dendritic cells including epidermal Langerhans cells, granulocytes, mast cells), "support" cells of the connective tissue (e.g., fibroblasts), and the terminal vasculature, especially postcapillary venules (Fig. 1). Schwann cells, the glial cells of the peripheral nervous system, also come into play as indirect actors in the multifaceted processes of neuroimmune interactions. As in the CNS, volume transmission contributes to the interrelation of the pleiotropic messengers.

"Local Talk." The "local talk" of the neuroimmune dialogue occurs within the local micromilieu. Different messengers and diverse messenger receptors are directly or indirectly involved in the communication between immune cells and nerve endings (Fig. 2). The interactions are characterized by changes in the release of neurotransmitters, which are caused or modulated by changes in the local levels of immune mediators (e.g., cytokines, prostaglandins). Released neurotransmitters influence the synthesis and release of cytokines from immune cells or support cells. Under pathophysiological conditions, there is a mutual buildup of the interplay between immune mediators and neuromessengers (Weihe & Schäfer, 1994; Weihe *et al.*, 1991b). The molecular scenario is complemented by the local production of neurotrophins and growth factors acting on pleiotropic receptors localized on heterogeneous immune cell populations, support cells, and neuroglial cells (Acheson & Lindsay, 1994; Jonakait, 1993; Leon, Buriani, Dal, Fabris, Romanello, Aloe, & Levi-Montalcini, 1994). These factors are crucial for the regulation of the innervation density and spatial connectivity in the lymphoid tissues and in the local neuroimmune territory. This issue has been recently addressed in transgenic animals overexpressing nerve growth factor

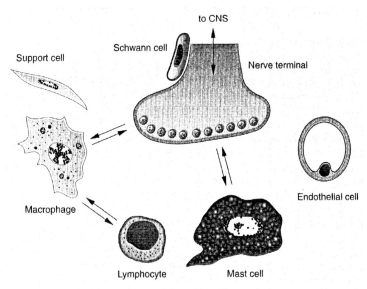

Figure 1.   Cellular network of the neuroimmune dialogue. Nerve fibers are in close contact with several types of immune and support cells within immune organs and peripheral tissues. The immune system and nervous system communicate at these interfaces directly, or indirectly, via support cells ("local talk"). Neural messengers (neuropeptides, classical transmitters) and immune messengers (cytokines) are released. In addition, peripheral nerve fibers have to be regarded as the starting point for "long-loop" mediation of immune signals to the CNS, and/ or as endpoints of "long-loop" signals from the CNS to the immune cells.

(Carlson, Albers, Beiting, Parish, Conner, & Davis, 1995; Carlson, Johnson, Parrish, & Cass, 1998). Depending on the body region, tissue compartments, and immune organ, the afferent sensory innervation and the efferent sympathetic noradrenergic and/or parasympathetic cholinergic innervation contribute to the interplay in combination or separately (Weihe & Hartschuh, 1988; Weihe *et al.*, 1991a).

"Tele Talk."   Neuronal chains of the sympathetic and parasympathetic nervous system originating from the CNS are responsible for the "tele" regulation of the immune system by the CNS (Fig. 2). One means of "tele" communication from the immune system to the CNS is realized by cytokines circulating in the bloodstream and entering the brain at sites where the blood–brain barrier is reduced (see Besedovsky & del Rey, 1996). Cytokines influence neuroendocrine regulatory centers in the hypothalamus and activate the hypothalamic–pituitary–adrenal (HPA) axis and the hypothalamo-sympatho-adrenomedullary axis. In addition to this immunoendocrine loop, immune signaling to the CNS appears to follow a direct immunoneural loop along primary afferents (Dantzer, 1994; Weihe *et al.*, 1991a).

We have earlier proposed that peripheral terminals of primary afferent neurons function as "immunoceptors" (Weihe & Schäfer, 1994; Weihe *et al.*, 1991a,b, 1994a), a view that has been experimentally confirmed in more recent investigations (Fleshner, Goehler, Hermann, Relton, Maier, & Watkins, 1995, Licinio & Wong, 1997). The perception of immune signals by primary viscerosensory neurons of the vagal nerve, or by primary somato- or viscerosensory neurons of the dorsal root ganglia (spinal ganglia), is transferred to

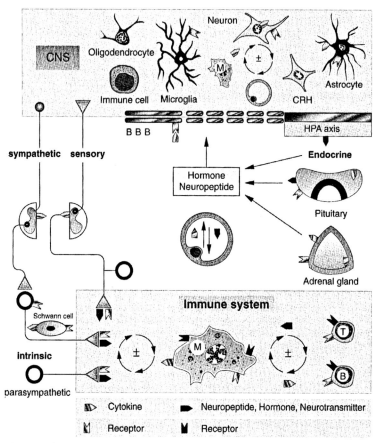

Figure 2. Neuro-immune-endocrine interactions. Schematic representation of the cells, molecules, and neural pathways for short- and long-loop neuro-immune-endocrine dialogue between the periphery and the CNS, and within the CNS. BBB, blood–brain barrier; HPA axis, hypothalamic–pituitary–adrenal axis; CRH, corticotropin-releasing hormone.

second-order neurons in the brain stem (vagal nerve) and the spinal cord (primary spinal afferents) (Fleshner *et al.*, 1995; Weihe *et al.*, 1991a). From there, the immune signaling is propagated via further neuronal chains to higher brain centers, particularly to the hypothalamus. It is assumed that the immune activation of these neural pathways also results in an activation of the HPA axis and of the hypothalamo-sympatho-adrenomedullary axis.

The long sensory loops that serve the "tele" communication from the immune system to the CNS may also be used antidromically as communication pathways from the CNS to the immune system (Weihe & Nohr, 1992; Weihe *et al.*, 1991a). Primary sensory neurons have a functional duality. They perceive and propagate signals to the CNS but also exert local effector functions in the periphery. This prompted us to argue for the theory that impulses from the brain activate terminals of primary afferents in the CNS and thus activate primary afferents antidromically, which could result in the peripheral release of immuno-modulatory neuropeptides (e.g., calcitonin gene-related peptide (CGRP), substance P (SP)) from sensory terminal in close spatial relationship with immune cells. This may cause priming, exacerbation, or suppression of immune reactions in the periphery, and could explain influences of the psyche on allergic reactions such as asthma (neuroimmune conditioning).

Furthermore, it can be anticipated that the net signaling generated in immune cells and adjacent support cells is perceived by the local endings of sensory, sympathetic, or para-sympathetic terminals directly or indirectly (by Schwann cells) and translated into short- or long-term changes of the gene expression in the respective neuronal cell bodies. Neurons altered in their metabolism in this way feed back into the local immune target milieu at their peripheral terminal. This provides the basis for protracted neuroimmune dialogues and neuroimmune homeostasis. In analogy to our arguing for antidromic activation of sensory efferent functions from the CNS, we propose the possibility that efferent sympathetic or parasympathetic neurons may be influenced antidromically, with the potential of antidromic signaling to the CNS along efferent pathways.

8.5. Principles of Local Neuroimmune Interactions in the CNS

Also within the CNS, there is evidence for mutual intercommunication between nerve endings and resident immunocompetent microglial cells of the CNS—the macrophages of the brain. This interaction is primarily of the "local talk" type but also involves long-loop neuronal networks between different brain areas (Hopkins & Rothwell, 1995; Kreutzberg, 1995; Travis, 1994). Microglial cells synthesize cytokines that can modify neurotransmission in the CNS. Neuronal activity regulates microglial function and cytokine production. Apparently, neuronal IFN-γ receptors are crucially involved (Neumann, Schmidt, Wilharm, Behrens, & Wekerle, 1997). Neuro-mast cell interactions in the brain may be much more important than currently believed, and could represent a causal link in CNS inflammatory and autoimmune disease. Astrocytes are also regarded as sources or targets of cytokines. They express cytokine receptors and neurotransmitter receptors for both classical transmit-ters and neuropeptide messengers. Microglial and astroglial cells can express neurotrophins and growth factors and are likely to interact with neurons and lymphocytes patrolling the CNS (Acheson & Lindsay, 1994). Endothelial cells express cytokine receptors and possibly also cytokines. Brain endothelial cells could be activated by circulating cytokines to produce cytokines, which when released abluminally influence iuxtaendothelial astrocytes, microglial cells, and possibly neurons (Weihe *et al.*, 1991a). Thus, brain endothelial cells may function as transducers of circulating immune signals to the brain parenchyma. The peripheral application of cytokine-inducing LPS or cytokines results in an activation of cytokine synthesis in brain resident cells (Bartholomew & Hoffman, 1993; Buttini & Boddeke, 1995; Rothwell & Hopkins, 1995), most likely microglial cells and/or perivascular macrophages. In the course of viral infections (e.g., influenza) or endotoxemia, cytokines produced in the periphery are likely to influence cytokine production in the CNS, by endocrine or neural transduction mechanism, resulting in the alteration of neuronal func-tions and sickness behavior. This may be independent of the induction of fever.

8.6. Molecular Anatomy of the Innervation of Immune Organs

The sympathetic, sensory, and parasympathetic innervation of primary and secondary lymphoid organs differ qualitatively, quantitatively, proportionally, and with respect to neuropeptide content. The innervation of the lymphoid organs is delivered along their blood vessels from where fibers branch off to form close spatial relationships with different subsets of immune cells (Madden & Felten, 1995; Weihe et al., 1991a). With few exceptions, the qualitative pattern of innervation in primates and rodents seems to be similar. However, the density of innervation of lymphoid organs in rhesus monkeys and humans is lower than in rats.

8.6.1. Primary Lymphoid Organs

Bone Marrow. Knowledge about the specific innervation of the bone marrow is sparse (Felten & Felten 1994; Weihe et al., 1991a). The presence of vascular and nonvascular fibers staining for CGRP and SP indicates that they are of primary sensory origin. Tyrosine hydroxylase (TH)- and neuropeptide Y (NPY)-positive fibers mark the sympathetic innervation (Imai, Tokunaga, Maeda, Kikkawa, & Hukuda, 1997). Thus, neuroimmune interactions at this level of the bone marrow are conceivable, especially influences of bone marrow innervation on the differentiation and mobilization of hematopoietic cells (Afan, Broome, Nicholls, Whetton, & Miyan, 1997; Benestad, Strom, Ole, Haug, & Nja, 1998). This is an area that deserves further investigation. There is as yet no evidence that the bone marrow is supplied by the parasympathetic nervous system. The bursa of Fabricius, the bone marrow equivalent of birds, receives a particularly rich innervation with fibers containing vasoactive intestinal polypeptide (VIP), which target B cells. Neuro-B-cell contacts in the bursa of Fabricius are frequent, but neuro-T-cell contacts are also formed (Zentel & Weihe, 1991; Zentel, Nohr, Albrecht, Jeurissen, Vainio, & Weihe, 1991). However, the origin and the classical transmitter content of the VIP innervation in birds are unclear.

Thymus. Based on double labeling experiments for neuropeptides and markers of the peripheral sympathetic innervation, such as TH or the vesicular monoamine transporter isoform 2 (VMAT2) (Erickson, Schäfer, Bonner, Eiden, & Weihe, 1996; Weihe, Schäfer, Erickson, & Eiden, 1994b), the presence of sensory CGRP/SP innervation and sympathetic noradrenergic NPY innervation in the mammalian thymus is suggested (Kendall & al-Shawaf, 1991; Kranz, Kendall, & von Gaudecker, 1997; Müller & Weihe, 1991; Weihe et al., 1989a). The thymic VIP innervation has been regarded to be cholinergic parasympathetic, although direct evidence for thymic innervation by fibers costaining for VIP and unequivocal markers of cholinergic innervation was absent. In fact, our current investigations with the recently established, novel, reliable, and highly sensitive marker of peripheral and central cholinergic innervation VAChT, the vesicular acetylcholine transporter (Schäfer, Weihe, Varoqui, Eiden & Erickson, 1994; Schäfer, Weihe, Erickson, & Eiden, 1995; Schäfer, Schütz, Weihe, & Eiden, 1997; Schäfer, Eiden, & Weihe, 1998a; Schäfer, Eiden, & Weihe, 1998b; Weihe, Tao-Cheng, Schäfer, Erickson, & Eiden, 1996), provide no evidence for the presence of cholinergic innervation in the mammalian thymus (Schäfer et al., 1998b). This is in contrast to previous reports claiming that cholinergic innervation of the thymus exists (Antonica, Magni, Mearini, & Paolocci, 1994; Antonica, Ayroldi, Magni, & Paolocci, 1996). The neuropeptidergic and catecholaminergic innervation of the thymus is concentrated in the capsule, in the interlobular septa, in paravascular spaces, and in the cortico-medullary boundary zone. SP and CGRP fibers frequently branch off from the perivascular space to run into the surrounding parenchyma. TH/VMAT2-positive noradrenergic fibers are more restricted to perivascular plexus. Paravascular mast cells and macrophages, but not T cells, are the main targets of sensory and sympathetic thymic innervation (Müller & Weihe, 1991). A few fibers containing opioid peptides are also encountered. Some cells of the thymus express neuropeptides (enkephalin, tachykinins, CGRP) or chromogranin A but their cellular identities and functions are not known. Like previously reported nonneuronal expression of oxytocin and vasopressin in the thymus, the presence of these molecules may reflect that the thymus has a neuroendocrine compartment, or that these molecules are expressed in thymic immune cells. These issues clearly have to be resolved in conjunction with the characterization of the neuropeptide receptor expression in identified thymic immune and thymoepithelial cells.

8.6.2. Secondary Lymphoid Organs

Lymph Nodes. Lymph nodes of different body regions receive sensory and sympa-thetic nerves that accompany blood vessels and lymphatic vessels, and branch off from paravascular spaces to enter the parenchyma (Fink & Weihe, 1988; Weihe et al., 1991a). CGRP/SP staining marks the sensory innervation and TH/NPY/VMAT2 staining delineates the sympathetic innervation. VIP, a cotransmitter candidate of acetylcholine innervation, is contained in a low number of nerves, which are most likely of sensory origin by staining of nerve fibers for the cholinergic parasympathetic marker. The presence of VIP innervation was taken as indirect evidence for a cholinergic innervation of lymph nodes because VIP is a typical neuropeptide cotransmitter in postganglionic parasympathetic cholinergic neurons. However, the coexistence of VIP and a reliable marker of peripheral cholinergic innervation has not yet been shown. Our current investigations with VAChT, the novel marker of cholinergic innervation, showed that cholinergic innervation is absent from lymph nodes (Schäfer et al., 1998b). Thus, the view of a parasympathetic cholinergic innervation of lymph nodes has to be reevaluated. Sensory CGRP/SP fibers running independent from the vasculature are more frequent than nonvascular sympathetic TH/NPY/VMAT2 fibers. Target cells of both sensory and sympathetic terminals include mast cells, dendritic cells, plasma cells, macrophages, and T cells (Fig. 3). Elimination of small-diameter C-fibers with capsaicin treatment modifies stimulated antibody production supporting the functional importance of neuro-plasma cell contacts. Germinal centers bearing B cells are regularly spared from any innervation. High endothelial venules do not appear to be innervated, excluding the possibility of direct neuronal control of lymphocyte migration through the wall of high endothelial venules. Afferent and efferent lymphatic vessels and arterial vessels are prominently innervated by sensory and sympathetic peptidergic fibers.

The presence of SP-containing innervation in conjunction with selective expression of the SP (NK-1) receptor gene in endothelial cells and paravascular macrophages, and perhaps mast cells, suggests that tachykinin-mediated immunomodulation in lymph nodes involves endothelial cells, mast cells, and macrophages but not T cells (Weihe et al., 1994a). In contrast to in vitro data providing evidence for NK-1 receptors in T cells, our in vivo studies indicate absence or very low abundance of NK-1 receptors in T or B cells. The cellular identity of NPY, VIP, and pituitary adenylate cyclase-activating polypeptide (PACAP) receptors (Garrido, Delgado, Martinez, Gomariz, & de la Fuente, 1996; Yada, Sakurada, Ishihara, Nakata, Shioda, Yaekura, Hamakawa, Yanagida, Kikuchi, & Oka, 1997) in lymph nodes under in vivo conditions is unclear. It remains to be elucidated to what

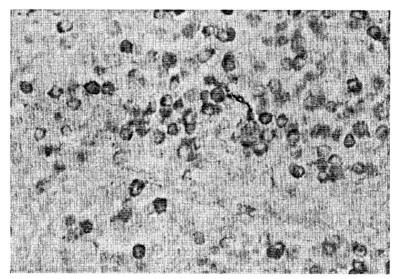

Figure 3.   Nerve fibers displaying CGRP immunoreactivity are in close contact with immune cells (plasma cells) in a lymph node of a rat. Plasma cells were visualized using antibodies against immunoglobulins.

extent the ample evidence of pleiotropic neuropeptide receptor expression in immune cell lines is matched by the *in vivo* situation, when immune cells are naturally exposed to pleiotropic paracrine control mechanisms of their activity by other cell types.

Retrograde tracing, combined with immunocytochemistry for opioid peptides, demonstrates a proenkephalin-derived enkephalinergic component of the sympathetic innervation to cervical lymph nodes, originating from the superior cervical ganglion that is segregated from the sympathetic NPY innervation (Romeo *et al.*, 1994). This is further evidence for a pathway- and transmitter-specific modulation of immune functions in lymph nodes. In addition, proenkephalin opioids are expressed in immunocytes of lymph nodes (most likely in both T cells and macrophages). LPS treatment upregulates the gene expression of proenkephalin (Behar *et al.*, 1994).

Spleen.   The sympathetic innervation predominates in the spleen (Felten & Felten, 1994; Weihe *et al.*, 1991a). Sensory CGRP/SP nerves are rare and cholinergic innervation is absent. The majority of splenic sympathetic fibers costoring noradrenaline and NPY are associated with the vasculature. Some varicose fibers, however, branch off and supply the iuxtavascular lymphoid parenchyma with synapselike target relationships to lymphocytes, and close spatial relationships to macrophages. B cells in the germinal centers are excluded from innervation. Opioidergic nerve fibers are species specific (Nohr, Michel, Fink, & Weihe, 1995). Nerve fibers containing proenkephalin derivatives occur in cow and pig but not in mouse, rat, guinea pig, or hamster. Whether opioidergic innervation is present in primate spleen is not yet known.

The expression of opioids in nonneuronal cells of the rodent spleen is indelible. In the mouse, the expression of proenkephalin mRNA is temporally coinduced with that of IL-2 by T-cell superantigen stimuli. The induced expression most likely localizes to T cells. IL-6 secretion is inhibited by electrical stimulation of spleen slices, an effect that is blocked by the opioid antagonist naloxone (Straub, Herrmann, Berkmiller, Frauenholz, Lang, Scholmerich, & Falk, 1997), indicating the possibility that opioids locally released from immunocytes may control IL-6 secretion in a paracrine manner because murine splenic innervation lacks an opioid component. Cytokines modulate splenic blood flow (Rogausch, del Rey, Kabiersch, & Besedovsky, 1995). The cytokine-induced modulation of blood flow in the rat spleen is reversed by naloxone (Bognar, Albrecht, Farasaty, Schmitt, Seidel, & Fuder, 1995). In the absence of opioidergic innervation (Nohr *et al.*, 1995) of rat splenic blood vessels, immunocytes of the spleen are the most likely cellular source of opioid synthesis and release, responsible for naloxone-reversible modulation of cytokine effects on blood flow. The inducible opioid expression in splenic immunocytes could be relevant in endotoxic shock. The superantigen- and LPS-induced cytokine expression in murine spleen appears to be altered by chemical sympathectomy. This is a further indication that the splenic innervation has functional significance during the immune response. The sympathetic/NPYergic reinnervation of splenic transplants depends on the age of the donor spleen and on the age of the host (Westermann, Michel, Lopez-Kostka, Bode, Rothkötter, Bette, Weihe, Straub, & Pabst, 1998).

Mucosal-Associated Lymphoid Tissue (MALT).   Virtually all mucosal tissues are endowed with a defense line of immunocytes, the MALT. Independent of the location, the innervation of the various MALT systems [gut-associated lymphoid tissue (Peyer's patches), bronchial-associated lymphoid tissue, nasal-associated lymphoid tissue, urogenital-associated lymphoid tissue, skin-associated lymphoid tissue] is stereotypically similar (Nohr & Weihe, 1991; Nohr, Buob, Gärtner, & Weihe, 1996; Weihe & Hartschuh, 1988; Weihe et al., 1991a; Weihe, Hartschuh, Schäfer, Romeo, & Eiden, 1998). The occurrence of nerve fibers is restricted to the outer margins of the lymphoid follicles, while germinal centers (B-cell areas) are regularly spared from innervation. This pattern is similar to that observed in lymph follicles of spleen, tonsils, and lymph nodes. As a rule, sensory fibers containing CGRP/SP, sympathetic noradrenergic NPY fibers, and cholinergic VIP fibers are associated with MALT. This pattern is relatively independent of species and also applies to human palatine tonsils (Weihe & Krekel, 1991). All three categories of innervation form close spatial relationships with mast cells, macrophages, and plasma cells and with the outer border of lymphoid follicles bearing T cells. Neuro-immunocyte and neuro-mast cell contacts are frequently encountered at sites distant from lymphoid follicles. In the skin, cholinergic/VIPergic neuro-mast cell contacts are less frequent than in visceral mucosal tissues and restricted to sweat gland and paravascular microterritories. In the skin, sensory peptidergic and sympathetic neuro-mast cell contacts prevail. The sensory peptidergic (CGRP) innervation of dendritic epidermal Langerhans cells seems to play a role in the modulation of antigen presentation by Langerhans cells to T cells (Hosoi, Murphy, Egan, Lerner, Grabbe, Asahina, & Granstein, 1993). This may be of relevance in cutaneous immunological reactions and in neurogenic inflammation, e.g., under the influence of ultraviolet radiation and sunburn (Gillardon, Moll, Michel, Benrath, Weihe, & Zimmermann, 1995)—with possible long term consequences for the development of a melanoma. Whether antigen-presenting dendritic cells in other locations are also modulated by peptidergic innervation is not known.

Disseminated Neuroimmune Connections.   In addition to the innervation of lymphoid organs and MALT, most organs and tissues exhibit areas where disseminated immunocytes and mast cells are targeted by nerve terminals, in particular in paravascular areas with special reference to postcapillary venules. Nerve terminals, mast cells, macrophages, and support cells form close spatial relationships. In the mucosal epithelial cells, local neuroendocrine cells, and in the synovia of joints, synovial lining cells enter the spatial scenario for multicellular direct and indirect neuroimmune communication. This dialogue involves multicellular synthesis and release of cytokines, growth factors/neurotrophins, and neuromessengers influencing each other. Such mixed paracrine interactions probably contribute to the local tissue homeostasis and are important factors in defense reactions, wound healing, and repair (Weihe *et al.*, 1991a, 1994a). Neuroimmune mechanisms underlying pathophysiological reactions in the respiratory, gastrointestinal, and urogenital tract, in joints, and in musculocutaneous tissues, as well as in the cardiovascular system, may be of particular importance. Neuro-macrophage and neuro-mast cell contacts in conjunction with neuro-endothelial contact may be key interfaces for disease-relevant neuroimmune dialogues. At these fronts, toxic substances from the environment meet the organism's first defense line and may break it by unbalancing the neuroimmune homeostasis. We postulate that environmental diseases are elicited by damage of the two main defense sensors of the organism, the nervous and the immune systems. Thus, ecological diseases are likely to involve disturbances of neuroimmune balance.

8.7. Pathophysiological Significance of Neuroimmune Interactions

Based on a variety of clinical and experimental observations, we assume, as outlined above, that a loss of neuroimmune homeostasis may be the common denominator in causing and chronifying diseases. Neuroimmune dysregulation could be a crucial pathophysiological principle in a variety of diseases (Table 3).

8.7.1. Peripheral Inflammation, Inflammatory Pain, Allergy

The sensory nervous system responds to peripheral inflammation with an upregulation of the expression of its key neuropeptide transmitters, SP and CGRP (Persson, Schäfer, Nohr, Ekström, Post, Nyberg, & Weihe, 1994; Weihe, Nohr, Schäfer, Persson, Ekström, Källström, Nyberg, & Post, 1995). The peripheral release of these transmitters in the inflamed tissue is enhanced and causes increased synthesis and local release of the cytokines IL-1 and TNFα (Weihe & Nohr, 1992; Weihe *et al.*, 1991b). These cytokines enhance the release of sensory neuropeptides. Endothelial cells, and possibly macrophages, react with upregulation of the expression of the NK-1 receptor on which SP exerts its main effects. Endothelial NK-1 receptors may further facilitate the immigration of leukocytes by upregulating the expression of the cell adhesion molecule ICAM. Taken together, these factors enhance inflammatory processes (Weihe *et al.*, 1994a) by entering into a vicious circle. This may be a key mechanism in triggering and sustaining chronification of inflammatory pain such as commonly observed in rheumatoid arthritis, chronic interstitial cystitis, chronic inflammatory bowel diseases such as sustained appendicitis, chronic pancreatitis, Crohn's disease, ulcerative colitis, and anal fissures (Büchler, Weihe, Friess, Malfertheiner, Bockman, Müller, Nohr, & Beger, 1992; Di Sebastiano, Fink, Weihe, Friess, Beger, & Büchler, 1995; Di Sebastiano, Fink, Weihe, Friess, Innocenti, Beger, & Büchler, 1997; Fink, Di Sebastiano, Büchler, Beger, & Weihe, 1994; Frieling & Strohmeyer, 1995; Hörsch, Kirsch, & Weihe, 1998; Watkins, Maier, & Goehler, 1995b; Weihe, 1998) (see Fig. 4).

Table 3
Diseases with Underlying Neuroimmune Mechanisms

Inflammation	Degenerative diseases
Arthritis	Alzheimer's disease
Crohn's disease	Parkinson's disease
Ulcerative colitis	Stroke
Chronic pancreatitis	Trauma, neural lesions
Appendicitis	Viral infection
Tonsillitis	Herpes
Cystitis	AIDS dementia
Autoimmune diseases in the CNS	Borna disease
Multiple sclerosis (experimental	Influenza
allergic encephalomyelitis)	Chronic fatigue syndrome
	Fibromyalgia

Peripheral inflammation also causes reactions in the CNS. In the spinal cord, genes encoding the precursors of endogenous opioids, prodynorphin and proenkephalin, are upregulated. This is probably a consequence of an increased spinal release of SP, CGRP, and glutamate from nociceptor terminals in the dorsal horn of primary afferents projecting from the inflamed tissue into the spinal cord (Persson *et al.*, 1994; Weihe, Millan, Höllt, Nohr, & Herz, 1989b). The spinal upregulation of opioid is seen as a mechanism of endogenous pain suppression. The spinal upregulation of NK-1 receptors is regarded as a mechanism to enhance spinal nociceptive neurotransmission resulting in increased pain (Schäfer, Nohr, Krause, & Weihe, 1993). Most interestingly, NK receptors even seem to be expressed on primary afferents themselves, implying autoactivation of immunoceptor signaling by SP (Stumm, Schäfer, & Weihe, 1997). The net effect of pain sensation results from the balance between pain enhancing and pain suppressing messengers and receptors along the ascending nociceptive and the descending inhibitory pain pathways. From inflamed tissues, both "immune signals" and nociceptive ("pain") signals are transmitted to the spinal cord, probably by an identical population of small-diameter (C, A-δ) afferents. Therefore, we introduced for these primary afferents the term *immunonociceptors* (Weihe *et al.*, 1994a). This terminology underlines the increasing knowledge of the involvement of neuroimmune interactions in inflammatory pain (Watkins *et al.*, 1995b; Weihe *et al.*, 1991b). There is mounting evidence that neuroimmune mechanisms within the CNS may be crucially

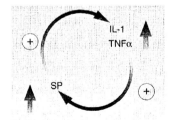

Figure 4. Chronification of inflammatory pain. As a vicious circle, inflammation and pain in the periphery become potentiated by inflammation-dependent mediators (cytokines such as TNFα or IL-1). These mediators, released from immune and other cell types, increase the production of proinflammatory substances (e.g., substance P) in sensory nociceptive nerve fibers. Additionally, substance P enhances the synthesis and release of cytokines from leukocytes and other local cell types involved in the ongoing inflammation. The outcome is a self enhancing process of chronic inflammation.

involved in the pathogenesis of chronic neuropathic pain, as suggested by our group (Weihe *et al.*, 1991a,b).

Consequently, it is an attractive strategy to treat inflammatory, and possibly neuropathic, pain by reestablishing the neuroimmune balance in the "local talk" and "tele talk" of the neuroimmune dialogue.

Direct clinical evidence for a neural contribution to chronic arthritis derives from the well-known fact that arthritis is reduced or disappears from joints that become functionally denervated by stroke or nerve lesions.

Similarly, in allergic reactions (asthma), neurodermitis, psoriasis, and chronic fatigue syndrome, neuroimmune factors are regarded to trigger or to sustain (chronify) disease (Moldofsky, 1995; Weihe & Hartschuh, 1988; Williams, Bienenstock, & Stead, 1995). Besides neuroimmune target relations, as described above, mast cells, macrophages, and T cells, as well as dendriticlike cells located in the peripheral nerves themselves, may come into play. They appear to be targeted by intranerve terminals. This constellation is of great importance in neuritis and neurotoxicological reactions (Purcell & Atterwill, 1995). Diabetic neuropathy may be explained by changes in the activity of neural immune components within peripheral sensory and autonomic nerves. Not surprisingly, AIDS patients often suffer from peripheral and central neuropathy. Here, a neuroimmune connection as a cause of the pain symptoms is obvious. Interestingly, certain tumors grow along and infiltrate into peripheral nerves, e.g., in pancreatic carcinoma, a process that is often accompanied by infiltration of inflammatory cells. Thus, neuroimmune mechanisms may also underlay tumor pain. One is tempted to use the term *neuroimmunooncology* or even *psychoneuroimmunooncology* with respect to the long antidromic loops between the nervous and immune systems outlined above.

8.7.2. Virus Infections, Autoimmune and Neurodegenerative Diseases of the CNS

Recent investigations reveal common neuroimmunological pathomechanisms underlying diverse inflammatory and degenerative diseases of the CNS (Morimoto, Hooper, Bornhorst, Corrisdeo, Bette, Fu, Schäfer, Koprowski, Weihe, & Dietzschold, 1996; Rausch, Heyes, Murray, Lendvay, Sharer, Ward, Rehm, Nohr, Weihe, & Eiden, 1994). Cytotoxic mediators (NO, peroxides) and cytokines, as well as chemokines, produced and released by resident macrophage/microglial cells or by invading immunocytes are likely to play a key role in the disease processes (Hooper, Bette, Morimoto, Weihe, Koprowski, & Dietzschold, 1998). Also, a growing number of cytokine receptors have been described to be expressed in resident brain cells which become regulated under pathophysiological conditions (Sternberg, 1997; Fig. 5). Growth factors/neurotrophins synthesized by astroglial cells or by neurons may have compensatory neuroprotective potential. Interestingly, a similar spectrum of neurotransmitter systems (opioids, CGRP, tachykinins) that are affected by peripheral inflammation or nerve injury are also regulated during viral infections of the CNS, neurodegeneration related to cerebral ischemia, and in experimental autoimmune disease (experimental allergic encephalitis). The regulation of somatostatin expression in the cortex seems to be characteristic for neuro-AIDS. Complement activation is another phenomenon common to both inflammatory and degenerative CNS diseases (Dietzschold *et al.*, 1995; Post, Salvati, Schäfer, Schwäble, Cini, Calabresi, Vaghi, Wong, & Weihe, 1996). Taken together, we propose that neuroimmune interactions within the brain contribute to regulatory mechanisms of neurodegeneration and neuroregeneration.

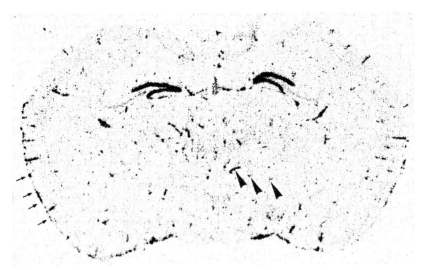

Figure 5. IL-1β receptor type I mRNA expression in the brain of a mouse intraperitoneally challenged with staphylococcal enterotoxin B (SEB). Note strong *in situ* hybridization signals in leptomenix (arrows), in the dentate gyrus, and in blood vessels (arrowheads).

8.8. Therapeutic Strategies Targeted to the Neuroimmune Connection

The main goal of a neuroimmune-oriented therapeutic strategy of chronic disease is the interruption of a disease-furthering vicious circle (Fig. 6). This can be achieved by blocking signals of the nervous system that evoke and sustain pathophysiological immune reactions. Thus, presynaptic inhibition of the release of proinflammatory and pronociceptive neuropeptides by opioids acting on peripheral opioid receptors on nociceptor endings may be one way to interrupt the circle under inflammatory pain conditions. By blocking peripheral postsynaptic NK receptors on endothelial cells or macrophages, the neural stimulus for immune cell immigration and increased cytokine production should be attenuated. Alternatively, the synthesis or release of proinflammatory/pronociceptive cytokines should be blocked, or the released molecules should be neutralized, immunopharmacologically or by antagonists. The combination of the two approaches may be even more efficient in disrupting the assumed viscous circle of neuroimmune dysregulation. In fact, the therapeutic trials with TNFα antibodies reduced inflammation and pain in rheumatoid arthritis and in Crohn's disease. Corticosteroids are not only anti-inflammatory but also regulate the expression of sensory neuropeptides believed to be crucial in the neurogenic inflammatory process (Weihe *et al.*, 1995). This very efficient drug, modulating both immune responses and neural messenger plasticity, supports the view that combined neuroimmune pharmacotherapy may be particularly effective.

In the CNS, neuroimmunological reactions in the cause of degenerative, inflammatory, and autoimmune neurological diseases (Alzheimer's disease, Parkinson's disease, stroke,

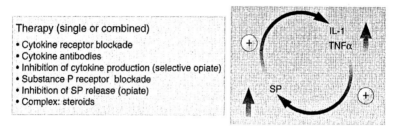

Figure 6. Neuroimmune therapy of chronic inflammatory pain.

multiple sclerosis) may offer a target for neuroimmune therapy. In particular, the activation of cytotoxic molecules in the microglia such as cytokines, nitrogen and oxygen intermediates, particularly peroxynitrite, complement factors, and others may be inhibited by a combined application of inhibitors of neuron hyperexcitation (NMDA receptor antagonists) and cytokine antagonists and/or free radical scavengers. As cyclooxygenase-2, the inducible rate-limiting enzyme of prostaglandin E_2 synthesis, is dramatically upregulated in neurons after experimental stroke and after experimental virus-induced encephalitis (Miettinen, Fusco, Yrjäneikki, Hirvonen, Roivainen, Närhi, Höckfelt, & Koistinaho, 1997; Morimoto et al., 1996), specific COX-2 inhibitors may also be beneficial (Appleton, Tomlinson, & Willoughby, 1994). The dramatic induction of CGRP in cortical neurons (Röhrenbeck, Bette, Hooper, Nyberg, Eiden, Dietzschold, & Weihe, 1999) and the changes of proenkephalin expression (Fu, Weihe, Zheng, Schäfer, Sheng, Corisdeo, Rauscher, Koprowski, & Dietzschold, 1993) after virus-induced encephalitis suggest CGRP and opioid agonists/antagonists as potential therapeutics in neurodegeneration and neuroinflammation.

In light of recently demonstrated upregulation of complement factors in brain resident cells during cerebral ischemia, multiple sclerosis, Alzheimer's disease, and virus-induced encephalitis, the inhibition of complement activation by novel complement inhibitors is another attractive therapeutic target.

Immunosuppressive therapy with diverse cytokines for the treatment of oncological diseases, in transplantation medicine, or in multiple sclerosis could exert short- and long-term side effects in the nervous system. It will be of paramount importance to recognize the underlying neuroimmune mechanisms of the therapy-induced behavioral effects that are very disturbing for the patients.

8.9. Summary, Conclusions, and Perspectives

On multiple intersections, the peripheral and central nervous system and the immune system are directly or indirectly connected to each other and to the endocrine system. Neuroimmune endocrine networks are characterized by anatomical, cellular, and molecular heterogeneity and complexity with tissue, organ-, and cell-specific patterns. The bidirectional communication between the nervous and immune systems occurs in short and long loops, also referred to as "local talk" and "tele talk." Neuroimmune dialogues are short term with acute release of messengers and long term with sustained changes in messenger and messenger receptor gene expression. Neuronal messengers (classical neurotransmitters and neuropeptides) act either directly on immune cells by binding on specific immune cell receptors or indirectly via activation of secondary messengers in support cells (e.g., growth factors/neurotrophins). In the CNS, neuronal activity and neuronal messengers seem to be capable to regulate microglial functions, and microglial mediators seem to influence neuronal functions. The expression of cytokine receptors on neurons seems to be limited. Therefore, cytokines appear to influence neuronal functions mostly indirectly, both in the periphery and in the CNS, by involving support cells, mast cells, and glial cells which produce neuroactive growth factors, neurotrophins, and a variety of bioactive molecules such as histamine and prostaglandins. The functional significance of the expression of neuromessengers (neuropeptides) in immune cells and of immune messengers (cytokines) in neurons are still a controversial matter. Opioids expressed in immune cells may serve autocrine, paracrine, or intracrine regulation. They may even participate in the regulation of the cell cycle and apoptosis. The sensory nervous system perceives immune and pain signals in the periphery and transmits this information to the CNS (immunonociceptor function). Vagal afferents have an especially potent immune signaling capacity to the brain. In most peripheral tissues, and both primary and secondary immune organs, sensory endings also exert an effector function. They release immunostimulatory and immunosuppressive neuropeptides and interact with the immunomodulatory messengers of the sympathetic terminals. A cholinergic parasympathetic neuroimmunomodulation appears to be restricted to MALT. The neural control of antigen presentation in dendritic cells (epidermal Langerhans cells) is apparently a domain of CGRP-containing sensory neurons. Neuroimmune dysregulation underlies a plethora of inflammatory, degenerative, autoimmune, and allergic diseases and also many diseases caused by environmental factors. The neuroimmune connection in pain is obvious. The recognition of neuroimmune factors in the pathogenesis of inflammatory and neuropathic pain led to the first trials to treat pain with neuroimmune therapeutics. A better understanding of the complex cellular and molecular mechanisms of neuroimmune interactions in health and disease opens far-reaching perspectives for innovative therapies, for long-term disease prevention, and for avoiding unpleasant side effects in ongoing and future pharmacotherapies.

ACKNOWLEDGMENTS. This work was supported by Volkswagen-Stiftung; Deutsche Forschungsgemeinschaft (DFG); BMFT; Kempkes-Stiftung, Marburg; and Naturwissenschaftliches Medizinisches Forschungszentrum (NMFZ), Universität Mainz. We would like to thank Prof. Dr. Helmut Beger (Universitätsklinikum Ulm), Prof. Dr. Markus Büchler (Universitätskliniken Bern), Prof. Dr. Bernhard Dietzschold (Jefferson University, Philadelphia), Dr. Pierluigi Di Sebastiano (Clinica Pierangeli, Universita di Chieti), Dr. Lee Eiden (Bethesda, NIMH), Prof. Dr. Bernhard Fleischer (Bernhard Nocht Institut für Tropenmedizin, Hamburg), Dr. Helmut Frieß (Universitätskliniken Bern), Prof. Dr. Michael Loos (Universität Mainz), Dr. Donatus Nohr (Universität Düsseldorf), Prof. Dr. Fred Nyberg (Uppsala Universität), Dr. Stephan Persson (Pharmacia/UpJohn, Uppsala), Prof. Dr. Claes Post (Pharmacia/UpJohn, Uppsala), and Dr. Wilhelm Schwaeble (University Leicester and Universität Marburg) for their contributions during different stages of this work. For technical assistance we acknowledge Martin Anlauf, Christian Brett, Candan Depboylu, Markus Gördes, Kristina Kappel, Friderike Kesten, Dagmar Schütz, Ralph Stumm, and Christian Ulke.

9 Neuroendocrine System and Immune Functions

Manfred Schedlowski and Robert J. Benschop

9.1. Introduction

There is abundant empirical evidence that immunocompetent cells are influenced by substances of the neuroendocrine system (reviewed by Ader, Felten, & Cohen, 1991; Besedovsky & del Rey, 1996; Madden & Felten, 1995). This chapter describes a selection of hormones, neurotransmitters, and neuropeptides that can be detected in peripheral blood and/or in the tissue and that are known to affect immunocompetent cells.

The effects of neuroendocrine substances on the immune system are analyzed in *in*

Manfred Schedlowski • Institute of Medical Psychology, Essen University Clinic, D-45122 Essen, Germany. Robert J. Benschop • Division of Basic Sciences, National Jewish Medical and Research Center, Denver, Colorado 80206.

vitro or *in vivo* experiments. In animal experiments, the neuroendocrine signals can be blocked via the administration of selective antagonists, or by surgical removal or chemical destruction of the endocrine organs under study. Such an approach usually yields unequivocal results on the role of a given substance on the immune system.

The effects of the different neuroendocrine parameters on the immune system are dependent on the species under investigation (e.g., humans, rodents), whether the effects are analyzed *in vitro* or *in vivo*, and the type of immune response investigated (e.g., cellular versus humoral). In addition, the effects of a certain hormone may differ, sometimes with the opposite effect, depending on the dosage (pharmacological versus physiological), the administration route (intraperitoneal, intravenous, intracerebroventricular), and the timing of administration (i.e. before or after antigen challenge). In view of this heterogeneity in research methods and immunological parameters investigated in animal and human experiments, this chapter will only exemplarily summarize empirical data that reflect the current state of the art of the interactions between the neuroendocrine and immune systems.

Substances produced by the neuroendocrine system serve as transmitters and can be grouped into hormones, neurotransmitters, and neuropeptides. Using this functional and topographical definition, hormones are produced by endocrine cells, and neurotransmitters are released from central and peripheral neurons. Neuropeptides can be detected in nerve tissues and are also produced by neurons (Fig. 1) (see also Chapter 3).

The nervous system can regulate immunological processes by autonomic nerve fibers, which are located in primary as well as in secondary lymphoid organs (see Chapter 8). In addition, circulating transmitter substances of the neuroendocrine system are able to influence immunocompetent cells via specific membrane receptors. Hormones, neurotransmitters, and neuropeptides can affect diverse processes in the immune system such as cellular development and differentiation, lymphocyte activation and proliferation, cell migration, production and release of cytokines, and the expression of cytokine receptors. An overview of the neuroendocrine factors with immunological relevance introduced in this chapter is presented in Table 1.

Three conditions must be fulfilled for a neuroendocrine–immune interaction:

1. Release and presence of endogenous substances in concentrations that allow receptor-mediated signal transduction on the target cell.
2. Presence of receptors in or on immunocompetent cells, which in turn induces processes in the cell affecting the ligand–receptor interaction of a target cell.
3. Reproducible effects of these ligand–receptor interactions on the migration and/ or functional capacity of immune cells in peripheral blood and/or in other compartments of the immune system.

Apart from intracellular cytoplasmic receptors for substances such as glucocorticoids and thyroid hormones, most receptors for the endocrine substances described in this chapter are expressed on the surface of leukocytes and/or lymphocytes. The binding of hormones on cell surface receptors induces an activation of intracellular and enzymatic systems, the so-called "second messengers" (Fig. 2; see Chapter 3). The second messengers include cyclic AMP, cyclic GMP, calcium, inositol phospholipids, tyrosine kinases, and arachidonic acid derivatives. This relatively low number of intracellular messengers is shared by a wide variety of cell surface receptors, suggesting a high level of regulation.

The receptors for most of the hormones, neurotransmitters, and neuropeptides are coupled to so-called G proteins (guanine nucleotide binding proteins). The G proteins are activated when the ligand binds to its membrane receptor. Activated G proteins (at least 12 are known at present) can in turn stimulate or inhibit the synthesis of second messenger

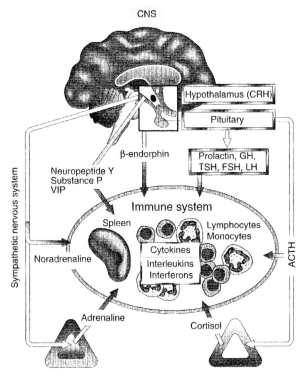

Figure 1. Schematic representation of neuroendocrine–immune interactions. CRH, corticotropin-releasing hormone; VIP, vasoactive intestinal peptide; GH, growth hormone; TSH, thyroid-stimulating hormone; FSH, follicle-stimulating hormone; LH, luteinizing hormone.

systems. The second messengers stimulate biochemical alterations within the cell, via the activation of specific protein kinases (third messengers) that phosphorylate other proteins. These so-called phosphoproteins act as a fourth messenger and induce physiological alterations within the cells, in particular, changes in membrane permeability, or induction of the synthesis of different proteins (Brown, 1994).

cAMP was the earliest second messenger to be discovered, and has been shown to regulate biochemical changes in a variety of target cells, including immunocompetent cells. For example, the stimulation of β-adrenoceptors induces an increased production of cAMP, whereas stimulation of α_2-adrenoceptors results in a decreased level of cAMP. It has been demonstrated that endogenous substances that activate the adenylate cyclase, and therefore increase the cAMP levels within the cell, inhibit lymphocyte functions.

9.2. Glucocorticoids

The synthesis and release of the glucocorticoids (e.g., cortisol and corticosterone) is stimulated by adrenocorticotropic hormone (ACTH) from the anterior pituitary gland (see

Table 1
Neuroendocrine Factors that Can Affect Immune Functions[a]

Hormones	Receptors in the immune system	Effects
Glucocorticoids	All immunocompetent cells	Inhibit cytokine production Inhibit T- and B-lymphocyte reactivity and NK activity "General" immunosuppression
Prolactin	T and B lymphocytes, NK cells	Stimulated T- and B-lymphocyte reactivity
Growth hormone	Thymocytes, mononuclear leukocytes	Stimulates T-lymphocyte reactivity and NK activity, increases thymus growth
Catecholamines (adrenaline/ noradrenaline)	β_2-Adrenoceptors on all lymphocyte subpopulations α-Adrenoceptors on PBL and NK cells (?)	Inhibit T-lymphocyte reactivity Stimulate NK activity Stimulate lymphocyte migration (in particular, NK cells)
β-Endorphin	Leukocytes and Lymphocytes	Stimulates lymphocyte reactivity and NK activity (also suppressive effects)
Substance P	T and B lymphocytes	Stimulates antibody production and lymphocyte proliferation
Vasoactive intestinal polypeptide	Monocytes, T and B lymphocytes	Stimulates lymphocyte migration (?)
Corticotropin-releasing hormone	Macrophages	Direct and indirect effects via the HPA axis and sympathetic nervous system
Adrenocorticotropic hormone	PBL, T and B lymphocytes	Stimulates/inhibits antibody production (in vitro) Inhibits IFNγ production (in vitro)
Enkephalin	Leukocytes and lymphocytes	Stimulates NK activity Inhibits antibody production, T-lymphocyte reactivity, and NK activity (in vitro)
Neuropeptide Y	Unknown	Inhibits NK activity (in vitro)
Thyroid-stimulating hormone	Phagocytes, B lymphocytes	Stimulates antibody production (in vivo and in vitro)
Follicle-stimulating hormone	Unknown	Stimulates T-lymphocyte reactivity (in vitro)
Luteinizing hormone	Unknown	Stimulates T- and B-lymphocyte reactivity and IL-1 and IL-2 production (in vitro)

[a]Effects are in vivo unless otherwise noted.

Chapter 3). Glucocorticoids that are released following stressful stimuli modulate carbohydrate metabolism, and generally have anti-inflammatory and immunosuppressive effects.

All immunological cell types express receptors for glucocorticoids. However, in contrast to hormones that bind to receptors located on the cell membrane, and that therefore need a signal transduction machinery, most of the steroid hormones are lipophilic and are able to pass through the cell membrane via diffusion. It is believed that highly developed organisms have a so called "carrier system" in the membrane, which transports glucocorticoids into the cell.

The effects of glucocorticoids in the target cells are mediated via a receptor molecule. This protein has binding sites for glucocorticoids and a high affinity to nucleic acids. When steroid hormones enter their target cells, they bind to intracellular receptors and form a so-called hormone–receptor complex, which regulates specific gene expression of proteins in

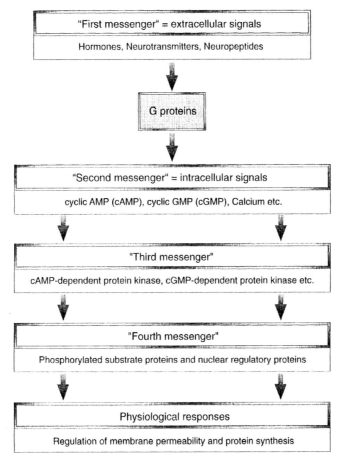

Figure 2. Schematic representation of the cascade of the first through fourth messenger systems leading to physiological responses in target cells. Once the first messengers activate their G proteins, biochemical changes occur in the target cells via a number of intracellular second messenger systems. These second messengers produce their actions through the activation of specific protein kinases, which are often referred to as third messengers. These protein kinases phosphorylate proteins that act as fourth messengers and induce physiological changes such as alterations in membrane permeability or protein synthesis. Redrawn from Brown (1994).

the nucleus of the target cell. The translation of mRNA for specific proteins is altered, which in turn leads to changes in the rate of synthesis of these proteins. This finally induces the metabolic actions of glucocorticoids (Brown, 1994).

The glucocorticoids are mainly produced by the cortex of the adrenal glands. In animal experiments, one can therefore test the functional relevance of corticosterone for the intact immune system via surgical removal of the adrenals. In adrenalectomized mice, numbers of T and B lymphocytes decrease in the spleen 24 h after surgical removal of the adrenals (Fig. 3). Adrenalectomy not only affects the cell numbers but also differentially alters the

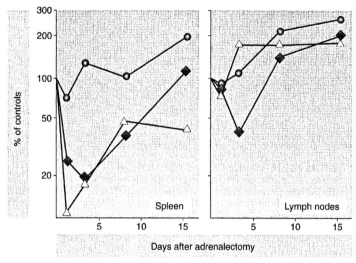

Figure 3. Numbers of B- and T-lymphocyte subpopulations in the spleen and in the lymph nodes, 1–15 days after surgical removal of the adrenal glands. Redrawn from Rocha (1985).

functional capacity of lymphocytes. After adrenalectomy, proliferation of T and B lymphocytes was increased in the spleen, but was decreased in cells taken from lymph nodes.

In animals and humans a continuous treatment with glucocorticoids induces immune suppression, inhibiting humoral as well as cellular immune functions. The action of glucocorticoids seems to mainly reflect an inhibition of the production and release of cytokines, which in turn leads to the suppression of immune responses (Fig. 4).

When rat splenocytes are incubated *in vitro* with pokeweed mitogen, they produce the cytokine interferon. However, when the synthetic glucocorticoid hydrocortisone is added to this culture, a dose-dependent suppression of interferon production is observed (Fig. 5). The highest level of the hydrocortisone (55.2 nM) completely inhibited interferon production by the cells. These suppressive effects of hydrocortisone could be completely reversed by adding RU-38486, a competitive synthetic antagonist for glucocorticoids, to the assay (Fig. 5).

Meanwhile, a number of *in vitro* studies demonstrate that glucocorticoids not only inhibit cytokine production, but also affect different cellular and humoral immune responses such as T- and B-lymphocyte reactivity, and natural killer cell (NK) activity. However, these immunosuppressive effects are only partly related to a diminished cytokine production. It is assumed that cortisol or corticosterone also affect adhesion of immunological effector cells, preventing the binding of these immunocompetent cells to the target cells. In addition, glucocorticoids are also able to affect the migration of lymphocytes.

The immunosuppressive effects of glucocorticoids are also demonstrated in a number of *in vivo* studies. For many years, glucocorticoids have been used as the standard therapeutic for the treatment of inflammation and for inducing immunosuppression in general. Glucocorticoids are commonly prescribed for the treatment of autoimmune diseases and for prevention of allograft rejection after organ transplantation. However, they do not neces-

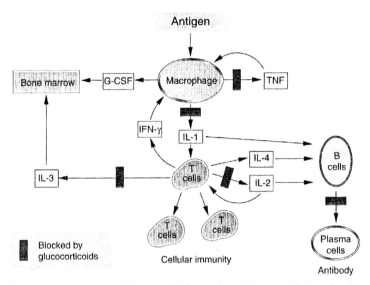

Figure 4. Schematic representation of glucocorticoid effects on the cytokine network. Redrawn from Munck and Guyre (1991).

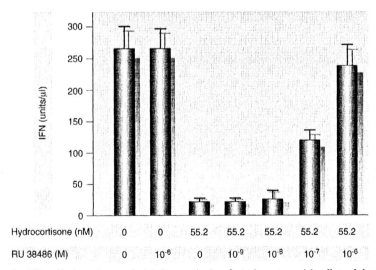

Figure 5. Effect of hydrocortisone on the interferon production of rat splenocytes, and the effects of glucocorticoid antagonist RU-38486 in different dosages on a glucocorticoid-induced inhibition of interferon production. Redrawn from Jiayi, Shikun, and Renabo (1989).

sarily always induce an inhibition of the immune response and their physiological role is far more complex than only a general immunosuppression induction. Many studies have used synthetic glucocorticoids, such as prednisone or dexamethasone. These substances have a much stronger action than corticosterone or cortisol. In addition, many studies applied pharmacological dosages, which are much higher than the concentrations of glucocorticoids found in the organism under normal (physiological) conditions.

For example, in rats, the surgical removal of the adrenals diminished T-lymphocyte reactivity. The administration of high (pharmacological) doses of glucocorticoids induced a further inhibition of the lymphocyte proliferation. However, when glucocorticoids were administered in lower, physiological dosages, a reversal of the reduction in T-cell function was observed (Fig. 6). It is assumed that different glucocorticoid receptors are responsible for these contrasting effects. More recently, it was also demonstrated *in vitro* that glucocorticoids in physiological concentrations accelerate anti-T-cell-receptor-induced lymphocyte proliferation, demonstrating that besides their inhibitory properties glucocorticoids also have stimulatory effects (Wiegers, Labeur, Stec, Klinkert, Holsboer, & Reul, 1995).

When cortisol was infused in physiological concentrations in humans, the numbers and function of NK lymphocytes remained unchanged. However, T cell numbers in peripheral blood significantly decreased after cortisol infusion (Tønnesen, Christensen, & Brinkløv), 1987). Similar cell-specific effects could be observed after oral or intravenous administration of dexamethasone. Ten hours after cortisol administration, CD4[+] cell numbers decreased, whereas NK cell numbers increased. However, the functional capacity of T lymphocytes and NK cells remained unaffected by the dexamethasone treatment (Chiapelli, Gormley, Gwirstman, Lowy, Nguyen, Nguyen, Esmail, Strober, & Weiner, 1992).

In addition to direct immunosuppressive effects, glucocorticoids in general seem to have important specific tasks within the immune system, in particular to prevent immune

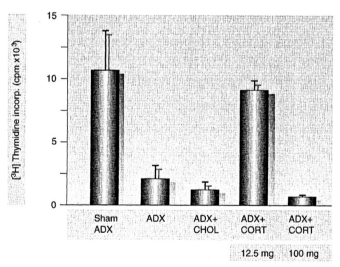

Figure 6.   Effect of glucocorticoid treatment on the T-cell reactivity of adrenalectomized rats. Redrawn from Wiegers, Croiset, Reul, Holsboer, and De Kloet (1993).

responses from "overshooting." In particular, the work of Besedovsky and colleagues illustrates these important immunomodulatory properties of this hormone. Their work clearly demonstrates the physiological role of glucocorticoids within the immune response, focusing on the bidirectional interaction between the hypothalamus–pituitary–adrenal axis and the immune system (see Chapter 11; Besedovsky & del Rey, 1996).

9.3. Pituitary Hormones: Prolactin and Growth Hormone

9.3.1. Hypophysectomy

An intact pituitary, in particular the production and synthesis of the pituitary hormones prolactin and growth hormone (GH), seems to be crucial for an undisturbed functioning of the immune system. In mice and rats the surgical removal of the pituitary induces a general immune deficiency, in particular a reduction in thymus and spleen weights, a reduction of lymphocyte numbers, a diminished primary and secondary antibody response to T-cell-dependent antigens, and a suppression of cytokine production and NK activity. These suppressive effects can be almost completely restored by administration of prolactin or GH, which are normally produced by the pituitary (Berczi & Nagy, 1991).

In order to analyze the effects of prolactin and GH on the immune response, one does not necessarily have to remove or destroy the pituitary. Recently, the effects of prolactin and GH has been analyzed in GH- or prolactin-deficient mice. These studies demonstrated that GH and prolactin can act differentially on the T lymphocytes of these mice. Whereas GH promotes early phase T-cell development in the thymus, prolactin seems to inhibit this development (Fig. 7).

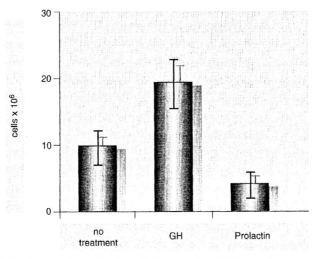

Figure 7.   Proliferation rate of T lymphocytes in the thymus in GH- and prolactin-deficient mice, after daily injection for 8 weeks of either 20 μg GH or 20 μg prolactin. Redrawn from Murphy, Durum, and Longo (1993).

9.3.2. Prolactin

The release of prolactin from the pituitary can be influenced, like many other hormones, by circadian rhythms, or short pulsatile rhythms, and is stimulated by reproductive cycles, stress, steroid hormones, neuropeptides, and cytokines. Prolactin is essential for initiating milk synthesis in the mammary gland, and also has many functions relating to growth, osmoregulation, fat and carbohydrate metabolism, reproduction, and parental behavior. However, the complex physiological role of prolactin is not well understood.

Prolactin can affect lymphocytes by binding to cell surface receptors. With approximately 360 receptors on T and B cells and 660 receptors on NK-cells, the number of prolactin receptors on lymphocytes in rats and humans is relatively small in comparison with liver cells or cells of the female breast (Kooijman, Hooghe-Peters, & Hooghe, 1996).

Direct *in vitro* effects of prolactin on the functional capacity of lymphocytes are very difficult to investigate, as fetal calf serum (in which lymphocytes are normally cultured) generally contains large amounts of prolactin. Studies that have overcome these difficulties demonstrated that prolactin reinforced the IL-2 production and proliferation of T lymphocytes. Conversely, small amounts of antibodies against prolactin inhibited T- and B-lymphocyte proliferation (Hartmann, Holaday, & Bernton, 1989).

Prolactin seems to have diverse effects on immunological functions *in vivo* which are mediated by different mechanisms. Mice that have been continuously injected with pharmacological concentrations of prolactin over 5 days, showed a dose-dependent increase in T- and B-cell reactivity in the spleen (Fig. 8). In contrast, when prolactin release from the pituitary was inhibited by bromocryptine, an inhibitor of prolactin secretion, the macrophage and T-cell reactivity in mice was diminished, and the mortality following infection with a bacterial pathogen increased. The administration of prolactin completely antagonized these immunosuppressive effects (Bernton, Meltzer, & Holaday, 1988).

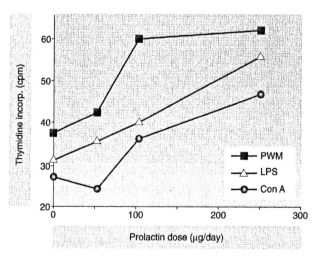

Figure 8. Mitogen (PWM, LPS, Con A)-induced proliferation of splenic lymphocytes in mice that were injected daily for 5 days with different concentrations of prolactin (0, 50, 100, or 250 μg). Redrawn from Bernton, Bryant, and Holaday (1991).

Besides these immunostimulating properties, prolactin seems to act indirectly as an antagonist on a glucocorticoid-induced immune suppression. The glucocorticoid-induced inhibition of lymphocyte proliferation in mice could be completely neutralized by an increased prolactin secretion. The exact mechanisms of the interactions between glucocorticoids, prolactin, and immune processes are not yet clear (Kelley & Dantzer, 1991; Matera, 1996).

9.3.3. Growth Hormone

The release of this pituitary hormone can be stimulated or inhibited by different stimuli similar to prolactin. GH is produced in somatotroph cells of the adenohypophysis, and promotes growth in almost all body cells (bone, muscle, brain, and heart tissue).

Receptors for GH have so far been detected on thymocytes and transformed lymphocytes, with approximately 7000 high-affinity receptors per cell on PBMC. GH receptor expression on the different lymphocyte subpopulations, such B, T, or NK cells, has not yet been analyzed in detail (Kooijman, Hooghe-Peters, & Hooghe, 1996).

Experimental data indicate that GH can influence a number of immunological functions, such as thymus functions, T- and NK-cell functions, cytokine production, phagocytosis, and hematopoiesis (Kooijman et al., 1996). This pituitary hormone seems to play an integrating and coordinating role within the immune response, in particular in thymus development and T-cell differentiation (Kelley, Arkins, & Lin, 1992).

Generally, GH seems to have a positive effect on DNA synthesis and increases T-cell activation and proliferation and B-cell function (proliferation and antibody production) in vitro (Kooijman et al., 1996).

In vivo, GH also appears to exert an immunoenhancing effect. In rodents, hypophysectomy induced an inhibition of NK activity. This reduction in the functional capacity of NK cells could be normalized by the administration of GH (see above). In parallel, an injection of GH in rats increased NK activity (Kelley, 1991).

Effects of GH on cellular immune function have also been observed in humans. One to two hours after a bolus injection of GH, an induction of monocyte migration and granulocytosis have been observed in young healthy volunteers (Wiedermann, Reinisch, Kähler, Geisen, Zilian, Herold, & Braunsteiner, 1992). However, other investigators could not observe any short-term effects of GH administration on the circulation of lymphocyte subpopulations, NK-activity, or the production of a number of cytokines such as IL-1, IL-2, IL-6, IL-8, TNFα, TNFβ, or IFNγ (Kappel, Hansen, Diamant, Jorgensen, Gyhrs, & Pedersen, 1993).

The effects of prolonged GH administration could be observed in children with deficient GH production, who were treated with GH for therapeutic reasons (Kooijman et al., 1996). These patients had a diminished NK activity in comparison with normal healthy controls. In parallel to the studies mentioned above, short-term treatment with GH did not affect the functional capacity of NK cells. However, regular GH treatment over a period of 3 or 6 months induced significant increases in the NK activity of these young patients (Fig. 9).

Prolactin can affect a number of immune functions and experimental data demonstrate that lymphocytes can synthesize and release prolactin and GH similarly to immunoreactive proteins. Thus, some authors concluded that GH and prolactin directly and/or indirectly modulate the immune response to infection or stress, and focus on the essential role of these two pituitary hormones in the maintenance of the homeostasis of the immune system (Kelley, 1991; Kelley & Dantzer, 1991; Kelley et al., 1992).

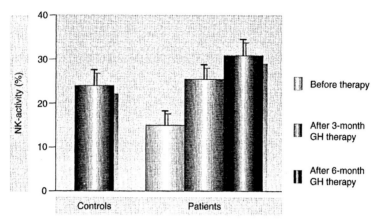

Figure 9. Activity of NK cells in children with deficient GH production. Patients received a weekly bolus injection with GH. Redrawn from Bozzola, Cisternino, Larizza, Maghnie, Moretta, Valtorta, Schimpff, and Severi (1989).

9.4. Catecholamines

The adrenal medulla is innervated by nerve fibers of the sympathetic nervous system, and releases the catecholamines adrenaline and noradrenaline during psychological or physical stress situations. Noradrenaline, however, is not only produced and released by the adrenal medulla. Most circulating noradrenaline is released as a neurotransmitter by postsynaptic neurons of the sympathetic nervous system. Both catecholamines are important regulatory factors in stress situations. Adrenaline increases heart rate and blood sugar levels, thus increasing the amount of work the muscles can perform. Noradrenaline acts to increase blood pressure, and to constrict blood vessels. Both catecholamines are important to prepare the organism for a "fight-or-flight" response in acute stress situations.

Adrenaline was the first hormone to be isolated from tissue, rapidly followed by chemical characterization and industrial production. The adrenal glands were discovered in 1563 by the Italian physician Bartolomeo Eustachi (1520–1574). The observation that injection of adrenal gland preparations caused a strong increase in blood pressure, led to great interest within the medical profession. In 1898, Otto von Fürth was the first to isolate the bioactive component from animal tissue and called this partly purified product *suprarenin*. Three years later, Takamine and Aldrich independently isolated the responsible component in crystalline form. The substance was named *adrenaline* by Takamine, and Aldrich found the correct formula ($C_9H_{13}NO_3$). During experiments in 1907, a by-product in the synthesis of adrenaline was identified. This substance, which became commercially available as "arterenol" in 1908, was in fact noradrenaline, which would formally be discovered and isolated from tissue 40 years later. Because the effects of arterenol were less pronounced than those achieved by adrenaline, production was canceled in 1910.

The production of adrenaline and its commercial availability led to a number of investigations testing its effects on white blood corpuscles. In 1904, Loeper and Crouzon were the first to describe the pronounced leukocytosis (two- to threefold increase) after subcutaneous injection of adrenaline (1 mg in humans). Frey (1914) reported that the response to adrenaline injections was biphasic in both animals and humans. The first phase of the leukocytosis lasted about 30 min and was predominantly characterized by an increase in lymphocyte numbers; during the second phase, granulocyte increases were observed. These observations were confirmed by other investigators, although it was noted that the lymphocytosis during the first phase was very consistent, but that there was more interindividual variation in the increase in granulocytes (Hoefer & Herzfeld, 1921; Walterhöfer, 1933). These and other observations have led to the general agreement that the response to adrenaline comprises an initial lymphocytosis (maximal response within 30 min), followed by a polymorph leukocytosis with relative lymphopenia (with a maximal response after 2–4 h; Samuels, 1951). Although the acute occurrence of catecholamine-induced leukocytosis seemed to be an acknowledged fact that had quickly taken a central position in hematological research, not everybody felt that such studies were useful. For instance, Ziegler (1924) wrote that "studies regarding displacement or distribution leukocytosis lack a sufficient scientific basis and should be rejected as misleading" (Benschop, Rodriguez-Feuerhahn, & Schedlowski, 1996b).

These early observations fell into oblivion for a long time. Two observations, however, brought catecholamines back into the focus of psychoneuroimmunological research. First, neuroanatomical studies provided evidence for a functional connection between the sympathetic nervous system and the immune system by demonstrating a noradrenergic innervation of primary and secondary lymphatic organs (Felten, Felten, Bellinger, Carlson, Ackerman, Madden, Olschowski, & Livnat, 1987; see Chapter 8). Second, exercise and acute psychological stress, which are primarily associated with an increased adrenaline and noradrenaline release, induce pronounced alterations in cellular immune functions (see Chapters 15 and 18).

The effects of catecholamines are mediated via adrenoceptors, which can be separated into four categories (α_1, α_2, β_1, and β_2) based on their different sensitivities to certain agonists. These receptors are present on many different cell types throughout the body. In particular, β_2-adrenoceptors have been identified on various lymphocyte subpopulations (Landmann, 1992; Mills & Dimsdale, 1993). Noradrenaline predominantly activates α- and β_1-adrenoceptors, and is a weak stimulator of β_2-adrenoceptors, whereas adrenaline is a strong stimulator of β_1-, and β_2-adrenoceptors.

Peripheral human blood lymphocytes, for example, express adrenoceptors of the β_2 type with varying levels of expression on different lymphocyte subpopulations. β_2-Adrenoceptors are expressed on B and T lymphocytes and NK cells, with the highest receptor density on B and NK cells, followed by CD8$^+$ and CD4$^+$ cells (Landmann, 1992). The existence of α-adrenoceptors on human peripheral blood lymphocytes has been demonstrated only in a few studies (Jetschmann, Benschop, Jacobs, Kemper, Oberbeck, Schmidt, & Schedlowski, 1997; Mills & Dimsdale, 1993).

Similar to the surgical removal of the adrenals and pituitary, the influence of noradrenaline release from sympathetic nerve terminals on functions of the immune system can be precisely analyzed via denervation strategies. In rats and mice, injection of the neurotoxin 6-hydroxydopamine leads to a destruction of noradrenergic nerve fibers of the sympathetic nervous system in the periphery and diminishes the noradrenaline concentration in lymphatic tissue (spleen or lymph nodes) by approximately 90%. This chemical denervation leads to a number of alterations in the immune system. After injection with 6-hydroxydopamine in mice, the T-cell reactivity in lymph nodes and spleen was significantly diminished 3 days to 2 weeks after the treatment. However, the proliferation rate of B lymphocytes was increased in the lymph nodes by this treatment whereas it remained unaffected in the spleen. Corresponding to these changes, B-cell numbers increased in the lymph node, whereas T-cell numbers decreased. These data indicate that chemical sympathectomy suppresses cell-mediated responses, may enhance antibody production, and modulate lymphocyte migration (Madden & Felten, 1995).

A further indication of the functional meaning of the noradrenergic innervation of the spleen comes from studies in which the spleen has been surgically denervated. Surgical denervation almost completely inhibits the noradrenaline supply to the spleen. In a following experiment, only the control animals with intact connection between sympathetic nervous system and the spleen demonstrated a stress-induced inhibition of T-cell reactivity in the spleen (Fig. 10). Denervated animals showed no such stress response, a clear evidence for the role of the sympathetic nervous system as a mediator between the brain and immune system.

In vitro studies demonstrate that catecholamines can modulate all aspects of an immune response. However, the effects of catecholamines on cells of the immune system are diverse and complex and cannot be characterized as a simple enhancement or inhibition, as each cell type seems to be differentially affected by catecholamines in a unique time frame. For example, when human peripheral blood lymphocytes are preincubated *in vitro* with adrenaline, there is a concentration-dependent increase in the activity of NK cells (Fig. 11). These concentration dependent effects of adrenaline can be blocked by the nonselective β-adrenergic antagonist propranolol. However, the exact opposite effects have been observed when adrenaline was added directly into the assay. The functional capacity of NK cells is then inhibited by adrenaline. These effects could also be antagonized by propranolol, indicating that adrenaline acts via β-adrenoceptors in both cases. Similar contrasting effects have been observed in cytokine production of mononuclear cells. Short-

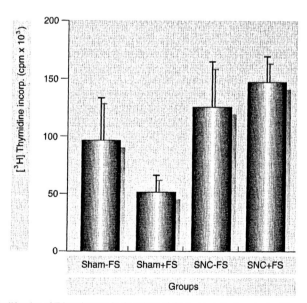

Figure 10. Proliferation of T lymphocytes in the spleen of sham-operated rats and rats with surgical denervation of the spleen, which either did (+FS) or did not (−FS) receive electric footshock. Redrawn form Wan, Vriend, Wetmore, Gartner, Greenberg, and Nance (1993).

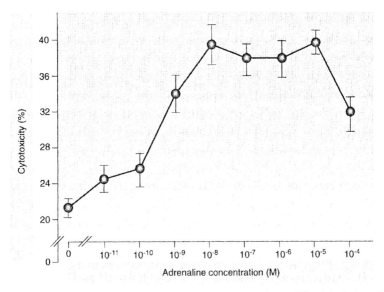

Figure 11. Effects of preincubation with adrenaline in different concentrations on NK activity. Redrawn from Hellstrand, Hermodsson, and Strannegard (1985).

term preexposure of these cells to adrenaline inhibited LPS-induced production of TNFα, whereas preexposure for 24 h resulted in an increased TNF-α release. Moreover, LPS-induced IL-10 release was increased by adrenaline via a combined effect of α- and β-adrenoceptors (van der Poll, Coyle, Barbosa, Braxton, & Lowry, 1996).

The physiological meaning of these *in vitro* effects of catecholamines on cells of the immune system, however, can only be analyzed in *in vivo* experiments. When rats were continuously (up to 4 days) treated with the synthetic catecholamine isoproterenol via a subcutaneously implanted osmotic minipump, differential effects on lymphocyte numbers in peripheral blood and in the spleen could be observed. Whereas NK cell numbers in the spleen and in peripheral blood increased following isoproterenol treatment, $CD8^+$ lymphocyte numbers in the spleen decreased. In parallel, there was a reduced reactivity of T lymphocytes in the spleen. Because isoproterenol selectively acts via β-adrenoceptors, it is most likely that these effects are mediated via β-adrenoceptors.

However, there is empirical evidence that the effects of catecholamines on immune functions are also mediated or regulated via α-adrenoceptors. In one experiment, rats received subcutaneously implanted capsules containing either adrenaline or noradrenaline. In this way, each catecholamine could be continuously administered for up to 20 h. When noradrenaline was given, a 50% decrease in T-cell activity in peripheral blood was observed. In contrast, adrenaline did not induce any effects on immune cells. However, when the animals were pre-treated with the β-blocker propranolol, blocking the β-adrenoceptors, both catecholamines induced a pronounced reduction of T-cell reactivity in peripheral blood of the animals, which was apparently mediated via α-adrenoceptors (Felsner, Hofer, Rinner, Porta, Korsatko, & Schauenstein, 1995).

The precise mechanisms and meaning of the *in vivo* effects of catecholamines on lymphocyte numbers and functions, in the different compartments of the immune system, are presently unclear. However, experimental data in animal studies have so far demonstrated a direct β-, and also α-adrenoceptor-mediated modulation of humoral and cellular immune function.

Investigations in humans demonstrated that both adrenaline and noradrenaline can affect lymphocyte subpopulations, in particular NK cell numbers and functions. In one study, healthy volunteers were subcutaneously injected with either NaCl, adrenaline, or noradrenaline. Adrenaline and noradrenaline administration induced pronounced increases in NK cell numbers (Fig. 12). In parallel to these increases in cell numbers, there were marked increases in the activity of NK lymphocytes after administration of both catecholamines (Schedlowski, Falk, Rohne, Wagner, Jacobs, Tewes, & Schmidt, 1993a). More recently, it could be demonstrated that the increase in NK cell numbers on adrenaline and noradrenaline infusion is inhibitable via a blockade of β_2-, but not β_1-adrenoceptors (Schedlowski, Hosch, Oberbeck, Benschop, Jacobs, Raab, & Schmidt, 1996).

Catecholamines are able to modulate a number of immunological functions. The *in vivo* effects of adrenaline on cellular immune functions in humans are meanwhile quite well documented. However, several fundamental questions remain to be solved. Future research activities should elucidate the cellular mechanism of catecholamine-induced leukocytosis. The exact migration pathways of the affected cell population, such as NK cells, after catecholamine administration need to be analyzed. In particular, the different molecules involved in these activation processes (e.g., adhesion structures) must be identified, and the intracellular mechanisms of how catecholamines alter these cellular immune functions need to be clarified (Benschop, Schedlowski, Wienecke, Jacobs, & Schmidt, 1997).

The first observations indicating that these catecholaminergic connections between the nervous system and the immune system are of clinical relevance are emerging. In rats, an infection with toxoplasmosis is normally not lethal. However, when rats were treated with high concentrations of propranolol, two-thirds of the treated rats died within 4 weeks of infection. This clearly indicates that stimulation of β-adrenoceptors plays a crucial role in the immunological defense against pathogens (Benedetto, Folgore, & Galdiero, 1993). Vice versa, immunosuppressive effects obtained by blockade of β-adrenoceptors might be particularly helpful in the treatment of autoimmune diseases. Experimentally induced arthritis in animals is often used as a model for rheumatoid arthritis in humans. Rats with experimentally induced arthritis were pretreated with a nonselective β-blocker. This treat-

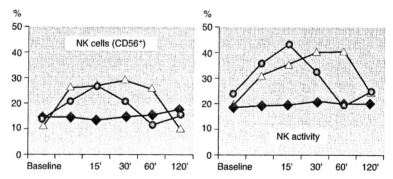

Figure 12. Healthy male volunteers were subcutaneously injected with either NaCl (△), adrenaline (●), or noradrenaline (◆). Blood was drawn before (baseline) and 5, 15, 30, 60, and 120 min after injection, in order to analyze NK cell numbers and functions. Redrawn from Schedlowski *et al.* (1993a).

ment not only delayed the onset of the disease, but also significantly decreased the disease symptoms relative to the controls (Levine, Coderre, Helms, & Basbaum, 1988).

Empirical evidence for the biological meaning of these catecholaminergic connections also comes from human experiments. In patients with rheumatoid arthritis, for example, the number and function of β-adrenoceptors on peripheral blood lymphocytes are altered (Kuis, de Jong-de Vos van Steenwijk, Sinnema, Kavelaars, Prakken, Helders, & Heijnen, 1996). In patients with juvenile chronic arthritis, α-adrenoceptors were found on mononuclear cells which normally do not express these receptors (Heijnen, Rouppe, van der Voort, Wolfraat, van der Net, Kuis, & Kavelaars, 1996). Changes such as these could have consequences for the responsiveness of lymphoid cells to catecholamines and could result in hypo- or hyperresponsiveness of the immune system, potentially with pathophysiological outcome (Benschop et al., 1996b).

9.5. β-Endorphin

β-Endorphin belongs to the opioid peptides. It consists of 31 amino acids and similar to ACTH is derived from the enzymatic cleavage of the large precursor molecule proopio-melanocortin (POMC). β-Endorphin is produced by the pituitary, CNS, and various other tissues, and has pronounced morphinelike activity (Brown, 1994). β-Endorphin binds to opioid receptors on the cell membrane (see also Chapter 10). To date, four opioid receptors and a so called nonopioid binding site have been identified. Leukocytes and lymphocytes express receptors for endogenous opioids. However, the exact subtype, number, and affinity of these receptors on the different lymphocyte subpopulations have not been analyzed.

β-Endorphin is able to modulate a number of immune functions, including effects on the primary and secondary antibody response, and in vitro effects on lymphocyte prolifera-tion, NK activity, and cytokine production. However, stimulatory as well as inhibitory effects of β-endorphin on immune responses have been reported. When β-endorphin was added to a mixed lymphocyte culture, the activity of lymphocytes significantly increased following 5 days of incubation. These stimulatory effects were observed independently of whether β-endorphin was added at the onset of the assay or after 4 days of incubation (Fig. 13). Another study showed that β-endorphin induced opposite effects in a dose-dependent fashion; high concentrations increased and low concentrations decreased NK activity of human lymphocytes (Williamson, Knight, Lightman, & Hobbs, 1987).

More recent in vitro studies demonstrated that β-endorphin stimulates the production and release of cytokines from mouse CD4$^+$ cells. The stimulation of a pure T-cell culture (CD4$^+$ cells) with β-endorphin induced a threefold increase of IL-2, IL-4, and IFNγ production. Because a five-amino-acid-smaller peptide fragment (β-endorphin 6–31), which only binds to nonopioid receptors, exerts comparable effects, the increase in cytokine production was thus modulated via nonopioid binding (van den Bergh, Dobber, Ramlal, Rozing, & Nagelkerken, 1994).

Only a few studies investigated the effects of β-endorphin on cellular and humoral immune functions in vivo. In parallel to the in vitro data, these experiments report stimula-tory as well as inhibitory effects of β-endorphin on lymphocyte proliferation. In rats the infusion of β-endorphin for 3 h induced a concentration-dependent increase (up to 37%) in the activity of T lymphocytes in the spleen. This stimulation of T cell activity could be blocked by naloxone, an opioid antagonist (Kusnecov, Husband, King, & Smith, 1987). More recently, it was demonstrated in rats that β-endorphin exerts a tonic inhibitory effect on the proliferative response of splenocytes that was mediated at a central as well as

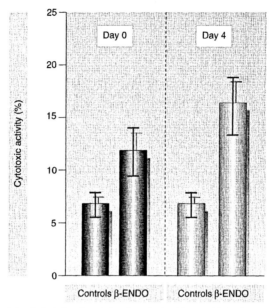

Figure 13.   Cytotoxicity of splenic lymphocytes in mice. β-Endorphin was added at the onset of the assay or after 4 days. Redrawn from Carr and Klimpel (1986).

peripheral level (Panerai, Manfredi, Granucci, & Sacerdote, 1995). This is in accordance with other reports showing inhibitory properties of β-endorphin, such as an inhibition of B-lymphocyte activation and NK activity. Different reasons are discussed for these opposite effects (Millar, Hough, Mazorow, & Gootenberg, 1990):

1. The immunomodulatory effects of β-endorphin seem to be concentration dependent: High concentrations appear to activate, whereas low concentrations inhibit, lymphocyte responses.
2. It is believed that the effects of opioid peptides on the immune response are highly dependent on the activation status of the lymphocytes. In this way, opioids would inhibit functions of activated lymphocytes, but stimulate resting lymphocytes.
3. This dependency of the activation status of the lymphocytes may explain other observations where β-endorphin did not induce any effects in 50% of the blood donors. In addition, it is assumed that there are large interindividual differences in the expression of opioid receptors on peripheral blood cells. Moreover, some investigators differentiate between opioid binding sites of β-endorphin, for example on NK cells, where stimulation of these binding sites increases NK activity, whereas stimulation of nonopioid receptors decreases NK activity.

Recently, it has been suggested that β-endorphin has a general immunosuppressive role, in particular under pathological conditions, as β-endorphin concentrations are elevated when the immune system is depressed (e.g., HIV infection) and decreased when the immune system is activated (e.g., rheumatoid arthritis) (Panerai & Sacerdote, 1997).

In addition to exerting direct effects on immunological functions, β-endorphin also is produced by activated lymphocytes (see Chapter 10). Therefore, β-endorphin seems to be a "fine tuning" immunomodulator that can maintain the homeostatic balance of the immune system (Heijnen, Kavelaars, & Bernards, 1991).

9.6. Substance P

The neuropeptide substance P belongs to the family of tachykinine peptides, consists of 11 amino acids, and is produced by sensory neurons. The primary function of substance P appears to be the transmission of pain signals in the spinal cord, but it also functions as a transmitter in the sensory pathways of the gastrointestinal and cardiovascular systems. Pathophysiologically, substance P is involved in a number of inflammatory processes, such as arthritis, asthma, and inflammatory gut diseases.

Concerning the immunological properties of substance P, research has focused on its effects on the production and release of cytokines and their receptors on the one hand, and on the other hand, on the molecular characterization of substance P receptors on cells of the immune system (McGillis, Mitsuhashi, & Payan, 1991; see also Chapters 8 and 10).

Lymphocytes express neurokinin-1 receptors that have a high affinity for substance P. However, substance P apparently can also activate monocytes and B cells via a not yet characterized nonneurokinin receptor. In addition, a receptor-independent activation of T cells by substance P has been described (Kavelaars, Jeurissen, & Heijnen, 1994).

In vitro experiments predominantly demonstrate stimulating effects of substance P on the function of T and B lymphocytes, monocyte functions, and hematopoiesis. An increased mitogen-induced (Con A/PHA) proliferation rate of T lymphocytes was observed, and B lymphocytes showed both heightened proliferation and IgM, IgA, and IgG production when substance P was added to the assay. In addition, substance P seems to play a role as a costimulator during the activation and differentiation of B lymphocytes to antibody-producing plasma cells. Moreover, substance P is able to stimulate monocytes and CD4$^+$ cells for the release of IL-1 and IL-2, respectively (McGillis *et al.*, 1991).

Less is known about the *in vivo* role of substance P on immunological functions. In observations paralleling the *in vitro* data, an infusion of substance P in rats increased immunoglobulin secretion, and treatment with a substance P antagonist decreased the secretion of immunoglobulins by approximately 70% (Bost & Pascual, 1992). In addition, substance P seems to have a stimulatory effect on T-lymphocyte proliferation in lymphatic organs such as the spleen or Peyer's patches. In mice the substance P concentrations in peripheral blood were experimentally increased (> twofold) over a period of 7 days via a subcutaneously implanted osmotic minipump. This led to a pronounced increase in lymphocyte proliferation compared with the control mice (Scicchitano, Bienenstock, & Stanisz, 1988).

These data, together with neuroanatomical studies demonstrating innervation of the spleen by substance P-containing neurons (see Chapter 8), strongly indicate that substance P is another important candidate used by the nervous system to communicate with the immune system.

9.7. Vasoactive Intestinal Polypeptide

The neuropeptide VIP consists of 28 amino acids. It is found in nerve cells of the cerebral cortex, hypothalamus, amygdala, hippocampus, and spinal chord. VIP has a wide range of effects on the cardiovascular, respiratory and gastrointestinal systems and other visceral functions. It may act as an excitatory neurotransmitter as well as a neuromodulator in the CNS and is able to affect a number of immunological functions.

Peripheral mononuclear cells of the immune system recognize VIP and express 2000–4000 receptors for this neuropeptide, with a different receptor density on human monocytes and different lymphocyte subpopulations with approximately 2500 receptors on T lymphocytes (CD4$^+$ > CD8$^+$) and 600 receptors on B cells. In addition, the receptor number seems to vary within the different immunological organs (Bellinger, Lorton, Brouxhon, Felten, & Felten, 1996; Ottaway, 1991).

The *in vitro* effects of VIP on cellular and humoral immune functions seem to be inhibitory in every respect. VIP suppresses T-cell reactivity and activity of NK cells. Production of IgA by lymphocytes in the Peyer's patches is inhibited by VIP, although IgA production in the spleen and lymph nodes is increased. The immunomodulatory properties of VIP seem to be at least partially mediated through the effects on the production of cytokines, in particular proinflammatory cytokines (Ganea, 1996).

In vivo, VIP seems to particularly affect the migration of lymphocytes. In sheep, the infusion of VIP induced a 70% decrease in the migration of lymphocytes from the lymph nodes (Moore, Spruck, & Said, 1988). However, the mechanisms by which VIP influences the migration of lymphocytes are still unclear.

9.8. Corticotropin-Releasing Hormone

CRH is primarily synthesized in the paraventricular nucleus and the anterior periventricular nuclei of the hypothalamus. It coordinates the neuroendocrine, visceral, and adaptive behavioral responses to stress and is the starting signal of the activation of the hypothalamus–pituitary–adrenal axis. CRH elevates noradrenaline levels in the sympathetic nervous system and stimulates the release of adrenaline from the adrenal medulla. CRH also elevates blood glucose levels, and increases blood pressure and heart rate.

Receptors for CRH in the immune system could be identified on splenic macrophages, and are comparable in structure to CRH receptors in the pituitary and brain.

In mice, *in vitro* experiments showed an effect of CRH on the splenic NK-cell activity, militating at first glance for a direct receptor-modulated mechanism (Fig. 14a). However, this stimulating CRH effect could be blocked by administration of the opioid antagonist naloxone, a clear indication of the involvement of endogenous opioids. Indeed, the administration of β-endorphin to the assay induced an increase in NK activity similar to the effect achieved with CRH administration. This effect could not, however, be repeated by adding ACTH to the assay (Fig. 14b). An additional experiment demonstrated that the CRH-induced increase in NK activity was absent when all macrophages in the assay had been neutralized. This finding suggests that CRH induced macrophages to release β-endorphin and that this macrophage-derived β-endorphin increased the functional capacity of NK cells via opioid receptors on the NK cells.

The mechanisms of how CRH indirectly affects functions of the immune system can be analyzed in *in vivo* studies. When rats were given an intracerebroventricular injection of CRH a significant reduction in the activity of NK cells in the spleen was observed 1 h later. However, subcutaneous injection of CRH did not exert any effects on NK activity. Further analysis of this phenomenon demonstrated that the central application of CRH led to an

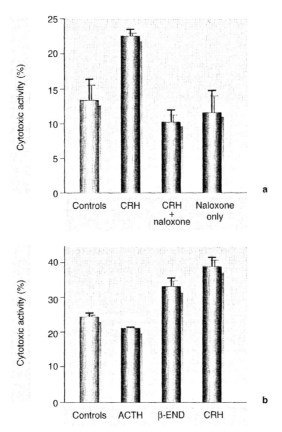

Figure 14. Mouse splenic NK activity following administration of either CRH, CRH + naloxone, or naloxone only (a); and following application of ACTH, β-endorphin (β-END), or CRH (b). Redrawn from Carr *et al.* (1990).

increased plasma concentration of noradrenaline in these animals, a well-known effect of CRH. Blockade with the β-adrenoceptor antagonist propranolol or chemical sympathectomy with 6-hydroxydopamine, and also a neutralization of the plasma noradrenaline levels antagonized the CRH-induced reduction of NK activity. These data demonstrate that CRH is able to indirectly modulate cellular immune functions via the sympathetic nervous system (Irwin, Hauger, Jones, Provencio, & Britton, 1990a).

Another pathway through which CRH can modulate immune functions seems to be the activation of the pituitary–adrenal axis and the release of glucocorticoids. Long-term intracerebroventricular CRH administration suppressed T- and B-splenocyte proliferation in normal but not in adrenalectomized rats (Labeur, Arzt, Wiegers, Holsboer, & Reul, 1995).

More recently, stimulatory as well as inhibitory effects of CRH on the inflammatory response have been described, which seem to be mediated by peripheral as well as central CRH (Karalis, Muglia, Bae, Hilderbrand, & Majzoub, 1997).

9.9. Further Hormones and Neuropeptides with Immunological Competence (ACTH, Enkephalin, Neuropeptide Y, Thyrotropin, FSH, LH)

Besides the above mentioned neuroendocrine substances, a number of endocrine factors are able to modulate functions of the immune system. Similar to β-endorphin, ACTH is spliced from the precursor molecule POMC and acts to stimulate the synthesis and release of glucocorticoid hormones in the adrenal cortex. ACTH receptors could be identified on B and T lymphocytes of the rat. ACTH seems to particularly affect B-cell functions, whereby stimulatory as well as inhibitory effects on antibody production could be observed. It is believed that the effects of ACTH on humoral immune functions are mediated via CD4+ cells and production of cytokines, as ACTH has been shown to diminish IFNγ production (Heijnen et al., 1991; see Chapter 10).

Enkephalins (Met-enkephalin and Leu-enkephalin) belong to the endogenous opioids. They are produced and released from the pituitary and adrenals and act via opioid receptors on the target cell. In vitro, Met-enkephalin suppressed the antibody response, but stimulated T-cell proliferation and NK activity. In vivo, Met-enkephalin increased NK activity in mice and in parallel reduced the number of experimentally induced lung metastases in these animals (Plotnikoff, Murgo, Faith, & Wybran, 1991; see Chapter 10).

Neuropeptide Y (NPY) consists of 36 amino acids and occurs in the brain in higher concentration than any other neuropeptide, with highest levels in the hypothalamus and limbic system. NPY has diverse effects on the cardiovascular and gastrointestinal systems and influences body temperature regulation. In the periphery, NPY is often colocalized with the classical neurotransmitter noradrenaline and is able to amplify noradrenergic effects. Nerve fibers containing NPY could be identified in lymphatic organs of the rat, in close vicinity to macrophages and lymphocytes. However, direct effects of this neuropeptide on the functioning of the immune system are as yet unknown. Concentration-dependent inhibitory effects of NPY on the NK activity of human peripheral blood lymphocytes could be observed in vitro, which could be neutralized by administration of an NPY antagonist.

Thyroid-stimulating hormone (TSH) is produced in thyrotroph cells of the adenohypophysis. TSH stimulated the synthesis and release of thyroxine and triiodothyronine from the thyroid. Data regarding the effects of TSH on immunological functions are rare in comparison with other pituitary hormones. TSH receptors seem to be predominantly expressed on B cells and phagocytes. TSH stimulated humoral immune functions, such as in vivo antibody production, and amplified the T-cell-dependent and -independent in vitro antibody response (Smith, Harbour, Hughes, Kent, Ebaugh, Jazayeri, & Meyer, 1991). More recently, TSH has been suggested to be a key immunoregulatory mediator in the intestine (Wang, Whetsell, & Klein, 1997).

The gonadotropic hormone-follicle stimulating hormone (FSH) and luteinizing hormone (LH) are produced in the gonadotroph cells of the adenohypophysis. Their immunomodulatory properties are not yet well understood. FSH and LH increase the proliferation rate of human T lymphocytes, and LH stimulates the activity of splenic T and B cells and induces the production of IL-1 and IL-2 in mice (Athreya, Pletcher, Zulian, Weiner, & Williams, 1993; Smith et al., 1991).

9.10. Summary

This chapter exemplarily described data demonstrating that hormones, neurotransmitters, and neuropeptides affect functions of the immune system. These observations are an important basis for the analysis of the complex interactions between the nervous, endocrine, and immune systems. From a more clinical point of view, these endocrine substances or the blockade of their specific receptors could one day be used specifically to modulate processes of the immune system. These manipulations could then influence the susceptibility to certain diseases or disease outcome. Whether and to what extent this basic knowledge can be introduced into humans is not foreseen at present.

10   Opioid Peptide Production by the Immune System

Cobi J. Heijnen and Annemieke Kavelaars

Cobi J. Heijnen and Annemieke Kavelaars   •   Department of Immunology, University Hospital for Children and Youth, "Het Wilhelmina Kinderziekenhuis", 3501 CA Utrecht, The Netherlands.

10.1. Introduction

The nervous, endocrine, and immune systems are intimately connected through mutually acting regulatory systems. The immune system is highly specialized in its recognition of and reaction to specific antigens from internal as well as external sources. The brain is specialized in recognition and response to internal and external emotional and physical stimuli. Both systems have their own characteristic responses to these stimuli. However, the response patterns of the two systems are partly overlapping, that is, they both produce mediators such as cytokines and neuropeptides that can act as communication pathways for the two systems. Moreover, the immune system expresses receptors or binding sites for every neurotransmitter or hormone tested so far, thereby enabling the immune system to recognize and to respond to neuroendocrine signals (Blalock & Smith, 1985).

During the last decade there has been an intensive debate about the problem of whether or not cells of the immune system can synthesize, produce, and secrete neuropeptides. Blalock and colleagues were the first to show that the immune system is capable of doing so by demonstrating the production of neuropeptides belonging to the proopiomelanocortin (POMC) family (see below) (Smith & Blalock, 1981). Since then a lot of other groups have confirmed and extended these data. By now the production of several peptides has been clearly shown; the list includes ACTH, endorphins, enkephalins, growth hormone, insulin-like growth factor (IGF-1), corticotropin-releasing hormone (CRH), thyroid-releasing hormone (TRH), vasoactive intestinal peptide (VIP), somatostatin, oxytocin, vasopressin (AVP), and substance P (Geenen, Legros, Franchimont, Baudrihaye, Defresne, & Boniver, 1986; Heijnen, Kavelaars, & Ballieux, 1991; Pascual & Bost, 1990). More recently, it has become evident that cells of the immune system can also produce hormones and neurotransmitters of nonprotein nature such as glucocorticoids and catecholamines.

In this chapter we will discuss the regulation of expression of opioid peptides in the immune system and their function as autocrine and paracrine mediators.

10.2. Regulation of Production and Secretion of Opioid Peptides in the Neuroendocrine System

10.2.1. POMC in the Pituitary

The POMC gene consists of three exons (100, 150, and 830 nucleotides) separated by two large introns (3.5–4 and 2–3 kb). The 5'-untranslated part of the mRNA is encoded by the first exon and part of the second one. The signal peptide and a portion of the N-terminal peptide are also encoded by the second exon, whereas exon 3 encodes all of the peptides with known biological activity (Cochet, Chang, & Cohen, 1982; Takahashi, Hakamata, Watanabe, Kikuno, Miyata, & Numa, 1983).

Expression of the POMC gene in pituitary cells is regulated primarily by CRH. CRH is the principal orchestrator of the neuroendocrine response to stress, as evidenced by the fact that it controls activation of the hypothalamic–pituitary–adrenal axis. In addition, CRH neurons also activate noradrenergic neurons ending in the hypothalamus, leading to activation of the autonomic nervous system. Finally, CRH acts as the pivotal integrator of the cytokine-mediated feedback signaling pathway of the immune system to the brain (Berkenbosch, Van Oers, del Rey, Tilders, & Besedovsky, 1987). In response to a stress stimulus or an immune-mediated cytokine stimulus, the 41-amino-acid peptide is secreted by the hypothalamus and transported via the portal circulation to the pituitary where it induces synthesis and production of the POMC precursor molecule. AVP released from AVP-secreting neurons in the hypothalamus, enhances the activity of CRH to induce the POMC precursor.

Corticotropic cells of the anterior pituitary and in some animals (e.g., the rat) also in the melanotrophic cells of the intermediate lobe, synthesize the precursor molecule POMC (Cochet *et al.*, 1982). After removal of the N-terminal signal sequence, the POMC molecule is posttranslationally processed by a series of enzymatic events to a number of peptide hormones, e.g., ACTH, β-LPH, and β-endorphin. The N-terminal fragment gives rise to γ-MSH. The midportion of the precursor encodes ACTH(1–39), which can be further cleaved into α-MSH (*N*-acetyl-ACTH 1–13) and CLIP (ACTH 18–39). The C-terminal part of POMC comprises β-LPH. Cleavage of β-LPH releases γ-LPH (β-LPH 1–58), which can be processed into β-MSH (β-LPH 41–58), and β-endorphin (β-LPH 61–91). A specific endopeptidase releases γ-endorphin (β-endorphin 1–17). Carboxypeptidase activity processes γ-endorphin into α-endorphin (β-endorphin 1–16) (Burbach, Loeber, Verhoef, Wiegant, De Kloet, & De Wied, 1980) (see Fig. 1).

10.2.2. Extrapituitary Sites of POMC Production

Brain POMC-derived peptides are synthesized mainly in cell bodies of the arcuate nucleus, from which nerve fibers innervate the median eminence, amygdala, preoptic area,

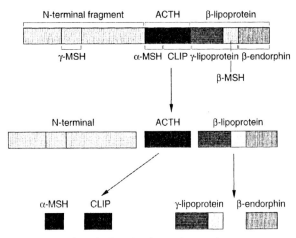

Figure 1. Proopiomelanocortin processing.

ventromedial nuclei, reticular formation, striatum and hippocampus. Furthermore, β-endorphin-positive neurons have been shown to project to the portal capillaries of the median eminence and to the ependymal layer of the third ventricle, suggesting that POMC-derived peptides are secreted into the cerebrospinal fluid.

Production of POMC-derived peptides also takes place in a number of peripheral tissues. β-Endorphin has been detected in extracts of adrenal medulla, thyroid gland, lung, duodenum, and pancreas. Moreover, POMC mRNA has been found in ovary, uterus, and testis. However, the major form of POMC mRNA in testis is approximately 200 bases shorter than that in pituitary and hypothalamus. Similar POMC-like mRNA was found in some parts of the brain outside the hypothalamus, e.g., the amygdala.

10.2.3. PEA in the Central Nervous System and Adrenal Gland

Proenkephalin A (PEA) encodes a precursor protein (preproenkephalin A) that is enzymatically processed into the pentapeptides Met-enkephalin (Menk) and Leu-enkephalin. Other larger cleavage products are also known, such as peptides E, I, F, amidorphin, Bam-20P, and Bam-22P (Saravia, Ase, Aloyz, Kleid, Ines, Vida, Nahmod, & Vindrola, 1993). The peptides have opioid activity and bind to delta-, mu-, and kappa-type opioid receptors. The precursor protein also encodes synenkephalins, but these peptides have no opioid activity and will not be discussed in the context of this chapter.

Menk is ubiquitously expressed in the body. It is produced in the adrenal gland, in the pituitary, and in brain regions, especially the striatum (Pittius, Kley, Loeffler, & Hollt, 1985). Like the endorphin family, the enkephalins function as neurotransmitters and neurohormones and are important in analgesia and in the responses to stress (Howlett & Rees, 1986).

PEA is also abundantly expressed in embryonic mesenchymal tissues during differentiation into cartilage, bone, dermis, kidney tubules, and choroid of the eyes. The expression of PEA mRNA drops as soon as the differentiation process stops. It is not yet clear whether PEA peptides actually regulate growth and development of mesodermal tissues, although the clear transient nature of PEA expression suggests that PEA peptides play an important role in development.

The processing of PEA has been studied extensively in the adrenal medulla. In chromaffin cells of the adrenal medulla, the majority of the PEA products consist of peptides in the size range of 3–5 kDA, like peptide E, peptide F, and peptide I. Less than 20% of the enkephalin sequences present in these cells are found as free pentapeptide. In contrast, in the central nervous system the majority of the enkephalin is found in the form of pentapeptide. Processing of PEA requires two different enzymatic activities: a trypsinlike enzyme to cleave dibasic amino acid sequences, and a carboxypeptidase B-like enzyme to remove C-terminal basic amino acids.

10.2.4. POMC and PEA Peptide Expression in Nonvertebrates

Unicellular organisms produce molecules that are similar to peptide hormones of vertebrates, despite the absence of structures that are associated with endocrine or neuroendocrine systems.

In the protozoan *Tetrahymena pyriformis*, a molecule was detected that is similar to POMC in vertebrates. This molecule contains both β-endorphin and ACTH immunoreactivity and was shown to be similar in size to vertebrate POMC by means of HPLC techniques. Moreover, ACTH- and β-endorphin-like material was shown to have the classical biological activity of opioid peptides (Roth, LeRoith, Collier, Weaver, Watkinson, Cleland, & Glick, 1985).

POMC or PEA-related peptides (immunoreactivity) have also been shown to be present in multicellar organisms, such as in neural elements of the annelid earthworm, the mollusks pond snail (*Lymnea stagnalis*), *Mytilus edulis*, and octopus, in the insect silkworm (*Bombyx mori*), and in the protochordate sea squirt. In the insect *Drosophila melanogaster*, ACTH-like material was found in neural as well as reproductive organs. In addition, specific hybridization of mouse POMC cDNA to restriction fragments of *Drosophila* DNA from neural tissue has been detected (Roth *et al.*, 1985).

The similarity in the structure and biological activity of these peptides suggests that peptide hormones are stable elements with a long evolutionary history as intercellular messengers (Roth *et al.*, 1985). It may be worthwhile considering that these peptides functioned long before the nervous and neuroendocrine systems were present as separate organ systems.

10.3. Opioid Peptides in the Immune System

10.3.1. POMC Gene Expression in Cells of the Immune System

In 1986, two groups showed independently the presence of a full-length mRNA for POMC in virus-activated mouse splenocytes and in unstimulated splenic macrophages, respectively, by means of Northern blot analysis (Westly, Kleiss, Kelley, Wong, & Yuen, 1986; Lolait, Clements, Markwick, Cheng, McNally, Smith & Funder, 1986). However, later on it appeared that human leukocytes contain several POMC mRNA species of various lengths. In many experiments, mRNAs were detectable that were 200 bp shorter, which resembles the length of the POMC transcripts found in testicular, placental, and cerebral POMC mRNA (Lolait *et al.*, 1986; Westly *et al.*, 1986). Stimulation of murine lymphocytes with CRH results in the expression of two truncated POMC transcripts. These transcripts lack exons 1 and 2, but contain either part or all of exon 3. It should be noted that exon 3 encodes all of the biologically active peptides (Galin, LeBoeuf, & Blalock, 1991). Stephanou and colleagues demonstrated POMC transcripts both larger and smaller than the 1.2-kb pituitary species, including 0.8-, 1.2-, 1.5-, and 9.5-kb transcripts in human peripheral blood T cells after stimulation with IL-2 (Stephanou, Fitzharris, Knight, & Lightman, 1991). EBV-transformed B cells only contain the 9.5- and 0.8-kb transcripts. Circulating neutrophils only contain the 9.5-kb transcript (Stephanou *et al.*, 1991).

In addition, Lyon and colleagues demonstrated the presence of full-length mRNA for POMC in rat splenocytes after polyclonal activation with the mitogen Con A (Lyons & Blalock, 1997). Using an RNase protection and primer-extension techniques, these authors showed that splenocytes can use the same translation initiation sites as pituitary cells do.

10.3.2. Presence of POMC-Derived Peptides in Immune Cells

Blalock and Smith were the first to demonstrate the presence of the POMC-derived peptides ACTH and β-endorphin in cells of the immune system (Smith & Blalock, 1981). They showed that leukocytes can produce POMC peptides *de novo* after appropriate stimulation and the peptides can be detected in culture supernatants. The amino acid sequence of lymphocyte-derived ACTH has been shown to be identical to that of pituitary ACTH (Smith, Galin, LeBoeuf, Coppenhaver, Harbour, & Blalock, 1990). In terms of bioactivity, antigenicity, molecular weight, and retention time on RP-HPLC, leukocyte-derived ACTH and endorphins appear to be identical to their pituitary counterpart.

Two types of stimuli are capable of inducing POMC-derived peptide production in cells of the immune system: immune stimulators and hypothalamic releasing factors such as CRH and AVP, which are classically associated with POMC production in the pituitary. Malignant transformation also induces the expression of the POMC gene with subsequent production of POMC-derived peptides (Oates, Allaway, Armstrong, Boyajian, Kehr, & Prabhakar, 1988). Using Northern blot analysis, POMC mRNA was observed in a T-lymphocyte line derived from a patient with lymphoma (Buzzetti, McLoughlin, Lavender, Clark, & Rees, 1989).

10.3.3. Neuroendocrine and Immune Mediators Induce POMC Peptides

T- and B-cell mitogens such as PHA, Con A, LPS, anti-CD3 and anti-IgM antibodies as well as recall antigens like *Candida albicans* are efficient stimulators of POMC peptide production (Heijnen *et al.*, 1991; Weihe, Nohr, Michel, Muller, Zentel, Fink, & Krekel, 1991). These stimuli have in common that they induce the secretion of various cytokines like IL-1, IL-2, and IL-6. When instead of the abovementioned stimuli, only cytokines are added to peripheral blood cells, the cells also start to produce substances such as β-endorphin or ACTH (Heijnen *et al.*, 1991). The latter finding may imply that POMC peptide production is, like most immune products, regulated by cytokines and that POMC-derived peptides are induced after every antigenic or mitogenic stimulus.

Pituitary cells are also capable of responding to cytokines like IL-1 and IL-2 with the production of POMC-derived peptides (Brown, Smith, & Blalock, 1987). IL-1 can exert its effect either by a direct binding to pituitary cells or by indirectly enhancing the secretion of CRH in the hypothalamus (Brown *et al.*, 1987; Berkenbosch *et al.*, 1987).

POMC expression in the immune system not only can be induced by classical immune stimuli, but also by neuroendocrine mediators. Our data have demonstrated that catecholamines trigger the release of endorphin from leukocytes. Catecholamines are mediators of the autonomic nervous system and can trigger β_2-adrenergic receptors on lymphocytes, resulting in increased β-endorphin secretion (Kavelaars, Ballieux, & Heijnen, 1990b). This means that theoretically, a stress stimulus not only will result in β-endorphin release from the pituitary, but also will evoke the secretion of β-endorphin by leukocytes.

The neuroendocrine mediator CRH can also induce POMC peptide expression in the immune system. In addition, AVP enhances the CRH-induced secretion of POMC-derived peptides by the immune cells. This is another example of the similarity of POMC regulation in both the pituitary and the immune system.

It appeared that CRH exerts its action by the induction of IL-1 in monocytes or macrophages, thereby enabling especially human B cells to produce β-endorphin. Therefore, the secretory roles of CRH and IL-1 in the immune system are opposite those of the brain, because CRH can stimulate IL-1 production in human monocytes (Kavelaars, Ballieux, & Heijnen, 1989; Kavelaars, Berkenbosch, Croiset, Ballieux, & Heijnen, 1990c; Singh & Leu, 1990). Systemic administration of CRH in rats also induces splenic and lymph node cells to secrete β-endorphin and at the same time increases the secretion of IL-1 in macrophages (Kavelaars *et al.*, 1989). Moreover, increased plasma levels of IL-1 have been found in humans after intravenous administration of CRH.

Glucocorticoids are capable of suppressing the production of β-endorphin, as evidenced by their inhibition of CRH-induced IL-1 production (Kavelaars, Ballieux, & Heijnen, 1990a). Taken together, these results support the concept that neuropeptides and glucocorticoids are mediating a reciprocal modulation of both neuroendocrine and immunological activities.

10.3.4. CRH in the Immune System

Although low levels of hypothalamus-derived CRH are present in the peripheral circulation, the immune system can also produce this mediator. Stephanou and co-workers were the first to show that CRH can be synthesized by cells of the immune system (Stephanou, Jessop, Knight, & Lightman, 1990). *In situ* hybridization demonstrated mRNA in unstimulated normal peripheral blood T and B cells as well as neutrophils. The product does not seem to be identical to hypothalamus-derived CRH 1–41, as it elutes earlier on RP-HPLC, which indicates that the molecules are larger. Moreover, it appears to be more labile than CRH 1–41. Northern blot analysis showed the presence of an mRNA corresponding to 1.7 kb in comparison with the 1.5-kb mRNA that is associated with human hypothalamic CRH 1–41 (Stephanou *et al.*, 1990). These elegant data indicate that although the molecules closely resemble each other, they are not identical.

The presence of CRH mRNA and immunoreactive CRH has been demonstrated in human leukocytes and rat thymus, spleen, and synovial tissue (Aird, Clevenger, Prystowski, & Redei, 1993; Redei, 1992). In the thymus, the epithelial cells as well as the thymocytes express the CRH message. In this context it is interesting that thymic epithelial cells also express vasopressin mRNA, a phenomenon that mirrors the situation in the hypothalamus (Geenen *et al.*, 1986).

Culturing of thymocytes or spleen cells leads to the release of about 50 pg/ml CRH in the culture medium. Stimulation with the lipoxygenase pathway inhibitor nordihydroguaiaretic acid (NDGA) leads to a severalfold increase in CRH mRNA and the secretion of CRH, a mechanism that finds its counterpart in the hypothalamus (Aird *et al.*, 1993). Secretion of CRH by cells of the immune system also occurs when cells have taken up the peptide from the peripheral circulation into secretory granules by an endocytotic mechanism (Aird *et al.*, 1993). Splenic adherent cells can secrete CRH after stimulation *in vitro* with NDGA but do not express CRH mRNA (Aird *et al.*, 1993).

There is a clear difference in the regulation of CRH production in the hypothalamus and the immune system. It is well-known that cytokines, especially IL-1, play a key role in the upregulation of the expression of CRH in the hypothalamus. However, IL-1 does not influence the thymic and splenic secretory response of CRH. In contrast, CRH induces IL-1 in immune cells.

Immunoreactive CRH expression has also been demonstrated in the joints and surrounding tissues of rats with adjuvant-induced arthritis (Crofford, Sano, Karalis, Webster, Goldmutz, Chrousos, & Wilder, 1992). As in the hypothalamus, glucocorticoids can inhibit CRH expression in the joints. The magnitude of CRH expression matches the severity of the inflammation (Crofford *et al.*, 1992). In this respect it is of interest to note that the Lewis/N rat, which is highly susceptible to induction of arthritis and has a defective CRH production in the hypothalamus, shows a clear increase in iCRH in the inflamed joint, despite the absence of a CRH increase in the hypothalamus (Crofford *et al.*, 1992; Sternberg, Hill, Chrousos, Kamiliaris, Listwak, Gold, & Wilder, 1989a).

Following adrenalectomy the level of CRH in the thymus increases (Redei, 1992). iCRH expression in the inflamed joint of the rat also decreases after administration of glucocorticoids (Crofford *et al.*, 1992). These results suggest that CRH expression is under the control of glucocorticoids. However, it may also be that the level of CRH expression parallels the cytokine expression and is determined by the intensity of the immune response, i.e., the inflammatory response.

10.3.5. Receptor and Function of CRH

The immune system can specifically recognize CRH. There are saturable binding sites for [^{125}I]-CRH found on resident macrophages in the spleen of rats (Webster, Tracey, Jutila, Wolfe, & de Souza, 1990). Synthetic CRH 1–41 has been shown to increase thymocyte proliferation. In this respect it is an interesting question whether expression of iCRH in the thymus in association with the coexpression of AVP might play a role in the selection process of antigen-reactive cells in the thymus. CRH also enhances mitogen-induced proliferation of lymphocytes (Singh & Leu, 1990). It does so, apparently, by increasing the expression of the IL-2 receptor on T cells as well as by enhancing the production of IL-1 and IL-2 (Singh, 1989; Singh & Leu, 1990). It is not known whether the enhancing effect of CRH on T cells occurs by an indirect effect via activation of monocytes or macrophages.

It is of interest to know whether iCRH, produced by the immune system and derived from an mRNA of slightly different length than that found in the hypothalamus, also acts as a proinflammatory mediator in the immune system. CRH, isolated from rat thymic extract, was shown to have the classical bioactivity of stimulating ACTH secretion in a culture of dispersed rat anterior pituitary cells (Redei, 1992). Therefore, we may assume that immune-derived CRH, the expression of which is upregulated during immune activation, is one of the mediators that enhances proliferation/clonal expansion of T cells in an inflammatory region. Because iCRH has been shown to be rather labile and plasma concentrations are low in view of the K_d of CRH binding to the receptor, iCRH may have solely an autocrine or paracrine effect on immune cells. This view is supported by the results of Karalis and colleagues, who demonstrated that systemic immunoneutralization of CRH in rats that were suffering from an aseptic inflammation, suppressed the inflammatory response by causing a decrease in the exudate volume and extravasated leukocytes. Also in this case the increase in iCRH expression could only be observed at the site of inflammation and not in the peripheral circulation (Karalis, Sano, Redwine, Listwak, Wilder, & Chrousos, 1991).

10.3.6. POMC Processing in the Immune System

As earlier discussed, the POMC precursor molecule is enzymatically cleaved into several neuropeptides. Stimulation of human peripheral blood cells with CRH (or IL-1) or Newcastle disease virus gives rise to the secretion of β-endorphin (1–31) and ACTH(1–39). However, when cells are cultured with LPS, one can detect the presence of ACTH 1–25 and β-endorphin 1–17 and 1–16, which are γ- and α-endorphin, respectively. These results were obtained using antisera specific for the various enzymatic cleavage products, and the data have been confirmed on the level of amino acid sequencing (Harbour, Smith, & Blalock, 1987; Smith et al., 1990).

10.3.7. Production of Enkephalins in the Immune System

During the last decade it has become clear that PEA mRNA can be found in most cells of the immune system, including Con-A stimulated CD4$^+$ T cells, CD4$^+$ thymocytes, B lymphocytes, as well as monocytes and selected lines of T cells, macrophages, and mast cells (Kamphuis, Kavelaars, Brooimans, Kuis, Zegers, & Heijnen, 1997; Kamphuis,

Eriksson, Kavelaars, Zijlstra, van de Pol, Kuis, & Heijnen, 1998; Kuis, Villiger, Laser, & Lotz, 1991; Linner, Beyer, & Sharp, 1991; Martin, Prystowski, & Angeletti, 1987; Rosen, Behar, Abramsky, & Ovadia, 1989). In thymocytes and some T-cell lines, the prepro-enkephalin (PPE) mRNA is not expressed constitutively; in freshly isolated T cells of some donors, the message is expressed constitutively (Kuis *et al.*, 1991).

Recently, we have demonstrated that cytokines can enhance PPE mRNA expression in human peripheral blood cells. Interestingly, especially Th2-type cytokines like IL-4, IL-10, and IL-6 and Th3 cytokines like TGFβ are potent enhancers of PPE mRNA expression in T cells, B cells, and monocytes (Kamphuis *et al.*, 1997, 1998).

Culture of PBMC with IL-4 or IL-10 results not only in enhanced expression of PPE mRNA, but also in the presence of immunoreactive Menk in the cytoplasm of PBMC. In contrast, Th1 cytokines are only weak inducers of PPE mRNA expression and do not result in detectable production of the protein. Stimulation of PBMC with IL-4 does not lead to secretion of the fully processed Menk peptide into the culture supernatant. However, larger precursor molecules containing Menk appear to be secreted after culture of PBMC with IL-4, as Menk was present after enzymatic treatment of the supernatant.

PEA mRNA expression can be induced or markedly enhanced in T cells or thymocytes by stimulation with Con A. B cells are capable of expressing PEA mRNA after stimulation with LPS (Rosen *et al.*, 1989). In contrast to the production of β-endorphin, IL-1 suppresses the Con A-induced expression of PEA mRNA in T cells. The glucocorticoid analogue dexamethasone also suppresses the expression of PEA (Behar, Ovadia, Polakiewicz, Abramsky, & Rosen, 1991) (see Fig. 2). The actions of IL-1 and dexamethasone in the brain, however, are permissive with respect to PEA mRNA expression, which underlines again that the regulatory mechanisms of neuropeptide production by the immune system and the brain can be opposite.

One of the physiological regulators of PEA mRNA expression may be the PEA peptides themselves. When the delta opioid receptors are triggered on cells, one can observe an up- or downregulation of the PEA mRNA, depending on the dose of the delta opioid receptor agonist (Linner, Quist, & Sharp, 1995).

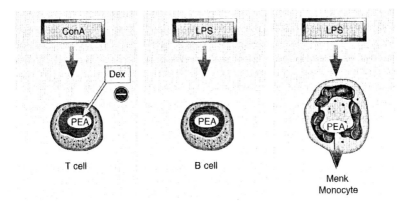

Figure 2. Schematic representation of the regulation of PEA and Menk production in mononuclear peripheral blood cells.

10.3.8. Processing of PEA in the Immune System

Menk produced by the immune system or other organs can be broken down by specific endopeptidases. An example of these peptidases is CD10 (common acute lymphoblastic leukemia antigen), which is expressed by normal lymphoid progenitor cells, polymorpho-nuclear cells, and nonhemopoietic cells (Shipp, Stefano, D'Adamio, Switzer, Howard, Sinisterra, Scharrer, & Reinherz, 1990). Cell-bound endopeptidases may therefore reflect another regulatory system for Menk expression in the immune system.

Interestingly, T cells can express PEA mRNA and produce large precursor molecules, although they cannot process it into the smaller pentapeptide Menk (Kamphuis *et al.*, 1997, 1998; Kuis *et al.*, 1991). Monocytes apparently have the full proteolytic machinery to process the endogenous preproenkephalin.

10.3.9. Intracellular Messenger Molecules Involved in Opioid Peptide Expression in the Immune System

There exist at least two different intracellular signaling pathways that regulate the production of POMC-derived peptides like β-endorphin. Activation of two protein kinases seems to play a key role in this process. The cAMP-dependent protein kinase (PKA), which is stimulated by ligands that increase intracellular cAMP levels, can induce the production of β-endorphin which becomes demonstrable after 18 h. Typical examples of the dominant activation of the PKA pathway are activation of monocytes and B cells by CRH (or IL-1) and LPS (Heijnen *et al.*, 1991; Kavelaars, Ballieux, & Heijnen, 1991).

Another pivotal protein kinase, PKC, that phosphorylates many important transcription factors is also involved in the production of β-endorphin. Ligands that use this activation pathway such as anti-CD3 or the direct activator of PKC, phorbol esters, give rise to a rapid production of β-endorphin (within 3 h) (Heijnen *et al.*, 1991; Kavelaars *et al.*, 1991). Figure 3 is a schematic representation of the pathways involved in the regulation of β-endorphin secretion by leukocytes.

With respect to the enhanced production of PPE mRNA and PEA products, cAMP can mimic the effect of IL-4 and TGFβ. Both cytokines are known inducers of cAMP, and subsequent activation of PKA. The effect of IL-4 on PPE mRNA can be inhibited by the addition of a PKA inhibitor, suggesting that the cAMP–PKA pathway can be used to

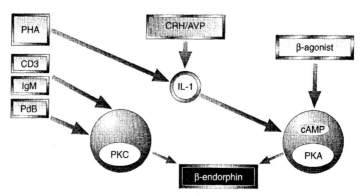

Figure 3. Schematic representation of the regulation of β-endorphin secretion by cells of the immune system.

activate PPE gene expression in leukocytes. The cAMP–PKA pathway is also the major pathway for induction of enkephalin production in the adrenal medullary chromaffin cell.

10.4. Opioid Receptors

The response of the primitive immune system of invertebrates to antigenic challenges appears to be sensitive to modulation by opioid peptides. Receptor-mediated opioid signals facilitate directed movement and aggregation of hemocytes or immunocytes from the mollusk *Mytilus edulis* and the insect *Leucophora maderae*. The opioid response in these organisms can also be "saturated," because after the immunocytes have become fully activated in the presence of exogenous endorphins or under stressful conditions, they no longer respond to opioids (Stefano, Cadet, & Scharrer, 1989).

Pharmacological data obtained from studying neuronal opioid receptors demonstrate the existence of at least three opioid receptor subtypes (Zukin & Zukin, 1981). A clear characterization of opioid receptors on cells of the immune system is still lacking, but it has been shown by several researchers that immune cells express mu- and delta-type opioid receptors, based on distinct binding and affinity profiles of selective agonists and antagonists.

As early as 1979, Wybran and colleagues described that Menk could enhance the binding of sheep erythrocytes to T cells (T-cell rosetting). They also showed that human peripheral blood T cells expressed specific opioid receptors, as the response could be blocked by the opiate receptor antagonist naloxone (Wybran, Appelboom, Famaey, & Govaerts, 1979). Since then, receptors for Menk have been studied more extensively, resulting in the demonstration that specific delta-type opioid receptors are involved, which are also present in the brain (Linner *et al.*, 1995).

A number of studies have shown that opioids affect practically every *in vitro* function measured, such as antibody production, NK cell activity, macrophage and polymorphonuclear functions, graft rejection, and T-cell proliferation (Faith, Jurgo, Clinkscales, & Plotnikoff, 1987; Rowland & Tokuda, 1989). Interestingly, opioids often have a dose-dependent effect on these immunological functions, being enhancing when applied in low doses and suppressive when used at higher doses (Rowland & Tokuda, 1989). In addition, when the immune response is vigorous, opioids usually suppress the response, whereas they enhance a weak immune response (Rowland & Tokuda, 1989). This dual effect of the peptides make them an important mediator in the regulation of immune responses (see Chapter 9).

For example, β-endorphin can dose-dependently enhance or suppress the production of antiherpes virus antibodies *in vitro*. The secondary antibody response to tetanus toxoid *in vitro* can also be modulated by β-endorphin (Heijnen *et al.*, 1991). β-Endorphin inhibits the antibody response in donors with a high baseline response (without endorphin), whereas stimulation of the antibody response was observed when the baseline response was low.

The primary antibody response of human peripheral blood cells can also be enhanced by β-endorphin (Heijnen, Bevers, Kavelaars, & Ballieux, 1985). Conflicting results have been reported on the influence of β-endorphin on T-lymphocyte proliferation. β-Endorphin can enhance as well as suppress mitogen-induced proliferation, depending on the lymphoid organ or species used for the experiments (Gilman, Schwarz, Milner, Bloom, & Feldman, 1982; Heijnen, Zijlstra, Kavelaars, Croiset, & Ballieux, 1987).

β-Endorphin has also been shown to modulate NK cell function. In this case the dose–response curve for β-endorphin has an inverted U-shape (Kay, Allen, & Morley, 1982).

Taken together with the above results, one can conclude that β-endorphin can modulate the immune response, but the direction of the modulatory effect is variable. One of the factors may be the intensity of the immune response, as has been shown in the antibody response to tetanus toxoid (Heijnen *et al.*, 1991). Another factor may be that β-endorphin can be cleaved *in vitro* into γ- and α-endorphin. These smaller peptides often have a totally different effect on the immune response.

In view of the widespread effects of opioids on the immune system, it is conceivable that opioids interfere with early metabolic events in leukocyte activation, thus affecting multiple cellular functions. One of these early events is the modulation of protein phosphorylation, which is one of the most general mechanisms of regulating cellular processes. One such example is the ability of β-endorphin to modulate the phosphorylating capacity of PKC, resulting in a change in the phosphorylation of the CD3 antigen–receptor complex of human T cells (Kavelaars, Eggen, De Graan, Gispen, & Heijnen, 1990d).

10.4.1. Function of Opioid Peptides Produced by the Immune System

The endogenous pentapeptides Menk and Leu-enkephalin trigger inflammatory responses in low vertebrate organisms such as *Mytilus edulis* by inducing morphological changes, directed migration, and aggregation of hemocytes (Stefano *et al.*, 1989).

When rat thymocytes are stimulated with Con A, the cells start to proliferate. During this stage of proliferation, the thymocytes also express PEA mRNA and produce Menk. Elegant *in vitro* studies have shown that in the mouse, the endogenous thymocyte Menk can inhibit its own proliferation. The latter was demonstrated by adding delta opioid receptor antagonists to the cultures of thymocytes or by neutralization of the Menk by anti-Menk antibodies. Moreover, inhibition of the production of PEA products by addition of an antisense oligonucleotide that binds to the translation initiation site, and thereby inhibits translation of PPE mRNA, also resulted in an enhanced thymocyte proliferation (Linner *et al.*, 1995).

In the human immune system, a similar approach has demonstrated that PEA products result in enhanced proliferation of peripheral blood T cells and decreased cytokine production by monocytes. It is not known which processing products of PEA are essential in the regulation of the immune response. The effect of the antisense oligonucleotide on monocyte cytokine production can be restored by addition of Menk. In contrast, for the T cells addition of Menk did not "reconstitute" the effect of the antisense oligonucleotide. This is especially interesting in view of the observation that T cells do not produce the fully processed peptide Menk and suggests that other processing products of PEA are involved in the endogenous effect of PEA on T-cell proliferation (Kamphuis *et al.*, 1997, 1998).

10.4.2. Function of POMC and PEA Peptides in the Control of Pain in Inflamed Areas

Pain can be dampened by various mechanisms, one of which is the central nervous system. Impulses from peripheral nerves are integrated in the dorsal horn of the spinal cord and transmitted centrally, resulting in pain perception and response. However, Stein and colleagues have shown that intrinsic modulation of nociception can also occur at peripheral terminals of afferent nerves. They demonstrated that opioid peptides, produced by immune cells at the site of inflammation, can mediate analgesia (Stein, Gramsch, & Herz, 1990a; Stein, Hassan, Prezwlocki, Gramsch, Peter, & Herz, 1990b). Moreover, inflammatory mediators like IL-1 and CRH, which are also locally produced by the immune system during inflammation (Harbuz, Rees, Eckland, Jessop, Brewerton, & Lightman, 1992), upregulate the opioid peptide production by immune cells and cells present in the synovial lining. Immunoneutralization of the peptides by specific antibodies or immunosuppression abrogated the analgesic effects. One important issue in this context is that peripheral opioid receptors are upregulated in the inflamed area. IL-1 and other cytokines have been shown to have potent anti-inflammatory effects in inflamed areas but not in noninflamed areas, indicating that upregulation of opioid receptors at the site of inflammation is a necessary event. These results demonstrate again the intertwining of the immune system and the neuroendocrine system and support the hypothesis that opioid peptide production by the immune system is physiologically relevant and may have clinical implications (see Fig. 4).

It will be clear by now that the conclusion is warranted that the immune system, like many other organ systems, can produce peptides that were originally thought to be exclusively produced by the neuroendocrine system. The peptides were likely present very early in evolution and have been well conserved. The immune system, however, did not evolve as an endocrine system that functions by secreting high levels of peptide hormone into the circulation. Most of the peptides discussed herein affect the immune system or neuronal systems mainly by a paracrine action. Therefore, one will never detect large quantities of the peptides in the supernatants of cell cultures. One has to keep in mind, however, that the immune system is not a fixed organ; most of the cells from lymphoid organs are traveling constantly in the circulation and to other organs. Some stimuli do not even give rise to secretion of the peptides, as the mRNA lacks the signal peptide. In this case the peptides are induced intracellularly, where they may be active as intracellular messengers. The length of the mRNAs is often, but not always, identical to that of the mRNA found in the neuroendocrine system. However, on the protein level the molecules resemble each other closely. Some authors question or deny the presence of substances such as POMC

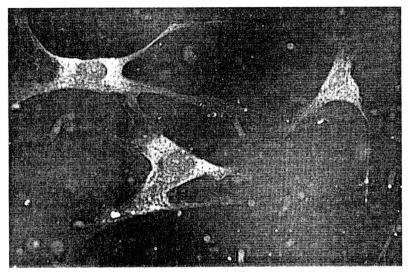

Figure 4.   Menk expression in fibroblasts in the synovial lining of a joint.

mRNA or POMC peptides in cells of the immune system, especially in peripheral blood leukocytes (Sharp & Linner, 1993; Van Woudenberg, Metzelaar, Van der Kleij, De Wied, Burbach, & Wiegant, 1993). However, in view of the large body of positive data from very carefully performed experiments, one has to conclude that leukocytes, like many other cells, synthesize and secrete POMC peptides, which may function in an autocrine and/or paracrine way. It is our expectation that in the coming decade these peptides will be viewed as important messengers between the immune system and the neuroendocrine system, functioning at the site of inflammation, as well as molecules that assist in the growth and differentiation of immune cells during development of the organism.

11   The Immune–Neuroendocrine Network

Hugo O. Besedovsky and Adriana del Rey

11.1. Introduction

A remarkable advance in medical sciences during the last part of this century has been based on the implementation of refined techniques that allowed a deep analysis of cellular processes at molecular levels. This analytical approach has not, however, been proportionally complemented by studies into how cellular and molecular mechanisms are physiologically integrated and controlled in a whole organism. A clear example is provided by the

Hugo O. Besedovsky and Adriana del Rey • Institute for Normal and Pathological Physiology and Immunophysiology, Marburg University, D-35037 Marburg, Germany.

evolution of our present knowledge of the immune system. We now know the structure of the main molecules that recognize internal and external antigens, and how a large diversity in these structures arises. The cell subtypes that participate in an immune response, how these cells are activated and interact, the main immune-derived soluble mediators involved in these processes, and the operation of well-programmed autoregulatory signals have been clarified to a large extent. However, the understanding of the immune system from "inside" has raised essential questions about its physiology, such as: What are the limits of the autonomy that immune cells display? How do immune cells and their products interact with integrative neuroendocrine agencies and with other bodily systems? What are the consequences of such interactions for the control of the activity of the immune system itself? To what extent are other physiological systems affected by the complex process of an immune response in an intact organism? In our view, studies that would contribute to answer these questions constitute the basis of immunophysiology. In this chapter, we shall discuss some concepts, methodological approaches and data related to interactions between the immune system and neuroendocrine mechanisms, mostly based on examples derived from our own work. Particular emphasis is put on immune interactions with the hypothalamus–pituitary–adrenal (HPA) axis and the central and autonomic nervous systems.

11.2. The Structure of the Network

To understand better how difficult it is to analyze the complexity of immune neuroendocrine interactions, three main aspects should be considered:

1. *The complexity of the immune system.* The immune system is built up of a large variety of interacting cells and molecules. The amount of information that it can process and the different responses that it can generate probably make the immune system the most complex body system after the central nervous system (CNS).

2. *The diffuse limits between physiological and pathological conditions.* Because, under natural conditions, changes in the activity of the immune system are often linked to disease, this system operates at a diffuse border between physiological and pathophysiological processes.

3. *The feedback interactions between the immune and nervous systems under basal conditions.* The immune system and the CNS are both constantly in operation. This implies that functional immune–neuroendocrine interactions must also be constantly established.

These three aspects show that multiple possibilities for interactions between immune and neuroendocrine mechanisms do exist. Thus, the communication between the immune and neuroendocrine systems can be best viewed as a functional network of interactions. In the following, a conceptual model of how this network may operate is proposed. However, it is necessary to state that some of the points raised are still hypothetical and may have to be modified according to experimental results.

Experimental data show that, after recognition of nonself and modified self antigens, activated immune cells convey messages to the brain or structures under brain control. Several messengers released by immune cells that mediate immune–CNS communication have already been identified and it is predicted that other such messengers will be found. Some of these cell products, such as cytokines, certain complement fractions, immunoglobulins, histamine, serotonin, mediators of inflammation, thymic hormones, putative immunologically derived "classical hormones," are possible mediators of immune–CNS communication. Depending on the type of antigenic stimulus, multiple possible combinations of subsets of lymphoid and accessory cells and their soluble products would occur. Thus, one can conceive a code based on combinations of soluble messengers that could inform the CNS about the type of immune response in operation. Also, information about the site of an ongoing immune response could be transmitted by stimulated nerve fibers in the "strategically" located lymphoid organs, or in tissues where the immune response takes place. The existence of an afferent pathway from the immune system to the CNS implies, as we have previously suggested, that the immune system is a receptor-sensorial organ.

After receiving signals from the immune system, the brain or structures under brain control may or may not respond by emitting regulatory signals. If brain intervention does occur, which may depend on homeostatic programs and other inputs into the CNS, the regulatory outcome will be the result of neuroendocrine signals and autoregulatory immune signals. It is known that hormones, neurotransmitters, and neuropeptides are capable of affecting the operation of the immune system. These agents could act directly by binding to specific receptors on defined subsets of immune cells, or they could affect more complex autoregulatory mechanisms of the immune system, e.g., cytokine production, suppressor–helper cell relationship, and idiotypic and antibody class feedback. This concept of immunoregulation provides a broader framework contrasting with the concept of a self-contained, self-monitored immune system.

In the following, we provide selected examples from our own work that support the concepts mentioned above. As a whole, the evidence obtained shows that:

1. Activation of immune cells affects neuroendocrine mechanisms that can, in turn, influence immune cell activity.
2. Products from immune cells mediate neuroendocrine effects during the immune response.

11.3. Immune Processes Can Affect Neuroendocrine Mechanisms

It has been known for a long time that neuroendocrine derangements occur during the course of infective, inflammatory, and neoplastic processes. Such derangements could be directly caused by microorganisms and neoplastic cells and/or their products, or by substances released following tissue injury. Thus, in an experimental approach to the problem of detecting effects intrinsically related to the activation of the immune system, the first difficulty encountered was how to distinguish these effects from those of the disease itself and the stress of sickness, and from the procedures used to stimulate the immune system. Several strategies and/or criteria have been used to circumvent this problem:

1. To trigger immune responses with innocuous, noninfective, nonneoplastic antigens and to search for changes in endocrine, central, and peripheral neural activity occurring in parallel with activation of the immune system.
2. To show that stimulation of immune cells with agents that cause diseases in which neuroendocrine alterations are noticed, induces the production of mediators capable of eliciting similar neuroendocrine changes in healthy individuals.
3. To confirm that neuroendocrine alterations occurring during diseases that involve the immune system, are not expressed in individuals lacking the particular type of immune cells involved.

4. To establish that neuroendocrine changes occurring after administration of infective and neoplastic agents that stimulate the immune system, are detectable before
the overt onset of the disease. This criterion is only indicative of immune-mediated
neuroendocrine changes.

These research strategies have been used in the past not only in our laboratory but also
by other research groups. In the following, examples of our work are used to illustrate the
points mentioned above.

The capacity of the immune system to affect the activity of the pituitary–adrenal axis
is, in this context, probably the most extensively studied phenomenon. Increased glucocorticoid blood levels have been observed during the course of specific immune responses to
different innocuous antigens (Besedovsky, Sorkin, & Mueller, 1975). These changes are
only detected when immune responses are intense enough to reach a given threshold. The
levels of corticosteroids in blood that are attained during immunization can interfere with
the response to other unrelated antigens (Besedovsky, del Rey, & Sorkin, 1979). Profound
increases in adrenocorticotropic hormone (ACTH) and corticosterone blood levels are also
observed after inoculation of Newcastle disease virus (NDV) in mice, a virus that produces a
mild disease in rodents. Already 2 h after injection of this virus, the levels of ACTH and
corticosterone in blood are increased severalfold (Fig. 1). This effect could be caused by the
virus itself, by the stress of the infection, or by the release of immune-derived cytokines. The
last alternative has proved to be correct. As can be seen in Fig. 2, coculture of peripheral
blood leukocytes with NDV results in the induction of a factor that stimulates the pituitary–
adrenal axis in normal, noninfected animals. This effect is totally neutralized when the
active material is incubated with anti-IL-1 serum. This was the first evidence that IL-1 can
mediate the increase in glucocorticoid levels following the administration of a natural
infective agent (Besedovsky, del Rey, Sorkin, & Dinarello, 1986).

It has long been known that bacterial endotoxins can stimulate the pituitary–adrenal
axis. This increase in glucocorticoid levels has protective effects during septic shock. Only
recently has it been shown that endotoxin stimulation of the pituitary–adrenal axis is the
result of products from activated macrophages, as depletion of macrophages interferes with
the increase in glucocorticoid levels in blood induced by LPS (Fig. 3). This endocrine

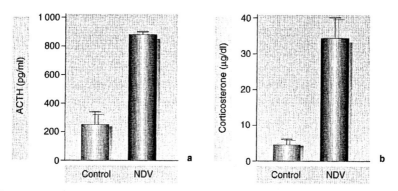

Figure 1. Injection of Newcastle disease virus (NDV) into mice induces an increase in the concentration of
adrenocorticotropic hormone (ACTH) and corticosterone in plasma. Control animals received an injection of the
vehicle alone. Mice were killed 2 h after the injection, and plasma levels of ACTH (a) and corticosterone (b),
the main glucocorticoid in the mouse, were determined. From Besedovsky and del Rey (1992).

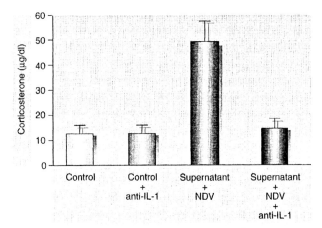

Figure 2. IL-1 mediates the glucocorticoid-increasing activity of supernatants from human peripheral blood leukocytes cocultured with NDV. NDV was added to cultures of human peripheral blood leukocytes. One fraction of the supernatant of these cultures was incubated with an anti-IL-1 antiserum, and another fraction was incubated for the same time but without further treatment. These supernatants as well as the culture medium alone and culture medium containing the anti-IL-1 antiserum were injected into rats. Two hours later, animals were killed and corticosterone levels determined. Anti-IL-1 treatment completely abolished the glucocorticoid-increasing activity of the culture supernatant. From Besedovsky *et al.* (1986).

response is predominantly elicited by IL-1, since administration of an IL-1 receptor antagonist also abrogates the increase in adrenocortical hormones following LPS administration.

An early increase in glucocorticoid blood levels is also observed following syngeneic tumor transplantation (Norman, Besedovsky, Schardt & del Rey, 1988). At least in one model this effect is mediated by factors derived from immune cells. A biphasic early and

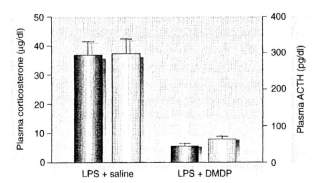

Figure 3. Macrophage depletion inhibits the increase in blood ACTH and corticosterone levels induced by endotoxin in the rat. One group of rats was depleted of macrophages by treatment with DMDP-containing liposomes, and a control group received an injection of physiological saline. Five days later, endotoxin (LPS) was injected into both groups of rats and ACTH (black bars) and corticosterone (white bars) levels in blood were determined after 2 h. From de Rijk *et al.* (1991).

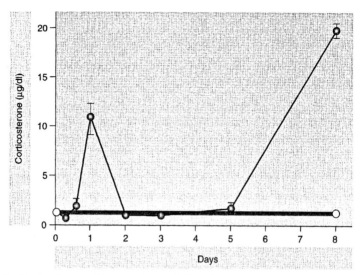

Figure 4. Injection of tumor cells induces an increase in blood glucocorticoid levels. EL-4 lymphoma cells were injected into mice and blood corticosterone levels were determined at different times (upper curve). Mice that received an injection of the vehicle alone served as control. From Norman *et al.* (1988).

late increase in glucocorticoid levels is observed following administration of EL-4 lymphoma cells (Fig. 4). The early eight- to tenfold increase in corticosterone blood levels occurs within 24 h after tumor cell transplantation, well before the tumor becomes detectable. Such changes in glucocorticoid levels do not occur in T-cell deficient recipients receiving either allogeneic EL-4 cells or syngeneic fibrosarcoma cells. Furthermore, the increase in corticosterone levels has been shown to be produced by a host-derived factor present in the ascitic fluid that, on transference to a normal host, stimulates the pituitary–adrenal axis (Norman *et al.*, 1988).

Not all types of immune responses result in stimulation of the pituitary–adrenal axis. During skin graft rejection, corticosterone blood levels are lower than in animals bearing autografts (Besedovsky, Sorkin, & Keller, 1978). It should also be considered that other endocrine mechanisms involving pituitary hormones different from ACTH are affected during an immune response (for review see Besedovsky & del Rey, 1996).

Autonomic responses also occur during activation of the immune system. We have observed decreased sympathetic activity in the spleen, as evaluated by noradrenaline content and turnover rates, during the immune response. A decrease in noradrenaline content of the spleen is observed 3 days after inoculation of sheep red blood cells (SRBC), i.e., preceding the peak of the immune response. In immunologically high responder animals, this decrease is maintained over several days (del Rey, Besedovsky, Sorkin, Da Prada, & Bondiolotti, 1982). A long lasting decrease in noradrenaline content in the spleen was also observed during the immune response to other antigens. This evidence indicates that sympathetic activity is reduced during certain immune responses.

From these data, it cannot be concluded that the CNS is directly involved in all of the peripheral effects discussed above. However, the observed changes in the endocrine and autonomic nervous systems reflect, even if only indirectly, the existence of afferent pathways in immune–brain communication. Because the hypothalamus integrates most endocrine and autonomic functions, the possibility was considered that hypothalamic neurons receive signals derived from the immune system. We studied the immune response and the rate of firing of individual hypothalamic neurons in the same animals at various intervals after injection of SRBC or TNP-hemocyanin (Besedovsky, Sorkin, Felix, & Haas, 1977). During the immune response to TNP-hemocyanin, the frequency of firing of rat hypothalamic neurons (ventromedial nucleus) was increased on day 2, a time close to the peak number of direct plaque-forming cells (PFC). Animals stimulated with SRBC showed no PFC and no changes in the frequency of firing on day 1, but on day 5 PFC in the spleen were maximal and there was a severalfold increase in the firing rate of the ventromedial neurons.

As mentioned before, the hypothalamus is an endocrine organ that is subject to hormonal feedback signals. Therefore, it cannot be determined whether a change in the electrical activity of hypothalamic neurons is causally related to the reception of immune messages or to the emission of endocrine signals. Furthermore, some hypothalamic neurons are controlled by neurotransmitters produced by neurons that follow anatomically well-identified pathways, connecting the hypothalamus with distant parts of the brain. Aminergic neurons projecting to the hypothalamus are important controllers of hypothalamic neuronal activity. On this basis, catecholamine turnover rates in different regions of the CNS during the immune response to SRBC were studied (Besedovsky, del Rey, Sorkin, Da Prada, Burri, & Honegger, 1983). In immunologically high-responder rats, a marked decrease in hypothalamic noradrenaline turnover rate compared with saline-injected controls was noted 4 days after antigen administration, whereas immunologically low-responder animals had turnover rates almost equal to those of the controls (Fig. 5). This effect seems to be specific, as no changes in dopamine turnover in the CNS were observed during the immune response.

The endocrine and autonomic changes and the alterations in electrical activity and

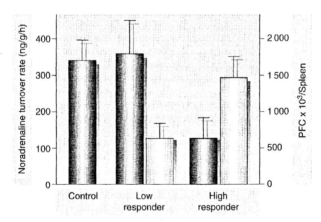

Figure 5. Decreased noradrenaline synthesis in the hypothalamus of rats during the immune response. Physiological saline was injected into a group of rats (control), and another group of animals received sheep red blood cells. The immune response in the spleen [expressed as plaque- forming cells (PFC), white bars] was determined 5 days later, and animals were divided into immunologically "high" and "low" responders. Noradrenaline synthesis (expressed here as turnover rate, black bars) in the hypothalamus of these animals was determined simultaneously. From Besedovsky *et al.* (1983).

turnover rate of catecholamines in the brain after activation of the immune system constituted clear evidence of the reception of immune-derived signals by the CNS. However, the nature of these signals needed to be investigated, and in recent years evidence has accumulated about the identity of at least some of the messengers involved.

11.4. Products of Immunocompetent Cells Can Affect Neuroendocrine Functions

Our prediction was that, among other agents, cytokines could serve as mediators for immune–brain communication. The first approach was to study *in vivo* neuroendocrine effects of supernatants from cultures of immune cells stimulated with mitogens or antigens *in vitro*. On injection into normal animals, these cytokine-containing supernatants stimulated ACTH and corticosterone output (Besedovsky, del Rey, & Sorkin, 1981) and decreased the noradrenaline content of the hypothalamus (Besedovsky *et al.*, 1983), thus mimicking two of the neuroendocrine changes observed following *in vivo* immunization. Products with glucocorticoid-increasing activity [glucocorticoid-increasing factors (GIF)] derive from allogeneic and/or mitogenic stimulation of human or murine lymphocytes. One million lymphocytes stimulated in culture produce enough GIF to induce a severalfold increase in blood ACTH and corticosterone levels on introduction into normal animals. Neither IL-2 nor IFNγ is responsible for this activity. Later, we showed that both natural purified and recombinant human IL-1β can also stimulate ACTH and corticosterone output in mice and rats, and that IL-1 is the most likely mediator of the glucocorticoid changes induced by certain viruses (Besedovsky *et al.*, 1986). This effect of IL-1 seems to be rather specific, as it is not paralleled by changes in other "stress hormones," such as somatotropin, prolactin, and αMSH. The levels of catecholamines in blood are only modestly increased following IL-1 injection (Berkenbosch, de Goeij, del Rey, & Besedovsky, 1989). It has also been shown that *in vivo* release of corticotropin-releasing factor is an obligatory step in IL-1-induced increase in corticosterone levels (Berkenbosch, Van Oers, del Rey, Tilders, & Besedovsky, 1987; for review see Besedovsky & del Rey, 1996).

Apart from IL-1β, other cytokines also stimulate the pituitary–adrenal axis (for review see Besedovsky & del Rey, 1996). We have performed comparative studies on the capacity of IL-1β, IL-6, and TNFα to stimulate the pituitary–adrenal axis. These studies showed that, on a molecular weight basis, IL-1β is more potent than TNFα and IL-6 (Besedovsky, del Rey, Klusman, Furukawa, Monge-Arditi, & Kabiersch, 1991). Figure 6 shows an example of these studies.

Following this primary evidence that cytokines can affect the HPA axis, the effects of cytokines on other neuroendocrine mechanisms have been studied. For example, the pituitary–thyroid axis, the pituitary–gonadal axis, and growth hormone secretion can be affected by IL-1, and/or IL-6, and/or TNFα, and/or IFNγ (for review see Besedovsky & del Rey, 1996). Immune cytokines can also affect brain neurotransmitters; IL-1β, for example, stimulates noradrenaline synthesis in the hypothalamus and hippocampus (Kabiersch, del Rey, Honegger, & Besedovsky, 1988).

The routes that peripheral cytokines follow to affect CNS mechanisms are at present under intense investigation. Cytokines or secondarily induced mediators could be transported through the blood–brain barrier, or act primarily in regions such as the circumventricular organs, which lack this barrier. Alternatively, immune-derived products could act on peripheral nerve fibers present in lymphoid organs.

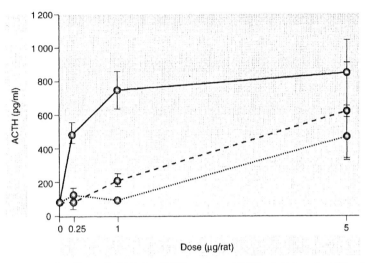

Figure 6. Comparative effects of IL-1β, IL-6, and TNFα on plasma ACTH levels. Rats received different doses of IL-1β (upper curve), IL-6 (middle curve), and TNFα (lower curve). Plasma ACTH levels were determined 1 h after injection. From Besedovsky *et al.* (1991).

Considering that not long ago, only a few immune messengers able to affect neuro-endocrine mechanisms were known, the examples mentioned above suggest that other messengers with these properties will be found in the near future.

11.5. The Biological Relevance of the Immune–Neuroendocrine Network

11.5.1. The Immunoregulatory Effect of Hormones and Neurotransmitters Depends on the Type and Stage of Activation of Immune Cells

The existence of neuroendocrine changes during the stimulation of the immune system strongly suggests the involvement of these mechanisms in immunoregulation (Besedovsky & Sorkin, 1977). However, a main conceptual difficulty is to understand how nonspecific agents such as hormones and neurotransmitters could contribute to the control of a highly specific mechanism such as the immune response to a given antigen. We have earlier postulated that such a problem could be circumvented if resting and activated immune cells were to have different sensitivities to neuroendocrine signals. An example of an endocrine effect that is predominantly exerted on resting immune cells is the well-known inhibitory action of glucocorticoids; this effect is readily noticed prior to and at the time of antigenic challenge, but not, or only marginally, during the course of the immune response. Antigen presentation and the release of certain cytokines necessary for the control of immune cell activation, proliferation, and differentiation, are sensitive to glucocorticoids. However, because these early events precede the increase in the levels of endogenous glucocorticoids,

the onset of the immune response will not be affected and clonal expansion, a main ampli-
ficatory mechanism, can be initiated. Interestingly, it has been reported that IL-1 protects
specific T-helper cells (CD4$^+$) but not suppressor and cytotoxic (CD8$^+$) T cells from
inhibition by glucocorticoids. As the immune response proceeds, the stimulation of the HPA
axis will result in inhibition of the production of certain cytokines (for review see Besedovsky & del Rey, 1996). In this way, the consequences of increased glucocorticoid levels for
already activated cells with high affinity for the antigen may be different from those for
accessory cells or resting or low-affinity lymphocytes. As the latter are more sensitive to
glucocorticoids than activated lymphocytes, this could explain the lack of response to an
antigen administered at a time when the levels of glucocorticoids are increased as a
consequence of the immune response to another unrelated antigen (Besedovsky *et al.*,
1979). Collectively, these data brought us to postulate that the function of a cytokine–
glucocorticoid circuit could be to prevent the excessive expansion of cells that are recruited
under the polyclonal influence of cytokines.

There are also examples of potential discriminatory effects of other neuroendocrine
mediators, such as insulin and β-adrenergic and muscarinic agonists, that predominantly
affect activated immune cells because they express more of the corresponding receptors
than resting cells (for review see Besedovsky & del Rey, 1996). Thus, the increase in the
sensitivity to a neuroendocrine messenger of the immune cells activated by a given antigen,
would allow the involvement of this messenger in the specific control of the immune
response that is triggered.

11.5.2. The Biological Significance of the Immune–Neuroendocrine Network during Pathological Conditions

The examples we have chosen to discuss here describe interactions between immune
mechanisms and the HPA axis under pathological conditions. There is reasonable proof that
normal bidirectional communication between these two systems can play a protective role
whereas disruption of this communication can lead to predisposition to, or aggravation of,
the disease. It is known that glucocorticoids in blood reach very high levels during sepsis.
This increase is protective for the host, as administration of sublethal doses of endotoxins
such as LPS into adrenalectomized animals results in high mortality. We now know that in
this model of sepsis, the stimulatory effect of LPS on glucocorticoid output is mediated by
cytokines, most likely IL-1, released by macrophages (de Rijk, van Rooijen, Besedovsky,
del Rey, & Berkenbosch, 1991; Fig. 3). Cytokines that stimulate the pituitary–adrenal axis
mediate the inhibition of the inflammatory reaction that is observed soon after tumor cell
inoculation. One example is the previously described effect of injection of EL-4 lymphoma
cells (Norman *et al.*, 1988). Twenty four hours after injection of these cells, an inhibition of
the inflammatory reaction occurs in parallel to the immune-mediated increase in glucocorticoid blood levels. This anti-inflammatory effect is not observed in adrenalectomized
animals.

As mentioned above, the immune–HPA axis circuit may contribute to the control of
excessive cumulative expansion of immune cell mass and activity. Disturbances in this
circuit may contribute to the expression of lymphoproliferative and autoimmune diseases in
genetically predisposed individuals, especially as these pathological events are the product
of multiple factors that progressively disrupt immune cell homeostasis. An example of the
moderating action of the pituitary–adrenal axis is observed in Lewis female rats, which are
susceptible to experimentally induced arthritis and encephalomyelitis. These rats have a
defective HPA axis that responds to cytokines like IL-1 less intensively than that of other

rat strains. A clinical counterpart of these studies shows that patients with rheumatoid arthritis and multiple sclerosis have a disturbed HPA axis. Also, in chickens with spontaneous autoimmune thyroiditis, the expression of the disease correlates with a deficient operation of the cytokine-mediated stimulation of the HPA axis (for review Besedovsky & del Rey, 1996).

The immune-mediated glucocorticoid output may not necessarily be beneficial for the host. We have demonstrated, for example, that IL-1 mediates the stimulation of the HPA axis after virus inoculation (Besedovsky *et al.*, 1986, Fig. 2). The resulting increase in glucocorticoid levels may contribute to the immunosuppression observed during viral diseases, which often results in the appearance of opportunistic superinfections.

11.5.3. The Immune–Neuroendocrine Network Plays a Relevant Role Not Only in Immunoregulation but also in the Control of Other Physiological Functions

Because of the multiplicity of the mechanisms involved, it is obvious that the operation of immune–neuroendocrine circuits will have consequences not only for immunoregulation but also for other physiological mechanisms, such as the control of metabolic processes, thermoregulation, food intake, and sleep patterns. In fact, evidence obtained during pathological conditions suggests that cytokines released during immune responses mediate alterations in these mechanisms by affecting neuroendocrine functions. Furthermore, it is also well known that during certain infectious and neoplastic processes, profound metabolic derangements occur (for review see Besedovsky & del Rey, 1996). However, evidence should be provided that these derangements are mediated by endogenous immune-derived products rather than by the agents (e.g., microorganisms, neoplastic cells) that primarily cause the disease. One piece of evidence that the immune response itself can also affect metabolic processes is the observation that 5 days after administration of SRBC to rats, i.e., at the peak of the immune response to this innocuous antigen, a decrease in glucose blood levels occurs. IL-1β is a likely candidate to mediate this effect, as administration of nanogram amounts of the cytokine into mice results in a reduction (40–50%) of glucose blood levels. This effect is long lasting, the animals remaining hypoglycemic for 8 to 12 h (del Rey & Besedovsky, 1989). The mechanism of action of IL-1β is only partially understood, but the hypoglycemia induced by this cytokine is dissociable from possible insulin secretagogue effects and is partially mediated at the level of the CNS (del Rey & Besedovsky, 1989). IL-6 and TNFα do not induce hypoglycemia, at least in the same range of concentrations as that used for IL-1.

11.6. Overview

Self antigens present in most cell membranes and bodily fluids constitute biological markers of cell and tissue constancy and integrity. Microorganism and foreign bodies, modification of self antigens as well as the expression of neoantigens disturb this constancy and represent a threat to the individual. The adaptive function of the immune response is to eliminate these disturbances. Because of this property, the immune response belongs to the category of homeostatic responses that maintain the *milieu interieur*. An extension of this integrationist view of the immune system includes the proposal that the immune response, as other homeostatic mechanisms, is under neuroendocrine control. This proposal implies that immune cells should be able not only to receive neuroendocrine signals but also to emit signals capable of affecting endocrine, autonomic, and CNS mechanisms. Increasing evidence supports this concept. On this basis, we have proposed that the degree of activity of such a complex network of functional interactions is changed by stimuli acting at the level of either the immune or the nervous system. Also, networks between locally released hormones and neurotransmitters and cytokines may operate at organ and tissue levels (Rogausch, del Rey, Kabiersch, Reschke, Örtel, & Besedovsky, 1997).

In our opinion, this view of immune–neuroendocrine interactions as a network has several conceptual advantages: The degree of activity of the network can be changed by stimuli acting at or generated from any of its components, e.g., antigens at the level of the immune system, and psychosocial stimuli at the level of the CNS. Furthermore, the tonic functioning of the network would influence all of its components and keep them informed about their mutual functional states. Finally, a change in the actual state of the network, such as during activation of the immune system, may have consequences beyond immunoregulation and affect general homeostasis. Homeostatic mechanisms that operate under basal conditions may differ qualitatively or quantitatively from those that are active during pathological states involving the immune system. In fact, the setpoint for the regulation of essential variables may need to be adjusted during the course of these states and potent immune-derived messengers may mediate these adjustments.

There is probably a nonhomeostatic, or even antihomeostatic, dimension to the operation of the network, as immune–neuroendocrine interactions could contribute to the aggravation of certain diseases. Immune cells, by producing exaggerated amounts of certain cytokines and other products, may induce neuroendocrine and metabolic derangements that lead for example to the lethal course of certain infective diseases. This process, which contrasts with homeostasis, may have the evolutionary aim of eliminating sick individuals who, although having a normal immune system, cannot cope with a particular infective agent. These individuals could become vectors of transmissible diseases that might occasionally compromise the survival of the species. According to this view, the immune system may contribute to evolution not only by positive protective mechanisms but also by active mechanisms of negative self-selection. The timely administration of appropriate cytokine antagonists or blockers, a therapeutic strategy being developed in the last few years, may contribute to alleviate immune-mediated deleterious effects.

12   Sickness Behavior
A Neuroimmune-Based Response to Infectious Disease

Robert Dantzer

Robert Dantzer • INSERM U394, 33077 Bordeaux Cedex, France.

12.1. Introduction

Nonspecific symptoms of infection and inflammation include fever and profound physiological and behavioral changes. Sick individuals experience weakness, malaise, listlessness,
and inability to concentrate. They become depressed and lethargic, show little interest in
their surroundings, and stop eating and drinking. This constellation of non-specific symptoms is collectively referred to as *sickness behavior*. Because of their commonality, sickness
symptoms are frequently ignored by physicians. They are often considered as an uncomfort-
able, but rather banal, component of the pathogen-induced debilitative process.

A few years ago, Hart (1988) proposed that the behavioral symptoms of sickness represent, together with the fever response, a highly organized strategy of the organism to fight
infection. This view was built on the already recognized role of fever in the host response to
pathogens (Kluger, 1979). In physiological terms, fever corresponds to a new homeostatic
state that is characterized by a raised setpoint of body temperature regulation. A feverish
individual feels cold at usual thermoneutral environments. Therefore, the individual not
only seeks warmer temperatures but also tries to enhance heat production (increased
thermogenesis) and reduce heat loss (decreased thermolysis). The higher body temperature
that is achieved during fever stimulates proliferation of immune cells and is unfavorable for
the growth of many bacterial and viral pathogens. In addition, the reduction of zinc and iron
plasma levels that occurs during fever decreases the availability of these vital elements for
growth and multiplication of microorganisms. The adaptive nature of the fever response is
apparent from studies showing that organisms infected with a bacterium or virus and are
unable to mount an appropriate fever response (when they are kept in a cold environment or
are treated with an antipyretic drug) have a lower survival rate than organisms that develop a
normal fever (Kluger, 1979).

The amount of energy that is required to increase body temperature during the febrile
process is quite high, as the metabolic rate in humans needs to be increased by 13% for a
raise of 1°C in body temperature. Because of this high metabolic cost of fever, there is little
room for activities other than those favoring heat production (e.g., shivering) and minimizing thermal losses (e.g., rest, curl up posture, piloerection).

The molecular signals that mediate the local immune reaction are also responsible for
ensuring the necessary synchrony between metabolic, physiological, and behavioral components of the systemic response to infection. These signals are proinflammatory cytokines,
such as IL-1, IL-6, TNFα, and interferons, which are released by activated monocytes and
macrophages during the course of an infection.

The focus of the present chapter is exclusively on the effects of cytokines on behavior.
The general biology of cytokines and their receptors is briefly treated, followed by a detailed
discussion of the evidence showing that cytokines have potent effects on the central nervous
system, which are responsible for the development of sickness behavior.

12.2. Biology of Cytokines and Cytokine Receptors

Cytokines are soluble mediators that are released by activated immune cells in contact
with an antigen. These factors play a key role not only in the regulation and differentiation of
the cells responding to the antigen, but also in the interactions between immune and
nonimmune cells during infection and inflammation (Baumann & Gauldie, 1994; Hamblin,
1993; see Chapter 2).

The main cytokines that are important in the acute inflammatory response are IL-1, IL-6, and TNF. These cytokines are synthesized by many cell types including activated monocytes and macrophages. A potent activator of these cells is lipopolysaccharide (LPS), the biologically active fragment of the membrane of gram-negative bacteria.

There are several members of the IL-1 family, each coded by a different gene (Dinarello, 1996). IL-1α and IL-1β behave as agonists of IL-1 receptors and share similar biological activities despite a relatively low homology (27%) at the amino acid level. IL-1α is mainly released in a membrane-associated form. IL-1β is synthesized as an inactive precursor form, and requires proteolytic cleavage to produce the mature biologically active protein. The enzyme responsible for this cleavage is interleukin-1β-converting enzyme. IL-1ra, an endogenous antagonist of IL-1 receptors, competes with the binding of both forms of IL-1 to their receptors and is able to block most of the biological effects of IL-1.

Two subtypes of IL-1 receptors have been cloned: an 80-kDa IL-1 receptor (type I IL-1 receptor) and a 68-kDa receptor (type II IL-1 receptor) (Dinarello, 1996). These two receptor types belong to the superfamily of immunoglobulins. The type II IL-1 receptor is structurally similar to the type I IL-1 receptor in its extracellular and transmembrane domains but possesses a shorter cytoplasmic domain. Although both receptors bind IL-1 ligands with similar affinity, there is evidence that most of the biological activities of IL-1 on immune and nonimmune cells are mediated by the type I IL-1 receptor, with the type II IL-1 receptor serving as a decoy protein that can be shed from target cells on binding of agonists. Although IL-1 receptors were initially believed to function as monomeric receptors, recent evidence indicates that they actually form a complex with an IL-1 receptor accessory protein (IL-1RAcP). This subunit of the IL-1 receptor does not bind IL-1 by itself, but increases the affinity of the type I IL-1 receptor for IL-1α and IL-1β. IL-1ra prevents the formation of a complex between the type I IL-1 receptor and the IL-1RAcP.

Like IL-1, TNFα is produced mainly by activated macrophages and monocytes. This cytokine has many biological activities in common with IL-1 and is actually a strong inducer of IL-1. There are two types of TNF receptors: a low-affinity 55-kDa type 1 receptor (TNFR1) and a high-affinity 75-kDa type 2 receptor. The extracellular regions are similar for both TNF receptors but the intracellular domains are entirely unrelated, suggesting different signaling mechanisms and functions. The common view is that activation of TNFR1 triggers a process of programmed cell death (apoptosis) whereas activation of TNFR2 induces the transcription of genes involved in cellular activation. TNF binds as a trimer to its receptors, which also undergo trimerization. Most cell types express both TNFR1 and TNFR2, although expression of one type of receptor usually predominates.

IL-6 is produced by a wide variety of cell types and is easily inducible by IL-1. Its receptor consists of two chains. The β chain, known as gp130, is responsible for intracellular signaling and is a member of the hematopoietin receptor family.

The interferons are members of a large family of proteins that are made by a wide variety of cells in response to viral infection. They all act on a common receptor.

All of the proinflammatory cytokines mainly exist in an inducible form, as they are not made in the absence of antigenic stimulation. As already mentioned, LPS is a strong inducer of IL-1 and TNF-α both *in vitro* and *in vivo*. Proinflammatory cytokines are characterized by pleiotropism (active on a wide variety of cell types) and redundancy (possessing many biological activities in common).

Studies of the biological activity of cytokines *in vivo* have been made possible by the availability of recombinant technology. Administration of recombinant human cytokines represents the most usual way of studying the specific biological activity of each of the proinflammatory cytokines. Although human recombinant IL-1 and IL-6 are fully active in laboratory animals, the same does not apply to TNFα and IFNα. TNFα behaves as a partial agonist in the mouse system, only binding to the 55-kDa TNF receptor. Recombinant human IFNα is a poor agonist of the IFNα receptor in the rat and mouse systems and homologous IFNα should be used instead. Although most recombinant mouse cytokines are easily available, the same does not apply to rat cytokines. Administration of LPS has the advantage of inducing the release of endogenous cytokines of which the individual role can then be studied, using appropriate tools. Antagonists of cytokines used for pharmacological studies include antibodies to cytokines, antibodies to cytokine receptors, soluble receptors, and various cytokine antagonists.

The creation of transgenic mice in which a specific cytokine or cytokine receptor gene is either overexpressed or "knocked out" provides an invaluable tool for the *in vivo* investigation of biological functions of cytokines. However, the redundancy of the cytokine network often complicates the interpretation of the results obtained with these animal models.

12.3. Cytokines and Cytokine Receptors in the Central Nervous System

The expression of cytokines in the brain can be studied at the protein level with immunohistochemical techniques and at the mRNA level with polymerase chain reaction after reverse transcription (RT-PCR) or with *in situ* hybridization. Because of the very low abundance of cytokine transcripts on most cells, the latter technique is close to its limit of sensitivity and has to be used carefully to characterize the neuroanatomical distribution of cytokine mRNA.

Using these techniques, IL-1α and IL-1β have been found to be expressed in the brain in two pools: a constitutive form, which appears to be mainly present in neurons, and an inducible form, which is expressed mainly in brain macrophages and microglial cells (Hopkins & Rothwell, 1995). Peripheral administration of LPS induces the expression of IL-1 transcripts in discrete areas of the mouse and rat brain (Fig. 1) (Layé, Parnet, Goujon, & Dantzer, 1994). The same treatment induces the expression of immunoreactive IL-1β in perivascular and meningeal macrophages and in microglia throughout the brain parenchyma (Buttini & Boddeke, 1995; Van Dam, Brouns, Louisse, & Berkenbosch, 1992). Constitutive expression of IL-1ra mRNA has been described in the rat brain. It is induced in the brain in response to LPS and it peaks later than IL-1 (Licinio, Wong, & Gold, 1991).

Equilibrium binding of recombinant human [^{125}I]IL-1α by brain membrane homogenates and brain slices has revealed the presence of binding sites for IL-1 in the mouse but not the rat brain (Haour, Ban, Baran, Millon, & Fillion, 1990; Takao, Tracey, Mitchell, & De Souza, 1990). These binding sites are located almost exclusively in the dentate gyrus of the hippocampus and the choroid plexus. They are also found in the anterior pituitary. *In situ* hybridization identified the type I IL-1 receptor mRNA in several brain structures including the anterior olfactory nucleus, thalamus, hypothalamus, amygdala, and cerebellum (Cunningham, Wada, Carter, Tracey, Battey, & De Souza, 1992). Double labeling revealed that the signal for the type I IL-1 receptor is mainly localized in neuronal cells, endothelial cells, and epithelial cells of the choroid plexus and ventricles. RT-PCR techniques have allowed identification of both type I and type II IL-1 receptor mRNAs in various brain regions in the mouse (Parnet, Amindari, Wu, Brunke-Reese, Goujon, Weyhenmeyer, Dantzer, & Kelley, 1994). Neuronal cell lines derived from chromaffin cells express only the type I IL-1

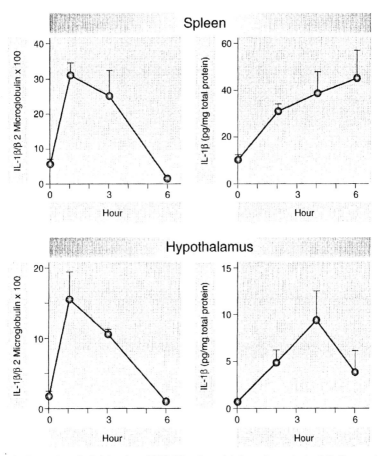

Figure 1. Intraperitoneal administration of LPS (10 μg/mouse) induces the synthesis of IL-1β not only in lymphoid tissues such as the spleen (top graphs) but also in the brain (bottom graphs). The left graphs represent tissue levels of IL-1β mRNA, as determined by polymerase chain reaction after reverse transcription of total RNA extracted from the spleen and hypothalamus of mice injected with LPS and killed just before or at different times after. IL-1β mRNA levels are expressed as percentage of mRNA of β_2-microglobulin, a housekeeping gene. Each value is the mean of three to five independent replications. The right graphs represent tissue levels of IL-1β, as determined by a specific ELISA. IL-1β levels are expressed as picograms per milligram of total proteins. Each value represents the mean of five to six independent replications.

receptor. Immunocytochemistry techniques with rat monoclonal antibodies directed against the different types of IL-1 receptors in the mouse system have confirmed that both type I and type II IL-1 receptors are present in the hippocampus, cerebellum, and choroid plexus, whereas only the type II IL-1 receptor is found in the paraventricular nucleus of the hypothalamus (French, Chizzonite, Dantzer, Parnet, Bluthé, VanHoy, Zachary, & Kelley, 1999). In the anterior pituitary, both types of IL-1 receptors are present only on those pituitary cells that synthesize and release growth hormone (French, Zachary, Dantzer,

Frawley, Chizzonite, Parnet, & Kelley, 1996; Parnet, Brunke, Goujon, Demotes-Mainard, Biragyn, Arkins, Dantzer, & Kelley, 1993).

TNFα-like immunoreactive neurons have been identified in the mouse brain under basal conditions (Breder, Tsujimoto, Terano, Scott, & Saper, 1993). These neurons have a widespread distribution and innervate in particular those brain areas involved in autonomic, neuroendocrine, and behavioral functions. In response to systemic LPS, TNFα is rapidly induced in the brain, as shown by RT-PCR (Gatti & Bartfai, 1993; Layé et al., 1994) and in situ hybridization (Breder, Hazuka, Ghayur, Klug, Huginin, Yasuda, Teng, & Saper, 1994). There is an early phase of induction that occurs mainly in perivascular cells and neurons in the circumventricular areas and cells in the meninges. This early phase is followed by a later phase of induction that involves neurons of pericircumventricular nuclei (paraventricular and arcuate nucleus of the hypothalamus, nucleus of the solitary tract). Binding sites for murine TNFα have been identified in the rodent brain, with substantial binding in the brain stem, cortex, cerebellum, thalamus, and basal ganglia (Kinouchi, Brown, Pasternak, & Donner, 1991). The cellular localization of brain TNF receptors is still uncertain. Rat neurons, microglial cells, and cerebrovascular endothelial cells express both TNFR1 and TNFR2, whereas astrocytes and oligodendrocytes predominantly express TNFR1 (Dopp, Mackenzie-Graham, Otero, & Merrill, 1997).

Although IL-6 is present in the pituitary in basal and stimulated conditions, its expression in the brain has been more difficult to detect, reflecting the low abundance of transcripts in this organ. RT-PCR studies revealed that IL-6 mRNA increases after systemic LPS injection, with a time course of induction that is delayed relative to TNFα and IL-1, suggesting that the production of IL-6 is mediated by these cytokines (Layé et al., 1994). The IL-6 receptor appears to be expressed in the anterior pituitary gland and brain (Schöbitz, Voorhuis, & De Kloet, 1992). In situ hybridization performed on rat brain revealed positive signals in the dentate gyrus of the hippocampus, the habenulae, the dorsomedial and ventromedial regions of the hypothalamus, the internal capsule, the piriform cortex, and the optic tract.

12.4. Assessment of Sickness Behavior

Most behaviorists prefer not to study the behavior of sick animals whereas pathologists are rarely interested in the behavior of sick animals. This explains why there has been for years no standard way of assessing sickness behavior in a quantitative way in animals.

Sickness was first studied in a very indirect manner by using the conditioned taste aversion (CTA) paradigm (see also Chapter 23). This paradigm is based on the association that animals form between the taste of the food or drink they have ingested and a subsequent episode of illness (Garcia, Hankins, & Rusiniak, 1974). In a typical experiment, rats are trained to drink their daily allocation of water during a 30-min presentation of the water bottle. On the day of conditioning, they are presented with a solution of saccharin instead of water and they are subsequently injected with a toxic agent. After recovery, they are presented with the saccharin solution that was paired with the poisoning episode, either alone or concurrently with water. Conditioned animals refrain from drinking the saccharin solution and the amount of saccharin drunk is an indirect measure of the intensity of the previously experienced sickness.

Although the CTA procedure is often used to demonstrate sickness-inducing properties of chemical substances, there is evidence that drugs of abuse, such as amphetamine and morphine, produce CTA at doses that are self-administered and therefore considered as rewarding (Goudie, 1987). This paradox somewhat limits the usefulness of this procedure for detecting the sickness-inducing properties of substances.

Another way of assessing sickness is to look for alterations of innate and learned behavior. For example, the sickness developed by morphine-dependent rats on challenge with naloxone has been measured in a very precise way by disruption of operant responding in animals trained to press a lever for food in a Skinner box (Babbini, Galardi, & Bartoletti, 1972). This measure is much more sensitive and specific than the physical symptoms that are often used to assess withdrawal (e.g., body weight and wet dog shakes). The choice of the behavioral endpoints to be used as indicators of cytokine-induced sickness can be guided by what is already known about the nature of fever symptoms (Hart, 1988). The reduction in body care activities leading to the scruffy-looking hair coat that is characteristic of sick animals is not easy to quantify. Changes in general activity, feeding behavior, and social interactions are easier to assess using automated recording of behavior or direct observation.

12.5. Cytokines Induce Sickness Behavior

The first evidence in favor of a possible role of cytokines in sickness behavior came from clinical trials with purified or recombinant cytokines in the treatment of intractable viral diseases and cancer. Patients injected with these molecules were observed to develop not only flulike symptoms but also, on repetition of injections, acute psychotic episodes characterized by depression or excitation. Fortunately, these symptoms spontaneously regressed on cessation of treatment. Toxicological studies carried out on laboratory animals confirmed the neurotropic activity of recombinant cytokines such as IL-1β and TNFα. Animals injected acutely or chronically with these molecules usually appear lethargic, anorexic, and withdrawn from their environment (Kent, Bluthé, Kelley, & Dantzer, 1992b). The same effect is observed when animals are injected with LPS.

More objective studies of the sickness-inducing properties of cytokines have been based on the CTA paradigm and the observation of changes in locomotor activity, social activities, and food intake (Dantzer, Bluthé, Kent, & Goodall, 1993). In the CTA paradigm, systemic injection of rats with LPS after presentation of the new taste solution resulted in a decreased intake of this solution when it was presented later. Intraperitoneal (i.p.) or intracerebroventricular (i.c.v.) injections of IL-1β had the same effect (Tazi, Dantzer, Crestani, & Le Moal, 1988; Tazi, Crestani, & Dantzer, 1990). CTAs have also been established to TNFα. However, treatment with IFNα failed to induce CTA (Segall & Crnic, 1990). The reasons for these differences between IFNα and other proinflammatory cytokines are still unclear.

Adult animals presented with juveniles of the same species spontaneously investigate these social stimuli. This investigatory behavior remains at a high level as long as unfamiliar juveniles are presented. Systemic administration of LPS, IL-1β, and TNFα to the adult animal decreases the duration of social investigation (Bluthé, Crestani, Kelley, & Dantzer, 1992a; Bluthé, Dantzer, & Kelley, 1992b; Bluthé, Pawlowski, Suarez, Parnet, Pittman, Kelley, & Dantzer, 1994a). These effects are slow to appear, not developing before 2–4 h after injection. They last for 4–6 h and recovery is usually complete by 24 h. A typical dose–response curve for the effects of IL-1β on social behavior is shown in Fig. 2.

Systemic administration of IL-1 or TNFα consistently suppresses feeding and drinking. This effect has been observed using various measurements of food and water intake in ad libitum as well as deprived conditions (Kent, Bret-Dibat, Kelley, & Dantzer, 1995; Plata-Salaman, 1996a). A reduction in food intake can occur as a result of different mechanisms,

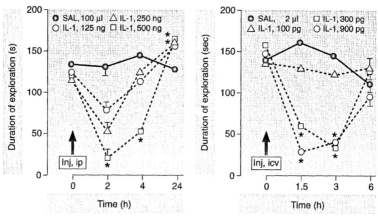

Figure 2. Effects of i.p. and i.c.v. IL-1β on social exploration in mice. Individually housed male mice were presented with a juvenile conspecific at time 0 and the duration of olfactory investigation of this social stimulus was assessed during a 4-min test. They were then injected with IL-1β and presented again with another juvenile at different time intervals. For clarity of representation, the SEM is presented only for the last point of each curve. *p < 0.05 compared with value at time 0.

so that all anorexic cytokines do not necessarily share the same mode of action. For example, low doses of IL-1β reduced both meal size and meal duration whereas higher doses also decreased meal frequency. In contrast, interferon reduced both meal size and meal duration but not meal frequency, whereas IL-8 decreased meal size exclusively (Plata-Salaman, 1996a). A direct comparison of the effects of administration of IL-1β on social exploration and food-motivated behavior in rats revealed that the time course of action differs according to the behavioral endpoint, as decreases in food intake typically develop within 1 h after injection (Fig. 3) (Kent et al., 1995).

An important component of sickness behavior is the increased somnolence that occurs during infectious episodes. The evidence in favor of a role of cytokines in this phenomenon came originally from the observation that the endogenous molecules responsible for the increased sleepiness induced by sleep deprivation in experimental animals are muramyl peptides, i.e., bacterial cell wall products (Krueger et al., 1982). Intravenous injection of IL-1β or TNFα in rabbits induces dose-related increases in the amount of time spent in slow-wave sleep (Krueger & Majde, 1994). In rats, the effect of IL-1β on the architecture of sleep is more complex and depends on the circadian phase and the dose used (Opp, Obál, & Krueger, 1991). For instance, low doses of IL-1β increased slow-wave sleep during the light period whereas higher doses increased wakefulness. In view of the potent somnogenic effects of cytokines, it could be argued that the behaviorally depressing effects of cytokines reflect nothing else than the intrusion of sleeping episodes in the time budget of sick animals. However, this does not appear to be the case, as LPS, at a dose that decreased the rate of responding in rats tested in an operant conditioning procedure in which the presentation of food was contingent on the intrusion of an operant lever in the cage irrespective of their behavior, did not alter food intake (Aubert, Vega, Dantzer, & Goodall, 1995b). In the same manner, observation of IL-1β-treated rats revealed that when they were exposed to a juvenile, they still responded to the juvenile when it came in contact but did not follow it.

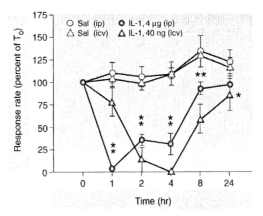

Figure 3. Effects of i.p. and i.c.v. IL-1β on food-motivated behavior in rats. Adult male rats were deprived to 85% of their free-feeding body weight before being trained to press a bar for a 45-mg food pellet in a Skinner box on a fixed ratio 10 schedule (i.e., one food pellet for every ten presses). Each rat served as its own control. Each test session lasted 5 min. Animals were injected with IL-1β i.p. (4 μg/rat, $n = 5$) or i.c.v. (40 ng/rat, $n = 5$) and physiological saline (1 ml i.p., 1 ml i.c.v.). Separate groups of rats were used for i.p. and i.c.v. injections. All injections were given immediately after the first session (time 0). Data are expressed as percentage of preinjection response rate. Vertical bars represent mean + SEM. *$p < 0.05$; **$p < 0.01$ compared with respective control values. From Kent *et al.* (1995).

One of the cardinal symptoms of inflammation is pain. Because of the key role played by cytokines in inflammation, their effects on pain sensitivity have been assessed using different experimental paradigms. Hyperalgesia in response to IL-1β has been observed in four different model systems: the rabbit isolated ear perfusion model (Schweizer, Feige, Fontana, Muller, & Dinarello, 1988), the paw pressure test in rats (Ferreira, Lorenzetti, Bristow, & Poole, 1988), the tail-flick test, and the formalin test (Wiertelak, Smith, Furness, Mooney-Heiberger, Mayr, Maier, & Watkins, 1994). In the rabbit isolated ear perfusion model, IL-1 enhanced the blood pressure effects induced by acetylcholine, which is typical of proalgic compounds. Intraplantar injection of IL-1 also resulted in an increased sensitivity to paw pressure (Fukuoka, Kawatani, Hisamitsu, & Takeshige, 1994). This effect was not restricted to the site of injection, as it was also observed in the contralateral paw. A similar hyperalgesic effect was also observed in the tail-flick test and the formalin test. Rats injected with i.p. LPS or IL-1β displayed prolonged hyperalgesia, which developed within 5–10 min following injection and lasted for at least 1 h. Although all of these results are strongly suggestive of an enhancing influence of cytokines on pain sensitivity, analgesia following cytokine injection has also been observed using the hot plate test and the phenylquinone writhing test. A possible explanation for these contradictory effects is the difference in time of testing after injection rather than the technique used to assess pain sensitivity. In a systematic study of the time course of the modulatory influence of LPS on pain sensitivity, using the hot plate test, LPS was found to first increase pain sensitivity and this effect was followed by an analgesic response beginning at 2 h and disappearing at 30 h after administration (Yirmiya, Rosen, Donchin, & Ovadia, 1994). These analgesic effects were certainly mediated by the release of endogenous opioids, as they were blocked by administration of naltrexone.

The effects of cytokines on nociception can also be observed when cytokines are ad-

ministered centrally. In accordance with the effects observed at the periphery, i.c.v. administration of IL-1β produced thermal hyperalgesia in rats (Watkins, Wiertelak, Goehler, Smith,
Martin, & Maier, 1994b). These effects appear to be of central origin, as administration of
the same dose of IL-1β at the periphery was totally ineffective. However, hyperalgesia in
response to central injections of cytokines is not always constant. Analgesia can also be
obtained when IL-1 and TNFα are administered intracerebroventricularly or into specific
brain nuclei. Whether hyperalgesic or analgesic effects develop depends on the dose and the
exact site of administration of the cytokine under study. For example, Sacerdote, Bianchi,
Ricciardi-Castagnoli, and Panerai (1992) reported that doses of IL-1β greater than 5–10 ng
were ineffective in inducing analgesia; Oka, Oka, Hosoi, Aou, and Hori (1995) observed
that IL-1β was hyperalgesic when it was injected into the preoptic nuclei of the hypothalamus whereas it was analgesic when injected into the ventromedial nucleus.

12.6. Motivational Aspects of Sickness Behavior

Sickness behavior is usually viewed by physicians as the result of debilitation and
physical weakness that inevitably occur in an organism whose total resources are engaged
in a defensive process against pathogens. An alternative hypothesis is that sickness behavior
is the expression of a highly organized strategy that is critical to the survival of the organism.
If this is the case, then it follows that sick individuals should be able to reorganize their
behavior depending on its consequences and the internal and external constraints to which
they are exposed. This flexibility is characteristic of what psychologists call *motivated
behavior*. Motivation can be defined as a central state that reorganizes perception and action.
In order to escape a potential threat, a fearful individual must be attentive to everything that
can occur in his environment to be able to exhibit at the right time the most appropriate
defensive behavioral pattern that is available in his behavioral repertoire. In other words, a
motivational state does not trigger an unflexible behavioral pattern. On the contrary, it un-
couples action from stimulus conditions and enables the selection of the appropriate strategy
depending on the eliciting situation.

The first evidence that sickness behavior is the expression of a motivational state rather
than a consequence of weakness was provided by Miller (1964). Working on the mechanisms of thirst, he was struck by the observation that thirsty rats injected with endotoxin
stopped bar pressing for water but, when given water, drank it although to a lesser extent
than normally. This effect was not specific to the thirst motivation, as the endotoxin
treatment also reduced the amount of food eaten and even blocked responding in rats trained
to press a bar for the rewarding effects of electrical stimulation in the lateral hypothalamus
(self-stimulation). Interestingly enough, when rats were trained to turn off an aversive electrical stimulation in this brain area, endotoxin also reduced the rate of responding, but to a
lesser extent than bar pressing for a rewarding brain stimulation. More to the point, however,
was the observation that rats that were placed in a rotating drum that they could stop for brief
periods by pressing a lever increased their response rate in response to endotoxin treatment.

The mere fact that endotoxin treatment decreased or increased behavioral output, depending on its consequences, gives strong support to the motivational interpretation of the
behavioral effects of such a treatment. The problem, however, with the results reported by
Miller (1964) is that they are anecdotal, no report of the original data having ever been
published.

To test the motivational interpretation of sickness behavior, Aubert and colleagues
(Aubert, Goodall, & Dantzer, 1995a; Aubert, Goodall, Dantzer, & Gheusi, 1997a; Aubert,

Kelley, & Dantzer, 1997b; Aubert, personal communication) engaged more recently in an extensive series of experiments. In one experiment (Aubert, personal communication), rats were trained to work for food on a progressive ratio schedule. In the schedule that was chosen for this particular experiment, rats had to press the lever once to get the first pellet, twice to get the second pellet, four times to get the third pellet, eight times for the fourth pellet, and so on. This means that food became more and more difficult to obtain with time. Normally, animals stop pressing the lever when the ratio becomes too high relative to its motivation for food. If food motivation is decreased, the achieved ratio is smaller whereas if food motivation increases, the achieved ratio is larger. When well-trained rats had to work for food in a cold environment (4–5°C), some of them increased their response rate whereas other animals responded less. These changes correspond to two different thermoregulatory strategies, consisting of either increasing energy intake to produce more heat, or stopping lever pressing and adopting a curled posture to decrease energy expenditure and minimize thermal losses. The important result is that when rats were injected with a small dose of IL-1β directly into the lateral ventricle of the brain, they responded to this immune signal in exactly the same way as when they were exposed to cold, e.g., those individuals that enhanced their response rate in the cold enhanced their response rate following IL-1β and vice versa.

Another important characteristic of a motivational state is that it competes with other motivational states for behavioral output. It is normally not possible to search for food and court a potential sexual partner at the same time because the behavioral patterns of foraging and courtship are not compatible with each other. The normal expression of behavior therefore requires a hierarchical structure of motivational states that is updated contingent on urgencies that occur in the internal and external milieus. When an infection occurs, the sick individual is at a life-or-death juncture and its physiology and behavior must be altered so as to overcome the disease. However, this is a relatively long-term process that needs to give way to more urgent needs when necessary. To make an analogy, it is obvious that if a sick woman lying in her bed hears a fire alarm ringing in her house and sees flames and smoke coming out of the basement, she will momentarily overcome her sickness behavior to try to escape danger. In motivational terms, fear competes with sickness and, in behavioral terms, fear motivated behavior overrides sickness behavior. An example of this competition between sickness and other motivational states is given by the observation that the depressing effects of LPS on behavior of lactating mice were abrogated when LPS-treated mothers had to retrieve their pups after they had been removed from the nest and scattered throughout the home cage (Aubert et al., 1997a). Additionally, LPS-treated mothers also engaged in nest building when the motivation to build a nest was increased by placing the mother with her litter at an ambient temperature of 6°C, whereas nest building did not occur when the animals were tested at 22°C.

From an adaptive point of view, the anorexic effect of cytokines is difficult to reconcile with the pyrogenic activity of these molecules. The decrease in food intake that accompanies fever appears to be inconsistent with the enhanced energy requirement of thermogenesis. To resolve this paradox, it has been proposed that cytokine-induced anorexia spares energy required for foraging and prevents a weakened organism from running the risk of being exposed to a predator during the search of food. From this perspective, it can be predicted that cytokines should be more effective in suppressing the foraging than the consummatory components of food intake. This appears to be the case, for, as previously mentioned, LPS- or IL-1-treated animals stopped pressing a lever for food but still ate the food pellets that were delivered independently of their behavior.

Food intake can be affected by cytokines not only in a quantitative but also in a qualitative way. It has been amply demonstrated that when rats are given the opportunity to select components of their diet, their selection pattern reflects the organism's nutritional and energetic requirements. To determine whether this selection pattern was altered during sickness, rats were submitted to a dietary self-selection protocol in which they had free access to carbohydrate, protein, and fat diets for 4 h a day (Aubert *et al.*, 1995a). After a 10-day habituation to this regimen, they were injected with LPS or IL-1β. Under the effect of this treatment, they decreased their total food intake but reorganized their self-selection pattern so as to ingest relatively more carbohydrate and less protein, whereas fat intake remained unchanged. This change in macronutrient intake contrasted with the increased fat intake that occurs in rats exposed to cold. Although eating fat would be a better way for feverish animals to fulfill their increased energy requirements, it would not be of much use, as cytokines have profound metabolic effects resulting, among others, in increased lipolysis and hypertriglyceridemia. Under these conditions, an increased intake of fat would actually be counterproductive, as it would further enhance hyperlipidemia without positively contributing to lipid metabolism.

12.7. Mechanisms of the Behavioral Effects of Cytokines

12.7.1. Role of Endogenous Cytokines

Proinflammatory cytokines act in a cascade fashion, each cytokine being able to induce its synthesis and the synthesis of other cytokines. Because of the pleiotropism and redundancy of the cytokine network, it is important to determine which cytokine contributes to sickness behavior and whether different cytokines mediate different components of sickness behavior. In order to answer these questions, it is not sufficient to test the effect of different cytokines on various behavioral patterns. It is also necessary to block the production or action of the specific cytokine that is under investigation. As pointed out earlier, administration of LPS induces the synthesis and release of a wide number of proinflammatory cytokines. In rats injected with LPS, depression of social exploration was attenuated by i.p. administration of the specific antagonist of IL-1 receptors, IL-1ra (Fig. 4) (Bluthé *et al.*, 1992b). However, the same pretreatment had no effect on the LPS-induced decrease in food motivation (Kent *et al.*, 1995). These results indicate that peripheral IL-1 is the main mediator of the social component of sickness behavior but that other cytokines mediate the alterations in food intake that occur in sick individuals. Concerning the possible involvement of IL-1 in the effects of other cytokines on social behavior, the observation that injection of TNFα mimicked the depressing effect of IL-1β on social exploration and synergized with IL-1β to induce sickness behavior led to the investigation of the effect of IL-1ra on the behavioral effect of TNFα (Bluthé *et al.*, 1994a). Pretreatment with IL-1ra abrogated the behavioral effect of TNFα, which can be interpreted to suggest that TNFα acts on social exploration by inducing the production of IL-1. The importance of IL-1β in the sickness-induced decrease in social behavior must not, however, be overestimated. Mice in which the gene coding for the type I IL-1 receptor had been deleted by homologous recombination no longer displayed a decrease in social investigation in response to IL-1β but were still fully responsive to LPS (Dantzer, Bluthé, Gheusi, Cremona, Layé, Parnet, & Kelley, 1998). In the same manner, mice lacking the gene coding for interleukin-1β-converting enzyme, responsible for processing the biologically inactive precursor of IL-1β into the biologically active mature form, were still found to be sensitive to the depressing effects of systemic LPS on food intake (Burgess, Gheusi, Yao, Johnson, Dantzer, & Kelley, 1998).

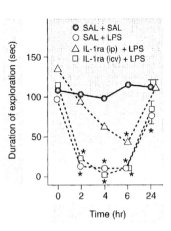

Figure 4. Effects of IL-1ra on LPS-induced depression of social exploration in rats. Rats were injected with either saline or IL-1ra followed by saline or LPS. Four experimental groups were used: saline (i.p.) + saline (i.p.) ($n = 5$); saline (i.p.) + LPS (250 μg/kg i.p.) ($n = 6$); IL-1ra (3 mg/rat i.p.) + LPS (250 μg/kg i.p.) ($n = 5$); IL-1ra (60 μg/rat i.c.v.) + LPS (250 mg/kg i.p.) ($n = 7$). Injections were administered immediately after the first session, in which their interest toward a juvenile conspecific was measured, and animals were tested again after 2, 4, 6, and 24 h. SEM is shown only for the last point. $*p < 0.05$ compared with values at time 0. From Bluthé *et al.* (1992b).

12.7.2. Central versus Peripheral Sites of Action of Cytokines on Behavior

Because cytokines are released locally in the body during the course of infection or inflammation, the question of how these molecules act on the brain has long been the subject of much debate. Cytokines are relatively big proteins (about 150 amino acids for IL-1α and IL-1β) and hydrophilic, and therefore cannot cross the blood–brain barrier. For this reason, they are believed to act on those brain sites that lack a blood–brain barrier, known as circumventricular organs because of their spatial location close to the brain ventricles. There, cytokines would trigger the synthesis and release of prostaglandins of the E series, which would then freely diffuse to the target neuronal cells. This mode of action has been proposed to account for both the corticotropic and pyrogenic effects of cytokines (Katsuura, Arimura, Koves, & Gottschall, 1990; Stitt, 1985).

The problem with this hypothesis is that it does not account for the previously mentioned observation that cytokines are present with their receptors in the brain and their local expression is modulated by peripheral immune stimuli. There are several reasons to propose that central cytokines are responsible for the behavioral effects of peripherally released cytokines. The most obvious one is that much lower doses of IL-1β or TNFα are needed, when injected directly into the lateral ventricle of the brain or specific brain structures, to depress feeding behavior and social exploration than the doses that are required when the same cytokine is injected at the periphery (Kent *et al.*, 1992b). The ratio is usually 1 to 100 or 1000 (Figs. 2 and 3). Another finding favoring the possibility of a central site of action is the demonstration of the blockade of the behavioral effects of peripherally injected IL-1β by a central injection of the specific antagonist of IL-1 receptors, IL-1ra (Kent, Bluthé, Dantzer, Hardwick, Kelley, Rothwell, & Vannice, 1992a). In this experiment, rats were pretreated with an i.c.v. dose of IL-1ra sufficient to block the depressing effects of i.c.v. IL-1β on social exploration and food-motivated behavior. This pretreatment completely abrogated the reduction of social exploration induced by i.p. administration of IL-1β and attenuated the decrease in response rate that occurred in IL-1β-treated rats trained to get their food by pressing a lever in a Skinner box (Fig. 5). These results indicate that IL-1 administered at the periphery acts in the brain via classical IL-1 receptors, but they do

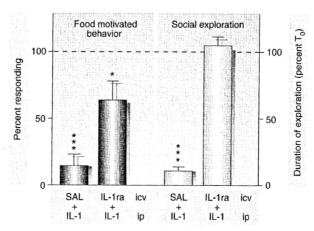

Figure 5. Differential effects of blockade of central IL-1 receptors on the effects of IL-1β on food-motivated behavior and social exploration in rats. IL-1ra or saline was injected intracerebroventricularly into rats trained to press a lever for food on a fixed ratio 10 schedule (24 μg of IL-1ra /rat) or into rats presented with a juvenile conspecific (4 μg of IL-1ra/rat). This injection was followed by i.p. IL-1β (4 μg/rat) or saline. Injections were given immediately after the first test session and animals were tested again 1 h (food-motivated behavior) or 2 h (social exploration) later. The figure represents percentage of variation with regard to baseline values. *p < 0.05; ***p < 0.001. Note that pretreatment with IL-1ra blocked the effects of IL-1β on social exploration but only partially attenuated the effects of this cytokine on food-motivated behavior. From Kent et al. (1992a).

not allow a determination of where the centrally acting IL-1 comes from, i.e., whether it is imported from the periphery or produced locally in the brain. The demonstration that peripheral administration of LPS induces the expression of cytokines in the brain (Gatti & Bartfai, 1993; Layé et al., 1994) is strongly suggestive of the second possibility.

Based on these findings, Kent et al. (1992b) proposed that cytokines released at the periphery activate primary sensory neurons and that this neuronal activation conveys the peripheral immune message to the brain, where it results in the synthesis and release of a functional pool of cytokines. In accordance with this hypothesis, i.p. LPS was found to increase the levels of sensory neuropeptides (substance P, neurokinin A, and calcitonin gene-related peptide) in the spinal cord (Bret-Dibat, Kent, Couraud, Creminon, & Dantzer, 1994) and induce the expression of the cellular immediate-early gene c-fos in various areas of the brain (Wan, Wetmore, Sorensen, Greenberg, & Nance, 1994). Inducible expression of the c-fos gene has been validated as a widely applicable marker for neural systems activated by a variety of extracellular stimuli. The protein product Fos of this gene interacts with nuclear proteins to act as a transcription factor. Fos immunoreactive neurons were identified in the primary projection area of the afferent branches of the vagus nerve, represented by the nucleus tractus solitarius, and secondary projection areas such as the parabrachial nucleus, the paraventricular nucleus, and the supraoptic nucleus of the hypothalamus. Furthermore, transection of the vagus nerve at the subdiaphragmatic level, which eliminates afferent neurons originating from the liver and the gastrointestinal tract, abrogated LPS-induced Fos immunoreactivity in these brain areas.

The functional nature of this vagal communication pathway between the immune system and the brain was evidenced by the demonstration that section of the vagus nerve abrogated LPS-induced hyperalgesia in rats (Watkins, Wiertelak, Goehler, Mooney-

Heiberger, Martinez, Furness, Smith, & Maier, 1994a) as well as LPS- and IL-1β-induced decreases in social exploration (Bluthé, Walter, Parnet, Layé, Lestage, Verrier, Poole, Stenning, Kelley, & Dantzer, 1994b) and food-motivated behavior (Bret-Dibat, Bluthé, Kent, Kelley, & Dantzer, 1995) in rats and mice (Fig. 6). The vagus nerve transmits to the brain not only the signals that are responsible for the behavioral effects of cytokines but also those that are responsible for fever and pituitary–adrenal activation, as vagotomy has been shown to block the two latter responses to LPS (Fleshner, Goehler, Hermann, Relton, Maier, & Watkins, 1995; Watkins, Goehler, Relton, Tartaglia, Silbert, Martin, & Maier, 1995a). In accordance with such a wide range of effects of vagotomy on the brain actions of peripheral immune stimuli, this surgical procedure was found to result in the abrogation of the induction of IL-1β in the brain at the mRNA and protein levels in response to peripherally injected LPS (Layé, Bluthé, Kent, Combe, Médina, Parnet, Kelley, & Dantzer, 1995).

Although all of these findings are strikingly suggestive of a neural pathway mediating the effects of peripheral cytokines in the brain, other interpretations are possible. In particular, section of the vagus nerve might alter in a nonspecific way the sensitivity of circumventricular organs to circulating cytokines because of the close anatomo-functional relationships between one of these circumventricular organs, the area postrema, and the primary projection area of the vagus nerve, the nucleus tractus solitarius. Another possibility is that the neural reorganization that takes place in the projection areas of the vagus nerve decreases the sensitivity of brain cell targets to cytokines. If the consequences of vagotomy on the behavioral effects of cytokines occur by such mechanisms, vagotomized animals should be less sensitive to the effects of cytokines that are administered by other routes of injection than the intraperitoneal route. However, section of the vagus nerve did not attenuate the behavioral effects of IL-1β when this cytokine was injected into the lateral cerebral ventricle of the brain (Bluthé, Michaud, Kelley, & Dantzer, 1996a), or by

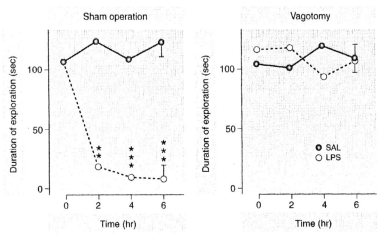

Figure 6.   Effects of vagotomy on LPS-induced decreases in duration of investigation. LPS (1.25 mg/kg) or physiological saline was injected into vagotomized and sham-operated rats immediately after the first behavioral session (time 0) and the same animals were tested again with different juveniles 2, 4, and 6 h later ($n = 4$ for each experimental group, except LPS-treated sham animals for which $n = 3$). **$p < 0.01$, ***$p < 0.001$ compared with saline values. From Bluthé *et al.*, 1994b).

systemic routes other than intraperitoneally (Bluthé, Michaud, Kelley, & Dantzer, 1996b). These results are important because they confirm the role of vagal afferent nerves in the transmission of the immune message from the periphery to the brain and they show that the vagus nerve conveys specific information concerning cytokines injected into the abdominal cavity. The mechanisms that are responsible for the transformation of the immune message into a neural message at the periphery, and the transduction of this neural message back into an immune message in the central nervous system still need to be worked out (Dantzer, 1994), but they provide a fascinating example of communication between the brain and the periphery (Fig. 7).

12.7.3. Role of Corticotropin-Releasing Factor in the Behavioral Effects of Cytokines

Cytokines have potent activating effects on the pituitary–adrenal system that are mediated via the release of CRF. The possible involvement of this hypothalamic neuropeptide in the neural effects of IL-1 has been assessed by i.c.v. administration of CRF antiserum or a peptide known as α-helical CRF(9–41) and behaving as an antagonist of CRF receptors [ahCRF(9–41)].

Immunoneutralization of endogenous CRF in the brain attenuated the anorexic effects of IL-1ß (Uehara, Sekiya, Takasugi, Namiki, & Arimura, 1989). The same treatment blocked the reduction of immobility induced by i.c.v. IL-1β in rats forced to swim in a confined space (Del Cerro & Borrell, 1990). Central adminstration of ahCRF(9–41) was also able to prevent

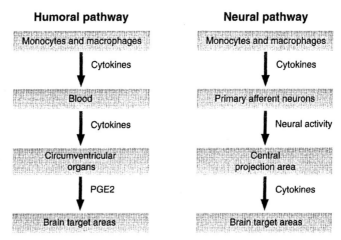

Figure 7. Possible communication pathways between the immune system and the brain. In the humoral interpretation of the effects of cytokines on brain functions, cytokines are released into the general circulation by activated monocytes and macrophages. They reach circumventricular organs where they induce the synthesis and release of prostaglandins, which then freely diffuse into the brain parenchyma to reach cell targets mediating the various neural effects of cytokines. In the neural interpretation of the effects of cytokines on brain functions, locally released cytokines activate peripheral afferent nerves. The subsequent changes in firing rate are transmitted to the brain where they induce the synthesis and release of cytokines by resident macrophages and microglial cells. The cell targets for the cytokines that are produced in the brain are still unknown.

the IL-1β-induced reduction of exploratory behavior of mice placed in a multicompartment chamber (Dunn, Antoon, & Chapman, 1991).

Although these findings can be interpreted to suggest that brain CRF mediates the behavioral effects of IL-1, there is evidence that contradicts this conclusion. In particular, the decrease in food-motivated behavior that was induced in rats by i.p. injection of IL-1β was not altered by i.c.v. administration of either ahCRF(9–41) or CRF itself (Bluthé, Dantzer, & Kelley, 1989). When IL-1β was injected centrally, ahCRF(9–41) did not alter the peak effect but facilitated the return to baseline (Bluthé *et al.*, 1992a).

The exact factors that are responsible for these differences are unknown. However, it is clear that the involvement of CRF in the behavioral effects of IL-1 is not a general phenomenon.

12.7.4. Role of Prostaglandins in the Behavioral Effects of Cytokines

IL-1 and other proinflammatory cytokines are potent inducers of the COX-2 gene that codes for the inducible form of cyclooxygenase, a key enzyme in the synthesis of prostaglandins. Administration of cyclooxygenase inhibitors such as indomethacin, at doses that abolish synthesis of prostaglandins, attenuates the pyrogenic and anorexic effects of IL-1. Pretreatment with indomethacin or piroxicam blocked the depressing effects of i.p. administration of IL-1β on food-motivated behavior in rats and social exploration in mice (Crestani, Seguy, & Dantzer, 1991). Whether these effects are mediated peripherally or centrally still needs to be elucidated. Central administration of ibuprofen failed to block the anorexia induced by centrally administered IL-1β although it attenuated the increase in body temperature caused by central IL-1β (Shimizu, Uekara, Shimomura, & Kobayashi, 1991). In contrast, i.p. injection of ibuprofen partially blocked the anorexic effects of i.c.v. IL-1β.

In rats continuously infused with recombinant murine IL-1α, piroxicam completely inhibited the stimulation of drinking behavior, but had no effect on the reduction in eating activity and locomotor activity induced by this cytokine (Otterness, Golden, Seymour, Eksra, & Daumy, 1991). Pretreatment with indomethacin had no effect on the depression of general activity and food intake induced in mice by peripheral injection of IFNα (Crnic & Segall, 1992).

In the face of these contradictory results, there is clearly a need for further studies on the role of prostaglandins in the behavioral effects of cytokines.

12.7.5. Role of Nitric Oxide in the Behavioral Effects of Cytokines

The sustained vasodilatation and hypotension induced by IL-1 and other proinflammatory cytokines are mediated by the local synthesis and release of nitric oxide (NO) via induction of an NO synthase in both endothelial and vascular smooth muscle cells. In addition to its potent vasodilatory activity, NO acts as an effector molecule in immunological reactions, and as a neurotransmitter in the central and peripheral nervous system.

Administration of agents that block synthesis of NO from L-arginine attenuated the dramatic fall in blood pressure that occurs in septic shock or in response to exogenous cytokines. In the same manner, inhibition of activity of NO synthase attenuated the somnogenic effects of cytokines. However, the same treatment had no effect on fever.

To test whether NO production is also involved in the behavioral effects of IL-1β, mice were pretreated with various doses of *N*-nitro-L-arginine methyl ester (NAME), a selective inhibitor of the brain and endothelial NO synthases (Bluthé, Sparber, & Dantzer, 1992c). Administration of a high dose of the antagonist potentiated the depressing effects of IL-1β

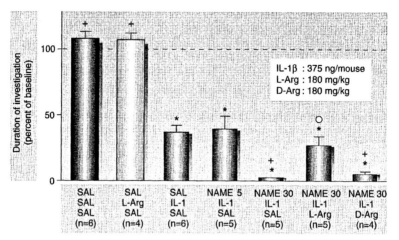

Figure 8. Potentiation of IL-1β-induced changes in social exploration by pretreatment with NAME, an inhibitor of NO synthase, and the effects of L-arginine or D-arginine. Each column represents the mean variation measured 4 h after injection of IL-1β. NAME was injected at a dose of 5 or 30 mg/kg. Previous experiments had established that these doses of L-NAME had no effect on social exploration. The number of mice in each group is given in parentheses. $*p < 0.05$ compared with saline; $+p < 0.05$ compared with IL-1; $\circ$, $p < 0.05$ compared with IL-1 plus NAME group. From Bluthé et al. (1992c).

on social exploration whereas a lower dose had no effect. This potentiation was the result of inhibition of the synthesis of NO, as it was attenuated by L-arginine but not by D-arginine (Fig. 8). Administration of the nitroarginine derivative had no effect on the weight loss induced by IL-1β. Although these results indicate that NO production might have a protective role on the effects of IL-1 on brain functions, they need to be complemented by further studies assessing the effects of IL-1 on brain NO synthases.

12.7.6. Other Mediators of the Behavioral Effects of Cytokines

Cholecystokinin (CCK) is a peptide that plays an important role in the regulation of food appetite and gastric emptying. CCK is released by the small intestine during feeding. It acts on CCK receptors in the gut, which stimulate centers in the brain that control feeding behavior via vagal nerve efferents, resulting in cessation of food consumption. CCK receptors are of two types: type A, which is located primarily in the gut, and type B, located mainly in the brain. The anorexic effects of CCK appear to be mediated via type A receptors.

Because the anorexic effects of IL-1 resemble those of CCK, the role of CCK in IL-1-induced anorexia has been investigated indirectly by measuring the effects of IL-1 on plasma levels of CCK, and directly, by administering a CCK-A receptor antagonist (L364,718) before injection of IL-1α. Administration of IL-1α significantly increased plasma levels of CCK and decreased food intake and gastric emptying in rats. The latter effects were partially blocked by pretreatment with L364,718 (Daun & McCarthy, 1993).

The results of experiments using CCK receptor antagonists are not all that clear. In a systematic series of studies carried out to assess the role of CCK in the behavioral effects of cytokines, peripheral or central administration of a CCK-A or CCK-B receptor antagonist did not alter the depressing effects of LPS and IL-1β on social investigation and food-motivated behavior in mice (Bluthé, Michaud, Kelley, & Dantzer, 1997). The same negative results were obtained when mice and rats were pretreated with capsaicin, a neurotoxic drug that depletes substance P and selectively destroys C-fiber afferents (Bret-Dibat, Creminon, Couraud, Kelley, & Dantzer, 1997). These results indicate that the behavioral effects of cytokines are unlikely to be mediated by a single sensory neuropeptide.

Several other peptides have been putatively involved in the behavioral actions of cytokines. It is particularly the case for leptin, a feeding inhibitory factor, and neuropeptide Y (NPY), a feeding stimulatory factor. Leptin is a product of the ob gene that is expressed specifically in white adipose tisue and plays a key role in energy regulation. Because administration of TNF, IL-1, and LPS increases serum leptin levels and leptin mRNA expression in adipocytes, it has been proposed that this peptide mediates the anorexic and cachexic effects of proinflammatory cytokines (Sarraf, Frederich, Turner, Ma, Jaskowiak, River, Flies, Lowell, Fraker, & Alexander, 1997). With leptin inhibiting the synthesis and release of NPY, whereas IL-1β opposes the stimulating effects of NPY, it is possible that the interactions between leptin, NPY, and IL-1 represent a critical interface mechanism for the regulation of feeding in health and disease (Plata-Salaman, 1996b).

12.8. Opposition of the Behavioral Effects of Cytokines by Cryogens

Concurrent with their effects on brain functions, cytokines trigger neurotransmitter mechanisms that are part of a feedback loop regulating the neurotropic effects of cytokines and promoting recovery. Because these neurotransmitters were originally identified on the basis of their ability to oppose the pyrogenic effects of cytokines, they were named *endogenous antipyrogens* or *cryogens* (Kluger, 1991).

The neuropeptide vasopressin (VP) is a key factor in regulation of water metabolism. It is produced by magnocellular hypothalamic neurons and accumulates in the terminals of these neurons in the posterior pituitary. It is released in the general circulation and acts as an antidiuretic hormone in conditions of water deprivation and in response to fever. VP can also act as a neurotransmitter in the brain: It is present particularly in neurons of which the cell bodies are located in the bed nucleus of the stria terminalis (BNST) and the terminals project to the lateral septum. This vasopressinergic pathway is highly sensitive to circulating androgens in rodents. Castration leads to a dramatic decrease in the content of VP mRNA in the BNST neuronal cell bodies and to a reduction in immunoreactive VP in the terminal areas of the septum. This androgen-dependent pathway is also activated during fever.

Central administration of VP attenuated the depressing effects of centrally injected IL-1β on social exploration. Conversely, central injection of an antagonist of vasopressin receptors, which has no biological activity on its own but prevents endogenously released vasopressin from reaching its receptors, sensitized rats to the behavioral effects of IL-1. The latter results are important as they suggest that endogenous VP plays a physiological modulatory role in the behavioral effects of IL-1. To determine whether this phenomenon is mediated by an androgen-dependent or -independent vasopressinergic pathway, castrated male rats were compared with intact male rats in their sensitivity to the modulatory role of the VP receptor antagonist. Castration by itself potentiated the depressing effects of IL-1β on social exploration. Central administration of VP was more effective in attenuating the behavioral effects of IL-1 in castrated than in intact male rats and, conversely, i.c.v. administration of the VP receptor antagonist was no longer active in potentiating the behavioral effects of IL-1 in castrated male rats lacking vasopressinergic innervation of the lateral septum (Dantzer, Bluthé, & Kelley, 1991).

The mechanisms by which the brain vasopressinergic system is activated by cytokines and the way in which vasopressin interacts with the effect of cytokines on their target cells remain to be elucidated.

The potent activating effects of proinflammatory cytokines on pituitary–adrenal activity have already been mentioned. Glucocorticoids represent another class of key molecules in the regulation of sickness behavior. Adrenalectomy is accompanied by an increased sensitivity to the depressing effects of IL-1β and LPS on social exploration. This effect was mimicked by acute administration of the antiglucocorticoid antagonist RU-38486 to intact mice and it was abrogated by the implantation of a corticosterone pellet in adrenalectomized mice (Goujon, Parnet, Aubert, Goodall, & Dantzer, 1995a).

Adrenalectomized animals implanted with a corticosterone pellet have constant levels of corticosterone but are unable to respond to administration of IL-1 by enhanced pituitary–adrenal activity. This phasic response to IL-1 appears to be important in regulating the behavioral effects of cytokines, as the protection offered by a corticosterone pellet that ensured plasma levels of corticosterone intermediate between normal and stress levels was effective only against low but not high doses of IL-1β.

The effects of glucocorticoids on the behavioral activity of IL-1 and LPS are likely to be mediated by both peripheral and central mechanisms. Adrenalectomized mice were also more sensitive to the depressing effects of a central injection of IL-1β on social exploration, and i.c.v. administration of the glucocorticoid receptor antagonist RU-38486 mimicked this effect (Goujon, Parnet, Cremona, & Dantzer, 1995b). The mechanism underlying these effects is represented by the downregulation of glucocorticoids on the synthesis and release of cytokines by both peripheral immune cells and brain cells and by the upregulation of the type II IL-1 receptor, which acts as a decoy target (Goujon, Parnet, Layé, Combe, & Dantzer, 1996).

12.9. Pathophysiological Implications of the Behavioral Effects of Cytokines

The demonstration that the immune system is able to influence behavior and mental states has important implications on our understanding of the relationships between psychological factors and disease. In the case of cancer for example, such psychological features as the feelings of hopelessness and helplessness that are commonly associated with the onset and progression of the disease might be secondary to the brain effects of factors released from immune or tumor cells during the early stage of the tumoral process. The same possibility applies to the relationship between psychological factors and autoimmune diseases. The possible causal role of cytokines in the mental and behavioral symptoms that occur in various pathological conditions has hardly been investigated, except in a few cases, such as infection and fever, cachexia, AIDS dementia complex, chronic fatigue syndrome, and depression.

There is ample evidence that proinflammatory cytokines are responsible for the development of subjective and behavioral symptoms of sickness during infection with a bacterial or viral pathogen. For instance, patients treated with IFNα showed fever, anorexia, fatigue, headache, myalgia, and arthralgia. These symptoms culminated in lethargy and withdrawal from the surroundings. The same symptoms were observed in volunteers injected with low doses of LPS.

The possibility that the release of cytokines accounts for more subtle changes in cognition and performance has been assessed by Smith, Tyrrell, Coyle, and Higgins (1988). On the basis of earlier work showing that infection with upper respiratory tract viruses decreased the efficiency with which psychomotor tasks were performed, these researchers injected IFNα into volunteers of both sexes. Volunteers injected with the larger dose were significantly slower at responding in a reaction time task when they were uncertain when the target stimulus would appear. Simultaneously, they displayed hyperthermia and experienced feelings of illness. However, they were not impaired on a pursuit tracking task or syntactic reasoning task. These effects were similar to the alterations in performance observed in patients with influenza. The possibility that proinflammatory cytokines have relatively specific effects on cognitive processes has been further investigated in animal models. IL-1β, but not IL-6, impaired the learning of a spatial navigation task in rats (Oitzl, van Oers, Schöbitz, & De Kloet, 1993). However, the possible interference of sickness-induced altered performance with the task requirements was not systematically examined in this report. A similar deficit in spatial learning was observed in mice injected with IL-1β or infected with the pathogenic agent *Legionella pneumophilia* (Gibertini, Newton, Friedman, & Klein, 1995). Interference of cytokines with formation of new memories was also demonstrated in an autoshaping task in which rats learned to press a lever that was introduced into the cage before food delivery (Aubert *et al.*, 1995b). These effects of cytokines were independent of their pyrogenic activity, as they were observed whether body temperature increased or decreased in response to the treatment under study.

There has been much speculation on the possible pathogenic role of cytokines in the chronic fatigue syndrome (CFS). Always feeling tired is a common complaint in patients afflicted with a viral infection and represents the core symptom of the so-called postviral fatigue syndrome. CFS patients feel the same but in the absence of any persistent viral infection (Kendell, 1991). Their symptoms are real, pervasive, and often incapacitating. The fact that a substantial proportion of these patients fulfill criteria for major depression and other psychiatric illness does not facilitate the classification of this disorder. Whatever the case, and in view of the similarities between the subjective effects of cytokines and the symptoms reported by CFS patients, many researchers have looked for possible hyperproduction of cytokines in this condition. Elevated plasma levels of IL-2 and IL-6 have been reported in one study of CSF patients, but this result has not been found in other studies. Such inconsistent results can be easily explained by technical problems associated with the detection of cytokines in biological fluids and the poor correlation between plasma levels of cytokines and local activity of these mediators. A better way of assessing peripheral cytokine function is to study the ability of peripheral blood mononuclear cells (PBMC) to produce cytokines when put in culture and stimulated with LPS. Using such a strategy, Chao, Janoff, and Hu (1991) reported a hyperproduction of IL-6 and TNFα and a lower production of the inhibitory cytokine TGFβ in CFS patients. However, it is important to note that abnormalities, if any, of the cytokine network are not necessarily present at the periphery but might instead preferably affect cytokines that are expressed in the central nervous system. Although this possibility is much more difficult to put to test, it has inspired quite elaborate ad hoc speculation (Goldstein, 1993).

Besides their common occurrence in CFS patients, lack of energy and loss of interest are very frequent in depressed patients. These symptoms are actually incorporated in the basic description of depressive episodes. The 10th revision of the *International Classifica-*

tion of Disease begins, "the subject suffers from a lowering of mood, reduction of energy, and decrease in activity. Capacity for enjoyment, interest and concentration are impaired, and marked tiredness after even minimum effort is common." The possibility that activation of peripheral blood monocytes and T lymphocytes plays a role in the pathophysiology of major depression has been proposed by Maes, Smith, and Scharpe (1995b). In addition to the evidence pointing out the profound effects of cytokines on behavior and the hypothalamic–pituitary–adrenal axis, this hypothesis is based on the observation of an increased production of cytokines by monocytes and T lymphocytes of depressed patients. For example, patients with major depression have elevated plasma levels of acute-phase proteins and increased concentrations of IL-6 and its soluble receptor, with a close relationship between IL-6 levels and acute-phase proteins. However, more research is needed before a role of immune products in the pathogenesis of depressive symptoms can be accepted. The observed immune alterations might be a trait rather than a state marker of depression, as they persist even when depressive symptoms regress. In addition, the possible contribution of antidepressant treatment to the changes in immune functions observed in depressed patients remains to be established.

At the experimental level, the hypothesis of common mechanisms mediating the brain effects of cytokines and symptomatology of depression prompted a renewal of interest in the investigation of analogies between immune stressors and other stressors. The pattern of neurochemical changes that is observed in animals subjected to immunological challenge is reminiscent of that provoked by stressors (Dunn, Powell, Meitin, & Small, 1989; Zalcman, Shanks, & Anisman, 1991). In the same manner, the behavioral consequences of immunological activation share some similarity with those induced by stressors. This similarity is found even when the actions of cytokines in relatively specific models of animal depression are considered. For example, antigenic stimulation with sheep red blood cells decreased self-stimulation from the nucleus accumbens but not the substantia nigra in rats, at the time of the peak immune response, and this effect was similar to what is observed following exposure to inescapable electric shock (Zacharko, Zalcman, Macneil, Andrews, Mendalla, & Anisman, 1997). In the same manner, administration of LPS reduced saccharin preference in rats (Yirmiya, 1996). These similarities encouraged researchers to test the effects of antidepressant treatment on the behavioral effects of cytokines and, reciprocally, the effects of treatments blocking cytokine actions on animal models of depression. Chronic but not acute treatment with imipramine abrogated the LPS-induced decrease in saccharin preference (Yirmiya, 1996), whereas i.c.v. administration of IL-1ra blocked the impairing effects of acute exposure to a stressor on the ability to learn an escape response (Maier & Watkins, 1995). Although there is still a long way to go, these findings are strongly supportive of the hypothesis of common mechanisms between cytokine-induced sickness and some forms of depression.

Cachexia and anorexia are prominent features of chronic infection and cancer. PBMC isolated from cancer patients release more TNFα in response to stimulation than cells from healthy controls. Elevated serum levels of TNFα have also been reported in patients with a variety of malignancies. The possible involvement of TNFα in the metabolic changes occurring in malnutritional status has been deduced from the demonstration of an enhanced release of this cytokine by stimulated PBMC from short-term acutely starved subjects and nonstimulated PBMC from patients afflicted with anorexia nervosa. Anorexia, weight loss and wasting are important causes of morbidity in AIDS patients. The pathophysiology of the AIDS wasting syndrome includes three components: impaired nutrient intake, decreased nutrient absorption, and metabolic alterations (Grunfeld & Kotler, 1992). The possibility that cytokines play a key role in the AIDS wasting syndrome has been proposed on the basis of two different sets of findings: (1) HIV infection is accompanied by activation of several cytokine genes and (2) many characteristic features of the AIDS wasting syndrome are similar to those that are seen in other chronic infections and cancer, and that are associated with cytokines. Infection and activation of CD4$^+$ lymphocytes and macrophages by HIV are associated with the production of a range of cytokines, especially TNFα. Elevated levels of this cytokine are found in the plasma of HIV-infected patients. In addition, PBMC of HIV-infected individuals produce higher levels of TNFα. Other cytokines, including IL-1, IL-6, and TGFβ, are also hyperexpressed in HIV-1 infection.

HIV-1 is able to enter the central nervous system and infect macrophages and microglial cells (Vintners & Anders, 1990). However, the small number of infected cells contrasts with the extensive amount of functional alterations, of which the most well known is AIDS-related dementia. Many AIDS patients present a primary neuropsychiatric disorder, including slowness in thinking, forgetfulness, cognitive changes, drowsiness, weakness, and difficulties in concentrating. These neurological abnormalities occur in as many as 80% of all AIDS patients, cannot generally be attributed to opportunistic infections, and may even precede most other clinical signs of infection with HIV. Because HIV-1 is well known to induce the synthesis of TNFα, IL-1, IL-6, and IL-2 from microglia, it has been proposed that these cytokines contribute to the neurological symptoms observed in AIDS patients. Examination of brains from HIV-1 infected patients by immunocytochemistry revealed higher expression of IL-1 in endothelial cells and TNFα in macrophages and microglial cells. TGFβ has also been identified in HIV-1-infected brain and appears to be localized exclusively to tissue areas with pathological abnormalities (Vitkovic, 1997).

12.10. Conclusion

Sufficient evidence is now available to accept the concept that cytokines are interpreted by the brain as internal signals of sickness. Sickness can actually be considered as a motivation, that is, a central state that organizes perception and action in face of the threat of infection by pathogen microorganisms. A sick individual does not have the same priorities as a well one, and this reorganization of priorities is mediated by the effects of cytokines on a number of peripheral and central targets. The elucidation of the mechanisms that are involved in these effects should give new insight into the way sickness and recovery processes are represented in the brain.

13 Stress Effects on Immune Function in Rodents

Melissa K. Demetrikopoulos, Steven E. Keller,
and Steven J. Schleifer

13.1. Introduction

There is a vast literature examining the effects of stress on immune function in rodents. This chapter will provide an overview of this literature by describing the immunological changes

Melissa K. Demetrikopoulos • Institute for Biomedical Philosophy, Chamblee, Georgia 30341.   Steven E. Keller • Department of Psychiatry and Neuroscience, UMDNJ New Jersey Medical School, Newark, New Jersey 07103.   Steven J. Schleifer • Department of Psychiatry, UMDNJ New Jersey Medical School, Newark, New Jersey 07103.

that occur in response to stress and the mechanisms by which these changes are believed to occur. It begins with a description of various rodent models used to study stress effects of immunity to provide an appreciation for the types and breadth of experiments likely to be performed in this field. This is followed by more specific examples of stress-induced immunological changes. The final section of the chapter explores neuronal mechanisms by which this occurs.

13.2. Rodent Models Employed to Study Stress Effects on Immunity

Much research has examined stress effects on immunity using small mammalian, particularly rodent, subjects. Rodents are often chosen as subjects for this area of research because it is possible to develop easily employable stress models that can be utilized to understand mechanisms of stress-induced changes in immunity. In fact, there are several stress models that have been commonly utilized to study these phenomena. For purposes of simplification, these can be divided into three main categories, namely, physical stressors, social stressors, and psychological stressors. The reader should keep in mind that this distinction is somewhat arbitrary, as most of the models employed would be likely to contain aspects of each of these categories.

13.2.1. Physical Stressors

Probably all stress models have some aspect of a physical stressor. This is certainly the case in any of the models utilizing electrical shock or other physical stimulus regardless of the other aspects of the stressor given. However, most early studies relied almost exclusively on large-magnitude physical stressors such as 18-h intermittent tail shock, or forced cold swim testing. These were utilized initially because it was not known whether any stressor would have an immune effect. As the field of psychoneuroimmunology evolved, the magnitude of the physical stressors could be reduced and social and psychological stressors were introduced. Attempts have been made to control for the physical stressors that often accompany the various social and psychological stressors in order to describe those aspects of the stressor experience that account for the psychological components per se. Physical stressors that are most relevant and currently commonly employed in psychoneuroimmunological research tend to be more ethologically valid. These are likely to be encountered by humans and animals alike and include sleep deprivation, water or food restrictions, malnourishment, and physical exertion. Often, outside the laboratory, several of these physical stressors may be occurring simultaneously with social or psychological factors. Studying these factors individually in a controlled manner should help to determine the contribution of each during stressful periods and suggest possible interventions. These interventions would be designed to reduce the effects of a particular stressor on health by careful attention to the maintenance of a suitable physical environment during times of psychological stress such as death of a loved one, divorce, job loss, or even final examinations.

Another physical stressor commonly employed to study psychoneuroimmunological phenomena is surgical stress. While performed to improve the health of the individual, the surgery itself provides a stressor of very large magnitude that may have important immunological consequences. The stress of the surgery has many components such as anesthesia, tissue damage, blood loss, physical disturbance of organs, and postsurgical pain. Again, studying the contribution of these various components may suggest possible interventions aimed at maximizing health outcomes. This may be of critical importance for severely immunocompromised individuals and surgical treatment of immune dysfunction such as that occurring with cancer.

13.2.2. Social Stressors

Because rodents are social animals, it is possible to use a change in their social environment as an experimentally induced stressor. Social stressors are predominantly psychological stressors but they are perhaps a distinct enough class of stressors to warrant a separate classification. In addition, social stressors may have a minor physical stressor component, such as moving and handling the subjects by the investigator as well as differential physical relationships with cage mates as a result of the social manipulation. For example, the subject would receive differential physical contact with cage mates such as grooming and dominance displays.

A commonly employed social stressor is maternal separation. This is often scheduled prior to weaning for a short period of time. The level of physical distress varies depending on whether the dam is removed from the pups or individual pups are removed. When the dam is removed, the pups are likely to be less disturbed as they do not need to be handled by the investigator and are still in the company of their siblings, which provides some level of social support and a degree of physical warmth. However, both models contain the physical stress of temporary lack of feeding opportunity and a reduction in warmth provided by the dam. These manipulations have been shown to have both long- and short-term immunological consequences, which may even be evident in these subjects when measured during adulthood.

Another, more chronic, social stressor commonly employed is differential housing conditions. Rats prefer to be group housed and individual housing per se is demonstrated to be stressful. Although group housing is seemingly preferable, beyond a certain limit animals also experience crowding as a stressor. Thus, both isolation and crowding have been shown to be immunomodulatory. The effect of crowding is more severe when there is limited access to food and water but is still evident when there is ample access.

Yet another social stressor is an intruder. In such a model, a new individual is moved into an established cage. The social dynamics of this model can be quite complex such that the interaction that ensues when the intruder is introduced depends on the characteristics of both the resident and the intruder. For example, utilizing a female dam with pups as the resident will provide a substantially different social environment for an intruding young adult male than for a sexually receptive female. Clearly there is an almost endless number of combinations of social pairings that can occur with this paradigm. Studying different combinations allows one to model the role of various social conditions on immunity.

13.2.3. Psychological Stressors

Recent psychoneuroimmunological studies have primarily considered psychological stressors. As previously mentioned, social stressors may be thought of as a special class of psychological stressors. Other psychological stressors involve learned responses originally linked to physical stressors such as conditioned place aversion. This model allows one to utilize the associative aspects of the conditioned situation in the absence of a true physical stressor. The physical stressor can be dissected out of the equation and the contribution of the psychological distress can be assessed. Another psychological stress component commonly utilized is the controllability of a shock, with differential effects found for control-lable versus uncontrollable shocks. In this model, subjects receiving the controllable shocks serve as controls for their yoked partners which have no control over the shock. Because the psychological aspects would account for any differential effects found between these two groups, the contribution of psychological factors on immunological functioning can best be examined.

13.3. Stress-Induced Changes in Immunity

The interaction between stress and immunological changes has been well documented and has been implicated in a variety of human illnesses. However, given the variety of factors that necessarily confound investigations utilizing human subjects, including ethical considerations, it is more difficult to isolate specific aspects of the effects of stress on immunity in humans, as well as to investigate mechanisms of such effects. Therefore, animal models, particularly rodent models, have been instrumental in assessing the effects of stress on both *in vitro* and *in vivo* immune responses. This section will examine the effects of stress on lymphoid cells *in vitro* and describe studies that have examined *in vivo* immune responses. An enormous literature has explored these aspects of stress-induced changes in immunity, and illustrative examples will therefore be presented. These examples were chosen to demonstrate the pervasiveness of psychoneuroimmunological phenomena across a wide variety of experimental settings.

13.3.1. Effects on Lymphoid Cells

In vitro measures of immunity have been widely utilized to study the effects of stressors such as life events, examinations, and bereavement in human subjects. Studies in animal subjects have been conducted to elucidate the specific effects on lymphoid cells (Table 1) and the mechanisms by which these changes may occur (see Table 3). This literature has been reviewed (Keller, Schleifer, & Demetrikopoulos, 1991; Weiss & Sundar, 1992).

The polyclonal activators of lymphocytes are widely used to assess stress effects on immunity because they are independent of antigen presentation, can be measured repeat-edly, and are an indication of the functioning of a substantial proportion of lymphoid cells.

Table 1
Effects of Stress on *In Vitro* Immune Responses

Stressor	Effect	Study
Intermale aggression		Hardy *et al.* (1990)
High physical contact	Decreased T-cell activity in submissive mice (proliferation and IL-2 production)	
Low physical contact	Increased T-cell activity in dominant mice	
Signaled shock	Decreased T-cell proliferation	Cunnick *et al.* (1990)
High-pressure chamber		Shibata *et al.* (1991)
Without fragrance	Decreased plaque-forming cell activity	
With fragrance	Blockade of stress-induced suppression	
Exercise (undernourished mice)	Increased proliferation	Filteau *et al.* (1992)
Immobilization	Decreased NK activity	Shimizu *et al.* (1996)

Many investigators have also examined natural killer (NK) cell activity following stressor exposure utilizing either mutated or virally infected cells as targets.

Hardy, Quay, Livnat, and Ader (1990) examined both T- and B-cell proliferation as well as NK-cell activity utilizing the naturally occurring stressor intermale aggression. Male mice that are individually housed will demonstrate aggression when placed together. The investigators examined the immunological variables under two forms of encounter that result in substantially different degrees of actual physical contact between the males in the establishment of a dominance relationship. Subjects were rated as dominant or submissive in the encounters and the results were examined using these two groups in comparison with control mice. In the high physical contact condition, the submissive mice showed reduced T-cell activity as measured by both proliferation and IL-2 production relative to dominant and witness mice. However, in the low physical contact condition, the submissive mice did not differ from the controls whereas the dominant mice demonstrated increased T-cell activity by these criteria. NK activity was not affected in either condition. This study demonstrates several important concepts. First is the importance of proper controls. If the investigators had not included the various control groups, they would have been unable to differentiate physical contact conditions, because in both cases the dominant animals had functional values higher than those of submissive animals. Second, they were able to demonstrate that the physical stress of fighting differentially affects the influence of the establishment of dominance on immune function. This has important implications as it suggests that it is not the social ranking that is of primary importance, but rather how that ranking was determined. Third, they demonstrated that while T-cell function was affected by these manipulations, NK activity was not. This demonstrates the importance of measuring several immunological variables at the same time so as to get a better approximation of the immunological status of the individual. And finally, they demonstrated that psychosocial factors can be immunoenhancing as well as immunosuppressing.

Cunnick, Lysle, Armfield, and Rabin (1991b) utilized a physical stressor, signaled shock, to examine possible mechanisms involved in stress-induced alterations in lymphocyte proliferation. They found that stress decreased T-cell proliferation following mitogen stimulation. However, there was no effect on IL-2 production, the percentage of IL-2 receptor-positive T cells, or T-cell subsets. Further, the mitogenic suppression was not alleviated by the addition of IL-2 *in vitro*. The stressed subjects, however, showed no calcium ionophore-induced proliferation, suggesting a biochemical mechanism of suppression. This study demonstrates that there are various mechanisms responsible for suppression of mitogen-induced T-cell proliferation. It is therefore important to know the specific step in the activation sequence that is being affected by the stressor.

Shibata, Fujiwara, Iwamoto, Matsuoka, and Yokoyama (1991) conducted an intriguing experiment examining the effects of olfactory stimuli on the progression of stress-induced immunological changes. They exposed mice to a high-pressure chamber for 1 h per day for 2 days and then continuously exposed the subjects to a woody fragrance for 4 days. The high-pressure stress suppressed plaque-forming cell activity but this suppression was blocked by fragrance exposure. Further, the effect of the fragrance was blocked by pretreatment with 2% procaine onto the olfactory cells. The authors suggest that because the limbic system is innervated by the olfactory bulb, it is possible that the fragrance worked by blocking stress-induced activation of the limbic system. While this hypothesis goes well beyond the data presented, it may have implications for procedures such as aromatherapy. This paper illustrates the wide range of factors that may be involved in modulating stress-induced immunological changes. The relationship between the olfactory and immune systems is further described in Chapter 22.

A large literature on the effects of exercise on immune function (see Chapter 18) demonstrates both immunoenhancement and immunosuppression depending on various experimental factors. Filteau, Menzies, Kaido, O'Grady, Gelderd, and Hall (1992) examined the immunological effects of exercise in undernourished mice. Their study was designed to serve as a model of anorexia nervosa patients who regularly exercise. They hypothesized that exercise could prevent the immunosuppression caused by undernourishment. Lymphocyte proliferative responses to lipopolysaccharide were increased by exercise, although there was no change in the response to the mitogens concanavalin A (Con A) or phytohemagglutinin (PHA). Additionally, they found that serum corticosterone was decreased and hypothalamic norepinephrine (NE) was increased by exercise. They suggested that exercise may affect undernourishment-induced immune changes by a stress-induced immunoenhancement. This experiment is instructive in that it demonstrates differential effects of the experimental manipulation on different mitogenic responses.

While the above studies predominantly address T- and B-cell function, there have also been many studies of the effects of stress on NK activity. We have found that NK-cell activity is suppressed by blood sampling techniques 1 h to 2 days postsampling (Fig. 1). This suggests that use of repeated sampling must take into consideration prior treatments such that time is allowed for a return to baseline levels of functioning before further experimental manipulations occur. Also, NK activity changes rapidly following stressor exposure and persists for an extended period of time. A possible mechanism for stress effects on NK-cell activity was suggested by Shimizu, Kaizuka, Hori, and Nakane (1996). They demonstrated that immobilization stress-induced suppression of splenic NK cell activity led to increased splenic NE levels. Both the change in NK activity and in NE levels could be attenuated by denervation of the splenic sympathetic nerve, suggesting that sympathetic splenic innervation contributes the stress-induced changes in NK activity. However, it has also been suggested that stress-induced NK suppression can be mediated through an opiate pathway (see Section 13.4.3).

13.3.2. Effects on *In Vivo* Immune Responses and Disease Models

Studies of stress effects on immunity that utilize *in vitro* techniques such as NK-cell activity and the proliferative response to mitogen are important in that they demonstrate the functional capacity of various immune cells under standardized conditions. However, interpretation of results is limited because they are obtained under artificial conditions outside the experimental subject. Therefore, many investigators utilize *in vivo* measures of immune function such as response to an antigenic, viral, or bacterial challenge (see Table 2). Clinical models can also be measures such as the progression of experimentally induced cancers. These *in vivo* measures have the advantage of being able to assess immunocompetence in terms of the response of the whole organism. This is especially important because the complexity of the immune system is such that *in vitro* functioning of isolated populations of cells may not accurately reflect the true immunocompetence of the experimental subject. Obviously, many of these challenges can only be examined in experimental animals such as rodents, as one would not be able to, nor want to, experimentally produce illness in humans. Rodent models utilizing both *in vivo* and *in vitro* immune measures may also be useful in identifying *in vitro* measures, which can be obtained from human subjects, that approximate clinically relevant *in vivo* responses. Nevertheless, it is important to appreciate the limitations in generalizing across species, as findings in rodents are not necessarily applicable to humans (some stress paradigms cannot even be replicated across strains within rat or mouse species).

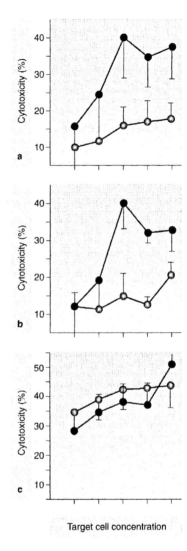

Figure 1. Time kinetic of changes in NK activity following blood sampling in the rat. Blood samples were taken 1 h (a), 2 days (b), or 7 days (c) following the baseline blood sampling. The baseline probes are shown by the filled circles, and the second probes by the open circles.

Target cell concentration

Antigen challenge is commonly used to assess immune function *in vivo*. While there are many advantages to *in vivo* assessment, there are many challenges as well. Various parametric manipulations complicate the interpretability of findings from these studies, as the effects of stress on the immune response are dependent on the timing of the stressor, antigen exposure, and immunological assessment. The authors have previously reviewed studies demonstrating that there is a critical period following antigen challenge when stress will affect immunological changes (Keller *et al.*, 1991). Once these parametric considerations are established, a model can be used to assess various mechanisms involved in stress effects on immune function. For example, Irwin (1993) found that central administration of

Table 2
Effects of Stress on *In Vivo* Immune Responses

Stressor	Effect	Study
Restraint stress	Decreased cellular immunity	Hermann *et al.* (1994)
	Decreased macrophage antimycobacterial activity	Brown and Zwilling (1994)
Forced swim	Increased metastatic colonization	Ben-Eliyahu *et al.* (1991)
Premature maternal separation	Increased mortality following opportunistic infection	Ackerman *et al.* (1988)
CRH administration	Decreased antigen processing	Irwin (1993)

corticotropin-releasing hormone (CRH) in rats 20 min prior to keyhole limpet hemocyanin (KLH) antigen challenge suppressed the IgG response. The author suggested that CRH is important in modifying initial antigen processing through a centrally mediated pathway, as peripheral CRH administration was ineffective and central administration of a CRH antagonist could block the central CRH effect.

Viral challenge has also been used to assess stress–immune interactions. Several experiments from Sheridan's laboratory (e.g., Hermann, Beck, Tovar, Malarkey, Allen, & Sheridan, 1994) have examined the effect of restraint stress on influenza viral challenge. They found that the inflammatory response in the lung was reduced by stress as were various measures of cellular immunity. Using DBA/2 mice, they demonstrated that both glucocorticoids and catecholamines are involved in the stress-induced mediation of the viral infection such that glucocorticoids affect cell trafficking and inflammation while catecholamines affect virus-specific effector cells. However, the pathogenesis of the infection varied according to the strain of inbred mouse used. This series of experiments demonstrates the usefulness of a well-characterized model in exploring possible mechanisms of stress-induced immune changes. They attempted to characterize the various immune changes that occurred, then to assess the phenomena across inbred strains, and finally to explore possible mediating CNS variables that could account for the phenomena.

Zwilling's laboratory (e.g., Brown & Zwilling, 1994) has utilized a similar approach to examine the effects of stress following bacterial challenge. They found that restraint stress suppressed the antimycobacterial activity of macrophages in a strain-specific manner. This appears to be related to corticosterone activity, as it could be duplicated by exogenous corticosterone and blocked by pretreatment with the receptor antagonist RU486.

Taken together, the above studies point to the importance of specifying the model employed in examining stress–immune interactions. Seemingly contradictory findings in the literature may be related to experimental parameters involving stressor exposure or immunological challenge as well as species or strain differences. These differences may ultimately provide insight into underlying mechanisms in stress-induced immune changes.

An often overlooked measure of *in vivo* immunity is population data containing information about susceptibility to infection or immune-mediated premature death. For example, Ackerman, Keller, Schleifer, Shindledecker, Camerino, Hofer, Weiner, and Stein (1988) found that adult male rats that had experienced premature maternal separation were more likely to die following an opportunistic infection that contaminated the colony. It is likely that other investigators have found similar group differences following laboratory contaminations. However, such data are often not reported.

An approach of potential clinical relevance to *in vivo* immune function involves the experimental induction of cancer (reviewed in Demetrikopoulos, Siegel, Schleifer, Obedi,

& Keller, 1994). Forced swim has been used as a physical stressor in conjunction with tumor studies. Ben-Eliyahu, Yirmiya, Liebeskind, Taylor, and Gale (1991) examined the effects of forced swim stress on metastases of the syngeneic MADB106 tumor in Fischer 344 rats. (To prevent the immune system from responding to the tumor as an invading pathogen, the injected tumor must be syngeneic to the host). A 45 g/kg weight was attached to the animal's tail prior to forced swim in 37°C water. The animals swam five times for 3 min with a 3-min rest period between exposures. The investigators found that stressed animals had increased numbers of lung metastases and decreased NK-cell activity compared with home cage controls (Fig. 2). These stress-induced changes were not attenuated by pretreatment with the opiate antagonist naloxone. Furthermore, effects were time-related in that increased metastases were only evident if the swim stressor was administered 1 h prior to injection of the tumor cells, and no effect was seen if the stressor was administered 24 h prior to, or 24 h after, tumor injection. Other studies have used similar approaches and are described in Section 13.4.3, as stress effects on tumor development may be partially mediated, in some cases, by a central opiate system.

It is likely that rodent models using *in vivo* measures of stress-induced changes in immune function will become better characterized, and more commonly utilized because their importance in determining the effects of stress on immune status in a whole functioning organism is obvious.

13.4. Mechanisms for Stress Effects on Immunity

A variety of pathways through which stressful conditions may produce their effects on immunity have been investigated. The classical stress response (associated with the pioneering work of Selye) consists of the release of corticotropin-releasing factor (CRF) from the hypothalamus, adrenocorticotropic hormone (ACTH) from the anterior pituitary, and glucocorticoid from the adrenal cortex. Therefore, some individuals have suggested that stress-induced immunomodulation is simply an elaborate way to measure the well-documented effects of glucocorticoids on the immune system. However, while this system

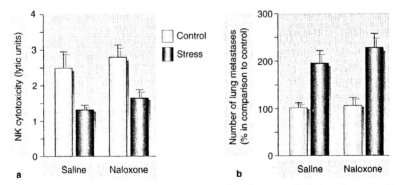

Figure 2. Effect of stress on (a) NK activity and (b) the number of metastases in the lung. Stress exposition produced a significant decrease in NK activity, paralleled by a twofold increase in the number of lung metastases. This effect is independent of opioid mechanisms, as naloxone administration produced no alterations. From Ben-Eliyahu *et al.* (1991).

is clearly an important contributor to the phenomena under consideration, it by no means provides a complete explanation for stress effects on immunity. Equally important contributions are made by the sympathetic nervous system and the opioid system. Further, central mechanisms are beginning to be explored such as neuroanatomical pathways involving the periaqueductal gray (PAG).

13.4.1. Hypothalamic–Pituitary–Adrenal Influences

Perhaps some of the earliest studies directly examining CNS control of immunity employed hypothalamic lesions (the following investigations are reviewed in Locke, Ader, Besedovsky, Hall, Solomon, & Stom, 1985). For example, in 1958 Filipp and Szentivanyi demonstrated that lesions of the tuberal hypothalamus were protective against anaphylactic shock. In another historical paper published in 1963, Korneva and Khai demonstrated that the anterior hypothalamus was important for production of antibodies. We also reviewed the more recent research involving the effect of hypothalamic lesions on immune function (Keller *et al.*, 1991). In 1980 Keller, Stein, Camerino, Schleifer, and Sherman published a study on the effects of bilateral anterior hypothalamic lesions in Hartley strain guinea pigs. This study found both *in vitro* and *in vivo* immunosuppression: Anterior hypothalamic-lesioned guinea pigs had reduced cutaneous tuberculin reactions and suppressed proliferative responses to PHA. Similarly, a series of studies from Roszman's laboratory in the early 1980s demonstrated immunosuppression in rats with anterior hypothalamic lesions. Korneva and Lesnikov (1985) examined the effects of lesions in the anterior and posterior hypothalamus in mice by measuring the colony-forming activity of hematopoietic stem cells. In this case, anterior hypothalamic lesions had no effect but posterior hypothalamic lesions produced decreased colony formation. Although these early hypothalamic lesion studies were somewhat nonspecific in their neuroanatomical methodologies, they were useful in demonstrating direct CNS effects on the immune system.

The role of the hypothalamus in immune function has been examined further by investigating the role of CRH in immune function. In a series of papers, Irwin demonstrated that centrally administered CRH reduces lymphocyte proliferation, cytotoxic activity of NK cells, and antibody production (reviewed in Irwin, 1993). This would suggest a mechanism of hypothalamic activation that may explain several of the early hypothalamic lesion studies.

Because pituitary hormones have been shown to be immunomodulatory, the role of the pituitary as an adrenal-independent regulator of stress effects on immunity has been explored. The effect of hypophysectomy on stress-induced immunomodulation was examined in Sprague–Dawley rats (Keller, Schleifer, Liotta, Bond, Farhoddy, & Stein, 1988). This study found a pituitary-dependent stress-induced lymphopenia and stress-induced splenic NK activity suppression. However, stress-induced suppression of peripheral blood mitogen response was exaggerated by hypophysectomy. These findings suggest a complex role for the pituitary in stress-induced immunomodulation such that the pituitary may be involved in both suppressive and compensatory effects on immune measures.

A similar experiment assessed the role of the adrenals in stress-induced immunomodulation (Keller, Weiss, Schleifer, Miller, & Stein, 1983). Adrenalectomy was performed and various immune measures taken following stressor exposure. While stress-induced lymphopenia was blocked by adrenalectomy, an adrenal-independent suppression of lymphocyte stimulation was shown. This study demonstrated that not all stress-induced changes in immunity can be reduced to the simple effect of corticosterone on immune cells. The role of the adrenals was explored further by Brown and Zwilling (1994) in an examination of

Table 3
Mechanisms for Effects of Stress on Immune Responses

Mediator	Effect	Study
Hypothalamus		
Tuberal hypothalamus lesion	Protective against anaphylactic shock	Filipp and Szentivanyi (1958)
Anterior hypothalamus	Antibody production	Kornevea and Khai (1963)
Anterior hypothalamus lesion	Decreased tuberculin reactions	Keller *et al.* (1980)
	Suppressed PHA response	
Posterior hypothalamus lesion	Decreased colony formation	Kornevea and Lesnikov (1985)
Central CRH administration	Decreased lymphocyte proliferation	Irwin (1993)
	Decreased NK activity	
Pituitary		
Hypophysectomy	Protective against stress-induced lymphopenia and splenic NK suppression but not mitogen response	Keller *et al.* (1988)
Adrenal		
Adrenalectomy	Protect against stress-induced lymphopenia but not mitogen response	Keller *et al.* (1983)
Glucocorticoid antagonist	Blocked stress-induced changes in macrophage function	Brown and Zwilling (1994)
Type II adrenal receptor	Leukocyte distribution	Dhabhar *et al.* (1996)
Catecholamine		
Adrenalectomy and β-adrenergic antagonist	Blocked stress-induced T-cell suppression in blood	Cunnick *et al.* (1990)
	Blocked stress-induced T-cell suppression in spleen	
β-adrenergic antagonist	Blocked morphine-induced T-cell suppression in spleen	Fecho *et al.* (1993)
Opiate		
Opiate-dependent foot shock stress	Suppressed NK activity and T-cell proliferation	Shavit (1991)
	Reduced survival of tumor-implanted subjects	
Morphine	Blocked surgery-induced increases in metastasis	Page *et al.* (1993)
Morphine and ganglionic blocker	Suppressed lymphocyte proliferation	Flores *et al.* (1996)
	Blocked morphine-induced suppression	
Serotonin		
Active serotonin	Immunoinhibition	Devoino *et al.* (1988)
Dopamine		
D1 agonist	Increased lymphocyte response in central amygdala	Nistico *et al.* (1994)
	Decreased in nucleus accumbens and CA1	
Dopaminergic lesion	Suppressed *in vivo* challenge to SRBC	Devoino *et al.* (1997)

antimycobacterial activity of macrophages. They demonstrated that stress effects could be blocked with a glucocorticoid receptor antagonist. They also demonstrated an adrenal-independent phenomenon, namely, a naturally resistant strain of mice was not affected by hypothalamic–pituitary–adrenal (HPA) activation or exogenous corticosterone treatment. Taken together, these studies suggest that the adrenals are indeed important for stress-induced immunosuppression, although other factors are also important. Dhabhar, Miller,

McEwen, and Spencer (1996) demonstrated that stress-induced changes in leukocyte distribution that have been shown to be mediated through corticosterone are likely to be acting through the type II adrenal steroid receptor. An excellent review of corticosteroid effects on immunity was recently published (McEwen, Biron, Brunson, Bulloch, Chambers, Dhabhar, Goldfarb, Kitson, Miller, Spencer, & Weiss, 1997).

13.4.2. Sympathetic Nervous System Influences

The complexity of sympathetic nervous system influences on immunity is evident from the several reviews of this system in the second edition of *Psychoneuroimmunology* (Ader, Felten, & Cohen, 1991). The innervation of immunological organs by the sympathetic nervous system provides an important mediating mechanism between the CNS and the immune system (see Chapter 8). In addition to the direct innervation of lymphoid organs, there is much evidence for adrenergic receptors on various immunologically important cells such as T and B cells. The presence of these receptors provides further evidence for communication between the sympathetic nervous system and the immune system.

These effects have been extensively examined by *in vitro* and *in vivo* exposure to catecholamines as well as by lesioning of the sympathetic innervation of immunological organs. *In vitro* studies suggest that, in general, stimulation of β-adrenergic receptors leads to an inhibition of immunological functioning whereas α-adrenergic receptor simulation leads to potentiation of immunological functioning. *In vivo* administration of epinephrine causes changes in lymphocyte trafficking and may affect antibody formation depending on the presentation schedule employed. Another *in vivo* approach extensively studied employs sympathetic denervation. However, this literature is fraught with seemingly inconsistent data reflecting the wide range of experimental models employed, such that there have been reports of immunoenhancement, of immunosuppression, and of no change in immune function. Ultimately these data can probably be resolved by a close scrutiny of the effects of species, subspecies, age, immune compartments assessed, and specific sympathectomy protocols. Furthermore, examination of the contribution of catecholamines by the use of splenic denervation is complicated by the fact that other organ systems, notably the adrenal, may increase their catecholamine production as a compensatory mechanism. A thorough examination of this literature is presented in the previously mentioned reviews, which the reader is encouraged to examine.

Another approach to examining the role of the sympathetic nervous system on immune function *in vivo* involves stimulation of an afferent pathway. Irwin, Hauger, and Brown (1992a) utilized lateral ventricular microinjection of CRH to stimulate sympathetic outflow in order to examine changes in immune function. They demonstrated that the increased sympathetic activity led to a suppression of splenic NK activity. Furthermore, this phenomenon was evident in older but not younger rats. They proposed that increased incidence of cancer in older organisms may be related to changes in CNS regulation of the sympathetic nervous system.

As previously mentioned, the adrenal represents another potential catecholamine source. Cunnick, Lysle, Kucinski, and Rabin (1990) examined the role of the adrenals and a ß-adrenergic antagonist on T-cell function in order to explore possible mechanisms for stress (foot shock)-induced suppression. They demonstrated that adrenalectomy blocked stress-induced T-cell suppression in peripheral blood, but not spleen; the β-adrenergic antagonist blocked this effect in the spleen, but not peripheral blood. In a related study, Fecho, Dykstra, and Lysle (1993) demonstrated that β-adrenergic antagonists are able to block morphine-induced suppression of splenic, but not peripheral blood, T-cell function. These experiments demonstrate that there are likely to be distinct mechanisms for adrenergic modulation of peripheral blood and splenic populations of lymphocytes. Both studies also suggest that the changes in splenic T-cell function are the result of peripheral effects of catecholamines.

13.4.3. Opiate Influences

Yehuda Shavit's laboratory has conducted a series of experiments examining the effects of opiates on NK activity (Shavit, 1991). In these studies, they take advantage of a model of stress analgesia developed by Lewis, Cannon, and Liebeskind (1980) in which inescapable foot shock stress analgesia may have an opiate or nonopiate mechanism depending on its temporal parameters. Sprague–Dawley rats were exposed to either brief continuous foot shock for 3 min, prolonged intermittent foot shock for 30 min (1-s pulses delivered every 5 s), or no shock (handled controls). Both stress conditions produced long-lasting analgesia; however, only the prolonged intermittent foot shock stress-induced analgesia was blocked by naloxone. Thus, differential parametric manipulations of the same stressor allow for differential mechanisms of stress analgesia and provided a model for studying an opiate- versus nonopiate-mediated stress analgesia. The presence of both an opiate and a nonopiate mechanism of stress analgesia was supported further by the findings of cross-tolerance between morphine- and naloxone-sensitive, but not naloxone-insensitive, forms of stress analgesia.

In 1983 Shavit, together with Lewis and Liebeskind, began to examine opiate foot shock stress effects on immune function (Shavit, Lewis, Terman, Gale, and Liebeskind, 1983). It had previously been shown that opiate antagonists injected can inhibit tumor growth and prolong survival time in animals given tumors. This group demonstrated immunosuppression by the opiate form of foot shock stress such that there were reduced Con A and PHA stimulation levels, suppressed NK lytic ability, and reduced median survival time of tumor-implanted subjects. Furthermore, all of these effects could be blocked by pretreatment with naltrexone (Fig. 3). None of the immune measures were affected by the nonopiate form of foot shock stress. Shavit et al. (1983) further demonstrated that morphine mimicked the endogenous opiate stress-induced response and produced a dose-dependent suppression of NK-cell activity.

Shavit and Liebeskind (1986) next determined that opiate-induced immunosuppression was centrally mediated. They administered morphine directly into the lateral ventricle and gave a morphine analogue (N-methyl-morphine), which cannot cross the blood–brain barrier, systemically. The intraventricular administration of morphine produced a dose dependent decrease in the cytotoxic activity of splenic NK cells that could be blocked by systemic naltrexone. Additionally, intraventricular administration of naltrexone could block the suppressive effects of systemic morphine administration on NK-cell lytic activity. Systemic administration of N-methyl-morphine had no effect on NK activity.

Shavit, Martin, Yirmiya, Ben-Eliyahu, Terman, Weiner, Gale, and Liebeskind (1987) examined the effects of a single exposure to the opiate stress paradigm, or a single morphine injection on NK activity. They examined possible gender effects, dose response, time course, blockage by naltrexone, and possible tolerance development using cells derived from spleen, bone marrow, and peripheral blood. Exposure to a single session of opiate stress suppressed splenic NK activity transiently for 3 h, with no effect seen 24 h poststress. Furthermore, a single systemic dose of morphine suppressed NK activity in all three immune compartments, suggesting that their previous findings were not related to redistribution of lymphocytes into different immune compartments. (An excellent review of this literature was published by Shavit, 1991.)

Although the above studies suggest that morphine produces immunosuppressive

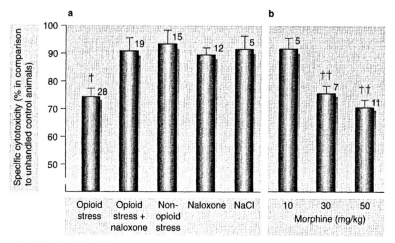

Figure 3. NK cell activity in the spleen. Animals underwent the opioid or nonopioid form of foot shock stress once a day for 4 days. Opioid stress reduced splenic NK activity, and this effect was blocked by the opioid antagonist naloxone. Morphine administration mimicked the endogenous opiate stress response. From Shavit *et al.* (1983).

effects, the role of opiates in immunosuppression is complex. For example, Page, Ben-Eliyahu, Yirmiya, and Liebeskind (1993) demonstrated that morphine administration blocked surgical stress-induced increases in metastatic colonization but did not affect colonization in a nonsurgery group, suggesting that morphine attenuated the colonization by blocking the experience of pain which may have otherwise led to immunosuppression and lung metastases. It may be somewhat surprising that the nonsurgery group was unaffected by the morphine, as other reports had previously demonstrated increased colonization in this group via an opiate-mediated stressor. However, another study demonstrated a dissociation between morphine and the opiate-mediated stressor effects on NK activity (Shavit, Depaulis, Martin, Terman, Pechnick, Zane, Gale, & Liebeskind, 1986). While the suppressive effect of morphine administration showed tolerance after repeated administration, the opiate-mediated stressor effect on NK activity did not. A recent study by Bayer's group (Flores, Dretchen, & Bayer, 1996) has suggested that the autonomic nervous system may act as a mediator for morphine-induced suppression of lymphocyte proliferation. They demonstrated that the ganglionic blocker chlorisondamine antagonized the effects of morphine while blockade of either the sympathetic or parasympathetic divisions of the autonomic nervous system failed to antagonize the effects of morphine. These studies suggest that the role of opiates in immunosuppression may involve different mechanisms for stress, pain, and pharmacological effects.

13.4.4. Other CNS Influences

In addition to the mediators described in the preceding sections, a variety of other potential neurochemical systems may be involved in stress-induced immune changes. This section will briefly review the possible contributions of the serotonergic and dopaminergic systems. The serotonergic system has been shown to play a role in stress-induced immunomodulation such that serotonin is increased in the solitary nucleus but decreased in the dorsal hippocampus following immunization (reviewed in Ader *et al.*, 1991). Further, Devoino, Morozova, and Cheido (1988) demonstrated that elevations of active serotonin produced inhibition of the immune response, while depression of active serotonin stimulated the immune response. Although much more needs to be learned about this system, these results may suggest that the raphe nucleus serotonergic system is an inhibitory system for immunomodulation that may be regulated by the HPA axis.

Nistico, Caroleo, Arbitrio, and Pulvirenti (1994) examined the neuroanatomically site-specific role of dopaminergic D1 receptors on immune function. There was no effect on NK activity in any of the sites tested. However, there was a site-specific effect of a D1 agonist on proliferative activity of splenocytes. They demonstrated that a D1 agonist microinfused into the central amygdala increased splenic mitogen response to Con A, while microinfusion into the nucleus accumbens or into the CA1 region of the hippocampus decreased splenic proliferative responses to both Con A and LPS. Additionally, microinfusion of the D1 agonist into the ventral tegmentum had no effect. The specificity of these findings was demonstrated by the blockade of the effects of the D1 agonist by systemic administration of a D1 antagonist. Devoino, Alperina, Galkina, and Ilyutchenok (1997) further explored the role of dopamine on immune function and demonstrated that lesions of the terminal regions of the nigrostriatal and mesolimbic dopaminergic systems suppressed *in vivo* immune response to sheep red blood cell challenge. Additionally, these regions showed high dopaminergic activity following immunization in intact subjects. Taken together, these studies suggest that the dopaminergic system may play a role in immune challenge.

13.4.5. The PAG as a Neuroanatomical Model for Stress-Induced Immunosuppression

The above sections described the role of various neurochemical systems on stress-induced immunosuppression. This section will explore the role of the PAG in stress-induced immunosuppression as an example of a model for suggesting various neuroanatomical pathways involved in stress-induced immunosuppression. Weber and Pert (1989) used neuroanatomical site-specific microinjections of morphine in order to determine a possible site for the central action of opiate-induced immunosuppression. Injection sites included the anterior hypothalamus, arcuate nucleus, ventral medial hypothalamus, medial thalamus, medial amygdala, dorsal hippocampus, and caudal ventral PAG. Fischer rats given morphine injections into the caudal ventral PAG showed decreased splenic NK activity that could be blocked with prior intraperitoneal injection of naltrexone. None of the other sites showed an effect on splenic NK activity. Because it had earlier been shown that endogenous opiates were released in the PAG during foot shock stress and that opiate-dependent foot shock stress produced decreased NK activity, it is possible that foot shock stress produces its effect through the PAG by enhancing the activity of endogenous opiate containing neurons in the spinoreticular pain pathways. Weber and Pert have suggested that opiate activity in the PAG may cause activation of the HPA axis through hypothalamic efferents or through an increased peripheral sympathetic output, either of which could affect NK activity.

Weber and Pert (1990) presented preliminary data suggesting that microinjection of the selective mu opiate agonist D-Ala^2-N-Me-Phe^4-Gly-enkephalin into the lateral ventricle decreased NK activity, whereas the selective kappa agonist S,S,-U50,488 or the high-affinity delta receptor agonist D-Pen^{25}- enkephalin had no effect. This suggests that the opiate-induced immunosuppression of splenic NK cells is specific for mu receptors. In additional preliminary data, Weber and Pert (1990) also demonstrated that electrical stimulation of the ventral PAG suppressed splenic NK activity, using intermittent 150-ms trains of biphasic pulses (50 μs in duration presented at 100 Hz and 500 μA) for 20 min. The most active area of suppression was in the caudal PAG lateral to the dorsal raphe. Suppression was also obtained by stimulation of the caudal periventricular gray matter extending from the level of the posterior hypothalamus to the red nucleus. Collectively, their data suggest that the immunosuppressive effect of opiate-dependent foot shock stress may be mediated through a central opiate system, through the caudal ventral PAG by way of mu receptors.

This proposed neuroanatomical pathway is only one of several through which the PAG may modulate the immune system. There are functional differences between subregions of the PAG as defined along both the rostral–caudal and dorsal–ventral axes. For example, the ventral and dorsal aspects of the PAG differ in terms of behavioral and neurochemical processes including: (1) ventral PAG stimulation produces mild behavioral changes such as head turning while dorsal PAG stimulation produces robust behavioral changes such as jumping and increased escape behavior; (2) ventral PAG stimulation produces an opiate-sensitive analgesia while dorsal PAG stimulation produces an opiate insensitive analgesia; and (3) ventral PAG stimulation produces sympathoinhibition while dorsal PAG stimulation produces sympathoexcitation. These differences suggest that ventral and dorsal PAG stimulation sites could affect immune function through different neuroanatomical pathways and neurotransmitter systems. We recently examined the effect of rostral–dorsal PAG stimulation on splenic and peripheral blood NK cells (Demetrikopoulos, Siegel, Schleifer, Obedi, & Keller, 1994). Bipolar electrical stimulation was delivered once a minute for 30 min to freely moving rats (0.4 mA, 62.5 Hz, 1 ms half-cycle duration). The stimulation was applied until a "flight" response was elicited (less than 1 s). The dorsal PAG stimulation did not alter splenic NK activity yet it suppressed peripheral blood NK activity (Fig. 4). Further, the blood NK suppression was not blocked by pretreatment with the opiate antagonist naltrexone. Taken with Weber and Pert's work, these studies suggest that the dorsal and ventral aspects of the PAG may have very different roles in stress-induced immunomodulation. This may reflect different distributions of various neurotransmitter systems between the dorsal and ventral aspects of the PAG neuropil. These regions have been shown to differ

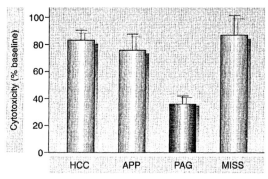

Figure 4.   Effect of periaqueductal gray (PAG) stimulation on peripheral blood NK activity in the rat. Some animals were left untreated in the home cage (home cage control, HCC), or placed in the experimental apparatus without PAG stimulation (apparatus, APP). In animals that received electrical stimulation, a division could be made between those where stimulation was located in the PAG (PAG), and those where stimulation occurred in areas other than the PAG (missed, MISS).

Table 4
Site Specificity of PAG-Induced Immunomodulation

Rostral–dorsal	Rostral–ventral	Caudal–dorsal	Caudal–ventral
Decreased blood NK activity[a]	Increased blood NK activity[b]	Decreased spleen NK activity[c]	Decreased spleen NK activity[f]
Increased tumor metastasis[b]	Decreased tumor metastasis[b]	Decreased antibody production to BSA[e]	Increased tumor metastatis[d]
		Increased delayed hypersensitivity reaction[e]	Decreased spleen and thymus T-cell proliferation[g]
		No effect on T-cell proliferation[c]	Decreased peritoneal macrophage activity[g]

[a]Demetrikopoulos *et al.* (1994).
[b]Demetrikopoulos and Zhang (1998).
[c]Lysle *et al.* (1996).
[d]Simon *et al.* (1980).
[e]Vlajkovic *et al.* (1994).
[f]Weber & Pert (1989).
[g]Weber *et al.* (1997).

with respect to neurotransmitters known to be important immunomodulators such as serotonin, opiates, and NE. Furthermore, these results suggest that there are different central mechanisms involved in the modification of splenic and peripheral blood NK function. This is consistent with earlier work showing different effects of stress on splenic and peripheral blood mitogen activity (Keller *et al.*, 1983).

There is a growing body of literature examining the effects of PAG stimulation on immunological function, which appear to be extremely site specific along both the rostral-caudal and dorsal–ventral axes (see Table 4). These studies were performed by either electrical stimulation or morphine administration at these sites. At least part of the effects at the caudal stimulation sites are likely to be related to activation of generalized stress effects as evidenced by increases in ACTH and cortisol which are correlated to the immune changes. The immunoenhancement and protection from tumor progression seen in the rostral–ventral PAG sites may be related to PAG pathways known to be important for aggressive behavior, which has been suggested to be immunoenhancing.

13.5. Conclusion

Research utilizing rodent subjects has provided insight into the mechanisms involved in stress-induced immunomodulation by employing a variety of models of various stressful situations. Although potentially complex interactions between the various stress-sensitive neuronal systems should be explored further, the neuroanatomical pathways and neuro-chemical mediators underlying these phenomena have begun to be elucidated.

14 Behavior/Immune Relationships in Nonhuman Primates

Mark L. Laudenslager and Julie M. Worlein

Mark L. Laudenslager • Behavioral Immunology Laboratory, Department of Psychiatry, University of Colorado Health Sciences Center, Denver 80220.   Julie M. Worlein • Regional Primate Research Center, University of Washington, Seattle, Washington, 98195.

14.1. Why Use Nonhuman Primate Models for Biobehavioral Studies?

Animal models are used in a wide variety of behavioral and biomedical research because they allow control of extraneous, confounding variables to a degree that is not possible with human subjects. The use of animal models also allows the researcher to perform experimental manipulations that, for ethical or practical reasons, are not practical in humans.

To serve as an adequate model for a human phenomenon, an animal model must meet several commonly accepted criteria, including similarities in etiology, pathophysiology, phenomenology, and effective treatments (physiological, pharmacological, or behavioral) for amelioration of negative outcomes (McKinney & Bunney, 1969). Nonhuman primates have been used extensively as models for human conditions such as grief and bereavement, social conflict and aggression, and early development and aging, as they fulfill many of these criteria. They are particularly useful in studies of development and the influence of early experience on later outcomes because they develop more rapidly than human children, thus supplying answers to developmental questions far more quickly than would be possible in humans.

Nonhuman primates offer a unique model for exploring neural–immune interactions. The nervous, endocrine, and immune systems of nonhuman primates are morphologically and physiologically similar to those of humans. Human and nonhuman primates share many phenotypic traits. Phenotypic similarities may be either analogous or homologous. Analogous phenotypes evolve independently in response to similar evolutionary selection pressures. Flight in birds and bats is an example of analogous similarities. However, analogous traits are not likely to arise from the same mechanistic processes. Conversely, homologous phenotypes evolve through genetic continuity from a common ancestor. The fact that humans and nonhuman primates share a common evolutionary history suggests that similarities in nervous, endocrine and immune systems are more likely to be homologous arising from comparable mechanisms. More importantly, nonhuman primates are complex social animals that form strong and lasting social bonds. The intricacy of nonhuman primate behavior also allows the role of individual differences in responses to be investigated.

14.2. Social Loss

The psychological and physical consequences associated with the experience of losing important individuals in one's life have been documented by personal experience, anecdotes, and large-scale epidemiological investigations. Regardless of the nature of the loss (i.e., romantic breakup, a brief or long-term separation, divorce, or death), humans feel sad, lonely, and quite often depressed. Medically, one might experience exacerbation of existing chronic illnesses, appearance of new or previously undetected illnesses, or more prolonged recovery from minor illnesses such as the common cold. There are epidemiological reports documenting increased morbidity and mortality in the year following the death of a spouse (for reviews, see Stroebe, Stroebe, & Hansson, 1993). Interestingly, the impact of loss and consequent grief on health tends to be greater for men than for women, and also greater in older populations.

A number of health behaviors may contribute to the relation between the experience of loss and altered health (Laudenslager & Reite, 1984). For example, changes in eating and sleep habits, increased use of psychoactive substances, and altered activity and exercise patterns often accompany a loss experience. All of these behaviors contribute significantly to alterations in immunoregulation and other homeostatic systems. Although the focus of this chapter is on immune-mediated phenomena, it is important to note that increased incidence of cardiovascular disease is also associated with recent loss experiences (see Stroebe *et al.*, 1993).

14.2.1. Primate Models of Social Loss

Through the use of well-defined animal models, it may be possible to disentangle some of these relationships. A number of studies have investigated acute and chronic behavioral and physiological impacts of maternal or peer separation on young nonhuman primates (Mineka & Suomi, 1978). Typically, studies have focused on young (4–6 month old) macaque monkeys housed with their mothers in social groups, and then removing the mothers from the group. In other studies, young monkeys have been removed from their social group or natal housing and placed in different housing without their mothers, so that they experienced not only loss but also isolation and exposure to a novel environment. The former situation may parallel the experiences of a young macaque in a natural setting more than the latter. For example, under free-ranging conditions, the first mating of a monkey's mother and the subsequent birth of a sibling produces significant conflict for the young monkey and its mother. During this time, the young monkey experiences frequent rejections by the mother, resulting in significant behavioral distress (e.g., Laudenslager, Rasmussen, Berman, Suomi, & Berger, 1993; Worlein, Eaton, Johnson, & Glick, 1988). Behavioral and physiological responses have also been observed following separation of young monkeys that had been peer-reared without mothers but that formed emotional bonds with each other (Mineka & Suomi, 1978).

Behaviorally, the response to separation, particularly maternal separation, is described as a two-step process comprising an initial agitation phase followed by a despair or depression phase (Kaufman & Rosenblum, 1967; Mineka & Suomi, 1978; Reite, Short, Seiler, & Pauley, 1981b). The agitation phase following maternal separation lasts 24–48 h. During this time, the young monkey vocalizes frequently and increases its vigilance, activity, and locomotion. Such behavior would be instrumental in attracting the mother to the isolated infant or young monkey if occurring in a naturalistic setting. These behaviors were first described as attachment behaviors (Bowlby, 1973), and more recently as coping behaviors (Levine, Wiener, Coe, Bayart, & Hayashi, 1987). After the initial period characterized by these agitation behaviors, the monkey withdraws. Activity declines, vocalization drops, and the young monkey may assume a withdrawn huddled posture while remaining cautiously aware of its surroundings. Social interactions and play behavior decline and are accompanied by a concomitant rise in ingestive behaviors (eating and drinking).

14.2.2. Some Physiological Changes following Loss in Nonhuman Primates

When the mother of a young macaque is removed from the social group, a number of behavioral and physiological changes ensue. The physiological responses associated with social separation have been characterized extensively. The agitation phase is associated with increased heart rate and body temperature, whereas the depression phase is characterized by altered sleep patterns (decreased total sleep, increased REM latency, reduced number of REM periods, and more frequent arousals), lower heart rates and body temperature, cardiac arrhythmia, and changes in circadian rhythms of heart rate and body temperature (Reite *et al.*, 1981b).

Furthermore, there are multiple indications of hypothalamic–pituitary activation (Levine & Wiener, 1988) including increased plasma cortisol (Laudenslager, Boccia, Berger, Gennaro-Ruggles, McFerran, & Reite, 1995), growth hormone (Laudenslager *et al.*, 1995), and prolactin (see Fig. 1). Changes in CNS biogenic amines and peripheral sympathetic activity have also been noted (Kraemer, Ebert, Schmidt, & McKinney, 1991). The magnitude of these physiological changes is related to a number of social and nonsocial factors such as the species under investigation, presence of peers or conspecifics, visual contact with the mother, and ability to touch the mother. In addition, physiological changes often covary with the magnitude of behaviors that reflect distress. Elevation in plasma cortisol during the first week after separation is positively correlated with vocalization and time spent in a withdrawn posture and negatively correlated with play and proximity to others (Laudenslager *et al.*, 1995). Because these hormonal and sympathetic alterations are exactly the changes expected to affect the regulation of the immune response, it is not surprising that altered immunoregulation also follows acute maternal or peer separation in nonhuman primates (Coe, 1993).

14.2.3. Immune Changes

In a landmark study, Bartrop, Lazarus, Luckhurst, Kiloh, and Penny (1977) provided credence to the notion that one potential mediator in increased morbidity and mortality associated with loss is altered immunoregulation. This cross-sectional study of recently bereaved spouses of accident victims observed that lymphocyte activation was depressed as early as 1–3 weeks after the loss event and remained low up to 6 weeks later compared with that of nonbereaved age- and sex-matched control subjects. Another study replicated these basic observations, but this time *prospectively* in spouses of breast cancer patients (Schleifer, Keller, Camerino, Thornton, & Stein, 1983). Although intriguing, these studies have not addressed a variety of issues, such as the mechanisms of immunomodulation, the role of confounding health behaviors, age, and social support, or the quantitative assessment of depression. Prior studies have attempted to address these questions in recently bereaved populations, but it is difficult to control for the many confounding issues that exist in human populations. Consequently, the relation between recent loss and immunoregulation remains somewhat indeterminate.

Demonstrating the validity of the primate model of loss for studying depressed lymphocyte activation as noted in bereaved humans (Bartrop *et al.*, 1977), Reite, Harbeck, and Hoffman (1981a) investigated lymphocyte activation by mitogens during a 2-week separation of 6-month-old pair-reared macaques. The results were the same as those seen in humans: The proliferative response to T-cell mitogens was decreased when the monkeys were separated, and returned to baseline when the pair was reunited. These results were extended to maternal separation in bonnet macaque monkeys in the absence of allomaternal care: Lymphocyte activation by B- and T-cell mitogens was reduced following separation (Laudenslager, Reite, & Harbeck, 1982). Neither of these studies occurred under the conditions of social group housing, however, and the subject populations were small. Therefore, another study was undertaken, this time with a large number of socially housed subjects, both bonnet and pigtailed macaques were included, as well as age- and species-matched control subjects to control for handling and cohort effects (Laudenslager, Held, Boccia, Reite, & Cohen, 1990). In this study, declines in lymphocyte activation were *not* observed in all subjects. The effects were limited to the young monkeys exhibiting the greatest behavioral distress after maternal separation. Monkeys that vocalized the most on the first day of separation and spent the greatest amount of time in huddled, withdrawn postures were the most likely to show indications of immunomodulation. These observations were consonant with those of another study also demonstrating individual differences in the response to social separation in nonhuman primates (Lewis, McKinney, Young, & Kraemer, 1976). Thus, not all monkeys are significantly affected by the absence of the mother from the social group. An important area for continued investigation is the determination of which monkeys are most vulnerable to social separation experiences: Is it related to dominance status, gender, the mother–infant relationship, parity, other older siblings present in the group, inherent biological reactivity, or other factors?

14.2.4. Modulation by Social Factors

Allomaternal and affiliative behaviors (i.e., grooming and other displays of physical proximity) play an important role in mitigating the negative effects of social stressors. Allomaternal care refers to maternal interactions of the young monkey with adults other than the mother (e.g., aunting). By studying species with different maternal styles, one can ask questions regarding the relative role of affiliative behaviors in response to social separation. For example, pigtailed macaque mothers are quite restrictive with their infants. Under normal circumstances in social groups, adult pigtailed females may maintain spatial separation from each other rather than sit in close proximity (Boccia, Laudenslager, & Reite, 1988). Thus, from an early age, pigtailed macaque infants interact infrequently with adults other than their mothers. When a mother is removed from the social group, the young monkey is unlikely to locate other adults (e.g., aunts) willing to provide either protection or social interaction. When it attempts to interact with other adults, the monkey may be punished or chased away. The resulting behavioral and physiological responses may be quite profound in this species. If, on the other hand, the young pigtailed monkey succeeds in receiving allomaternal care from other members of the group, the negative effects of maternal separation can be mitigated. For example, greater contact with other members of the social group reduces the magnitude of heart rate changes following maternal separation in pigtailed macaques (Caine & Reite, 1981).

Bonnet macaques display a far less restrictive pattern of maternal care. During quiet periods, individuals typically sit in close proximity in small clusters. Bonnet macaques have the opportunity to participate in a number of affiliative interactions with many members of their natal group. If a mother is removed from the social group, her offspring is likely to have alternate adults with which it can interact. These allomaternal caregivers may enclose the monkey, groom it, and provide protection. As a result, bonnet macaques show far fewer behavioral and biological changes during maternal separation than do pigtailed macaques.

When 6-month-old bonnet macaques are separated from their mothers in the presence of a preferred peer partner, there are no changes in the activity of NK cells. However, when young bonnet monkeys were separated from their mothers in the absence of a preferred peer partner (Boccia, Scanlan, Laudenslager, Berger, & Reite, 1997), cytotoxicity toward tumor targets declined as rapidly as 2 h after separation but returned to baseline after 1 week. These monkeys showed a similar pattern of changes in mitogen-induced activation of lymphocytes (Boccia et al., 1997). The magnitude of these changes was related to allomaternal and affiliative behaviors that the young monkeys experienced during the separation period. Thus, the more social affiliation received by a monkey, the higher the cytotoxicity of peripheral blood lymphocytes toward tumor cells. It is thought that the presence of these affiliative and allomaternal behaviors in nonhuman primates models similar influences of social support in humans. Certainly the presence of social support reduces the negative consequences of stressors on health in humans and is particularly important for patients in recovery from serious illnesses of many etiologies (Cohen, 1988) and even minor illnesses such as the common cold (Cohen, Doyle, Skoner, Rabin, & Gwaltney, 1997).

The impact of social affiliation on immune function is similar during other social challenges. Macaques establish dominance hierarchies in their social groups, and disruptions of the social structure are quite distressing because dominance status has to be reestablished each time a group is rearranged. During these periods, lymphocyte activation by mitogens is depressed and cardiovascular risk increases. However, some animals cope with this highly conflictual situation by increasing social affiliation. Monkeys with higher levels of social affiliation show less reduction in lymphocyte activation and NK-cell activity, and more rapid return of peripheral blood lymphocyte phenotypes to baseline values (Cohen, Kaplan, Cunnick, Manuck, & Rabin, 1992; Kaplan, Heise, Manuck, Shively, Cohen, Rabin, & Kasprowicz, 1991).

Reciprocal grooming is an important component of social interactions between non-human primates. A monkey's heart rate can fall as low during social grooming as it does during sleep. Grooming is a part of the reconciliation process among nonhuman primates following agonistic encounters. In adult monkeys, the elevation in heart rate associated with agonistic encounters or flight declines most rapidly when an individual receives social grooming (Boccia, Reite, & Laudenslager, 1989). Social grooming may alter sympathetic activity in other autonomically controlled systems and perhaps influence immunoregulatory processes to reduce the impact of the stressor. This was observed during social group formation in pigtailed macaques wherein the amount of grooming received, although related to dominance ranking in the group, was also related to the recovery of peripheral blood phenotypes after rearrangement (Gust, Gordon, Wilson, Brodie, Ahmed-Ansari, & McClure, 1996). The link between social affiliation and diminished changes in immune responses under social stressors is not surprising in light of these observations.

14.2.5. Social Separation in Adult Animals

Disruption of social bonds associated with removal from social groups can result in immunomodulation in fully developed animals. Experimentally removing juvenile rhesus macaques from their natal social groups and housing them in peer groups produced marked changes in behavioral and immune parameters as well as an increase in cortisol (Gordon, Gust, Wilson, Ahmed-Ansari, Brodie, & McClure, 1992; Gust, Gordon, Wilson, Brodie, Ahmed-Ansari, & McClure, 1992). Similar to young monkeys separated from their mothers, separated juveniles exhibited less play and more coo vocalizations than control juveniles remaining in their social groups. Separated juveniles also showed a rise in cortisol on the day of separation. Although cortisol levels returned to baseline by 3–4 weeks, separated individuals showed continued reductions in peripheral blood CD4$^+$ (T-helper) and CD8$^+$ (T-cytotoxic/suppressor) cells for up to 11 weeks after separation. It is likely that separation also produced dysregulation in other physiological systems, as weight gain in separated subjects was one-third of that exhibited by controls.

When adult female rhesus macaques were separated from their social group and housed either alone or with a preferred companion, they underwent changes similar to those seen in juveniles, namely, an increase in plasma cortisol concentrations, a decrease in CD4$^+$ and CD8$^+$ populations, and an increase in coo vocalizations. The presence of a preferred companion modulated these changes and was associated with a quicker return to baseline levels (Gust, Gordon, Brodie, & McClure, 1994). However, there were large individual differences in the degree to which the companion modulated the stress-associated changes in lymphocyte subsets. Although separation from social partners resulted in immuno-modulation for both male and female juveniles and for adult females, the effect was somewhat different for adult males. Removal from a large social group had no effect on cortisol, absolute number of T-helper and T-suppressor cells or B cells, or absolute number of white blood cells (Gust, Gordon, & Hambright, 1993). These differences in immunomodulation may reflect the different life histories of male and female macaques. Females remain in the social group in which they are born and maintain strong social bonds with maternal kin, whereas males immigrate to other troops as young adults.

In a laboratory situation, macaques can live to be over 30 years of age, with females exhibiting menopause at about 25 years of age. Older animals (>20 years of age) show significantly lower proliferative responses to mitogens and lower NK cytotoxicity than do younger animals (Erschler, Coe, Gravenstein, Schultz, Klopp, Meyer, & Houser, 1988). The age-related decrease is greater in males than in females. Older animals also tended to have lower antibody responses to tetanus vaccine. Because younger monkeys have been shown to benefit from affiliative behaviors during social challenges, it was thought that older animals might also benefit from the effects of social housing and increased social interaction. In a study to test this relationship, social housing did indeed enhance the well-being of older animals initially: Old individuals that were housed with one or more juveniles were more active than their counterparts that were housed alone or with one other aged individual. However, the immunological data showed that these animals had decreases both cross-sectionally and longitudinally in lymphocyte proliferation and NK activity. It is apparent that unregulated social interaction was physiologically very stressful for older animals. Housing the older monkey in a cage that allowed it to control the access of a single juvenile alleviated these effects. Under these conditions, lymphocyte proliferation or NK activity was not depressed. It is noteworthy that old macaques (>25 years of age) living under open corral conditions also engage in fewer social interactions than younger individuals (Worlein, unpublished observations). Thus, the ability of social affiliation and social interaction to mitigate stressor effects may not be constant but actually vary quite significantly across the life span of an individual.

14.2.6. Prediction of Increased Risk following Social Loss

Autonomic indicators, such as resting heart rate, may be useful predictors of risk for untoward behavioral and physiological responses to social stressors such as maternal separation (Boccia, Laudenslager, & Reite, 1994). This concept has received attention in studies of both human and nonhuman primates. Average daytime heart rate is related to the behavioral response to separation in 6-month-old pigtailed macaques. Rates of vocalization (e.g., agitation) and time spent in withdrawn, huddled postures (e.g., depression) were greatest in monkeys whose heart rates were above the median of average daytime baseline heart rate noted the week before separation. Higher heart rates may reflect more reactive physiological and behavioral patterns in these monkeys. One might predict that monkeys with higher heart rates at baseline would show greater immunological and endocrinological responses to separation because the same behaviors were correlated with greater changes in lymphocyte activation and plasma cortisol during maternal separation.

Human children with more reactive response patterns are at greater risk for immunological disease such as allergic disorders (Kagan, Snidman, Julia-Sellers, & Johnson, 1991). Studies have also demonstrated that humans with high cardiac sympathetic reactivity also have high hypothalamic–pituitary–adrenal reactivity and immune modulation during stress

(Cacioppo, 1994). Given the inherent variability in the response of monkeys to maternal and other forms of social separation (Lewis *et al.*, 1976), intrinsic differences in biological reactivity, as reflected by these autonomic parameters, may act as important markers for individuals at greater risk for untoward responses to social stressors.

14.2.7. Pharmacological Intervention in Nonhuman Primate Models of Social Loss

Animal models afford the opportunity to investigate the efficacy of pharmacological modification of responses to social stressors. Social separation experiences and rearing conditions alter CNS neurochemistry as indicated by changes in neurotransmitters and metabolites in cerebrospinal fluid (CSF) (Clarke, Hedeker, Ebert, Schmidt, McKinney, & Kraemer, 1996; Kraemer *et al.*, 1991). Levels of CSF norepinephrine and its metabolite, 3-methoxy-4-hydroxyphenylglycol, rise following maternal separation. To the extent that psychoactive agents may attenuate these changes in brain neurochemistry during separation, use of antidepressants may be important in the treatment of severe responses to loss that require clinical intervention. This may be particularly important when autonomic homeostasis is also disrupted. Tricyclic antidepressants attenuate some of the behavioral responses to social separation in nonhuman primates (Hrdina, von Kulmiz, & Stretch, 1979; Suomi, Seaman, Lewis, DeLizio, & McKinney, 1978). Continuous administration of a monoamine oxidase inhibitor, clorgyline, not only reduced the rise in plasma cortisol and prolactin associated with maternal separation (Fig. 1), but also diminished several behav-

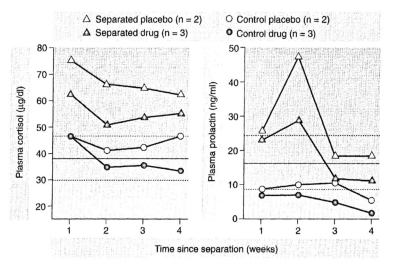

Figure 1. Plasma total cortisol and prolactin as determined by radioimmunoassay of samples obtained from pigtailed macaque infants that were separated from their mothers and from matched control subjects not experiencing separation. Samples (and matched controls) were obtained 1, 2, 3, and 4 weeks after separation. All subjects were implanted with devices that continuously administered either a monoamine oxidase inhibitor (clorgyline) or vehicle (placebo). Mean plasma cortisol and prolactin during baseline are indicated by the solid line; 1 S.E.M. is indicated by the dotted horizontal lines. Drug treatment attenuated the rise in plasma cortisol and prolactin. Unpublished observations of Laudenslager, Reite, and Boccia.

ioral, immunological, and cardiovascular changes in 6-month-old pigtailed macaques that were experiencing separations (Laudenslager, Reite, & Boccia, unpublished observations).

There is a potential downside to treatment of the young organism with these anti-depressants. The long-term consequences of treatment are poorly understood. If fluoxetine, a selective serotonergic reuptake inhibitor, is administered during social separation experiences in 8-month-old macaques, trafficking of lymphocytes into the CSF is affected as many as 2 years later (Coe, Hou, & Clarke, 1996). Furthermore, young monkeys that received antidepressant treatment (either fluoxetine or desipramine) during social separation experiences prior to 1 year of age, showed reduced specific antibody levels following tetanus immunization when immunized at 4 years of age (Laudenslager, Clarke, Kraemer, & Goldstein, 1997). The use of psychoactive substances for treatment of depression in young children needs to be carefully considered in light of the recent observations of their previously unknown effects on subsequent blood–brain barrier and immune regulation. Data suggest that other aspects of early rearing conditions may also have long-term influences on immune regulation.

14.3. Early Experience

Early social separation experiences, albeit brief, may have long-term behavioral and immunological consequences. The experience of a single 2-week maternal separation experience at 6 months of age, followed by reunion with the mother, affects subsequent behavioral responses to novelty in juvenile and adult monkeys (Capitanio, Rasmussen, Snyder, Laudenslager, & Reite, 1988; Capitanio & Reite, 1984). These monkeys are more timid in new situations, taking longer to explore novel objects or obtain preferred fruit under conditions of novelty. They have smaller social networks and fewer grooming partners as adults. These behavioral differences may be mediated partly through long-term changes in the mother–infant relationship following reunion with the mother. We have noted that pigtailed macaque mothers that have been separated briefly from their young tend to be more attentive and restrictive when they are reunited: They wean the young monkeys at a later age than if they had not been separated. They also maintain greater control over their physical proximity.

Immunologically, adult monkeys that have had early experiences with brief social separation have lower mitogen activation of lymphocytes than do adults that have not experienced these events (Coe, Lubach, Schneider, Dierschke, & Erschler, 1992b; Laudenslager, Capitanio, & Reite, 1985). In contrast, the influence of similar experiences on NK activity is quite different. Laudenslager, Berger, Boccia, and Reite (1996) studied NK activity of both bonnet and pigtailed macaques every 6 months after an early two week separation experience after which the mother was returned. Under resting, nonchallenged conditions, lymphocytes from previously separated monkeys showed *greater* lysis of tumor targets than lymphocytes from control animals.

14.3.1. Role of the Mother in Mediating Long-Term Effects

The preceding observations are reminiscent of the observations of Solomon, Levine, and Kraft (1968) in which rat pups experiencing early handling were more responsive to antigen challenge as adults. The changes in behavioral and physiological regulation in rats following early handling have been attributable to differential reactions of the mother to the pups that were handled. Handled pups received increased grooming from the mother after return to the nest (Smotherman, Brown, & Levine, 1977). Increased maternal care has been shown to affect the nature of hypothalamic–pituitary–adrenal responses to stressors (Liu, Diorio, Tannenbaum, Caldji, Francis, Freedman, Sharma, Pearson, Plotsky, & Meaney, 1997). Perhaps a part of the long-term differences observed in monkeys that experienced early separation may also be related to alteration in the relationship between the mother and the young monkey when the mother is returned.

Free-ranging macaque mothers typically resume mating activity when the young monkeys are 5–8 months old, before weaning is complete (Laudenslager *et al.*, 1993). This is a time of significant conflict for the young monkey because its mother frequently rejects and punishes it as she attempts to develop a consortship with a male. Antibody responses to prophylactic immunization with tetanus toxoid in yearlings were directly related to behavioral responses that occurred during the period of their mothers' mating activity. Six- to eight-month-old monkeys that showed high distress during their mothers' mating activity had lower antibody levels to immunization as yearlings.

14.3.2. Disease Risk and Early Rearing

Consonant results were obtained in a study of the relation between stressful early experiences and the course of infection following inoculation with simian immunodeficiency virus (SIV): The number of cage moves and separations from either mother or familiar peers was correlated with a shorter latency to clinical disease following inoculation (Capitanio & Lerche, 1991). These observations demonstrate the significant impact that early experiences may have on subsequent immune regulation and resistance to disease.

14.3.3. Hand-Rearing Experiences

Early rearing conditions can affect the developing immune system in rhesus macaques (for review see Coe, 1993). Infant monkeys that had been hand-reared by humans in a nursery had significantly higher lymphocyte responses to mitogen stimulation at 5–8 months of age than infants that had been reared by their mothers (even when the mother-reared infants had also been subjected to either daily or one 2-day maternal separation). Although levels were still elevated 1½ years later, by 2½ years of age only concanavalin A (Con A) responses were still significantly elevated while phytohemagglutinin (PHA) and pokeweed mitogen responses had returned to normal ranges. The nursery-reared infants also had a significantly higher ratio of T-helper (CD4$^+$) to T-cytotoxic/suppressor cells (CD8$^+$) at 6, 12, 18, and 24 months of age and consistently lower cytotoxic activity than infants that had been reared with their mothers (Lubach, Coe, & Erschler, 1995). Young monkeys that were weaned from their mothers early (at 6 months of age) had mitogen stimulation profiles that were intermediate between the profiles of nursery-reared infants and those of monkeys that were reared with their mothers until they were more than 1 year old, the age at which macaques are normally weaned from their mothers under free-ranging conditions.

Although some of these effects on immune profiles may have been related to the psychological impact of rearing without regulatory input from the mother (Hofer, 1994), infants raised in a nursery setting differ from mother-reared infants in a number of other ways. First, infants raised in the nursery were fed human infant formula and were larger and heavier than mother-reared infants at all developmental time points (Coe *et al.*, 1992b). Second, infants raised in the nursery were more prone to gastrointestinal infections than infants raised with their mothers (Coe, 1993). Nursery-reared monkeys may have been exposed to a different range of infectious agents because of their close exposure to humans during rearing as opposed to a macaque mother. Although there is no doubt that early rearing conditions can have long-term effects on immune profiles in macaques, it should be noted that the true functional significance of these differences is not entirely clear, especially in light of the fact that antibody responses to vaccination with three different antigens were no different in nursery-reared than in mother-reared infants (Coe et al., 1992b).

14.4. Social Conflict and Immunity

14.4.1. Social Group Formation

Immune functioning is affected not only by social separation but also by any type of social situation that increases conflict between animals. In studies of immune responses during social group formation, unfamiliar monkeys were placed together in a common housing area (Gust, Gordon, Wilson, Brodie, Ahmed-Ansari, & McClure, 1991). While the animals were establishing a new social structure and dominance hierarchy, there were increased levels of aggression, but the levels commonly decreased to baseline within a few days. Social instability and aggression received were correlated with a variety of physiological and behavioral consequences, which were exacerbated in individuals that attained a lower dominance rank. Both T-helper and T-suppressor cells were depressed in all individuals for the first 24 h after the group was formed. However, these decreases were longer lasting in females that attained low social rank, and they were still apparent even 9 weeks later. Immunological measures were not highly correlated with aggression received, but were associated with grooming, an affiliative behavior. Higher-ranking individuals received more grooming and also recovered from the effects of group formation more rapidly. Similar effects have been seen in male macaques following frequent social group reorganization (Kaplan et al., 1991).

14.4.2. Return of Individuals to Familiar Groups

Formation of groups of unfamiliar individuals is not the only experimental manipulation that can result in increased levels of aggression in macaques. Return of individuals to a familiar group also results in increased aggression directed toward the familiar intruder. When familiar adult male rhesus macaques were released into their group after a 1-year absence, they were threatened by the majority of the resident group members. These intruder males showed significant increases in cortisol and significant decreases in absolute numbers of T-helper and T-suppressor lymphocytes. Changes in physiological parameters were significantly correlated with the amount of aggression the monkeys received. Individuals that received more aggression showed a greater increase in cortisol and a greater decrease in T lymphocytes (Gust et al., 1993).

A similar pattern of behavioral and physiological responses was seen in juvenile animals that were returned to their social group after shorter absences (11–18 weeks) (Gordon & Gust, 1993). After their return to the social group, the animals experienced aggression and showed significantly higher cortisol levels and significantly lower T-helper and T-suppressor lymphocytes. There were also significant correlations between increases in cortisol and decreases in T-cell subsets within subjects as well as significant correlations between aggression and T-cell subsets and cortisol. Therefore, it is likely that immune changes were related to cortisol, which increased in response to aggression received.

14.4.3. Constantly Changing Social Conditions

In a study of the cardiovascular and immune consequences of long-term exposure to unstable social conditions, cynomolgus macaque males were subjected to monthly reorganizations of their social groups. After 26 months of social instability, the subjects showed lower responses to mitogen stimulation than did males that were housed in stable social groups. Subjects that showed more affiliation and were less aggressive had higher levels of lymphocyte response to Con A and PHA at the end of the study. NK activity was higher in all animals that showed higher levels of affiliation, even the more aggressive males. Interestingly, even in the stable social groups, animals that showed more affiliative behavior had higher levels of Con A-stimulated T-lymphocyte mitogenesis (Cohen *et al.*, 1992; Kaplan *et al.*, 1991). As with nursery rearing data, the functional significance of these differences has not yet been fully explored. The primary IgG response to tetanus toxoid immunization was higher in subordinate animals (those receiving more aggression) at the beginning of the 26-month study (Cunnick, Cohen, Rabin, Carpenter, Manuck, & Kaplan, 1991a). When secondary response to tetanus toxoid was measured 9 months later, animals that had undergone repeated social reorganization had a greater IgG response regardless of rank. These observations emphasize the need for any immune panel to include some measure of *in vivo* immune function (Maier & Laudenslager, 1988) so that the functional significance of these changes can be ascertained.

In a second study, male cynomolgus monkeys were reorganized four times over a period of 5 months. As with the longer study, reorganizations resulted in immunomodulation, increased lymphocyte counts, and decreased lymphocyte proliferation. These effects were especially prevalent in individuals that showed high levels of fear. However, as with the longer study, a functional measure of immune competence (antibodies to herpes B virus) was not negatively affected by the stressor (Line, Kaplan, Heise, Hilliard, Cohen, Rabin, & Manuck, 1996). However, more recently, Capitanio and Lerche (1998) found that housing relocations and social separations occurring in the 90-day period prior to inoculation with SIV and 30 days after were associated with decreased survival. There appear to be clear indications of greater health risk associated with unstable social conditions.

14.4.4. Competition for Restricted Resources

Another situation that increases conflict in nonhuman primates resulting in immunomodulation is competition for a restricted resource. When groups of macaques were denied access to water for 18–24 h, the animals displayed increased agonistic behavior and competition for the water spout when access was restored. However, behavioral and immunological changes varied according to species (Boccia, Laudenslager, Broussard, & Hijazi, 1992). There were clear differences between pigtailed and bonnet macaques in immune and endocrine measures both at baseline (before experimental manipulation) and in response to experimental manipulation. At baseline, pigtailed macaques showed lower lymphocyte activation in response to mitogens, higher NK activity, lower percentages of lymphocytes and higher percentages of segmented neutrophils in plasma, and lower total WBC than did bonnet macaques. After the experimental manipulation, only pigtailed macaques showed an immunological change, that is, significantly increased NK cytotoxicity and decreased proliferative responses to mitogens compared with preexperimental values. These changes may be related to the fact that agonistic behaviors were consistently higher in pigtailed macaques. Although immune measures were not directly influenced by the dominance status of the individual, there was a relation between the number of times a given individual was displaced from the water spout during the test and the proportion of change in responses to PHA and Con A. Animals that received more displacements also showed a greater decrease in mitogen stimulation.

14.4.5. Aggression under Free-Ranging Conditions

Physiological correlates of increased aggression are not unique to laboratory settings. Increased rates of aggression have also been associated with altered numbers of lymphocytes and higher levels of cortisol in baboons in free-ranging situations. When an extremely aggressive adult male baboon joined a troop of baboons, the troop experienced increased levels of aggression, mostly instigated by the immigrating male and disproportionately aimed at females (Alberts, Sapolsky, & Altmann, 1992). Physiological profiles of troop members taken 2 weeks later showed that levels of cortisol were higher than before the male joined the troop, and these levels were significantly higher in animals that were the focus of his aggressive behavior. Interestingly, the immigrating male himself had high levels of cortisol. Mean numbers of lymphocytes in blood of resident troop members were lower after the male's immigration and negatively correlated with plasma cortisol.

14.5. Other Laboratory Experiences

A few studies have investigated immune responses in macaques subjected to stressors unique to laboratory environments. Vervet monkeys were inoculated with an antigen, bovine serum albumin (BSA), and subjected to a variety of stressful but harmless manipulations (housing in small cages, a 21-min cycle of noise, lights, and silence, dropping housing cages 15 cm at random intervals, and an irregular feeding protocol) (Hill, Greer, & Felsenfeld, 1967). Control animals were not disturbed except for the venipuncture necessary to obtain blood samples. Cortisol levels initially rose in the stress-exposed group, but fell to the level of control animals during week 4 of the protocol. At 5 and 6 weeks after the beginning of the experiment, the stressed group showed significantly lower levels of antibodies to the BSA. The process of restraint, which frequently occurs in a laboratory environment, is capable of changing relative numbers of lymphocytes, monocytes, and neutrophils in addition to increasing plasma levels of cortisol and β-endorphin and reducing NK activity (Morrow-Tesch, Mcglone, & Norman, 1993).

Administration of medications routinely used for chemical restraint, such as ketamine, has been found to affect the generation of specific antibodies in laboratory rats (Lockwood, Siebert, Laudenslager, Watkins, & Maier, 1993) and alter patterns of lymphocyte activation and natural cytotoxicity for several days in macaques (Worlein, Berger, & Laudenslager, 1995). These results emphasize the need for careful handling of experimental subjects in a laboratory setting.

14.6. Summary

The relations between social experiences and physiological well-being are summarized in Fig. 2. Each social challenge a nonhuman primate experiences is associated with behavioral and physiological outcomes (Pathway 1). These outcomes are modulated by both intrinsic (Pathway 2) and extrinsic (Pathway 3) factors. Some of the intrinsic factors

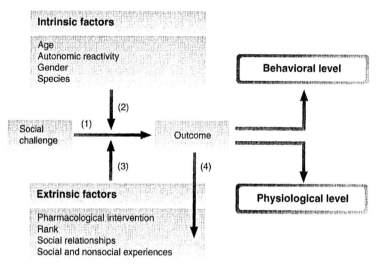

Figure 2. Summary of factors contributing to differences in outcomes following social challenge in nonhuman primates.

discussed include age, inherent reactivity of the autonomic nervous system, gender, and species. Extrinsic factors are rank, prior social and nonsocial history, and social relationships. Thus, experience with any social or nonsocial challenge is likely to affect responses to subsequent challenges (Pathway 4). Many of the behavioral and physiological outcomes can be modified pharmacologically. Finally, study of the role of social interactions and social relationships has indicated that affiliative behaviors such as grooming or simple proximity can mitigate many negative consequences of social challenge.

Nonhuman primates provide a unique opportunity to study models of many processes, both normal and pathological, in humans. Although this chapter has focused primarily on social separation and conflict, there are many other situations in which nonhuman primates can provide important information regarding the human condition.

14.7. Methodological Notes on Observing Behavior

Watching monkeys would seem to be a pleasant and entertaining enterprise. However, when it comes to identifying discrete observable events among behaviors occurring in a constant stream of actions, the task is not as simple as it first appears. Behavioral research depends on the researcher's ability to construct a behavioral taxonomy for the behaviors of interest within the constraints of both the experiment and the species under investigation. This involves dividing behaviors into discrete categories and providing labels for those categories. Before assembling a behavioral taxonomy and selecting a sampling method, the researcher must choose a strategy on the basis of at least three factors: the purpose of the study, the setting in which it will be conducted (laboratory, semi-free-ranging, or field), and the species and number of animals to be studied. For accurate statistical analysis, the observed behaviors must be mutually exclusive (no two behaviors can occur at the same time) and exhaustive (no time can pass without a behavior being scored). To be exhaustive, a taxonomy typically includes a category labeled "other" for all behaviors not otherwise specified. It should be noted that even inactivity (not engaging in any particular behavior) can be an important behavioral classification, for there may be processes that actively inhibit a subject's behavioral response. Therefore, an "inactive" or "passive" category should be included in any behavioral taxonomy.

Having selected a behavioral taxonomy, the researcher must decide how to collect behavioral data. There are several commonly used sampling methods (Altmann, 1974). In *focal animal* sampling the observer records all behaviors of one animal for a specified period of time. Similarly, in *focal subgroup* sampling the observer records all behaviors of a selected subgroup, such as a mother–infant pair, for a specified period. In *specific behavior* sampling the observer focuses on the whole group of animals and records all occurrences of a specific behavior within the group. This method is effective for studying behaviors that occur infrequently. In *scan* sampling the observer records the behavior of each group member at a specified time. With all of these methods, the order in which subjects are sampled during each observation period must be randomized to avoid bias. The time of day that observations are collected also should be randomized because there are likely to be diurnal variations in behavior.

The duration of observation sessions depends on the behaviors to be studied. If the behaviors occur frequently, sampling sessions may be brief. Conversely, if the behaviors of interest occur infrequently, sampling periods must be extended. Within an observation period, behavior can be sampled either continuously (called *real-time sampling*) or at predetermined intervals (called *instantaneous sampling*). Real-time sampling is most feasible when used in conjunction with a computer-based data collection system (Noldus, 1991). Instantaneous sampling is done by dividing the observation period into intervals, which are signaled by a timer, and recording the behavior of the focal animal or focal subgroup at the instant the timer sounds. A third method, called *modified frequency sampling*, permits a behavior to be scored only once during a time interval, no matter how many times it occurs. This method is not recommended because neither true frequencies nor proportions of time engaged in behavioral states can be obtained from the data.

Behavioral observations can be recorded on a computer in real time or on simple check sheets. If more than one observer will be observing behavior during a study, the percentage of agreement or correlation between two observers watching the same interactions and behaviors (*interobserver reliability*) becomes an important issue. Most published studies report interobserver reliability scores of 85% or better. A more precise assessment of interobserver reliability is the kappa score (Cohen, 1960), defined as "the proportion of interobserver agreement after chance agreement is removed from consideration."

From this brief overview one can see that the observation of behavior is not as simple as it might seem and is subject to a number of methodological difficulties. For a more detailed description of the issues associated with behavioral observations, see Sackett (1978).

ACKNOWLEDGMENTS. The preparation of this chapter was supported in part by the United States Public Health Services, National Institutes of Mental Health research grants MH37373 (M.L.L.), MH19514 (M.L.L.), and MH45045 (M.L.L.), NIMH Institutional Postdoctoral Training Grant MH15442 (J.W.), and NIH grant RR00166 to the Washington Regional Primate Research Center. We would like to thank Kate Elias for her editorial assistance.

15   Acute Psychological Stress

Robert J. Benschop and Manfred Schedlowski

15.1. Introduction

Organisms survive by maintaining homeostasis, a complex, dynamic, and harmonious equilibrium. This equilibrium is constantly challenged or threatened by disturbing factors or stressors. For survival, organisms have to actively react with an adaptational or stress response, which should lead to the preservation or reestablishment of the steady state. Stressors are qualitatively and quantitatively very diverse in nature. They are either external or internal in origin and may be caused by psychological (e.g., feelings of fear) or physiological (e.g., exercise-induced glucose shortage) events.

In contemporary literature, stress responses are discussed in the context of whether the stressor is physical or psychological in origin. In our modern daily life, these components are for the most part difficult to separate. Feelings of anxiety may very well lead to a physiological response (muscular tension) and physical activity is often dependent on

Robert J. Benschop • Division of Basic Sciences, National Jewish Medical and Research Center, Denver, Colorado 80206.   Manfred Schedlowski • Institute of Medical Psychology, Essen University Clinic, D-45122 Essen, Germany.

psychological events. In this chapter the effects of acute psychological stress on components of the immune system will be discussed.

In contrast to chronic stress (discussed in Chapter 16), acute psychological stressors usually fall within a time frame of 30 to 60 min and require our immediate attention. Two main forms of psychological stress, emotional and mental stress, are distinguished. Emotional stressors induce feelings of fear, joy, anxiety, anger, and so on, whereas mental stressors require more rational cognitive processes, and lack strong emotional components. Both types of stress are experimentally used to evoke an acute response in psychoneuroimmunological research.

15.2. Methodological Aspects

In contrast to research on the effects of chronic stress on the immune system, studies on the effects of acute stressful situations can be performed in the laboratory. This has several advantages over field studies. First, different subjects can be exposed to the same standardized situation. Another advantage of these laboratory studies is that the physiological changes induced by these stressors can be measured relatively easily; cardiovascular measurements can be taken continuously and blood can be drawn at designated time points to determine endocrine parameters and components of the immune system. From a methodological point of view, the inclusion of a control (i.e., nonstress) group is very important. Although no major changes are to be expected in a control group, it enables the investigators to distinguish between stressor-induced changes and accidental changes (e.g., related to circadian rhythms). Several models are being used to induce psychological stress, and an example of the most important ones will be described. A point of criticism regarding these laboratory studies is that they are artificial and do not have much ecological value. Although this is more true for mental stressors than for emotional stressors, some experimenters have attempted to counteract this argument by making the experimental models more lifelike, as we will see later on.

The effects of acute stress on cardiovascular and endocrine parameters have long been subjects of study. Given the role of the immune system in disease prevention and control, considerable attention has been directed lately to the question of whether and to what extent the immune system is sensitive/susceptible to short-term mental and/or emotional processes. In this context, the effects of acute stress on immunological variables such as changes in WBC numbers, lymphocyte subset distribution, and cellular immune functions (e.g., lymphocyte proliferation and NK cell activity) are typically investigated.

15.3. Acute Psychological Stress and the Immune System

15.3.1. Emotional Stress

Studies dating back to the 1930s yielded evidence for an effect of emotional stress on the immune system: Farris reported considerable changes in lymphocyte numbers in subjects who had to take an important exam (Farris, 1938) (Fig. 1). The increase in cell numbers was clearly associated with an altered emotional state, as all subjects reported feelings of anxiety, nervousness, excitedness, and fear. The "personality" of the examiner also seemed to be important; instructors who were strict and known by reputation to fail students, undoubtedly had a more marked effect on lymphocyte numbers.

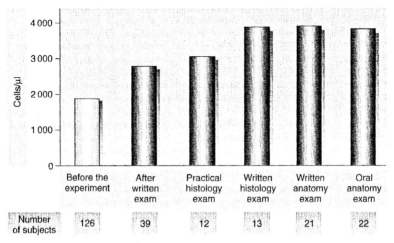

Figure 1.   The effects of several different exams on lymphocyte numbers. From left to right, the exams became progressively more difficult. Based on Farris (1938).

Exam stress is not the best example of an acute stressor, because it usually stretches over a period of days or weeks. Early studies such as this one, however, indicate the longstanding interest in the effects of psychological stress, and its potential to induce changes in the body, particularly in the immune system.

In other studies, the effect of acute emotional stress on the immune system was investigated using a parachute jump as a stressor. Pilot studies had clearly established that this stress model represented a well-controllable, standardized situation yielding reproducible results. Every subject undergoes the same routine procedure, including the check of the equipment, boarding of the airplane, climbing to jump altitude, jump, opening of the parachute, and landing. In addition, the amount of stress induced in such a field setting surpasses that induced in the laboratory. In one study (Schedlowski, Jacobs, Stratmann, Richter, Hädicke, Tewes, Wagner, & Schmidt, 1993b), physiological changes were monitored in subjects before, during, and after their first parachute jump. In these so-called tandem-parachute jumps, the subject is secured in front of an experienced tandem instructor (Fig. 2). In this manner, the subject is merely a passenger with no active role in the actual jump procedure, thus minimizing the effect of physical activity. As might be expected, state anxiety increased in this situation, and peak heart rates of 160 bpm were monitored at exit of the plane. Lymphocyte subset analyses revealed large increases, predominantly in the number of NK cells (CD16$^+$ and CD56$^+$) and NK cell activity (Fig. 3). The immunological variables were also monitored 1 h after the jump, to determine the longevity of the induced stress. It was found that at this time, the number of NK cells was lower than at the beginning of the experiment, indicating a biphasic effect on NK cells (Fig. 3). The same pattern of change was observed for the activity of NK cells. In a follow-up experiment, an additional blood sample was taken 4 h after the jump; at this time point, both NK cell number and activity had returned to baseline (Benschop, Jacobs, Sommer, Schuermeyer, Raab, Schmidt, & Schedlowski, 1996a).

Speech stress models take advantage of the fact that most people dislike speaking in

Figure 2. Parachute jumping as a model of emotional stress. The test subject is secured in front of an experienced tandem-master.

public and want to avoid a confrontation. This model can be categorized under mental as well as under emotional stress, although the emotional part probably is the more important one. In the experimental model in the laboratory, role-play is used in which the experimental subjects can be accused of shoplifting and have to defend themselves before a jury or they may be asked to present themselves at a job application. Usually, the situation is intensified by videotaping the whole procedure and telling the subjects that these tapes will later be analyzed by experts in behavior. Although this model has strong effects on stress hormones (cortisol and adrenaline), to date only few data are available for changes in the immune system. Preliminary data from our laboratory reveal significant increases in the number of

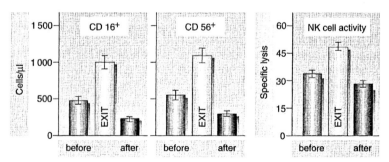

Figure 3. Effects of a first-time parachute jump on the number of NK cells (CD16+ and CD56+) in the circulation and on NK cell activity. From Schedlowski *et al.* (1993b).

NK cells following a 5-min speech. One hour after the speech, the number of NK cells had returned to baseline.

The abovementioned studies all induce nonspecific emotional stress. In an attempt to study the effects of specific emotions in the laboratory, subjects were asked to relive a particularly sad or joyful moment. This led to a decrease in lymphocyte proliferation to concanavalin A and phytohemagglutinin, with larger effects of negative emotions (Knapp, Levy, Giorgi, Black, Fox, & Heeren, 1992). There are, however, very few studies in this area, mainly because it is rather difficult to investigate the effect of one isolated emotion on the immune system.

15.3.2. Mental Stress

The "Stroop-color-word conflict task" is a classical example of a laboratory stressor used to induce mental stress. In this test, subjects have to name the color of letters forming various words. This sounds easy, but the presented words are names of different colors and the whole procedure aims to confuse the subject. For example, the word *red* is written in yellow letters and the subject has to say *yellow*. The words are presented at high speed to increase the rate of mistakes.

In a study by Landmann, Mueller, Perini, Wesp, Erne, and Buehler (1984), heart rate and blood pressure were both increased during the test (+8 bpm and +13 mm Hg, respectively). Furthermore, significant increases were found in the numbers of NK and B cells. Unfortunately, no control group was included, but the results were interesting enough to prompt a follow-up experiment. In another study (Bachen, Manuck, Marsland, Cohen, Malkoff, Muldoon, & Rabin, 1992), the same stressor was used and changes in a control group were also measured. These investigators reported similar increases in heart rate and blood pressure in the stress group as Landmann *et al.*(1984), but did not find an increase in B cells. Likewise, the number of NK cells (characterized by CD16$^+$ and CD56$^+$) increased by approximately 35%, which was slightly less than reported in the first study. Lymphocyte proliferation in the stress group to the mitogen phytohemagglutinin was slightly decreased following the stressor as compared with the control group.

A second classical model for mental stress is mental arithmetic. Here, subjects have to serially subtract a small number (e.g., 7, 13) from a larger number (e.g., 2905) for approximately 10 min. In this process, they have to make as few errors as possible and are paced by a metronome and/or further pressured by coaching comments (e.g., "go faster," "try and concentrate"). Naliboff, Benton, Solomon, Morley, Fahey, Bloom, Makinodan, and Gilmore (1991) studied the effects of this stressor on the immune system in 23 female subjects. The subjects' impressions of their feelings in this situation were measured by mood checklists; feelings of stress, anger, and anxiety clearly increased during the task (Fig. 4a), as did heart rate (+20 bpm) and blood pressure (+22 mm Hg). No changes were observed in the number of T (CD3$^+$, CD4$^+$) and B (CD20$^+$) lymphocytes, whereas the number of cells expressing NK-cell-associated markers (i.e., CD8, CD16, CD56, and CD57) was increased by mental arithmetic by 60 to 100% (Fig. 4b). NK cell activity was also tested and increased significantly.

Because these two stress models have been criticized for a lack of representing relevant everyday situations, a more complex stress model for a population of high-school teachers was contrived (Brosschot, Benschop, Godaert, de Smet, Olff, Heijnen, & Ballieux, 1992). In this study, the teachers had to solve a three-dimensional puzzle under time pressure, but did not know that the puzzle was not completely solvable. Subsequently, the teachers had to explain their solution to another subject, who actually was a confederate of the experimen-

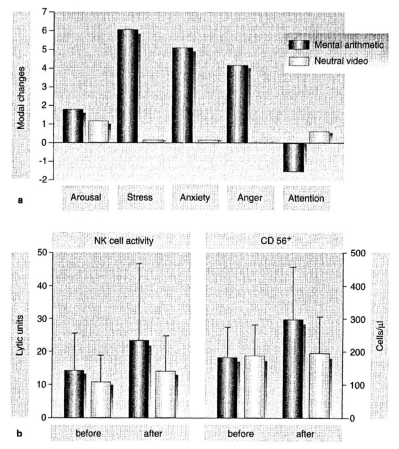

Figure 4. Mental arithmetic induces feelings of stress, anxiety, and anger, whereas a neutral stimulus (video) does not affect subjective mood (a). In association with this, the number of CD56+ lymphocytes increases, as does the NK cell activity (b). Based on data from Naliboff *et al.* (1991).

ter. Because the puzzle was unsolvable and the performance of the confederate did not improve despite the efforts of the teacher, they were faced with an impossible situation and were frustrated in what they are supposed to do best: teach. Indeed, the teachers reported feelings of tenseness and irritation, which indicates the induction of mild psychological stress. As evidence for physiological arousal, heart rate and blood pressure were elevated. The stressor also increased the numbers of CD8+, CD57+, and CD16+ cells. Because CD8 is expressed on subsets of both T and NK cells, two-color fluorescence stainings were performed, to determine whether this rise in CD8+ cells was related to an increase in NK or in T cells. The results demonstrated that only the number of cells expressing CD57 (i.e., NK cells) increased, but that the number of CD8+ cells lacking CD57 did not change (Fig. 5). T- and B-lymphocyte proliferation to phytohemagglutinin and pokeweed mitogen did not

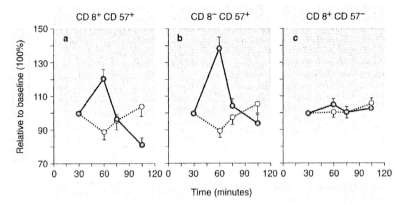

Figure 5. The effects of an interpersonal stressor on lymphocyte subset numbers. Because CD8 is expressed both on a subpopulation of T (CD3$^+$) cells and on a subpopulation of NK (CD3$^-$) cells, two-color fluorescence analyses were performed to separate these two subsets. Only cells that express the NK cell marker CD57 increase after the stressor (either CD57$^+$CD8$^+$ or CD57$^+$CD8$^-$); the number of CD8$^+$ cells that are negative for CD57 (CD8$^+$CD57$^-$) do not increase. From Brosschot *et al.* (1992).

change in this situation. Analysis of two more blood samples, which were taken 15 and 45 min after the stressor, revealed that the number of NK cells had already returned to baseline levels 15 min after termination of the stressor. Thus, although mental stress can induce marked increases in NK cell number, the effects are relatively short-lived.

Summarizing, these results demonstrate that both mental and emotional stressors have a positive effect on heart rate and blood pressure. In addition, it is clear that the immune system is sensitive to acute stress as well. Across studies, stress-induced increases in the number of lymphocytes are being reported, the majority of which are NK cells as evidenced by the increase in cells expressing cell-surface molecules such as CD2, CD8, CD16, CD56, or CD57. Along with the change in NK cell number, the total NK cell activity increases. The duration of the effect seems to depend on the type and/or intensity of the stressor: as described, a parachute jump induces a decrease in NK cell number and activity 1 h after the jump, whereas after mental stress cell number returns to baseline already 15 min after the stressor.

In contrast, lymphocyte proliferation, which mainly tests the capacity of T and B cells to divide after mitogenic stimulation, did not change or was even slightly decreased in the experiments described. The altered composition of lymphocyte subsets during psychological stress could be one of the reasons for the observed decrease; while the number of NK cells increases, the relative number of T and B cells decreases, and as a consequence fewer cells will proliferate in the proliferation assay.

15.4. Mechanisms of Stress-Induced Immunological Alterations

The effects of psychological stress on the immune system spurred investigators to look for the factors responsible for these changes. Emotional and mental stress causes the release of various hormones and neurotransmitters. Because lymphocytes express receptors for many of these factors (see Chapter 9), it could very well be that one or several of these substances are responsible for the changes in the immune system through direct binding to their respective receptors on lymphocytes. A great deal has been learned about stress-induced hormone cascades and their effects on the immune system from animal studies. The results of animal studies may serve as a theoretical basis on which to design human stress studies, although, from an ethical perspective alone, the experimental possibilities in studying these cascades in humans are limited.

15.4.1. Indirect Evidence

To get an indication of the factors that are important in influencing the immune system in humans, we may consider changes in other physiological systems. In one study the changes in cardiovascular variables positively correlated with the changes in NK cell number (Benschop, Godaert, Geenen, Brosschot, de Smet, Olff, Heijnen, & Ballieux, 1995). A positive relationship may be an indication for a common factor causing effects in both systems. Although correlations do not prove cause-and-effect relationships, they may be the basis for developing new hypotheses.

Traditionally, there is much more fundamental knowledge of the consequences of stress on regulation of the cardiovascular system than of the immune system. It has long been known that changes in blood pressure and heart rate positively correlate with changes in plasma levels of catecholamines (Lovallo, Pincomb, Brackett, & Wilson, 1990; Nestel, 1969). Release of catecholamines is a sign of CNS activation. This activation leads to the release of neurotransmitters at nerve endings in the tissue and to the production and release of hormones. We can therefore hypothesize that acute psychological stressors activate the sympathetic nervous system, leading to increases in the concentration of plasma cate-cholamines. This in turn induces changes in the cardiovascular and the immune systems. This effect of catecholamines can be systemic (i.e., in the circulation) or more local. Indeed, nerves belonging to the autonomic nervous system are found in almost all lymphoid organs (e.g., spleen, bone marrow, lymph nodes, thymus), with endings in close proximity to lymphoid cells (Felten and Felten, 1991) (see Chapter 8). The production of these substances may directly or indirectly be responsible for the changes in the immune system. In the parachute jump study discussed earlier, a positive correlation was found between the number of NK cells, measured immediately after a parachute jump, and plasma levels of noradrenaline before the jump (Schedlowski *et al.*, 1993b). Noradrenaline is one of the important neurotransmitters released at nerve endings. Thus, these findings demonstrate that the CNS is important in inducing changes in the immune system.

15.4.2. Direct Evidence

As mentioned earlier, correlations can only be used as indicators for relationships, but do not prove cause and effect. The positive relationship between noradrenaline and the number of NK cells does not necessarily mean that noradrenaline causes the changes in NK cells. It can be argued that noradrenaline is only one factor in a chain of events and causes the release of an unknown factor X that is actually responsible for the observed changes. It is also possible that a factor released simultaneously with noradrenaline is causing the NK cell changes. To prove the involvement of noradrenaline, a different approach is required.

Based on the results just described, follow-up studies were performed, hypothesizing that factors released after activation of the sympathetic nervous system are responsible for immunological changes. In animal studies, physical or biochemical denervation can be used to identify the neural pathways causing changes in the immune system. Obviously, such an approach is not suitable for human studies and a different experimental approach is required. In studies involving humans, a pharmacological approach seems indicated, in which either the release of the substance under investigation is inhibited or the binding of this substance to its receptor is blocked. Because hormones and neurotransmitters have to bind to their specific receptors to exert an effect, the participation and importance of a particular substance can be tested by inhibition of hormone–receptor binding with blocking drugs.

This basic pharmacological knowledge was used in further studies, in which it was hypothesized that the release of catecholamines and the subsequent binding to their receptors explain the effects of acute psychological stress on the immune system. To test this hypothesis, subjects were treated with a drug (propranolol) that binds to the β-adrenoceptors on all target cells in the body and thus inhibits the effects of adrenaline and noradrenaline (Fig. 6). These medicated subjects, as well as a group of subjects who had swallowed a placebo, were exposed either to a mental stressor (Benschop, Nieuwenhuis, Tromp, Godaert, Ballieux, & van Doornen, 1994b) or to an emotional stressor (Benschop et al., 1996a) and the physiological changes were monitored.

To induce mental stress a set of two tasks of the "active coping" type was chosen. These tasks are known to activate the sympathetic nervous system and to evoke cardiovascular reactivity. The first task is the Tone Avoidance reaction time task. Subjects have to react to a stimulus (an "X") that flares up irregularly in one of the corners of a computer screen. Subjects have to respond as fast as possible to this stimulus by pressing the button opposite to this corner on their response panel. Incorrect or too slow responses are punished

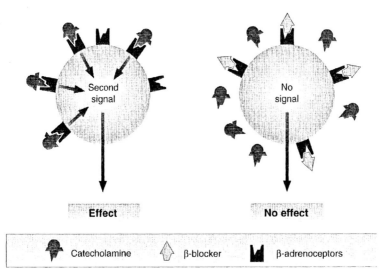

Figure 6.   Mechanism of action of β-adrenoceptor blocking drugs. Normally, binding of adrenaline to its receptor induces the generation of second messengers, which leads to downstream effects. β-Adrenoceptor blocking drugs a.k.a. β-blockers bind to the same receptor, but do not induce an intracellular signal. This occupation of the β-adrenoceptors prevents the binding of the natural ligand adrenaline and thus inhibits any downstream effects normally caused by adrenaline.

with a loud noise burst. The second task is a Memory Search reaction time task. Subjects have to memorize a set of four letters that is presented on a video screen at the start of the task. Subsequently, sets of one to four letters are presented in which either none (50%) or one (50%) of the memorized letters is present. Subjects have to respond as quickly as possible by pressing either the "present" or "not present" button. In both tasks the subjects received continuous feedback on the screen about their performance. If the subject starts to perform better, the tasks are automatically made more demanding. Motivation in this experiment was kept high by a combination of feedback, competition, and punishment.

In the placebo group, both heart rate and systolic blood pressure increased (18 bpm and 28 mm Hg, respectively), indicating activation of the sympathetic nervous system. An increase in the number of NK cells (CD56$^+$) was also observed in the placebo-treated subjects, paralleled by an increase in NK cell activity (Fig. 7). These increases in immunological variables and heart rate were completely inhibited in subjects treated with the β-adrenoceptor blocking drug propranolol. This demonstrates that binding of either adrenaline or noradrenaline to β-adrenoceptors is an essential step in stress-induced changes in NK cells.

Similar conclusions regarding the mechanisms of NK cell changes can be drawn from the results obtained with the emotional stress model (Benschop et al., 1996a). Here the already described parachute jump paradigm was used. Heart rate responses were markedly reduced in subjects treated with a ß-blocker. The adrenaline and noradrenaline concentrations in these subjects after the jump were increased to a similar extent as observed in the placebo-treated subjects (Fig. 8). This demonstrates that blockade of the β-adrenoceptors does not influence the endocrine stress-response, but only inhibits the binding of these catecholamines to their receptors. The increase in NK cell number and activity was completely abolished in subjects treated with the blocking drug (Fig. 8). This clearly indicates that changes in NK cell number are caused by the interaction of catecholamines with their receptors. Stimulation of β-adrenoceptors is thus a crucial step in the mobilization of NK cells.

15.5. From Psychological Stress to Biological Mechanisms

Using β-adrenoceptor blocking drugs, the central role of β-adrenoceptors in eliciting changes in NK cell number and activity was identified. To prove that the catecholamines adrenaline and/or noradrenaline are causing these changes, we have to leave the stress models and take a different approach. To demonstrate cause-and-effect relationships, investigators directly applied either adrenaline or noradrenaline to human volunteers and monitored the effects on the immune system. Such infusion protocols allow for additional experimental manipulation, such as dose–response relationships, and have the advantage that as a model, they are rather clean. Application of a given substance usually will result in the concentration increase of only the injected substance, whereas psychological stressors cause changes in the levels of various hormones and neurotransmitters.

In one of these studies, a single injection of adrenaline resulted in a transient increase in the absolute number of circulating lymphocytes (Crary, Hauser, Borysenko, Kutz, Hoban, Ault, Weiner, & Benson, 1983b). Analysis of the different subsets revealed that the percentage of NK cells was significantly elevated (Fig. 9). In parallel, the authors found a decreased percentage of CD4$^+$ lymphocytes, although the absolute number of CD4$^+$ cells did not change. These results have been confirmed by other groups and all studies indicate that injection of adrenaline influences lymphocyte circulation, most notably NK cells.

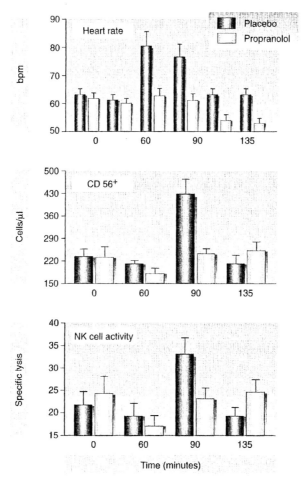

Figure 7.   Effect of β-blockade on the effects of a series of two mental tasks on heart rate, number of CD56$^+$ cells, and NK cell activity. The stress-induced increases in these variables seen in the placebo group were completely absent in the propranolol group. Adapted from Benschop *et al.* (1994b).

Similar observations were made after the injection of noradrenaline, although a much higher dose of noradrenaline is needed.

Additional studies show that this redistribution of lymphocytes involves the activation of β_2-adrenoceptors, but not the β_1 ones. In a similar design as described for the β-blockade studies of mental and emotional stressors, subjects who had received either placebo, propranolol, or bisoprolol were infused with adrenaline (Schedlowski, Hosch, Oberbeck, Benschop, Jacobs, Raab, & Schmidt, 1996). Propranolol blocks both β_1- and β_2-adreno-ceptors, whereas bisoprolol specifically binds to β_1-adrenoceptors and does not affect the binding of catecholamines to β_2-adrenoceptors. The increase in the number of NK cells induced by the adrenaline infusion could only be inhibited when both β-adrenoceptor

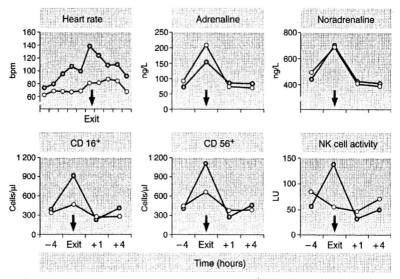

Figure 8. Effect of β-blockade on changes in heart rate, catecholamines, NK cell number (CD16+ and CD56+), and NK cell activity, induced by an emotional stressor (first-time parachute jump). While the endocrine stress-response is unaltered, the increase in heart rate, NK cell number, and activity clearly is abrogated. Adapted from Benschop *et al.* (1996b).

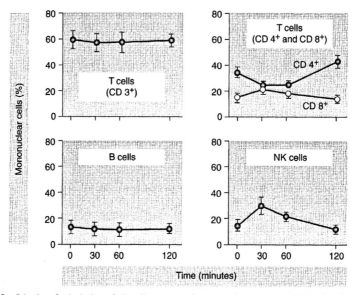

Figure 9. Injection of a single dose of adrenaline causes an increase in the percentage of NK cells and a decrease in the percentage of CD4+ cells. Adapted from Crary *et al.* (1983b).

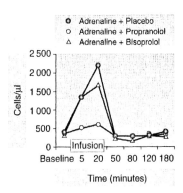

Figure 10. Effects of adrenaline infusion on NK cell number in the circulation of healthy volunteers treated with different drugs. In placebo-treated subjects, the NK cell number increases by 600%. Blockade of both β_1- and β_2-adrenoceptors by propranolol prevents this increase. Specific blockade of the β_1-adrenoceptors with bisoprolol does not inhibit the effects of adrenaline on the increase in NK cell number. Adapted from Schedlowski *et al.* (1996b).

subtypes were blocked. When only the β_1-adrenoceptors were blocked, the increase in NK cell number was similar to that observed in the placebo group (Fig. 10).

15.6. The Biological Meaning of Stress-Induced Immune Changes: "Good" or "Bad"?

The studies described above clearly demonstrate that acute psychological stressors can transiently influence the circulation of lymphocytes (predominantly NK cells) in the body. Additional experiments revealed the pivotal role of the catecholamines adrenaline and noradrenaline in these processes, which involves specific binding of catecholamines to their receptors. Why do catecholamines influence lymphocyte circulation and what is the biological meaning of the changes in NK cell circulation? The involvement of catecholamines implicates a broader picture: The stress-induced increase in NK cell number could be seen in the context of a general physiological response (fight–flight), as described by Cannon (1929). These responses to stressors are vital for evolutionary survival and are characterized by the activation of systems that are necessary for the defense of an organism, while other activities are being put "on hold." It is known that catecholamines reduce most immune functions *in vitro*, including NK cell activity (Kammer, 1988). This implies that immune functions are temporarily deactivated for the duration of a stress reaction in favor of other energy-demanding events. Intuitively, the effects observed during stress seem to be in contradiction to this reasoning: It is clear that after an increase in catecholamine concentration, a higher number of NK cells is present in the circulation, and that, as a result, the total NK cell activity is higher.

At this point, it is important to realize that the alteration in NK cell number is not so much a function of the immune system, but merely indicates that stress (in this case in the form of catecholamines) induces the recruitment and redistribution of these cells. It does not seem likely that this effect on NK cell circulation is an artifact of nature. Therefore, the real question is whether the organism could benefit from these changes. It is clear from many *in vitro* experiments that NK cell activity is greatly inhibited in the presence of catecholamines. The immunosuppressive effects of adrenaline *in vivo*, however, are short-lasting, because it is rapidly broken down by enzymes in the liver (half-life in the circulation <2 min). Interestingly, *in vitro* experiments have also shown that NK cell activity can be increased for up to 24 h after a short preincubation with cAMP-inducing agents (Kendall & Targan, 1980). This means that short exposure to adrenaline has important consequences for immune functioning after stress, i.e., when adrenaline concentrations have returned to baseline.

Originally, the goal of activating certain processes under acute stress was to eliminate threats and minimize physical damage, using fight or flight strategies. For humans, this original meaning is mostly lost in modern daily life. In a phylogenetically earlier stage, however, one of the consequences of an acute stressor could be tissue damage (e.g., wounds), with an associated danger of infection. At these damaged sites, tissue factors are being released, which attract NK cells and granulocytes. Because higher numbers of these cells circulate shortly after an acute stressor, more cells can be attracted, thus reducing the risk for infections. This hypothesis is supported by the fact that stress specifically induces the recruitment of cells of the first line of immune defense (i.e., NK cells and granulocytes) and not of the more specialized T and B lymphocytes. In addition, adrenaline can "prime" NK cells, rendering them more active once the peak concentrations of adrenaline are over (i.e., on termination of the stressful situation) such that they represent a more effective immune defense.

Are these stress-induced changes in the immune system "good" or "bad"? From the previous paragraph, we can conclude that the temporary increase in NK cells is a normal feature of physiological arousal. Moreover, the immune system could potentially benefit from the effects of adrenaline. Therefore, the effect of acute stress on NK cell number can be perceived as "good" (although "normal" or "useful" would be better terms). Nonresponsiveness, on the other hand, can be considered as an insufficient, thus undesirable ("bad"), response. Research in this area demonstrated that in chronically stressed subjects, acute psychological stress did not induce an increase in NK cell number (Benschop, Smet, Brosschot, Godaert, de Geenen, Olff, Heijnen, & Ballieux, 1994a). This observation suggests that the sensitivity of the β-adrenergic system is decreased in chronically stressed persons. Chronic stress can be regarded as a sequence of successive acute stressors. In order to prevent overstimulation, the sensitivity of the β-adrenergic system is downregulated to protect the organism from excessive effects of catecholaminergic stress. This is necessary because certain physiologically relevant processes are put on hold in the presence of high levels of adrenaline. The sensitivity of the system can be altered by downregulation of the number of β-adrenoceptors, or by lowering the affinity of these receptors for adrenaline. Indeed, prolonged exposure *in vivo* to agonistic drugs like terbutaline leads to a reduced number of β-adrenoceptors on lymphocytes (Aarons, Nies, Gerber, & Molinoff, 1983). Also, in patients with congestive heart failure (a disease state of increased sympathetic activity), the exercise-induced increase in adrenaline levels and lymphocyte numbers is markedly reduced or even absent (Maisel, Knowlton, Fowler, Rearden, Ziegler, Motulsky, Insel, & Michel, 1990b). This indicates that a state of chronic sympathetic stimulation (e.g., chronic stress) can modify the physiological responses to novel demanding situations.

15.7. Conclusions

The results described in this chapter clearly demonstrate that certain parts of the immune system are extremely responsive to psychological stress, or, to be more precise, to the hormones that are released in these situations. This again demonstrates the existence and importance of intricate relationships between the brain and the immune system.

16   Concepts and Models of Immunological Change during Prolonged Stress

Christopher L. Coe

16.1. Introduction

An understanding of the relationship between stress and immunity is critical for students and researchers in psychoneuroimmunology (PNI), because this type of immune alteration directly addresses several controversial tenets of the discipline. Over the last three decades we have learned a remarkable amount about how immune responses change during and after stressful events, but there is still considerable debate over the normal versus pathological nature of these shifts in immunity. For example, should these stress-induced changes be

Christopher L. Coe • Department of Psychology, University of Wisconsin, Madison, Wisconsin 53706.

considered adaptive or maladaptive? What types of life events are aversive and disruptive enough to elicit a sufficiently large and prolonged immune impairment to actually result in an immune-related disease? This chapter reviews some important concepts and findings that will help to answer these questions.

A discussion about stress-associated changes in immunity highlights both the resilience and vulnerability of the immune system, and the sometimes delicate balance between wellness and illness. Fortunately, most immune alterations induced by stressors usually do not result in overt disease because of the many redundant backup responses built into the immune system and because compensatory processes ensure that immune alterations are typically short-lived. Cognitive and emotional changes that facilitate coping with negative events are importantly involved in this compensatory adjustment. On the other hand, some stress-induced changes in immune responses have been found to create windows of opportunity for certain pathogens that might normally have been resisted and contained. Further, there are at least two phases in the life span, childhood and old age, when environmental and psychosocial perturbations may have greater consequences for health, because the immune system is not likely to recover as quickly and in some cases as completely.

16.2. Historical Overview: What Is Stress?

At the top of the list of controversial issues is the term *stress*. One of the founding fathers of stress research, Hans Selye, then a young Canadian physician in the 1930s, suggested that the word had already been so overused and applied in so many different contexts that a better one was needed to describe the body's reactions to negative and traumatic stimuli (Selye, 1936, 1956). He proposed *general adaptation syndrome* (GAS), but this label never gained popularity or universal acceptance. For PNI, however, it is of considerable historical interest that Selye and others (Mora, Amtman, & Hoffman, 1926) had discovered that shifts in immunity were a significant component of the stress or GAS response. Along with increases in endocrine activity and ulceration of the gastrointestinal tract, Selye noted that stressed animals and human patients showed similar immune alterations following various traumas and diseases:

- Lymphoid tissues tended to decrease in size and undergo structural alterations (i.e., thymolymphatic involution).
- The profile of white blood cells (leukocytes) in circulation changed (i.e., neutrophilia, lymphocytopenia, eosinopenia).

Because these changes appeared to be so common and they occurred after such diverse provocation, Selye concluded that these physiological changes were the *nonspecific*, global response of the body to any challenge, physical or psychological. This characterization built on observations of earlier researchers in the nineteenth century, and can even be traced back to the ancient Greek and medieval notion of *body humors* and their relationship to well-being and disease. From a scientific perspective, though, most refer back to the French physiologist Claude Bernard (1879). Bernard described how the body is buffeted by many environmental forces, but has regulatory processes to ensure the correct balance of the internal milieu (milieu intérieur). This idea was developed further in the twentieth century by U.S. scientist Walter Cannon (1929), who championed the term *homeostasis* to characterize the maintenance of normal physiological setpoints and detailed the adjustments mediated by the autonomic nervous system (ANS) during arousing and stressful times. During the activation phase, which he described as underlying behavioral components of the *fight-or-flight response*, the body's physiology is dominated by the actions of the sympathetic nervous system (SNS). Even today, we are still exploring the ramifications of this observation as many attribute the mediation of stress-related changes in immunity to the innervation of lymphoid tissue by the ANS, and to the effects of adrenaline and noradrenaline released into the bloodstream.

The topic of physiological mediation is discussed in more detail later, but at this point it is also important to emphasize that many researchers consider an appreciation of the cognitive and emotional reactions to be equally essential for an understanding of stress. Indeed, some have been opposed to describing stress on the basis of physiology alone, and have advocated that stress, at least in humans, is better understood as a psychological concept. This mind/body dichotomy has led to confusion in the scientific literature, because stress is sometimes defined and/or quantified by its emotional and cognitive antecedents or consequences—anxiety, dysphoria, hostility, negative outlook, or poor attention and performance. To facilitate our discussion, the emotional responses associated with stressful events will be considered as affective signs of "distress," and we will reserve the term *stress* for the physical responses of the body. Following this logical scheme, the precipitating events will be labeled as *stressors* to distinguish the initiating stimuli from the stress response itself.

It should be emphasized further that the debate over terminology emanates in part from the correct realization that personality and other psychological variables figure prominently in determining how one will respond in stressful situations, and whether or not the stressor will even elicit a physiological response. To describe the complex evaluative processes that occur at the cognitive level, researchers have often used the term *appraisal*. Richard Lazarus (1966) was one of the pioneering thinkers on this topic and advocated that how we look at a negative event, with respect to harm done and harm that may occur later, will determine the subsequent course of responses and actions. Our personal appraisal of the event guides the behavioral response (i.e., approach or avoidance) and influences the possible need for intrapsychic adjustments (i.e., denial, anxiety) that are involved in shaping our physiological response to stressful events. Even in extreme situations, such as when confronted with the impending death of a sick child, psychological defense mechanisms such as denial have been found to be capable of blocking signs of stress physiology in the parents (Wolff, Friedman, Hofer, & Mason, 1964). To facilitate the discourse for this chapter, these complicated cognitive and emotional processes will be grouped together and described simply as a component of *coping*. This process is important to consider, because when coping is adequate and successful, it may obviate the need for a bodily response to stressors. Put even more strongly, the effectiveness of coping may be gauged by the absence of stress physiology in situations that we would normally deem as stressful. Cognitive and behavioral adjustments may also account for why some immune alterations resolve and disappear over time in chronically stressful situations. One way to prevent immune alterations from persisting and causing disease is by changing the psychological meaning of the situation.

Many other perspectives have also proved to be important for predicting how people will react in negative situations. Within the field of PNI, there has been considerable interest in those personality variables that account for differences in stress reactivity. One influential theorist in this area was social psychologist Albert Bandura, who promulgated the view that a sense of *self-efficacy* or a confidence in one's ability to control and master situations, strongly influences both our behavioral and physiological reactions (Bandura, 1977). Many other personality traits, and ways of looking at the world and our coping resources, have been shown to be useful for explaining the magnitude and duration of immune alterations and their ramifications for health. Significant on this list are (1) an optimistic versus pessimistic outlook, (2) an internal or external locus of control, (3) a stoic and fatalistic acceptance of adversity, and (4) an ability to express emotion or to utilize social support (Cohen & Wills, 1985; Levy, Herberman, Maluish, Schlien, & Lippman, 1985; Rodin, 1986; Scheier & Carver, 1985; Schmale, 1972).

Some enduring traits have also been posited as contributing to a "disease-causing personality" or conversely, a more "hardy" makeup (Friedman & Booth-Kewley, 1987; Kobasa, 1979). From this perspective, certain personality types might be predisposing to different immune-related diseases, similar in many ways to the idea of an ulcer-prone personality (Weiner, Thaler, Reiser, & Mirsky, 1957), or to the association between a Type A personality and heart disease (Friedman & Rosenman, 1974). Rudy Moos and George Solomon (1965) were among the first to advocate that there might be a personality profile typical for individuals with autoimmune diseases such as rheumatoid arthritis. More recently, a Type C personality—denoting unexpressive individuals who are not fully cognizant of their emotions—has been posited as a risk factor for cancer morbidity and mortality (Temoshok & Dreher, 1992). Using the same type of perspective, one study of 913 medical students from Johns Hopkins University also argued that "closeness to family members" at the time of graduation between 1948 and 1964 was a significant predictor of which 48 members of the class would develop cancer several decades later (Thomas, Duszynski, & Shaffer, 1979). In a sense, many of these purported relationships are descendent from the concept of *somatization* in psychiatry, i.e., emotions and unresolved conflicts were believed to impact specific systems of the body (Alexander, 1939). One challenge for PNI will be to determine whether variation in stress reactivity and stress-associated changes in immunity could contribute to this type of long-term effect. Further resolution is important because among the lay public it is commonly believed to be proven; many patients formulate decisions about medical treatment on the basis of a faith in the relationship between psychological factors and immunity (Cousins, 1976).

16.3. Assessing Stress and Immunity

Even this brief introduction reveals the many challenges potentially facing any intrepid investigator contemplating a study of stress and immunity in humans. Beyond the formidable methodological hurdles involved in assessing immunity, which have already been reviewed in other chapters, one must endeavor to quantify the complex psychological aspects of stressful situations. In addition, the experiment must typically include an accepted index of stress physiology to verify autonomic or endocrine activation (see Fig. 1). First, we will consider some of the immune issues and the solutions found by different investigators, and then discuss a few of the assessment approaches used for assessing other physiological and psychological aspects of stress.

Unlike animal studies, where one may have access to samples from lymphoid tissues (e.g., thymus, spleen), research on humans usually relies on cells collected from the bloodstream. Tremendous insights have been gained from these blood-derived leukocytes, although it must be acknowledged that they represent only about 2% of the total cellular pool and are sometimes not representative of the functioning of cells located in lymphoid tissue. While not always an important methodological issue, if one is interested in just documenting a general effect of a stressor on immunity, the site of cell origin can be very

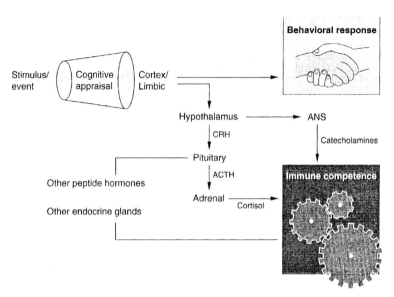

Figure 1. Flow diagram illustrating the brain, autonomic nervous system, and endocrine pathways through which stressful events must act in order to impinge on the immune system. Cognitive processes and behavioral responses can augment the immunological effects or obviate the need for physiological activation.

germane for certain diseases. For example, in the case of asthma, cell activity in lung tissue is much more critical than the activational state of cells circulating in the blood.

16.3.1. Immune Measures from Blood Leukocytes

Once the blood sample has been obtained, studies of stress and PNI in humans typically have used the standard measures and assays employed throughout the field. These have included the following.

Determining the Leukocyte Profile. During the stress response there is often an increased number and percentage of neutrophils (which may be elevated enough to raise the total WBC), and a decrease in the percentage and/or number of lymphocytes. For example, even 6 years after a leak from a nuclear reactor at Three Mile Island, higher numbers of neutrophils along with fewer B, cytotoxic, and NK lymphocytes were found in distressed residents of the surrounding community (McKinnon, Weisse, Reynolds, & Baum, 1989).

Enumerating Lymphocyte Subsets. Using monoclonal antibodies to mark cell surface proteins, different lymphocyte populations can be quantified via flow cytometry. For example, in stressful situations it has often been found that $CD4^+$ cells decline relatively more than $CD8^+$ cells, usually reported in old-fashioned terms as a decrease in the helper-to-suppressor lymphocyte ratio. Separated and divorced women were found to have a lower percentage of $CD4^+$ and NK cells, below married women, along with a lower blastogenic response to stimulation with mitogens as well as higher Epstein–Barr (EB) antibody titers

(Kiecolt-Glaser, Fisher, Ogrocki, Stout, Speicher, & Glaser, 1987a). Divorced men were also found to have lower CD4/CD8 ratios and higher EB antibody.

Lymphocyte Proliferation. This measure assesses the ability of lymphocytes to proliferate *in vitro*, usually following stimulation with plant proteins (mitogens). Stressors typically decrease proliferative responses, although the effect can become variable during chronic stress, and proliferation can sometimes increase over time. Stresslike cell margination and decreases in lymphocyte proliferative responses have been reported in U.S. and Russian astronauts during space flight (Kimzey, 1975; Taylor, 1993). For 4 months postpartum, mothers of preterm, low-birth-weight infants had lower lymphocyte proliferative responses than did mothers of term infants, although it should be noted that these differences could not be attributed specifically to anxiety and depression scores, and their NK activity did not differ (Gennaro, Fehder, Nuamah, Campbell, & Douglas, 1997).

Lymphocyte Cytolytic Activity. Another *in vitro* assay commonly used to evaluate lymphocytes involves a test of their ability to kill virally infected or mutagenic target cells. Cytolytic activity is one of the most reliable immune parameters to decrease during sustained stress, although there is often a transient increase in lytic activity during the first hours of acute stress as a result of a brief rise in the number of NK cells in circulation. In one study, blood samples were obtained from a devastated Florida community between 1 and 4 months after Hurricane Andrew (Ironson, Wynings, Schneiderman, Baum, Rodriguez, Breenwood, Benight, Antoni, LaPerriere, Huang, Klimas, & Fletcher, 1997). Relative to laboratory controls, impacted individuals had fewer T lymphocytes, and lower lytic activity, despite more NK cells. The magnitude of NK changes was associated with the level of property damage and posttraumatic stress disorder (PTSD) symptoms. Similar immune changes were found at 1, 2, and 4 months after an earthquake in southern California (Solomon, Segerstrom, Grohr, Kemeney, & Fahey, 1997). Lowest T cell and NK numbers, proliferative, and cytolytic responses were found at the 4-month time point.

Cytokine Production. Recent studies often include assays of soluble immune products, such as the interleukins and interferon. Most indicate a change in the synthesis and response to cytokines during stress, often a decrease, although some proinflammatory cytokines have been reported to increase during stressful situations. School exams have been the most systematically studied stressful life event in this regard. Changes in lymphocyte subsets, proliferative, and cytolytic responses have been found to occur reliably in many types of students, including high school, undergraduate, nursing, and medical. Changes in cytokine release and cytokine receptors on cells have also been found, e.g., a decreased percentage of cells positive for IL-2R and increased IL-2 production by Con A-stimulated blast cells (Glaser, Kennedy, Lafuse, Bonneau, Speicher, Hillhouse, & Kiecolt-Glaser, 1990).

Asthma has long been purported to be a psychosomatic illness, but it has not been easy to show that stress actually aggravates asthma through immune mechanisms. One notable difference in asthmatic high-school students was that their cells released more IL-5 during exam week, whereas the mitogen-stimulated cells from controls released relatively more IL-2 (Kang, Coe, McCarthy, Jarjour, Kelley, Rodriguez, & Busse, 1997).

16.3.2. Immune Measures from Saliva

Beyond the many immune measures that can be obtained from blood-derived cells, there have been a number of attempts to find noninvasive ways to monitor immune competence in human studies of PNI. With regard to stress, the alternative body fluid of choice has often been saliva. Sensitive and biologically meaningful measures of immunity in saliva would be ideal because they would complement assays of cortisol, which diffuses readily into saliva. Moreover, the assessments could be made as regularly and frequently as needed. Despite dozens of studies and many interesting observations, a cautionary note for future work is still warranted because of the continuing concern that salivary measures of immunity can be affected by flow rate, diet, and other factors in the mouth including dental cavities and gum disease. Some salivary measures used in previous studies of stress are discussed next.

Salivary IgA.   Notwithstanding the concern about flow rate, it is clear that stressors decrease the absolute quantity of secretory IgA (sIgA) in the mouth. Diverse events, from exercise to the inhibition of one's psychological needs for power, have been reported to lower sIgA titers (McClelland, Alexander, & Marks, 1982). Conversely, reducing arousal through relaxation and meditation can increase both salivation and sIgA levels (see Chapter 19).

Specific Antibody.   As a refinement of the total IgA measure, investigators have assessed IgA specifically directed against certain antigens (Stone, Cox, Valdimarsdottir, & Neale, 1987). It can be used as a measure of herpes simplex reactivation or to verify exposure to certain disease pathogens. Children raised in orphanages suffer from a psychological neglect that affects many aspects of development, including physiological systems (Holden, 1996). A high percentage of 3-year-olds raised in Romanian orphanages expressed salivary antibody against *Haemophilus influenzae*, indicative of exposure to bacterial meningitis (55%), as well as high titers of salivary antibody and other proteins against herpes simplex (39% of children), suggestive of poor immune containment of the virus (Coe, Carlson, Gunnar, Lubach, Dragomir, Macovie, & Scripcaru, 1997).

Cytokine Levels.   While salivary cytokine levels have been used only infrequently in PNI research, cytokine release from other mucosal surfaces has been a valuable tool in clinical immunology studies. We can expect to see more reports as we better understand the function of different cytokines secreted into saliva.

16.3.3. *In Vivo* Assessments of Stress and Immunity

Many researchers interested in the effect of stress on immunity have also utilized other creative approaches to carry out *in vivo* assessments of immune competence. These alternative strategies have been especially important in beginning to address the health relevance and clinical practice implications of stress-induced immunomodulation. Four successful approaches are discussed next.

Reactivation of Latent Viral Infection.   Considerable interest has focused on the herpesviruses (Glaser, Kiecolt-Glaser, Speicher, & Holliday, 1985a) (see Chapter 27). In general, stress studies have found decreased cellular containment of the virus, resulting in partial virus reactivation and an increase in virus-specific antibody titers. Recrudescence of latent herpesviruses has been suggested by an increase in antibody titers during the week of school exams. Elderly people tend to have high antibody titers to herpesviruses, especially when stressed by the demands of caregiving for an incapacitated relative. In contrast, 1 month of relaxation therapy was shown to lower antibody levels presumably indicative of better viral containment (Kiecolt-Glaser, Glaser, Williger, Stout, Messick, Sheppard, Ricker, Romisher, Briner, Bonnell, & Donnerberg, 1985).

Antibody Response to Vaccination.   Vaccine studies have also been important in demonstrating that the degree of immune suppression induced by stressful events can be of clinical significance (Glaser, Kiecolt-Glaser, Malarkey, & Sheridan, 1998). Typically, antibody responses to the vaccine are reduced, the transition from primary to secondary antibody response may be delayed, and additional booster vaccinations may be required for high antibody titers to be attained. For example, the level of psychological distress during the school year, as reflected by the Profile of Mood States (POMS), influenced the number of booster injections required to elicit an efficacious antibody response to a three-part hepatitis vaccine protocol in medical students (Glaser, Kiecolt-Glaser, Bonneau, Malarkey, Kennedy, & Hughes, 1992).

Induction of Disease.   Although not a commonly employed approach for obvious ethical reasons, the unique studies of controlled infection with cold and influenza viruses are widely cited for directly demonstrating a link between stress and upper respiratory illnesses (Cohen, Tyrrell, & Smith, 1991). Following a controlled exposure to rhinoviruses, the likelihood of experiencing cold symptoms was correlated with the number of major stressful life events in the past year and scores on the Perceived Stress Scale.

PNI in Patient Populations.   An alternative approach tackles the question of stress, immunity, and disease progression in ill rather than healthy individuals. In this type of study, immune measures are usually correlated prospectively with disease morbidity and mortality. This strategy has been successfully employed to demonstrate effects of stress and other psychological processes in cancer, AIDS, and some autoimmune conditions.

Breast Cancer.   Studies of cancer patients indicate that low NK activity is associated with fatigue, difficulty in expressing emotions, lower levels of social support, and tends to be predictive of reduced longevity (Levy *et al.*, 1985; Levy, Herberman, Lippman, & d'Angelo, 1987). In turn, provision of social support has been reported to increase longevity (Spiegel, 1992).

Skin Cancer.   Histopathological measures including mitotic rate of cells at tumor site and number of lymphocytes at deepest invasion point were correlated with ability to express emotions, verbal articulateness, and tumor progression 3 years later (Temoshok & Fox, 1984).

Autoimmune.   Stressful life events have been associated with the onset and aggravation of symptoms in many illnesses, including rheumatoid arthritis and multiple sclerosis, in retrospective studies (Grant, Brown, Harris, McDonald, Patterson, & Trimble, 1989; Heisel, 1972). Differences in the way leukocytes from individuals with rheumatoid arthritis respond to adrenergic and corticoid stimulation has also been linked to disease pathology and could exacerbate the effects of stressors (Kuis, de Jong-de Vos van Steenwijk, Sinnema, Kavelaars, Prakken, Helders, & Heijnen, 1996).

AIDS.   Many studies have suggested that psychological factors, including stress-related emotions and exposure to stressful incidents, can influence the numbers of T lymphocytes and NK cells in individuals with AIDS (reviewed in O'Leary, 1990).

16.3.4. Psychological Assessment

As emphasized earlier in the Historical Overview section, however, a study of stress and immunity would be incomplete without giving equivalent attention to the psychological

Table 1

Psychological and (Nonimmune) Physiological Measures that Have Been Used
in Studies of Stress and Immunity, in Addition to Descriptive Measures
of the Negative Event Itself (e.g., Major Life Events, Daily Hassles)

Psychological	Physiological
Perceived stress	Endocrine activation (e.g., cortisol)
Negative emotions (dysphoria)	Sympathetic activation (e.g., urinary NE)
Hostility and anger	Cardiovascular (e.g., BP and reactivity)
Optimisim/pessimism	Brain activation (e.g., EEG)
Self-efficacy	Neurotransmitters (e.g., NE, DA, 5-HT)
Self-discrepancy	Neuromodulators (e.g., substance P, neuropeptide Y)
Stoicism/fatalism	Proinflammatory substances (e.g., histamine,
Social networks and support	superoxides, prostaglandins)

domain and the many (nonimmune) physiological systems involved in mediating stress-induced changes (Fig. 1). Because a complete review of psychological variables would be prohibitively long, a number of the ones commonly considered in PNI studies are highlighted in Table 1 (see also Chapter 5). Depending on one's interest and the goals of the study, each researcher must decide whether and how to characterize psychological dimensions of the stressful situation, and how best to assess aspects of an individual's perceptions and reactions. When the study involves a major life event, such as loss of a loved one, a detailed psychological assessment may not seem as critical, because we can all agree that this would be a severe stressor. Holmes and Rahe (1967) suggested that we could categorize events on the basis of the amount of life adjustment required. But many PNI studies have involved more benign negative events—such as school exams—where individual reactions vary considerably. Some have advocated that these commonplace events, often described as daily hassles, ultimately have a greater summative impact on psychological outlook and health (Dohrenwend & Shrout, 1985; Lazarus & Folkman, 1989).

Even when a situation would seem to be obviously aversive—for instance, being informed of a disease diagnosis, such as HIV status—there can be considerable individual variation in the nature of the immune response to receiving this type of bad news (e.g., Ironson, Laperriere, Antoni, O'Hearn, Schneiderman, Klimas, & Fletcher, 1990) (see Chapter 25). In this example, it is immediately evident then that more information on the psychological reaction must be obtained, especially if one wants to develop good predictors of why stress-induced immune alterations vary so much across individuals. We know that a person's psychological reaction can shape the type of immune reaction that occurs during the acute response. Our research group found, for example, that it was aversive for dysphoric and anxious students to write about discrepant aspects of their personal ideals and expectations, resulting in lower NK activity, whereas this cognitive conflict, emotional distress, and NK response to writing were not seen in psychologically well adjusted students (Strauman, Lemieux, & Coe, 1993). In studies involving stressors of a longer duration, many personality variables have been shown to be important, including attributional style, internal or external locus of control, as well as the prevailing way in which an individual looks at the world and outcomes in the future. Are we generally optimistic or pessimistic? Do we accept adversity with a stoic and fatalistic attitude, or alternatively seek out solutions actively and use coping strategies effectively to reduce stress reactions? In particular, PNI studies have suggested that an ability to express emotions and to benefit from social companionship and support are important in buffering us against the effects of stress.

Assessments of expressiveness and sociality may also serve as a prognostic markers of longevity in individuals with terminal illnesses (Spiegel, 1992).

16.3.5. Physiological Assessments in Studies of Stress and Immunity

We are still not able to rank the intervening physiological steps linking psychological processes and immunity in order of importance, at least not in a hierarchy that would receive universal endorsement, despite studies on this topic. As a consequence of research on stress in animals and experiments on *in vitro* stress-associated substances, we can offer many likely candidates, but outside of the laboratory in the real world it is usually not immediately evident which stress-responsive system is the primary mediator (see Table 1). In keeping with the Selye tradition, much interest has been focused on endocrine activation, especially on the hypothalamic–pituitary–adrenal (HPA) axis. Certainly, opiate hormones—endorphin from the pituitary—and cortisol from the adrenal gland have the potential to be immunomodulatory. Nevertheless, while these hormones are elevated acutely in stressful situations, it is often difficult to show that the HPA axis remains activated after repeated or chronic stress (Ursin, Baude, & Levine, 1978). Further research may still implicate the endocrine system though, because the lingering effect of sustained stressors on hormones is often manifest in more subtle ways, such as by a disturbance of the diurnal rhythm. Many immune responses tend to rise in the afternoon and evening as HPA hormones decline (Kronful, Nair, Zhang, Hill, & Brown, 1997); thus, if stress interferes with the daily hormone decline to the nocturnal nadir, it may be of great health consequence over time.

While our abiding interest in the HPA axis reflects the well-documented, immunomodulatory potential of corticosteroids, many studies of chronic stress continue to remind us of the prophetic implications of Cannon's pioneering research. Even in situations where it seems that adrenocortical activity has subsided to normal, sympathetic activation may persist. This sensitivity of the SNS has been well-documented in aversive work environments in demanding situations requiring sustained vigilant performance, as well as accompanying the lingering arousal and challenges after environmental disasters. The fact that injections of norepinephrine (NE) and epinephrine (E) in humans and animals have been shown to affect lymphocyte profiles in blood and to inhibit proliferative and cytolytic responses in a manner mimicking the effects of stressors supports the conclusion that the SNS is involved in mediating many long-term effects on immunity (Benschop, Rodriguez-Feuerhahn, & Schedlowski, 1996b). An important lesson one quickly learns in PNI is an appreciation for the complexity of the body and a realization that a perturbation in one system usually has cascading effects on others.

16.4. Models of Stress and Immunity

Understanding the ways in which immunity can change following acute stressors lasting from a few minutes to several hours is complex enough, but longer-lasting perturbations present some unique analytic challenges. One important consideration is that there are likely to be changes in the magnitude or direction of the effect over time. Studies in animals have indicated that stress-induced physiological alterations do not progress continuously in a linear manner. One widely cited study of daily stress in mice documented a biphasic change in lymphocyte proliferation, first inhibited for 3 weeks, then rising above normal for the remaining 3 weeks of the study (Monjan & Collector, 1977). Selye anticipated this problem in his discussions of the GAS, and suggested that there were three phases: (1) an alarm reaction, (2) a stage of resistance, and (3) a stage of exhaustion. There is a general agreement about the meaning of the first stage, but in human studies of PNI, it has been extremely difficult to apply Selye's conceptualization of the resistance and exhaustion stages. Indeed, with the possible exception of severe physical stressors, such as starvation and trauma, it is not clear that they are directly applicable. Despite a subjective sense that one may be in the "exhaustion phase," as Norman Cousins concluded for himself prior to initiating his successful psychological battle with cancer, humans rarely reach this physiological point within the realm of typical life stressors (Cousins, 1976, p. 1459).

One reason we do not proceed through this triadic sequence pertains to a question raised at the outset: Do the adaptive and maladaptive aspects of stress-induced immunomodulation change over time? Although somewhat of an oversimplification, the initial shift in immunity during the alarm phase can be described as beneficial, a component of the global shift from an anabolic to catabolic state (from energy storing to energy releasing in order to support the fight–flight response of Cannon). Resources and energy are diverted from many vegetative functions—including growth, reproduction, and immunity—to favor brain and muscle functions, presumably an adaptive response to meet the exigencies and demands of the threatening situation. Even within the immune system, one can point to several seemingly adaptive aspects of the acute response: (1) a massive mobilization of the phagocytic neutrophils into the bloodstream in anticipation of trauma and (2) a homing of lymphocytes into bone marrow and lymph nodes where they may be spared certain damaging effects of stress hormones, recruited to stimulate new cell growth, or be well-situated for encounters with antigen-presenting cells. Although adaptive in the short run, maintenance of these immune changes and the catabolic energy state can become costly and maladaptive over time. Within the immune system, the changes associated with prolonged stress will eventually retard the production of new cells, and stress hormones can activate cell death programs, the sum of which begins to take a toll (manifest as a lymphocytopenia). Sustained elevations in stress hormones have also been shown to prevent cells from trafficking to the appropriate locations or to inhibit the cellular response to chemokine attractants, a prerequisite for adherence at sites of inflammation and infection. As time progresses, across days and weeks, there may also be a reduced ability to make antibody and sometimes a disruption of the T-cell regulatory support necessary to ensure immunoglobulin class switching (from M to G in a secondary antibody response). With regard to inflammatory conditions, there may be an increase in proinflammatory mediators coincident with a hampering of those cellular processes that normally prevent overreactions. The latter changes help to explain the seemingly paradoxical observation that stressors sometimes increase cutaneous hypersensitivity responses to recall antigens, even though this might seem antithetical to the immune-inhibiting aspects of the acute stress response.

Commonly Employed Conceptual Models

Figure 2 presents in cartoon fashion some of the conceptual models that have been employed to explain the more sustained influences of stress and other environmental and psychological variables on immune responses and immune-related disease. The first model illustrates the idea that an external or psychic stimulus can directly affect our physiology in a potent enough manner to initiate a pathological disease. There are a number of examples of this direct model in the general field of stress research, including the link between stomach acid secretion and ulceration of the gut, or the stress-associated increases in cholesterol and fatty acids that aggravate hypertension and cardiovascular disease. Within PNI, examples

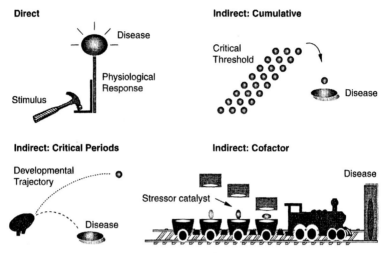

Figure 2.   Direct and indirect models commonly used to explain the relationship between stress, immunity, and disease. Direct and immediate effects on physiology may result in pathology, but more typically the influence is indirect. Negative effects may accumulate in an incremental manner (Cumulative); perturbations may alter development or the setpoints at which immune responses become established (Critical Periods); alternatively, stressors may serve a catalytic or permissive function, facilitating the actions of other variables (Cofactor).

of such direct and immediate effects of stress *causing* immunological disease have been somewhat more difficult to demonstrate, although a disruption of cell trafficking to sites of infection might provide one example. It has also been posited that stressors could impair the surveillance processes involved in detecting a metastatic cell migrating through the bloodstream. More typically, though, stress researchers have proposed indirect models in which recurrent or prolonged stressors exert their disease-promoting influences in a gradual manner over time.

Indirect Models.   One indirect model is based on the idea that small alterations in immunity do not usually result in illness by themselves, unless the host is simultaneously exposed to a disease-causing agent. That is, a bacterial or viral pathogen is the actual vector of infectious disease; the stressful event is influential because it compromises the host's ability to mount an effective immune response. A specific example might be the observation that entering new school settings is stressful and can increase exposure to viruses, such as EBV, but only those students experiencing the greatest distress will be likely to succumb and develop mononucleosis (Kasl, Evans, & Niederman, 1979). While this type of vulnerability model is relatively easy to demonstrate in animal experiments, it is likely that alternative explanations incorporating even more indirect relationships occur more frequently in humans. It is possible that the long-term negative consequences actually result from an accrual of many small alterations over time, in which case the effect on health might appear to be delayed from the first antecedent event and emerge only if the events were recurrent. Boyce, Chesterman, Martin, Folkman, Cohen, and Wara (1996) used this type of conceptual model to explain why only a subset of young children starting kindergarten in San Francisco had more frequent respiratory illness, and the higher disease incidence emerged only after an earthquake also occurred later during the school year. Presumably with this type of indirect model, physiological changes may be evident after each stressful episode, but only the aggregate toll has eventual health implications (Fig. 2, upper right).

Cumulative or wear-and-tear models have become more popular recently, in conjunction with the idea that the regulatory setpoints of the body may become imbalanced over time (sometimes described as an *allostatic load* model) (McEwen & Stellar, 1993; Seeman, Singer, Rowe, Horwitz, & McEwen, 1997). This model is particularly applicable to the developing individual, and there is increasing evidence that psychological and environmental factors can affect the developmental trajectory of the immune system (Fig. 2, lower left). Stressful perturbations may affect the rate of maturation, as well as change the setpoints at which certain responses are established in the adult. Because this type of effect appears to be more common in the young infants, when even a single event may appear to have lasting effects, it is probably important to add the concept of *critical periods* of vulnerability to the cumulative model. It can also be argued that old age should be viewed as another period in the life span when the effects of stress are magnified by intrinsic age-related changes in immunity.

Finally, under the heading of indirect models, we must consider the situation where stressful events are not the initiating cause of disease, nor even exert the predominant effect, but should just be considered as one *cofactor*, catalyzing or augmenting the actions of other variables (Fig. 2, lower right). For example, researchers have had difficulty showing that stress alone increases the occurrence of herpes lesions. However, recurrences of genital herpes were associated with the depression subscale of the POMS, and thus stressors could act in an indirect manner through dysphoria (Kemeny, Cohen, Zegans, & Conant, 1989). Here one might also list those studies showing an influence of psychological factors on an illness with a strong genetic basis, possibly accelerating or worsening the expression and perception of symptoms in a sick individual. For example, it has been reported that stressful family functioning may influence the age of asthma onset in children genetically prone to this respiratory condition (Mrazek, Klinnert, Mrazek, & Macey, 1991). In addition, retrospective surveys of patient populations suggest that stressful life events should be viewed as a "catalytic cofactor" often associated with the first clinical expression and progression of autoimmune disease and cancer (Grant *et al.*, 1989; Heisel, 1972; Jacobs & Charles, 1980; Levy *et al.*, 1985; Temoshok & Fox, 1984). As these longer-duration effects become better understood, it should be possible to determine to what degree stress is involved in tilting the balance of immunity toward health or illness.

16.5. Recovery or Disease following Altered Immunity

It is important to consider one other temporal factor, which concerns the issue of recovery. When highlighting changes in immunity that occur during stressful events, it is often not emphasized enough that most immune changes resolve without evoking disease. It is important to know that lymphocyte proliferation and cytolytic activity are decreased in bereaved individuals for 1–2 months after loss (Bartrop, Luckhurst, Lazarus, Kiloh, & Penny, 1977; Irwin, Daniels, Smith, Bloom, & Weiner, 1987e), but equally significant that a return to baseline responses was found 1 year later, presumably reflecting a successful resolution of the grieving process (Schleifer, Keller, Camerino, Thornton, & Stein, 1983). In addition, while it is often stated that the immune changes caused by bereavement are

associated with higher morbidity and mortality during the year after loss, it is less emphasized that relatively few of the illnesses are immune related (Klerman & Izen, 1977;
Maddison & Viola, 1968).

In our research examining immune changes associated with academic exams, we were
struck by several aspects of this apparent resiliency. First, when assessing healthy and
asthmatic adolescents during exam week, we found it was possible for the students to
undergo rather marked changes in immunity without any clinically significant deterioration
in lung function (Kang, Coe, McCarthy, & Ershler, 1996). Second, when the students were
evaluated 2 weeks after the school exam period, most of the stress-induced deviations in
proliferative and cytolytic activity had been restored to normal. Excessive production of
superoxides by neutrophils was the one alteration to linger at the 2-week time point.

Further research on the process of recovery and restoration may provide us with new
insights about the relationship between stress and disease. Figure 3 illustrates a number of
possible paths of physiological change that might occur during sustained stressors, which
may explain why certain immune alterations resolve and others progress toward pathology
or a worsening of disease symptomatology. The first line conveys a quick and successful
recovery from a stress-induced immune alteration. In contrast, the second connotes a
progressive deterioration in immunity that eventually crosses the threshold for disease. The
third line illustrates a point from the previous section. Even a smaller decline in immunity
could become consequential, if there was a coincidental exposure to a pathogen, or if one
had a genetic predisposition for a certain disease.

Recurrent experiences with the same stressor or with multiple stressors could also
hamper recovery, lowering immunity further and compounding the risk for disease. Even if
the events do not actually reoccur, emotional and memory processes could serve to reelicit
the physiological responses. This explanation might account for the immune findings reported in a study of the reaction to a disaster scene involving exposure to trauma and death
(Delahanty, Dougall, Craig, Jenkins, & Baum, 1997). NK activity was assessed in professional workers at 2 and 6 months after a plane crash. Those who had witnessed trauma
directly at the crash site had higher NK responses than did either controls or more

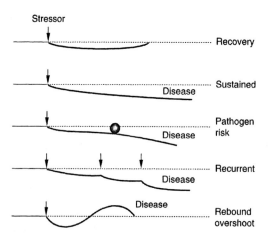

Figure 3. Following a stress-related
change in immunity, we can hypothesize
a number of potential paths leading to
recovery or disease. Immune alterations
may persist and worsen reaching a critical
threshold. Alternatively, an opportunistic
pathogen may take advantage of the compromised host, or repeated stressors may
have a summative effect. Finally, some
stress-induced pathology may be related
to a compensatory rebound above the
original baseline, possibly important in
inflammatory disorders.

experienced morgue workers. The findings of increased NK activity suggest recurrent acute activation, especially as they were correlated with a higher level of intrusive thoughts in the crash site workers, whereas the similarity between controls and morgue workers was interpreted as indicative of more effective coping in those individuals acclimated to witnessing traumatic injury.

These different modes of recovery could be anticipated from the prior discussion about the importance of the psychological dimensions of stress, but there is one other recovery pattern that has received almost no attention. Following a period of immune suppression, there may be a compensatory overshoot that aggravates symptomatology or causes a disease to emerge during the phase of the recovery. This unique profile of physiological adaptation is illustrated on the last line as a rise above baseline following the initial poststress inhibition. Although not normally discussed in the context of stress physiology, some immune changes associated with pregnancy and birth could exemplify this biphasic model. Pregnant women with autoimmune diseases may experience a temporary alleviation of symptoms during the third trimester, mediated in part by the normal pregnancy-associated suppression of certain cellular immune responses. However, the illnesses often return with renewed and sometimes even greater vigor following birth. This type of rebound effect can also occur on a much more rapid scale. A specific example is the dynamic parasympathetic rebound after a period of sympathetic activation, which might have important implications for a number of diseases including asthma. In this example, it would make more sense to attribute stress-associated changes in lung function to the rebound phase, rather than to the prior activational release of adrenaline or corticosteroids. Both of these stress-related compounds are actually used clinically to dilate the lungs and inhibit inflammation in the airways.

While these conceptual models are important to consider when reading reports on stress and immunity, in most cases it has not been practical for investigators to assess the full temporal course of psychological and physiological change. The cost and difficulty of conducting immune assays would typically preclude the serial measures needed for such a longitudinal study. Most studies have instead assessed immunity or health at a specific time point. In addition, in some cases investigators have taken advantage of naturalistic events or environmental disasters, opportunities that do not readily permit prospective analyses. Often in these studies it was necessary to use a cross-sectional design, with comparison values obtained from a healthy control group matched on certain criteria, rather than a pre- and postdesign that would allow for temporal analyses.

16.6. Vulnerable Points in the Life Span

In concluding an earlier review of the literature on stress and immunity, two other authors hypothesized that "distress-related immunosuppression may have its most important health consequences in those individuals whose immune function is already impaired" (Kiecolt-Glaser & Glaser, 1991, p. 864). This idea is particularly applicable in a developmental model, because immune responses are not as competent in children and the elderly. While stress-induced changes in immunity can occur at any point in the life span, these alterations may be of most consequence for the very young and old. Morbidity and mortality in both historical and contemporary disease epidemics continually remind us of the importance of age when discussing immunity. Many immune responses are still maturing in infancy and childhood and thus stressful perturbations may be able to alter the trajectory of

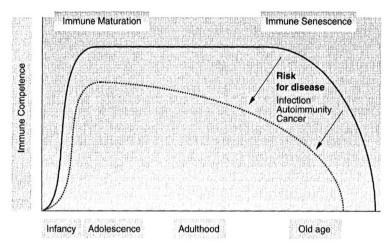

Figure 4. Stressful environmental and psychological events may alter the developmental trajectory of the immune system, reducing immune competence and increasing the risk for immune-related diseases. In old age these stress-related changes may influence the occurrence and progression of immune senescence.

development (Fig. 4). Certainly, we know that severe physical challenges to the body, such as malnutrition, have that potential to undermine immune development in children (Chandra & Newberne, 1977).

16.6.1. Pediatric Studies

Support for this view about the potential importance of developmental PNI can be found in many studies.

Retrospective Assessment of Pediatric Patients. Major stressful life events have been reported to occur more frequently during the 1–2 years preceding illness, including in pediatric cases of leukemia (Greene & Miller, 1958; Jacobs & Charles, 1980). Family difficulties and childhood disturbances have also been associated with an earlier age of disease onset in children genetically prone to asthma (Mrazek *et al.*, 1991).

Upper Respiratory Infection. Quality of family functioning has been associated with the incidence of streptococcal infection in children (Meyer & Haggerty, 1962) as well as the duration and severity of viral respiratory infections (Boyce, Jensen, Cassel, Collier, Smith, & Ramey, 1977).

Initiating School. Lymphocyte subsets were found to change during the first week after children started kindergarten (Boyce, Chesterman, Martin, Folkman, Cohen, & Wara, 1993). Behavioral adjustment problems were associated with changes in numbers of CD4[+] and CD19[+] cells. Later in the school year, children who had had more difficulty adjusting, and had shown larger cell subset changes, experienced more upper respiratory infections when an earthquake occurred 6 weeks after the start of school.

Psychosocial Deprivation.   Children raised in orphanages present with a different immune profile indicative of greater pathogen exposure and lower immune competence. Besides exposure to diseases such as bacterial meningitis, higher levels of salivary sIgA suggested greater expression of Herpes simplex (Coe *et al.*, 1997).

16.6.2. Geriatric Studies

In a complementary manner, we know that many immune responses undergo a progressive decline in old age. This process of immune senescence appears to be quite susceptible to influence by stressful events. Because of the intrinsic tendency for an age-related decline at the end of the life span, perturbations may exert a more lasting effect or push immune competence below a critical threshold. It has also been suggested that immune reactivity to acute stressors may change in the elderly, at least as reflected by changes in NK cell numbers (Naliboff, Benton, Solomon, Morley, Fahey, Bloom, Makinodan, & Gilmore, 1991). Two examples of stress-related immune studies in the elderly are highlighted below.

Stress and Vaccine Efficacy.   Many old individuals fail to mount an effective antibody response to vaccines. Sustained caregiving for spouses with Alzheimer's was shown to impair the response to influenza vaccination, in conjunction with decreased cellular immune responses and elevated antibody to herpesviruses (Kiecolt-Glaser, Glaser, Gravenstein, Malarkey, & Sheridan, 1996).

Wound Healing.   In addition to effects on immunity, the effort and psychological demands of caregiving have been found to retard wound healing of a small skin biopsy site in the elderly (Kiecolt-Glaser, Marucha, Malarkey, Mercado, & Glaser, 1995).

16.7. Summary

As reviewed in this chapter, significant changes in immunity can linger for several days to months after negative and threatening life events, especially if the psychological distress persists. Recurrent stressful episodes or cognitive processes that re-create the emotional disturbance associated with the original event—intrusive thoughts in the case of PTSD— also seem to contribute to the maintenance of immune changes over prolonged periods of time. In contrast, when cognitive and emotional processes facilitate effective coping, immunological alterations tend to subside. The latter type of psychological adjustment was used to account for the absence of immune alterations during a 2-month period of confinement, investigated in order to simulate the psychosocial challenges of prolonged space flights (Schmitt, Peres, Sonnenfeld, Tkackzuk, Arquier, Mauco, & Ohayon, 1995). High morale and team spirit, which resulted in the four volunteers not perceiving the enforced isolation as stressful, presumably accounted for the absence of immunological change. Similarly, when demands subside after a discrete stressful event, the immune modulation usually resolves fairly quickly: in the case of school exams in about 2 weeks (Kang *et al.*, 1996).

When more lasting stress-induced immune alterations persist, they appear to be superficially similar to those first seen during the day after stress onset. That is, most investigators report signs of cell margination, a decreased percentage and/or number of T-cell subsets (especially CD4$^+$ cells), and a reduction in proliferative responses. However, there is one dramatic difference between acute and chronic stressors, NK activity is the measure most reliably decreased during prolonged stress. In contrast, during the first hours after a stressor, there is a transient rise in NK cells in the bloodstream, which creates the appearance of increased cytolytic activity for a few hours. Studies of prolonged stress have also indicated that assessing antibody responses to latent viruses can provide a good index of immune competence. Because these antibody titers usually rise rather than decrease in stressful situations, it has been argued that the change in antibody should be interpreted as reflecting an impairment in cellular immunity, necessitating a humoral response.

A number of investigators have commented on the similarity between immune changes associated with stress and depression (Irwin, Patterson, Smith, Caldwell, Brown, Gillin, & Grant, 1990c). Presumably the overlap reflects a common effect of the physiological mediators in both conditions, such as HPA activation. While it remains difficult to rank the major mediators of stress-induced immunomodulation in order of importance, the preponderance of stress research will likely continue to focus on the HPA axis and the ANS. Signs of this endocrine or sympathetic activation are usually critical for verifying that a state of stress existed. From a historical point of view, it is interesting to mention again that immune alterations were initially included as one of the three criteria of the stressed state (along with endocrine and gastrointestinal measures).

In this chapter we have emphasized cognitive and emotional factors as the primary psychological mediators of the long-term response to stressful situations, but behavioral and lifestyle variables are also likely to be involved in prolonged immune alterations. Health-promoting and disease-causing behavior (e.g., exercise, diet, smoking, drug intake) probably contribute to the course and duration of sustained immune changes. For example, given that stressful events are known to cause a lasting increase in the release of superoxides by neutrophils and monocytes (Kang et al., 1996), it is possible that a diet high or low in antioxidants could influence this stress-associated response. Somewhat surprisingly, there have been few studies of interventions exclusively focused on trying to block the immune alterations associated with stress.

We have also touched on two other aspects of the relationship between stress and immunity that warrant brief reiteration. Notwithstanding the importance of stress-induced immune changes, or maybe because they are so potentially significant, there seem to be many processes that mitigate against their effects on health. The overlapping and redundant functions of many immune cells often appear to compensate for a decline in the capabilities of any one arm of immunity. In addition, we have also suggested that psychological processes related to coping should be considered as an essential, protective reaction of the host, helping to ensure against the maladaptive cost of maintaining a stressed state. Even a psychic defense, such as denial, can serve to downregulate stress physiology when the situation is too overwhelming (Wolff et al., 1964). We need to learn more about those homeostatic processes that facilitate a return to normalcy, especially during the recovery phase. One recovery pattern in need of more study is the poststress, physiological rebound that can sometimes push the body above the original baseline. Allan Munck and others have written about the actions of corticosteroids in this way, suggesting that their immune-inhibiting properties are related to the fact that they can overcompensate in containing immune activation (Munck, Guyre, & Holbrook, 1984).

Finally, we have suggested that the effects of stress on immunity may be of special importance during certain phases of the life span when the immune system is undergoing developmental change. Stressful perturbations at these times are more likely to exert immune alterations that might result in immune-related disease in the young and old. Many of the seminal studies in PNI were conducted with this perspective in mind and showed that stressors had a more profound and long-lasting impact when they occurred in immature animals (Ader, 1983; Solomon, Levine, & Kraft, 1968; for review of primate studies see Coe, 1993, or Chapter 14). At other points in the life span, especially during the prime of young adulthood, it may be possible to withstand rather large changes in certain immune responses without experiencing any disease sequelae. When discussing stress and immuno-modulation, it is just as important to emphasize this resilience as to highlight our potential vulnerability. To understand resilience, it will be critical to put as much emphasis on understanding the complex psychological domain as deciphering the intricate connections of the underlying physiological substrate. Moreover, cognitive and behavioral interventions will likely be more effective for health promotion than discovering medical treatments to repair damage done to our physiological systems by stress.

Further Reading on Stress and Immunity

Cohen, S., & Williamson, G. M. (1991). Stress and infectious disease in humans. *Psychological Bulletin, 109*, 5–24.

Cohen, S., Kessler, R. C., & Gordon, L. U. (1995). *Measuring stress.* New York: Oxford University Press.

Jemmott, J. B., & Locke, S. E. (1984). Psychosocial factors, immunologic mediation, and human susceptibility to infectious diseases: How much do we know? *Psychological Bulletin, 78*, 78–108.

Kiecolt-Glaser, J. K., & Glaser, R. (1991). Stress and the immune function in humans. In R. Ader, D. L. Felten, & N. Cohen (Eds.), *Psychoneuroimmunology* (2nd ed., pp. 849–857). San Diego: Academic Press.

O'Leary, A. (1990). Stress, emotion, and human immune function. *Psychological Bulletin, 108*, 363–382.

17 Does Psychological Depression Cause Immune Suppression in Humans?

Michael Irwin and Elliot Friedman

17.1. Introduction

Considerable evidence suggests that major depression places people at increased risk for physical morbidity and mortality. The mechanisms responsible for the decline of well-being in persons with major depression are not known, but recent research has hypothesized that immune alterations might be one pathway through which depression influences physical health. However, disparate results have been obtained in the study of immune function in depressed patients, and a general consensus that depression is associated with immune alterations has not been reached.

Michael Irwin   •   Departments of Psychiatry, San Diego VA Medical Center and University of California, San Diego, California 92161.   Elliot Friedman   •   Department of Psychology, Williams College, Williamstown, Massachusetts 01267.

Studies of immunity in depression have been primarily concerned with the evaluation of enumerative (numbers of circulating white blood cells or lymphocyte subsets) and/or *in vitro* correlates of cellular immune function (lymphocyte responses to mitogens and NK activity). The results of these studies are inconsistent. Moreover, the conclusions of three comprehensive reviews summarizing and evaluating the disparate results from the several dozen studies that have been conducted since 1978 have not been in agreement (Herbert & Cohen, 1993a; Stein, Miller, & Trestman, 1991; Weisse, 1992).

The reasons for these discrepant results in studies examining the link between major depression and immune function are not clear. It is possible that the heterogeneity of major depression results in differences in immune function in some depressed subjects but not in others. However, before any conclusions can be generated with confidence about the presence or absence of an association between major depression and immunity, it is essential to first evaluate the methodologies of the studies. Because diagnostic heterogeneity, medication use, age, and gender could affect immune function, the influence of these factors on the interpretation of results deserves attention. Consequently, the present chapter focuses on those studies that have evaluated samples of depressed patients who fulfilled diagnostic criteria (either DSM-IV or Research Diagnostic Criteria) for major depression, were assessed while psychotropic medication free for at least a week, and were compared with healthy age- and sex-matched comparison controls. In addition, because of the recognized variability of functional immune assays and the need for interassay controls, this chapter will only consider those studies that have been conducted using a matched subject–control pair design and/or laboratory controls with one or more cryopreserved standards.

17.2. Methodological Issues

Even in series of studies restricted to those that are carefully controlled as described above, the discrepant results found may be related to differences in sample characteristics. The following subsections will discuss the implications of age, gender, hospitalization stress, life stress, comorbidity, and severity of depressive symptoms on immune variability and how such subject variability might contribute to different immunological results in the study of subjects with major depression.

17.2.1. Age

Immune decrements occur in aged individuals (Irwin, Patterson, Smith, Caldwell, Brown, Gillin, & Grant, 1990c), and depressed patients who are older may be more vulnerable to the effects of depression (Irwin, Lacher, & Caldwell, 1992b). Schleifer, Keller, Bond, Cohen, and Stein (1989) found that depressed people show decreased proliferative responses to mitogens with advancing age, while controls appeared to show increases. Furthermore, meta-analytic techniques have demonstrated that the effect of depression-related reduction of immune function is reliably stronger in older clinically depressed people than in young subjects (Herbert & Cohen, 1993a). However, in most studies, subjects were on the average middle-aged, and it has not been convincingly demonstrated that differences in the ages of various study samples have resulted in the discrepant findings.

17.2.2. Gender

Gender appears to have a striking influence on the depression-related reduction of NK activity. Many studies have focused only on men and typically found a reduction of NK activity (Irwin, Daniels, Bloom, Smith, & Weiner, 1987a; Irwin, Smith, & Gillin, 1987b;

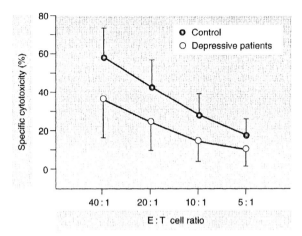

Figure 1.   Natural killer cell activity in male depressed subjects (n = 19) and male control subjects (n = 15). From Irwin *et al.* (1987b).

Irwin, Caldwell, Smith, Brown, Schuckit, & Gillin, 1990b; Irwin, Brown, Patterson, Hauger, Mascovich, & Grant, 1991) (Fig. 1), whereas the largest study that was conducted using predominantly female subjects found no difference in cellular immunity (Schleifer *et al.*, 1989). Evans, Folds, Petitto, Golden, Pedersen, Corrigan, Gilmore, Silva, Quade, and Ozer (1992) have separately compared gender differences for NK activity and NK cell counts, and reported that depressed men exhibit a significant reduction in these immune measures. Female depressed patients have levels of NK activity and NK numbers similar to those of female controls, suggesting the possibility that major depression has differential effects on NK cell function in male and female subjects. Indeed, increased NK activity was reported in female depressives, which was speculated to reflect direct effects of progesterone, or possibly progesterone-mediated antagonism of glucocorticoid suppression on cellular immune function (Miller, Asnis, Lackner, Halbreich, & Norin, 1991).

17.2.3. Hospitalization Stress

Hospitalization stress has been proposed to exacerbate the relationship between depression and immunity, although this relationship may represent the increased severity of depressive symptoms in hospitalized depressives. However, hospitalization status was not found to statistically contribute to mitogen-induced lymphocyte proliferation or NK activity (Schleifer *et al.*, 1989), and schizophrenic patients undergoing the stress of hospitalization failed to show a reduction of cellular immune function, even though a group of similarly hospitalized depressed inpatients did show such a reduction of immunity (Caldwell, Irwin, & Lohr, 1991). Together, these data suggest that the effects of depression on immunity are not likely to be related to nonspecific effects of hospitalization.

17.2.4. Life Stress

Considerable evidence has found that life stress can alter immune function. Depressed patients often undergo increased numbers of life events secondary to their affective illness, and depression-related reduction of cellular immunity could possibly reflect the effects of life stress, not depression (Irwin *et al.*, 1990a). Conversely, about 10% of control subjects who volunteer for clinical research and are free of any psychiatric diagnosis are found to be undergoing severe life stress as identified by structured life event interviews (Irwin *et al.*, 1990a). Failure to assess life stress in depressed patients and controls could lead either to the inclusion of depressed patients who show a reduction of immunity secondary to life stress, or alternatively to the enrollment of stressed "controls" who show a stress-related reduction of immune function.

To evaluate the relative contribution of stress to depression-related immune decreases, as well as to strengthen the methodological rationale for assessing life stress by structured interviews in controls, one study of immune function in depression has rated severity of life stress in depressed patients and controls. Both depression and life stress have separate but not additive effects on NK activity (Fig. 2) (Irwin *et al.*, 1990a). In other words, depressed patients who were or were not severely stressed had a similar reduction of NK activity. Psychiatric controls who are undergoing stress also show reduced immunity similar to depressed patients, even though these "controls" reported no significant depressive symptoms. This is important for the design of further clinical studies, as inclusion of such stressed persons as "normal controls" in the comparison group could potentially obscure immunological differences between the control and depressed groups.

The influence of life stress on immune function in major depression may depend on cognitive variables, such as how life events are perceived by the depressed individuals. One study examined the role of locus of control, the degree to which individuals believe that they have control over their conditions ("internal" locus of control) or that their condition is more influenced by others ("external" locus of control), in depression- and stress-related changes in immune function. NK activity was found to be significantly lower in adults with major depression compared with healthy controls, results that support previous studies. But, in addition, NK activity was significantly lower in patients determined to have an external locus of control than in patients with an internal locus (Raynaert, Janne, Bosly, Staquet, Zdanowicz, Vause, Chatelain, & Lejeuene, 1995). Although this study did not assess levels of life stress, these findings suggest that the immunological consequences of life stress in depression may be mediated in part by stable cognitive and coping styles in the subjects.

Research on the immunological effects of acute laboratory stress in depressed patients and nondepressed controls has also supported this alternative hypothesis. In one study,

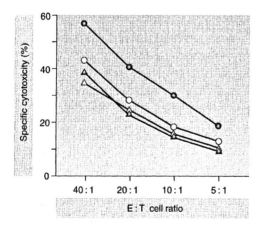

Figure 2. Natural killer cell activity across four effector-to-target cell ratios in control subjects with low ($n = 28$; ●) and high stress ($n = 8$; ▲) and in depressed patients with low ($n = 14$; ○) and high stress ($n = 22$; △). From Irwin *et al.* (1990c).

circulating numbers of NK cells were found to be increased in depressed patients compared with dysthymic and control subjects. After a mathematical challenge, however, circulating numbers of NK cells increased in all subjects to similar degrees, suggesting that the depressed subjects were not differentially sensitive to the effects of acute stress. Interestingly, the authors also reported a correspondence between NK numbers and "emotion-focused" coping styles in the depressed patients, indicating further that variables related to the perception of both day-to-day and traumatic life stressors may contribute to the immunological impact of stress (Ravindran, Griffeths, Merali, & Anisman, 1996).

17.2.5. Substance Abuse Comorbidity

Increased use of tobacco and/or alcohol might worsen depression-related immune changes; depressed patients increase their tobacco smoking, and over 30% escalate their alcohol drinking during a depressive episode. Despite the potential importance of alcohol and tobacco smoking to immune alterations in depression, evaluation of the contribution of either substance to altered immunity during depression is essentially limited to a few studies. In one study (Irwin *et al.*, 1990a), smoking was found to be a strong predictor of the increased number of circulating neutrophils and lymphocytes in the depressives, but cigarette use did not affect the association between depression and NK activity. More recent evidence suggests that smoking and depression interact to produce a graded decrement of NK activity. Depressed subjects who are smokers show a greater decline of NK activity than depressed nonsmokers, but there appear to be no differences in lytic activity between smoker and nonsmoker comparison controls (Jung & Irwin, 1998). In contrast, Fischler, Bocken, DeWaele, and Thielemans (1990) found that depressed patients who are smokers have higher levels of lytic activity than tobacco smoking controls.

The effects of current alcohol intake and past long-term alcohol abuse on immune function in depressed patients have also been evaluated (Irwin *et al.*, 1990b). While alcohol consumption in the 6 months prior to immune assessment was a significant predictor of total leukocyte and neutrophil cell counts similar to the effects of smoking, histories of recent alcohol use were not associated with NK activity (Irwin *et al.*, 1990b). Past heavy alcohol consumption, as reflected by histories of alcohol abuse and alcohol dependence, did, however, produce a graded decrement of NK activity beyond the effects of depression alone (Fig. 3). Nevertheless, depression had an effect on NK activity independent of the effect of past alcoholism.

17.2.6. Psychiatric Comorbidity

In addition to careful assessment of substance comorbidity, psychoimmunological studies of psychiatric patients with affective disorders need to evaluate comorbidity for other psychiatric disorders, and the potential influence of this comorbidity on immune function. For example, Andreoli, Keller, Rabaeus, Zaugg, Garrone, and Taban (1992) investigated the contribution of the additional presence of panic disorder to the variability of lymphocyte proliferation in depressed patients, and found that depressives with simultaneous panic disorder had a greater number of T cells and PHA mitogen responses than depressed patients without comorbid panic disorder.

17.2.7. Severity of Depressive Symptoms

Severity of depressive symptoms is a correlate of immunity, and may be a more important determinant of altered immunity than depression diagnostic *classification* per se.

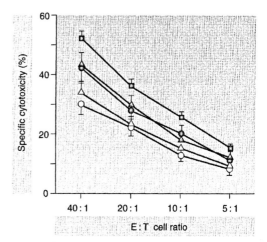

Figure 3. Natural killer cell activity across four effector to target cell ratios in control subjects ($n = 50$; ■), primary alcoholics ($n = 19$; ●), primary alcoholics with secondary depression ($n = 9$; ○), depressed patients ($n = 18$; ▲) and depressed patients with histories of alcohol abuse ($n = 26$; △). From Irwin *et al.* (1990b).

For example, Schleifer *et al.* (1989) reported that depressed patients had levels of lymphocyte proliferation similar to those of controls, yet severity of depressive symptoms was associated with suppression of mitogen proliferation. Furthermore, Schleifer, Keller, Meyerson, Raskin, Davis, and Stein (1984) suggested that ambulatory depressives, because of their modest symptom severity, have similar levels of immune function as controls. In our studies of NK activity in depression, severity of depressive symptoms has been reliably associated with a reduction of NK activity (Irwin *et al.*, 1990a,b, 1987b), a finding that has been replicated by other groups and strengthened by the meta-analysis reported by Herbert and Cohen (1993a). Finally, Irwin and colleagues used a longitudinal case–control design and found that NK activity increases when symptom severity resolves following clinical psychopharmacological treatment (Fig. 4) (Irwin *et al.*, 1992b).

Other measures of immune dysfunction, including elevated numbers of circulating NK cells and monocytes, and increased production of IL-2 and IFN-γ, have also been shown to return to control values in depressed patients after treatment with psychotropic medication (Ravindran, Griffeths, Merali, & Anisman, 1995), and the extent of the decrease toward control levels was correlated with the degree of clinical improvement (Seidel, Arolt, Hunstiger, Rink, Behnisch, & Kirchner, 1996a,b). In addition, biochemical analysis of intracellular calcium concentrations in lymphocytes revealed a 50% decline in responsiveness to mitogenic stimulation in cells isolated from depressed patients compared with those taken from healthy controls. This difference was abolished following successful treatment with interpersonal psychotherapy (Aldenhoff, Dumais-Huber, Fritzsche, Sulger, & Vollmayr, 1997).

To extend these findings regarding symptom severity, and evaluate the effects of diagnostic subtype on immune function, Hickie, Hickie, Lloyd, Silove, and Wakefield (1993) assessed delayed-type hypersensitivity (DTH) response and PHA-induced lymphocyte proliferation in 57 patients with major depression versus age- and gender-matched controls. Patients with melancholic depression had a higher rate of abnormally reduced DTH induration diameter compared with both nonmelancholic patients and age- and gender-matched controls; a difference that could not be accounted for by age, severity of depressive symptoms, hospitalization status, or weight loss. A similar pattern of results was

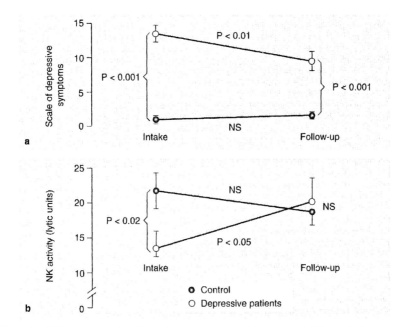

Figure 4. (a) Change in severity of depressive symptoms from intake to 6-month follow-up in the depressed patients ($n = 20$) and control subjects ($n = 20$). (b) Change in NK activity from intake to 6-month follow-up in the depressed patients ($n = 20$) and control subjects ($n = 20$). From Irwin *et al.* (1992b).

found for *in vitro* assessment of immunity by PHA lymphocyte proliferation, in that melancholic depressed patients showed impaired blastogenic response relative to those with nonmelancholic depressive disorder.

Finally, in an effort to broaden immunological assessment and to explore whether mediators important in the regulation of cellular immune responses are altered in melancholic depression, production of IL-1, IL-6, and soluble IL-2 receptors in PHA culture supernatants has been determined in depressed patients with melancholic subtype (Maes, Bosmans, Suy, Vandervost, DeJonckheere, & Raus, 1991b; Maes, Scharpe, Meltzer, Bosmans, Suy, Calabrese, & Cosyns, 1993). Depressed melancholic patients showed higher expression of IL-1, IL-6, and IL-2 receptors. These patients also showed increased circulating levels of IL-6 (Maes, Meltzer, Bosmans, Bergmans, Vandoolaeghe, Ranjan, & Desnyder, 1995a; Maes, Bosmans, De Jongh, Kenis, Vandoolaeghe, & Neels, 1997), IL-1 receptor antagonist (Maes, Vandoolaeghe, Ranjan, Bosmans, Bergmans, & Desynder, 1995c, Maes *et al.*, 1997), and soluble IL-6 and IL-2 receptors (Maes *et al.*, 1995a). Furthermore, dexamethasone suppressed IL-1 and IL-2 receptor production in controls, but failed to alter responses in depressives. These findings, together with evidence of an increased number of pan-T, pan-B, and T-suppressor/cytotoxic cells, and elevated levels of a variety of nonspecific mediators of inflammation in melancholics, have suggested to Maes and colleagues that melancholics exhibit a systemic immune activation (Maes, 1995; Maes, Lambrechts, Bosmans, Jacobs, Suy, Vandervorst, DeJonckheere, Minner, & Raus, 1992).

Contrary to the hypotheses that blunted lymphocyte responses or NK activity are mediated through a deficit in T-helper cells and/or lymphokine stimulation (Schleifer *et al.*,

1989), or via inhibition by various neuroendocrine signals (Irwin *et al.*, 1991), Maes *et al.* (1992) hypothesize that lower *ex vivo* mitogen responses may be related to immune activation *in vivo*, which leads to increased *in vivo* production of soluble IL-2 receptors inducing a state of relative deficiency of IL-2, necessary for proliferation.

17.3. Mechanisms of Depression-Related Immune Alterations

17.3.1. Neuroendocrine–Immune Interactions in Depression

Adrenal Axis. The secretion of corticosteroid has long been considered the mechanism of stress-induced and/or depression-related suppression of immune function (see Chapters 9 and 16). Specific intracytoplasmic corticosteroid receptors have been identified in normal human lymphocytes, and these receptors bind corticosteroids and appear to play a role in the regulation of cellular function through modulation of cyclic AMP levels. Indeed, *in vitro* studies have demonstrated that glucocorticoids inhibit IL-1 and IL-2 production at the level of cytokine gene transcription, with resulting suppression of lymphocyte responses to mitogenic stimulation and NK cell activity (antibody-dependent cytotoxicity is relatively refractory to glucocorticoids).

Glucocorticoids also have numerous *in vivo* effects on immune function. Serum levels of IgG, IgA, and IgM are suppressed by pharmacological doses of glucocorticoids *in vivo*. In addition, glucocorticoids affect the distribution of circulating lymphocytes, potently inhibit cellular cytotoxicity, and suppress mitogen-induced T-cell proliferation *in vivo*. B-cell proliferation is relatively resistant to glucocorticoids. Together these data demonstrate that glucocorticoids pharmacologically modulate cellular and humoral immune responses.

The role of physiologic elevations of glucocorticoids *in vivo* in mediating alterations of immune function following stress has not been as conclusively demonstrated as the pharmacological effects of glucocorticoids. For example, in rats, acute administration of either forced immobilization (Irwin & Hauger, 1988) or audiogenic stress (Irwin, Segal, Hauger, & Smith, 1989) produces an activation of adrenal steroid secretion, but does not alter cellular immune function. Likewise with repeated exposure to the stressor, pituitary adrenal activation is dissociated from a reduction in cytotoxicity (Irwin & Hauger, 1988). Furthermore, stress-induced suppression of lymphocyte function and/or NK activity following inescapable aversive stress occurs in either adrenalectomized (Keller, Weiss, Schleifer, Miller, & Stein, 1983) or hypophysectomized animals (Keller, Schleifer, Liotta, Bond, Farhoody, & Stein, 1988). Finally, antagonism of stress-induced activation of the pituitary adrenal axis by the peripheral preadministration of an antiserum to corticotropin-releasing hormone (CRH) fails to alter stress-induced suppression of NK activity, even though the release of ACTH and corticosterone is inhibited (Irwin, Vale, & Rivier, 1990d) (see Chapter 13).

Consistent with these animal studies, clinical research has found no relationship between adrenocortical activity and immunity in depressed patients and in stressed persons. In depressed patients, decreased lymphocyte responses to mitogens were not correlated with circulating levels of adrenocortical hormones, increased excretion rates of urinary free cortisol, or dexamethasone nonsuppression. Furthermore, in bereavement, in which a reduction of NK activity has been demonstrated, this immunological change occurs even in subjects who had plasma cortisol levels comparable to control subjects (Irwin, Daniels, Risch, Bloom, & Weiner, 1988a). Miller and colleagues (Miller *et al.*, 1991) also found no association between hypercortisolemia and NK activity in depressed patients. In contrast to these findings, Maes, Bosmans, Suy, Minner, and Raus (1991a) found that baseline cortisol, urinary free cortisol, and postdexamethasone β-endorphin values explained up to 45% of the variance in lymphocyte proliferation, although in these studies lymphocyte function was inversely related to baseline cortisol, but positively correlated with urinary free cortisol excretion.

It is alternatively hypothesized that lymphocytes of depressed patients may be even less sensitive to the effect of glucocorticoids. Following chronic exposure to glucocorticoids, adrenal steroid receptors expressed in immune tissues exhibit downregulation. Furthermore, dexamethasone-induced inhibition of T-cell blastogenesis is inversely correlated with plasma concentrations of cortisol, suggesting the possibility that elevated activity of the adrenal axis downregulates the lymphocyte glucocorticoid receptor axis in depressed patients, and decreases the sensitivity of lymphocytes to pharmacological as well as physiological concentrations of glucocorticoids.

Autonomic Nervous System. Anatomical studies have revealed an extensive presence of autonomic fibers in both primary and secondary lymphoid organs, innervating both the vasculature and the parenchyma of the tissues, serving as one pathway for communication from the brain to cells of the immune system (see Chapter 8). Immunohistochemical studies of splenic tissue have demonstrated that nerve fibers containing norepinephrine, substance P, or neuropeptide Y branch into the parenchyma where lymphocytes (primarily T cells) reside. These postganglionic sympathetic fibers are not only adjacent to T cells but, at the electron microscopic level, end in synapse-like contacts with lymphocytes in the spleen.

Lymphocytes have been found to receive signals from the sympathetic neurons by adrenoceptor binding of norepinephrine, epinephrine, and dopamine (see Chapter 9). These β-adrenoceptors are linked to adenylate cyclase and have a functional role in the modulation of cellular immunity. *In vitro* incubation of lymphocytes with varying concentrations of either norepinephrine or epinephrine decreases NK activity and mitogenic responses, an effect that is reversed by a β-adrenoceptor antagonist. Additional *in vitro* studies have found that neuropeptide Y, a sympathetic neurotransmitter present in peripheral sympathetic nerves, also acts to inhibit NK activity.

In vivo studies have shown that the sympathetic nervous system has a role in the modulation of immune function. Either surgical denervation of the spleen or chemical sympathectomy using the neurotoxin 6-hydroxydopamine produces an augmented antibody response to thymus-dependent antigens such as sheep red blood cells, an enhanced plaque forming cell response to thymus-independent antigens, increased macrophage phagocytosis, and enhanced responsiveness of T and B cells to mitogen stimulation that is strain specific. Conversely, infusion of adrenergic agonists such as isoproterenol in rats and in humans results in a downregulation of β-adrenoceptors in circulating mononuclear cells, and a dose-dependent, transient decrement in mitogen responses (Murray, Polizzi, Harris, Wilson, Michel, & Maisel, 1993) that is likely to be mediated by a redistribution of circulating lymphocyte subpopulations via β-adrenoceptor mechanisms. In addition to sympathetic-induced alterations of nonspecific immune function, sympathetic neurotransmitters also suppress antigen-specific lymphocyte proliferation by inhibiting macrophage antigen processing/presentation, and indirectly, T-helper responses (Heilig, Irwin, Grewal, & Sercarz, 1993).

Acute stress-induced elevation of sympathetic activity is one pathway that likely plays a role in stress-induced suppression of cellular and humoral immune responses. For example, in animals exposed to aversive stress, or central doses of neuropeptides such as CRH that activate sympathetic outflow, suppression of cellular immunity or *in vivo* specific antibody responses is completely antagonized by either autonomic blockade (Irwin, Hauger, Brown, & Britton, 1988b), chemical sympathectomy, or β-adrenoceptor antagonism (Irwin, Hauger, Jones, Provencio, & Britton, 1990a).

Chronic elevation of sympathetic tone has also been found to mediate a reduction of immunity. For example, in animals, induction of a chronic hyperadrenergic state (experimental congestive heart failure, or 2-week infusion of the β-adrenoceptor agonist isoproterenol) reduces NK activity, *in vivo* antibody responses, and lymphocyte proliferation (Waltman, Irwin, Harris, & Maisel, 1992). Similar to findings following acute stress, the immunosuppressive effect of chronic, increased sympathetic tone is completely antagonized by β-adrenoceptor blockade. In humans, chronic elevated sympathetic tone, as reflected by circulating concentrations of neuropeptide Y, may also contribute to the modulation of immune function during severe life stress and depression (Irwin *et al.*, 1991). Neuropeptide Y is present in peripheral sympathetic nerves and is released following emotional stress potentiating the effects of vasoactive catecholamines and other pressor substances. Irwin *et al.* (1991) have shown that plasma concentrations of neuropeptide Y are elevated in depressed patients as well as in aged individuals and persons undergoing severe Alzheimer caregiver stress. Furthermore, activation of the sympathetic nervous system and release of neuropeptide Y is associated with a reduction of natural cytotoxicity in depression and life stress (Fig. 5). Additional findings also support the hypothesis that elevated sympathetic activity in depression is associated with immune alterations. In depressed patients, excretion of 3-methoxy-4-hydroxyphenylglycol (MHPG) has been used as an index of total body noradrenergic turnover or sympathetic activity, and MHPG excretion was inversely related with lymphocyte proliferative responses in depressed patients (Maes, Bosmans, Suy, Minner, & Raus, 1989). Finally, Hickie, Hickie, Bennett, Wakefield, Silove, Mitchell, and Lloyd (1995) recently reported that in those melancholic patients with suppressed DTH responses, levels of urinary epinephrine were also significantly elevated compared with patients without suppressed responses. Together, these data suggest that elevated sympathetic tone in patients with major depressive disorder and/or in persons undergoing life stress is inversely correlated with cellular immune function.

17.3.2. Behavioral Factors

Sleep. A link between sleep and immune related activity has been demonstrated by the ability of immune mediators to alter sleep activity in animals (Krueger & Karnovsky, 1987) (see Chapter 27). In addition to influence of immune response modifiers on sleep, it is also possible that sleep processes and sleep–wake activity may mediate changes of cellular immunity, possibly by producing changes in the nocturnal secretion of immune response modifiers such as interleukins. Indeed, Moldofsky, Lue, Eisen, Keystone, and Gorczynski (1986) have reported a temporal association between the onset of slow wave sleep and the secretion of IL-1 and IL-2, two lymphokines that are known to stimulate or activate NK cells. Other data in normal controls have shown that sleep deprivation alters immunity. Palmblad, Petrini, Wasserman, and Akerstedt (1979) found that 48 h of sleep deprivation reduced lymphocyte proliferative responses to PHA relative to baseline levels, and Moldofsky and colleagues (Moldofsky, Lue, Davidson, & Gorczynski, 1989b) found a decline in NK activity following 40 hours of wakefulness. In contrast, Dinges, Douglas, Zaugg, Campbell, McMann, Whitehouse, Orne, Kapoor, Icaza, and Orne (1994) reported significant increases in WBC counts and NK activity during and after 64 h of sleep deprivation, although many of these effects may be attributable to the stress associated with profound sleep loss. Finally, our group has shown that the loss of even modest amounts of sleep can

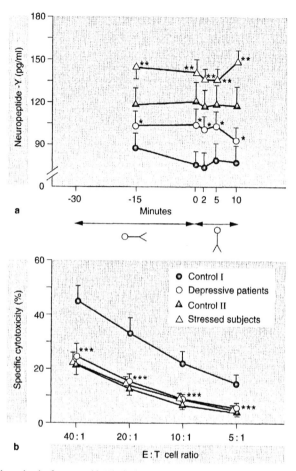

Figure 5.   (a) Plasma levels of neuropeptide Y in the four groups. (b) NK activity across the effector-to-target cell ratios in the four groups. From Irwin *et al.* (1991).

impair immune responses. In healthy subjects, late-night partial sleep deprivation (sleep time 11 PM to 3 AM) resulted in 30% declines in NK (Irwin, Mascovich, Gillin, Willoughby, Pike, & Smith, 1994), and early partial sleep deprivation (sleep time 3 AM to 7 AM) produced significant reductions in NK and lymphokine-activated killer (LAK) activity, as well as IL-2 production by stimulated lymphocytes (Irwin, McClintick, Costlow, Fortner, White, & Gillin, 1996). Interestingly, while NK and LAK activity returned to normal levels after a full night's sleep, IL-2 production remained suppressed.

The effects of sleep deprivation on immune function suggest that insomnia and sleep disturbance that occur during life stress and depression may be associated with depression-related alterations of immunity, especially given the prevalence of sleep abnormalities in depressed individuals (Benca, Obermeyer, Thisted, & Gillin, 1992). To test whether sleep disturbance accounts for the reliable association between depressive symptoms and reduced immunity, correlational analyses have been conducted. Insomnia is negatively correlated with NK activity in bereaved women, individuals undergoing other severe life stress, patients with major depression, and patients comorbid for depression and alcoholism (Cover & Irwin, 1994; Irwin *et al.*, 1987a,b, 1990a,b). Together, these data are consistent with the hypothesis that specific symptom profiles contribute to the reduction of NK activity found in depression, and further implicate severity of insomnia in depression-related suppression of NK activity.

Insomnia or subjective sleep disturbance is characterized objectively by disturbances of several sleep EEG continuity measures, including increased sleep latency and decreased total sleep time and sleep efficiency. To extend the correlational findings between insomnia and NK activity, Irwin and colleagues predicted that EEG measures of sleep continuity would also be associated with NK activity (Irwin *et al.*, 1994). Indeed, in depressed patients, EEG measures of total sleep time, sleep efficiency, duration of non-REM sleep and Stage 2 sleep were positively correlated with lytic activity. Furthermore, these relationships between sleep amounts and continuity measures were independent of the presence of a mood disorder, because similar correlations between lytic activity and total sleep time, sleep efficiency, and duration of non-REM sleep were found in nondepressed control subjects. It appears that sleep amounts and/or quality, whether assessed by objective or subjective methods, are associated with immune function.

Further studies are needed to examine the mechanisms underlying an association between sleep disturbance and reduced NK activity in depression, although sleep-related changes in lymphokines or neuroendocrine signals are hypothesized. For example, a decrease in amounts of slow-wave sleep occurs in depression, and Moldofsky *et al.* (1989b) have reported a temporal association between the onset of slow-wave sleep and the secretion of IL-1 and IL-2, two lymphokines that are known to stimulate or activate NK cells. Alternatively, sleep is associated with decreased sympathetic outflow, and disturbance of total sleep time or loss of sleep efficiency has been proposed to elevate sympathetic activity and in turn reduce cellular immunity.

Activity.   Changes in physical activity in depressed patients might also contribute to the reduction of cellular immunity (see Chapter 18). Assessment of psychomotor retardation and/or agitation using the Hamilton Depression Scale has shown that ratings of retardation negatively correlate with values of NK activity (Cover & Irwin, 1994), whereas agitation is positively associated with lymphocyte proliferative responses. Separate from the study of depressed patients, a reduction of physical activity such as induced bed rest or immobilization in humans has been found to produce a reduction of immune function, as measured by decreased antibody response to specific antigens. Conversely, acute physical exercise acutely increases NK activity (Murray *et al.*, 1993), as well as circulating levels of interferon and IL-1, and exercise training of sedentary subjects increases basal levels of cytotoxicity compared with nonexercise controls. Finally, given the possible role of NK cells in immune surveillance, it is of interest that increased levels of physical activity are associated with a decreased incidence of certain tumors in animals.

17.4. Health Implications of Depression-Related Immune Alterations

Depressed patients have been found to show a higher incidence and higher titers of herpes simplex virus antibodies, as well as increased incidence of herpes simplex. In one recent study, patients diagnosed with clinical depression exhibited significant impairment of herpes zoster virus-specific cellular immune responses compared with age-matched controls (Irwin, Costlow, Williams, Artin, Levin, Hayward, Chan, Stinson, & Oxman, 1998). Importantly, while control responses showed age-related declines in these virus-specific responses, responses in the depressed subjects were impaired at all ages measured (Fig. 6). The extent to which impairment of disease-specific immune responses mediates increased incidence of viral illness in depressed individuals remains unclear, but these data support this likelihood, especially given the vulnerability of the depressed population to viral infection.

The possibility that depression may be linked to disease vulnerability is supported by experiments demonstrating a relationship between stress and illness. Cohen *et al.* (1991) found that psychological stress was associated in a dose response manner with an increased risk for acute infectious respiratory illness. Importantly, this association reflected increased rates of infection rather than an increased frequency of symptoms after infection. Nonetheless, the role of the immune system in mediating illness vulnerability in this study was not clear, as measures of cellular immune function were not obtained. Other studies, but not all, have found that stressful events increase the risk of verified acute respiratory illness. However, again no study has yet delineated a causal chain showing that severe life stress or a particular psychological state such as depression produced an immunological response that in turn resulted in an altered clinical outcome.

The findings of increased cancer incidence in clinically depressed patients are consistent with a "null or weak relationship" (Fox, 1989). While several prospective follow-up studies have demonstrated a relationship between affective disorders and clinical depression and increased cancer morbidity and mortality, particularly in male patients over age 40 who had primary diagnoses of affective disorders, small case numbers indicate that these results should be cautiously interpreted. Likewise, epidemiological studies on the role of depression in cancer have been inconclusive. Depressed mood, as measured by the Minnesota Multiphasic Personality Inventory depression scale, was found to be a significant risk factor for cancer incidence and mortality over 20 years of follow-up. Although subsequent

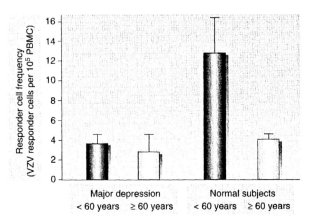

Figure 6.   Varicella-zoster virus responder cell frequency in depressed subjects < 60 and ⩾ 60 years of age, and in normal subjects < 60 and ⩾ 60 years of age; mean ± SD. (Depressed subjects < 60 years: $n = 7$; mean age 41.4 ± 8.0 years, range 32–56. Depressed subjects ⩾ 60 years: $n = 4$; 68.5 ± 8.2 years, range 60–77. Normal subjects < 60 years: $n = 7$; 40.1 ± 9.3 years, range 28–57. Normal subjects ⩾ 60 years: $n = 35$; 71.2 ± 6.7 years, range 60–80.)

studies have failed to replicate the observation that depression is associated with an increase in the relative risk of cancer, future studies are needed to evaluate whether smoking or alcohol use interacts with depressive symptoms to increase the risk of cancer. For example, some data show an increased risk of cancer for persons with depressed mood who are smokers compared with persons who never smoked without depressed mood. Furthermore, there was a dose response effect of cigarette smoking on the depressed mood–cancer relationship in the smokers (Linkins & Comstock, 1990).

The promise of recent investigations has suggested that psychiatric interventions that reduce psychological distress may have beneficial effects on cancer survival. Spiegel, Bloom, Kraemer, and Gottheil (1989) found that group psychotherapy of breast cancer patients improved quality of life and extended life expectancy for patients with metastatic breast cancer, and Fawzy, Kemeney, Fawzy, Elashoff, Morton, Cousins, and Fahey (1990) showed that other types of cancer patients improve both psychologically as well as immunologically with psychosocial group therapy. The psychoneuroimmunological link between depression, psychological stress, and the outcome of immune-related medical disorders will be refined by further research that addresses the role of depressive symptoms such as sleep disturbance in the modulation of immunity, and defines the neurobiological paths that underlie the interaction between the brain, behavior, and immunity.

ACKNOWLEDGMENTS. This work was supported by VA Merit Review and National Institutes of Mental Health Grants (MH-44275-05 and MH-46867-01) to M.I. Additional support was provided in part by a General Clinical Research Center Grant (M01 RR00827) and a Mental Health Clinical Research Center Grant (MH-30914). Portions of this chapter may have been published previously.

18   Exercise and Immune Functions

Bente Klarlund Pedersen

Bente Klarlund Pedersen   •   Copenhagen Muscle Research Center, Department of Infectious Diseases, Rigshospitalet, 2200 Copenhagen N, Denmark.

18.1. Introduction

The increasing popularity of regular exercise is probably the result of increased public awareness regarding the beneficial effects of exercise on physical and emotional well-being. The positive effects of regular exercise are obviously a complex interaction between physiological and psychological effects. For example, it has been shown that exercise can lower blood pressure, improve the lipid profile, and decrease sensitivity to pain (Thoren, Floras, Hoffmann, & Seals, 1990).

Many individuals claim that regular exercise increases resistance to infections such as the common cold (Fitzgerald, 1988; Nash, 1987). On the other hand, there have also been anecdotal reports from athletes and their coaches that hard training is associated with increased respiratory tract infections (Fitzgerald, 1988). There is epidemiological evidence supporting the anecdotal impression (Nieman & Henson, 1994), and it has become clear that the function of the immune system is enhanced by moderate physical activity and may be somewhat responsible for exercise-related reduction in illness. In contrast, it has been repeatedly shown that intense exercise causes inhibition of the function of the immune system in the recovery phase following intense exercise (Hoffman-Goetz & Pedersen, 1994).

Today much research in exercise immunology is stimulated by the acceptance of exercise as a tool to study the immune system. Exercise can be employed as a model for temporary immunosuppression occurring after severe physical stress. The exercise stress model can be easily manipulated experimentally, and allows for the study of interactions between the nervous system, the endocrine system, and the immune system. Furthermore, eccentric exercise that is associated with muscle damage may represent a model of the acute-phase response to local injury.

This chapter provides an overview of the effects of acute and chronic exercise on the immune system and discusses the clinical significance of these findings.

18.2. Acute Exercise and the Cellular Immune System

18.2.1. Leukocyte Subpopulations

There are several consistent patterns that emerge regarding leukocyte subpopulations in the blood. The neutrophil concentrations increase during exercise and continue to increase postexercise (McCarthy & Dale, 1988). The lymphocyte concentration increases during exercise and falls below prevalues following intense long-duration exercise (McCarthy & Dale, 1988). The increased lymphocyte concentration is related to recruitment of all lymphocyte subpopulations to the blood. Thus, CD4$^+$ T cells, CD8$^+$ T cells, CD19$^+$ B cells, CD16$^+$ natural killer (NK) cells, and CD56$^+$ NK cells increase during exercise. Simultaneously, the CD4/CD8 ratio decreases, because the CD8 count increases more than the CD4 count. The percentage of CD4$^+$ cells declines primarily because of the fact that NK cells increase more than any other lymphocyte subpopulation. Accordingly, the relative fraction of lymphocyte subpopulations changes.

18.2.2. NK Cells

NK cells are a heterogeneous population of cells (see Chapters 1 and 2). By definition, they are CD3$^-$ large granular lymphocytes that express characteristic NK cell markers such as CD16 and CD56.

NK cells are recruited to the circulation within a few minutes after the onset of exercise and other stressors, and NK cells are more sensitive to stress stimuli than any other subpopulation (Pedersen, Kappel, Klokker, Nielsen, & Secher, 1994). Results from the area of stress immunology thus indicate that the rapidly increasing NK cell response in relation to infections probably also includes immediate recruitment of NK cells to the circulation and the site of infection. Because the concentration of NK cells increases more than any other lymphocyte subpopulation, this means that the NK cell percentage of blood mononuclear cells (BMNC) increases. Using the *in vitro* assay measuring the NK cell activity (lysis per fixed number of BMNC), the NK cell lysis increases as a consequence of increased proportion of cells mediating non-MHC-restricted cytotoxicity. During exercise the NK cell activity on a per NK cell basis does not change in some exercise models (Nieman, Miller, Henson, Warren, Gusewitch, Johnson, Davis, Butterworth, & Nehlsen Cannarella, 1993b; Palmø, Asp, Daugaard, Richter, Klokker, & Pedersen, 1995), but is reduced in relation to very intense exercise (Nielsen, Secher, Kappel, Hanel, & Pedersen, 1996a).

Following intense exercise of long duration, the concentration of NK cells declines below prevalues. Furthermore, the NK and LAK cell activity (lysis per fixed number of BMNC) decreases. The NK cell concentration and the NK cell activity are maximally suppressed 2 to 4 h after exercise.

18.2.3. NK Cells at Exercise of Different Duration and Intensity

NK cells have been studied in various models, including laboratory and field studies, studies on running, cycling, and rowing, studies on concentric, eccentric, or combined concentric and eccentric exercise, and exercise lasting from a few minutes to several hours. This has recently been reviewed (Pedersen & Nielsen, 1997).

In general, the NK cell activity is increased immediately after or during both moderate and intense exercise of a few minutes' duration. The intensity, more than the duration of exercise, is responsible for the degree of increment in the number of NK cells. If the exercise has lasted for a long time and has been very intense (e.g., a triathlon race), only a modest increase in NK cells is found postexercise (Rohde, MacLean, Hartkopp, & Pedersen, 1996).

The NK cell count and the NK cell activity are suppressed only following intense exercise of a certain duration, and at least 1 h seems to be a critical duration of exercise in terms of postexercise suppression of natural immunity. Initial fitness level or sex does not influence the magnitude of exercise-induced changes in NK cells (Brahmi, Thomas, Park, & Dowdeswell, 1985).

18.2.4. Antibody Production

The secretory immune system of mucosal tissues such as the upper respiratory tract is considered to be the first barrier to colonization by pathogenic microorganisms causing upper respiratory tract infections (URTI) (Dufaux & Order, 1989b; Mackinnon & Hooper, 1994). Although IgA constitutes only 10–15% of the total immunoglobulin in serum, it is the predominant immunoglobulin class in mucosal secretions, and the level of IgA in mucosal fluids correlates more closely with resistance to URTI than serum antibodies (Liew, Russell, Appleyard, Brand, & Beale, 1984) (see Chapters 2 and 19).

Tomasi, Trudeau, Czerwinski, and Erredge (1982) reported suppressed levels of salivary IgA in cross-country skiers after a race. This finding was confirmed by a 70% decrease in salivary IgA for several hours after 2 h of intense ergometer cycling (Mackinnon, Chick, van As, & Tomasi, 1987). Decreased salivary IgA was also found after swimming (Tharp & Barnes, 1990), marathon running (Muns, Liesen, Riedel, & Bergman,

1989), and incremental treadmill running to exhaustion (McDowell, Hughes, Hughes, Housh, Housh, & Johnson, 1992). In contrast, submaximal exercise of VO_{2max} had no effect on salivary IgA (Housh, Johnson, Housh, Evans, & Tharp, 1991; McDowell, Chaloa, Housh, Tharp, & Johnson, 1991).

In order to study the mechanism behind the suppression of immunoglobulins, a plaque-forming cell assay has been used. This assay allows identification of the individual immunoglobulin-secreting cells of blood. Stimulation of cells with pokeweed mitogen, IL-2 and Epstein–Barr virus resulted in significantly decreased numbers of IgG-, IgA-, and IgM-secreting blood cells during, as well as 2 h postexercise. The percentage of B cells among BMNC does not change in relation to exercise, suggesting that the suppression of immunoglobulin-secreting cells is not related to changes in numbers of B cells. Purified B cells produce plaques only after stimulation with Epstein–Barr virus and in these cultures no exercise-induced suppression was found. The addition of indomethacin to IL-2-stimulated cultures of BMNC partly reversed the postexercise suppressed B-cell function. Therefore, it was concluded that the exercise-induced suppression of the plaque-forming cell response was partly mediated by monocytes (Tvede, Heilmann, Halkjaer Kristensen, & Pedersen, 1989).

18.2.5. Neutrophil Function

Neutrophils represent 50 to 60% of the total circulating leukocyte pool (Chapters 1 and 2). They are considered to be part of the innate immune system, and are essential for host defense. However, they are also involved in the pathology of various inflammatory conditions. Therefore, neutrophils can be considered to be a double-edged sword, as suggested by Smith (1994).

One of the more pronounced effects of physical activity on immune parameters is the prolonged neutrocytosis following acute long-term exercise (McCarthy & Dale, 1988). Short bouts of vigorous exercise lasting from approximately 10 s to half an hour (e.g., sprinting, rowing, or gymnastics), sustained strenuous exercise lasting a few hours (e.g., marathon races), or intermittent vigorous exercise (e.g., U.S. football or soccer) can all induce an immediate leukocytosis (McCarthy & Dale, 1988). At the end of short-term exercise lasting up to half an hour, the leukocyte count usually falls quickly, approaching its normal levels within half an hour. After sustained exercise, lasting 2½ to 3½ h, the leukocyte count returns slowly and is often elevated above preexercise levels at 24 h. In contrast, after intermittent exercise over 1½ h or sustained exercise of 1 h, the leukocyte count continues to rise for 1 to 4 h, before falling slowly (Smith, 1994; Smith, Telford, Mason, & Weidermann, 1990; Weidemann, Smith, Gray, McKenzie, Pyne, Kolbuch Braddon, & Telford, 1992).

Regarding the function of neutrophils, exercise has both short- and long-term effects. The neutrophil response to infections includes adherence, chemotaxis, phagocytosis, oxidative burst, degranulation, and microbial killing. In general, moderate exercise boosts neutrophil functions, including chemotaxis, phagocytosis and the oxidative burst, whereas extreme exercise suppresses these functions with the exception of chemotaxis and degranulation, which are not affected (Brines, Hoffmann-Goetz, & Pedersen, 1996; Nieman, 1994a,b; Ortega, Collazos, Maynar, Barriga, & De la Fuente, 1993; Smith et al., 1990; Smith, McKenzie, Telford, & Weidemann, 1992). In the study by Smith et al. (1990), neutrophil killing capacity was enhanced for at least 6 h following 1 h of moderate ergometer cycle exercise. In another study, immediately following a 20-km race, neutrophils from runners were less able to ingest bacteria, an effect that lasted for 3 days (Nieman and Henson, 1994).

In a comparative study of male distance runners, triathletes, and untrained controls, progressive exercise to exhaustion doubled neutrophil phagocytic capacity in samples taken up to 24 h after exercise. Superoxide production in phorbol myristate acetate-stimulated cells fell slightly immediately after exercise but then increased 1 h later and remained elevated for 24 h; there were significant changes in chemotactic capacity or random migration (Hack, Strobel, Rau, & Weicker, 1992). These results were confirmed by the same group in both trained and untrained individuals (Hack, Strobel, Weiss, & Weicker, 1994). Phagocytic activity increased immediately after exercise but no changes in adherence or bactericidal activity were seen in untrained individuals.

Several studies have shown that exercise activates the release of neutrophil granule constituents into the circulation, indicating direct neutrophil activation *in vivo*. Elevated concentrations of neutrophil elastase have been detected in plasma after exercise (Busse, Anderson, Hanson, & Folts, 1980; Camus, Pincemail, Ledent, Juchmes Ferir, Lamy, Deby Dupont, & Deby, 1992; Dufaux & Order, 1989a; Hansen, Wilsgard, & Osterud, 1991; Kokot, Schaefer, Teschner, Gilge, Plass, & Heidland, 1988).

18.2.6. *In Vivo* Immunological Methods

While significant changes in the concentration and functional activity of immune parameters have been observed, this may not necessarily lead to a higher incidence of infections and illness. *In vitro* tests may not always provide accurate assessments of systemic immune responses, epidemiological studies may reflect concomitant conditions that have their own separate effects on immune function, and procedures involving animals may not represent reasonable models of human exercise.

It is possible that the use of *in vivo* immunological methods may provide a tool to study clinically relevant immune changes. The systemic *in vivo* function of the immune system can be assessed using at least two different principles. The cellular immune response can be evaluated by measuring the delayed hypersensitivity response to a recall antigen introduced to the skin. In that case an absent response (anergy) or a poor response indicates an impairment of the cellular immune response. The humoral immune response can be evaluated through the measurement of the specific antibody titers following vaccination. Low specific titers following antigen challenge would indicate a defective humoral response.

Bruunsgaard, Galbo, Halkjaer-Kristensen, Johansen, MacLean, and Pedersen (1997) investigated whether an *in vivo* impairment of cell-mediated immunity and specific antibody production could be demonstrated after intense exercise of long duration (triathlon race). The cellular immune system was evaluated by the skin test response to seven recall antigens: two toxoids (tetanus and diphtheria), three bacterial (streptococcus, tuberculin, and proteus), and two fungal (candida and trichophyton). The humoral immune system was evaluated by the antibody response to pneumococcal polysaccharide vaccine, which is generally considered to be T cell independent, and two toxoids (tetanus and diphtheria) that are dependent on T cells. The subjects had received tetanus and diphtheria vaccinations previously during their life, but not pneumococcal vaccination.

The protocol included 22 male triathletes, who performed one-half of an ironman. Vaccinations were given after the race and the skin test was applied. Specific antibody titers were measured before and 2 weeks after the race. The skin test was read 48 h after application. Eleven nonexercising triathletes and 22 moderately trained men were used as controls.

The cumulated skin test responses (sum of the diameters of indurations and number of positive skin spots) were significantly lower in the ironman group compared with the nonexercising triathletes and the moderately trained subjects, whereas no differences were found between the two latter groups. In addition, no differences were found between the groups regarding specific antibody titers against diphtheria, tetanus toxoid, or six pneumococcal antigens.

Thus, *in vivo* cell-mediated immunity was impaired in the first days after prolonged high-intensity exercise. The delayed-type hypersensitivity reaction is a complex immunological reaction involving several different cell types and chemical mediators (see Chapter 2). When antigens are injected into the skin, a typical reaction consisting of erythema and induration is seen after 48 h. Histologically, an exudation of monocytes and polymorphonuclear cells is seen. The latter soon migrate out of the lesion and mononuclear cells are left. Memory T cells recognize the antigen-presenting cells and are stimulated to blast transformation and secretion of cytokines. If a Th1 response is generated, the cytokines will attract cytoxic T cells, NK cells, and macrophages. The finding of a decreased cellular immunity following prolonged intense exercise may be a result of a decrease in the accumulation of cells or a result of a functional impairment including decreased production of IL-2 and IFNγ as suggested by Weinstock, Konig, Harnischmacher, Keul, Berg, and Northoff (1997).

The diverse findings regarding influence of strenuous exercise on *in vivo* cellular and humoral responses may be that primarily unspecific host responses are influenced by exercise, whereas the specificity of the immune system is not altered. In that case, the transient decrease in cellular immune response may be of importance in weakening the first line of defense and increasing the risk of acute infections such as URTI, and may have no effect on the generation of a specific immune response during the following days and thus may have no effect on clinical recovery from disease. Alternatively, the time factor may be taken into consideration.

18.3. Exercise and Cytokines

Cytokines are a group of low-molecular-weight regulatory proteins secreted by white blood cells, and a variety of other cells in the body, in response to a number of inducing stimuli (see Chapter 2). Cytokines generally function as intercellular messenger molecules that evoke particular biological activities after binding to a receptor on a responsive target cell. In the acute-phase response, IL-1, IL-6, and TNFα have been most extensively studied (see Chapter 12).

A study by Cannon and Kluger (1983) was the first to suggest that exercise induced a cytokine response. They showed that plasma obtained from human subjects after exercise and injected intraperitoneally into rats, elevated rat rectal temperature and depressed plasma iron and zinc concentrations. Plasma obtained prior to exercise failed to produce these responses. The pyrogenic component was heat-denaturable and had an apparent molecular mass of 14 kDa. Human mononuclear leukocytes obtained after exercise and incubated *in vitro* released a factor into the medium that also elevated body temperature in rats and reduced trace metal concentrations. These results suggested that endogenous pyrogen was released during exercise.

In 1986 two studies were published on exercise and IL-1 (Cannon, Evans, Hughes, Meredith, & Dinarello, 1986; Evans, Meredith, Cannon, Dinarello, Frontera, Hughes, Jones, & Knuttgen, 1986). Evans *et al.* (1986) studied the effects of 45 min of eccentric bicycle exercise on IL-1 activity in the plasma of four male runners and five untrained men. IL-1 activity was measured by the ability of Sephadex G-50 chromatographed plasma fractions to increase thymocyte proliferation, and results were expressed as percent basal thymocyte proliferation. Plasma IL-1 activities increased 3 h after exercise in all five untrained men, whereas there was no increase in the trained individuals. Interestingly, the highly trained endurance runners tended to have higher IL-1 activity prior to exercise, but these values did not increase further in response to exercise. Furthermore, the patterns of change in plasma creatine kinase (CK) activities between trained and untrained men were strikingly different. Compared with the untrained men, preexercise CK levels were significantly higher in the runners. In the untrained men, CK levels increased progressively for 5 days afterwards, reaching a peak value that was 33 times greater than the baseline level. In contrast, the increase in the plasma CK level in trained subjects was significant only at 24 h after the exercise bout, reaching a maximum value 2.3 times greater than the preexercise level.

In the study by Cannon *et al.* (1986), IL-1 activity was measured by the thymocyte proliferation bioassay in plasma 3 h after bicycle exercise for 1 h. The IL-1 effect was neutralized by an antibody to IL-1. However, although initial findings indicated a correlation between CK and IL-1 levels (Evans *et al.*, 1986), this study (Cannon *et al.*, 1986) found no significant correlation between CK level and IL-1 activity in plasma.

These initial studies thus showed that systemic elevations of cytokines occur in serum after strenuous exercise. The identity of the observed cytokine was, however, uncertain. Recently, Bagby, Crouch, and Shepherd (1996) suggested that although IL-1 was believed to be the cytokine responsible for the exercise-induced plasma activities in the above mentioned studies, the possibility exists that cytokines other than IL-1 were measured. The latter studies were conducted prior to the availability of recombinant IL-1 proteins. Furthermore, the thymocyte proliferation bioassay also detects IL-6. Therefore, the possibility exists that the cytokine responsible for the activity measured in the thymocyte bioassay or for the fever-inducing properties of plasma was IL-6 and not IL-1. As pointed out by Bagby *et al.* (1996), a number of studies have failed to detect significantly elevated levels of IL-1 in plasma. However, as IL-6 was found to be enhanced in some studies, Northoff and Berg (1991) suggested that IL-6 might be involved in the generation of the acute-phase response after exercise. They tested 17 participants of a marathon run. The serum was examined using the 7TD1 assay, which is very sensitive for IL-6, but negative for other cytokines. In 15 of the 17 runners, there was a significant increase of IL-6 measurable immediately after exercise, which returned in most cases to baseline 24 h later.

The effect of concentric exercise on cytokine plasma levels and cytokine pre-mRNA in BMNC has been investigated (Ullum, Haahr, Diamant, Palmo, Halkjaer-Kristensen, & Pedersen, 1994). A small (less than twofold) but significant increase in plasma IL-6 was found during exercise using an ELISA kit. The plasma levels of IL-1α, IL-1β, IL-6, and TNFα were below the detection limit in most subjects. Although pre-mRNA for IL-1α, IL-1β, IL-6, and TNFα could be detected in BMNC using nuclear run-off analysis, the amounts did not change in relation to exercise. Increased cytokine levels after exercise have been found mainly in relation to eccentric exercise or combinations of eccentric and concentric exercise.

Bruunsgaard *et al.* (1997) performed a study to test the hypothesis that the exercise-induced increase in cytokine levels was associated with muscle damage. Healthy young males performed two bouts of bicycle exercise; the first trial consisted of 30 min concentric exercise and the second, 30 min eccentric exercise. The CK level increased almost 50-fold 4 days after eccentric exercise, whereas there was no change in CK level in relation to concentric exercise. The IL-6 level increased 5-fold in relation to eccentric exercise, but not in relation to concentric exercise. Thus, this study supports the hypothesis that the post-exercise cytokine production is related to skeletal muscle damage. Recently, we have found a 50-fold increase in plasma levels of IL-6 and IL-1 receptor antagonist following a marathon (Ostrowski *et al.*, unpublished data). Thus, several studies have identified an exercise-induced increase in IL-6 activity in response to exercise. In addition, a significant correlation was found between IL-6 and CK in an eccentric exercise study (Bruunsgaard *et al.*, 1997).

One study clearly shows that the cytokine cascade is activated as a consequence of exercise (Sprenger, Jacobs, Nain, Gressner, Prinz, Wesemann, & Gemsa, 1992). It is very interesting that although exercise has a minimal effect on most circulating cytokine values, several cytokines can be measured in the urine. This indicates that some cytokines are produced in response to exercise, but rapidly removed from the circulation. Although most studies fail to detect circulating cytokines other than IL-6 in the plasma, the presence of multiple cytokines in urine following exercise shows that the expression of a broad spectrum of cytokines in response to exercise is possible. Thus, exercise may alter the local production of several cytokines and these may have important local functions, e.g., in the local acute response. Increased cytokine level after exercise is primarily found after eccentric muscle contractions. The critical question remains, however, whether a causal relationship exists, e.g., does muscle damage cause cytokine production or vice versa?

18.4. Neuroendocrinological Regulation

Studies where either hormones were infused, hormone receptors were blocked by drugs, or hormone production was inhibited by epidural blockade, in relation to physical stress, contribute to our understanding of the mechanisms of action. Based on these studies, a model on the possible roles of stress hormones in exercise-induced immune changes can be proposed (see Chapters 9 and 15).

18.4.1. Catecholamines

During exercise, adrenaline is released from the adrenal medulla and noradrenaline is released from the sympathetic nerve terminals. Arterial plasma concentrations of adrenaline and noradrenaline increase almost linearly with duration of dynamic exercise and exponentially with intensity, when it is expressed relative to the individual's maximal oxygen uptake (Kjaer, 1989). The expression of β-adrenoceptors on T, B, and NK cells, macrophages, and neutrophils in numerous species provides the molecular basis for these cells to be targets for catecholamine signaling (Madden & Felten, 1995).

It was previously reported that proliferation, antibody production, and cytotoxic activity could be reduced by β-adrenergic stimulation, and in some cases enhanced by α-adrenergic stimulation (Madden & Felten, 1995). From such studies, elevation of intracellular cAMP was viewed as a way to limit lymphocyte proliferation and effector function, thereby suppressing the immune response. However, researchers from several laboratories have been unable to replicate findings of reduced proliferative responses with doses of catecholamines other than pharmacological (Butler, Kelly, O'Malley, & Pidgeon, 1983; Dufaux, Order, Geyer, & Hollmann, 1984). The *in vitro* effect of adrenaline on NK cell activity has been examined by Katz, Zaytoun, and Fauci (1982), who demonstrated that human NK cell activity was inhibited by the addition of cAMP inducers directly to target and effector cells in a ^{51}Cr-release assay. More complex effects were reported by Hellstrand, Hermodsson, and Strannegard (1985). They found that pretreatment of lymphocytes with low concentrations of adrenaline (10^{-7} to 10^{-9} M) followed by removal of the drug increased NK cell activity. In contrast, direct addition of adrenaline (10^{-6} M) to the lymphocyte–target cell mixture inhibited the NK cell activity. In a study by Kappel, Tvede, Galbo, Haahr, Kjaer, Linstow, Klarlund, and Pedersen (1991), adrenaline was present during preincubation of mononuclear cells as well as in the NK cell assay. The adrenaline concentration (3.8×10^{-9} M) was comparable to adrenaline concentrations measured during exercise. We were unable to demonstrate any *in vitro* effects of adrenaline on NK cells isolated before, during, or after adrenaline infusion. Therefore, these results indicate that adrenaline may act by redistributing BMNC subsets, rather than directly influencing the activity of the individual NK cells.

The numbers of adrenergic receptors on the individual lymphocyte subpopulations may determine to what degree the cells are mobilized in response to catecholamines. In accordance with this hypothesis, it has been shown that different subpopulations of BMNC have different numbers of adrenergic receptors (Khan, Sansoni, Silverman, Engleman, & Melmon, 1986; Maisel, Harris, Rearden, & Michel, 1990a; Rabin, Moyna, Kusnecov, Zhou, & Shurin, 1996; van Tits and Graafsma, 1991). NK cells contain the highest number of adrenergic receptors, with CD4$^+$ lymphocytes having the lowest number. B and CD8$^+$ lymphocytes are intermediate between NK cells and CD4$^+$ lymphocytes (Rabin *et al.*, 1996). Dynamic exercise upregulates the β-adrenergic density, but only on NK cells (Maisel *et al.*, 1990a). Interestingly, NK cells are more sensitive to exercise and other stressors than any other subpopulation. CD4$^+$ cells are less sensitive and CD8$^+$ and B cells are intermediate (Hoffman-Goetz & Pedersen, 1994). Thus, a correlation exists between numbers of adrenergic receptors on lymphocyte subpopulations and their sensitivity to exercise.

In humans, a single adrenaline injection induced transient increases in the number of circulating blood lymphocytes and monocytes and decreased the response to T-cell mitogens (Crary, Borysenko, Sutherland, Kutz, Borysenko, & Benson, 1983a). Selective administration of adrenaline to obtain plasma concentrations comparable to those obtained during concentric cycling for 1 h at 75% of VO$_{2max}$, mimicked the exercise-induced effect on BMNC subsets, NK cell activity, LAK cell activity, and the lymphocyte proliferative response (Kappel *et al.*, 1991; Tønnesen, Christensen, & Brinkløv, 1987; Tvede, Kappel, Klarlund, Duhn, Halkjaer-Kristensen, Kjaer, Galbo, & Pedersen, 1994). However, adrenaline infusion caused a significantly smaller increase in neutrophil concentrations than that observed following exercise (Kappel *et al.*, 1991; Tvede *et al.*, 1994).

Ahlborg and Ahlborg (1970) showed that after administration of propranolol, exercise resulted in practically no increase in lymphocyte concentration. β$_{1+2}$-receptor blockade more than β$_1$-blockade inhibited head-up tilt-induced lymphocytosis (Klokker, Secher, Olesen, Madsen, Warberg, & Pedersen, 1997) and the increase in the number of NK cells is in accordance with the finding that primarily β$_2$-receptors are expressed on lymphocytes (Benschop, Oostveen, Heijnen, & Ballieux, 1993). β-Receptor blockade did not abolish the head-up tilt-induced neutrocytosis, which is in agreement with previous findings showing that adrenaline infusion caused a smaller increase in neutrophil concentration than the exercise-induced increase (Kappel *et al.*, 1991; Tvede *et al.*, 1994). The effect of noradrenaline on recruitment of lymphocytes to the blood resembles that of adrenaline, although noradrenaline was a much weaker stimulator of the immune system (Kappel *et al.*, unpublished data).

There are several pieces of evidence showing that adrenaline is responsible for the recruitment of NK cells to the blood during physical exercise and other physical stress forms. The experimental basis included the findings that (1) adrenaline infusion mimicked the exercise-induced effect especially on NK and LAK cells, (2) the β-adrenergic receptors are upregulated on NK cells during exercise, (3) β-receptor blockade abolished lymphocytosis during exercise and the increase in NK cell number during head-up tilt, and (4) β$_2$-

receptor agonists induce selective detachment of NK cells from endothelial cells. The latter results strongly support the hypothesis that epinephrine mediates recruitment of NK cells from the marginating pool in blood vessels during stress.

18.4.2. Growth Hormone

Plasma levels of pituitary hormones increase in response to both exercise duration and intensity, although the growth hormone (GH) response is more related to the peak exercise intensity than to duration of exercise or total work output (Kjaer, 1992; Vanhelder, Radomski, & Goode, 1984).

Addressing the question of a possible role of GH in mediating acute exercise-induced immune changes, Kappel, Hansen, Diamant, Jorgensen, Gyhrs, and Pedersen (1993) injected GH into humans to obtain blood concentrations of GH comparable with those observed during exercise. An intravenous bolus injection of GH had no effect on BMNC subsets, NK cell activity, cytokine production, or lymphocyte function, but induced a highly significant neutrocytosis (Kappel *et al.*, 1993). Therefore, GH does not seem to play a major role in exercise-induced recruitment of lymphocytes to the blood. However, epinephrine and GH in conjunction are probably responsible for the recruitment of neutrophils to the blood during physical stress.

18.4.3. β-Endorphin

The role of β-endorphin in exercise-induced immunomodulation has been difficult to establish. However, the available data indicate that β-endorphin is not responsible for the immediate recruitment of NK cells to the blood during acute exercise, but likely responsible for increased NK cell activity during chronic stress. This is based on the fact that NK cells are recruited to the blood immediately after the onset of exercise and even at exercise of very low intensity. In contrast, the concentrations of β-endorphin increase only on exercise at high intensity and long duration, which make it unlikely that β-endorphin plays a major immunomodulatory role in the immediate recruitment of NK cells to the blood. Furthermore, although Fiatarone, Morley, Bloom, Benton, Makinodan, and Solomon (1988) reported that naloxone blocked the exercise-induced increase in NK cell activity, the differences were, although statistically significant, small, and there was no effect of β-endorphin receptor blockade on the number of cells. It has been found that β-endorphin receptor blockade during head-up tilt and inhibition of afferent nerves from working muscles, which subsequently cause an increase in β-endorphin, did not inhibit the recruitment of NK cells to the blood. This further supports the idea that β-endorphin does not play a role in the immediate recruitment of NK cells during acute exercise. The hypothesis that β-endorphin is important in maintaining increased NK cell activity during chronic stress is primarily based on studies showing that voluntary chronic exercise augments *in vivo* natural immunity (Jonsdottir, Asea, Hoffmann, Dahlgren, Andersson, Hellstrand, & Thoren, 1996a). In addition, β-endorphin and dynorphin infusion in hypertensive rats (Hoffmann, Terenius, & Thoren, 1996; Jonsdottir, Johansson, Asea, Hellstrand, Thoren, & Hoffmann, 1996b; Persson, Jonsdottir, Thoren, Post, Nyberg, & Hoffmann, 1993) and β-endorphin but not dynorphin infusion in normal rats, increased the NK cell activity (Jonsdottir *et al.*, 1996b).

18.4.4. Cortisol

The concentration of cortisol increases only in relation to exercise of long duration (Galbo, 1983). Thus, short-term exercise does not increase the cortisol concentration in

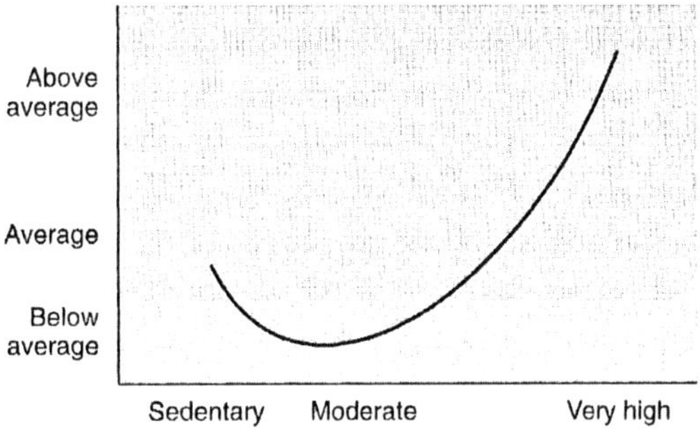

Figure 1. "J"-shaped model of relationship between varying amounts of exercise and risk of upper respiratory tract infection. This model suggests that moderate exercise may lower risk of respiratory infection while excessive amounts may increase risk.

plasma, and only minor changes in the concentration of plasma cortisol were described in relation to acute time-limited exercise stress of 1 h (Galbo, 1983).

Based on these studies, we propose a model (Fig. 1) of the possible roles of stress hormones in mediating exercise-related immune changes: Adrenaline and, to a lesser degree, noradrenaline are responsible for acute exercise effects on lymphocyte subpopulations, NK and LAK cell activities. The increase in catecholamines and GH mediates the acute effects on neutrophils, whereas cortisol exerts its effects within a time lag of at least 2 h and contributes to maintain the lymphopenia and neutrocytosis only after long-term exercise. The role of β-endorphin is less clear, but we do not believe that β-endorphin plays an important role in the immediate recruitment of NK cells to the blood. We do hypothesize that β-endorphin is important in chronic stress. Stress hormones do not seem to be responsible for the exercise-induced increase in cytokines. This hypothesis is an extension of the hypothesis previously suggested by McCarthy and Dale (1988) that the immediate leukocytosis during exercise is attributable to elevated catecholamine levels, whereas the delayed neutrophilia is the result of raised cortisol levels.

18.5. Glutamine—A Metabolic Link between Muscle and the Immune System?

It has generally been accepted that cells of the immune system obtain their energy by metabolism of glucose. However, it has been established that glutamine is also an important fuel for lymphocytes and macrophages. Several lines of evidence suggest that glutamine is used at a very high rate by these cells, even when they are quiescent (Newsholme, 1994). It has been proposed that the glutamine pathway in lymphocytes may be under external regulation, associated partly with the supply of glutamine itself (Ardawi & Newsholme, 1984).

Skeletal muscle is the major tissue involved in glutamine production and known to release glutamine into the bloodstream at a high rate. Thus, it has been suggested that the skeletal muscle plays a vital role in maintenance of the rate of the key process of glutamine utilization in the immune cells. Consequently, the activity of the skeletal muscle may directly influence the immune system. It has been hypothesized (the so-called "glutamine hypothesis") that under intense physical exercise, or in relation to surgery, trauma, burn, and sepsis, the demands on muscle and other organs for glutamine are such that the lymphoid system may be forced into a glutamine debt, which temporarily affects its function. Thus, factors that directly or indirectly influence glutamine synthesis or release could theoretically influence the function of lymphocytes and monocytes (Newsholme, 1990, 1994). Following intense long-term exercise and other physical stress disorders, the glutamine concentration in plasma declines.

Clearly, optimal lymphocyte proliferation is dependent on the presence of glutamine, although there are no published data showing that glutamine supplementation restores impaired immune function after exercise. The critical question therefore is not whether concomitant decreased plasma glutamine concentration and lymphocyte function occur following intense exercise, but whether a causal relationship exists. Two recent placebo-controlled glutamine intervention studies (Rohde et al., unpublished data) prevented the postexercise decline in plasma glutamine without influencing postexercise immunosuppression. Thus, these studies did not support the hypothesis that postexercise immunosuppression is caused by a decrease in the plasma glutamine concentration.

18.6. Repetitive Bouts of Acute Exercise

Little is known about effects of repeated bouts of exercise on the cellular immune system and cytokine production. Hoffman-Goetz, Simpson, Cipp, Arumugam, and Houston (1990) studied the effect of repeated bouts of submaximal cycle ergometry exercise on changes in the percentage of BMNC subpopulations. Healthy volunteers exercised daily for 1 h at 65% of VO_{2max}. The increase in percentage of NK cells to five repetitive bouts of cycling over 5 days did not differ from that elicited by the first bout.

In another study the effects of two bouts of exhaustive cycle ergometry exercise, lasting 12.9 and 13.2 min, respectively, and separated by 1 h were investigated (Field, Gougeon, & Marliss, 1991). A significant increase in total leukocytes (2-fold), neutrophils (1.9-fold), and lymphocytes (2.3-fold) occurred during the first exercise bout. The concentrations of leukocytes and neutrophils increased to the same level during both experiments, but the concentrations 1 h after the second bout were higher than those 1 h after the first bout. The concentration of lymphocytes increased less during the second bout compared with the first bout. One hour after the second bout, the lymphocyte concentration decreased below baseline. This suppression was similar to that developed following the first bout.

Nielsen, Secher, and Pedersen (1996b) investigated the effects on the immune system of 6 min of "all-out" ergometer rowing over 2 days (two × three bouts) in elite male oarsmen. Compared with levels at rest, the first bout of exercise increased the concentration of leukocytes (twofold), neutrocytes (twofold), and lymphocytes (twofold). During the last bout, even higher levels were observed for leukocytes (threefold), neutrocytes (threefold), and lymphocytes (fourfold). During the recovery periods, all values were at or above the level at rest, and elevated concentrations of leukocytes, neutrocytes, lymphocytes, and NK cell activity were also noted on the day following the last bout.

A recent study by Rohde et al. (unpublished data) examined the effect of bicycle exercise lasting 1 h, 45 min, and 30 min with intervals of 2 h. The lymphocyte concentration and the mitogen-induced proliferative response declined 2 h after each bout of exercise, whereas the LAK cell activity declined 2 h after the third bout. During the last bout of exercise, significantly higher levels of leukocytes, neutrophils, lymphocytes, and several lymphocyte subpopulations were observed.

The divergence between the studies mentioned above may reflect that enhancement of the immune system depends on the intensity of exercise. The study by Nielsen et al. (1996b) shows that intense exercise with large muscle groups of short duration provokes higher immune responses during repetitive bouts of exercise. The study by Rohde *et al.* (unpublished data) reveals that immunosuppression may occur with repeated bouts of exercise, when the model employed is just at the critical border in duration and intensity in terms of postexercise immunosuppression.

18.7. Chronic Exercise

In contrast to the large number of studies on the immune response to acute exercise, much less is known concerning the effect of physical conditioning or training on immune function. This largely reflects the difficulties in separating fitness effects from the actual physical exercise. Thus, the changes induced by intense physical exercise may last at least 24 h, and even moderate acute exercise induces significant immune changes for several hours. As it is not easy to persuade athletes to abstain from their normal training program even for just a day, it may be difficult to obtain results on true "resting levels." The influence of chronic exercise has been studied in both animal and human models, the latter including longitudinal as well as cross-sectional studies.

18.7.1. Cross-Sectional Studies

A good indicator of chronic exercise as a lifestyle factor is to study resting levels of the immune system in untrained controls versus athletes who have been competing for several years. Two studies have been conducted on competitive male cyclists. In order to eliminate the effects of acute exercise in these studies, none of the subjects were allowed to exercise 20 h prior to blood sampling. All subjects belonged to the elite Danish cycling group; they had been active in sports for a median of 4 years and had trained a medium of 20,000 km per year. In the first study, the NK cell activity was measured in 27 highly trained cyclists and in 15 age- and sex-matched untrained controls. Median NK cell activity was 38.1% in the trained group compared with significantly lower levels (30.3%) in the untrained group, along with a lower percentage of CD16$^+$ NK cells in the untrained group (Pedersen, Tvede, Christensen, Klarlund, Kragbak, & Halkjaer Kristensen, 1989). An additional study (Tvede, Steensberg, Baslund, Halkjaer Kristensen, & Pedersen, 1991) examined the cellular immune system in 29 cyclists and 15 controls. Measures were taken under resting conditions, during a period of low-intensity training (winter). Fifteen cyclists and ten controls were reexamined during a period of high-intensity training (summer). The NK cell activity was significantly elevated in the trained group, during both low-intensity and high-intensity training. During low-intensity training, the increased NK cell activity in trained subjects correlated with an increased percentage of NK cells. However, during high-intensity training, NK cell function in the trained subjects was increased, despite a comparable number of circulating NK cells in trained subjects and controls.

The mechanisms of this enhanced activity might be secondary to differences in NK cell numbers. The results suggested that the NK cells were activated in trained subjects during high intensity training and that this may lead to an adjustment of the number of CD16$^+$ cells in circulation by some unknown mechanism. In these studies, other lymphocyte subpopulations and the lymphocyte proliferative responses did not differ among trained and untrained subjects.

A study by Nieman, Buckley, Henson, Warren, Suttles, Ahle, Simandle, Fagoaga, and Nehlsen Cannarella (1995) supported the findings that athletes have increased NK cell activity. Twenty-two marathon runners who had completed at least seven marathons and had been training for marathon race events for at least 4 years, were compared with a group of 18 sedentary controls. Despite a large difference between groups on VO_{2max}, percent body fat, and physical activity, only the NK cell activity (but not lymphocyte numbers or mitogen-induced lymphocyte proliferation) significantly differed between the groups (higher among the marathoners).

Nieman, Henson, Gusewitch, Warren, Dotson, Butterworth, and Nehlsen Cannarella (1993a) compared resting levels of several immune parameters in a group of 12 highly conditioned and 30 sedentary elderly women (67–85 years of age). The NK cell activity and mitogen-induced proliferative responses were significantly elevated in the highly conditioned group compared with the inactive group.

Although not all cross-sectional studies could confirm elevated NK activity after chronic exercise, most experiments demonstrated increased NK activity in trained subjects.

18.7.2. Longitudinal Studies

The effect of chronic exercise has been studied in longitudinal designs. This is advantageous as the studies use randomization, thus in principle excluding confounding factors. The disadvantage is that the majority of studies investigate the effect on the immune system of at most 16 weeks of training, whereas the cross-sectional studies reflect many years of training. All studies, however, show significant effects on VO_{2max} as a result of training.

Nieman et al. (1993a) found no influence on NK cell activity or any other immune parameters for 30 elderly women randomized into a 12-week walking program. However, Crist, Mackinnon, Thompson, Atterbom, and Egan (1989) reported that treadmill exercise for 16 weeks enhanced the NK cell activity in elderly women. Nieman, Nehlsen Cannarella, Markoff, Balk Lamberton, Yang, Chritton, Lee, and Arabatzis (1990b) found that 15 weeks of walking enhanced the NK cell activity in moderately obese, previously inactive women. For 18 patients with rheumatoid arthritis allocated to an 8-week cycling program, chronic exercise had no effect on the NK cell activity, lymphocyte proliferative responses, concentrations or proportions of lymphocyte subpopulations, or cytokine production (Barnes, Forster, Fleshner, Ahanotu, Laudenslager, Mazzeo, Maier, & Lal, 1991). Furthermore, exercise had no influence on erythrocyte sedimentation rate, C-reactive protein, or number of swollen and tender joints. However, using a visual analogue scale it was shown that the patients who had performed the exercise program had less pain, less morning stiffness, and were less fatigued.

NK cell activity is increased in athletes versus nonathletes, although in randomized, longitudinal studies chronic training does not show a consistent effect on NK cell activity. This may be explained as follows: In order to become a successful athlete, good health is an absolute necessity. Thus, the level of natural immunity may select who will be an athlete and who will not, and the cross-sectional studies may in theory simply reflect this selection. However, whereas the longitudinal training studies in humans fail to show consistent effects on natural immunity, most animal studies have shown that chronic exercise enhances resting levels of natural cytotoxic activity. The latter indicates a true relationship, not just an association, between NK cell activity and chronic exercise. Thus, a high level of resting natural immunity exists in trained individuals. How do we then explain that elite athletes as a group are more prone to URTI than their sedentary counterparts? The most likely explanation is that the increased susceptibility to infections is the result of postexercise suppression of NK and B cell functions, thus providing an opportunity for pathogens (Brines et al., 1996). In a review on the regulation of neutrophil function during exercise, Pyne (1994) suggests that repetitive high-intensity training sessions by elite athletes may leave a significant proportion of their circulating neutrophils in a chronically refractory state. This may also explain the observation that elite athletes, as a group, are more susceptible to infections (Nieman, 1996).

18.8. Exercise and Aging

As pointed out by Mazzeo (1996), it is important to know how the elderly can respond and adapt to stress imposed by an acute bout of exercise. This is not only important from a mechanistic standpoint, but there are also significant clinical implications as infectious diseases can be more debilitating in this age group, and because intense exercise suppresses the immune system. In theory, however, chronic exercise training may enhance resting levels of the immune system, and thus abrogate age-related immunodeficiency.

Fiatarone, Morley, Bloom, Benton, Solomon, and Makinodan (1989) examined the effect of one bout of maximal cycling in young (average age 30) and old (average age 71) women. They found no difference among the groups in the NK cell activity measured at rest or in response to acute exercise.

Crist et al. (1989) studied two groups of elderly women (mean age 72). One group participated in a program of physical training, the other group served as nonexercising controls. The resting level of NK cell activity was elevated in the trained individuals. This study thus supports the finding of increased resting levels of NK cell activity in trained versus untrained young athletes. Interestingly, in response to acute exercise, the increase in NK cell activity was greater in trained individuals compared with untrained.

Nieman et al. (1993a) studied the basic mitogenic responses of PHA-stimulated T cells in elderly women (average age 73) who were in regular training for long-distance running. These elite elderly subjects showed an increased proliferative response and NK cell activity compared with sedentary controls. Surprisingly, however, a 12-week training program of 30 previously sedentary women did not result in a significant increase in T cell and NK cell function, despite an increase in cardiorespiratory capacity.

Because animal studies in mice and rats have yielded inconsistent results, it is not possible to conclude at this stage whether an endurance training program may alter age-related decline in immune function. The major reason for this uncertainty is related to the scarcity of data addressing the issue of exercise and immune function in the elderly. There is especially a lack of human studies. The available data suggest that although age-related decline in immune function can be retarded, the greatest effect will be seen only in very highly conditioned subjects (Brines et al., 1996).

18.9. Upper Respiratory Tract Infections

A number of studies have found that resting levels in the immune system are not impaired in trained versus untrained subjects (Nieman, 1996). However, based on anecdotal information a general feeling has been that while regular training promotes resistance to URTI, severe exertion, especially when coupled with mental stress, places athletes at increased risk for URTI (Fitzgerald, 1991; Nieman, 1994b) (see Chapter 19).

Peters and Bateman (1983) carried out a prospective study of the incidence of symptoms of URTI in 150 randomly selected runners who took part in the 1982 Two Oceans Marathon in Cape Town, and compared this with the incidence in individually matched controls who did not run. Runners were questioned on the day before and 2 weeks after the race. Symptoms of URTI occurred in 33.3% of runners compared with 15.3% of controls, and were most common in those who achieved the faster race times. Similarly, Nieman, Johanssen, and Lee (1989) studied the incidence of infectious episodes in 273 runners during a 2-month training period prior to a 5-km, 10-km, or half-marathon race. In addition, the effect of the race experience on infectious episodes was studied. Only 6.8% of the runners preparing for the half-marathon race reported becoming sick versus 17.95% of the 5- and 10-km runners. This study showed a trend that runners with a more serious commitment to regular exercise had less infectious episodes.

The largest epidemiological study on exercise and URTI was performed by Nieman, Johanssen, Lee, and Arabatzis (1990a) who researched the incidence of URTI in a group of 2311 marathon runners who took part in the 1987 Los Angeles Marathon race. Some 12.9% of the participants reported an infectious episode during the week following the race, in comparison with only 2.2% of similarly experienced runners who had applied but did not participate (for reasons other than sickness). Controlling for important demographic and training data by using logistic regression, it was demonstrated that the odds were 6 to 1 in favor of sickness for the marathon race participants versus the nonparticipating runners.

There are a lack of studies comparing URTI in large groups of moderately active individuals. However, two randomized experimental trials using small numbers of subjects have provided important data in support of the view that moderate physical activity may reduce URTI symptomatology. A study on 36 women who performed a 15-week walking program (45 min of walking, 5 days a week) found significantly fewer days with URTI symptoms during the 15-week study than in the control group (5.1 versus 10.8 days). The number of separate URTI did not vary between groups, but the number of URTI symptom days per incident was lower in the exercising group (Nieman *et al.*, 1990b). In a randomized, controlled study of elderly women, (aged 67–85), the incidence of the common cold during a 12-week period in the fall was the lowest in highly conditioned subjects who exercised moderately each day for about 1.5 h (8%), compared with elderly subjects who walked 40 min 5 times a week (21% incidence) and a sedentary control group (50%) (Nieman *et al.*, 1993a). This study clearly showed that elderly women not engaging in cardiorespiratory exercise are more likely to experience URTI during the fall season than those who do exercise regularly.

Based on the abovementioned epidemiological studies, a relationship between exercise and URTI has been modeled in the form of the "J" curve (Fig. 1). This model suggests that while moderate exercise training may decrease risk of URTI below that for sedentary individuals, risk may rise above average during periods of excessive amounts of high-intensity exercise (Nieman, 1994b).

18.10. Exercise and Cancer

Based on roughly 40 epidemiological studies, there is accumulating substantial evidence for a protective role of exercise in colon cancer and breast cancer risk (Pedersen & Clemmensen, 1997). Although there is the same trend for prostate cancer, it is too premature to draw strong conclusions regarding a potential effect of exercise (Pedersen & Clemmensen, 1997) (see Chapter 24).

The role of exercise in tumorigenesis is probably based on multiple and overlapping mechanisms of action. These include altered peristalsis, bile acid composition, diets, prostaglandins, and cytokines. Exercise is known to induce dramatic changes in hormonal factors and may thereby exert its effect on tumor growth. The exercise-induced effect on sex hormones may directly influence the tumor growth of some cancers such as breast cancer and prostate cancer. Experimental evidence also suggests that estrogens may influence tumor growth by natural immune suppression (Hanna & Schneider, 1983). Certainly, exercise induces dramatic changes in neuroendocrine factors, which again cause alterations in the cellular immune system, in particular in natural immunity. There is, however, a clear need for well-designed investigations to address the putative interactions among exercise, natural immunity, and cancer. A common assumption is that because chronic training enhances natural immunity, the mechanism for reduced tumor growth is enhanced natural immunity. This approach has, however, not been experimentally invalidated. Animal data showed that pretreatment of mice with an antibody (anti-asialo-GM1) completely abolished the exercise effects on *in vitro* cytotoxicity, and partially abrogated the *in vivo* cytotoxicity effect, as well as the tumor count (MacNeil & Hoffman-Goetz, 1993). Although these findings suggest a link between exercise training and reduced metastasis, mediated through natural immune factors, this interpretation is strongly influenced by the specific tumor model used, exercise characteristics, and the timing of the exercise.

Regarding the effects of acute exercise on the immune response, it has been shown that natural immunity is enhanced during moderate exercise and severe exercise. However, the numbers and function of cells mediating cytotoxic activity against tumor target cells are suppressed after intense, long-term exercise. In accordance with the immune surveillance theory, it therefore was to be expected that moderate exercise protected against malignancy whereas exhaustive exercise was linked with increased cancer risk. There are limited data to support this theory presently. Although the majority of epidemiological studies show that physical activity protects against cancer, there are as yet no published data on increased cancer risk in association with exhaustive training.

18.11. The "Open Window" Hypothesis

In summary, the immune system is enhanced during moderate and severe exercise, and only intense long-duration exercise is followed by immunodepression. The latter includes suppressed concentration of lymphocytes, suppressed non-MHC-restricted cytotoxicity (NK and LAK cell activity), and secretory IgA in mucosa. During the time of immunodepression referred to as the "open window," microbacterial agents, especially viruses, may invade the host and infections may be established. One reason for the "overtraining effect" seen in elite athletes could be that this window of opportunism for pathogens is longer and the degree of immunosuppression more pronounced (Fig. 2).

Based on the available data, a model on exercise duration and intensity is suggested. NK cells and other lymphocytes are recruited to the blood from a pool. This pool may be located in various organs such as the spleen, lungs, bone marrow, and lymph nodes. The number of cells that enter the circulation is determined by the intensity of the stimulus. If it has been a long lasting and very intense stimulus, the total concentration of lymphocytes declines. This may in part be ascribed to a redistribution of lymphocytes to organs. Whether or not the "open window" in the immune system occurs is thus dependent on the intensity and duration of exercise, and somewhat dependent on recovery periods. It remains to be shown whether prophylactic treatment with amino acid combinations, antioxidant, prostaglandin inhibitors, or other nutrients will diminish or abolish postexercise suppression,

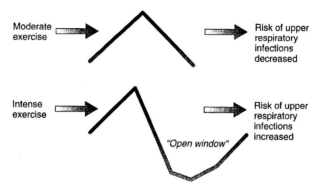

Figure 2. Schematic representation of the effects of moderate and high-intensity exercise for a fixed period of time. The model includes the effects on lymphocyte number, natural killer and lymphokine-activated killer cell activities, and antibody production. During intense exercise, the immune system is enhanced, but is followed by a period of immunosuppression. It is hypothesized that this represents an "open window" of opportunity for pathogens.

thus closing the "open window." It has recently been suggested that neutrophils serve as a last line of defense. During the "open window" when lymphoid cells are immunosuppressed, neutrophils are being mobilized to plug the gaps. The removal of this backup system following extreme activity would be compatible with the propensity of "overtrained" individuals to develop URTI (Brines *et al.*, 1996).

ACKNOWLEDGMENT. This work was supported by the National Research Foundation (#504-14).

19   Biobehavioral Influences on Respiratory Immunity

Joachim Kugler

19.1. Infections of the Upper Respiratory System: A Challenge for Public Health

The upper respiratory system of humans consists of the nose, paranasal sinus, oral cavity, pharynx, and larynx. The upper respiratory tract forms an important entrance for pathogens and respiratory allergens, for the following reasons:

1. An adult breathes about 12,000 liters of air per day.

Joachim Kugler • Department of Health Sciences/Public Health, Medical School Dresden, D-01309 Dresden, Germany.

2. Solid and fluid food passes through the oral cavity and pharynx, which is the common part of the respiratory and digestive tract.
3. The moist and warm milieu in the upper respiratory tract offers an ideal climate for microorganisms.

Such an enormous exposure to pathogens and respiratory allergens leads to a high incidence of upper respiratory tract disease. Acute infections are typical for the upper respiratory tract, although diseases of the teeth (e.g., caries) and gum (e.g., periodontosis), allergic reactions (e.g., hay fever), chronic inflammatory diseases (e.g., chronic tonsillitis), and neoplastic diseases are also prevalent. Among the acute infections, the common cold is the most frequent disease of the upper respiratory tract (Gwaltney, 1990):

1. On average, an adult suffers two to four times, and children six to eight times per year from episodes of the common cold.
2. The common cold is the most frequent cause for absenteeism from work and school.
3. Suffering from a common cold is one of the frequent reasons for visiting a doctor.

The common cold is a viral disease. After a short incubation period, it causes symptoms for 5 to 7 days if no complications occur. The disease is characterized by three stages:

1. During the incubation period the patient suffers from a reduction of general well-being, subfebrile temperature, irritation of the nose and throat, and sneezing.
2. During the catarrhal period the patient suffers from serous secretion and obstruction of the nasal breathing, impairment of olfaction, and a further reduction of general well-being.
3. During the mucopurulent period the patient suffers from adhesive secretions, improvement of olfaction, and improvement of local and general symptoms.

In 30–35% of cases, rhinoviruses are the cause of the common cold. More than 100 types of this virus have been described, which makes the development of an efficient vaccine nearly impossible. Moreover, the common cold can be induced by a variety of other viruses, such as influenza, parainfluenza, and coronaviruses. In one-third of patient cases, no viral cause can be detected.

A causal therapy for the common cold is currently not available. Nonetheless, the high incidence of infections of the upper respiratory tract prompts the consumption of medica-

Table 1
**Mortality of Acute Upper Respiratory Tract Infections
and Acute Bronchitis/Bronchiolitis
in Germany, Japan, and the United States[a]**

	Germany	USA	Japan
Acute upper respiratory tract infections			
Deaths per year	307	264	828
Deaths per year per 100,000 persons	0.4	0.1	0.7
Acute bronchitis/acute bronchiolitis			
Deaths per year	428	633	1418
Deaths per year per 100,000 persons	0.5	0.3	1.1

[a]From World Health Organisation Statistics Annual (1994).

tions against the common cold that are rated as "not efficient" or "questionable" by pharmacologists. It has been estimated that one of every eight medication packages sold in Germany is related to symptoms of the common cold.

Acute infections of the upper respiratory tract may also result in a dramatic disease course, which is potentially fatal (Table 1). Such developments include acute croup and pseudocroup syndromes in children, bacterial superinfections, as well as cardiac irregularities induced by viral infections in adults. Taking acute bronchitis also into account, it becomes evident that acute infections of the upper respiratory tract pose a significant but minor mortality risk in industrialized countries, especially for babies, children, and the elderly.

19.2. Biobehavioral Stress and Susceptibility to Infections of the Upper Respiratory System

The inoculation of pathogens is a necessary but not a sufficient condition for the manifestation of infectious disease. The onset and severity of infection are modulated by the immune function of the inoculated organism.

More than 100 years ago, Rudolf Virchow, a pioneer of the study of infectious diseases, studied the typhus epidemic in Upper Slesina. He emphasized the relationship between social living conditions and the outbreak of infectious disease:

> There is no doubt, that the typhus epidemic was only possible because of the living conditions especially poverty and low hygiene in Upper Slesina. If those conditions were removed I am convinced that the epidemic typhus would not come back. (1868, translated by the author.)

With improvements in socioeconomic conditions in industrialized countries, the issue of biobehavioral stress as a factor for susceptibility to infections has come into focus. Retrospective epidemiological studies showed that patients with a higher incidence of upper respiratory infection tended to complain more about everyday stress than people with a low incidence. However, an evaluation of such findings is difficult, as it cannot be determined whether stress is a causal factor for the susceptibility to upper respiratory infection, or if reports of stress events form a post-hoc explanation from the subjective view of the patient. To tackle this issue more decisively an experimental virological approach, using prospective designs, is more appropriate. This makes it necessary to study the incidence and course of disease following experimental virus inoculation, instead of observing natural infections. However, such methodology requires a sophisticated experimental setup, i.e., the volunteers have to be isolated so as to exclude natural infections. Furthermore, it requires that volunteers face the risk of an infection, which leads to ethical constraints. It is for these reasons that experiments examining the impact of stress on the onset of virological disease are rarely conducted in humans (Jemmott & Locke, 1984).

Nevertheless, a paradigm for this type of research is provided by a seminal study conducted by Cohen, Tyrrell, and Smith (1991). Four hundred and twenty volunteers were studied in a specialized research institution, the Medical Research Council's Common Cold Unit. The average subject age was 33.6 years (range between 18 and 54 years). They suffered from no chronic or acute disease, were not taking medication, and were rated as healthy in the medical entrance examination. During the entrance examination the stress index was assessed. The index consisted of critical life events within the year prior to the study, and present strain and negative mood. Next, volunteers were isolated so as to exclude natural infections of the upper respiratory tract. After 2 days, volunteers were inoculated with various types of virus in low dose into the nose: rhinovirus type 2 ($n = 86$), rhinovirus type 9 ($n = 122$), rhinovirus type 14 ($n = 92$), RSV ($n = 40$), coronavirus type 229E ($n = 54$), or saline solution ($n = 26$). The dose was chosen so that it resembled the level achieved by air transmission between patients with upper respiratory infections. The volunteers were medically examined on each of the following days, especially with regard to symptoms of infection of the upper respiratory tract.

The results showed a linear dose–response association between stress and susceptibility to infections (Fig. 1a,b). Volunteers with high stress ratings demonstrated higher risk of contracting clinical symptoms of an acute rhinitis ("common cold") than volunteers with low stress ratings. If secretory virus identification was used as a criterion of infection, volunteers showing high stress ratings had a nearly sixfold increase compared with volunteers scoring low on the stress rating. Furthermore, the relationship between stress and susceptibility to infection could be demonstrated for each type of virus.

Such findings lead to the issue of which physiological systems mediate the effects of stress on infection risk. One position is that biobehavioral stress reduces immune function, which may lead to an increased susceptibility to infections of the upper respiratory system. Accordingly, the local immune system of the upper respiratory tract may play an important role.

19.3. sIgA: An Indicator for the Local Immune System of the Upper Respiratory System

Since the 1920s, immunologists have distinguished local immune systems from the systemic immune system. Besredka (1926) at the Pasteur Institute in Paris revealed that certain pathogens causing anthrax or dysentery are only pathogenic if they are inoculated via the skin or intestines. Antibody titers in serum are not a valid indicator of immunity toward these pathogens. It followed that parenteral immunization with vaccines is not necessarily better than oral or subcutaneous immunization, and in some cases even worse.

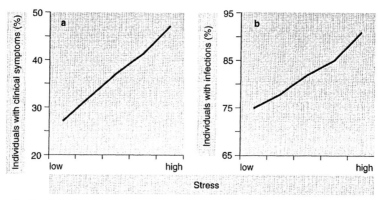

Figure 1.   Association between perceived stress and (a) frequency of clinical symptoms and (b) frequency of infections. Modified from Cohen *et al.* (1991).

Their early credo can be summarized as "every pathogen has its cell, every cell its immunity."

The upper respiratory tract can be differentiated into an inborn and an acquired immune system. The inborn immune system forms the genetically determined first line of defense, which is uniform toward all antigens. The acquired immune system provides an antigen-specific immune response, which is determined by previous exposure of the organism to the particular antigen.

The inborn immune system consists of biochemical factors that aid pathogen elimination (Roitt & Lehner, 1980). These include:

1. Lysozyme, which reacts with the cell wall of bacteria like staphylococcus
2. Peroxidase and lactoferrin, which inhibit bacterial growth

The inborn immune system also implements mechanical factors, such as:

1. Washing and cleaning by saliva secretion, inducing swallowing of microorganisms into the stomach where they are destroyed by the low pH of the gastric juice
2. Transport of foreign substances or microorganisms by the cilial system of the tracheas

The acquired immune system includes the local mucosa-associated immune function. Local immune systems are formed by lymphoid tissues that are connected to the body surface, while the systemic immune system comprises lymphoid tissue of the spleen and lymph nodes, and its circulating components in blood and lymph. They are called *local* because they are associated with certain parts of the body. A skin-associated lymphoid tissue can be distinguished from mucosa-associated lymphoid tissue. The local immune system of the upper respiratory tract is located in the mucosa and submucosa. Local immune systems are described as secretory, because their important components like antibodies or immune cells can be found in external secretions (i.e., tear fluid and saliva). The distinction between local and systemic is not exclusive because a local immune reaction is usually accompanied by a systemic response and a systemic immune response can be triggered by a local one.

Secretory IgA is the predominant humoral component of the mucosa-associated lymphoid tissue (Table 2). It is evident that different parts of the mucosa-associated lymphoid tissue are structurally similar. If one part is exposed to an antigen, specific antibodies can be found in different sections of the mucosa associated lymphoid tissue. For example, a food-dependent antibody can be found in tears but not in serum (Mestecky,

Table 2
Percentage of
Immunoglobulin-Producing Granulocytes
in Mucosa-Associated Lymph Tissues[a]

Tissue	IgA	IgM	IgG	IgD
Salivary glands	77%	7.2%	5.8%	9.7%
Parotid	91	3.0	3.7	2.5
Jejunum	81	17	2.6	1
Ileum	83	11	5.0	1
Colon	90	6	4.2	1
Lactating breast	68	13	16.0	2.4

[a]From Hanson and Brandtzaeg (1980).

1993). Although the mucosa-associated lymphoid tissue also secretes other immunoglobulins, research in psychoneuroimmunology has focused mainly on the modulation of IgA secretion.

19.3.1. Synthesis and Secretion of sIgA

IgA plays an important role in the secretory immune system. While only about 15% of immunoglobulins in the circulating blood belong to class A, they form the majority of the secreted immunoglobulins, e.g., in saliva IgA represents more than 80% of the total.

Secretory IgA differs in various physicochemical properties from serum IgA. It has a higher molecular weight and a higher sedimentation and diffusion coefficient. Serum IgA has typically a monomeric structure; secretory IgA, including salivary IgA (sIgA), has a dimeric structure consisting of two IgA monomers, a J chain, and a secretory component (Fig. 2).

An IgA monomer consists of four chains, two heavy alpha chains and two lighter kappa or lambda chains. The J chain serves to bridge the IgA monomers. Both IgA monomers and the J chain are built and secreted by plasma cells in the submucosa of the local lymphoid tissue or in exocrine glands.

The secretory component is a glycoprotein that is bound by disulfide bridges to both IgA monomers. This allows the transport of the immunoglobulin through the epithelium and protects it from proteolysis.

The secretion of IgA is schematized in Fig. 3. After intake, antigens pass through the mucosal epithelium and come into contact with B lymphocytes in the submucosa, which are triggered to produce IgA. Intracellularly, two IgA monomers are connected by a J chain. The secreted IgA dimer receives the secretory component while passing back through the mucosal epithelium. Whether plasma cells in the submucosa are also capable of producing IgA monomers, as observed in the serum, remains to be determined.

The secretion of IgA differs according to body fluid type (Table 3). About 20–30% of the total daily production of IgA can be found in serum, compared with about 70% of the total daily production of IgG. Various components of the digestive system, including saliva, are the major sites containing IgA.

The secretion of IgA takes place from the major salivary glands (parotid gland, sublingual gland, submandibular gland), minor salivary glands (such as the palatine gland

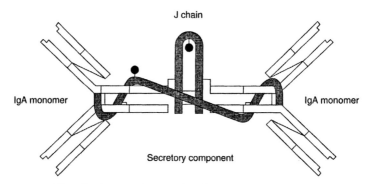

Figure 2.   Structure of secretory IgA. From Roitt and Lehner (1980).

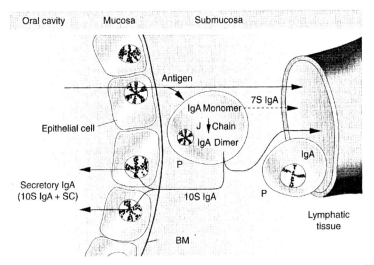

Figure 3.   Synthesis, transport, and secretion of IgA in the oral cavity. From Roitt and Lehner (1980).

and buccal gland), and directly from lymphocytes of the mucosa (Fig. 4). Considering total quantity the parotid gland does not have the most important role, for although its watery protein poor secretion forms the main fluid component of saliva, the concentration of immunoglobulin in parotid secretion is markedly lower than in whole saliva (Roitt & Lehner, 1980; Table 4). In serum, IgG shows the highest concentration, while those of IgA and IgM distantly follow. However, in whole saliva, the IgA concentration is substantially higher than that of other immunoglobulins. The differences between body fluids can be expressed by the IgG:IgA ratio, which is 5.7 for the serum, but only 0.07 for whole saliva.

19.3.2. Function of sIgA

sIgA is involved in protection from infection. Its main functions include:

1. Immune exclusion, i.e., the intrusion of the antigen into the mucosa is prevented (mucosa block)
2. Immune elimination, i.e., elimination of antigens is facilitated by leukocytes and macrophages

Table 3
Estimates of Daily IgA Production in an Adult Weighing 70 kg[a]

Tissue	IgA	Tissue	IgA
Blood	1295–2100 mg/day	Nasal secretion	45 mg/day
Saliva	100–200 mg/day	Small intestine	2100–5234 mg/day
Tear fluid	1–6 mg/day	Urine	1–3 mg/day
	Total	3598–9158 mg/day	

[a]From Mestecky *et al.* (1986).

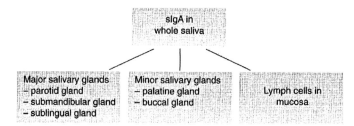

Figure 4. Loci of secretion of secretory (salivary) IgA (sIgA). From Roitt and Lehner (1980).

Among other functions the following mechanisms can be distinguished (Mestecky, Russell, Jackson, & Brown, 1986):

1. Agglutination of antigens
2. Inhibition of bacterial adherence to epithelial cells
3. Neutralization of viruses, toxins, and enzymes
4. Augmentation of natural immune functions, like lactoferrin and lactoperoxidase
5. Antigen-dependent cell-mediated cytotoxicity by macrophages and leukocytes
6. Suppression of inflammatory reactions

sIgA is only one factor in local immunity. In patients with complete or selective genetically induced IgA Mangel syndrome, there is an increased secretion of IgM and lysozyme (Mestecky *et al.*, 1986). This partly compensates the theoretically expected increased susceptibility to infection incurred following IgA downregulation.

The relevance of sIgA for the protection of the upper respiratory tract from infection is underlined by epidemiological studies. In a meta-analysis of nine studies, Jemmott and McClelland (1989) concluded that the level of IgA secretion indicates the vulnerability toward infections of the upper respiratory tract. Low secretion of sIgA is associated with a higher susceptibility to upper respiratory infections.

The secretion of IgA has a great interindividual variability (e.g., Kugler, Hess, & Haake, 1992). The secretion of sIgA under resting conditions depends on a variety of factors, such as age, flow rate of saliva, caries activity, dental state, nutritional state, and mucosal changes.

Changes in secretion can also be related to either production or mobilization of IgA. As for other immunoglobulins, changes in production of IgA require days to weeks. However,

Table 4
Concentrations (mg/100ml) of IgA, IgG, and IgM
in Serum, Parotid Fluid, and Whole Saliva[a]

Tissue	IgA	IgG	IgM	IgG:IgA
Serum	220	1250	80	5.7
Parotid	4.0	0.04	0.04	0.009
Whole saliva	19.4	1.4	0.2	0.07

[a]From Roitt and Lehner (1980).

the secretion of stored IgA is not constant over time. A short-term mobilization of IgA can take place within minutes.

One aim of psychoneuroimmunological research in this area is to describe how biobehavioral influences such as stress or relaxation impact on the secretion of sIgA.

19.4. Biobehavioral Stress and sIgA

One foundation of psychoneuroimmunological research was based on the general adaptation syndrome. Considering Selye's general adaptation theory, it was proposed that psychosocial stress triggers the hypothalamus–pituitary–adrenal axis, leading to an increased secretion of glucocorticoids. Corticosteroid release was posited to induce an immunosuppression, thus increasing susceptibility to infection (Fig. 5). In the case of disease, emotional disorders such as anxiety and depression reduce the adaptation resources to stress, and thus a vicious cycle between stress and disease may occur.

However, this model is not completely generalizable, as glucocorticoids produce marked immunosuppression in the systemic immune system, but not in the local immune system of the respiratory tract. The modification was simplified by proposing that stress lowers the secretion of sIgA, thus leading to increased susceptibility to upper respiratory tract infections.

However, this general hypothesis of stress-induced immunosuppression is not compatible with the body of empirical evidence provided by studies on IgA (Kugler, 1994; Van Rood, Bogaards, Goulmy, & van Houwelingen, 1993). The newly proposed model of adaptive immunomodulation takes into account differential psychological variables, such as a comparison between acute and chronic stress, and relaxation (Fig. 6).

19.4.1. Effects of Relaxation and Imagery

Relaxation and imagery techniques have a long tradition in psychotherapy. Several studies demonstrated that mental relaxation has a significant positive effect on the secretion of IgA (Kugler, 1994). Effects could be seen as early as after 20 min of relaxation, but also after a long-lasting relaxation training over several weeks.

Green and Green (1987) studied the influence of five relaxation techniques—Benson's relaxation response, guided imagery, massage, lying down, or finger touching without massage—on the concentration of sIgA (Table 5). Fifty college students were assigned to these five treatment groups. Each treatment lasted approximately 20 min. It was found that Benson's relaxation response, guided imagery, and massage produced significant increases in sIgA concentration.

Jasnoski and Kugler (1987) examined immune function following two relaxation techniques: (1) progressive muscle relaxation and focused breathing and (2) progressive muscle relaxation, focused breathing, and guided imagery. Subjects undergoing these techniques were compared with a group who worked on a vigilance task consisting of a tone discrimination. The 28 college students rated their mood and provided a saliva sample

Model I : Stress-induced immunosuppression

Figure 5. Model I: Association between stress, salivary IgA (sIgA), and risk for acute upper respiratory tract infections (URTI).

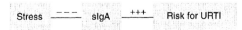

Model II : Adaptive immunomodulation

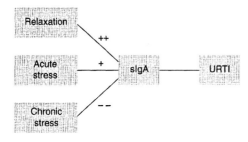

Figure 6. Model II: Modified association between relaxation, acute respiratory chronic stress, salivary IgA (sIgA), and acute upper respiratory tract infections (URTI).

before and after each 60-min treatment. The results showed that both relaxation techniques induced mental relaxation and increased the concentration of sIgA compared with the vigilance task.

It presently remains equivocal whether the impact of imagery on immune functions is beneficial for IgA secretion. However, it is evident that mental relaxation is capable of increasing IgA secretion.

19.4.2. Effects of Acute Stress

As suggested by the work of Selye and Cannon, the effects of stress on the organism depend on its duration. Stress episodes of only a few minutes or hours are usually accompanied by flight–fight responses, leading to an activation of the sympathetic nervous system, which is associated with a marked release of catecholamines and increase in sympathetic tone. In contrast, chronic stress is associated with a withdrawal response, which is related to activation of the hypothalamus–pituitary–adrenal axis, leading to an increase of glucocorticoid release.

In a number of studies, a significant increase in IgA secretion was recorded after acute stress (Kugler, 1994). A study examining 17 professional soccer coaches before, during, and after a game found a significant increase in sIgA secretion during the halftime break and at the end of the game (Kugler, Reintjes, Tewes, & Schedlowski, 1996; Fig. 7). An hour after the game, the concentration of sIgA was comparable to that 2 h prior to the game.

Additionally, studies on air traffic controllers reveal a significant increase in IgA

Table 5
Effects of Various Relaxation Techniques
on the Secretion of sIgA[a]

Technique	Effect on sIgA concentration[b]
Benson's relaxation response	↑
Guided imagery	↑
Massage	↑
Lying down	↔
Finger touching	↔

[a]From Green and Green (1989).
[b]↑, significant increase; ↔, unchanged.

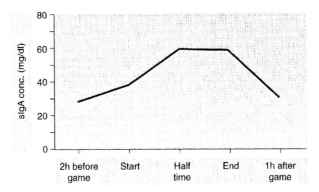

Figure 7.   Concentration of sIgA in soccer coaches during a game. From Kugler *et al.* (1996).

secretion after work shifts of 4 h compared with control situations (Zeier, Brauchli, & Joller-Jemelka, 1996). Furthermore, periods of negative mood in everyday life, which can be interpreted as an indicator of stress, have been shown to be accompanied by higher IgA secretion (Evans, Bistow, Hucklebridge, Clow, & Walters, 1993).

However, the results are not homogeneous. In several studies, no significant change could be seen after acute stress. This may be related to difficulty defining acute stress, because besides duration, the intensity of the stressor may also be important. It is nonetheless noteworthy that there is little evidence for a decrease in IgA secretion after acute stress.

19.4.3. Effects of Chronic Stress

If chronic stress is defined as lasting longer than 24 h, there is a nearly homogeneous body of evidence that chronic stress is associated with a decrease in sIgA secretion (Kugler, 1994; Van Rood *et al.*, 1993).

Jemmott, Borysenko, Borysenko, McClelland, Chapman, Meyer, and Benson (1983) studied dental students over an academic year with regard to stress rating and sIgA secretion. It was found that in exam periods in November, April, and June, stress rating increased while sIgA secretion decreased compared with exam-free periods in September and July (Fig. 8).

Furthermore, McClelland, Alexander, and Marks (1982) studied perceived stress and concentration of sIgA in male prisoners. Participants who suffered from high stress had a lower concentration of sIgA compared with participants who rated stress low.

19.4.4. Personality Traits and sIgA

Biobehavioral stress and mental relaxation are defined by subjective rating. Contrary to physical stress, biobehavioral stress is not completely defined by stimulus features like duration and intensity of stress. For example, an academic exam can be appraised as a stress or as a challenge. The degree of perceived stress and relaxation can be seen as an interactive process between conditions of a specific situation and personality traits. It is for these reasons that one approach in psychoneuroimmunology is concerned with relationships

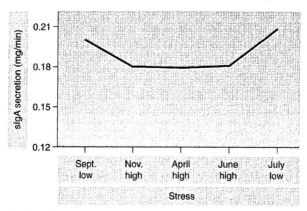

Figure 8. Secretion of sIgA in dental students during an academic year. From Jemmott *et al.* (1983).

between personality traits that play an important role in coping, stress, and parameters of the immune system.

With regard to IgA, McClelland and colleagues have investigated the role of the "inhibited power motive" (Jemmott & McClelland, 1989). The latter is characterized as the need to have power over other persons, and a high active inhibition which results in a high ego control not to display the power motive. From motivation theory, the "relaxed affiliation motive" may be seen to be contrary to the "inhibited power motive": it can be characterized as the need to congregate with others, and is associated with low inhibition so that the affiliation motive is included in planning overt behavior. Studies showed that persons high in "inhibited power motive" had a significantly lower sIgA secretion of sIgA than persons high in "relaxed affiliation motive."

Research on the personality trait "neuroticism," which can be seen as a measure for maladaptive coping with stress, showed that persons with high neuroticism ratings have a lower secretion of IgA than persons with lower ratings in neuroticism (Hennig, 1994).

On the whole, the studies using the personality trait approach indicate that persons with personality traits that are dysfunctional for stress coping have a lower secretion of IgA.

19.4.5. How Are the Biopsychological Effects on IgA Secretion Mediated?

The regulation of the mucosa-associated lymphoid tissue is currently the subject of intensive research (Stead, Tomioka, Pezzati, Marshall, Croitoru, Perdue, Stanisz, & Bienenstock, 1991). It can be assumed that the tissue receives both sympathetic and parasympathetic input. The effects of sympathetic and parasympathetic activation are not antagonistic on the salivary gland target tissue (Dole & Role, 1991); both result in an increased secretion of saliva. Differences are mainly related to the primary location: Parasympathetic activation leads mainly to an increased secretion from the parotid, which increases the serous part in whole saliva. Sympathetic activation increases mainly the mucous secretion of the small salivary glands.

Both mental relaxation and acute biobehavioral stress result in an increased secretion of IgA via storage depletion. It can be assumed that both mechanisms induce similar short-term effects. The intake and transport of IgA into the epithelium and its secretion are active processes, which may be influenced by direct neuronal stimulation as well as by neuropeptides like cholecystokinin (Stead *et al.*, 1991).

Chronic stress or personality traits that are dysfunctional for stress coping may lead to a constant increase in the depletion of stored IgA, which may be only partly compensated by the *de novo* synthesis of IgA by plasma cells. It is of great interest that some studies show that plasma cells in the mucosa are influenced by the central nervous system. Besides direct nerve–plasma cell interaction, it has been proposed that such modulation may be effected by neurotransmitters and neuropeptides such as somatostatin, norepinephrine, and substance P (Stead *et al.*, 1991).

19.5. Summary and Future Perspectives

The upper respiratory system is the main entrance for pathogens and allergens into the organism. Among the diseases of the upper respiratory system, the common cold is of utmost importance because of its high incidence and the enormous cost it imposes on health care systems. Experimental studies in virology indicate that biobehavioral stress increases the susceptibility to upper respiratory infections. One issue in psychoneuroimmunology is to study whether the local immune system is influenced by biobehavioral stress and mental relaxation. Studies examining the modulation of IgA as a predominant humoral factor of the local immune system have been introduced. Studies in this field are not compatible with a model of general stress-induced immunosuppression. On the contrary, acute stress and mental relaxation lead to an increased secretion of sIgA, whereas chronic stress and personality traits that make coping with stress difficult, are related to a decreased secretion of IgA. How these different effects on the local immune system are mediated remains an open question.

One issue for future research is the application of psychoneuroimmunological results for prevention and therapy of infectious diseases. It seems necessary to demonstrate in intervention studies that changes in stress coping and personality traits are accompanied by changes in the function of the local immune system, and that this leads to a decreased susceptibility to infections, or a shorter and less intensive course of disease.

20   Effects of Psychosocial Interventions on the Immune System

Margaret E. Kemeny and Gregory Miller

20.1. Introduction

Studies that determine whether psychosocial interventions can impact on the immune system are critical to the field of psychoneuroimmunology for three primary reasons. First, an intervention study design can determine whether a psychosocial independent variable can affect an immunological dependent variable. Thus, causality can be established. In contrast, data from cross-sectional studies of psychological factors and immune processes

Margaret E. Kemeny   •   Departments of Psychology and Psychiatry and Biobehavioral Sciences, University of California–Los Angeles, Los Angeles, California 90095.   Gregory Miller   •   Western Psychiatric Institute and Clinic, Pittsburgh, Pennsylavnia 15213.

cannot be used to infer a causal relationship. Even longitudinal studies that demonstrate that a psychological variable can predict changes in an immune variable over time are weak tests of a causal relationship. The question of "what causes what" has become an increasingly important issue as new findings indicate that immune products, such as cytokines, can enter the brain and alter cognitive and affective states (Maier & Watkins, 1998). In other words, under certain conditions, the relationship between psychological states and immune processes may be a function of immunological processes acting on the brain and behavior. Thus, intervention studies are crucial in establishing that psychosocial factors can in fact affect the immune system.

Second, intervention studies provide evidence of the clinical utility of relationships between psychosocial factors and the immune system. For example, it is possible that psychological factors contribute to immune impairment and disease progression in patients with a particular disease. However, many other factors may also influence biological processes relevant to the progress of that disease. As a result, it is possible that a psychosocial intervention cannot significantly modify these processes, because of the other factors that may drive those processes. Moreover, a relationship between certain psychosocial factors and immunity may be a function of "trait"- rather than "state"-like processes. For example, depression may be correlated with immune status because of genetic factors, or early life experiences that have had stable effects on both. An intervention to alleviate the "state" of depression may not be capable of altering this stable relationship.

Third, intervention studies can be used to address applied issues in medicine and health policy. It is very important to determine whether or not interventions can impact the immune system and health of individuals. If they do, then this information can be used to support changes in the health care delivery system that will promote the incorporation of psychological interventions into the medical care of patients. Well-conducted intervention studies, depending on their results, could influence the reimbursement policies of health insurance companies. In addition, clinical trials could point to interventions that are not beneficial, which could also have an important impact on health care delivery and insurance policies.

This chapter will review the findings of studies that have tested the impact of a psychosocial intervention on the immune system in either healthy individuals or those with a defined immunologically related illness. Only studies that are methodologically sound are included. The following criteria were used to select studies in order to focus on those with the best chance of providing solid information about the impact of interventions on the immune system:

1. Participants were randomly assigned to either an intervention or a control condition. We included studies in which each subject participated in both an intervention and a control condition, as long as the ordering of the conditions was randomly selected or counterbalanced.
2. The design included pre- and postintervention measures of immune parameters.
3. The intervention could extend over consecutive weeks or it could be a brief manipulation of psychological state that was induced over a short period of time. Thus, experimental studies that manipulated key aspects of a psychotherapeutic intervention were included.
4. Intervention studies that measured health status, but not immune parameters, as outcomes were not included.

Intervention studies in healthy individuals are reviewed first, followed by studies of individuals with cancer, HIV, and other diseases. Within each study population, studies are organized by the type of intervention used. The following overall intervention types were included:

1. Individual stress management techniques (e.g., hypnosis, relaxation, imagery, biofeedback)
2. Cognitive–behavioral stress management
3. Supportive–expressive/experiential therapy
4. Other (e.g., conditioning)

Table 1 lists all of the studies reviewed in this chapter organized by the study population (healthy, HIV positive, cancer, other disease) and by the intervention types above. For each study, the table includes the sample studied, the design used, the type of intervention selected, the points at which immune parameters were assessed, the outcomes measured, and the central findings.

20.2. Psychosocial Interventions in Healthy Individuals

20.2.1. Hypnosis

Hypnosis involves the induction of a state of focused attention or awareness (Olness, 1993; Spiegel & Spiegel, 1987). It is also considered to be a state of high suggestibility. The hypnotic state can be induced by another, or individuals can learn to induce this state themselves. Hypnosis is used in a variety of medical contexts to manage chronic pain, aid in recovery from medical procedures, and facilitate adjustment to a chronic illness. In most medical or biological studies of hypnosis, hypnosis is used in an attempt to induce a relaxed state or to alter a specific bodily function.

The earliest published clinical trials of the effects of a psychosocial intervention on the immune system used hypnosis. In these studies, hypersensitivity reactions on the skin were the primary immune outcome. These tests involve placing an antigen under the skin and measuring the size of the wheal that forms either immediately [immediate-type hypersensitivity (ITH) test] or 12 to 24 h later in the delayed-type hypersensitivity (DTH) test. The wheal is measured in terms of the induration (the size of the bump) and the erythema (the redness). At some point prior to, during, or immediately following the placing of the antigen, individuals are hypnotized and given specific suggestions to either increase or decrease the size of the wheal.

There are two interesting aspects of these studies. First, an entire *in vivo* immune response is measured rather than an isolated *in vitro* response. Thus, all of the processes that are involved in the development of the wheal in response to the antigen may come into play. *In vitro* assays isolate immune cells from their biological context. In addition, a direction of effect is explicitly specified as part of most hypnosis interventions. In other interventions (e.g., relaxation, support groups), the immune system itself is not the focus of the intervention and a specific direction of effect is not explicitly stated to the participants. Thus, most hypnosis studies are based on a theoretical premise that it is possible to manipulate the direction of an immune response via neuroendocrine pathways.

Including the first study published in 1963 by Black, we have found nine studies that used a hypnotic intervention to impact a hypersensitivity response in a healthy sample (and met our study criteria). All of these studies utilized highly hypnotizable individuals as part or all of their sample. Most of the studies included conditions that involved increasing the wheal size as well as conditions that involved suppressing the wheal size. In some cases, more specific suggestions were also given (the arm that should have the increased wheal). Three of the nine studies found no differences in erythema and induration across conditions (Beahrs, Harris, & Hilgard, 1970; Locke, Ransil, Covino, Toczydolwski, Lohse, Dvorak,

Table 1

Categorization of Studies According to Sample and Intervention Type[a,b]

Randomized clinical trials with nonclinical populations
Supportive–expressive interventions

Study	Sample	Design	Intervention	Assessments	Outcomes	Central findings
Booth et al. (1996)	38 healthy medical students	RCT RA to disclosure intervention or control group	Ss wrote about traumatic experiences for 20 min on 4 consecutive days Controls wrote about trivial matters	Preintervention Postintervention 1 week postintervention 2 weeks postintervention	T_h numbers T_c numbers Lymphocyte numbers Neutrophil numbers Erythrocytes Platelets	Time × Group interaction for T_h, T_c, and lymphocyte numbers: Although the two groups showed no immunological differences at preintervention, at postintervention, disclosers showed lower T_h, T_c, and total lymphocyte numbers than controls No group differences on other measures
Esterling et al. (1994)	57 healthy undergrads	RCT RA to written disclosure intervention, verbal disclosure intervention, or control group	Ss in verbal disclosure group discussed traumas into a tape recorder for 20 min on three occasions Ss in written disclosure group wrote about traumas with same format Controls wrote about trivial matters	Preintervention Postintervention	Abs to EBV	Time × group interaction for Abs to EBV: Both verbal and written disclosers showed a decreased in Ab titers to EBV from pre- to postintervention while controls showed no change. Verbal disclosers showed a larger decrease than written disclosers High disclosers showed greater decreases in Abs than low disclosers
Lutgendorf et al. (1994)	76 healthy undergrads	RCT RA to verbal disclosure intervention or control group	Ss discussed traumas with experimenter for 20 min on 3 days	Preintervention Postintervention	Abs to EBV	No group differences in Abs to EBV Greater involvement in treatment and decreases in avoidance of topic predicted decreased in Abs to EBV

Study	Sample	Design	Intervention	Time points	Measures	Results
Pennebaker et al. (1988b)	50 healthy undergrads	RCT RA to disclosure intervention or control group	Ss wrote about traumatic experiences for 20 min on 4 consecutive days Controls wrote about trivial matters	Preintervention Postintervention 6 weeks postintervention	PHA Con A Physicians visits	Time × Group interactions for PHA: Disclosers showed no change in proliferative response to PHA from pre- to postintervention; controls showed a decrease. No group differences for Con A. Time × Group interaction for physician visits: Disclosers showed a decrease in physician visits from pre- to postintervention while controls showed an increase. High disclosers showed larger PHA responses than low disclosers
Petrie et al. (1995)	40 medical students	RCT RA to disclousre intervention or control group	Ss wrote about traumatic experiences for 20 min on 4 consecutive days Controls wrote about trivial matters	Postintervention 1 month postintervention 4 months postintervention 6 months postintervention	Abs to hepatitis B vaccine T_h numbers T_c numbers NK numbers Basophil numbers NKCA	Time × Group interaction for Abs to hepatitis B: Disclosers had higher Ab titers to hepatitis B vaccine than controls at 4- and 6-month follow-ups. No group differences on other measures

[a]A study was coded as randomized controlled trial (RCT) if (1) it included intervention and control conditions and subjects were randomly assigned to participate in one or the other; (2) it included intervention and control conditions and subjects were randomly assigned to participate in both a randomly selected, counterbalanced order; or (c) it included multiple intervention groups that were instructed to alter immune responses in opposite directions (e.g., one group was instructed to suppress a response and another to enhance it) and subjects were randomly assigned to those conditions.

[b]Abbreviations used: Abs, antibody titers; Ags, antigens; BFB, biofeedback; CD3mAb, lymphocyte proliferative response to stimulation with CD3 monoclonal antibody; CMV, cytomegalovirus; Con A, lymphocyte proliferative response to concanavalin A; DCP, diphenylcyclopropenone; DNCB, dinitrochlorobenzene; DTH, delayed-type hypersensitivity response; EBV, Epstein–Barr virus; HH, highly hypnotizable; HHV-6, human herpesvirus type 6; HSV, herpes simplex virus; HW, homework; IFN-augmented NKCA, interferon-α-augmented natural killer cell activity; ITH, immediate-type hypersensitivity response; LGL, large granular lymphocytes; LH, low hypnotizable; Mϕ, macrophage/monocyte; MH, medium hypnotizable; MLR, mixed lymphocyte response to mitogen; NK, natural killer cells (CD16⁺ or CD56⁺ or CD16⁺/CD56⁺); NKCA, natural killer cell activity; PHA, lymphocyte proliferative response to phytohemagglutinin; PWM, lymphocyte proliferativ response to pokeweed mitogen; RA, randomly assigned; RCT, randomized controlled trial; RT, relaxation training; sIgA, salivary IgA; T_c, suppressor/cytotoxic T lymphocyte (CD3⁺/CD8⁺); T_h, helper/inducer T lymphocyte (CD3⁺/CD4⁺); T_h/T_c ratio, ratio of helper/inducer T lymphocyte to suppressor/cytotoxic T lymphocyte (CD4⁺/CD8⁺); T_i, inducer T lymphocyte (CD45⁺/CD4⁺); T_t, total T lymphocyte (CD3⁺); VZ, varicella zoster; WBC + DIFF, white blood cell count plus differential.

(continued)

Table 1 *(Continued)*

Randomized clinical trials with nonclinical populations

Conditioning interventions

Study	Sample	Design	Intervention	Assessments	Outcomes	Central findings
Booth et al. (1995)	15 adults	RCT Ss received a conditioning intervention and a placebo intervention	Ss met the experimenter individually on 10 days. On the first 6 days liquid droplets were paired with an allergen injection in one arm, and droplets with saline injection on the other arm. On days 7 and 8 these pairings were covertly switched between arms	Postintervention	Induration Titration gradient Titration endpoint following ITH challenge	No changes in induration, titration gradient, or titration endpoint were observed when the droplet–allergen and droplet–saline pairings were covertly switched between arms
Buske-Kirschbaum et al. (1992)	24 healthy undergrads	RCT RA to conditioning intervention or one of three control groups	Intervention Ss met individually with experimenter on 5 consecutive days. On first 4 days sherbet was paired with epinephrine injection; on day 5 sherbet was paired with saline injection. Controls received five trials of sherbet + saline injection, sherbet + epinephrine injection, or sherbet + epinephrine injection with long interval between	Preintervention Postintervention (Assessments conducted at all five sessions)	NKCA Epinephrine	Time × Group interaction for NKCA: Ss who received the conditioning intervention showed an increase in NKCA when presented with the sherbet + saline injection pairing, while Ss in the three control groups showed no changes in NKCA when presented with this pairing. Epinephrine levels were not elevated among intervention subjects the day of the sherbet + saline injection pairing, indicating that sherbet presentation did not alter NKCA by increasing circulating epinephrine

Kirschbaum et al. (1992), Study 1	14 healthy adults	RCT RA to conditioning intervention or a control group	Intervention Ss met individually with experimenter on 5 consecutive days. On first 4 days sherbet sweet was paired with epinephrine injection; on day 5 sherbet was paired with saline injection. Controls received five trials of sherbet sweet + saline injection	Preintervention Postintervention (Assessments conducted at sessions one and five)	NKCA numbers NK numbers Epinephrine	Time × Group interaction for NKCA: Ss who received the conditioning intervention showed an increase in NKCA when presented with the sherbet + saline injection pairing, while Ss in the control groups showed no changes in NKCA when presented with this pairing. Among Ss who received the conditioning intervention, NK numbers increased after epinephrine injection at session 1 but did not change following the sherbet + saline pairing. Epinephrine levels were not elevated among intervention subjects the day of the sherbet + saline pairing, indicating that sherbet presentation did not alter NKCA by increasing circulating epinephrine
Kirschbaum et al. (1992), Study 2	6 adult males	RCT RA to conditioning intervention or a control group	Intervention Ss met individually with experimenter on 4 consecutive days. On first 3 days a bitter drink was paired with epinephrine injection; on day 4 drink was paired with saline injection. Controls received four trials of bitter drink + saline injection	Preintervention 5 min postintervention 45 min postintervention (Assessments conducted at sessions one and four)	NKCA NK numbers Epinephrine	Among Ss who received the conditioning intervention, NKCA and NK numbers did not change following the bitter drink + saline injection presentation

(continued)

Table 1 (*Continued*)

Study	Sample	Design	Intervention	Assessment	Outcomes	Central findings
Kirschbaum *et al.* (1992), Study 3	7 adult males	RCT RA to conditioning intervention or a control group	Intervention Ss met individually with experimenter on 5 consecutive days. On first 4 days sherbet sweet was paired with epinephrine injection; on day 5 it was paired with saline. No control group was used.	Preintervention 5 min postintervention 45 min post intervention (Assessments conducted at session one and four)	NKCA NK numbers Epinephrine Epinephrine-stimulated NVCKA	Among Ss who received the conditioning interventions, NKCA and NK numbers did not change following the bitter drink + saline injection presentation Epinephrine-stimulated NKCA was weaker at session five than at session one, suggesting that the NK cells of intervention Ss had become less responsive to epinephrine over the course of the experiment
Smith & McDaniel (1983)	7 adults	RCT Ss received a conditioning intervention and a placebo intervention	Ss met with experimenter individually on 7 days. On the first 5 days an allergen injection was given in one arm and a saline injection in the other arm. On days 6 and 7 the pairings were covertly switched between arms	Postintervention	Erythema Induration 24 and 48 h after DTH challenge	Ss showed smaller erythema and less induration in response to DTH challenge on days when the allergen and saline injections had been covertly switched between arms, suggesting conditioned suppression of DTH response

Randomized clinical trials with nonclinical populations

Brief hypnotic interventions (< 4 sessions) with suggestions to alter immune response

Study	Sample	Design	Intervention	Assessment	Outcomes	Central findings
Beahrs *et al.* (1970)	5 undergrads; HH	RCT RA to suppress ITH (Study 1) or DTH (Studies 2 & 3) response in one arm and not alter it in other arm	Ss were hypnotized briefly and given posthypnotic suggestion to suppress ITH/DTH response in one arm and not alter it in other	Study 1: 24 hours postintervention Study 2: 5, 10, 15 min post-intervention Study 3: 5, 10, 15 min post-intervention	Induration Erythema following ITH/DTH challenge	No differences in wheal induration or erythema as a function of posthypnotic suggestion

Black (1963)	18 adults with history of psycho-somatic illness; 7 HH and 11 MH	RCT Ss participated in a control session followed by a hypnosis session Separate control studies done to assess effects of time and multiple injections on outcomes	Ss were hypnotized and instructed to suppress ITH response	Midintervention following ITH challenge	Induration Skin temperature following ITH challenge	50% of Ss showed a reduction in induration and skin temperature during the hypnosis session, compared with the control session
Locke et al. (1987)	12 healthy adults; HH	RCT RA to one of hypnosis interventions, each of which enhanced, suppressed, or did not alter DTH response in one arm and did opposite in other arm	Ss met with experimenter individually six times for hypnosis. At final meeting instructed to enhance, suppress, or not alter DTH response HW given	24 h post-intervention/ DTH challenge 48 h post-intervention/ DTH challenge	Induration 24 and 48 h following DTH challenge	No differences in induration as a function of instructions to suppress, enhance, or not alter response
Locke et al. (1994)	24 healthy adults; HH	RCT Ss participated in three sessions of hypnosis and an assessment-only control session in random order	Ss had four individual sessions 1 week apart. Each of the three hypnosis sessions had a different posthypnotic suggestion: relax, suppress DTH response, enhance DTH response HW given	24 h post-intervention/ DTH challenge 48 h post-intervention/ DTH challenge (Assessments conducted at all four sessions)	Induration 24 and 48 h following DTH challenge	No differences in induration across the three hypnosis sessions or the control session, despite 0.80 statistical power to detect 0.91-mm difference in induration

(continued)

Table 1 (*Continued*)

Study	Sample	Design	Intervention	Assessments	Outcomes	Central findings
Olness *et al.* (1989)	60 healthy children	RCT RA to one of two self-hypnosis interventions or a control group Session 1: Ss learned about the immune system and practiced relaxation Session 2: Ss received self-hypnosis intervention	Ss listened to audiotapes that promoted self-hypnosis. The tapes either gave specific instructions to increase sIgA and sIgG levels or nonspecific suggestions to do so	Session 1: preintervention Session 2: pre- and postintervention	sIgA sIgG	Time × Group interaction for sIgA: Ss who were given specific instructions to increase immunoglobulin levels showed increases in sIgA from pre- to postintervention during session 2 while the other groups showed no change There were no changes in sIgG from pre- to postintervention in any of the groups Hypnotizability was unrelated to immune response
Rider and Achtenberg (1989)	30 healthy adults	RCT RA to RT + imagery intervention that encouraged either circulating lymphocyte numbers or circulating neutrophil numbers	Ss met in groups on two occasions and received RT + immune imagery focusing on either lymphocytes or neutrophils Given HW Ss met individually with experimenter for third session, where they engaged in 20 min of tape-assisted imagery	Preintervention Postintervention (Assessment was conducted at third sessions only)	Lymphocyte numbers Neutrophil numbers	Ss who were instructed to increase circulating lymphocyte numbers showed a decrease in lymphocytes from pre- to postintervention and no change in neutrophils Ss who were instructed to increase circulating neutrophil numbers showed a decrease in neutrophils from pre- to postintervention and no change in lymphocytes
G. R. Smith *et al.* (1992), Study 1	28 healthy adults; HH	RCT RA to RT + imagery + hypnosis intervention that instructed Ss to either suppress or enhance DTH response	Ss met in groups for 2 h on 3 consecutive days and were taught RT, immunological imagery, and hypnosis HW given	24 h post-intervention/ DTH challenge 48 h post-intervention/ DTH challenge	PHA Induration 24 and 48 h after DTH challenge	Ss instructed to enhance DTH response showed greater induration at 48 h postchallenge than Ss instructed to suppress response. No group differences in induration at 24 h postchallenge No group differences in proliferative response to PHA at any point

G. R. Smith et al. (1992), Study 2	45 healthy adults; HH	RCT	RA to RT + imagery + hypnosis intervention that instructed Ss to either suppress, enhance, or not alter DTH response	Ss met in groups for 2 h on 3 consecutive days and were taught RT, immunological imagery, and hypnosis HW given	Preintervention 24 h post-intervention/DTH challenge 48 h post-intervention/DTH challenge	Con A VZ Induration 24 and 48 h after DTH challenge	Time × Group interaction for induration: Ss instructed to enhance DTH response showed greater increases in induration from preintervention to 24 h postintervention/challenge than controls or Ss instructed to suppress the response. No group differences at 48 h postintervention/challenge No group differences in proliferative response to Con A or VZ at any point
Zachariae et al. (1989)	18 undergrads (8 intervention Ss were HH; 10 controls were not tested for hypnotizability)	RCT	RA to hypnosis intervention that instructed Ss to suppress ITH and DTH responses in one arm and increase those responses in the other arm	Ss hypnotized on one occasion for 7 min. Imaged leukocytes attacking Ag in one arm and not attacking Ag in other arm HW given	ITH: preintervention; postintervention/following ITH challenge DTH: preintervention 48 h postintervention/DTH challenge	Induration Erythema following ITH challenge and 48 h after DTH challenge	Time × Group interaction for ITH erythema: Ss reduced erythema response to ITH challenge when instructed to do so, but were not able to increase erythema to ITH No changes in induration to ITH challenge Ss reduced/increased induration and erythema to DTH challenge when instructed to do so. Control group showed no changes
Zachariae & Bjerring (1993)	20 adults; HH	RCT	All Ss were sensitized to skin Ags DCP and DNCB, then RA to hypnosis + imagery intervention that instructed them to suppress or enhance DTH reponse	Ss hypnotized once for 30 min. Imaged suppressed DTH response to DCP and enhanced DTH response to DNCB or vice versa HW given	6 weeks postintervention	Induration Erythema 48 h after DTH challenge	No group differences in induration Ss instructed to enhance DCP response and suppress DCNB showed larger erythema to DCP than to DNCB. Ss instructed to do the opposite showed no erythema differences between DCP, DNCB

(continued)

Table 1 *(Continued)*

Randomized clinical trials with nonclinical populations

Relaxation training interventions

Study	Sample	Design	Intervention	Assessments	Outcomes	Central findings
Green & Green (1987)	50 undergrads	RCT RA to one of four relaxation interventions or control group	Ss met in groups on one occasion and received a 20-min intervention that consisted of either RT, back massage, guided imagery, or lying down quietly	Preintervention Postintervention	sIgA	Ss in each of the four relaxation interventions showed significant pre- to postintervention increases in sIgA while control Ss showed no changes over the same period.
Green et al. (1988)	40 adults	RCT RA to one of three RT interventions or a wait-list control group	Ss met in groups for an initial session and were taught one of three relaxation strategies: relaxation while standing, relaxation while sitting, or imagery. They were instructed to practice the intervention 20 min daily over the next 3 weeks	Preintervention Postintervention	IgA levels IgG levels IgM levels sIgA levels	No IgA, IgG, or IgM differences between control group and relaxation groups at postintervention, and none of the groups showed significant changes in these measures over the course of the intervention Time × Group interaction for sIgA: Ss in the relaxation groups showed greater increases in sIgA from pre- to postintervention than control Ss No difference between the three relaxation interventions on any of the immune outcomes at any assessment point sIgA increased after 20 min of relaxation practice, but other measures did not change

Hall et al. (1992, 1993)	45 high school and college students	RCT RA to one of two relaxation interventions or control group	Ss in relaxation intervention received either brief training or extended training Brief: Ss trained in RT + immune imagery at session one, performed relaxation by themselves at session two Extended: Ss received four individual RT + imagery training sessions, followed by same protocol as Ss in brief intervention HW given	Preintervention Postintervention (Assessments conducted at all sessions)	Neutrophil adherenece WBC + DIFF	Neither of the intervention groups altered neutrophil adherence in a reliable fashion from pre- to postintervention during a single session Ss in the extended intervention showed a decrease in neutrophil adherence from session five to session six, while brief intervention and control Ss showed no change during the corresponding sessions No group differences on WBC + DIFF Immunological changes were not associated with hypnotic ability or with changes in pulse or skin temperature
Jasnoski & Kugler (1987)	28 undergrads	RCT RA to one of two relaxation interventions or a mildly amusing control task	Ss met in groups on one occasion and received 60 min of either RT + focused breathing, RT + focused breathing ± immune imagery, or a mildly stressful control task	Preintervention Postintervention	sIgA	Ss in each of the four relaxation interventions showed significant pre- to postintervention increases in sIgA while control Ss showed no changes over the same period No differences in sIgA change between two relaxation groups

(continued)

Table 1 (*Continued*)

Study	Sample	Design	Intervention	Assessments	Outcomes	Central findings
Kiecolt-Glaser et al. (1985)	45 healthy older adults	RCT RA to RT intervention, social contact intervention, or control group	Ss in both interventions met alone with experimenter for 45 min, three times a week for 1 month. RT group received relaxation training, social contact group received conversation	Preintervention Postintervention 30 days postintervention	PHA PWM NKCA Abs to HSV	Time × Group interaction for NKCA: RT intervention group showed a significant increase in NKCA from pre- to postintervention, which returned to preintervention levels at the 30-day follow-up. The social contact group and the control group showed no change in NKCA over time Time × Group interaction for Abs to HSV: RT intervention group showed a significant decrease in Abs to HSV from pre- to postintervention which was maintained at the 30 day follow-up. The social contact group and the control group showed no change in Abs to HSV over time No group differences on PHA or PWM at any point
Kiecolt-Glaser et al. (1986)	34 healthy medical students	RCT RA to RT + self-hypnosis intervention or control group	Ss met two or three times weekly in groups for 3-week period before final exams. Taught self-hypnosis, RT, and imagery HW given	Preintervention Postintervention (final day of exam period)	T_h numbers T_c numbers T_h/T_c ratio NKCA	No group differences on outcome measures at pre- or postintervention More frequent relaxation practice associated with greater increases in T_h cells from pre- to postintervention

McGrady et al. (1992)	31 healthy adults	RCT RA to RT/BFB intervention or control group	Ss had eight 30-min sessions, four of which provided RT in groups and four of which provided BFB individually	Preintervention Postintervention	Con A PHA WBC + DIFF	Time × Group interaction for PHA: Intervention group showed a significant increase in the proliferative response to PHA from pre- to postintervention while the control group showed no change Time × Group interaction for WBCs and neutrophils: Intervention group showed a significant decrease in WBCs and neutrophils from pre- to postintervention; control group showed no change No group differences for Con A, eosinophils, basophils, or Mφ
Peavey et al. (1985)	16 healthy adults	RCT RA to RT/BFB intervention or control group	Ss met alone with experimenter twice a week for 60 min until they had achieved criterion levels of muscle tension and relaxation	Preintervention Postintervention	Neutrophil phagocytic capacity WBC + DIFF	Time × Group interaction for phagocytic capacity: Intervention group showed greater phagocytic capacity than controls at postintervention, despite no group differences at baseline No group differences on WBC or DIFF at either timepoint
Rider et al. (1990)	45 healthy undergrads	RCT RA to one of two relaxation interventions or control group	Ss were given imagery audiotapes + entrainment music and instructed to practice with them three times a week for 6 weeks. One group received immune imagery and the other nonspecific imagery	Preintervention Midintervention Postintervention (All three assessments conducted before and after a relaxation exercise in the laboratory)	sIgA	At all three assessments Ss in the relaxation groups showed a significantly larger increase in sIgA from pre- to posttask than Ss in the control group The RT + immune imagery group showed larger sIgA increases from pre- to posttask than the RT + nonspecific imagery group at mid- and post-intervention assessments

(continued)

Table 1 *(Continued)*

Study	Sample	Design	Intervention	Assessments	Outcomes	Central findings
Ruzyla-Smith *et al.* (1995)	55 healthy undergrads; 29 HH and 26 LH	RCT Ss RA to hypnosis intervention, floating relaxation intervention, or wait-list control group	Ss in the hypnosis group were hypnotized for 20 min on two occasions, during which they were directed to image their immune system Ss in the floating relaxation intervention spent two 60-min sessions "floating effortlessly" in a fiberglass tank filled with a solution of water and 20% Epsom salts	Preintervention Post-session 1 Post-session 2	T_h numbers T_c numbers T_t numbers T_h/T_c ratio B numbers	Time × Group interaction for B numbers: Ss in the hypnosis condition showed significant increases in B cell numbers at the post treatment assessments, compared with Ss in the floating relaxation and the control groups Time × Group interaction for T_h numbers: Ss in the hypnosis condition showed significant increases in T_h numbers at the posttreatment assessments, compared with Ss in the floating relaxation and the control groups Time × Group interaction for T_t numbers: HH Ss in the hypnosis condition showed significant increases in T_t number at the posttreatment assessments, compared with HH Ss in the floating relaxation group. HH Ss in the control group showed higher T_t numbers than HH Ss in the floating relaxation group
Whitehouse *et al.* (1996)	35 healthy medical students	RCT RA to self-hypnosis intervention or control group	Ss met weekly in groups for 90 min of self-hypnosis and RT HW given	Preintervention Late intervention Postintervention 3 weeks postintervetion	T_h numbers T_c numbers T_t numbers B numbers Mϕ numbers G numbers NK numbers Con A PHA PWM	No group differences on the immunological outcome measures at any of the assessment periods Frequency of home practice was unrelated to immunological outcome measures

Study	Sample	Design	Intervention	Assessment	Outcome measures	Results
Zachariae et al. (1994), Study 1	30 healthy undergrads; 15 HH and 15 LH	RCT RA to relaxation intervention, relaxation + imagery intervention, or control group	Ss met alone with experimenter for three 60-min sessions, and received RT or RT with immunological imagery HW given	Preintervention Postintervention (Assessments conducted at all three sessions) Note: At sessions 2 and 3, control Ss did not receive the preintervention assessment	Mϕ chemotaxis Con A PHA PWM	The RT group and the RT + imagery group showed decreases in Mϕ chemotaxis and the proliferative response to Con A, PHA, and PWM from pre- to postintervention at al three sessions. The control group showed no pre to post changes in these measures during session 1, but such comparison data are not available for sessions 2 and 3 HH showed greater decreases on all outcome measures from pre- to postintervention than LH
Zachariae et al. (1994), Study 2	30 healthy undergrads; 15 HH and 15 LH	RCT RA to one of two relaxation interventions or control group	Ss met alone with experimenter for three 60-min sessions, and received RT or RT with immune system imagery HW given	Preintervention Postintervention 60 min postintervention (Assessments conducted at all three sessions)	NKCA	No group differences in NKCA change from pre- to postintervention HH showed greater decreases in NKCA from pre- to postintervention than LH

(continued)

Table 1 (*Continued*)

Randomized clinical trials in HIV infection
Cognitive–behavioral stress management intervention

Study	Sample	Design	Intervention	Assessments	Outcomes	Central findings
Antoni et al. (1991a)	47 men unaware of their HIV serostatus at baseline, all of whom were asymptomatic at study entry	RCT RA to stress management intervention or control group before notification of serostatus	Ss met in groups twice a week for 10 weeks. Taught RT, imagery, stress management skills, cognitive restructuring HW given	Week 6 of intervention (72 h prior to serostatus notification) Week 7 of intervention (1 week after serostatus notification)	T_h numbers NK numbers PHA PWM NKCA	Group × Time interaction for T_h numbers and PHA: Among seropositive Ss, the intervention produced a significant increase in T_h numbers and proliferation of PHA from pre- to postnotification. Controls showed no change Group × Time interaction for NKCA and NK numbers: Among seropositive and seronegative Ss, the intervention produced no changes in NKCA and NK numbers from pre- to postnotification while controls showed significant decreases No differences in PWM More frequent home relaxation practice was associated with higher T_h numbers and NK numbers at pre- and postnotification.
Coates et al. (1989)	64 HIV seropositive males, all of whom were asymptomatic at study entry	RCT RA to stress management intervention or waitlist control group	Ss met for eight 2-hr sessions and one full-day session and were taught stress management, health behavior change, and RT	Preintervention Postintervention	T_h numbers T_h/T_c ratio IgA levels NKCA Con A Abs to CMV Abs to Candida	There were no immunological differences between the stress management group and the control group at either pre- or postintervention

Study	Sample	Design	Intervention	Timing	Immune measures	Results
Esterling et al. (1992)	65 men unaware of the HIV serostatus at baseline, all of whom were asymptomatic at study entry	RCT. RS to stress management intervention, aerobic exercise intervention, or control group before notification of serostatus	Ss in stress management met in groups twice a week for 10 weeks. Taught RT, imagery, stress management skills, cognitive restructuring. Ss in aerobic exercise met in groups for 45 min, three times a week for 10 weeks and engaged in stationary bicycle ergometry. HW given	Preintervention; Weeks 5, 6, 7, 8, during intervention	Abs to EBV; Abs to HHV-6	Group × Time interaction for Abs to EBV: Ss in the intervention groups showed declines in Abs to EBV over the course of the intervention, while Ss in the control group showed no changes. Group × Time interaction for Abs to HHV-6: Ss in the stress management group showed declines in Abs to HHV-6 over the course of the intervention, while Ss in the exercise and control groups showed no changes. These effects were independent of total IgG levels and polyclonal B cell activation
Lutgendorf et al. (1997)	26 HIV seropositive males with mildly progressed HIV infection and no signs of AIDS	RCT. RA to stress management intervention or wait-list control group	Ss met in groups twice a week for 10 weeks. Taught RT, anger management and coping skills, cognitive restructuring. Educated about physiology of stress and emotion. HW given	Preintervention; Postintervention	T_h numbers; T_c number; Abs to HSV-1; Abs to HSV-2	Group × Time interaction for Abs to HSV-2: The intervention group experienced a significant decline in Abs to HSV-2 from pre- to post-intervention. Controls showed no change. No group differences emerged for T_h numbers, T_c numbers, or Abs to HSV-1. Reductions in dysphoria over the course of the study were associated with declines in Abs to HSV-2 at posttreatment

(continued)

Table 1 (*Continued*)

Study	Sample	Design	Intervention	Assessments	Outcomes	Central findings
Mulder et al. (1995)	26 HIV sero-positive males, all of whom were asympto-matic at study entry	RCT RA to stress management intervention or an experiential group therapy intervention	Ss met once a week in groups for 15 weeks	Measured collected at 3 month intervals throughout the study and 2-year follow-up	T_h numbers CD3mAb	No groups differences in T_h numbers or CD3mAb at postintervention or follow-up Neither group showed immune changes from pre- to postintervention
			Randomized clinical trials in HIV infection Relaxation training intervention			
Eller (1995)	69 HIV sero-positive adults at all stages of the disease	RCT RA to RT intervention, guided imagery intervention, or a control group	Ss received tape that contained either a 12-min relaxation intervention or a 22-min guided imagery intervention. They practiced with it daily for 6 weeks	Preintervention Postintervention	T_h numbers T_c numbers T_h/T_c ratio NK numbers	Time × Group intervention for T_h numbers: Ss in the RT intervention showed a significant increase in T_h numbers from pre- to post-intervention, while imagery Ss and controls showed no changes None of the groups showed changes in T_c numbers, T_h/T_c ratio, or NK numbers
Taylor (1995)	10 HIV sero-positive males, all of whom were asympto-matic but had T_h counts < 400 at study entry	RTC RA to RT intervention or wait-list control group	Ss attended hourlong sessions twice a week for 10 weeks. Ss trained in RT, BFB, and immune system imagery HW given	Preintervention Postintervention 1 month postintervention	T_h numbers	Time × Group interaction for T_h numbers: Ss in the intervention group showed an increase in T_h numbers from pre- to postinter-vention relative to controls Neither group exhibited changes at follow-up

Exercise training interventions

Randomized clinical trials in cancer
Cognitive–behavioral stress management interventions

Study	Subjects	Design	Intervention	Timing	Measures	Results
LaPerriere et al. (1990)	50 men unaware of their HIV serostatus at baseline	RCT. RA to exercise intervention or control group before serostatus notification	Ss met in groups for 45 min, three times a week for 10 weeks and engaged in stationary bicycle ergometry	Midintervention (72 h before serostatus notification) Midintervention (7 days after serostatus notification)	T_h numbers NK numbers NKCA	Time × Group interaction for NK numbers: Seropositive Ss in the control group showed a decrease in NK numbers from the pre- to postnotification, while seronegative Ss and seropositive Ss in the intervention showed no change in NK numbers over the same time period. No group differences or changes over time in T_h numbers or NKCA
Fawzy et al. (1990b, 1993)	61 adults with stage I or II malignant melanoma	RCT. RA to multicomponent psychiatric intervention or control group	SS met weekly in groups for 90 min for 6 weeks. Ss received cancer education, stress management training, cancer-related problem-solving skills, and psychological support	Preintervention Postintervention 6 months postintervention 6 eyars postintervention	T_h number T_c numbers T_h/T_c ratio LGL NK numbers NKCA IFN-augmented NKCA T cell activation markers Recurrence Mortality	Time × Group interaction for T_h numbers, NK numbers, LGL, NKCA: Ss in the intervention group showed a significant increase in T_h numbers, NK, numbers, LGL, NKCA from preintervention to 6 months postintervention, while control Ss showed no immunological changes. Most of these changes were not observable at the postinterventional assessment Ss in the intervention experienced lower rates of melanoma recurrence and lower rates of melanoma mortality at the 6-year assessment than control Ss Group differences in morbidity and mortality were not explained by immunological change

(continued)

Table 1 (*Continued*)

Study	Sample	Design	Intervention	Assessments	Outcomes	Central findings
			Randomized clinical trials in cancer			
			Relaxation training interventions			
Gruber et al. (1993)	13 stage 1 breast cancer patients	RCT RA to relaxation intervention or wait-list control group	Ss met weekly in groups for 9 weeks, and then monthly for the following 16 weeks. Taught RT, BFB, and imagery HW given	Preintervention (three samples, one per week) Midintervention (nine samples, one per week) Postintervention (three samples, one per month)	IgA levels IgG levels IgM levels Con A MLR NKCA WBC + DIFF	Collapsing across nine samples taken during the study, Ss in RT showed larger proliferative responses to Con A and MLR, higher NKCA, and higher levels of IgG, WBCs, and total lymphocytes than controls IgM levels and proliferation to Con A and MLR increased from pre- to postintervention No changes on other outcome measures
			Randomized clinical trials in asthma, hypersensitivities, and autoimmune disease			
			Relaxation training and hypnosis interventions			
Fry et al. (1964), Study 1	18 adults with asthma or hay fever	RCT RA to hypnosis intervention or control group	Ss met individually with experimenter three times and received hypnosis and the suggestion that they would suppress response to ITH challenge	Preintervention Postintervention	Induration Erythema following ITH challenge	Time × Group interaction for induration and erythema: Ss in the intervention group showed a significant decrease in erythema and induration from pre- to postintervention at two of the four allergen doses. Controls showed no changes in erythema or induration

Study	Sample	Design	Intervention	Assessment	Measures	Results
Fry et al. (1964), Study 2	29 adults with asthma or hay fever	RCT RA to one of three hypnosis interventions One intervention group was instructed to suppress ITH response in the right arm, another to suppress it in both arms, and the third was told nothing	Ss met in groups with experimenter three times and received hypnosis	Preintervention Postintervention	Induration Erythema following ITH challenge	None of the groups showed pre- to postintervention changes in erythema or induration
Laidlaw et al. (1994)	5 adults with asthma, untrained in hypnosis	RCT Ss participated in an assessment session, followed by three sessions of hyonosis and three control sessions in random order	During the intervention sessions, hypnotic induction was maintained for the 10 min between Ag injection and measurement of induration and erythema. During the control sessions, Ss engaged in a verbal task for the 10 min between Ag injection and measurement of induration and erythema	Postintervention (following ITH challenge) (Assessments conducted at all sessions)	Induration Erythema following ITH challenge	Ss showed reduced erythem during hypnosis sessions compared with control sessions Induration did not vary as a function of the intervention
Levine et al. (1966)	10 healthy undergrads 10 adults with allergies 10 adults with chronic hives	RCT RA to hypnosis intervention that instructed Ss to suppress ITH response in one arm and either enhance reponse or not alter it in other arm	Ss met with experimental individually for two sessions Session 1: Ss hypnotized, awakened but not given instructions to alter response Session 2: Ss hypnotized and instructed to suppress, enhance, or not alter ITH	Session 1: Post-intervention (following ITH challenge) Session 2: Midintervention (following ITH challenge)	Induration Erythema following ITH challenge	No changes in erythema or induration from control session to hypnosis + suggestion session, and no difference as a function of instructions to suppress, enhance, or not alter response No differences in erythema or induration between three groups of Ss

(continued)

Table 1 (*Continued*)

Study	Sample	Design	Intervention	Assessments	Outcomes	Central findings
Randomized clinical trials in asthma, hypersensitivities, and autoimmune disease						
Cognitive–behavioral stress management interventions						
O'Leary *et al.* (1988)	33 women with rheumatoid arthritis	RCT RA to stress management intervention or educational control condition	Ss met in groups once a week for 5 weeks. Taught relaxation, imagery, coping, effectiveness, goal setting	Preintervention Postintervention	T_h/T_c ratio	There were no immunological differences between intervention and control groups postintervention. Increases in pain management self-efficacy correlated with increases in T_h/T_c ratio

Arndt, & Franke, 1987; Locke, Ransil, Zachariae, Molay, Tollins, Covino, & Danforth, 1994). The remainder found mixed support for study hypotheses. One study found a difference across conditions when wheal size was measured at 48 h but not at 24 h (Smith, Conger, O'Rourke, Steele, Charlton, & Smith, 1992) However, in a separate study, the same investigators found differences across conditions at 24 h and not at 48 h. Another group found differences for one antigen but not for another (Zachariae & Bjerring, 1993). In a separate study, these investigators found that hypnosis could reduce or increase DTH wheal size (Zachariae, Bjerring, & Arendt-Nielsen, 1989). However, when ITH was examined, they found reduced wheal size but not increased wheal size with the matched suggestion. The earliest study in this area found the desired effect; namely, a reduction in induration following a hypnotic suggestion of suppression, but in only 50% of subjects (Black, 1963a).

A few studies have examined the effects of hypnosis on immunological processes other than the hypersensitivity response. Olness, Culbert, and Uden (1989) recruited 60 healthy children and randomly assigned them to conditions that involved self-hypnosis and directed suggestions or control conditions. They found that those who were given a specific instruction to increase salivary levels of IgA or sIgG showed increased IgA levels but no changes in sIgG levels. No changes were seen in the control conditions. Rider and Achtenberg (1989) examined the effects of hypnosis on circulating levels of lymphocytes and neutrophils. Those instructed to show an increase in lymphocyte number showed a decrease in these cells and no change in neutrophils. Those instructed to show an increase in neutrophils showed a decrease in neutrophils but no change in lymphocytes. These findings are interesting because they show a certain kind of specificity of effect (in terms of cell type) despite the fact that the direction of the effect differed from that suggested. One study of hypnotic suggestion regarding DTH wheal size also measured the proliferative response to mitogens following hypnosis. They found no effects on this immune function tested *in vitro* (G. R. Smith *et al.*, 1992).

Conclusion.   The results of studies examining the relationship between hypnotic suggestion and immune responses remain mixed. There are no obvious differences in study design, sample, or other factors when successful and unsuccessful studies are compared. However, some of the successful studies included imagery with hypnosis (G. R. Smith *et al.*, 1992; Zachariae *et al.*, 1989; Zachariae & Bjerring, 1993). The most well designed study to date, from a methodological standpoint, yielded no significant differences despite the use of three hypnosis sessions (Locke *et al.*, 1994). It is important to note that nonimmunological processes could explain significant findings in this area. For example, changes in blood flow to the local area, as a result of hypnosis, could cause the observed changes in wheal size even if the immune system was not affected directly. It may be useful to include more molecular-level techniques to examine the effects of hypnotic suggestion on the hypersensitivity response. For example, a biopsy of the local tissue could determine whether immunological processes differ at the site of the antigen placement following hypnosis, which are not reflected in differences in wheal size.

20.2.2. Relaxation

There is a large literature documenting the effects of relaxation training on psychological states and on physiological functioning. For example, relaxation training has been shown to reduce sympathetic nervous system activity and cortisol levels (Hoffman, Benson, Arns, Stainbrook, Landsberg, Young, & Gill, 1982; McGrady, Woerner, Bernal, & Higgins, 1987). A variety of approaches have been used to induce a relaxed state. Progressive muscle relaxation involves the suggestion to reduce muscle tension in each muscle system individually until the entire body is relaxed. The *relaxation response* (a term coined by Herbert

Benson; see Benson, 1993) involves selecting a meaningful focus word, sitting quietly and relaxing one's muscles, breathing slowly, repeating the focus word, and assuming a passive attitude. Relaxation can also be induced with other techniques including hypnosis, biofeedback, and meditation.

In the case of biofeedback, special instruments are used to convey to the subject the activity of a physiological function, often one that is under autonomic nervous system control (Schwartz & Schwartz, 1993). These signals are believed to allow the subject to make changes in these systems. For example, electromyographic (EMG) biofeedback measures muscle tension and is used to induce a relaxed state.

Imagery can also be used to induce a relaxed state of mind. The client creates a vivid mental image of a relaxed state, or a context or cue, that would induce relaxation. In addition, imagery is often used in a highly specific way to attempt to change bodily processes. For example, some therapists have their cancer patients imagine that immune cells are attacking their tumors (Simonton, Mathews-Simonton, & Creighton, 1978). There are no published clinical trials that have evaluated the efficacy of this approach in terms of its effects on cancer progression. In fact, there is insufficient evidence that the immune system plays a significant role in the progression of most forms of human tumors. Therefore, it is unclear whether tumor progression would be altered even if imagery could affect the immune system in such a specific way.

Since the mid-1980s, studies have been conducted to determine if the benefits of relaxation can be extended to the immune system. To date, 13 clinical trials have been conducted with healthy individuals to determine if relaxation training can impact the immune system. A variety of relaxation interventions were investigated, including single modalities or combinations of the following: progressive muscle relaxation, back massage, guided imagery for the purposes of relaxation, lying down quietly, relaxation while standing, relaxation while sitting, focused breathing, biofeedback for relaxation, audiotapes including music, and floating in salt water. Some interventions were conducted individually and some were conducted in groups.

Four studies used salivary levels of IgA as the outcome measure, of which all found significant increases following the relaxation intervention, and no differences in the control group (Green & Green, 1987; Green, Green, & Santoro, 1988; Jasnoski & Kugler, 1987; Rider, Achtenberg, Lawlis, Goven, Toledo, & Butler, 1990) (see Chapter 19). However, these findings are difficult to interpret as it is known that hydration can affect salivary levels of IgA and that anxiety can reduce levels of saliva. In one of these studies (Green et al., 1988), serum levels of immunoglobulins were also measured, and no effect of relaxation training was found.

Studies investigating the effects of interventions on the numbers of various immunological cells are quite contradictory. For example, two groups of investigators found no effect on the number of WBCs, or the WBC differential, following either one or four sessions of relaxation plus imagery training (Hall, Minnes, Tosi, & Olness, 1992a; Hall, Minnes, & Olness, 1993), or two sessions per week of individual relaxation training and biofeedback (Peavey, Lawlis, & Goven, 1985). However, another group found a decrease in WBCs and neutrophils but no effect on eosinophils, basophils, or monocytes, following eight 30-min sessions (four group sessions of relaxation training and four individual sessions of biofeedback; McGrady, Conran, Dickey, Garman, Farris, & Schumann-Brzezinski, 1992). In contrast, consistent null effects were found when more specific cell types were evaluated. No effects on CD4 helper or CD8 cytotoxic T cells, B cells, or NK cells were found in studies using either relaxation training, floating in salt water, or self-hypnosis (Kiecolt-Glaser, Glaser, Strain, Stout, Tarr, Holliday, & Speicher, 1986; Ruzyla-

Smith, Barabasz, Barabasz, & Warner, 1995; Whitehouse, Dinges, Orne, Keller, Bates, Bauer, Morahan, Haupt, Carlin, Bloom, Zaugg, & Orne, 1996). However, Kiecolt-Glaser *et al.* (1986) did find that relaxation training prior to exams for medical students was associated with increased levels of CD4 helper T cells, but only in those participants who practiced the relaxation at home.

Studies examining the proliferative capacity of lymphocytes following mitogenic stimulation have generated contradictory results. Weekly sessions of relaxation training for 4 weeks (Kiecolt-Glaser, Glaser, Williger, Stout, Messick, Sheppard, Ricker, Romisher, Briner, Bonnell, & Donnerberg, 1985) and weekly sessions of relaxation and self-hypnosis (Whitehouse *et al.*, 1996) had no impact, while eight 30-min sessions of either relaxation or biofeedback increased this function (McGrady *et al.*, 1992). Relaxation training alone or with imagery in three 60-min sessions resulted in a decrease in the proliferative response, as well as a decrease in monocyte chemotaxis (Zachariae, Hansen, Andersen, & Jinquan, 1994). Other immune functions have been shown to be affected in single studies [i.e., neutrophil phagocytic capacity, NK cell activity (NKCA), antibody to herpes simplex virus (HSV)]. However, NKCA was not affected by relaxation training in medical students prior to their exam period (Kiecolt-Glaser *et al.*, 1986).

Conclusion.   Overall, the findings are inconsistent in terms of the exact nature of immunological results. Part of the difficulty in getting an overall picture is related to the very different immune parameters examined across studies. Clearly, relaxation training alone, and in combination with other techniques, has the capacity to affect immune parameters over the short term. Only two studies investigated the stability of effects once the training was complete. One study found no immediate or long-term effects (Whitehouse *et al.*, 1996). The other found that the immediate effect on antibody to HSV was retained 1 month later, but the NKCA changes were not (Kiecolt-Glaser *et al.*, 1985). In the latter study, 45 healthy older adults were randomly assigned to participate in either relaxation training three times per week for a month, social contact with a college student over the same time frame, or a nonintervention control condition. As shown in Fig. 1, NKCA increased from pre- to postintervention, but the effect was not maintained at the 1-month follow-up point. In addition, antibody to HSV decreased from pre- to postintervention, indicating greater immunological control of this latent virus. These effects were maintained at the follow-up point.

20.2.3. Supportive–Expressive Therapy Interventions

The notion that adverse health consequences can result from the inhibition of psychologically significant thoughts and feelings originated with the Greek physician Galen in the second century (Siegel, 1968). This view has prompted research over nearly five decades, showing that inhibition is associated with lower resting levels of sympathetic nervous system activity and more pronounced sympathetic reactivity to psychosocial stressors (e.g., see Buck, 1979; Gross & Levenson, 1993, 1997; Jones, 1935, 1950). It has also spawned efforts to determine whether psychological interventions aimed at helping people to disclose previously inhibited material can decrease physiological arousal, and thereby improve health.

Most of these efforts have come in the form of experiments wherein healthy subjects are randomly assigned to describe their thoughts and feelings around either a previously undisclosed personal trauma, or a trivial, emotionally neutral topic, such as the contents of their closet (e.g., see Pennebaker & Beall, 1986). Several recent experiments have used

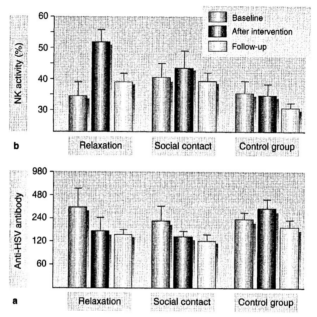

Figure 1. (a) Alteration of herpes simplex antibody titers over time in relaxation, social contact, and control groups. (b)Alteration of NK cell activity over time ($\pm$ SEM) in relaxation, social contact, and control groups, using a target-to-effector cell ratio of 40:1. Both from Kiecolt-Glaser *et al.* (1985).

immunological parameters as outcomes. For example, Pennebaker, Kiecolt-Glaser, and Glaser (1988b) assigned 50 undergraduates to write an essay that described either trivial matters or the most traumatic and upsetting experience of their life. Over the course of 4 days of writing, subjects who wrote about traumatic life events showed higher blastogenic responses to the mitogen PHA than controls. Disclosure subjects also exhibited a decline in health center utilization in the 6 months following the study, relative to controls. Booth, Petrie, and Pennebaker (1997) had 38 medical students write about either a traumatic experience or trivial matters, for 20 min on 4 consecutive days. Whereas subjects in the disclosure condition showed no changes in CD4$^+$ T lymphocyte, CD8$^+$ T lymphocyte, and total lymphocyte numbers from baseline to posttreatment, subjects who wrote about trivial issues exhibited increases in each of these subsets over that time period. Esterling, Antoni, Fletcher, Marguiles, and Schneiderman (1994) randomly assigned 57 undergraduates to spend three 20-min sessions either (1) writing about a traumatic personal experience, (2) discussing a traumatic personal experience with an experimenter, or (3) writing about trivial matters. Both verbal and written disclosers showed a significant decrease in antibody titers to Epstein–Barr virus (EBV) from pre- to postintervention, while controls showed no change. Verbal disclosers showed a larger decrease in antibody titers than written disclosers. Lutgendorf, Antoni, Kuman, and Schneiderman (1994) had undergraduates discuss a traumatic life experience with an experimenter for 20 min on three separate occasions. Subjects who completed the disclosure intervention showed no posttreatment differences in antibody titers to EBV compared with subjects assigned to an assessment-only control condition.

Although the studies described above document that disclosing a traumatic life event can influence specific immune functions and decrease health center utilization, they do not shed light on the issue of whether disclosure interventions have direct implications for physical health. To address this issue, Petrie, Booth, Pennebaker, Davison, and Thomas (1995) had 40 medical students write essays that described either trivial matters or the most traumatic and upsetting experience of their life. Subjects wrote for 20 min on 4 consecutive days. The day after subjects completed their final writing task, they were given a vaccination against hepatitis B virus, with booster vaccinations 1 and 4 months later. Blood samples were collected 1, 4, and 6 months later to assess antibody titers to hepatitis B, leukocyte subsets, and NKCA.

Compared with controls, subjects who participated in the disclosure intervention exhibited higher antibody titers to hepatitis B throughout the course of the follow-up period. These findings are illustrated in Fig. 2. Differences between the groups were most prominent at the 4- and 6-month assessments, suggesting that the benefits of the disclosure intervention may be maintained over a relatively lengthy period of time, even in the absence of any continued intervention. This was not the case for other immune measures collected during the study, however. Aside from small differences in $CD4^+$ T lymphocyte and basophil numbers during the writing period, control and intervention subjects did not evidence any differences in leukocyte subset numbers or NKCA during the course of the study.

Conclusion. Experimental studies assessing the efficacy of supportive–expressive interventions have been almost uniformly successful. Compared with controls, subjects who spend as few as three 20-min sessions writing about personally significant events show improved cellular immune control of latent EBV (Esterling *et al.*, 1994), larger blastogenic responses to PHA (Pennebaker, Kiecolt-Glaser, & Glaser, 1988b), and what appear to be

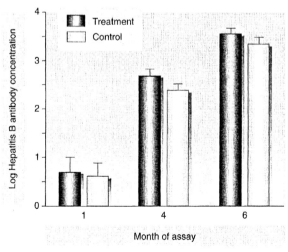

Figure 2.   Mean (± SEM) log hepatitis B concentration in treatment and control groups. From Petrie *et al.* (1995).

clinically meaningful benefits, such as fewer physician visits (Pennebaker, Kiecolt-Glaser, & Glaser, 1988b) and higher antibody titers following a hepatitis B vaccination series (Petrie *et al.*, 1995). These benefits are maintained for up to 6 months in the absence of continued intervention (Petrie *et al.*, 1995). Confronting traumatic events through writing appears to be superior to doing so through verbal disclosure (Esterling *et al.*, 1994; Lutgendorf *et al.*, 1994). The relatively poor performance of verbal disclosure paradigms in these studies is unlikely to be indicative of how efficacious a trial of psychotherapy might be in modulating immunity and/or health. Rather, this performance probably reflects the limited benefits one can achieve from spending three 20-min sessions disclosing highly personal thoughts and feelings to an unfamiliar experimenter, who is not trained in psychotherapy.

The interpretation of these studies is complicated to some extent, however, by the fact that most conclusions have been drawn on the basis of between-subjects comparisons. One alternative to this is to use within-subjects comparisons. From this perspective, many of the available studies on disclosure actually show that control subjects exhibit *declines* in immune function and/or *increases* in the frequency of health center utilization over the course of the study, whereas disclosure subjects often show small changes on health outcome measures (e.g., see Booth *et al.*, 1996; Greenberg, Wortman, & Stone, 1996; Pennebaker & Beall, 1986; Pennebaker *et al.*, 1988b; Petrie *et al.*, 1995). Cole (1999, p. 4) argues that this occurs because control subjects are inadvertently forced to inhibit the expression of personally significant thoughts and feelings. He writes:

> It is important to note that most disclosure studies recruit participants with the explicit understanding that they may be asked to discuss their thoughts and feelings about a traumatic personal experience. Upon commencement of the study, participants in the experimental group do just that, whereas those assigned to the control group find themselves writing objectively and dispassionately about a subjectively trivial topic. The structure of the control group's experience therefore closely resembles other psychological inhibition paradigms in which subjectively significant events are first primed and participants are then asked to suppress the experience or expression of these events.

To examine this hypothesis, Cole (1997) asked 110 undergraduates to write about a previously undisclosed traumatic event or a trivial matter for 30 min on 3 consecutive days. Half of the participants were warned in advance that they might be asked to disclose a personally significant event. Consistent with Cole's hypothesis about the health-damaging effects of psychological inhibition, subjects in the "thwarted-disclosure" condition (i.e., those who were warned that they might be required to disclose a personal trauma but were instructed to write about trivial matters) exhibited a significant increase in illness incidence in the months following the study. None of the other groups displayed significant changes in illness incidence over the course of the study, including those subjects who participated in the standard disclosure paradigm used in previous studies (i.e., subjects who were warned that they might be required to disclose a personal trauma and did so during the intervention). Although these findings are preliminary and await replication, they suggest that disclosure experiments may be revealing important health-damaging effects of psychological inhibition.

20.2.4. Classical Conditioning

Classical conditioning has been shown to affect a wide array of behavioral and physiological processes. In a set of pioneering studies, Ader and Cohen demonstrated that the immune response could be classically conditioned (see Ader & Cohen, 1981, and

Chapter 23). While these immune effects are usually small, they have been shown to be clinically significant. In one study, mice genetically susceptible to a form of systemic lupus erythematosus (SLE) were classically conditioned with an immunosuppressive drug. These mice showed retarded age-dependent onset of, and survival from, the disease (Ader & Cohen, 1982).

Five studies have attempted to demonstrate these effects in humans. One group of investigators has conducted four studies in which healthy subjects were randomly assigned to a conditioning intervention or one of a number of control conditions (Buske-Kirschbaum, Kirschbaum, Stierle, Lehnert, & Hellhammer, 1992; Kirschbaum, Jabaaij, Buske-Kirschbaum, Hennig, Blom, Dorst, Bauch, DiPauli, Schmitz, Ballieux, & Hellhammer, 1992). In three of the studies, an injection of epinephrine [the unconditioned stimulus (UCS)] was paired with eating sherbet [the conditioned stimulus (CS)] on 4 consecutive days. On day 5, sherbet was paired with a saline injection. The outcome measure [conditioned response (CR)] was NKCA, as research has shown that epinephrine can result in a rapid increase in the activity of NK cells. Those assigned to the conditioning sequence showed the expected increase in NKCA when presented with the sherbet–saline injection pairing. Subjects in the control conditions, who received nonconditioning variants on the above pairings over 5 days, did not show changes in NKCA when presented with the sherbet–saline pairing on day 5.

Additional analyses clarified potential biological mechanisms of these effects. The number of NK cells was not affected by conditioning, so it is likely that effects were related to the activity of NK cells. Also, epinephrine levels were not elevated following the sherbet–saline pairing, indicating that the increase in NKCA on day 5 was not the result of an increase in circulating epinephrine. Also, in one of their studies they found that epinephrine-stimulated NKCA was weaker on day 5, suggesting that the NK cells may have become less responsive to epinephrine following conditioning. These effects were not replicated when the UCS was a bitter drink.

Smith and McDaniel (1983) conducted a similar study using an allergen injection as the UCS, and the arm in which the injection was given as the CS. On the first 5 days the allergen injection was given in one arm and the saline injection in the other. On days 6 and 7 the injections were covertly switched. The investigators found differences in wheal size on these days. Specifically, the DTH response was smaller on the arm that had received saline injections for 5 days.

Conclusion.   The findings of these studies are fairly consistent in that immune effects in the expected direction are seen in most studies of conditioning. However, it is unclear whether these effects are related to conditioning per se or to other factors, such as conscious expectancy or carryover effects.

20.3. Psychosocial Interventions in Cancer

There is a long history of the use of support groups for cancer patients to alleviate the distress associated with a cancer diagnosis and, in some cases, with the aim of improving prognosis. The pivotal study conducted by Spiegel *et al.* (1989) supports the notion that a specific type of support group may in fact impact the course of disease in women with metastatic breast cancer. He showed that women randomly assigned to receive supportive–expressive group therapy, once a week for a year, lived an average of 18 months longer than women assigned to the control condition. However, these investigators were unable to determine the mediating mechanisms that explained their mortality effect. Was the effect related to the impact of a psychological change on the neuroendocrine and/or immune status of the patients? Or did the women who received the group intervention change their behavior in some way that promoted health, such as improving their nutrition? There is a great deal of interest in whether this group effect, and others like it, are effective because of their impact on relevant immune processes (see Chapter 24).

20.3.1. Relaxation Training

There has been one study of the effects of relaxation training in cancer patients. Gruber, Hersh, Hall, Waletzky, Kunz, Carpenter, Kverno, and Weiss (1993) randomly assigned 13 stage I breast cancer patients to either a relaxation intervention or a wait-list control group. The patients met weekly for 9 weeks and then monthly for 16 weeks. The intervention involved relaxation training, biofeedback, and imagery. Patients were instructed to practice techniques at home. Blood was drawn and immune parameters measured on a weekly basis with three baseline samples prior to the start of the intervention, nine samples over the course of the intervention, and three samples (one per month) during the postintervention follow-up period. The intervention group showed greater NKCA across the 9 weeks of the intervention period compared with the control group. In addition, the relaxation group had higher proliferative responses to mitogenic stimulation, higher levels of IgG, WBC, and total lymphocytes than controls during the intervention period. However, only IgM levels and the proliferative response showed an increase from pre- to postintervention.

20.3.2. Cognitive–Behavioral Stress Management

Cognitive–behavioral therapy (CBT) is a well-studied psychotherapeutic approach that can successfully treat a variety of psychological conditions. In addition, there are now studies evaluating its effectiveness in a wide array of medical problems. A central assumption of CBT is that awareness of behavior is critical (Meichenbaum, 1977). Awareness of one's own internal dialogue and its effect on behavior is an essential ingredient of change. A major focus of CBT is to try new dialogue with the help of the therapist. Patterns of response to the environment are believed to change when cognitions change and elicit coping behaviors (Meichenbaum, 1977). For example, in rational–emotive therapy, a variant of CBT (Ellis & Grieger, 1977), clients examine and change irrational beliefs (e.g., I should be perfect). This adaptive shift in cognitions is called *cognitive restructuring*.

Cognitive–behavioral group therapy usually includes cognitive restructuring, coping skills training, relaxation training, and systematic problem solving. Problem solving involves teaching methods to analyze a problem, find a range of solutions, evaluate the solutions, and implement the best solution in the real world (Rose, 1993). For example, a patient can present a problem and the group can go through the problem-solving stages together as they discuss solutions. Individual successes can be reinforced by the therapist and other group members. In addition, various social skills that facilitate problem solving may be taught in the group (e.g., assertiveness). A central feature of the CBT approach is monitoring behavior. Responses to stressful encounters and self-defeating, or otherwise problematic, thoughts are documented (using diaries for example). Social support is an essential part of most group approaches, and may enhance the effectiveness of the CBT techniques.

Fawzy and colleagues conducted the only published clinical trial of psychosocial intervention in cancer patients, examining the effects on the immune system and cancer progression (Fawzy, Cousins, Fawzy, Kemeny, Elashoff & Morton, 1990a; Fawzy, Kemeny, Fawzy, Elashoff, Morton, Cousins, & Fahey, 1990b; Fawzy, Fawzy, Hyun, Elashoff, Guthrie, Fahey, & Morton, 1993). Sixty-one patients with stage I or II malignant melanoma were randomly assigned to receive either a form of cognitive–behavioral group therapy once a week for 6 weeks, or no psychosocial intervention. The patients in the group intervention received health education, training in enhancing problem-solving skills, stress management, and social support. All patients in the study were assessed psychologically, and had blood samples drawn at baseline before the first group meeting, at the end of the 6-week intervention period, and 6 months later. In addition, patients were followed for 5–6 years from baseline to assess the number of recurrences of melanoma and deaths caused by melanoma. Immunological testing focused on the NK cell system, as it has been shown to play a role in controlling tumor metastases in some animal models. The number of various subtypes of NK cells (CD16[+], CD56[+], CD57[+]) was assessed as well as NKCA. In addition, the ability of IFNα to augment the cytotoxic capacity of NK cells was assessed.

The intervention group showed a significant decrease in negative mood, and an increase in the reported use of active behavioral coping strategies over the 6-month initial follow-up period, compared with the control group (Fawzy et al., 1990a). In addition, the 6-week intervention was shown to enhance various aspects of the NK cell system (Fawzy et al., 1990b). First, an increase in the percentage of NK cells was found at the 6-week point in the intervention patients. For some NK subtypes, however, this increase was not detectable until the 6-month assessment point, suggesting the possibility of a delayed benefit of the intervention. In addition, the ability of NKCA to be augmented by IFNα was enhanced at the 6-month point in intervention patients (see Fig. 3). Thus, NK cells may have become more responsive to cytokine signals that regulate their activity following the 6-week intervention. An inspection of individual changes in these immune parameters among the members of the intervention condition demonstrated that a majority of them showed some increase in NK cell number or function. Subgroup analyses indicated that those subjects most likely to show increases in NK cell number or function were those who showed a decrease in depression and anxiety over the 6 months, and an increase in anger.

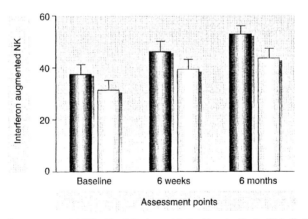

Figure 3.   Interferon-α-augmented NK cell activity (percent lysis of target cells at a 25:1 effector–target cell ratio) at baseline, 6 weeks, and 6 months in the intervention group (light bars; n = 17) and control group (dark bars; n = 16). From Fawzy et al. (1990b).

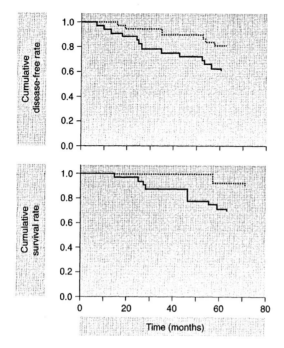

Figure 4. (Top) Survival to recurrence; (bottom) survival to death. Solid line represents the control group, dotted line the experimental group. From Fawzy *et al.* (1993).

The intervention also had a significant impact on disease course and mortality (Fawzy *et al.*, 1993). Survival analyses demonstrate that patients assigned to the intervention condition experienced a greater disease-free interval, and greater survival than the control patients (Fig. 4). Specifically, 10 of 34 patients assigned to the control group died during the follow-up period, while 3 of 34 intervention group patients died (dropouts were retained in this analysis). A number of factors were examined as potential confounders, including age, gender, depth of the initial melanoma lesion (Breslow depth), and site of original tumor. Adjusting for the one factor that related to melanoma recurrence and survival, Breslow depth, analyses showed that treatment continued to significantly predict recurrence and survival.

It has been argued that health differences between intervention and control groups may be related not only to the benefits of the intervention, but to the negative psychological consequences of random assignment to a control group. In order to address this issue, the 30 original stage I control patients were matched to patients from a large melanoma database according to gender, Breslow depth, and state of disease at initial diagnosis. The survival curves comparing this matched group with the control group were not statistically different. Thus, the effects are more likely attributable to some form of benefit obtained as part of the intervention condition.

This is the only study that has found psychological, immunological, and clinically meaningful health benefits of a psychological intervention. Therefore, we will provide additional detail on the nature of the intervention. Groups of seven to ten patients met for 1½ hours a week for 6 weeks. The groups were led by an experienced psychiatrist and by Norman Cousins, who is well known for his view that positive psychological states can improve health. The research team called the intervention a structured psychoeducational group intervention. However, many of the elements of the actual process overlap with the cognitive–behavioral approach described in more depth earlier in this chapter.

The approach used in this study has four components. First, the patients are given health education, such as cancer prevention strategies (e.g., information on nutrition, ways to avoid sun exposure). The second component is enhancement of illness-related problem solving skills, which was modeled after Project Omega (Cohen, Cullen, & Martin, 1982). It involves presenting specific problems that cancer patients encounter and teaching active methods of coping with the problems. Active behavioral and cognitive coping is contrasted with avoidance methods. More active approaches are learned, practiced, and applied to problems presented by group members. To illustrate specific problems and effective methods of coping with them, a series of pictures is shown in the group. The pictures illustrate possible problem areas such as isolation and loneliness, the physician–patient relationship, and changes in body image. As part of the discussion of active coping, talks are given on the promotion of feelings of hopefulness, and mobilizing coping resources.

The third component of this intervention is stress management. Patients are taught how to assess their own sources and levels of stress and are also taught a variety of relaxation techniques, including guided imagery and self-hypnosis. As part of the stress management component, patients are also instructed in ways to develop more positive appraisals of stressors. The final component is social support, which was mobilized from among group members and leaders.

20.4. Psychosocial Interventions in HIV Infection

There has been considerable interest in developing psychosocial interventions for people with HIV infection (see Chapter 25). Part of this interest stems from the fact that HIV infection has a highly variable clinical course. Some patients remain in the asymptomatic "clinical" latency period for many years, while others rapidly advance to full-blown AIDS after they become infected. Viral strain is a significant predictor of this variability. In addition, a number of host cofactors are thought to be responsible for this variability, including the genetic characteristics of the host immune system, alcohol and drug use, poor nutrition, and coinfection with other pathogens (see Saah, Hoover, He, Kingsley, & Phair, 1994). Psychosocial factors have also been hypothesized to serve as cofactors, as patients who are infected with HIV must contend with the significant psychological, social, medical, legal, and occupational burdens associated with the disease. To the extent that the resulting distress is associated with physiological changes that could impact HIV replication, or host containment of the virus, these burdens may lead to a more rapid onset of AIDS-related morbidity and mortality. There is increasing evidence in well-controlled longitudinal studies that psychosocial factors, such as depression, negative expectancies, and rejection sensitivity, can predict immunological and clinical evidence of HIV progression (see Cole & Kemeny, 1997).

20.4.1. Relaxation Training

Several attempts have been made to slow the course of HIV infection by teaching patients stress-reducing relaxation techniques. For example, Eller (1995) had 69 HIV seropositive individuals, at varying stages of the disease, practice either progressive muscle relaxation (PMR) or guided imagery daily for 6 weeks. Instruction was given through audiotapes. Compared with a wait-list control group, patients assigned to the PMR intervention showed increased CD4$^+$ T lymphocyte numbers at posttreatment. No posttreatment immunological differences emerged between patients in the guided imagery group and wait-list controls. Taylor (1995) randomly assigned 10 men to a wait-list control condition or to 20 sessions of a relaxation intervention, which included PMR, EMG biofeedback, meditation training, and hypnosis training. All patients were asymptomatic at study entry, but had CD4$^+$ T lymphocyte levels below 400 per mm^3. The intervention group showed a significant increase in CD4$^+$ T lymphocyte numbers from pretreatment to 1 month posttreatment, while controls experienced a decline in CD4 numbers over the same time frame.

20.4.2. Cognitive–Behavioral Stress Management

Most of the psychosocial interventions in the context of HIV infection have employed multifaceted, cognitive–behavioral stress management approaches. For example, in one study (Coates, McKusick, Kuno, & Stites, 1989), 64 asymptomatic males were enrolled in an 8-week intervention comprised of relaxation training, health habit change, and stress management skills. Compared with those who were randomized to a wait-list control group, individuals in the intervention showed a reduction in high-risk sexual behavior, but no differences in enumerative or functional measures of immunity. Mulder, Antoni, Emmelkamp, and Veugelers (1995) studied cognitive–behavioral stress management training versus experiential group therapy in a sample of 26 asymptomatic males. No differences in immune parameters were apparent after 15 weeks of treatment or at a 2-year follow-up. However, patients who showed the largest reductions in psychological distress over the course of treatment tended to experience less rapid declines in CD4$^+$ T lymphocyte levels at follow-up. Lutgendorf, Antoni, Ironson, Klimas, Kumar, Starr, McCabe, Cleven, Fletcher, and Schneiderman (1997) assigned 26 men with symptomatic, mildly progressed HIV infection, but no signs of AIDS, to a 10-week trial of cognitive–behavioral stress management or to a wait-list control condition. Patients in the intervention group exhibited a significant decline in antibody titers to HSV-2 from pre- to posttreatment, while control patients showed no changes. Declines in antibody titers to HSV-2 were most prominent among patients who showed a reduction in dysphoric mood during treatment. Neither the intervention group nor the control group showed changes in CD4$^+$ or CD8$^+$ T lymphocyte numbers over the course of the study.

To determine whether a psychosocial intervention could buffer patients from the psychological and immunological changes associated with HIV serostatus notification, Antoni, Baggett, Ironson, August, LaPerriere, Klimas, Schneiderman, and Fletcher (1991a) asked 47 healthy gay men who were unaware of their HIV serostatus to participate in either a cognitive–behavioral stress management intervention or a wait-list control group. Patients randomized to the stress management intervention met in groups twice a week for 10 weeks. One session each week was devoted to PMR training, while the other focused on cognitive restructuring and education regarding stress management, HIV infection, and HIV-risk reduction behaviors. Five weeks into the study, blood samples were collected to determine patients' HIV serostatus and provide prenotification assessments of immune function. Three days after the blood collection had taken place, patients met individually with a clinical social worker and were notified of their HIV serostatus. Blood samples were collected again 1 week after this meeting to assess patients' immunological responses to serostatus notification.

Figure 5 illustrates the central immunological findings from this study. Relative to seropositive controls, seropositive patients who participated in the stress management

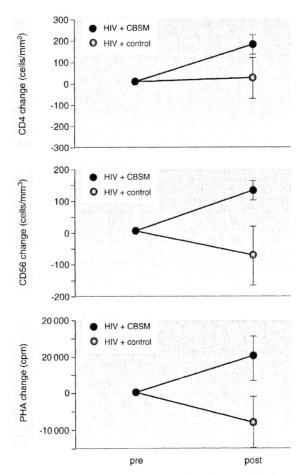

Figure 5.   Helper-inducer (CD4) and NK (CD56) cell counts and lymphocyte proliferative responses to phyto-hemagglutinin (PHA), expressed as counts per minute at pre- and postnotification in the seropositive cognitive–behavioral stress management (HIV + CBSM) and HIV + control groups. From Antoni *et al.* (1991a).

intervention exhibited pre- to postnotification increases in CD4$^+$ T lymphocyte numbers, NK cell numbers, NKCA, and the proliferative response to PHA. These buffering effects appeared to be mediated primarily by the PMR training, as the frequency of relaxation practice during the pre- to postnotification period was strongly correlated with immunological change. In a separate report (Ironson, Friedman, Klimas, Antoni, Fletcher, LaPerriere, Simoneau, & Schneiderman, 1994), adherence to treatment (number of groups attended and relaxation practice at home) predicted 2-year disease progression.

LaPerriere, Antoni, Schneiderman, Ironson, Klimas, Caralis, and Fletcher (1990) used a similar design to determine whether a social exercise intervention could buffer patients from the immunological effects of HIV serostatus notification. Compared with patients in a wait-list control condition, seropositive patients who participated in 30 sessions of stationary bicycle ergometry with other HIV seropositive men showed a decrease in NK cell numbers from pre- to postnotification. However, the social exercise intervention did not yield any changes in CD4$^+$ T lymphocyte levels among seropositive patients. Esterling, Antoni, Schneiderman, Carver, LaPerriere, Klimas, Ironson, and Fletcher (1992) combined data from the Antoni *et al.* (1991a) and LaPerriere *et al.* (1990) trials and found greater pre- to postnotification decreases in antibody titers to EBV among seropositive patients who had participated in either the exercise intervention or the cognitive–behavioral stress management intervention, relative to seropositive controls. Patients in the stress management intervention also exhibited declines in antibody titers to human herpesvirus-6 over the course of the intervention, while patients in the control and exercise conditions showed no change.

     Conclusion. Psychosocial intervention trials in the context of HIV infection have yielded mixed results. Cognitive–behavioral stress management treatments appear to buffer patients from the immediate immunological sequelae of serostatus notification (Antoni *et al.*, 1991a; Esterling *et al.*, 1992). Both cognitive–behavioral and relaxation training interventions appear to improve cellular immune function among symptomatic patients with mildly progressed disease (Eller, 1995; Lutgendorf *et al.*, 1997). However, these approaches have not proven efficacious among asymptomatic patients in the early stages of HIV infection (Coates *et al.*, 1989; Mulder *et al.*, 1995). The reasons for this disparity are not clear. Patients in the midst of major life transitions, such as HIV seroconversion, or the onset of more advanced disease, may derive the most benefit from psychosocial interventions because they are experiencing high levels of distress during those transitions. This appears to be particularly true if the intervention is successful at reducing dysphoria or increasing participation in distress-reducing relaxation exercises at home. CBT during the symptomatic phase may be useful in that it provides individuals with active tools for dealing with the problems they confront at this stage of illness. CBT also allows individuals to share their experiences with others undergoing the same transition. Asymptomatic patients who are in the prolonged "clinical" latency period of HIV infection might find these approaches unhelpful as they struggle to maintain normal life activities and not actively think about the status of their infection. This would explain why cognitive–behavioral stress management interventions have proven ineffective at modulating immunity among asymptomatic patients (Coates *et al.*, 1989; Mulder *et al.*, 1995).

     It is important to note that although psychosocial interventions have been shown to influence several immunological parameters in patients with HIV/AIDS, it is not clear whether these changes in immunity are of sufficient magnitude, or are maintained for a long enough period of time, to slow progression of the disease. To address this problem, future intervention trials will need to follow patients for more extended periods of time and supplement biological markers of disease progression (e.g., CD4$^+$ T lymphocyte numbers) with time-to-endpoint analyses of clinical markers of disease progression, such as AIDS onset and AIDS-specific mortality (see Cole & Kemeny, 1997).

20.5. Psychosocial Interventions in Other Clinical Populations

     While the psychoneuroimmunology literature is replete with examples of psychosocial interventions for patients with cancer and HIV/AIDS, only a handful of studies have examined whether such interventions might influence the course of other immunologically mediated or resisted diseases. Studies of psychosocial interventions and immune processes in such groups are sparse as well.

20.5.1. Hypnosis

Four studies have evaluated the effects of hypnosis on patients with asthma, allergies, or chronic hives. Laidlaw, Richardson, Booth, and Large (1994) recruited five asthmatics to participate in a series of hypnosis sessions during which they were asked to control their ITH reactions to a histamine prick. All subjects were untrained in hypnosis. Compared with control sessions, subjects showed decreased erythema in response to the histamine prick, but no difference in induration size. Levine, Geer, and Kost (1966) studied 30 subjects; 10 of the subjects were healthy, and the remainder had ragweed allergies or hives. They were randomly assigned to a hypnosis session that instructed them to suppress the ITH response in one arm and enhance the response in the other arm, or to a session that instructed them to suppress the response in one arm and not alter the response in the other. Hypnotic suggestions did not influence the subjects' ITH responses to any of three dilutions of a histamine prick. Fry, Mason, and Pearson (1964) studied 18 patients with asthma or hay fever. They found reduced ITH responses among patients who had undergone hypnosis with direct suggestions to abolish skin reactions. In a second study by the same investigators (Fry *et al.*, 1964), 29 individuals with asthma or hay fever were recruited. The same procedure was undertaken except that one group was instructed to suppress the ITH response in the right arm, the second group was instructed to suppress the response in both arms, and the third group was given no instructions. There were no pre- to postintervention changes in wheal size.

20.5.2. Cognitive–Behavioral Stress Management

One study evaluated the effects of cognitive–behavioral stress management on immune processes in patients with rheumatoid arthritis. O'Leary, Shoor, Lorig, and Holman (1988) recruited 33 women with rheumatoid arthritis to either a control condition, or a cognitive–behavioral stress management program that involved relaxation, imagery, goal setting, self-reinforcement, and facilitation of coping. After 5 weeks of the intervention there were no significant differences in $CD4^+/CD8^+$ lymphocyte ratios in the intervention group compared with controls. However, those women who reported increased self-efficacy in managing pain showed an increase in the $CD4^+/CD8^+$ ratio pre- to postintervention.

Conclusion. In studies of individuals with asthma and hypersensitivities, mixed findings have emerged, with some studies showing that patients are able to modify ITH responses following direct suggestions under hypnosis (Fry *et al.*, 1964, Study 1), and others failing to find such effects (Fry *et al.*, 1964; Study 2; Laidlaw *et al.*, 1994; Levine *et al.*, 1966). Some of the variability in outcome may stem from methodological differences across studies. Laidlaw and colleagues have shown that a number of generally unmeasured variables (e.g., ambient temperature, mood, day of the histamine prick) can account for substantial amounts of variance in subjects' skin responses (Laidlaw *et al.*, 1994). It will therefore be important for investigators to consider these issues when planning future studies in this area. At this point, however, too few data are available to warrant any substantive conclusions about the efficacy of hypnotic interventions among patients with asthma and hypersensitivity disorders.

Only one study has been conducted in patients with autoimmune diseases. The study by O'Leary *et al.* (1988) found no main effect of the intervention although immunological changes in a subgroup of patients were observed. This is an important area that deserves further investigation.

20.6. Conclusion

Five primary classes of interventions have been evaluated for effects on the immune system: cognitive–behavioral stress management, supportive–expressive therapy including disclosure experiments, relaxation training (including imagery and biofeedback), hypnosis, and conditioning. In most categories, approximately two-thirds of the studies showed a change in at least one immune parameter (on the low end, about half of CBT and hypnosis studies showed a change; on the high end, about 80% of the relaxation studies showed immune changes, although many of these studies only found effects for sIgA). Often, certain immune parameters are affected while others are not. There is no consistent pattern with regard to immune parameters being more vulnerable to modification by intervention.

A critical question is whether there is a central ingredient that accounts for the success of certain interventions. There is very little information available to address this question. Based on the studies reviewed, a few interesting possibilities emerge. In a study of CBT in HIV-positive individuals, those who showed a greater decrease in dysphoric mood following the intervention were more likely to show changes in antibody titers to a latent virus (Lutgendorf *et al.*, 1997). A study of melanoma patients found that a form of cognitive–behavioral group therapy impacted mood, coping, NK cells, recurrence rate, and mortality (Fawzy *et al.*, 1990a, 1990b, 1993). The investigators found that those individuals most likely to show beneficial changes in immune parameters were those who showed a decrease in depression and anxiety and an increase in anger. Thus, decreases in negative moods such as depression and anxiety may be important mediators of some intervention effects.

It is also possible that certain affective responses are enhanced during therapy and are also involved in beneficial physiological changes (as in the anger findings in the study by Fawzy *et al.*, 1990b). This is one of the few studies to suggest a specificity in terms of the affective responses associated with immunity with some responses showing positive associations and some showing negative relationships (see also Futterman, Kemeny, Shapiro, & Fahey, 1994). It is important to note that Ekman and Levenson (Ekman *et al.*, 1982) have shown different patterns of autonomic nervous system arousal with distinctive negative emotion states. Thus, it is possible that interventions that reduce the mood states of depression and anxiety may have beneficial immunological effects. However, interventions that enhance feelings of anger may also be beneficial. Alternatively, the anger findings may really represent changes in the level of assertiveness or emotional expressiveness. In this case, the findings would be consistent with the results of Pennebaker who found that those individuals who were most likely to benefit from disclosing their thoughts and feelings regarding a trauma, were those who used a moderate level of negative emotion words during the disclosure process (Pennebaker, 1997). Low levels and very high levels of negative emotion words were not associated with health benefit. Also, these findings are consistent with a small group of longitudinal studies that suggest that confronting a traumatic event, and experiencing emotional responses to it, may predict beneficial immunological changes, at least in the context of HIV infection (Bower, Kemeny, Taylor & Fahey, 1998; see Kemeny & Gruenewald, 1999 for a discussion of specific affective and cognitive correlates of immune status).

Another possible mediator of intervention effects involves imagining the desired outcome. The hypnosis literature is quite mixed, with many very well conducted studies showing no effects, and others showing effects under limited conditions. However, it appeared that those studies that combined hypnosis with imagery were more likely to show positive effects. Despite the widespread interest in imagery in cancer patients, there are no controlled clinical trials of imagery in cancer. However, imagery is used in a variety of contexts to bolster behavioral effects such as in sports performance. There is also some evidence that neuronal activity while imaging a given action is quite similar to the neuronal response observed while actually engaging in the action.

It is also possible that emotional support mediates the effects of interventions that involve others. Social support has been shown to be associated with decreased morbidity and mortality in a large body of epidemiological studies (Cohen & Syme, 1984). In addition, a growing number of studies demonstrate what appear to be beneficial changes in specific immune parameters with low levels of loneliness, marital satisfaction, and emotional support (e.g., Kiecolt-Glaser et al., 1984c). Beneficial changes in neuroendocrine products such as low levels of cortisol and epinephrine, which are associated with social support, may underlie these changes (Seeman, Berkman, Blazer, & Rowe, 1994). However, the relationship between social support and immunity and health is not always positive. A number of studies are emerging that indicate the possible detrimental aspects of social support (Miller, Kemeny, Taylor, & Visscher, 1997).

Another possible mediating mechanism is positive expectancies. Many of the interventions focus on the expectations of the client. For example, hypnosis is the most direct type of expectancy manipulation. In this case, the expectancy regards the immune system specifically. Imagery is also a direct manipulation of expectancy, either in terms of response to stress (imagery directed at a relaxed rather than stressed state) or imagery directed at the immune system or health. In cognitive–behavioral group therapy, clients are provided with the expectation that they can learn to deal with a life-threatening illness (cognitive restructuring of negative beliefs), and in some cases the intervention includes the promotion of positive beliefs about future health. Even some human conditioning studies are designed in such a way that an expectation about the outcome is manipulated (expectancy could be considered a confounding factor in these studies). In addition, biofeedback increases expectations that a specific physiological system can be manipulated. There is now increasing evidence that generalized positive expectancies and disease-specific positive expectancies are associated with better health over time (e.g., Reed, Kemeny, Taylor, Wang, & Visscher, 1994).

Other possible mediators that relate to specific types of interventions include stressor appraisal, active coping, modification of negative thoughts, increases in self-efficacy, control, or mastery (Antoni et al., 1991a). There is evidence in both human and animal studies that control can buffer the effects of stress on the immune system. Even perceived control in the absence of "real" control has protected subjects from the effects of loud noise stress on the immune system (Sieber, Rodin, Larson, Ortega, Cummings, Levy, Whiteside, & Herberman, 1992).

It may be useful to consider the possibility that there is a factor that is interfering with the potential benefit of some of the interventions that have been studied. Many interventions for patient populations take place in groups. While group activity is a major key to success for many techniques, such as CBT, there is one disadvantage to the group format in a medically ill population. If the patients in the group are ill, then group members may be exposed to the worsening of others' conditions or even the death of members. In a group process that lasts for an extended period of time, there is time to deal with such a traumatic experience. However, in short-term group interventions, such experiences may result in significant dysphoria and apprehension, as well as potentially bolstering group members' negative expectancies about their own future health. These psychological states could then have negative biological consequences or interfere with the benefits of the group experience. Future studies of longer-term group interventions, or more individually oriented approaches, might be useful.

There are a number of limitations to the literature on intervention effects on immune processes. First, even among studies that show changes pre- to postintervention, the sizes of the effects are relatively small. While they clearly demonstrate that a change in psychological state can affect the immune system, it is not clear whether these effects would have clinical implications. Would the effects of relaxation in an elderly population be sufficient to decrease their vulnerabilities to colds or influenza, for example? Only one study addressed this issue directly. Fawzy *et al.* (1990b, 1993) demonstrated that a 6-week cognitive–behavioral intervention impacted the NK cell system, as well as recurrence rate and mortality. However, changes in the NK cell system, as measured in this study, were not predictive of recurrence and mortality effects. Thus, while the immune system was affected by the intervention, it is not clear that these changes mediated the changes in health. Other physiological systems, or other aspects of the immune system, may be responsible. Alternatively, changes in behavior, such as reductions in sun exposure as a result of the health education component of the intervention, may explain these findings. Thus, the health effects of immune changes induced by interventions remain unknown. Studies are needed that measure psychological factors and manipulations, immune processes, and health, and the interrelationships among the three (see Kemeny, 1994, for a discussion of this issue). It is important to note, however, that even a small immune effect could impact a disease process. For example, in the studies of classical conditioning of mice, a relatively small immunological effect induced by the conditioning resulted in lower mortality in an autoimmune model (Ader & Cohen, 1982).

A second limitation of the studies is the restricted range of measures of immune processes utilized. Many have focused on salivary IgA, which is difficult to interpret because of hydration effects. Others have focused on the distribution of cell types, e.g., number of lymphocytes, number of $CD4^+$ T cells, which may be difficult to relate to health in nonclinical populations. There are other approaches that could be taken in this area of research. One approach is to evaluate the production of cytokines, as they play such a pivotal role in immune regulation. Cytokine levels, expression of cytokine receptors, and receptivity of lymphocytes to cytokine signals can be assessed *in vitro*. Also, expression of receptors for neuropeptides and neurohormones on lymphocytes might illuminate the links between interventions, neuroendocrine processes, and immune function. Also, in studies of DTH reactions, biopsies of the area could be obtained, and the local tissue could be examined for lymphocyte type and receptor expression.

Another limitation of studies is the lack of long-term follow-up. While short-term changes in immune parameters have been observed, it is unclear whether these effects will be maintained over time. Long-term maintenance is crucial if we are to expect health benefits of these interventions. A few studies did demonstrate long-term effects. Petrie *et al.* (1995) found an increase in antibody to the hepatitis B vaccine administered 1 and 4 months after a 4-day disclosure experiment. In fact, effects were stronger at 4 and 6 months than at the initial point. However, a similar disclosure study found that increases in proliferative response to PHA were not maintained over time (Pennebaker *et al.*, 1988b). One study of relaxation in the elderly showed that the immunological effects were not maintained after 1 month (Kiecolt-Glaser *et al.*, 1985).

Overall, these studies provide good initial indications that the immune system can be impacted by cognitive manipulations, stress management techniques, and disclosure processes. However, the long-term effects and health benefits of these processes remain to be determined. It may turn out that in most cases short-term interventions cannot generate long-term physiological and health benefits. If that is the case, then more intensive and longer-term interventions will need to be designed and evaluated.

21   Sleep and Immune Functions

Jan Born

Jan Born • Clinical Research Group–Clinical Neuroendocrinology, Lübeck Medical University, D-23538 Lübeck, Germany.

21.1. Introduction

Sleep is a fundamental biological need. Humans spend more than one-fourth of their life asleep, suggesting that this organismic state is of superior adaptive significance. In fact, total loss of sleep for a period of several weeks is lethal. Also, less extended periods of sleep deprivation may be damaging, and are popularly believed to lead to increased vulnerability to infection. It is a common claim, for example, that one caught a cold because of too little sleep the days before. As common as this belief, on the other hand, is the opinion that sleep helps healing and promotes the recovery from infectious disease. This view is further strengthened by the fact that fatigue and the desire for sleep is often distinctly increased during infectious diseases. Thus, a bidirectional interaction between the process of sleep and immune function seems evident. Such an interaction was recognized by scientists much earlier in this century (e.g., von Economo, 1930). However, only in the last 25 years have more thorough and extensive research efforts been made to outline some of the essential features of this interaction in animals and humans. The two key questions were (1) whether immune processes alter sleep and (2) whether sleep influences immune competence. Also, how such influences are transferred was the aim of these investigations (Fig. 1).

21.2. Sleep

Sleep is an organismic state characterized by behavioral inactivity and a reduced capability to process environmental stimuli. A distinct mode of neurophysiological regulation during sleep is indicated by electroencephalographic recordings of neocortical neuron activity as well as by recordings of vegetative and endocrine activity.

With reference to the signs of reduced behavioral activity and sensory intake during sleep, it has been assumed that this process primarily serves to permit recovery of the organism from the stress of the wake phase. In fact, sleep and stress represent extremely

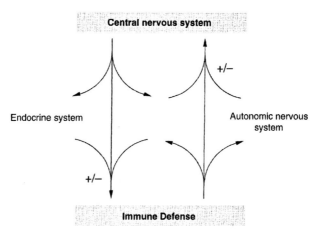

Figure 1.   Central nervous sleep influences immune function, and immune responses influence sleep. The effects can be either direct or mediated via the endocrine and autonomic nervous systems.

opposite states along several dimensions: While during stress the organism reacts to strong exogenous stimuli, during sleep the organism is almost exclusively regulated by endogenous rhythms. The reaction to a stressor is associated with catabolic metabolism and a maximum expenditure of energies, whereas sleep enhances anabolic metabolism thereby saving energies and supporting repair functions.

Reduced energy consumption during sleep as compared with wakefulness has been consistently confirmed in studies examining central nervous activation as well as activation of peripheral organ systems by measures of local blood flow, oxygen consumption, and temperature. Nevertheless, deactivation and the saving of energies describes just one aspect of sleep. The regulation of central nervous processes, vegetative and endocrine activity during sleep—as will be described in detail below—appears to be too complex to simply serve to deactivate the organism. Hence, current sleep research is in great part guided by the question of whether and to what extent sleep regulation, in addition, supports functions of repair and memory formation. It may well be that the acute processing of environmental stimuli during sleep is suppressed for the sake of an undisturbed consolidation of memories from stressful events during the wake period. The memory consequently would optimize the organism's adaptation to a stressor at future reexposures. This view bears some resemblance to Freud's view on sleep and dreams, and it may not only hold for psychological memory in response to psychological stressors but also for immunological memory formation in response to antigen challenge.

21.2.1. Polysomnographic Recording of Sleep

In the laboratory, sleep is commonly determined by means of polysomnography. This includes the continuous recording of electroencephalographic activity (EEG), electro-oculographic activity (EOG), and electromyographic activity (EMG). The recordings are used to discriminate between sleep and wakefulness and—within sleep—among five stages of sleep: sleep stages 1 to 4 and rapid eye movement (REM) sleep. The criteria commonly used for the classification of sleep stages in humans are those of Rechtschaffen and Kales (1968) (Fig. 2). According to this sleep scoring system, a subject is still awake as long as the EEG shows predominant activity in the alpha rhythm (8–12 Hz) and the EMG indicates enhanced muscle tension. The transition to sleep stage 1 is scored as soon as the EEG-alpha rhythm is replaced by irregular oscillations with higher frequency and lower amplitude and the EMG activity decreases. Sleep stage 2 begins as soon as sleep spindles and K-complexes appear in the EEG. Sleep spindles represent phasic events of fast oscillations (12–14 Hz), which like K-complexes do not last much longer than 0.5 s. Both events may occur as rarely as once every min during stage 2 sleep, but typically are more frequent. Sleep stages 3 and 4 are characterized by the emergence of very slow (<3 Hz), high-amplitude (>75 µV) oscillations in the EEG, termed *slow waves*. When slow waves are present more than 20% of the time, sleep stage 3 is scored, and when they occupy more than 50% of the time, sleep stage 4 is scored. Together, sleep stages 3 and 4 constitute slow-wave sleep (SWS). At the transition to REM sleep, the EEG activity switches to a pattern characterized by relatively fast oscillations of mixed frequency and low voltage. This pattern is reminiscent of the EEG during attention in awake subjects. Nevertheless, REM sleep can be readily distinguished because of the episodic occurrence of REMs and the concurrent tonic suppression of EMG activity.

By assigning a certain sleep stage to each successive 30-s period of polysomnographic recordings, a hypnogram can be constructed, which reflects the dynamic temporal organization of sleep over the entire night (Fig. 3). In the beginning of sleep, young healthy subjects

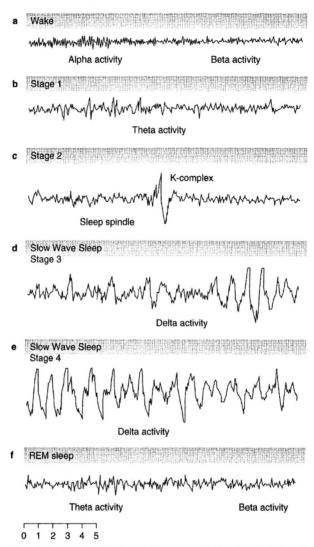

Figure 2. EEG signs of sleep. During wakefulness alpha waves prevail. Stage 1 sleep is indicated by fading alpha activity and the advent of sharp waves. During stage 2 sleep, sleep spindles and K-complexes occur. Slow-wave sleep is characterized by high-amplitude delta activity (< 3 Hz). During REM sleep fast oscillations with low amplitude (beta activity) predominate, interrupted by intervals of theta activity.

typically proceed quickly from stage 1 sleep to SWS. Then, after a period of about 30 min in SWS they return to lighter sleep (stage 2 or stage 1) in order to enter a period of REM sleep. This sequence of stages constitutes a non-REM–REM-sleep cycle, which normally takes about 100 min. The total night includes three to five such cycles. Across the cycles the time in SWS gradually decreases while the duration of the respective REM sleep epochs

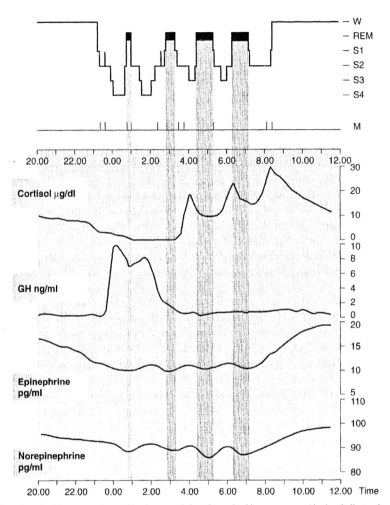

Figure 3. (Top) Representative profile of nocturnal sleep from a healthy young man. Abscissa indicates time of day, ordinate indicates sleep stage (W, wake; REM, rapid eye movement sleep; S1, S2, S3, S4, sleep stages 1–4; M, movements). S3 + S4 represents the time in slow-wave sleep (SWS). Note that the first half of sleep is dominated by SWS, the second half by extended epochs of REM sleep. (Bottom) Profiles of plasma concentrations of cortisol, growth hormone (GH), epinephrine, and norepinephrine. Minimum plasma concentrations of cortisol and maximum concentrations of GH occur during early sleep. Catecholamine concentrations are reduced during sleep and, in particular, during REM sleep.

increases from cycle to cycle. Thus, SWS is typically predominant during the first half of nocturnal sleep whereas REM sleep dominates the second half.

The circadian sleep–wake rhythm is primarily controlled by hypothalamic structures (Moore-Ede, Sulzman, & Fuller, 1982). Experimental animals with lesions of the nucleus suprachiasmaticus of the hypothalamus display fragmented sleep evenly distributed across the 24-h cycle. Additional circadian oscillators, probably localized in the ventromedial hypothalamus, control rhythms of REM sleep, temperature, and possibly also adrenal cortisol release. These hypothalamic structures interact with brain stem structures responsible for the generation of non-REM and REM sleep. In the dorsal nucleus raphe, neurons have been identified with firing patterns closely linked to the cyclic shift between non-REM and REM sleep (Hobson & Stickgold, 1995).

21.2.2. Autonomic Nervous Activity during Sleep

Aside from the changes in central nervous activity, non-REM and REM sleep are accompanied by a discrete regulation of activity within the autonomic nervous system. During sleep and, in particular, during non-REM sleep, sympathetic activity is decreased while parasympathetic output appears to be enhanced, compared with wakefulness. This is reflected, for example, by a distinctly reduced cardiovascular activity, reduced heart rate and cardiac output, and reduced blood pressure during SWS compared with wakefulness (Mancia & Zanchetti, 1980). Also, plasma concentrations of norepinephrine and epinephrine decrease at the transition to nocturnal sleep and increase again around the time of awakening in the morning (Dodt, Breckling, Derad, Fehm, & Born, 1997). Compared with non-REM sleep, REM sleep seems to be accompanied by a further depression of tonic sympathetic outflow associated with the persistence of increased tonic parasympathetic activity. Accordingly, in humans plasma catecholamine concentrations during REM sleep have been found to be even lower than during non-REM sleep and SWS. But, curiously, heart rate and blood pressure are increased during REM sleep to values comparable with those during wakefulness. Moreover, the tonic depression of sympathetic outflow during REM sleep is more or less frequently interrupted by phasic sympathetic discharges, causing an increased variability and instability of autonomic regulation during this stage of sleep.

21.2.3. Endocrine Activity during Sleep

Sleep is associated also with a rather specific regulation of endocrine activity constituting a prominent feature of this state of consciousness. Characteristic time courses during sleep have been documented for the release of hormones from the pituitary including adrenocorticotropin (ACTH) and growth hormone (GH), which are both known to also exert distinct influences on a variety of immune functions (Fig. 3). ACTH acts on immune functions mainly by stimulating the adrenal release of glucocorticoids, which at high concentrations generally suppress immune activation.

The episodic release of ACTH and cortisol from the pituitary–adrenal axis is at a minimum during the early part of nocturnal sleep when SWS is dominant. However, the release of these hormones begins to increase distinctly about 180–200 min following sleep onset so that plasma ACTH/cortisol concentrations during the second half of nocturnal sleep, which is dominated by REM sleep, are on average more than three- to fivefold higher than during the first half. This relationship between nocturnal sleep and pituitary–adrenal secretory activity reflects the influence of circadian oscillators entraining endocrine activity to the central nervous sleep–wake cycle. However, strong evidence has been provided that, in addition, the inhibition of pituitary–adrenal secretory activity during the first part of sleep reflects a sleep-dependent process (Born & Fehm, 1998). Episodes of pituitary–adrenal secretion during nocturnal sleep occur primarily during non-REM sleep while during REM sleep the release of ACTH/cortisol appears to be inactivated. Cortisol is known to suppress REM sleep, and at low concentrations, to increase SWS.

Compared with the nocturnal concentration of cortisol, concentrations of GH display a reversed pattern. They are at a maximum during the first hours of sleep and typically remain below the detection threshold during the late hours of sleep. The nocturnal GH peak usually coincides with the first extended epoch of SWS during the night. However, nocturnal GH release does not depend on SWS. Rather, sleep onset appears to be the triggering event. GH secretory episodes, like those of ACTH, normally fall into epochs of non-REM sleep.

21.2.4. Endocrine and Autonomic Nervous Activity: Mediators of Sleep–Immune Interactions?

The foregoing description underlines that sleep is not a mere central nervous event but also includes a specific regulation of activity within the autonomic nervous system and the endocrine system. During early sleep, for example, the inhibition of pituitary–adrenal secretory activity in conjunction with maximum release of GH represents an endocrine pattern that during wakefulness cannot be observed under any circumstances (the stress of wakefulness is typically associated with joint activation of the release of GH and ACTH/cortisol). The uniqueness of these patterns of endocrine and autonomic activity generated during sleep is suggestive for a particular functional significance, which could pertain to a specific control of immune processes. In the case of the secretory quiescence in the pituitary–adrenal system during early sleep, this period of suppressed cortisol release could help to establish defense mechanisms most sensitive to inhibitory effects of corticosteroids. However, which immune processes in particular benefit from neuroendocrine regulation during early sleep, as well as during sleep in general, is at present unknown.

Aside from their efferent actions, endocrine and autonomic signals feed back to the brain and in this way control central nervous sleep processes. Considering that endocrine activity such as the release of ACTH and cortisol, as well as autonomic activity is highly sensitive to signals from the immune system (i.e., to influences of cytokines), humoral and autonomic feedback signals presumably integrate messages from the immune system to the brain that interfere with the ongoing regulation of sleep and wakefulness. For example, several proinflammatory cytokines including IL-1β, IL-6 and IFNα have been demonstrated to be potent stimuli of the adrenal release of corticosteroids (Späth-Schwalbe, Porzolt, Digel, Born, Kloss, & Fehm, 1989; Späth-Schwalbe, Hansen, Schmidt, Schrezenmeier, Marshall, Burger, Fehm, & Born, 1998). The circulating glucocorticoids, in turn, acutely inhibit REM sleep. In contrast, enhancing effects of these cytokines on SWS in animals have been suspected to involve activation of the GHRH/GH axis (Krueger & Majde, 1994). However, in general, the mechanisms mediating immunological influences on sleep are poorly understood.

21.2.5. Species Specificity

It should be emphasized in this context that the phenomenon of sleep shows remarkable variations among different species. For example, when referring to sleep experiments in rats, it should be kept in mind that rats—unlike humans—sleep during the day, and the length of their non-REM–REM sleep cycle is distinctly shorter (about 15 min) than that in humans (90 to 100 min). Moreover, in rats the cyclic structure of sleep is frequently interrupted by periods of food intake, defecation, locomotion, and other behaviors. Considering that the immune system also shows distinct differences among species such as mice, rats, rabbits, and humans, it seems reasonable to doubt that findings in animals concerning the relationship between sleep and immune functions can be a priori generalized to humans.

This chapter, therefore, will also be a critical comparison of findings in animals and humans. Findings in humans will be highlighted although the majority of research on the sleep–immune issue has been done in animals.

21.3. Immune Functions during Undisturbed Sleep

Any systematic analysis of the interaction between the central nervous sleep process and immune function will begin with the state of immune functioning during regular nocturnal sleep. Is there a specific pattern of immune activity during nighttime sleep different from that during daytime wakefulness? While this is a fundamental question, it has to be kept in mind that a mere coincidence of immune changes in association with nocturnal sleep does not allow any conclusions to be drawn as to a causative role of sleep for these changes. Oscillators determining circadian rhythm may in parallel affect sleep–wake processes and immune functions. Whether a correlation observed between the sleep–wake rhythm and immune processes results from an influence of sleep on the immune system or, vice versa, an influence of immune factors on sleep has to be explored by experimental manipulation of either one of these processes. Nevertheless, temporal associations between sleep and immune processes as revealed under conditions of undisturbed regular nocturnal sleep represent the point of reference for such experiments.

21.3.1. White Blood Cell Counts

Most prominent changes are observed for white blood cell counts across the day (Haus, 1992). With regular sleep–wake schedules, the number of circulating leukocytes as well as granulocytes is at a maximum during the evening hours and declines during nocturnal sleep to reach a minimum during the early morning hours (see Fig. 10, dashed lines). A roughly similar dynamic with declining numbers of circulating cells across nighttime sleep has been revealed also for monocytes and the major lymphocyte subsets, including T-helper cells (CD4$^+$), T-suppressor cells (CD8$^+$), and B cells. In contrast, the number of NK cells as well as the cytotoxic activity of these cells reaches a minimum during the early hours of sleep, after which it steadily increases until the afternoon hours. The time course of NK cell numbers is a clear hint that cell counts within the various leukocyte subsets are subject to differential influences during nocturnal sleep.

The variations in blood cell counts across the sleep–wake cycle have been suspected to reflect an influence of glucocorticoid release on cell migration. Glucocorticoids are known to induce neutrophilia and to diminish lymphocyte counts. Minimum cortisol concentrations during the first half of the night, therefore, may contribute to the declining numbers of neutrophils across nighttime sleep. However, changes in cortisol would not explain the change in lymphocyte counts. Decreasing lymphocyte counts could result from reduced sympathetic activation during sleep (Ottaway & Husband, 1994). Yet, these mediating mechanisms remain to be confirmed experimentally.

A matter of controversial discussion is the functional significance of changes in white blood cell counts. Do decreasing numbers of leukocytes across nocturnal sleep reflect increased or decreased immune defense or neither? The decrease could result from margination of these cells sticking to the walls of the blood vessels or from an accumulation in the spleen. Alternatively, the cells may have completely left the blood compartment to enter extravascular spaces, lymphoid tissues, and the bone marrow. An accumulation of leukocytes and lymphocytes in lymphoid tissues could, in fact, speak for an enhanced cellular immune activity at these sites. However, the mechanism of leukocyte migration during nocturnal sleep is at present entirely obscure.

21.3.2. Cytokines

The state of immune defense during regular nocturnal sleep has also been described with regard to cytokine activity. Cytokines derived mainly from monocytes and macrophages, such as IL-1β, TNFα, and IL-6, as well as lymphocyte-derived cytokines such as IL-2 and IFNγ have been examined. Levels in serum for most of the cytokines are usually rather low and close to the detection limit of the assays available. This in part results from binding of the cytokines to soluble receptors. Nevertheless, the functional significance of cytokine levels in serum, which primarily exert paracrine actions in local tissues, presently remains obscure. An exception is IL-6, the concentration of which in systemic serum is considerably higher and which seems to also exert hormonelike actions via the bloodstream. However, the investigation of IL-6 concentrations across the 24-h cycle yielded inconsistent results. While in some studies enhanced IL-6 concentrations during sleep as compared with wakefulness were observed, this relation was not confirmed in several other studies (Born & Hansen, 1997).

Alternatively, the release of cytokines after *in vitro* stimulation of the sampled leukocytes with different mitogens has been assessed to indicate the state of immune defense during sleep. This method provides an estimate of cytokine production, i.e., of the capability of white blood cells to synthesize (and release) a certain cytokine on mitogen stimulation. In some experiments, production of TNFα and IL-1 was found to decrease during nocturnal sleep. However, because the primary source of these cytokines in blood are monocytes, this decrease could be attributed to the decline in the number of circulating monocytes across nocturnal sleep. Normalizing the production of IL-1 or TNFα to the number of monocytes stimulated *in vitro* with the mitogen did not reveal any change in the capability of monocytes to produce these cytokines during sleep (Born, Lange, Hansen, Mölle, & Fehm, 1997; see Fig. 11, ○). In contrast to these negative findings, the analysis of IL-2, which is predominantly produced by T-cell subsets, revealed rather consistent changes during nocturnal sleep. The production of this lymphokine (relative to the number of T-cells stimulated with mitogen) seems to be distinctly increased during nocturnal sleep compared with daytime wakefulness. Less consistent results were found in the evaluation of IFNγ produced mainly by the T-helper 1 cell (Th1) and T-suppressor cell subsets. Completely unknown at present is whether, and to what extent, the synthesis of cytokines derived from Th2 cells, such as IL-4 and IL-10, is also enhanced during nighttime sleep. Together, the available data suggest that during regular nocturnal sleep, the capability of T-cells to produce certain cytokines is enhanced while no such variation could be confirmed with regard to cytokines produced primarily by monocytes and macrophages. To obtain a more comprehensive picture, examination of cytokines characteristic for T-cell subsets seems promising. Also, examination of intracellular cytokine production (in combination with flow cytometry) across the 24-h cycle could prove a useful approach to distinguish those cells that actually produce a certain cytokine, thereby allowing for a clear-cut separation of changes in production and those related to redistribution and cell migration.

21.4. Regulation of Sleep by Immunological Factors

A basic and historically old view assumes that sleep is induced and maintained by some kind of factor that accumulates during wakefulness and the degradation and metabolization of which requires a particular type of organismic rest and brain state, i.e., sleep. Around the turn of this century, this view was strongly supported by fundamental experiments conducted by Legendre and Pieron (1913) in France and by Ishimori (1909) in Japan. Legendre and Pieron deprived dogs from sleep by activating and moving them for several days. When transferring cerebrospinal fluid from the sleep-deprived animals to normal dogs, the recipient dogs fell asleep for a prolonged period of time. From this observation they postulated that during sleep deprivation, the concentration of a "hypnotoxin" increases in cerebrospinal fluid, promoting sleep. However, at that time researchers could not identify the nature of the substance.

21.4.1. Infection and Sleep

A link between the putative sleep-promoting factors and immunological host defense mechanisms is suggested by several clinical observations. As early as the fourth century BC, Aristotle reported that during ongoing host defense responses, sleep is modified in humans. In some diseases, a disturbance of sleep–wake behavior is among the most prominent symptoms. Encephalitis lethargica, which is presumed to be a viral infection of the CNS, was epidemic at the beginning of this century and in more than 80% of the patients was associated with symptoms of severe hypersomnia or hyposomnia and changes in the distribution of sleep across the 24-h cycle. Mononucleosis resulting from infection with Epstein–Barr virus is associated with prolonged subjective sleepiness in about 11% of the patients. In a long-term study, Guilleminault and Mondini (1986) documented an increase in total sleep time and shortened sleep latency in these patients. These disturbancies of sleep did not improve with stimulant or antidepressive treatment.

Obvious increases in sleep time and SWS have been reported to be prominent symptoms in patients with African trypanosomiasis (sleeping sickness). The disease is induced by parasites (*Trypanosoma brucei*) that enter the brain. *In vitro*, they have been shown to stimulate the release of IFNγ from T cells, which in turn stimulates the release of IL-1β and TNFα from macrophages and astrocytes (Bentivoglio, Grassi-Zucconi, Olsson, & Kristensson, 1994). These cytokines are suspected to exert strong somnogenic influences (see below). However, more recent examinations failed to confirm an increase in total sleep time or SWS in these patients but indicated rather a severe disturbance of the distribution of sleep across the day, i.e., a primary disruption of the circadian sleep–wake cycle (Buguet, Bert, Tapie, Tabaraud, Doua, Lonsdorfer, Bogui, & Dumas, 1993). HIV is also associated with distinct changes in sleep. HIV infection has been reported to be accompanied by a characteristic increase in SWS dominating in the latter portions of sleep, even in patients still asymptomatic (Norman, Chediak, Kiel, & Cohn, 1990). However, subsequent studies failed to confirm this observation (e.g., Wiegand, Möller, Schreiber, Krieg, Fuchs, Wachter, & Holsboer, 1991). Also, it is nearly impossible to determine whether a change in the sleep pattern in these patients reflects a direct impact of the HIV virus or symptoms secondary to opportunistic infections.

Narcolepsy has been hypothesized to be mediated by an autoimmune process. Narcolepsy is an inheritable primary disorder of sleep characterized by excessive sleepiness and REM sleep-attacks associated with muscle atonia and cataplexy. Honda, Asaka, Tanaka, and Juji (1983) were the first to report that all of their Japanese narcoleptic patients tested carried the specific human leukocyte antigen (HLA) DR2, a finding that was later extended to Caucasians. The narcolepsy susceptibility marker was subsequently pinned down to the HLA-DQ allele *DQB1*0602*. The differences in HLA-DR alleles could be related to a changed production of cytokines, such as TNFα and IL-1β, in these patients (Hinze-Selch, Pollmächer, Wetter, Zhang, Lu, & Holsboer, 1994). Likewise, chronic fatigue syndrome has been considered to be of viral origin. The disease is characterized by complaints of daytime tiredness, myalgia, and disturbed, unrefreshing sleep. T-cell abnormalities have been described in these patients. However, their relevance for the sleep disturbance is unclear (Moldofsky, 1995).

Sleep alterations seen in conjunction with spontaneous infectious diseases stimulated the investigation of experimentally induced infections. A considerable number of these studies were conducted in rabbits by Toth and co-workers (Toth, 1995). After microbial inoculation, rabbits typically demonstrate an increase and a subsequent decrease in the amount and intensity of SWS (Fig. 4). The precise temporal pattern of this biphasic change in SWS varies substantially depending on the infecting microorganism. For example, gram-negative bacteria enhance sleep more quickly and for a much shorter period than do gram-positive bacteria or fungi. Sleep enhancements have also been observed following viral infection, for example following inoculation with influenza virus. After experimental infection with influenza virus, humans reported reduced sleep during the time of incubation, while sleep time increased with the appearance of clinical symptoms. Inoculation with rhinovirus subjectively increased the time spent asleep (Smith, 1992). Although in those experiments polysomnographical recordings of sleep were not performed, these studies are in line with those in animals suggesting that the occurrence of potential somnogenic effects of infections critically depends on the time after the infectious challenge.

21.4.2. Muramyl Peptides

The most direct evidence for an involvement of host defense mechanisms in the regulation of natural sleep was obtained by Pappenheimer and Karnovsky at Harvard Medical School. In seeking the "sleep factor" in goat cerebrospinal fluid, they detected a factor S that accumulated during sleep deprivation and when transferred to rats induced excessive non-REM sleep. Eventually, this group extracted over 10,000 brains from sleep-deprived rabbits and about 5000 liters of human urine to purify the putative sleep-promoting factor at quantities sufficient for amino acid analysis and mass spectrometry studies (Krueger & Karnovsky, 1995). Factor S was N-acetylglucosaminyl-1,6-anhydro-N-acetylmuramyl-

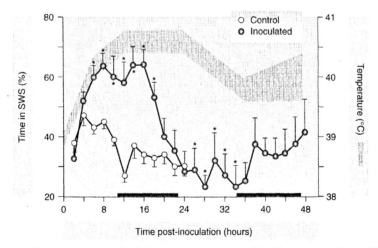

Figure 4. Sleep and temperature in *Staphylococcus aureus*-inoculated rabbits. The percentage of time in SWS in rabbits ($n = 16$) was evaluated for 24 h prior to (○) and 48 h after (●) the intravenous administration of 10^7 to 10^8 colony-forming units of *S. aureus*. To facilitate visual comparison, preinoculation data are superimposed over postinoculation values. Rectal temperature was measured prior to and every 6–12 h after inoculation (shaded area). Bars on abscissa indicate the lights-off period. Data indicate means ($\pm$ S.E.M.). *$p < 0.03$, relative to corresponding preinoculation values. From Toth (1995).

alanyl-glutamyl-diaminopimelyl-alanine, which belongs to the group of muramyl peptides (MP). The material derived from brain contained the same amino acids and amino sugars as that from urine (Krueger *et al.*, 1982). This outcome was surprising, as MPs are a component of bacterial cell walls and there is no known mammalian pathway for the synthesis of these substances. The findings led to the hypothesis that MPs are vitaminlike in that they are required, but not synthesized, by the mammalian host. Interestingly, muramyl dipeptide has been described as the simplest structure capable of replacing the mycobacterial component in Freund's complete adjuvant. Added to an antigen, Freund's adjuvant amplifies the immunization response. Based on this discovery, a number of synthetic MPs were developed, and the structural requirements of MPs for somnogenic activity could be refined (Krueger & Majde, 1990). A bacterial contribution to sleep regulation was further supported by findings of reduced sleep in decontaminated animals treated with antibiotics (Brown, Price, King, & Husband, 1990).

The question arises as to the mechanism by which MPs induce sleep. One assumption is that MPs act directly on the central nervous structures regulating sleep. Indeed, MPs appear to bind to serotonergic receptors. A prominent contribution of the serotonergic system originating from the dorsal nucleus raphe in the formation of sleep and SWS is well established. A direct action of MPs on brain structures is further supported by pharmacological studies indicating that the somnogenic effect of MPs is much stronger after intracerebroventricular administration than after intravenous administration.

21.4.3. Humoral Mediators of the Effects of MPs

Another mechanism by which MPs affect sleep derives from their potency to stimulate the release of various proinflammatory cytokines. This pathway has been extensively studied in animal models by Krueger's group at the University of Tennessee (Krueger, Takahashi, Kapas, Bredow, Roky, Guha-Thakurta, Novitsky, & Obál, 1995). Bacterial cell wall structures such as MPs as well as lipopolysaccharides (endotoxin) are sufficient and rather potent stimuli for the release of IL-1β, TNFα, and IL-6 from macrophages and monocytes. These cytokines are among the factors initializing the cascade of cellular immune defense. Their release, in turn, stimulates the proliferation of T cells and the production of T-cell-derived cytokines, such as IL-2 and IFNγ. Like MPs, IL-1β and TNFα were found to induce sleep and to markedly enhance SWS in animals following direct application to the brain and also when given intravenously (Fig. 5). Effects after administration of the cytokines emerged even more quickly than after MP administration, suggesting that the effects of MPs are for the greatest part mediated via release of IL-1β and TNFα.

An unresolved question is how IL-1β enters the brain to regulate sleep. In this context, various phenomena have been discussed such as carrier-mediated transport across the blood–brain barrier (BBB), the circumventricular organs which lack a BBB, as well as amplification mechanisms provided by endothelial cells at the BBB (Krueger & Majde, 1994) (see Chapter 12). Interestingly, the brain itself also produces IL-1 and many other cytokines. Astrocytes have been found to synthesize IL-1 as well as TNFα. Hence, rather than peripheral sources, the brain-borne cytokines could well play the more essential role in the promotion of sleep. Aside from IL-1 and TNFα, in the last few years several other cytokines have been demonstrated to be capable of inducing sleep in animals. Rather striking effects have been observed following administration of IFNα, which is most effectively induced by viral infectious agents in a great variety of cells in the body periphery and probably also in the CNS. IFNα and TNFα have been shown to induce the release of IL-1β, and through this action these cytokines may indirectly affect central nervous sleep.

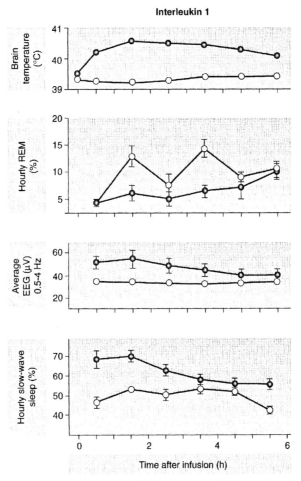

Figure 5. Brain temperature, hourly percentage of REM sleep, average EEG slow-wave amplitude, and hourly percentage of SWS in rabbits after intracerebroventricular injection of IL-1 (●) and of vehicle (artificial cerebrospinal fluid; ○). Means (± S.E.M.) are indicated. IL-1 induced fever, enhanced the duration of SWS and the amplitudes of EEG slow waves. REM sleep was inhibited. From Krueger (1990).

Brain receptors for proinflammatory cytokines such as IL-1β, TNFα, and IFNα have been detected in different structures some of which (e.g., the hypothalamus and hippocampus) are known to be involved in sleep regulation. The somnogenic action of IL-1β at these sites appears to involve secretion of growth hormone-releasing hormone (GHRH). Secretory activity of the GHRH–GH system is closely associated with the first episodes of SWS. GHRH-containing neurons are found in the arcuate nucleus and the ventromedial nucleus of the hypothalamus. While GHRH neurons of both the arcuate nucleus and the ventromedial nucleus project to the median eminence to control endocrine activity, neurons of the ventromedial nucleus also project to the basal forebrain. These neurons are probably involved in the regulation of sleep. Krueger *et al.* (1995) proposed that IL-1β triggers GHRH release from GHRH-containing neurons that, in turn, induces GH release from the pituitary gland and sleep via the basal forebrain. Consistent with this hypothesis is the observation that anti-GHRH antibodies block IL-1β-induced sleep as well as the associated fever response. The sleep promoting actions of IL-1β, TNFα, and IFNα could ultimately involve the synthesis of nitric oxide (NO). All of these substances as well as GHRH induce NO synthesis, the inhibition of which inhibits normal sleep as well as IL-1β-induced sleep. Interestingly, inhibition of NO does not abolish the fever response to IL-1β, indicating that these processes are mediated by different mechanisms. This dissociation is of considerable importance in understanding to what extent the proposed immunological mechanism could be relevant for the regulation of normal sleep, which is accompanied by a distinct decrease rather than increase in body temperature. Assuming an independence of the mechanisms inducing fever and sleep, the process promoting sleep during infection could be viewed as an amplification of the mechanisms regulating sleep under normal nonpathological conditions. This view is, in fact, challenging, as it implies that normal regulation of human consciousness is under control of microbes. The sleep-inducing mechanisms as proposed by Krueger and Majde (1994) and Krueger and Karnovsky (1995) are depicted in Fig. 6.

However, it has to be noted that the somnogenic activity of proinflammatory cytokines even in animals is much more complex than described above. Thus, in rats the sleep-enhancing effects of IL-1β depend on the time of administration and the dose (Opp, Obál, & Krueger, 1991). In those experiments, IL-1β exerted a distinct enhancing effect on sleep when given in the beginning of the active period, which is the night in rats. Given in the beginning of the light phase of the circadian cycle when rats sleep extensively, the cytokine rather inhibited sleep. Moreover, the promoting effect on sleep was restricted to lower doses of IL-1β, while the inhibitory effect was observed at high doses. The inhibitory effect has been ascribed to an activation of secretory activity within the hypothalamus–pituitary–adrenal system. Interestingly, a suppressing action on sleep has also been observed following administration of IL-10, the primary source of which are Th2 cells (Opp, Hughes, &

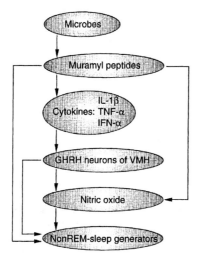

Figure 6. Schematic representation of the possible mediation of influences of muramyl peptides on sleep according to Krueger and Majde (1994). VMH, ventromedial hypothalamus.

Smith, 1994). IL-10 exerts inhibitory actions also on the production of IL-1β, TNFα, and IL-6 by macrophages and on NO synthesis.

21.4.4. Human Studies on the Effects of Endotoxin

Human studies have confirmed an influence of microbial products and cytokines on sleep. However, contrasting with the findings in rats, an inhibitory action of these substances prevailed. An accepted model to study the consequences of bacterial infections in humans is based on the induction of low-dose experimental endotoxemia. Endotoxins are lipopolysaccharides and are the dominant constituents of the outer cell membrane of gram-negative bacteria (Rietschel, Kirikae, Schade, Ulmer, Zähringer, Schreier, & Brade, 1994). Karacan, Wolff, Williams, Hursch, and Webb (1968) were the first to describe the effects of endotoxin (derived from *Salmonella abortus equi*) on sleep. In addition to endotoxin, that study also tested the effects of etiocholanolone (a pyrogenic steroid metabolite). Both substances increased body temperature to values close to 39°C, which were reached at about the onset of nocturnal sleep. Subjects complained of malaise and headache, and most of them had chills when temperature was rising. Sleep recordings indicated that both pyrogens increased the number and duration of awakenings, decreased the time in REM sleep as well as the time spent in stage 4 sleep.

Pollmächer and colleagues from the Max Planck Institute in Munich performed an elaborate series of studies that confirmed and extended these early results (Pollmächer, Schreiber, Gudewill, Trachsel, Galanos, & Holsboer, 1993; Pollmächer, Mullington, Korth, & Hinze-Selch, 1995; Korth, Mullington, Schreiber, & Pollmächer, 1996). In the first study, they evaluated influences of 0.4 ng/kg *Salmonella abortus* endotoxin on polysomnographic signs of sleep in 15 young healthy men. The antigen was administered at 19.00 h in the evening before sleep. Figure 7 plots the main changes in sleep, body temperature, heart rate, and plasma concentrations of cytokines and hormones. As expected, the bacterial antigen induced distinct increases in plasma concentrations of TNFα, IL-6, cortisol, and GH. Increases in body temperature were moderate and on average did not exceed 1°C. The time spent in SWS was not enhanced, but the time spent in the more shallow stage 2 sleep increased significantly after endotoxin administration. REM sleep and also time spent awake were decreased after endotoxin. Comparable changes were observed in depressed patients treated with endotoxin (Bauer, Hohagen, Bruns, Krieger, Lis, Riemann, & Berger, 1995). While plasma concentrations of IL-1β, TNFα, and IL-6 increased, REM sleep significantly decreased in these patients. The decrease in REM sleep can be attributed to the stimulation of cortisol release whose suppressing influence on this sleep stage is well documented.

Because observations in rats had indicated a dependence on the time of day for the somnogenic effects of IL-1β, Pollmächer's group also examined the effects of endotoxin on daytime sleep in humans. A placebo-controlled endotoxin challenge (0.8 ng/kg) was administered in ten volunteers at 9.00 h in the morning after a full night of sleep. The extent of host response as indicated by increases in body temperature and the release of cytokines and cortisol was similar to that observed in the first study. However, no differences were found between the effects of placebo and endotoxin in subjective sleepiness ratings made throughout the day, and also no differences were found in sleep onset latency, non-REM sleep stage 2, or any other sleep measures for any of the naps occurring during the day. The only sleep-related effect of endotoxin was a significant suppression of REM sleep in the first nap following endotoxin challenge (Korth, Mullington, Schreiber, & Pollmächer, 1996). Interestingly, higher peak responses of cortisol and IL-6 release to endotoxin were associ-

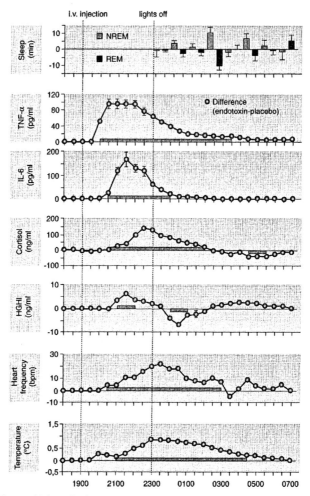

Figure 7. Influence of *Salmonella abortus equi* endotoxin in 25 healthy volunteers on non-REM and REM sleep, plasma concentrations of TNFα, IL-6, cortisol, and growth hormone (HGH), heart rate, and body temperature. The mean (± S.E.M.) differences relative to the effects of placebo are indicated. Data are collapsed across two subject samples receiving 0.4 ng/kg body weight (*n* = 15) and 0.8 ng/kg body weight endotoxin (*n* = 10). Asterisks and shaded areas indicate significant differences compared with the effects of placebo administration. From Pollmächer *et al.* (1995).

ated with reduced non-REM sleep during the first nap. While all of these findings indicate an impairment rather than improvement of sleep after host defense activation by endotoxin, there are preliminary data from one study indicating that very low nonpyrogenic concentrations of endotoxin, in parallel with the findings in animals, could enhance the time spent in SWS in humans (Mullington, Hermann, Holsboer, & Pollmächer, 1996a).

21.4.5. Human Studies on the Effects of Cytokines

First trials assessing the effects of cytokines on human sleep were conducted in cancer patients who were treated with high doses of TNFα, IFNα, or IFNγ. A common observation in these patients is a marked increase in lassitude and feelings of fatigue. In phase I clinical trials, similar observations have been made following administration of IL-1β. However, the extremely high doses used in those studies may not be comparable with endogenous release of these substances on normal infection. Consideration must also be given to the severe disease of the patients, which itself may induce sleepiness and sleep. Most importantly, those studies just examined subjective feelings of tiredness. It is well known from other diseases, such as major depression, that subjective tiredness is often a symptom of disturbed sleep. In fact, when evaluating the EEG activity in cancer patients treated with IFNα or IFNγ, we found signs of increased rather than decreased cortical arousal in these patients although they reported increased fatigue (Born, Spath-Schwalbe, Pietrowsky, Porzsolt, & Fehm, 1989).

Meanwhile, several studies have been undertaken in healthy humans to examine the influence of lower concentrations of cytokines on sleep. Two recent experiments assessed the effects of IL-6 and IFNα on sleep (Spath-Schwalbe *et al.*, 1998; Spath-Schwalbe, Perras, Fehm, & Born, 1999). IL-6 and IFNα were chosen because they play key roles in the mediation of the immune response to bacterial and viral infections. The substances were administered subcutaneously to healthy men at 19.00 h in the evening prior to the experimental night. The doses used (IL-6: 0.5 μg/kg body weight, IFNα: 1000 and 10,000 IU/kg body weight) were rather low and should induce increases in plasma concentrations mimicking those occurring during moderate bacterial and viral infections. IL-6 induced distinct feelings of tiredness in the men, as assessed by an inventory an hour before lights were turned off for sleep. Also, subjective self-reports indicated signs of sickness after IL-6. Overall, the cytokine did not affect SWS. However, a separate evaluation of the first and second half of nocturnal sleep revealed that it acutely suppressed SWS during the first half. This immediate inhibitory effect was compensated by a pronounced increase in the time in SWS during the second half of the night (Fig. 8). Thus, IL-6 shifted SWS from the first half of the night where it normally dominates, into the second half. REM sleep was reduced after IL-6, during the first as well as during the second half of the night. The effects on sleep were paralleled by a moderate increase in body temperature (averaging 1°C) and increased pituitary–adrenal release of ACTH/cortisol. The latter effect probably explains the loss of REM sleep after IL-6.

The essential effects of IFNα were strikingly similar to those of IL-6. Like IL-6, IFNα increased subjective signs of deactivation and fatigue. Moreover, IFNα acutely suppressed the time spent in SWS during the first half of the night, and the time in REM sleep across the entire night. Effects on REM sleep were restricted to the higher dose of IFNα. Contrasting with the effects of IL-6, the suppression of SWS following IFNα was accompanied by a distinct increase in GH release during early sleep. However, in parallel with the actions of IL-6, changes in sleep after IFNα were also associated with increased secretory activity within the pituitary–adrenal system. Part of the effects of IFNα could have been mediated by IL-6, the concentration of which in plasma was increased following the administration of IFNα.

Together, these findings indicate an acutely disturbing influence on sleep of the pro-inflammatory cytokines IL-6 and IFNα in healthy humans. With this outcome, the experiments, like those testing the effects of endotoxin in humans, warn against a premature generalization of psychoimmunological findings in mice, rats, and rabbits to humans. It

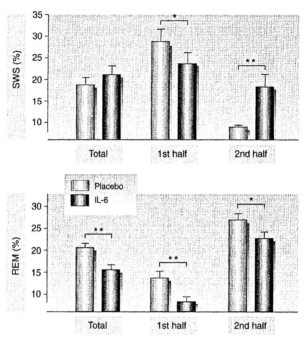

Figure 8. Mean (± S.E.M.) percent of time spent in SWS (top) and REM sleep (bottom) for the total nocturnal sleep period as well as separately for the first and second half of sleep after subcutaneous administration of IL-6 (0.5 μg/kg body weight; dark bars) and placebo (light bars) in 16 healthy men. *$p < 0.05$, **$p < 0.01$ for pairwise comparison between the effects of IL-6 and placebo. From Späth-Schwalbe *et al.* (1998).

could be argued that in rats a sleep promotion following cytokines such as IL-1 was observed mainly when the cytokine was administered in the beginning of the usual period of activity (i.e., the night) while an opposite effect can dominate when the cytokine is administered during the usual period of excessive sleep (i.e., the day). From that a corresponding dependence on the time of day could be expected also for the effects of cytokines on human sleep. Effects of cytokine administration on daytime sleep so far have not been investigated in humans. However, as mentioned above, endotoxin in humans failed to enhance sleep and SWS also when tested during the daytime, despite the fact that the microbial challenge at that time likewise induced distinct increases in plasma levels of proinflammatory cytokines (Korth *et al.*, 1996). Moreover, even if cytokines like IL-6 and IFN-α were found to increase SWS in naps over the day, such an effect would contradict a role for these cytokines in the regulation of normal sleep.

21.5. Regulation of Immune Functions by Sleep

In order to determine whether sleep causes changes in immune function, sleep must be manipulated experimentally. The most common approach is to totally deprive the organism of sleep for a certain time interval. In addition, a few experiments have examined influences of partial sleep deprivation of either the first or the second half of nocturnal sleep. Effects of the selective deprivation of certain sleep stages have not been examined, except in one study in rats yielding no effects after selective REM sleep deprivation (Benca, Kushida, Everson, Kalski, Bergmann, & Rechtschaffen, 1989). A general problem of sleep deprivation procedures is that with increasing time of deprivation, nonspecific effects of stress mediated by pituitary–adrenal secretion and activation of the sympathetic nervous system, can mask suspected effects of a mere lack of sleep. Stress effects seem to be even more disrupting in animals than humans, since the typical deprivation techniques applied to animals force the subject into a state of persistent activation, while participation in human sleep deprivation experiments is voluntary.

21.5.1. Animal Studies

The first systematic evaluation of the immunological consequences of total sleep deprivation in dogs was described by Manacaine (cited from Kleitman, 1963) at the end of the nineteenth century. She observed a decrease in white and red blood cells after a few days of sleep deprivation. Further extension of the deprivation period led to a dramatic decrease in body temperature before the animal died after about 1 week of sleep deprivation. The lethal effect of prolonged sleep deprivation has been subsequently confirmed in several animal studies. The deregulation of temperature homeostasis seems to be an essential factor precipitating this outcome. A most impressive series of experiments on this matter was conducted by Rechtschaffen and colleagues in Chicago (Benca *et al.*, 1989; Rechtschaffen, Bergmann, Everson, Kushida, & Gilliland, 1989) who deprived rats entirely from sleep for up to 32 days. While a fall in body temperature was consistently revealed in the end stage of the deprivation period, the changes in immunological measures (spleen cell counts, *in vitro* lymphocyte proliferation responses to mitogens, *in vitro* and *in vivo* plaque-forming cell responses to antigens) remained negligible. However, in an experimental replication of these studies, Everson (1993) reported that many of the rats that died after prolonged periods of sleep deprivation showed signs of infections and bacteremia. This observation, indeed, supports the view that prolonged sleep deprivation weakens host defense mechanisms.

As noted above, the extreme duration of the sleep deprivation employed in those studies represents a strong stressor that influences the immune system. The data, therefore, remain ambiguous as to a role of sleep in the regulation of immune function. However, complementing evidence for a supportive influence of sleep on immune defense derives also from studies that did not burden the animal with sleep deprivation. Most remarkable are the results from a study by Toth, Tolley, and Krueger (1993; Toth, 1995). They inoculated rabbits with different bacterial antigens (*Escherichia coli*, *S. aureus*), which led to a temporary increase in the time the animals spent asleep and, in particular, in SWS. Most important, however, was the correlation between sleep enhancement and survival rate revealed in this study. Long periods of enhanced sleep were associated with a more favorable prognosis and less severe clinical signs than were short periods of enhanced sleep followed by prolonged sleep suppression. Rabbits that eventually died exhibited less sleep than rabbits that survived the infection. Referring to studies by Norman *et al.* (1990) and Kubicki, Henkes, Terstegge, and Ruf (1988), Toth (1995) considered the changes observed in rabbits to bear some similarity with the picture in HIV-infected humans: Some of these individuals who are HIV seropositive but otherwise healthy appear to display excessive stage 4 SWS. However, sleep deteriorates and becomes disrupted as the disease progresses. Hence, good sleep may indicate efficient host defense. However, we must caution against concluding from the data of Toth and co-workers that enhanced SWS was the cause of an increased chance to survive infection in the rabbits. It could have been vice versa, that changes in SWS were a consequence of a more or less effective immune defense; i.e., compared with the surviving rabbits, the animals that eventually died developed more severe clinical symptoms in response to the antigen exposure which disrupted sleep and SWS.

The view of a beneficial influence of sleep on immune defense is further strengthened by a study performed at the University of Newcastle, Australia, by Brown, Pang, Husband, and King (1989). Those experimenters focused on the secondary immune response to infection with an influenza virus in mice. The mice were orally immunized with the virus, and challenged intranasally 1 week later. Some animals were deprived of sleep for a 7-h period immediately following challenge. Sleep was prevented by gently handling the mice or by softly tapping on the wall of the cage whenever an animal adopted a sleeping posture. This procedure was expected to minimize influences of stress. Three days after the challenge, virus clearance and virus specific antibody were determined in the lungs. Also, a control group of unimmunized mice was tested. Immunized mice that had normal sleep achieved total virus clearance within 3 days after challenge. Sleep deprivation in immunized mice totally abrogated this effect such that sleep-deprived animals behaved as though they had never been immunized (Fig. 9). Also, antibody concentrations in serum were reduced compared with the mice on regular sleep conditions. Interestingly, there was no difference in virus clearance and antibody concentration in unimmunized mice whether sleep deprived or not, indicating that sleep deprivation introduced immediately after the challenge did not itself have a direct effect on virus replication. Together, the study results show an optimizing effect of sleep on immune function. In addition, they suggest that the various stages of host defense responses are possibly not equally sensitive to the influence of sleep deprivation, as the detrimental effect of this procedure was significant only for the indicators of the secondary immune response but not for the primary response. However, the primary response usually develops rather slowly, with antigen concentrations peaking around 2 weeks after the primary challenge. Therefore, the measurement performed in this study 3

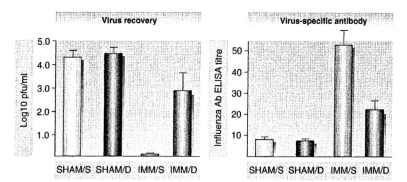

Figure 9.   Virus recovery in lungs (left) and virus-specific antibody in lung homogenate supernatants from mice at 3 days after intranasal challenge with influenza virus. Means ($\pm$ S.E.M.s) are indicated for four groups of five mice: SHAM/S, unimmunized normally sleeping mice; SHAM/D, unimmunized sleep-deprived mice; IMM/S, immunized normally sleeping mice; IMM/D, immunized sleep-deprived mice. Immunized normally sleeping mice differed from all other groups whereas immunized sleep-deprived mice did not significantly differ from the unimmunized mice. Modified from Brown *et al.* (1989).

days after the challenge may have been too early to reveal clear effects of sleep deprivation on the primary immune response.

21.5.2. Human Studies—White Blood Cell Counts

In experiments in humans, the duration of sleep deprivation introduced, with some exceptions, was in general less extreme. Mostly, effects on white blood cell counts and cytokines were studied. Surprisingly, results do not support the view of a generally impairing effect of sleep deprivation on host defense.

A most vigorous examination of the effects of a 64-h period of continuous wakefulness was conducted by Dinges, Douglas, Hamarman, Zaugg, and Kapoor (1995). With increasing time awake, the number of leukocytes steadily increased, which primarily reflected an increasing number of circulating granulocytes and monocytes. The number of NK cells, after a transient decrease, likewise increased toward the end of the vigil. Also, cytotoxic activity of NK cells increased during sleep deprivation. Consistent with other reports, the increase in NK cell counts during sleep deprivation developed more slowly. Moreover, following a night of recovery sleep, it did not immediately return to baseline levels, suggesting that these changes reflect a secondary effect of sleep deprivation that perhaps could be driven by slowly acting humoral factors. Overall, the results were interpreted in terms of an activation rather than suppression of nonspecific cellular immune defense during longer periods of sleep deprivation. While enhanced NK cell counts and NK cell activation have been likewise observed in response to different stressors, the increases in the number of granulocytes and monocytes during sleep deprivation cannot be easily explained by a stress effect.

Several sleep deprivation studies revealed changes diverging from those of Dinges's group. Depriving healthy young men from sleep just for one night, gross changes in the total number of leukocytes were not observed (Born et al., 1997). Likewise, Palmblad, Cantell, Strander, Froberg, Karlsson, Levi, Granstrom, and Unger (1976) failed to find any changes in leukocyte counts after a 76-h period of persistent wakefulness. These discrepancies can be attributed to methodological differences among the studies. First, just one night of sleep deprivation may be too short to induce discernible alterations in the distribution of white blood cells. Second, and more important, is the fact that changes in white blood cell counts during sleep deprivation ride on pronounced circadian oscillations in the numbers of these cells. This is illustrated in Fig. 10, which shows data from our own study that aimed to unravel the interaction between effects of sleep deprivation and circadian rhythm. For this purpose, blood cell counts were assessed in ten healthy men every 3 h over two consecutive 24-h cycles, starting at 8.00 h in the morning. In one condition, subjects were subjected to a regular wake–sleep schedule over the entire recording epoch. In the other condition, the men were deprived from sleep during the first night, which was followed by a night of recovery sleep (starting 23.00 h the following day). In both conditions, white blood cell counts and its major subsets displayed pronounced circadian rhythms. Acutely, sleep deprivation enhanced the number of circulating monocytes, NK cells, and lymphocytes including B cells, T-helper cells (CD4$^+$), T-suppressor cells (CD8$^+$), and activated T cells (HLA-DR$^+$). However, in the afternoon and evening following sleep deprivation, the numbers of NK cells, lymphocytes, as well as the numbers of all T-cell subsets measured were lower than during the regular wake–sleep condition.

Thus, compared with nocturnal wakefulness, sleep appeared to exert an acute suppressing effect on the numbers of these cells, thereby introducing distinct phase shifts in the circadian rhythms. Because of these shifts in phase, depending on the time of blood

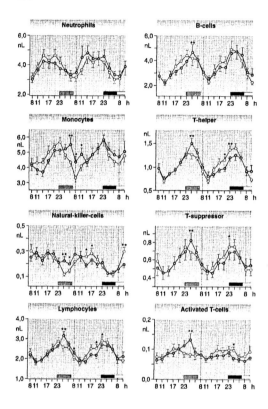

Figure 10. Mean (± S.E.M.) numbers of (left) circulating neutrophils, monocytes, natural killer cells, and total lymphocytes, and (right) of B cells, T-helper cells (CD4+), T-suppressor cells (CD8+), and activated T cells (HLA-DR+) during two 51-h experimental sessions (WS-WS and WW-WS). The WS-WS condition (dashed lines) included two regular wake–sleep cycles. During the WW-WS condition (solid lines) subjects were deprived of sleep during the first night and recovered sleep during the second night. Time in bed (horizontal bars) started at 23.00 h, and on the first night ended at 7.00 h. On the second night, subjects were allowed to sleep ad libitum, and stayed in bed until 11.00 h. $n = 10$. **$p < 0.01$, *$p < 0.05$, for pairwise comparisons between the effects of both conditions. From Born et al. (1997).

collection, cell counts during the condition of sleep deprivation could be either higher, comparable, or lower than during the condition of a regular wake–sleep cycle. For example, as shown in Fig. 10, the number of lymphocytes was significantly higher at 2.00 h during the night of sleep deprivation than during regular nocturnal wake–sleep conditions, comparable in both conditions between 11.00 and 17.00 h the following day and, thereafter, between 20.00 and 23.00 h in the evening, significantly lower in the condition of sleep deprivation. Given the obvious shifts in phase of the circadian rhythm after sleep deprivation, results obtained from studies in which blood was collected only once or twice a day, necessarily remain inconclusive. This accounts likewise for studies that, in addition to blood cell counts, measured proliferative functions of these cells after stimulation with different mitogens. Thus, signs of proliferative capabilities of lymphocytes after *in vitro* stimulation with phytohemagglutinin have been found to be increased, unchanged, or decreased under conditions of sleep deprivation as compared with regular sleep–wake conditions, in three different studies (Dinges *et al.*, 1994; Moldofsky, Lue, Davidson, Jephthalh-Ochola, Cara-yanniotis, & Gorczynski, 1989a; Moldofsky, Lue, Davidson, & Gorczynski, 1989b).

A further important result from the above described examination of the interaction between effects of sleep deprivation and circadian rhythm was that the average count across the total 24-h cycle was not influenced by sleep versus sleep deprivation for any of the leukocyte subsets evaluated. Hence, while the study revealed an acute suppressing influence of sleep on lymphocyte counts including all major subsets, this is apparently completely compensated during the following day.

To conclude from the acute suppression of lymphocyte counts that there is an impairing effect of sleep on immune function, is certainly premature. The changes in cell counts, in the first place, indicate a redistribution of these cells among different tissues. The direction of these migratory effects is at present unclear. But, it may well be that the decrease in immune cells in circulating blood is related to an accumulation of these cells in extravascular lymphoid tissues that could function to enhance processes of restoring an ongoing immune defense in these tissues.

21.5.3. Human Studies—Cytokines

Results concerning the effects of sleep and sleep deprivation on cytokines remain conflicting. The measurement of cytokine concentrations in plasma revealed ambiguous results, probably because the serum concentrations of most cytokines in healthy subjects are usually rather low and close to the threshold sensitivity of the assay. More consistent changes were observed when the production of cytokines after *in vitro* mitogen stimulation of cell samples was evaluated. The production of cytokines derived from monocytes and macrophages after stimulation with lipopolysaccharides such as IL-1β and TNFα was found to be increased following total as well as partial sleep deprivation. Notably, increases in these cytokines were also observed in aged persons and in certain psychiatric diseases, e.g., in schizophrenic patients, known to display severe disturbances of sleep. These findings were taken to suggest that not only acute but also subchronic states of sleep deprivation can lead to an activation of nonspecific (i.e., T cell independent) mechanisms of immune defense.

However, the influence of sleep and sleep deprivation on the production of IL-1β and TNFα disappears when cytokine production is measured relative to the number of mono-cytes stimulated with LPS (Fig. 11). Therefore, the decrease in the availability of these cytokines during sleep is very likely a result of a decreased number of circulating monocytes representing the main source of these cytokines. The production process per se does not seem to be affected by sleep (Born & Hansen, 1997).

A different picture evolved from the analysis of cytokines produced mainly by lymphocytes (Fig. 11). Sleep deprivation was found to suppress the production of IL-2 in several studies, and a similar effect was observed even with partial deprivation of just 3 to 4 h of sleep (McClintick, Costlow, Fortner, White, Gillin, & Irwin, 1994; Uthgenannt, Schoolmann, Pietrowsky, Fehm, & Born, 1995). An inhibiting effect of sleep deprivation and, vice versa, a relative stimulation of IL-2 production during nocturnal sleep have been confirmed also when normalizing the production of IL-2 to the number of stimulated T cells as an estimate of cells principally producing this cytokine. The effects concerning the production of IFNγ were less consistent. Some early studies indicated an increased produc-tion of IFNγ during sleep deprivation (e.g., Palmblad *et al.*, 1976). However, others failed. Curiously, increases in the production of IFNγ were also observed during recovery sleep after prolonged periods of wakefulness. Thus, the production of IFNγ is probably also affected by sleep and sleep deprivation. However, compared with the influences on IL-2, those on IFNγ appear to be inert, developing on a longer time scale and, perhaps, biphasically. Further research is needed to clarify the sensitivity of IFNγ production to the influences of sleep, and also of other T-cell products, like IL-10 and IL-4, which, so far, have not been thoroughly examined in the context of sleep deprivation. An additional examina-tion of these cytokines seems promising, as it could provide hints concerning a possibly

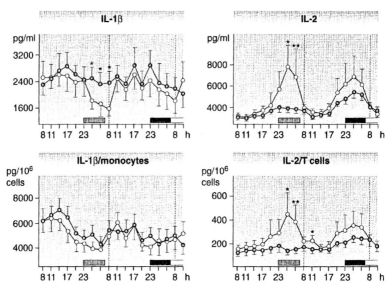

Figure 11. Mean (± S.E.M.) production of IL-1β (left) and of IL-2 (right) during two 51-h experimental sessions (WS-WS and WW-WS). The WS-WS condition (○) included two regular wake–sleep cycles. During the WW-WS condition (●) subjects were deprived of sleep during the first night and recovered sleep during the second night. The top panels show absolute cytokine production. In the bottom panels, production of IL-1β was determined relative to the number of monocytes stimulated with lipopolysaccharide and production of IL-2 was determined relative to the number of T cells stimulated with phytohemagglutinin. Time in bed (horizontal bars) started at 23.00 h, and on the first night ended at 7.00 h. On the second night, subjects were allowed to sleep ad libitum, and stayed in bed until 11.00 h. $n = 10$. $**p < 0.01$, $*p < 0.05$, for pairwise comparisons between the effects of both conditions. Modified from Born et al. (1997).

more selective influence of sleep on functions related to specific T-cell subsets. IL-4 and IL-10 are produced mainly by Th2 cells, considered to support preferentially antibody-dependent immune defense, while supportive actions of Th1 cells releasing IFNγ (in addition to IL-2) concentrate on cell-mediated immune responses via macrophages, NK cells, and cytotoxic T cells. Via the release of IFNγ and IL-10, respectively, Th1 and Th2 cells exert a mutual inhibitory effect on each other, thereby setting up a balance between Th1- and Th2-directed responses. Sleep and sleep deprivation could shift this balance to either one side or the other. Regarding this issue, the observed increase in the production of IL-2 during sleep remains inconclusive, because IL-2 is a product of a greater variety of T-cell subsets. Therefore, this increase in IL-2 suggests no more than a general facilitating influence of sleep on all T-cell-mediated mechanisms of immune defense.

The observation of an increasing influence of sleep on the production of IL-2 contrasts with findings reported by Brown, Husband, King, and Bull (1992). To gain a clearer understanding of the effect of sleep loss on the activated immune response, those experimenters tested healthy humans vaccinated with a *Salmonella typhi* mutant (Typh-Vax vaccine). Subjects were orally administered a capsule containing 109 live *S. typhi* mutant on the morning prior to the sleep deprivation night. Thereby, antigen challenge was separated from the next sleep period by more than 24 h. Seven days after the challenge, the sleep-deprived subjects showed a surge of IgA and IgG antibodies against *S. typhi* in saliva as well as enhanced mitogen-driven secretion of IgM. In another study by Mullington, Hermann, Holsboer, and Pollmächer (1996b), host response to endotoxin administered at 23.00 h did not depend on whether the subjects remained awake or slept the following night. These findings suggest that the immune response to microbial antigens like *S. typhi* and endotoxin is not affected or even enhanced by sleep loss, but in any case not impaired. However, this outcome may be restricted to a critical period, namely, immediately after initial challenge, where T-cell-independent mechanisms prevail. Whether sleep deprivation still proves to be beneficial (or just ineffective) when introduced at later stages of an ongoing immune process (with prevailing T cell activity) must be doubted considering the distinct suppressing influence of sleep loss observed on IL-2 production.

21.6. Concluding Remarks

This chapter summarized findings concerning the interaction between central nervous sleep and immune function. Together, the studies allow for preliminary conclusions only. This is related in part to the, by and large, small number of studies that have been devoted to this issue, and also to some serious methodological shortcomings in these studies. Nevertheless, the body of available data provides sufficient evidence for two major conclusions, namely, the immune system can affect central nervous sleep processes, and sleep affects immune functions.

Apart from these more general statements, research in this field has yielded a number of contradictory findings that are difficult to integrate. These discrepancies prevent any decision as to the central questions of whether immune activation *promotes* sleep and whether sleep *promotes* immune defense. A more complex interaction is possible, with an incompatibility of sleep and immune defense during the early stages of coping with infectious challenge and a mutually supportive influence during later stages of immuno-logical coping.

The most obvious discrepancies are between findings in animals, mostly rodents, and in humans. In animals, the stimulation with bacterial antigens, MPs, endotoxins, as well as the administration of various proinflammatory cytokines including IL-1β, TNFα, and IFNα, has been found to enhance sleep and in particular SWS. In humans, administration of endotoxin at pyrogenic doses impaired sleep. Correspondingly, the administration of IL-6 and IFNα led to an acute suppression of SWS and REM sleep, although subjects reported increased fatigue. Studies in animals supported an impairing influence of sleep deprivation on host defense mechanisms and which eventually can be lethal. In humans, sleep deprivation was found to enhance numbers of circulating monocytes, NK cells, lymphocytes and the production of specific antibodies against *S. typhi*, suggesting a facilitation of certain immune functions after sleep loss.

It is worth noting, however, that the conditions and parameters examined in the human and animal studies in most cases are not fully comparable. Therefore, the divergent outcomes from those studies, in the first instance, argue against a generalized and ubiquitous effect in that microbial antigens and cytokines exert an unconditioned promoting influence on sleep. Vice versa, the data exclude that sleep exerts a ubiquitous supportive effect on any aspect of immune functioning. For example, as mentioned above, sleep-enhancing effects of IL-1β in rats were observed primarily with tests taking place during the regular period of increased activity while testing during the usual hours of rest rather revealed an inhibiting effect on sleep. This latter finding indeed fits with the finding in humans of an impairing effect of IL-6 and IFNα on regular nocturnal sleep.

Rather than a ubiquitous enhancing influence, the immune system may exert differential effects on sleep depending on the stage in the cascade of defense reactions considered and, consequently, on the timing with reference to the initial immune challenge. For example, administration of IL-6 in humans acutely suppressed SWS during the first half of nocturnal sleep, but significantly enhanced SWS during the second half. This latter increase, in fact, fits findings in animals of an improving influence of proinflammatory cytokines on sleep. Hence, it could be that given earlier, IL-6 in humans increases SWS already during the first half of the night, and similarly may account for the effects of an infectious agent. Thus, while acute host defense appears to be accompanied by an immediate disturbing influence on sleep, later processes associated with prevailing activation of T and B cells, and with the formation of immunological memory, could still enhance sleep in humans.

Correspondingly, the detrimental influence of sleep deprivation on host defense could selectively pertain to later T and B cell interactions involved in the formation of specific antigens and immunological memory. This view would be consistent with findings in animals where after extended periods of sleep deprivation signs of infection increased. Also, with shorter periods of sleep deprivation immediately after challenge, the secondary immune response, as indicated by rates of virus clearance, was found to be impaired. In contrast, the primary immune response in unimmunized animals remained unchanged. In humans, a comparatively short period of sleep deprivation of one night introduced immediately after vaccination (with *S. typhi*) was found to even accelerate specific antigen formation. Also, acute sleep deprivation increased numbers of circulating monocytes, lymphocytes, and NK cells, as well as NK cell cytotoxicity. On the other hand, production of IL-2 by T cells was found to be suppressed during nocturnal sleep deprivation.

Based on these data, it seems tempting to propose that, in humans, initial stages of immune defense associated with prominent release of IL-1β, TNFα, and IL-6 from macrophages and monocytes, and strong NK cell activation, disturb sleep, whereas sleep loss conversely facilitates these early host defense processes. At later stages of host defense, the sleep–immune interaction may change into the opposite, i.e., sleep could facilitate the formation of specific antibodies and immunological memory resulting from more inert T and B cell interactions. The cytokine release associated with these processes, conversely, could promote sleep. However, as far as human studies are concerned, evidence has been provided almost exclusively for an incompatibility of sleep and acute host defense. Whether specific immunological processes subsequent to T cell activation are accompanied by sleep enhancement and, conversely, whether such processes are facilitated by sleep in humans, is at present mostly unknown. Thus, common sense currently remains confronted with the failure of scientific research to reveal any beneficial effect of human sleep on the recovery from disease.

ACKNOWLEDGMENTS. Thanks go to A. Otterbein and Dr. L. Marshall for their valuable help in preparing the manuscript.

22   Functional Relationship between the Olfactory and Immune Systems

Frank Eggert and Roman Ferstl

22.1. Introduction

The development of psychoneuroimmunology has changed our view of the immune system, which has been thought of as an autonomous system within the organism. Results from psychoneuroimmunology have brought us to the conclusion that the immune system interacts with the neural as well as the endocrine system. Therefore, the immune system is now understood as being an integrative part of the organism acting on other systems and being influenced by them.

The immune system of vertebrates is based on specific molecular mechanisms, allowing the identification of and response to potentially harmful pathogens. Because somatic recombination within the anticipatory immune system of vertebrates also gives rise to immunocompetent cells that react with self-epitopes, a mechanism of self-tolerance has evolved. Thus, the self–nonself discrimination is one of the primary necessities for the proper functioning of the immune system. The molecular basis of this self–nonself discrimination has been particularly analyzed in experimental and clinical transplantation medicine.

Studies have shown that the immunogenetic differences between organisms are also associated with odor differences between these organisms. Thus, in order to understand some basic features of the immune system of vertebrates, the behavioral ecology of

Frank Eggert and Roman Ferstl   •   Institute for Psychology, University of Kiel, D-24098 Kiel, Germany.

organisms needs also to be considered (Eggert, Luszyk, Westphal, Müller-Ruschholtz, & Ferstl, 1990).

A pivotal role of self–nonself discrimination in the immune system is played by trans-membrane glycoproteins, which are expressed on all somatic cells. These glycoproteins are coded by genes within one gene cluster, the major histocompatibility complex (MHC) (see Chapter 2). The proteins coded by genes within the MHC are involved in the deletion of developing autoimmunoreactive cell lines, which takes place in the thymus, thus establish-ing the necessary self-tolerance of the immune system. Furthermore, MHC proteins are also the proteins that present antigenic epitopes to the immunocompetent cells, in order to initiate an immune response.

The MHC shows a remarkable genetic diversity, as found in nearly all vertebrate spe-cies studied so far. The extraordinary genetic diversity shows up at four different levels:

1. There are a number of functionally equivalent loci within the MHC.
2. At each of these loci there is a remarkable allelic polymorphism.
3. The loci are codominantly expressed.
4. In natural populations there is a remarkable high level of heterozygosity at these loci.

This genetic diversity of the MHC makes it a primary candidate for a genetic basis of a marker of individuality. Markers of individuality have been demonstrated in a huge number of species, many of them using chemosensory signals in order to identify other members of the same species; preliminary results demonstrate genetically determined chemosensory markers of individuality, even in humans (Ferstl, Eggert, Westphal, Zavazava, & Müller-Ruchholtz, 1992). Because a marker of individuality has to be stable over longer times, a genetic basis of such a marker has been proposed (Beauchamp, Yamazaki, & Boyse, 1985).

In addition to the identification of individual species members, a signal based on the MHC would also allow the identification of different degrees of genetic relatedness (kin recognition). This information could then be used to avoid inbreeding (Brown & Eklund, 1994). These mechanisms of inbreeding avoidance could be involved in the maintenance of the genetic diversity of the MHC. However, in order to play a role in these behavioral mechanisms, the MHC has to be phenotypically expressed in a way that allows other members of the species to evaluate the specific MHC type present. Furthermore, differences in the MHC should also be associated with differential behavior, e.g., mating preferences between MHC-different organisms.

Research about the chemosensory expression of immunogenetic differences in the MHC thus is centered about two questions. First, are there signals associated with the MHC, and second, do these signals influence behavior in conspecifics? To answer these questions a closer look at the MHC is necessary.

22.2. The Major Histocompatibility Complex

The MHC codes for transmembrane glycoproteins which are the basis of the self–nonself discrimination of the immune system (Roser, Brown, & Singh, 1991; Srivastava, Ram, & Tyle, 1991). Initially, these proteins were studied because of their clinical relevance in transplantation medicine.

Based on differences in their function and their patterns of expression, two classes (I and II) of MHC proteins have been identified. Each of the MHC proteins is a dimer consisting of two polypeptides coded by genes of subclasses (a) and (b).

Class I proteins are expressed on the surface of somatic cells of adult vertebrates and their density differs in different tissues. In mice, these proteins are also expressed on erythrocytes. In contrast, class II proteins are only expressed on B lymphocytes and in some vertebrate species also on activated and virus-infected T lymphocytes.

The immunological function of MHC proteins, called *MHC restriction*, can be described as restricting antigen recognition by T lymphocytes. T lymphocytes recognize foreign antigens by specific T-cell receptors (TCRs), which recognize a foreign antigen only when presented by an MHC protein. T-helper cells (CD4$^+$), which support proliferation of B lymphocytes by the formation and sensitization of antibodies, recognize a foreign antigen only if it is bound to a class II protein on antigen-presenting macrophages. In contrast, cytotoxic T cells (CD8$^+$) recognize foreign antigen only if it is presented by an MHC class I protein (see Chapter 2). With these mechanisms the MHC is controlling to some extent immune recognition ability.

Besides their role in the recognition of foreign antigens, the MHC proteins also interact with products of the rearranged TCR genes in the developing thymus and restrict the diversity of antigen recognition sites of TCRs. This restriction gives rise to a selection of T-cell clones capable of recognizing foreign antigen if it is presented by MHC proteins, but incapable of recognizing self-epitopes.

The class I proteins are classically described as membrane-bound proteins but they can also be found in a soluble form in the body fluids. For example, this soluble form has been found in serum, lymph, and urine of rat inbred strains. Soluble class II proteins have only been found in some species, e.g., in mouse inbred strains, but not in rats. In humans, there are certain alleles (*A9; A23; A24*) expressed in large amounts in the soluble form. Genetic systems similar to the MHC have been found in all vertebrates studied so far. In mice the MHC is *H-2*; in rats, *RT1*; and in humans, *HLA* (Fig. 1).

Under normal conditions there is a high level of heterozygosity in the MHC genes, depending on a remarkable allelic polymorphism. Mice have approximately 100 different alleles at each locus and it is believed that the genetic diversity in other species is comparably high. The MHC is one of few examples of loci with dozens of different alleles. The genetic diversity in the MHC is further enhanced by the codominant expression of these loci and by differences in the patterns of expression. This high level of heterozygosity ensures a unique MHC type for individuals, monozygotic twins being the exception. Although the genetic diversity in the MHC has been extensively studied, the phylogenetic development

MHC
Major histocompatibility complex

- Gene cluster
- Codes for proteins involved in cell-cell interactions
- Proteins play a major role in immunological processes
  - Presentation of antigens
  - Induction of self-tolerance
- Basis of somatic self/non-self discrimination
- Has been found in all vertebrates studied so far
  - *H-2* = MHC of the mouse
  - *RT1* = MHC of the rat
  - *HLA* = MHC of humans

Figure 1. The major histocompatibility complex.

MHC
Major histocompatibility complex

- Remarkable genetic diversity
  - A number of functional loci
  - High level of allelic polymorphism
  - Codominant expression
  - High level of heterozygosity

- Selective forces
  - Balancing parasite- or pathogen-driven selection
  - Negative frequency-dependent selection
  - Overdominance

Figure 2.   Immune and population genetics of the MHC.

as well as the selection forces that maintain this high level of diversity are currently under debate.

It is generally assumed that the diversity in the MHC genes is maintained by a balancing parasite-driven selection (Potts & Wakeland, 1990) that includes two mechanisms: frequency-dependent selection by which rare alleles are favored, and overdominance by which heterozygosity is favored (Fig. 2).

A number of studies have meanwhile shown that the expression of odor signals in mice, rats, and humans is associated with the MHC. Therefore, it has been proposed that these mechanisms of natural selection have favored the development of a mechanism of sexual selection that is involved in the maintenance of the genetic diversity in the MHC (Potts & Wakeland, 1993; Fig. 3). Such a mechanism implies the existence of discriminative

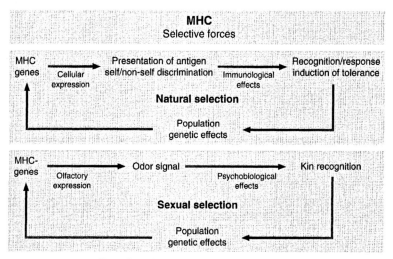

Figure 3.   Selective forces operating on the MHC.

stimuli that can be used to evaluate the MHC type of a given organism by its conspecifics. Such stimuli would not only give information about individuality but also about genetic relatedness (kin recognition). Such discriminative stimuli could be of enormous importance for the evolved social behavior of a species. Lewis Thomas (1975) was the first to propose that chemosensory signals associated with the MHC might function as discriminative stimuli. His proposal about the phylogenetic origin of the MHC led to the conclusion that the phylogenetically older function of MHC genes was to secure identity and integration of an organism.

A system that is phylogenetically based on cell–cell interactions in multicellular organisms allows for functions such as tissue differentiation and the self–nonself discrimination, especially the self–nonself discrimination that developed into the complex immune functions of recent vertebrates. Thomas (1975) postulated that this evolutionary origin of the MHC genes is still important and can be found in the recent function of providing each organism with an individual-specific chemosensory signal. This hypothesis has been supported empirically by a number of studies in which MHC-associated odor signals have been demonstrated in mice, rats, and humans.

22.3. MHC-Associated Odor Signals

The empirical proof that MHC-associated odor signals are present in mice was gained using genetically defined inbred strains. Inbred strains are animals that are genetically uniform and homozygous. A pair of inbred strains differing only with regard to a certain gene or part of a certain chromosome is called a pair of *congenic inbred strains*. A number of congenic inbred strains of mice are available; some of them differ only in the *H-2* (*H-2*-congenic inbred strains), whereas others do not differ in the *H-2* but in a number of other loci (non-*H-2*-different inbred strains).

Differences between *H-2*-congenic inbred strains can be traced back to the only genetic difference between them (the *H-2*). If one controls in an experimental setting other factors such as age, diet, and so forth, the difference observed between strains can be traced back to the genetic influences on the MHC. To examine whether the MHC in mice, rats, and humans is associated with specific odor signals, a number of different paradigms have been used (Fig. 4):

1. Y-maze and olfactometer: In both of these paradigms, animals are trained to distinguish the odors of two congenic inbred strains. The odor source could be either the whole animal, or just urine samples of these animals. The water-deprived animals are trained by reinforcing them for correct discrimination by a drop of water. Transfer-of-training tests are used to record their behavior to previously unknown stimuli, thus allowing study of the similarity between odor samples. The major difference between the two paradigms is the simultaneous presentation of odors in the Y-maze versus the successive presentation in the olfactometer.

2. Habituation–dishabituation paradigm: In this paradigm, rats are presented successively with nine odor samples recording their explorative behavior toward these samples. Following three presentations of water the first odor sample is presented three times, followed by three presentations of the second odor sample.

Olfactory expression of the MHC

Empirical evidence

Species studied	Odor source	Paradigm	Indicator species
	Animal, urine, feces	Y-maze	Mouse
		Mating preferences	Mouse
		Pregnancy block	Mouse
		Field study	Mouse
		Olfactometer	Rat
		Pair comparison	Humans
	Urine	Habituation-dishabituation	Rat
		Olfactometer	Rat
	Urine, person	Olfactometer	Rat
		Field study	Humans
		Experiment	Humans

Figure 4. Evidence of MHC-associated odors.

During the three presentations of the first odor sample the rats show habituation to this stimulus and do not explore it any further. If the second odor sample can be distinguished from the first one by the rats, dishabituation can be observed, otherwise the rats remain habituated until the last odor presentation.

3. Mating preferences: In this paradigm, mate choice of a male mouse (stud) is observed. The male mouse is presented with two female mice from two *H-2*-congenic inbred strains. Both female mice are in estrus. Mate choice is defined as successful copulation of the male with one of the females. After copulation the other female is removed and tested with a new male alone; only when copulation takes place is the trial scored valid.

4. Pregnancy-block paradigm: If an inseminated female mouse is removed from the male stud and presented with a new male, pregnancy is aborted in a higher proportion as if the female is presented with the same stud again. This so-called pregnancy block or Bruce effect is the result of induction of estrus by the foreign male and depends on chemosensory signals from this male. In the studies that

used this paradigm, a female mouse was inseminated. It is then presented with a new male (or its urine) from the same strain, or a male from an *H-2*-congenic strain. Differences in abortion rates indicate differences in the odor signals from the two males.

Based on observations during a breeding program of inbred strains, Yamazaki and his co-workers at the Monell Chemical Senses Center in Philadelphia (Yamazaki, Beauchamp, Bard, Boyse, & Thomas, 1991) demonstrated in a first series of experiments MHC-associated mating preferences in a number of inbred strains of mice. The chemosensory basis of this phenomenon was then examined in more detail. In a first attempt, mice were trained in a Y-maze to determine whether they are able to distinguish MHC-congenic strains by olfactory cues alone. As these experiments led to positive results, it was concluded that the MHC indeed is associated with specific odor cues that allow conspecifics to evaluate the MHC types present.

Such an association between the MHC and olfactory signals appears not to be restricted to mice alone. Richard Brown and his co-workers from Dalhousie University and from the University of Cambridge were able to show that MHC-associated odor signals are present in rats using the habituation–dishabituation paradigm with MHC-congenic inbred strains of rats (Brown, Singh, & Roser, 1987).

To demonstrate the existence of MHC-associated odors in humans requires a research strategy different from the one used with mice and rats, as of course no MHC-congenic groups are available in humans. In one of our experiments, a group of human subjects homogeneous with regard to the MHC was compared with an MHC-heterogeneous group. We examined whether in the first group a common odor cue is present, indicating that homogeneity with regard to the MHC is associated with homogeneity in the odor signals produced. Adopting experimental designs that have been used in research about concept formation, we developed a cumulative training and testing procedure. Rats were trained in an olfactometer to distinguish urine samples from the two groups, and previously unknown samples were tested in transfer-of-training tests. Stimulus generalization occurred after training with a number of different samples from the two groups, indicating the presence of a common odor cue within the samples from the homogeneous subjects. Replications of the experiments confirmed these results and led to the conclusion that MHC-associated odors are also present in humans (Ferstl *et al.*, 1992).

22.4. Olfactory Expression of the MHC

How MHC-associated odors are expressed is still a matter of discussion. Integrating the evidence so far, we assume that the primary gene products of the MHC from different tissues show up in the body fluids and are transported to the body surface where they undergo degradation. By means of a still unknown process the information of the specificity of the MHC proteins is transformed into specific profiles of volatile substances, giving rise to the odor signal (Fig. 5).

Studies with bone-marrow-transplanted *chimeras* identified the hematopoietic system as one of the tissues involved, but also showed that other tissues must be involved in the odor production. In immunological studies, soluble MHC proteins have been identified in different body fluids of different species. They are excreted into the urine where they undergo rapid degradation. First attempts to chemically characterize the profile of volatile substances constituting the odor signal were successful in showing specific differences in

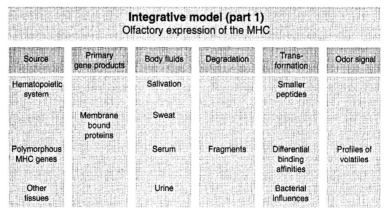

Figure 5. A model of the olfactory expression of the MHC.

these profiles in MHC-congenic strains of mice. However, the mechanism that links the degraded moieties of MHC proteins with the profiles of odorous substances remains unknown.

Using a number of approaches, investigators currently try to analyze the mechanisms including selective binding of odorant substances to MHC proteins, individual specificity of colonizing bacteria, and studies in which MHC proteins are incubated with bacteria occurring regularly on the surface of the human body. Besides these attempts, chemical analyses are used to characterize the odor signal in more detail in an effort to understand how these signals are generated.

One of the major problems of the chemical analysis, especially of human odor signals, is the degree of variability that can be observed in the profiles produced. Using urine samples gathered over 24 h, a relatively stable individual-specific profile is obtained. If these pooled samples are compared between subjects who are homogeneous with regard to the MHC, the analysis of MHC-specific components is possible by use of gas chromatography techniques.

The studies so far performed with this technique showed that only a limited number of substances are associated with the MHC type. A very similar pattern of results is obtained from studies with inbred strains of mice. Currently, studies are being conducted to analyze whether similar patterns found in urine can also be detected in other body fluids such as the perspiration of humans.

22.5. Psychobiological Effects of the MHC

The profiles of volatile substances constituting the odor signal are processed via the main olfactory system and include information about individuality and genetic relatedness. This information is used in controlling differential behavior which in turn is involved in the maintenance of the genetic diversity in the MHC. The second part of our integrative model is concerned with the psychobiological effects of these MHC-associated odors (Fig. 6).

Integrative model (part 2)
Psychobiological effects of the MHC

Odor signal	Peripheral processing	Central processing	Signal-information	Behavioral effects	Population-genetic effect
				Sexual selection	
			Individuality	Reciprocal altruism	Maintenance of polymorphism
Profiles of volatiles	Main olfactory system	CNS			especially
			Genetic relatedness	Genetic altruism	the high degree of heterozygosity
				Inbred avoidance	

Figure 6. A model of the psychobiological effects of the MHC.

Two species have so far been examined with regard to behavioral effects of MHC-associated odors. In mice, MHC-associated mating preferences have been demonstrated in inbred strains as well as in seminatural populations (Fig. 7), and MHC-associated pregnancy-block effects have been shown to occur in inbred strains. Furthermore, communal nesting in seminatural populations also appears to be MHC associated. Experimental data in humans are much harder to generate and to interpret (Fig. 8).

Statistical studies of the occurrence of heterozygotes and homozygotes in natural populations did not lead to any simple conclusion. However, a number of field studies conducted by us gave first evidence of associations between similarity in the MHC, hedonics of body odor, and differential social behavior in humans. These studies led to the following preliminary conclusions. First, there appears to be an association between the

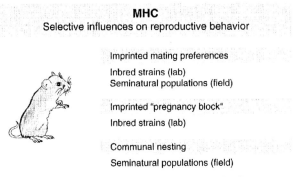

MHC
Selective influences on reproductive behavior

Imprinted mating preferences

Inbred strains (lab)
Seminatural populations (field)

Imprinted "pregnancy block"

Inbred strains (lab)

Communal nesting

Seminatural populations (field)

Figure 7. Psychobiological effects of the MHC in mice.

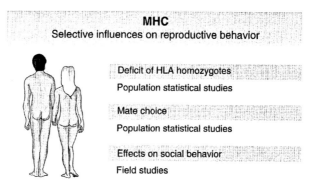

Figure 8.   Psychobiological effects of the MHC in humans.

concentration of soluble MHC proteins and the salience of body odor cues. A higher proportion of subjects with the HLA types *A9(A23,A24)* and *B15(B62,B63)* occurs in samples of subjects whose body odor was described as more salient than usual. Subjects with these HLA types also show an elevated concentration of soluble MHC proteins in their body fluids. Second, the concentration of soluble MHC proteins appears to be correlated with the reproductive cycle in women with the highest concentrations within the first half of the cycle. Third, the hedonics of the body odors of other subjects is similarly associated with the MHC. In same sex groups, body odors of similar subjects are rated more positive than body odors of dissimilar subjects. In different-sex groups, the reverse results were obtained. Fourth, the same pattern of results was obtained with regard to the degree of acquaintance or familiarity between subjects.

22.6. Conclusions

This brief overview has shown that a number of predictions from our integrated model have been empirically supported. One of the greatest problems presently unsolved is to understand how the specific odor profiles are linked to the specificity of the MHC proteins. It is largely unknown whether the behavioral effects demonstrated in mice and humans are also present in other species. The results from studies with mice are well documented, whereas the results from studies with humans need further refinement. Future research may demonstrate that MHC-associated odors occur in different species, and that the reproductive behavior of these species is linked to the MHC. If that is the case, basic features of the immune system influence social behavior, and social behavior is involved in the maintenance of a well-functioning immune system in vertebrates.

ACKNOWLEDGMENTS. Preparation of this manuscript, as well as the studies described in it, was supported by grants from the Volkswagen Foundation and the German Science Foundation.

23   Behavioral Conditioning of Immunity

Michael S. Exton, Maurice G. King,
and Alan J. Husband

23.1. What Is Behavioral Conditioning?

One established method of modifying immune responses is by behavioral conditioning (for reviews see Kusnecov, King, & Husband, 1989; Ader & Cohen, 1993). Conditioning procedures were initially described by Ivan Pavlov who observed that a dog can be made to salivate to an unnatural stimulus (e.g., a tone), if the neutral stimulus regularly preceded its food. In its simplest form, Pavlovian or classical conditioning, as the procedure is sometimes called, consists of presenting two stimuli in close temporal proximity to an organism. These stimuli are referred to as the *conditioned* and the *unconditioned stimulus*.

Michael S. Exton • Institute for Medical Psychology, Essen University Clinic, D-45122 Essen, Germany.
Maurice G. King • Institute for Behavioural Research in Health, Curtin University, Perth, Australia.
Alan J. Husband • Faculty of Veterinarian Sciences, The University of Sydney, Sydney, Australia.

1. The conditioned stimulus (CS) is usually an external sensory stimulus that is discrete and noticeable but also neutral and meaningless to begin with. For example, sounds, tastes, odors, flavors, cutaneous stimulation, and visual signals have been successfully used as CSs.
2. The unconditioned stimulus (UCS) reliably follows the CS shortly after its presentation. The UCS is usually an exteroceptive stimulus that elicits a particular physiological event (e.g., meat powder on a dog's tongue elicits salivation). The response elicited by the UCS is called the *unconditioned response* (UCR).

With further CS–UCS pairings, the dog will begin to salivate to the CS alone, i.e., without the meat powder. The conditioning may be tested by presenting the CS without the UCS. Because salivation was not unconditionally elicited by the CS, it is termed a *conditioned response* (CR). In summary, the repeated sounding of a tone (CS) to the dog in close temporal contiguity with food presentation (UCS) eventually enables the tone alone to elicit salivation (CR).

One important way in which Pavlovian conditioning differs from other forms of learning (e.g., trial-and-error learning) is that the UCR, the response that eventually provides the CR, is forced by the UCS. The conditioning is also controlled by the experimenter via the imposed signals (CS, UCS), not by the subject's choice of response as is the case in trial-and error learning.

23.2. Behavioral Conditioning of Immunity

The range of bodily responses that have been shown to be conditionable is extensive. Most readers are familiar with examples of simple nervous system reflexes in lower animals. But it is a mistake to stop there, as similar conditioning also occurs in humans. The list of conditionable responses includes certain drug effects, some hormone secretions, emotional and behavioral reactions such as fear, reactions to pathogens, illness, and some components of the immune response (e.g., human allergy). Here we concentrate on immunoconditioning, that ability to modify immune functions using a Pavlovian-based paradigm.

The first published studies of immunoconditioning were reported over 70 years ago by Metal'nikov and Chorine (1926) at the Pasteur Institute. They published the following three series of experiments in guinea pigs.

Series 1 demonstrated that leukocytosis could be conditioned to a neutral CS. The CS was either placing a warm bar on the subject's skin or lightly scratching it. The UCS was an injection of tapioca into the peritoneum. This elicits an elevation of leukocytes (UCR). After multiple pairings, the CS alone induced leukocytosis (CR).

Series 2 showed that the protective immune reaction to a specific disease (anthrax) could be conditioned to a neutral CS. The CS was either a warm bar placed on the skin or scratching the skin. The UCS was an injection of killed *Bacillus anthracoides* (anthrax), which also induces leukocytosis. After multiple CS–UCS pairings, the CS alone induced leukocytosis.

Series 3 revealed that the conditioned protection against one disease (anthrax) helps to protect against another disease (cholera). As in Series 2, the CS was trained to elicit leukocytosis by pairing it with anthrax administration. Subjects were then infected with cholera and presented with the CS alone. More conditioned animals survived cholera than nonconditioned controls.

The Metal'nikov series forces a conceptual broadening of strictly Pavlovian conditioning procedures. The overview is that the UCS makes the subject ill and engages the immune system and that this illness is subsequently elicited by an associated stimulus that in itself was benign. There follow several specific ways in which the above series extends the procedure of Pavlovian conditioning. (1) The traditional salivary reflex (UCR) studied by Pavlov is far less complex than the cascades of immunological events (UCR) arising from anthrax infection. (2) The UCS (injection) is operationally different from placing food powder on a sensory receptor such as the tongue. It has been argued that the UCS in the Metal'nikov series *is* the injection as this is what occurs in close temporal proximity to the CS. But this raises problems, for at the time of an injection, an i.p. dose of tapioca would not be perceptually different from an injection of anthrax or even a placebo control for that matter. (3) The UCS (anthrax) will make the subjects ill whereas meat powder does not. (4) The salivary reflex is fairly slow as reflexes go, i.e., a matter of seconds between UCS administration and the peak of the UCR, the consequence of the UCS that becomes conditioned. By comparison, the elevation of leukocytes against anthrax (Series 2) takes somewhat longer. This means that a long delay between CS and UCR is not inimical to immunoconditioning using Pavlovian procedures. Despite these difficulties in fitting the Metal'nikov series into the Pavlovian mold, Pavlov endorsed the series and cited them as examples of conditioning.

During the 1950s a further period of research activity in immunoconditioning followed in the USSR but these studies have been largely overlooked in the West. Like the Metal'nikov studies, the Russian experiments used Pavlovian procedures to form CS–CR associations between exteroceptive stimuli (e.g., scratching, warming the skin, odors, auditory stimuli) and alterations in immune parameters induced, for the most part, by injection of bacterial preparations, but which also included viral, malarial, and erythrocyte antigens.

23.3. Conditioned Taste Aversion and Immune Modulation

Behavioral conditioning of immunity gained a new impetus when it was shown that conditioned taste aversion (CTA), a variant of Pavlovian conditioning, could produce conditioned immunosuppression under certain circumstances (Ader & Cohen, 1975). These circumstances were made clearer by an astute observation during the course of experiments investigating CTA by Ader. Laboratory mice were exposed to the taste of saccharin in paired association with an injection of cyclophosphamide (CY), a substance that causes nausea but that is also commonly used in clinical practice to produce immunosuppression. When these mice were reexposed to saccharin alone (CS), the researchers obtained the expected result: The conditioned animals avoided saccharin because they had learned to associate it with the nausea produced by the drug. However, an unexpectedly high mortality was also observed and the cause of death was attributed to common animal house pathogens. The increased incidence of disease appeared to be accounted for by a deficiency in the immune response in the conditioned animals. Thus, it seemed that the pharmacological effects of CY, that is, immunosuppression, had been reenlisted in the animals exposed to the conditioning stimulus with which it had been originally paired (Fig. 1a,b). This marked the beginnings of a new surge in psychoimmune investigation into the role of conditioning in modifying immune reactivity.

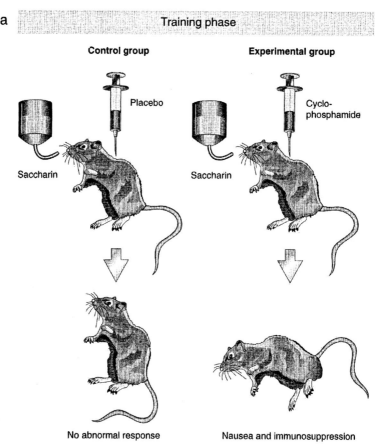

Figure 1. (a). In the training phase, animals that have been previously water deprived receive an injection following a short session of saccharin drinking. The control animals are injected with a placebo (UCS), which produces no physiological effect. The experimental group receive an injection of an immunomodulatory substance [UCS, e.g., cyclophosphamide (CY)], which in the case of CY produces gastrointestinal upset and suppression of certain immune functions. (b) In the test phase, the animals are again presented with saccharin. Animals in the control group drink a normal level of saccharin and display no abnormal physiological response. In contrast, the animals conditioned with CY avoid drinking saccharin after the initial taste (= conditioned taste aversion), concomitant with alterations in immune functions.

CTA

CTA was described by its discoverer, J. Garcia (see Garcia & Hankins, 1977), as an active dislike for the flavor of a particular food that can even spread to the place where the food was eaten. The dislike is usually provoked by ingesting something that upsets the gut axis, i.e., the animal becomes sick, in one sense or another. Olfactory, taste, and visual cues associated with the food and the place will elicit signs of aversion in animals. In humans, hearing or thinking about CTA elicits reports of nausea and facial expressions of loathing.

The defining characteristic of a CTA is that an otherwise harmless taste/odor/flavor is

b   Test phase

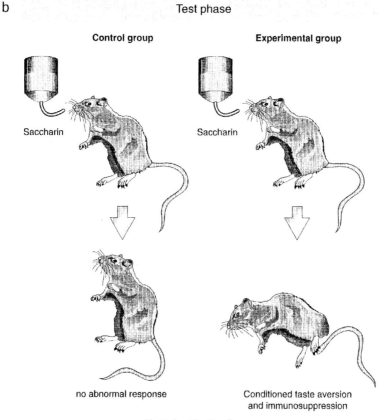

Figure 1.   (*Continued*)

avoided after it has become associated with another agent that inherently produces a gastrointestinal upset. However, the term *gastrointestinal upset* calls for sharper definition especially where immunomodulation is involved. The procedure usually is taken to mean that the organism ingests something and becomes "sick to the stomach." Equally this may occur via olfactory cues, the mucosal epithelia, and the lungs.

Although Garcia and Hankins (1977) couch their exposition in an evolutionary context, little use is made of the ethological data on "bait shyness." CTA serves a similar biological end to "bait shyness," whereby an animal samples a poisoned bait and consequently becomes slightly ill.

The UCS-induced illness can be of rapid onset or, as was also the case with Metal'nikov's Pavlovian series, even of delayed onset, e.g., illness onset may take up to 24 h and conditioning can still occur even with such a time lag. If the animal is subsequently presented with the bait, it will choose not to eat the rest of it. Note that in "bait shyness" the UCS is chosen, i.e., selected by the subject. In CTA, the position is unclear—the UCS can be either chosen or imposed.

458 BEHAVIORAL CONDITIONING OF IMMUNITY

However, in taste aversion, it is not usually the taste of the UCS that is important. In CTA the organism learns to avoid the disturbing stimulus by sampling associated odors, tastes, or flavors that are in themselves benign but that signal the agent that inherently produces the illness. Survival of the individual is clearly subserved by both of these processes, especially in rats, which are unable to regurgitate.

CTA is thus a variant of classical conditioning in which ingestion of a novel and benign taste, odor, or flavor (CS) is paired with an agent (UCS) that causes illness, viz., vomiting and/or nausea (UCR). Subsequently, the CS alone elicits aversion, which is measured by the reduction in the consumption of the CS. This has become known as CTA even though odors and flavors have also been used effectively as CSs. (For the sake of simplicity, we will use taste aversion in the generic sense to cover both odor and flavor CSs. In practice, it is technically difficult to avoid flavor when tastes or odors are used. Conditioned flavor aversion would seem to be a more appropriate label. However, unless otherwise specified, CTA will be used in the present context to cover taste, odor, or flavor CSs.)

CTA is most easily obtained with classes of stimuli that are readily associable, viz., flavor stimuli seem to be associated most readily with unpleasant gastrointestinal sensations. However, sounds and shapes can also be used as CSs though less economically in mammals. In such cases it is clearly not the taste of the CS that is averse.

The most commonly used taste CS has been weak saccharin in drinking water. Although rats occasionally exhibit some neophobia to saccharin, typically this passes after the initial experience. Subsequently, weak saccharin (0.1 to 1.0%) in drinking water is preferred to drinking water alone. In addition, saccharin has very little odor so that taste, not flavor, aversion would seem an appropriate label. Other effective taste CSs include milk and casein hydrolysate, and weak solutions of the following: sucrose, ethanol, vanilla, sodium chloride, and citric acid. Olfactory CSs have included the following: camphor, eucalyptus, lemongrass, peppermint, amyl acetate. The most commonly used UCSs that also have immune effects are CY, lithium chloride, α-melanocyte-stimulating hormone, lipopolysaccharide (LPS), albumin, poly I:C, and IL-1.

Although the phenomenon of CTA has been studied intensively since the mid-1950s, many of the operations producing CTAs have varied considerably during that time. Accordingly, in the present context some attempt must be made to clarify the procedures.

Garcia and Hankins (1977) writing on the origin of food aversion paradigms list ten propositions that bear on CTA. Not all of these are relevant to immunoconditioning via CTA. The following four are pertinent.

"*Proposition 1*: Food aversion learning. Acceptable food becomes aversive when it is followed by illness." If we recall the Metal'nikov series, there are at least two separate ways in which an animal in these circumstances may become ill. Substances like the emetic apomorphine will make an animal "sick to the stomach" without necessarily engaging the immune system. It seems that what the learning theory studies call for is an animal that becomes nauseated and recovers without necessarily activating its immune defenses. By contrast, other substances such as antilymphocyte serum will powerfully affect immunity without making the animal "sick to the stomach." The latter outcome is what the immunologists are primarily interested in. For an experiment where the hypothesis aims primarily at cognitive learning, a treated subject that is sleepy, febrile, and listless cannot be of much utility. Conversely, for an experiment where the hypothesis aims primarily at immunoconditioning, short-term nausea cannot be of much utility. The emphasis on CY by both learning theorists and the immunologists seems an unfortunate choice as both nausea and immunity are heavily involved.

"*Proposition 5*: Non-cognitive processing. 'The association between flavor and effect on food is often subconscious rather than fully reasoned and tends to pass into an automatic habit.' "

"*Proposition 6*: Long-delay learning. Aversions can be established with long intervals separating the flavor and the illness." This proposition is said to place CTA in a different category than Pavlovian conditioning. However, as we have seen from the Metal'nikov series, Pavlovian immunoconditioning is possible with long CS–UCR delays and Pavlov was aware of it.

"*Proposition 8*: Taste primacy. (In mammals) taste stimuli are favored as cues for illness." It should be added that novelty is also important. Novel taste/flavors seem to condition most readily to a food aversion. The speed of acquisition that characterizes CTA is probably dependent on this factor.

If the chosen UCS also impacts on the immune system, then some immune reactions may become conditioned along with the CTA. It is of considerable theoretical interest whether the immunoconditioning is (1) a by-product of the CTA or (2) whether two conditionings occur in parallel or (3) whether the CTA merely sets the context for the immunoconditioning process, which occurs independently of the CTA.

Historically, we can distinguish three phases in the study of CTA. In the mid-1950s, CTA was initially reported using X-irradiation as a UCS. Here, the CS was chosen by the subject and the UCS was imposed by the experimenter (cf. the strictly Pavlovian procedure in which both CS and UCS are imposed). Interest then shifted to studies of food aversions and these included some field studies (1960s) similar to "bait shyness." Here, both the CS and the UCS were selected by the subject. (However, hunger was used to motivate the subject.) In the third stage (1970s onward), the focus shifted to laboratory studies in which the UCS is usually injected, i.e., the CS is selected by the subject but the UCS is imposed by the experimenter. Such a focus is apparent with the development of conditioned immuno-modulation, with most paradigms using a CS selected by the subject associated with an injected UCS.

Nevertheless, subcategories can be identified within what is loosely referred to as CTA. It is pertinent that these be defined in terms of different actions of the UCS. Their relevance for immune conditioning is also noted.

Type I: In one of the earliest studies, CTA for saccharin (chosen) was produced using X-irradiation as the UCS (imposed). However, X-rays do not impact directly on the gastrointestinal system. It is presented here as a particular type because the gastrointestinal system is involved indirectly through radiation sickness, which includes nausea. Conditioned immunomodulation should be able to be produced in this way. There are two possible conditionings here. First, X-irradiation will directly reduce cell division, including lympho-cytes. This could be a UCR for simple Pavlovian immunoconditioning. In addition, the CTA related to radiation sickness might also be used to produce conditioned immunosuppression.

Type II: CTAs may be conditioned by external physical stimuli (imposed) that are only indirectly involved with gastrointestinal upset if at all, e.g., pairing voluntary saccharin drinking with mild footshock may produce a taste aversion for saccharin. It can be argued that this conditioning does not directly implicate the gastrointestinal system or illness. To date, conditioned immunomodulation has not been produced in this way.

Type III: CTAs can be produced by voluntary UCSs that taste bad but that do not produce gastrointestinal upset, e.g., brucine sulfate and bitter aloes. For instance, "The acceptable taste of saccharin signals that bitter aloes will follow"—a benign taste warns of a bitter taste, but there is no gastrointestinal upset involved. Learning theories have not focused on this type of CTA. To date, conditioned immunomodulation has not been reported in this way.

Type IV: CTAs can be produced by voluntary UCSs that taste bad and that produce gastrointestinal upsets, e.g., CY. However, learning studies are usually not primarily interested in the taste of the UCS per se. Besides making the animal nauseated, CY also engages the immune system. Conditioned immunomodulation can be produced in this way.

Type V: The most common voluntary UCSs for learning studies of CTA are those that do not taste bad but that produce gastrointestinal upset, e.g., lithium chloride. Both the learning theorists and the immunoconditioners are more concerned with the consequences of the UCS, i.e., the UCR. The former are usually interested in the nausea, the latter in immune activation. Immunoconditioning can be produced in this way.

Type VI: In allergy-prone individuals, conditioned aversions are not confined to gastrointestinal upset, e.g., food aversions. Allergic responses can also be conditioned using involuntary inhalation of or skin contact with allergens as UCSs. Conditioned immunomodulation can be produced in this way.

Type VII: At the other extreme are agents that produce "illnesses" that perturbate homeostasis *without* gastrointestinal upset, e.g., antilymphocyte serum. These can also act as UCSs in the taste aversion paradigm. They are of interest to immune conditioning especially if perturbation of homeostasis involves immune responses. It is probably inappropriate to attach the CTA label to this class of UCSs, as nausea and/or vomiting is not necessarily elicited by the UCS. Perhaps a different label should be applied, viz., conditioned sickness behavior, but up to the present that has not been the case and it is inappropriate to begin here and now. We will continue to refer to this as CTA. Conditioned immunomodulation can be produced by this method.

Type VIII: Subjects are injected with a UCS (e.g., emetic, pathogen, vaccine, drug, antigen) and this procedure does not involve ingesting the UCS. As the study of CTA became almost completely laboratory based in the 1970s, most experimenters preferred to inject the UCS. This deleted the element of taste/flavor of the UCS although the illness component of the UCS/UCR remained. In such experiments, sensation/perception of injection of the UCS varies little from one study to another. It is the consequences of the UCS—the UCR (nausea, immunity)—that seems more salient to the conditioning. It is the sick-making consequences of the injection, viz., the UCR, that are crucial to the conditioning. Since the 1970s the majority of studies of CTA, regardless of whether they are focused on learning issues or on immunomodulation, have been of this type. Conditioned immunomodulation has been produced in this way.

The above categorization reveals that both learning theorists and immunoconditioners have been primarily concerned with CS–UCR associations. One must add the further qualification: The former focus on CS–nausea and the latter on CS–immune responses. However, there are studies in which the UCR is mixed, where both nausea and immunity are jointly involved. An overview of such stimuli that induce both a behavioral and a conditioned immune response is shown in Table 1. One such UCS is CY and this is dealt with at length below as many reported studies have used saccharin–CY pairings.

23.4. The Relevance of CTA to Conditioned Immunomodulation

In the context of immunomodulation, the advantage of CTA-linked versus traditional Pavlovian procedures is that usually only a single CS–UCS pairing is sufficient for CTA. By contrast, multiple pairings are required in strict Pavlovian procedures.

If one compares the broader conditioning procedures used by Metal'nikov with those

Table 1
Overview of the Effectiveness of UCS and CS Modalities
in Producing Conditioned Taste Aversion and Conditioned Immunomodulation

	Type of CS	
Route of UCS Administration	Physical stimulus 1. Visual 2. Auditory 3. Tactile	Chemical stimulus 1. Aroma 2. Taste 3. Odor
Irradiation Ingestion Inhalation	Induces CTA and a weak conditioned immunomodulation	Induces strong CTA and moderate conditioned immunomodulation
Injection Infection	Induces effective CTA and conditioned immunomodulation	Induces most effective CTA and conditioned immunomodulation

of the contemporary researchers (e.g., Ader & Cohen, 1975; Kusnecov, Sivyer, King, Husband, Cripps, & Clancy, 1983), the procedures are very similar. The main difference is that the CS in Metal'nikov's series was never a novel taste but cutaneous stimulation. In the final analysis this may be what gives the CTA-linked procedures their characteristic rapid acquisition.

23.4.1. Humoral Immune Responses

To date, the majority of studies on CTA and immunity (particularly those conducted in the 1970s and 1980s) have used the immunopharmacological agent CY as the UCS. CY is a cytotoxic alkylating agent that on activation by liver microsomes, yields metabolites with potent immunosuppressive effects. Concomitantly, CY also causes considerable gastro-intestinal distress and this effect has been consistently reliable in promoting CTA. It is interesting to note that more than 30 papers had been published on CTA using CY as a UCS before it was realized that immune conditioning also occurred under those conditions. As was mentioned above, CY has the disadvantage of being a complex UCR (both nausea and immunosuppression). On the other hand, CY as a UCS does produce robust CTA and reliable conditioned immunosuppression.

In the original contemporary study of conditioned immunomodulation, Ader and Cohen (1975) hypothesized that the CS, by way of an association with CY, becomes endowed with immunosuppressive properties. In their first experiment, rats were given a paired presentation of saccharin drinking and an injection of 50 mg/kg CY. Three days later, the rats were injected with sheep red blood cells (SRBC) and 30 min later reexposed to the saccharin without any further CY. Six days after receiving SRBC antigen and CS reexposure, the rats were sacrificed and serum hemagglutinating antibody titers determined. The results revealed that rats consumed significantly less saccharin on reexposure, indicating the presence of CTA. More importantly, however, was the observation that these animals also showed lower anti-SRBC antibody titers compared with relevant controls. Ader and Cohen (1975) concluded that this decrement in antibody titer was attributable to the presence of the conditioned behavioral response to saccharin and proposed the term *behaviorally conditioned immunosuppression* to describe the effect. Replication of this outcome by independent laboratories using the same or similar paradigms served to establish the reliability of the phenomenon.

Nevertheless, a report that followed soon after the initial publication of Ader and

Cohen's results suggested that although the antibody response to a T-cell-dependent antigen (SRBC) could be conditioned, the response to a T-cell-independent antigen (*Brucella abortus*) could not (Wayner, Flannery, & Singer, 1978). Such results suggested that the conditioning of primary humoral antibody responses was restricted to T-cell modulation. This limitation was examined by Cohen, Ader, Green, and Bovbjerg (1979), using a similar paradigm to the original Ader and Cohen (1975) study. Mice were conditioned using a saccharin CS and CY UCS. However, animals were challenged 2 weeks following conditioning with 2,4,6-trinitrophenyl coupled to the thymus-independent carrier LPS. These researchers revealed that similar paradigms could produce conditioned suppression of antibody titers in response not only to SRBC, but also to a T-cell-independent antigen. Thus, CTA-linked immunosuppression is generalizable to having effects on both T and B cells.

The temporal contiguity of the CS and UCS is an important factor determining the effectiveness of CTA behavioral conditioning of immunity. Kusnecov, Husband, and King (1988) used the classic saccharin–CY pairing to examine alterations in both cellular and humoral immune functions. In the initial experiment, animals were presented with the gustatory CS, which was immediately followed by UCS administration. Rats were exposed to the CS 5 or 10 days following conditioning, and then sacrificed 24 h later. Conditioned animals displayed a reduction in the synthesis of IgM from splenocytes that were cultured with pokeweed mitogen. The same paradigm was then implemented in a further experiment, although in this investigation the animals were "backwardly conditioned." That is, CY was initially administered, followed by CS presentation 4 h later. These investigators found no evidence of CTA or conditioned immunomodulation using this technique. These results demonstrated that temporal contiguity between the CS and UCS is important in establishing conditioned alterations in immune functions. Whether such a rule is generalizable across numerous conditioned immunomodulation paradigms is not yet known.

The aforementioned studies have examined the conditionability of pharmacological effects. Such results may bear implications for the modulation of drug taking regimes. However, perhaps a more functionally relevant model is one that approximates an actual immune response, namely, by applying antigen as the UCS. Such experimentation more closely approximates the initial work completed in the three-part series by Metal'nikov and Chorine earlier this century. Little is known regarding the behaviorally conditioned effects of antigen. Nevertheless, some encouraging results have been demonstrated. Ader, Kelley, Moynihan, Grota and Cohen (1993) paired a gustatory CS with keyhole limpet hemocyanin (KLH) over five conditioning trials. Conditioned animals displayed no change in antibody production following CS readministration. However, when one conditioned group was re-presented with the CS plus low-dose (0.5 ng) KLH, the increase in IgG anti-KLH antibody was significantly stronger than in animals given the 0.5-ng dose of KLH without CS representation. Indeed, the conditioned effect approximated that in the positive control group, which received a 50-ng dose immunization of KLH without conditioning. The results demonstrated that CSs alone are not sufficient to reintroduce an antigen-invoked response. Nevertheless, a CS combined with a subphysiological dose of antigen may produce additive or even synergistic effects.

23.4.2. Cellular Immune Responses

In addition to humoral immunity, it is apparent that cellular aspects of the immune response are also affected by conditioning. Initially, the effectiveness of conditioned modulation of a graft-versus-host response (GvHR) was examined. This model is effected by injecting splenocytes from Lewis rats into a hind footpad of hybrid (Lewis × Norwegian) F_1 rats. Such a procedure results in an increased weight of the draining popliteal lymph node, indicative of an inflammatory response produced by the grafted cells recognizing the host as "foreign." Bovbjerg, Ader, and Cohen (1982b) showed that rats taste aversion conditioned with CY exhibit a reduction in the GvHR when challenged concomitantly with CS (saccharin) reexposure 48 days after the conditioning trial. In this experiment, conditioned rats were reexposed to saccharin three times on consecutive days, with a "reminder" dose of 10 mg/kg CY concomitant with the second reexposure being necessary to ensure a conditioned change in the GvHR. Thus, conditioned immunosuppression could reduce cellular inflammation. Moreover, this effect conformed to a major law of learning, in that graded amounts of saccharin reexposure prior to induction of the GvHR resulted in the extinction of the conditioned reduction of the inflammatory response (Bovbjerg, Ader, & Cohen, 1984).

An early example of Type VII/VIII CTA (see above) came from Kusnecov *et al.* (1983). The CS was a saccharin solution and the UCS employed was a single injection of antilymphocyte serum (ALS), a biological immunosuppressant. ALS suppresses lymphocyte proliferation and produces mild "serum sickness" but does not, at the dose used, produce gastrointestinal upset. In line with this, the CTA to saccharin, while significant, was not marked—the animals were wary of the saccharin CS rather than averse to it. In contrast, however, was the conditioned reduction of mixed lymphocyte culture reactivity among mesenteric lymph nodes from conditioned rats reexposed to the CS (Fig. 2). Surprisingly, the conditioned suppression of lymphocyte proliferation was far stronger than any previously reported (about one-third). These results helped establish the current thinking that CTA and conditioned immunosuppression are concomitant phenomena, but not necessarily interdependent for effectiveness.

In addition to mixed lymphocyte culture, behavioral conditioning can also modulate the proliferative response of lymphocytes to mitogen stimulation. The previously described study of Kusnecov *et al.* (1988) revealed that by pairing CY and saccharin, the suppression of splenocyte reactivity following *in vitro* mitogen stimulation could be suppressed by CS

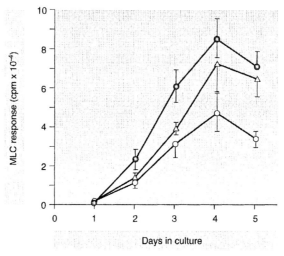

Figure 2. Proliferation of mesenteric lymph node cells in a mixed lymphocyte culture (MLC), 21 days after conditioning with saccharin and antilymphocyte serum (○), saccharin and placebo (●), and water and antilymphocyte serum (△). Plotted data represent mean ± S.E. From Kusnecov *et al.* (1983).

representation. This effect was also abrogated by implementing the backward conditioning technique.

Furthermore, it has been shown that sensitized mice first challenged with SRBC together with CY treatment initially demonstrate a depressed delayed-type hypersensitivity response, which is subsequently enhanced in response to future SRBC challenges. Bovbjerg, Cohen, and Ader (1987) paired consumption of a saccharin solution with CY administration. These researchers demonstrated that reexposure of conditioned mice to the CS on 3 consecutive days produced a conditioned enhancement of the delayed-type hypersensitivity response to SRBC injections subsequent to the initial challenge.

Therefore, it was deduced that the conditioning procedure can be implemented to enhance as well as suppress immune functions. More recent evidence supplemented this position, by using the CTA procedure with the immunopotentiating drug levamisole as a UCS (Husband, King, & Brown, 1987). This procedure not only induced a mild CTA but also produced a conditioned increase in the T-helper : T-suppressor subset ratio by selectively depressing the number of cytotoxic/suppressor T cells.

Canadian researchers examining conditioned increases in allograft-specific cytotoxic T-cell precursors provided a new slant on conditioned enhancement of immune processes. Focusing on the immune response to foreign tissue, they first tested the rejection by mice of a skin graft (Gorczynski, Macrae, & Kennedy, 1982). These investigators hypothesized that the mice could become conditioned by associating the neutral stimuli—handling, anesthesia—with a graft of nonself skin. Following three skin grafts (conditioning trials), all animals were sham grafted (CS reexposure). A majority of the conditioned mice displayed an increase in the number of precursors of cytotoxic T lymphocytes reacting against donor tissue. These results demonstrated that antigen is also an effective UCS in producing cellular immunomodulation.

23.4.3. Acute-Phase Reactions

In addition to humoral and cellular immune functioning, it is now certain that the homeostatic environment that influences immunocompetence is also alterable via conditioning. One arm of the immune reaction to pathogens is the production of the acute-phase response, incorporating immunological, neural, endocrinological, metabolic, and behavioral alterations. Each of these sets of responses is posited to optimize the host environment for pathogen elimination, and is effected by the release of cytokines (such as IL-1) from immunological and neural sources.

One particular pyrogen, LPS, has been commonly used in conditioning studies. The pairing of LPS with a saccharin solution produces a pronounced CTA on saccharin representation. Furthermore, a number of acute-phase reactions, such as fever, anorexia, sleep alterations, and hormonal and metabolic changes, are evidenced (Exton, Bull, King & Husband, 1995a,b; Janz, Green-Johnson, Murray, Vriend, Nance, Greenberg, & Dyck, 1996). Thus, in addition to the reenlistment of the cellular and humoral immune responses, host homeostasis, which is posited to serve as an optimal environment for immune effectiveness, is also conditionable.

23.5. Non-CTA Conditioning of Immune Functions

Although this chapter has focused on those studies examining the effectiveness of the CTA paradigm in producing conditioned immunomodulation, it must be noted that it is not restricted to this method. One major research direction has been to employ a novel environment (a cage other than the home cage; shock chamber) as the CS in combination with either a pharmacological or stress (electric shock) UCS.

Coussons, Dykstra, and Lysle (1992) used a novel environment (conditioning cage) as the CS paired with a subcutaneous injection of 15 mg/kg morphine. It was hypothesized that reexposure to the novel environmental cues would reenlist the changes in immune function that typically follow morphine administration. Indeed, conditioned animals displayed a suppression of immune functioning in both splenic and blood lymphocytes relative to controls. Environmental CSs are also associable with a nonpharmacological UCS. Environmental cues have the ability to reenlist changes in immune function after being paired with an aversive UCS (foot shock within a conditioning chamber) (Lysle, Cunnick, & Maslonek, 1991). In combination, these studies show that not only are environmental cues effective CSs, but that the immunological effect of both reinforcing (morphine) and aversive (electric shock) stimuli can be conditioned.

23.6. Mediating Mechanisms

Contemporary studies of the effects of conditioning on the immune system have extended the initial reports of this phenomenon by French and Soviet laboratories to show that various cellular and humoral immune responses can be modulated bidirectionally by environmental stimuli associated with immunomodulating events. Conditioned immunomodulation is not confined to CTA-linked procedures, and is of potential clinical utility (see Section 23.7). The nature of the mediating mechanisms involved are neither clear nor generally agreed upon. Nevertheless, these data are consistent with the presence of intimate communication between the CNS, the endocrine system, and the immune system.

23.6.1. Stress/Adrenal Steroid Mediation

It is important to point out that behavioral conditioning of immunity cannot be explained on the basis of stress-induced immunosuppression, as many studies have demonstrated its occurrence in the absence of pituitary–adrenal hormone involvement. Further, immunosuppression using a taste aversion model is not necessarily dependent on a noxious unconditioned stimulus. Kusnecov *et al.* (1989) used rabbit anti-rat lymphocyte serum instead of CY to induce immunosuppression without causing nauseating effects. Rats were exposed to oral saccharin paired with ALS and 14 days later were exposed to saccharin alone in the drinking water. After saccharin reexposure the immunosuppressive effects were reenlisted and the T-cell response of conditioned rats to antigen challenge was suppressed by 35% compared with controls. The fact that mild but significant CTA and conditioned immunosuppression were both achieved without an association between the conditioned stimulus and a noxious drug indicates that a true state of conditioning is achieved by this procedure that is not mediated via stress response pathways.

The hypothalamic–pituitary–adrenal (HPA) axis has been theorized to be the endogenous factor modulating conditioned immunosuppression, with the immune changes reflecting stress induction. CY causes considerable gastrointestinal illness and, presumably, it is for this reason that it may activate the HPA axis, resulting in high plasma levels of corticosteroids, an elevation that can be reenlisted on reexposure of rats to the CS in the taste aversion paradigm. It has been proposed that the consistently observed reduction in the immune responsiveness of rat CTA with CY is merely a consequence of a conditioned increase in circulating corticosteroids, which are in themselves immunosuppressive. However, considerable evidence has accumulated indicating that although the conditioned steroid response may accompany immunoconditioning, it does not drive conditioned alterations of immune function. It has been demonstrated from several laboratories that endogenous or exogenous elevation of plasma corticosterone in rats is not necessary for the production of conditioned immunosuppression (Ader, Cohen, & Grota, 1979; Exton, von Hörsten, Schult, Vöge, Strubel, Donath, Steinmüller, Seeliger, Nagel, Westermann, & Schedlowski, 1998; King, Husband, & Kusnecov, 1987; Roudebush & Bryant, 1991). Nevertheless, adrenalectomized mice taste aversion conditioned with CY do not show conditioned changes in the IgM plaque-forming cell response to SRBC (Gorczynski, Macrae, & Kennedy, 1984). Initially this may suggest possible mediation of conditioned immunosuppressive effects by adrenocorticoid steroids. However, it is well established that adrenalectomized animals incur a dramatic alteration in homeostasis, as the adrenal contains potentially immunomodulatory substances such as the catecholamines and enkephalins, and adrenalectomy results in significantly increased basal levels of adrenocorticotropin and β-endorphin.

In contrast, other studies have shown that adrenal steroids may *limit* rather than produce the conditioned changes in immune function. Such a position has been identified by pairing a distinct odor with administration of bovine serum albumin to guinea pigs (Peeke, Ellman, Dark, Salfi, & Reus, 1987). These researchers showed by attenuating plasma corticosterone levels prior to CS reexposure (via dexamethasone administration) that the conditioned increase of plasma histamine to bovine serum albumin, administered at the same time as CS reexposure, was enhanced. This finding suggests, at least with the conditioning model used, that levels of corticosterone in blood during CS reexposure may exert a slight inhibitory hold on the histamine modulating effect of the CS. However, because dexamethasone administration also blocks pituitary adrenocorticotropin and β-endorphin release, interpretation of the results solely in terms of corticosterone reductions is suspect.

While adrenal steroids are involved in some conditioning, steroids per se are not sufficient for conditioning. Attention now focuses on other routes by which neural influences impact on cells of the immune system, through small molecule neurotransmitters, hormones, and neuropeptides.

23.6.2. Other Possible Mediators

When examining the mechanisms of conditioned immunomodulation, three phases can be considered. First, what are the mechanisms by which the CNS recognizes the immune alterations induced by the UCS (afferent phase)? Second, what processes occur within the CNS enabling it to recognize the CS and potentiate the immune alterations (processing phase)? Third, what CNS/peripheral effector mechanisms produce the conditioned immunomodulation (efferent phase)?

Mediators of Non-CTA-Linked Immunomodulation.   Studies using environmental cues as the CS (Section 23.5) have revealed a number of important mediators of the efferent phase of conditioning. Researchers have primarily used receptor antagonists to delineate which substances are released either centrally or peripherally following CS representation to produce conditioned changes in immune function.

A number of factors are known to play a role in conditioned effects using environmental cues as the CS and an aversive stimulus (electric shock) or morphine as the UCS. The conditioned immunosuppression (commonly a reduction in splenocyte proliferation in response to mitogens, lowered splenic NK cell activity) observed following placement in the novel environment during the testing phase is at least partly induced by corticotropin-releasing hormone, endogenous opioids, and catecholamines. The heterogeneity of effector molecules most likely represents the various conditioning agents, stimuli, and paradigms that are implemented in behaviorally conditioned immunomodulation experiments.

Mediators of CTA-Linked Immunomodulation.   Little is presently known regarding the mechanisms of CTA-linked immunomodulation. Most evidence has been obtained from a single model by a U.S. research group. These researchers have implemented a camphor odor CS paired with poly I:C as the UCS. On subsequent CS re-presentation, an increase in splenic NK cell activity is observed (Hiramoto, Hiramoto, Solvason, & Ghanta, 1987). This group has used pharmacological antagonists and brain lesioning techniques to identify the afferent, processing, and efferent phases involved in producing the conditioned effect. As IFNβ is produced in response to poly I:C administration, Solvason, Ghanta, and Hiramoto (1993) infused IFNβ into the cisterna magna of the hypothalamus as the UCS, paired with camphor odor. As this method produced the previously described activation of splenic NK activity, it was argued that IFNβ must be the afferent communication molecule linking the immune response and CNS. However, such a conclusion may be premature, as the effects need to be supported by studies blocking the effect of IFNβ. Additionally, Ghanta, Rogers, Hsueh, Demissie, Lorden, Hiramoto, and Hiramoto (1994) demonstrated that the arcuate nucleus of the hypothalamus is essential in the acquisition of the CS–UCS association, but not the production of the conditioned increase in NK cell activity. Such effects were revealed by showing that lesioning the arcuate nucleus prior to the CS–UCS pairing blocked the conditioned effect, whereas a lesion induced after the pairing trial, but before the CS re-presentation had no effect on the conditioned response.

Direct neural innervation of lymphoid organs may be one possible efferent mechanism of conditioned changes of immune functions. A recent series of studies have examined the conditioned changes in immune functions using saccharin solution as the CS paired with i.p. 20 mg/kg cyclosporin A (Exton *et al.*, 1998). A major finding of this group is that on CS reexposure a conditioned reduction in the splenocyte proliferative response to mitogen stimulation is reduced, accompanied by a conditioned reduction in splenocyte cytokine (IL-2 and IFNγ) production. Nevertheless, when autonomic innervation of the spleen is surgically removed prior to conditioning, the conditioned suppression is abrogated. It appears that denervation of the spleen does not influence the afferent phase, as the conditioned animals display a normal CTA response. These results indicate that conditioned changes in immune functions within lymphoid organs can be effected via direct neural innervation. Thus, conditioned immunomodulation may not only be produced by hormones/neurotransmitters acting in an endocrine way, but also via the paracrine route of intimate neural–immune cell contact.

Despite these studies showing strong and replicable effects, the results are as yet far from generalizable across conditioning methodologies. Nevertheless, some insight is provided by studies examining the mechanisms of non-CTA-linked immunomodulation.

23.7. Biological Relevance and Practical Implications of Conditioned Immunomodulation

In studies of conditioned changes in immune function, the one criticism that has commonly been offered is that the effects were relatively small compared with actual drug/

antigen administration. Thus, for conditioning to have any biological relevance, it must be demonstrated that behaviorally conditioned alterations in immune functions affect disease outcome. Several animal studies have been able to produce such effects.

In the classic study of the effects of chemotherapeutic treatment with CY on survival rate among mice suffering from the autoimmune disease systemic lupus erythematosus (Ader & Cohen, 1982), it was observed that in animals given CY in paired association with saccharin, reexposure to saccharin alone significantly delayed mortality and onset of proteinuria compared with nonconditioned mice—the conditioned immunosuppression prevented the autoimmune response (Fig. 3). More specifically, among mice receiving weekly presentations of saccharin, the administration of 30 mg/kg CY after 50% of such presentations (C_{50} group) significantly delayed mortality rate and onset of proteinuria compared with nonconditioned mice that received an equal amount of CY but not together with saccharin (NC_{50} group). Moreover, it was found that the mortality rate of the NC_{50} group was comparable to that of mice that received CY treatment invariably after each presentation of saccharin (C_{100} group). These data are consistent with the model of conditioned immunosuppression, demonstrating that saccharin presentations that had occasionally been reinforced with CY treatment (NC_{50} group) exert a modulatory effect on autoimmune mechanisms when presented alone.

In an animal model of rheumatoid arthritis, a significant reduction in joint inflammation was observed after three consecutive daily re-exposures to saccharin/vanilla solution in rats previously conditioned by paired administration of the taste with CY (Klosterhalfen & Klosterhalfen, 1990). More specifically, this study revealed a significant reduction in the inflammatory response to Freund's complete adjuvant, administered to rats 30 min prior to

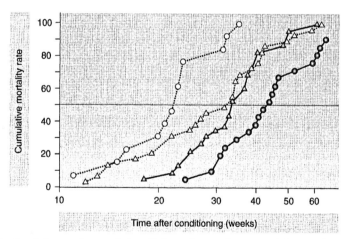

Figure 3. *Cumulative mortality rate of mice with the autoimmune disease systemic lupus erythematosus.* All animals received weekly a saccharin drinking solution administered via pipette. Control animals received only saccharin (○). A second group received saccharin and an injection of cyclophosphamide (two injections every 4 weeks), but the drug and saccharin were not paired (△). A third group received two injections of cyclophosphamide every 4 weeks, with the injection immediately following saccharin drinking (▲). The main conditioned group received a weekly injection of cyclophosphamide following the saccharin drinking (●). The conditioned mice survived significantly longer (27.6 ± 1.5 weeks) than nonconditioned animals (22.1 ± 1.7 weeks) receiving the same amount of drug. From Ader and Cohen (1982).

the first of three consecutive daily reexposures to saccharin/vanilla solution that was previously paired with CY. This finding demonstrates further the potential therapeutic efficacy of conditioned immunosuppression in the treatment of autoimmune disease.

Other studies have shown that conditioning can modify host resistance to tumor growth (Blom, Tamarkin, Shiber, & Nelson, 1995). Taste aversion conditioning of mice with CY and saccharin was followed by injection of the chemical carcinogen 9,10-dimethylbenz[a]-anthracene (DMBA). Conditioned immunosuppression following CS representation resulted in an increase in DMBA-induced tumor growth. In contrast, using conditioning to *stimulate* immune function can have beneficial effects on tumor progression. Using the camphor odor—poly I:C model described previously, Ghanta, Miura, Hiramoto, and Hiramoto (1988) examined whether the observed conditioned increase in NK cell activity may improve survival in tumor bearing mice. The conditioning paradigm, proposedly via the increase in NK cell activity, increased survival time of the animals by approximately 30%.

Recent studies have indicated that the rejection of transplanted organs (heterotopic heart allografts) may be inhibited by CTA using the immunosuppressive drug cyclosporin A as a UCS (Exton *et al.*, 1998). Cyclosporin A (a potent suppressor of T-cell IL-2 synthesis used clinically to prolong organ graft survival) was paired with the taste of saccharin, and the ability of saccharin reexposure alone to modify rejection of histoincompatible heart grafts was observed (Fig. 4a). Experimental groups were given three trials of cyclosporin paired with saccharin prior to transplantation. Rats were subsequently reexposed to saccharin alone, and on the third reexposure day an allogeneic heart was grafted heterotopically into the abdominal cavity. Saccharin was presented on the day of transplantation and every day thereafter until rejection of the graft (measured as the presence/lack of a palpable heartbeat). The experimental animals showed a significant prolongation of heart graft survival compared with control unconditioned groups, with this effect mirroring the prolongation produced by short-course cyclosporin treatment (Fig. 4b).

Thus, these animal studies indicate that conditioned effects on immunity are more than an interesting biological phenomenon and may have a powerful influence on the course of disease. However, for conditioning to have any practical clinical relevance, the immunological effects must be able to be produced in humans.

In this regard, CTA-like procedures may have potential for management of human immune-mediated allergic reactions such as hay fever. The symptoms of hay fever are caused by the release of a variety of substances from mast cells in the linings of the upper airways as part of an immune-mediated reaction to airborne allergens. Recently, studies have examined conditioned allergy in humans prone to perennial rhinitis (Gauci, Husband, Saxarra, & King, 1994). The studies showed that similar conditionings can occur in human subjects allergic to house dust mite or rye grass pollen who can learn to associate these symptoms with a novel taste sensation administered at the time of an induced allergic provocation. More specifically, challenge with house dust mite allergen (UCS) was paired with a novel colored–flavored drink (CS) in a single conditioning trial. The CR was levels of mast cell tryptase in nasal washings. On reexposure to the novel drink alone (CS), levels of mast cell tryptase (CR) released in conditioned subjects were significantly greater than in control subjects, i.e. after conditioning trials in which a novel-tasting soft drink was consumed at the time of the allergic attack, patients reexposed to the soft drink alone some days later displayed elevated levels of mast cell factors normally released during hay fever. This raises the important question of how much human allergy is triggered by the allergen and how much by associated stimuli.

Conditioned alterations in the activity of NK cells in the peripheral blood of humans

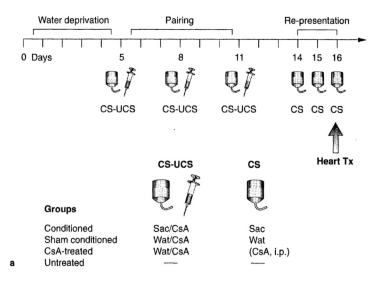

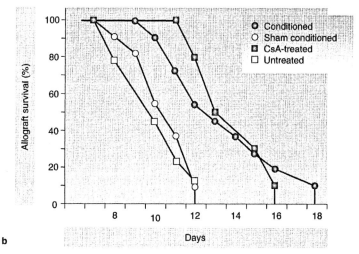

Figure 4. *Overview of the group design and conditioning protocol for examining the conditioned prolongation of heart transplantation survival.* (a) Following water deprivation, animals are administered three CS/UCS training days, receiving either saccharin (Sac) or water (Wat) as the CS, with cyclosporin A (CsA) administered as the UCS. On the third CS test day (day 16 of the experiment), rats have a heart allograft transplanted heterotopically into the peritoneum. The CS is re-presented every day thereafter, with the heart palpated to check for graft survival (with rejection defined as the last day of cardiac activity). (b) Rats conditioned with saccharin and cyclosporin A (●) have a longer surviving graft than animals sham conditioned with water and cyclosporin A (○) or untreated animals (□). Furthermore, the conditioned effect is the same as that produced by a short course of the drug itself (CsA-treated rats, ■). From Exton *et al.* (1998).

have also been demonstrated (Buske-Kirschbaum, Kirschbaum, Stierle, Lehnert, & Hellhammer, 1992). This study paired the oral administration of a sherbet sweet (CS) with an immediate subcutaneous injection of adrenaline (UCS), which produced a strong increase in blood NK cell activity 20 min later. Two days later, subjects in the conditioned group were once again given the sherbet sweet with an injection of saline. This procedure raised NK cell activity to a similar extent as adrenaline administration when compared with baseline values. However, such results are not easily reproduced. The same authors attempted to replicate the findings in collaboration with a Dutch research group (Kirschbaum, Jabaaij, Buske-Kirschbaum, Hennig, Blom, Dorst, Bauch, DiPauli, Schmitz, Ballieux, & Hellhammer, 1992). Despite adherence to the protocol as closely as possible, the Dutch group could not produce the conditioned increase in NK cell activity. These results allude to the potential difficulty in producing conditioned immunomodulation in humans. The ability to implement a CS that is novel to a human may prove difficult, and indeed the factors that encompass the conditioning procedure such as the visual and auditory environment may themselves become conditioned stimuli. Furthermore, a number of other factors such as what the subject has eaten, stress encountered, mood, and cognitive appraisal of the conditioning situation may all influence the conditioning process. It is these difficulties that must be overcome if conditioning is to move into a clinical sphere.

23.8. Summary

The phenomenon of immunoconditioning was first reported 70 years ago. Nonetheless, intensive investigation has only occurred in the last 20 years primarily using CTA-linked procedures. The present state of knowledge is that both strictly Pavlovian and taste aversion procedures produce robust and long-lasting effects, both immunosuppression and immunostimulation. Although we are not in a position to say precisely where conditioning actually takes place in the cascade of responses that constitute an immune response, what is clear is that not one but many components of the immune response are conditionable. The most likely candidates for mediation between CS and conditioned immune responses are direct neural–immune cell/organ connections and receptors on lymphocytes for soluble fractions.

A vast amount of careful basic research underpinning immunoconditioning has been carried out and replicated in several countries around the world. The question no longer is if the CNS affects the immune system. The question of how these effects are produced has been partially answered, although it remains relatively poorly understood. However, perhaps the ideal that the future can expect is some contribution to psychological and medical practice, this is proving elusive but some advances have been made.

ACKNOWLEDGMENT. This work was composed with the financial aid of an Alexander von Humboldt Postdoctoral Fellowship to Mike Exton.

24 Psychoneuroimmunology in Oncology

Dana H. Bovbjerg, Heiddis B. Valdimarsdottir, and Robert Zachariae

Dana H. Bovbjerg • Biobehavioral Medicine Program, Ruttenberg Cancer Center, Mount Sinai School of Medicine, New York, New York 10021.   Heiddis B. Valdimarsdottir • Memorial Sloan-Kettering Cancer Center, New York, New York 10021.   Robert Zachariae • Aarhus University, Risskov, Denmark.

24.1. Overview

Immunologists long viewed the immune system as self-contained, functioning autonomously in defense of the body. That view is no longer tenable. Evidence that the brain and immune system interact with each other is now overwhelming, as is amply documented in other chapters of this book. In this chapter, we explore potential implications of psychoneuroimmunology for oncology, focusing first on the "conventional view" that psychological influences on immune defenses may be the mechanism by which psychosocial factors may affect the development and/or progression of cancer (Bovbjerg, 1994). We then explore other ways in which psychoneuroimmunology may be relevant for oncology, including associations with infectious disease, and possible dysfunction in regulatory loops between the brain and immune system.

24.2. Psychoneuroimmunology and Cancer: The Conventional View

The conventional view of the importance of psychoneuroimmunology in oncology, prevalent in both the scientific literature (e.g., Andersen, Kiecolt-Glaser, & Glaser, 1994) and popular press (e.g., Chopra, 1989), is conceptually based on three largely independent lines of research (see Fig. 1). In the following sections, we consider each of these areas separately and then highlight those few studies that have attempted to bridge these different areas of research.

24.3. Psychosocial Factors Influence the Development and/or Progression of Cancer

Numerous studies have examined the possibility that psychosocial variables are associated with cancer incidence and cancer progression. Only a few representative studies

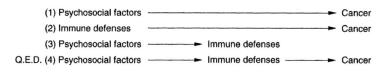

Figure 1. Psychoneuroimmunology and oncology: the conventional view. As indicated in this schematic, the conventional view is conceptually based on three essentially independent research literatures. As reviewed in the text, there are numerous conceptual issues and technical concerns within each of these literatures that highlight the need for additional research.

will be mentioned in this chapter as this literature has been extensively reviewed elsewhere (e.g., Holland, Breitbart, Jacobsen, Lederberg, Loscaizo, Massie, & McCorkle, 1998; Lewis, O'Sullivan, & Barrraclough, 1994). As will become clear below, research in this area is fraught with both conceptual and methodological challenges, which has slowed its progress. Despite the difficulties, there have been some striking results that continue to fire interest in this provocative area of research. Historically, the most common study designs employed in this literature have been retrospective, quasiprospective, and prospective.

24.3.1. Retrospective Studies

Retrospective studies have been the most common methodology to examine the psychosocial influence on cancer. In these studies, subjects diagnosed with, or being treated for, cancer are compared with healthy comparison subjects or individuals suffering from other diseases. Some retrospective studies have found support for the hypothesis that psychosocial factors, such as stressful life events and personality characteristics, are related to the development and/or progression of cancer.

Several studies have found that breast cancer patients retrospectively reported more stressful life events in the 5 or 6 years preceding their diagnosis compared with healthy women (Bremond, Kune, & Bahnson, 1986; Forsen, 1991). Ramirez, Graig, Watson, Fentiman, North, and Rubens (1989) found that patients with a recurrence of breast cancer reported more stressful life events than either patients who experienced no recurrence or breast cancer patients who were newly diagnosed. In contrast, Priestman, Priestman, and Bradshaw (1985) failed to find a difference in the number and severity of stressful life events between breast cancer patients, patients with benign disease, and healthy hospital personnel.

Using a potentially very powerful study design to control for possible genetic differences by assessing identical twins, Smith, Harrison, Ashworth, Montano, Davis, and Fefer (1984) failed to find differences in the number of stressful life events between the twin with cancer and the healthy twin.

In addition to life stressors, personality characteristics, particularly the concept of emotional repression, have received considerable research attention. For example, LeShan (1966) found the inability to express aggressive emotions in 64% of a group of cancer patients, compared with only 32% in a healthy control group. Similarly, Watson, Greer, Blake, and Shrapnell (1984) found that women with breast cancer suppressed their feelings of anger significantly more often than women without breast cancer. Kneier and Temoshok (1984) found that melanoma patients were more likely to repress their emotions than cardiovascular patients and healthy controls.

The methodological problems related to retrospective study designs have been discussed at length by Fox (1995). A critical issue is that it is impossible to demonstrate that the findings of such studies did not occur as a consequence, rather than as an antecedent, of the disease. Mood has been shown to affect memory and cognition, and experimental investigations have consistently found differential effects of induced moods on the accessibility of personal memories (Bower, 1981). Depressed patients have been found to recall specific negative events more easily than positive events (Moffit, Singer, Nelligan, Carlson, & Vyse, 1994), and it is therefore possible that the knowledge that one has cancer may not only affect the patient's mood, but also their recollection of past events and scores on personality questionnaires. Consistent with this possibility are the findings of Kreitler, Chaitchik, and Kreitler (1993) who reported no difference in emotional repression among breast cancer patients and noncancer patients who were assessed prior to surgery, at which time their cancer diagnosis was not yet known. After surgery, when the women had been informed of their diagnosis, emotional repression increased in the cancer group but not in the noncancer group. Although these methodological limitations make interpretation of positive findings difficult, retrospective studies serve a valuable purpose in the initial testing of novel hypotheses, which can then be evaluated in more time-consuming and expensive prospective studies.

24.3.2. Quasi-Prospective Studies

Quasi-prospective studies address a number of the methodological flaws in retrospective studies. In these studies, psychosocial variables assessed prior to biopsy are used to predict the diagnosis of cancer and/or the course of the disease. Such designs control for the psychosocial and emotional consequences of explicit knowledge that one has cancer, in investigation of differences between individuals with and without cancer. Quasi-prospective studies have also assessed psychosocial factors in patients already diagnosed, and examined the relations of such factors to the subsequent course of disease (prognosis).

Although not without exception, quasiprospective studies have found that psychosocial variables, such as emotional repression and stressful life events, predict which subjects will receive a cancer diagnosis. For example, Greer and Morris (1975) interviewed and administered a battery of psychological tests to women the day before their breast tumor biopsy. Women who were later diagnosed with breast cancer were found to express significantly less anger than patients who were found to have benign breast disease. Cooper, Cooper, and Faragher (1989) assessed the number and perception of stressful life events among women undergoing diagnosis for breast cancer. Although the women who were later found to have breast cancer reported fewer stressful life events than the women found to be cancer free, they rated the events as more distressing. This finding suggests that a person's perception of stressful life events may be a more important predictor of cancer development than simply the number of stressful life events they have experienced. Effects of psychosocial variables may also depend on interactions between several factors, which by themselves do not contribute significantly to the development of disease. In a study by Goodkin, Antoni, and Blaney (1986), for example, the diagnosis of invasive cervical cancer was found to be only modestly related to stress levels. This correlation, however, was substantially enhanced when other psychological factors were also taken into account (e.g., pessimism).

Quasi-prospective studies have also provided some evidence that psychosocial factors may affect the prognosis of patients who have been diagnosed with cancer. For example, a widely cited series of investigations by Greer and colleagues assessed psychological responses to breast cancer 3 months after mastectomy. Based on their responses, the patients were classified into four categories: denial, fighting spirit, stoic acceptance, and helplessness/ hopelessness. Recurrence-free survival was longer among patients who responded to their diagnosis with fighting spirit or denial than among patients who responded to their diagnosis with stoic acceptance or feelings of helplessness/hopelessness at a 5-year (Greer, Morris, & Pettingale, 1979), 10-year (Pettingale, Morris, Greer, & Haybittle, 1985), and 15-year follow-up (Greer, Morris, Pettingale, & Haybittle, 1990). Similar findings have been reported by other researchers, including Diclemente and Temoshok (1985), who found stoic acceptance and helplessness/hopelessness in malignant melanoma patients to be associated with subsequent disease progression.

Quasi-prospective designs can control for the effects of conscious knowledge that one has cancer, although the potential confounding effects of biological influences (e.g., the tumor itself or other factors that contributed to the carcinogenesis) on psychosocial factors remain an important issue for the interpretation of these results (Fox, 1995). Even the apparent control over conscious knowledge can be questioned. One must always be concerned that patients may be able to accurately predict their diagnosis before they receive explicit notification. Consistent with this possibility, Schwartz and Geyer (1984) found that both patients and their physicians were able to predict the outcome of a biopsy with a fairly high degree of accuracy. Of 23 patients who were found to have cancer, 65% of the patients and 74% of the physicians correctly predicted the outcome.

24.3.3. Prospective Studies

Prospective studies provide the strongest test of the role of psychosocial factors in cancer onset. In these studies psychosocial assessments are obtained from healthy individuals who are then followed for several years and those who develop cancer are compared with those who do not. Several such studies have examined the role of depression in cancer development. A study of 2020 male employees of Western Electric found that depression assessed 17 years earlier was associated with twice the risk of death from cancer (Shekelle, Raynor, Ostfeld, Garron, Bieliauskas, Liu, Maliza, & Paul, 1981). More recent studies have failed to find a relation between depression and cancer mortality and morbidity (Hahn & Petitti, 1988; Linkins & Comstock, 1990; Zonderman, Costa, & McCrae, 1989), and one study even found a significantly lower depression score in men who developed cancer (Dattore, Shontz, & Coyne, 1980).

Personality has also been reported to play a role in prospective studies of cancer development. Compared with individuals who did not develop cancer ($n = 125$), individuals subsequently diagnosed ($n = 75$) were more likely to repress their emotions, as measured by the MMPI obtained on average 55.6 months before clinical evidence of breast cancer (Dattore et al., 1980). In a recent large-scale study, women who subsequently developed breast cancer were more likely to score high on antiemotionality (i.e., did not trust their feelings or did not let their behavior be influenced by their emotions) than women who did not develop breast cancer during a 6 year follow-up interval, although 11 other personality traits (e.g., suppression of emotions, anger) were not found to be predictive of the development of breast cancer (Bleiker, van der Ploeg, Hendriks, & Ader, 1996). In a widely cited, but controversial, series of studies, Grossarth-Maticek and colleagues reported that individuals classified as Type 1 (characterized by feelings of hopelessness and depression, suppression of emotions, and high on rationality and antiemotionality) were more prone to develop cancer than other personality types (Eysenck, 1993; Grossarth-Maticek, 1980; Grossarth-Maticek, Siegrist, & Vetter, 1982b). Methodologies in this work have been severely criticized and further research is clearly needed (see Anonymous, 1993). Individuals who were characterized as Type A (hard-driving, time-oriented, impatient, and hostile) were found to be more likely to die of cancer than Type B individuals during the first 17-year follow-up in the Western Collaborative Group Study (Fox, Ragland, Brand, & Rosenman, 1987). Failures to find relations between particular personality factors and development of cancer have also been reported. For example, further analysis of the Western Collaborative Group Study indicated that the relationship between Type A personality and cancer development over the 22-year follow-up period may have been the result of other known risk factors (e.g., smoking) (Ragland, Brand, & Fox, 1992).

It is clear that the pattern of results in these studies is neither consistent nor simple. Considering the multiple steps involved in the development of cancer, and the potential complexity of psychological influences, it is noteworthy that any of the studies found relations between psychosocial factors and the development of cancer. There are a number of possible explanations for discrepant outcomes in this literature. First, because most of these studies were designed to examine the role of psychosocial factors (e.g., anger) thought to affect the etiology of other diseases (such as heart disease), and cancer was an incidental outcome, they may not have included the best measures to quantify variables theorized to play a role in cancer (e.g., repressive coping style). Second, the majority of the studies included multiple sites, stages, and types of cancer. This mixed pool of subjects may have diluted or distorted the findings, because psychosocial vulnerability may prove to be cancer-specific. Third, several of the studies did not control for other known risk factors (e.g., age, health behaviors) that could account for apparent relationships to psychosocial factors. Fourth, studies have rarely examined the possibility that psychosocial factors may interact with other more established risk factors for cancer, consideration of which may reveal psychosocial influences that would be otherwise missed. For example, Eysenck (1994) reported a synergistic interaction between family history of breast cancer and emotional distress. Women who had family histories of breast cancer and who also reported high levels of psychological distress were more likely to develop breast cancer than (1) women with and without a family history of breast cancer who reported low levels of distress and (2) women without a family history who reported high levels of psychological distress. Behavioral risk factors (i.e., smoking, drinking) have also been found to interact with psychosocial risk factors. In their studies with lung cancer, Eysenck and colleagues found that stress considered by itself did not contribute to lung cancer mortality. On the other hand, when stress was considered in conjunction with smoking, the odds of dying from lung cancer were increased considerably. These studies of interactions between psychosocial factors and other known risk factors suggest an important strategy for future research, which may yield more powerful and consistent results than have been seen to date.

24.3.4. Psychosocial Interventions

The evidence that psychosocial variables may affect the development and/or progression of cancer raises the possibility that psychosocial interventions may reduce the negative impact of psychological distress on cancer onset and/or progression. Supporting this possibility, there is now a growing literature showing that various types of psychosocial interventions, such as biofeedback and cognitive–behavioral therapy, can affect quality of life and psychosocial adjustment, and may possibly even improve survival time among cancer patients (Andersen, 1992; Spiegel, 1997) (see Chapter 20). In a well-known, controlled prospective study by Spiegel, Bloom, Kraemer, and Gottheil (1989), metastatic breast cancer patients met in weekly support groups for 1 year. At 12 months the patients in the intervention group showed significantly more vigor and significantly less fatigue, tension, or confusion. At a follow-up 11 years later, results showed a significant effect of intervention on survival time, with a mean survival of 36.3 months in the intervention group compared with 18.9 months in the control group. Similar results were found by Fawzy, Fawzy, and Hyun (1994) when they evaluated the immediate and long-term effects of a structured psychiatric intervention on psychosocial adjustment and coping in cancer patients. The patients who had stage I or II malignant melanoma were randomly assigned to an intervention group or to a control group. The intervention patients were seen in groups of seven to ten patients who met for 1½ h a week for 6 weeks. Following the intervention, the patients reported less distress as well as increases in active behavioral coping strategies. Most of these changes were seen for as long as 6 months following the intervention. At the 6-year follow-up (Fawzy *et al.*, 1994), 29% of the patients in the control group and only 9% of the patients in the intervention group had died. Although it is obviously important to await the results of ongoing replication studies, the results of these initial studies suggest that this approach may be a strong one for exploring psychosocial influences on cancer progression.

24.4. Immune Defenses Influence the Development and/or Progression of Cancer

Among the lay public, there is a widespread perception that immune defenses are important both in the prevention of cancer and in its cure. How else to explain Uncle Albert, who smoked all his life and never even had a cold, let alone cancer? Or Aunt Sue, who lived for 5 years after the doctors said she had 3 months? After all, haven't immune defenses against cancer been invoked by guru after guru on national television? Among the medical community, clinical experience of a tumor that spontaneously regressed, or a patient whose survival defied the odds, is considered anecdotal evidence. Such evidence, even when published, is viewed with considerable skepticism. After all, haven't all of those trumpeted theories—that the immune system is involved in surveillance against cancer, that magic bullets to cure cancer can be found among the armament of the immune system—repeatedly floundered when viewed in the harsh light of data? In our view, the truth lies somewhere between these two extremes. The huge scientific literature examining the interactions between cancer and the immune system is replete with examples of exciting findings from initial studies that have failed to fulfill their early promise. Despite the difficulties of this complex research topic, however, much quiet progress has been made, as reviewed in more detail elsewhere (e.g., Brittenden, Heys, Ross, & Eremin, 1996; Clark & Weiner, 1995; Schreiber, 1993; Shu, Plantz, Krauss, & Chang, 1997; Tanneberger & Hrelia, 1996). In this brief section, we focus on some of the critical concepts and continuing controversies in this rapidly developing area of research.

Any consideration of the importance of the immune system in defending against cancer must address at least four basic questions: (1) Can the immune system detect cancer cells? More precisely, are transformed cells antigenic? (2) Following detection, does the immune system initiate a response to cancer cells? More precisely, are transformed cells immunogenic? (3) Do the responses of the immune system to cancer cells affect the development of a tumor or its progression? (4) Are there interventions for augmenting immune responses to cancer that might be clinically effective?

Important to keep in mind when considering the research addressing these questions is the possibility that the answers will depend on the particular type of cancer under study. While there are clearly many common phenotypic features, differences in the molecular mechanisms, as well as other features specific to the particular tumor under study, are likely to affect the interactions with the immune system (Schreiber, 1993). The potential importance of specific features of particular types of tumors, or even the same tumor at different stages of progression (in interaction with specific features of the host), is clear from the huge literature on animal models of cancer. One can find studies with positive answers to each of the above questions, while other studies with other models, or the same model in another species (or even a different strain of the same species), have yielded negative results (Schreiber, 1993). One must be wary of generalizing too far from any particular animal model. Indeed, the relevance of animal models for our understanding of human cancers has been questioned by a number of investigators, who note the many differences between species; in mice, for example, most "spontaneous" tumors are induced by viruses and metastatic disease is rare (Schreiber, 1993).

24.4.1. Is There Evidence that the Immune System Can Detect Cancer Cells in Humans?

For cancers with a viral etiology (e.g., Burkitt's lymphoma), it has been clear for a long time that virus-associated determinants can serve as antigens, allowing recognition by the immune system (Schreiber, 1993). For tumors without a known viral etiology (e.g., the solid tumors responsible for the vast majority of cancer deaths), possible antigenic determinants are less obvious. There are, however, at least three ways in which distinctive proteins might arise in transformed cells, and thus serve as antigens for immune recognition (Greenberg & Riddell, 1992). First, tumor-specific proteins could arise by mutated oncogenes (e.g., p53). Second, specific proteins could arise following somatic mutations. Third, aberrant expression of various self-proteins (e.g., mucins) could distinguish transformed from normal cells. Considerable evidence now indicates that such antigens exist. In addition to the decades-old serological evidence of antitumor antibodies in some patients, recent research has revealed an increasing number of examples of T-cell recognition of specific tumor antigens, which have been characterized extensively using the tools of molecular biology (Maeurer & Lotze, 1997).

24.4.2. Is There Evidence that the Immune System Initiates Responses to Cancer Cells in Humans?

Immune responses to cancer have been reported for virtually every known effector mechanism, including antigen-specific responses of both antibody-producing B cells and cytotoxic T cells, as well as nonspecific responses by other cell types (Greenberg, 1991). The specific binding of antibody has been reported to affect cancer cells through several mechanisms, including complement-mediated cell lysis, antibody-dependent cellular cytotoxicity, and direct interference with cellular growth processes. Evidence for cytotoxic T-cell responses includes examples of classic MHC class I-restricted CD8[+] cell responses, as well as more recent evidence that certain T cells (e.g., gamma/delta subsets) can recognize some tumor determinants (e.g., differentially glycosylated mucins) independent of MHC class I expression (Maeurer & Lotze, 1997). There is also an extensive literature indicating that natural killer (NK) cells, operating through as yet not-well-established molecular recognition mechanisms, can detect and lyse some types of cancer cells, a reaction that is dramatically enhanced by the presence of cytokines (e.g., interferon) (Brittenden *et al.*, 1996). Macrophages have been reported not only to assist (as antigen-presenting cells) more specific responses, but also to have direct cytotoxic effects on some tumor cells (Greenberg, 1991).

Immune responses of one type or another to autologous tumor antigens have been convincingly documented in cancer patients, but such responses are typically weak (Brittenden *et al.*, 1996). Some investigators argue that the weakness of these responses suggests that immune defenses have little to do with cancer, while others argue that weak responses would be expected in individuals who have cancer, because immune surveillance mechanisms have already clearly failed in such cases. During the prolonged cellular expansion phase, before tumors become clinically detectable, selection pressures from immune defenses may result in the tumor-expression of various adaptive mechanisms for eluding the immune system (Greenberg, 1991; Shu *et al.*, 1997). For example, some tumors have been found to express very low levels of the MHC class I molecules necessary for targeting by cytotoxic T cells; others have been found to secrete factors that inhibit immune function;

while yet others have been found to shed tumor antigens and thus induce immune tolerance (Schreiber, 1993). In addition to such tumor-related factors, host factors have also been shown to play a role in patients' low levels of immune response to their tumors. Most obviously, patients' immune responses have been found to be suppressed by routine cancer treatments, including surgery and chemotherapy, the latter of which is known to have profound and lasting effects on the immune system. Other possible suppressive influences include exposure to carcinogens, concurrent infectious disease, and, of course, emotional distress (Ehrke, Mihich, Berd, & Mastrangelo, 1989; Lewis & McGee, 1992). Such suppressive effects may operate through a multitude of mechanisms, including direct influences on primary effector processes, as well as influences on cell–cell interactions (e.g., antigen presentation) and amplification steps (e.g., IL-2 secretion) known to be critical to the development of an effective immune response (Schreiber, 1993).

24.4.3. Do the Responses of the Immune System to Cancer Cells Affect the Development of a Tumor or Its Progression in Humans?

Although it is clearly evident that the immune system can recognize and destroy cancer cells, data that such effects play a major role in the prevention of cancer, as hypothesized by immune surveillance theories, are scant. In part, the paucity of such evidence reflects the difficulties inherent in conducting this type of research. Random assignment of people to have their immune defenses compromised is ethically indefensible. Prospective correlational studies of healthy individuals are impractical because the low incidence of cancer necessitates that large numbers of people be studied, while the slow development of clinically evident cancer requires years of follow-up. "Natural experiments" with immunosuppressed participants comprising congenital immune deficits, immunosuppressive disease (e.g., AIDS), or immunosuppressive treatments (e.g., following transplantation), are confounded by concurrent pathogenic processes and thus are difficult to interpret. Among those with immune deficits, there has been little evidence to suggest a generalized increased risk of the more common solid tumors (e.g., lung cancer, breast cancer), which are responsible for the bulk of cancer mortality, although increased risk of virally induced cancer has been clearly documented (Greenberg, 1991; Schreiber, 1993).

Prospective studies examining the relationships between immune measures and the progression of disease in patients with a diagnosis of cancer have been more common. There have been a number of reports of positive results. For example, a study of 90 women undergoing standard treatment for breast cancer found that lower T-cell responses to *in vitro* challenge by the mitogen PHA were prognostic of patients' risks of subsequently developing metastatic disease (Wiltschke, Krainer, Budinsky, Berger, Muller, Zeillinger, Speiser, Kubista, Eibl, & Zielinski, 1995). The results in the literature, however, have been mixed and are difficult to interpret, because investigators have often failed to control for even the most obvious confounding variables (e.g., subsets of participants receiving immunosuppressive chemotherapy). More compelling evidence of the potential clinical impact of immune defenses in cancer progression has come from recent therapeutic intervention trials (Scott & Cebon, 1997). Ongoing trials designed to enhance immune responses to tumor antigens using both nonspecific (e.g., interferon treatment) and specific approaches (e.g., tumor targeting with antibodies, vaccination with specific tumor antigens) suggest that the next century will see the development of effective immune treatments for cancer (Scott & Cebon, 1997).

24.5. Psychosocial Factors Affect Immune Defenses

As extensively reviewed by other contributors to this volume, studies have clearly established that psychosocial factors can affect the immune system. Both experimental and naturally occurring stressors have been shown to affect a number of endocrine, immune, and other biological parameters, which may be of importance regarding the risk of developing cancer. While acute experimental stressors generally are associated with short-term increases in NK cell activity (Zachariae, 1996), major life events, including surviving the death of a spouse and taking care of a sick relative, have generally been linked to diminished NK cell numbers and reduced NK cell activity, as indicated by a recent meta-analysis (Herbert & Cohen, 1993b). For example, Irwin, Daniels, Smith, Bloom, and Weiner (1987e) studied two groups of women whose husbands were either receiving treatment for or had died of lung cancer and compared both NK cell activity and depression scores with those of a control group. Impairments of NK cell activity found both in bereaved women and in the women whose husbands were receiving treatment were associated with the severity of depressive symptoms.

As reviewed in Chapter 20, numerous studies with healthy subjects have indicated that immune measures are affected by a variety of psychosocial interventions (Ironson, Antoni, & Lutgendorf, 1995), including relaxation (van Rood, Bogaards, Goulmy, & van Houwelingen, 1993), hypnosis (Zachariae, 1996), and self-disclosure (Christensen, Edwards, Wiebe, Benotsch, McKelvey, Andrews, & Lubaroff, 1996), as well as classical conditioning (Bovbjerg & Redd, 1992). There are also several studies on psychosocial interventions in cancer patients (Lewis *et al.*, 1994). To take one recent example, relaxation during chemotherapy for ovarian cancer was found to positively affect patients' lymphocyte counts (Lekander, Furst, Rotstein, Hursti, & Fredrikson, 1997).

24.6. The Effects of Psychosocial Factors on Cancer Are Mediated by Changes in Immune Defenses

As highlighted above, the data supporting the individual links in the chain of reasoning underlying the conventional view of psychoneuroimmunology and cancer are weaker than is often assumed. There is even less evidence to support the entire chain of reasoning, because only very few psychosocial studies of cancer have included immune variables in their investigations. One exception comes from the early work of Levy and Roberts (1992), who combined the study of the role of psychosocial factors in breast cancer prognosis with assessments of immune function. In one of their investigations, 90 breast cancer patients were assessed after surgery for psychosocial factors (e.g., fatigue, social support, adjustment) and NK cell activity, and the status of the women was then followed for several years, during which time 29 women had recurrences. Analyses indicated that both NK cell activity and psychosocial factors were individually related to risk of recurrence. However, variability in NK cell activity was not found to mediate the demonstrated relationship between the psychosocial variables and prognosis. Similar results arose when Fawzy and colleagues analyzed data on NK cell activity for the participants in their intervention study (described earlier) (Fawzy *et al.*, 1994). In addition to beneficial psychological effects and increased survival time, immune assessments indicated that the intervention patients had increases in the percentage of large granular lymphocytes (the NK cell phenotype), increases in NK cell numbers, and increases in IFNα-augmented NK activity. Statistical analyses indicated, however, that the intervention-induced immune changes were not found to mediate the intervention effects on the length of patients' survival.

24.7. The Chutzpah Hypothesis[1]

On the basis of the research findings reviewed here, a skeptic may be tempted to reject the conventional view of the role of psychoneuroimmunology in oncology. In our view, that response would be inappropriate for at least two reasons. First, the links in the chain of reasoning may be weak, but they are not without substance. Second, according to our chutzpah hypothesis (Bovbjerg, 1994), one of the reasons for the weakness of the evidence relating immune function to the development and/or progression of cancer may be the failure of immunologists to consider the influence of psychosocial variables on their immune measures. There is substantial evidence that the diagnosis of cancer and its treatment are powerful life stressors (Holland *et al.*, 1998). Even healthy individuals at risk because of a family history of cancer have been found to have higher levels of distress than individuals at normal risk (Valdimarsdottir & Bovbjerg, 1994). In light of the compelling evidence that emotional distress can influence immune function, one must be concerned that immune assessments collected without consideration of psychological effects may be misleading. Contradictory findings in previous studies of immunological defenses against cancer may reflect the fact that samples were collected under different stress conditions. Studies in the immunological literature rarely even mention the timing of blood sampling, which may be critical when one considers the obvious medical stressors typically confronting participants in such studies (e.g., cancer diagnosis). Consistent with the chutzpah hypothesis, we have found that the lower natural cytotoxic activity in individuals with family histories of cancer, previously reported by others, at least in part could be attributed to higher levels of emotional distress in these individuals (Bovbjerg & Valdimarsdottir, 1993). It should be emphasized that such findings do not rule out possible underlying heritable differences in immune function in individuals with family histories of cancer, but do suggest that taking into account the effects of psychological factors may help to clarify the relationships between immune function and cancer. These initial results suggest that the role of immune defenses in cancer might be clarified if immunological investigators routinely included assessments of psychosocial factors in their studies.

24.8. Psychoneuroimmunology and Cancer: Turning the Conventional View on Its Head

The conventional view of the role of psychoneuroimmunology in cancer is often presented as a one-directional, often monocausal, chain of reasoning that fails to take into account the fundamental bidirectional nature of psychoneuroimmunological connections (Zachariae, 1996). That is, one must recognize the possibility that the causal arrows of the "conventional view of psychoneuroimmunology in cancer" (Fig. 1) could all be reversed (Bovbjerg, 1994). See Fig. 2.

It has been argued that the prevailing "defense model of immunity," which focuses on

---

[1]For readers unfamiliar with the concept of chutzpah, it is perhaps best captured by the story of the boy who, having killed his parents, throws himself on the mercy of the court because he is an orphan.

(1) Cancer & treatment → Psychosocial factors
(2) Immune system → Psychosocial factors
(3) Cancer & treatment → Immune system
Q.E.D. (4) Cancer & treatment → Immune system → Psychosocial factors

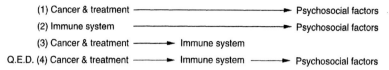

Figure 2. Psychoneuroimmunology and oncology: turning the conventional view on its head. As indicated in this schematic, there is evidence to suggest that the causal connections view between psychological factors, immune function, and cancer may operate in both directions. That is, analogous to the well-established effects of infectious disease, cancer and its treatment may affect the activity of the immune system, which in turn may affect the activity of the brain and thereby influence behavioral and psychological processes (e.g., illness behaviors).

the immune system as a relatively autonomous defense system, could represent an inadequate view of the role of the immune system in the living organism (Booth & Ashbridge, 1993). While psychoneuroimmunological research in oncology has focused primarily on the influence of psychological and behavioral factors on the immune system, other research has provided strong evidence that the activity of the immune system can affect behavioral and psychological processes, as has been reviewed in depth in Chapter 12. The possible impact of immunologically induced "sickness behavior" in cancer patients has yet to be examined. It is tempting to speculate that many of the common complaints of cancer patients (e.g., fatigue) may, at least in part, be the result of tumor- or treatment-related effects on immune processes that, in turn, affect brain activity. Increased understanding of such effects is important, not only in its own right, but also because some of these effects are likely to confound attempts to understand the impact of psychological factors on immune function in cancer. Prospective relationships among depression scores, immune function, and cancer progression, for example, may be obscured by patients' endorsement of the fatigue items on depression scales.

24.9. Psychoneuroimmunology and Cancer: Infectious Disease

In addition to affecting putative immune defenses against cancer, psychological influences on the immune system are likely to affect immune defenses that are undeniably involved in protecting us from infectious disease. The potential importance of psychoimmunology for infectious disease has been reviewed elsewhere in this book (see Chapter 27). For oncology, infectious disease is a critical problem, because it is a major source of morbidity and the number-one killer of cancer patients. In addition to the immunosuppressive effects that have been reported to be caused by the presence of the tumor itself, common treatments for cancer, such as chemotherapy, have powerful immunosuppressive consequences (Ehrke *et al.*, 1989). Psychological influences on immune function in cancer patients may have a particularly large impact on the risk of infection for at least two reasons. First, as noted earlier, distress levels are likely to be high in individuals who have been diagnosed with cancer and face aversive treatment procedures. Second, the impact of any given level of distress on immune defenses is likely to have a more profound impact on infectious disease in patients whose immune function is already compromised by other factors. Indeed, it is possible that psychological influences may synergize with underlying immune deficits to dramatically increase risk of infection, but as yet research data are scant. It is also important to recognize that once one gets beyond the assessment of antigen-specific aspects of

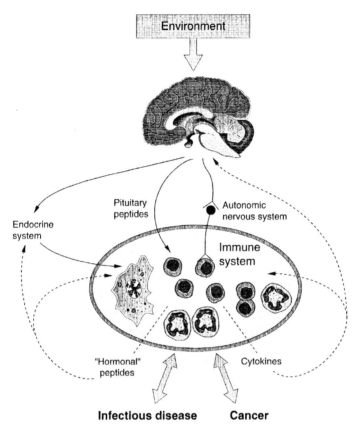

Figure 3. Psychoneuroimmunology and oncology: a schematic of interactions. As indicated in this diagram, there is now incontrovertible evidence of both efferent and afferent pathways by which the brain and immune system may interact. The existence of this "loop" provides the "hardwiring" necessary for the regulation of the immune system by the brain, the "software" for which is largely unknown. By analogy to other bodily systems, psychological and behavioral factors are thought to have their impact on the immune system by modulating the ongoing, day-to-day regulation of the immune system by the brain.

immune defenses, the measures that one would use in studies of infectious disease (e.g., cytokine secretion rates) would largely overlap with those typically used in cancer research.

24.10. Taking the "Psycho" out of Psychoneuroimmunology and Cancer

For other bodily systems (e.g., the cardiovascular system), fundamental regulatory processes (e.g., feedback and feedforward mechanisms) that operate independent of psychological processes have been clearly identified. For example, the CNS-mediated orthostatic response increases blood pressure when we stand up and keeps us from fainting (Weiner, 1992). Although the ways in which the brain may regulate the activity of the immune system are not yet well established, there is no reason to believe that feedback or feedforward regulatory strategies will not operate here as they do with other bodily systems. Indeed, there is considerable evidence that each of these types of regulatory processes modulates the activity of the immune system (Bovbjerg, Cohen, & Ader, 1982a). In addition, there is now incontrovertible evidence of the existence of both efferent (from the brain) and afferent (to the brain) pathways, the "loop" required for regulatory processes (see Fig. 2), as has been described in more detail in other chapters of this book. The possibility that cancer may be associated with dysfunction in such regulatory processes has received little research attention. Research in this area might suggest novel intervention strategies for cancer prevention and control.

24.11. Summary

In this chapter, we have first examined the conventional view of the role of psycho-neuroimmunology in oncology. According to this commonly held view, psychological effects on immune function provide the biological link by which psychosocial factors influence the development and/or progression of cancer. To date, few studies in the oncology literature have included both psychological and immune assessments. Fewer still have yet found evidence that psychological influences on immune function have an impact on the development or progression of cancer. The absence of such data is perhaps more understandable when one considers that the individual research literatures, exploring psychological influences on cancer and the involvement of immune defenses in cancer, do not provide strong guidance as to which measures are predictive of the development or progression of cancer. We have argued that immunologists' failure to consider the impact of psychological variables on their immune measures may be one of the reasons there has yet to be a strong consensus of evidence that particular immune defense mechanisms play a significant role in the development or progression of cancer. Recent excitement about the genetic basis of cancer should not blind us to the possible impact of immune defenses. "Either/or" thinking must be avoided; for example, the growing evidence of a link between cancer risk and mutations in specific susceptibility genes for cancer does not rule out the possibility that the penetrance of those genes may be modulated by polygenic inheritance of deficiencies in immune surveillance mechanisms, and the immunosuppression effects of stress. Future research should examine the possibility that cancer risk may be increased by interactions among a number of psychosocial, immune, and hereditary factors. We have also emphasized the important role that psychological influences on immune function may play in oncology, because these influences seem likely to increase the risk of infectious disease in cancer patients who not only have underlying immune deficits (e.g., from chemotherapy), but also are likely to experience high levels of distress. Finally, we have noted that interactions between the brain and immune system are likely to take place independent of psychological influences and that our understanding of the possible impact of such regulatory processes is in its infancy. A Chinese proverb says that even the longest journey must begin with a single step. Psychoneuroimmunology research in oncology has taken its first steps, but the journey has only begun.

ACKNOWLEDGMENTS. We thank Drs. Guy Montgomery and Joel Erblich, as well as Ms. Vida Petronis, for their comments on an earlier draft of this manuscript. We also acknowledge the financial support of the National Cancer Institute, the American Cancer Society, the U.S. Department of Defense, and the Martell Foundation. We are required to indicate that the content of the information contained in this report does not necessarily reflect the position or policy of the U.S. government.

25 Psychoneuroimmunology and HIV/AIDS

Neil Schneiderman, Michael Antoni, Gail Ironson,
Nancy Klimas, Barry Hurwitz, Mahendra Kumar,
Arthur LaPerriere, Kimberly Brownley,
and Mary Ann Fletcher

25.1. Introduction

The causative agent of the acquired immunodeficiency syndrome (AIDS) is the human immunodeficiency virus, type 1 (HIV-1). This is a retrovirus of the human T-cell leukemia/lymphoma line. An important feature of HIV spectrum disease, whose terminal stage is AIDS, is that it can have a long asymptomatic phase that may last as long as 10 to 15 years (Munoz, Wang, Good, Detels, Ginsberg, Kingsley, Phair, & Polk, 1988). This is followed by the appearance of constitutional signs and symptoms (Kaplan, Wofsky, & Volberding, 1987). Death usually ensues within 2 years (Lemp, Payne, Neal, Temelso, & Rutherford, 1990).

Neil Schneiderman, Michael Antoni, Gail Ironson, Nancy Klimas, Barry Hurwitz, Mahendra Kumur, Arthur LaPierre, Kimberly Brownley, and Mary Ann Fletcher • Department of Psychology, University of Miami, Coral Gables, Florida 33124-2040.

The first AIDS cases consisted of five gay men who were diagnosed in 1981 with Pneumocystis carinii pneumonia (Centers for Disease Control [CDC], 1981). It has been estimated that in 1980 there must have been approximately 100,000 individuals infected with HIV (Mann, 1990). By 1990 this number had increased 100-fold to 10 million HIV-infected people throughout the world (Mann, 1992). This included 6 million people in Africa, 1 million in Asia, 1 million in North America, 1 million in South America, and 500,000 in Europe. Some 60,485 cumulative cases of AIDS were recorded in Europe by 1991 (AIDS Feedback, 1991). By the end of the century, there will be more than 5 million AIDS cases reported worldwide (Chin & Mann, 1990).

Transmission of HIV requires the exchange of bodily fluids. In industrialized countries, the transmission of HIV has occurred primarily in homosexual men and injecting drug users although the prevalence of heterosexual transmission is increasing. Of 233,907 AIDS cases reported to the CDC in the United States by 1992, 25,947 cases (11%) occurred in women and 4051 cases (2%) in children (CDC, 1992). In contrast, in sub-Saharan Africa, more than 90% of HIV infections have been transmitted by heterosexual intercourse (Mann, Tarantola, & Netter, 1992). Despite the identification of risk factors for heterosexual HIV transmission, including contact with sex workers, multiple sexual partners, lack of male circumcision, and genital ulcers, the predominance of heterosexual transmission in sub-Saharan Africa remains poorly understood (Sandala, Lurie, Sunkutu, Chani, Hudes, & Hearst, 1995).

Most patients with primary HIV infection develop an acute mononucleosislike syndrome in the first few weeks after initial infection (Tindall & Cooper, 1991), which is associated with a high level of viremia (Daar, Moudgil, Meyer, & Ho, 1991). Although detectable viremia tends to decline markedly in the weeks after the acute syndrome subsides, the virus becomes widely disseminated during the early stage of infection. The decrease in viremia is temporally associated with the occurrence of an HIV-specific immune response that occurs 4 to 12 weeks after the onset of acute infection (Clerici, Berzofsky, Shearer, & Tacket, 1991). Viral replication does not appear to be completely curtailed, however, as it remains detectable in lymph nodes throughout the asymptomatic stages of HIV spectrum disease (Pantaleo, Graziosi, & Fauci, 1993).

Following the initial sequence of primary infection, viral dissemination, development of HIV-specific antibody production, and extensive curtailment of viral replication, a period of *clinical* but not *microbiological* latency occurs, which may last for many years. The progression of HIV disease continues during the clinical latency period; this is reflected in the depletion of T-helper/inducer cells (CD4) in peripheral blood and in the presence of HIV-infected lymphoid structures (Pantaleo *et al.*, 1993). Although the progressive decline of CD4$^+$ cells in peripheral blood is a major characteristic of HIV spectrum disease, it is conceivable that the high level of viremia in late-stage disease may in part reflect the recirculation of HIV particles removed from the constraint of lymph node entrapment (Pantaleo *et al.*, 1993). Nevertheless, it is the decline in CD4$^+$ cells destroyed by the virus that leaves the infected individual susceptible to opportunistic infections characteristic of AIDS. These opportunistic infections typically include Pneumocystis carinii pneumonia, cryptococcal meningitis, toxoplasmosis, candida esophagitis, and herpes simplex encephalitis (Kaplan *et al.*, 1987).

The recent advent of accurate viral load assessment has led to findings that viral burden provides a better predictor than CD4$^+$ cell count of subsequent CD4$^+$ cell count decline, AIDS development, and death in HIV-infected individuals (Mellors, Rinaldo, Gupta, White, Todd, & Kingsley, 1996). Modeling, however, indicates that CD4 and viral load markers have independent predictive value and that the markers together provide better prognostic discrimination than either one alone. Reductions in HIV-1 viral load correlate strongly with improved clinical outcome (O'Brien, Hartigan, Martin, & Esinhart, 1996). Theoretically, if the virus is completely suppressed, viral replication is arrested. Typically, if no measurable virus is found in peripheral blood, it is not found in the lymph nodes or semen either (Chun, Curruth, Finzi, & Shen, 1997). It appears that individuals who sustain unmeasurable viral loads for prolonged periods have no genotypic evidence of developing resistance, whereas patients with very low levels of measurable virus after antiviral therapy develop resistance over time (Deeks, Smith, Holodniy, & Kahn, 1997).

The amount of immune reconstitution and immune repertoire restoration after treatment depend on the degree of immunological damage at the time of intervention (Gaulton, Scobie, & Rosenzweig, 1997). A person with a relatively high CD4 count at the time of highly active antiretroviral therapy introduction will have a fuller immune repertoire in the periphery to draw from (i.e., somatic cells), a relatively undamaged thymus, and intact bone marrow. Such a person should theoretically be able to sustain a complete immune recovery. In contrast, a person in the late stage of disease, with a severely damaged thymus, and a severely depleted pool of somatic cells not only has an extremely limited somatic immune repertoire, but cannot draw naive cells from the bone marrow effectively. Despite increases in CD4 count, the cells are simply copies of whatever remained in the repertoire. Basically, the body replicates what is available, which is the depleted pool of somatic cells, and is unable to draw from the bone marrow, because of damage to the thymus and limits in extrathymic factors. If a person, for instance, had completely lost an immune response to cytomegalovirus (CMV), that individual might not be able to develop a new anti-CMV response (Jacobson, Kramer, Pavan, Owens, Pollard, & NIAID ACTG Protocol 266 Team, 1997). Thus, a patient who had a CD4 count of 40, reconstituted to 225 by therapy, may still be at higher risk for opportunistic infection than a person whose CD4 count had just fallen to 225 for the first time.

In our own work, we began by studying baseline immunological differences in asymptomatic gay men who wanted to learn their HIV serostatus (Klimas, Caralis, LaPerriere, Antoni, Ironson, Simoneau, Ashman, Schneiderman, & Fletcher, 1991). These men were initially tested 5 weeks before we drew blood for serostatus notification. We examined 45 men who subsequently turned out to be seronegative, 25 men who turned out to be seropositive, and 25 age- and gender-matched laboratory controls. The laboratory control condition was included because we suspected that enrolling high-risk gay men into a study, in which they would first learn about their serostatus, might be stressful to them and could thereby influence their immune status.

In the seronegative and control groups, the median value of CD4 counts was approximately 1000/mm³. In contrast, the median CD4 count in the seropositive group, which was asymptomatic, was 721/mm³. The median value observed in the asymptomatic seropositive group is consistent with the median CD4 count of 715 cells/mm³ found in the Multicenter AIDS Cohort Study (MACS) within 6 months of seroconversion (Giorgi & Detels, 1989). Because the MACS study found a drop to 626/mm³ after 1 year and a further drop to 530/mm³ for longer-term asymptomatic gay men, our seropositive group would appear to have been in an early stage of infection. In addition to lower CD4$^+$ cells and decreased numbers of the memory subset CD4$^+$CD29$^+$ in the seropositive as compared with the other two groups, our seropositive group also had lower proliferative responses to both phytohemagglutinin (PHA) and pokeweed mitogen (PWM) than the two seronegative groups.

We also observed some differences between the seronegative *study* group and the seronegative *laboratory control* group. The median ratio of T helper-inducer/T suppressor-cytotoxic (CD4$^+$/CD8$^+$) cells was approximately 1.3 in the seronegative group, which was significantly higher than the seropositive group but significantly lower than the laboratory control group. Both seropositive and seronegative study groups had significant elevations in the percentages of $CD8^+$ cells and the subsets $CD8^+12^+$ and $CD8^+12^-$ compared with the laboratory control group. Natural killer cell cytotoxicity (NKCC) and the mitogen response to PHA were lower in both study groups compared with the laboratory control group. The differences between the seronegative *study* group and the laboratory *control* group were consistent with a preliminary hypothesis that the decision to enter a study in which high-risk individuals were soon to learn their HIV serostatus was sufficiently stressful to influence the immune system.

The relationships among behavior, psychosocial variables, immunological status, and HIV spectrum disease are not yet understood, but there is reason to believe that such relationships exist. This chapter will examine these putative relationships within the contexts of psychosocial stressors, psychological states, disease progression, and behavioral interventions as secondary prevention.

25.2. Behavioral Stressors, Immune Function, and HIV

25.2.1. Stress, Hormones, and Immune Function

Exact functional relationships among behavioral stressors, immune function, and disease progression are almost completely unknown. This is especially the case for the relationships among behavioral stressors, immune status, and HIV progression. Relationships between various neurohormones and immune status, however, have been sufficiently documented to allow advancement of hypotheses relating psychosocial stressors on the one hand and immunomodulation on the other (see Chapter 9).

Different patterns of autonomic nervous system, neuroendocrine, and neuropeptide changes have been associated with the perception of psychosocial stressors with regard to the availability or nonavailability of coping responses (McCabe & Schneiderman, 1985). A specific pattern of autonomic activation, which occurs when coping responses are available and potentially adequate to meet stressful demands, is often associated with active coping. The sympathoadrenomedullary (SAM) system, activated during such active coping episodes, releases norepinephrine and epinephrine, and may promote increases in heart rate, cardiac contractility, and cardiac output. In contrast, another physiological response pattern appears to be dominant when coping responses are unavailable, as in those stressful situations defined as unpredictable, uncontrollable, or unrelenting. This pattern, associated with hypervigilance and lack of adequate coping resources, is characterized by behavioral inhibition as well as increases in norepinephrine and total peripheral resistance. Depending on the intensity, duration, and other characteristics of the stressor, the hypothalamic–pituitary–adrenocortical system (HPAC) may also become engaged. Activation of the HPAC system is associated with release of corticotropin-releasing hormone (CRH) from the hypothalamus, adrenocorticotropic hormone (ACTH) from the pituitary, and cortisol from the adrenal cortex.

The hormones of the HPAC and SAM systems are known to influence the immune system. Thus, CRH can inhibit NKCC (Pawlikowski, Zelazowski, Dohler, & Stepien, 1988); ACTH impairs the responsiveness of T (thymus-derived) lymphocytes to antigenic (CD3 antibody) and mitogenic stimuli such as concanavalin A (Kavelaars, Ballieux, & Heijnen, 1988); and corticosteroids impair or modify several components of cellular immunity including T lymphocytes (Cupps & Fauci, 1982), macrophages (Pavlidis &

Chirigos, 1980) and NKCC (Levy, Herberman, Lippman, & d'Angelo, 1987). Receptors for cortisol and ACTH have been established in lymphocytes and the interaction of cortisol receptors with appropriate levels of cortisol may inhibit cellular immune responses via changes in DNA and RNA synthesis and uptake (see Antoni, 1987 for review). Similarly, elevations in peripheral catecholamines have also been shown to influence the immune system. For example, sympathetic noradrenergic fibers innervate the vasculature as well as the parenchymal regions of lymphocytes and associated cells in several lymphoid organs in which nerve terminals are generally directed into zones of T lymphocytes (Felten, Felten, Carlson, Olschawka, & Livnat, 1985); activation causes the release of these lymphocytes into the peripheral circulation (see Chapter 8). The administration of β-adrenergic agonists has been associated with reduced NKCC (Katz, Zaytoun, & Fauci, 1982) and decreased T-lymphocyte proliferation (Plaut, 1987), which is consistent with the presence of β-adrenergic receptors found on lymphocytes.

Although catecholamines tend to reduce most immune functions in vitro, including NKCC (Bourne, Lichtenstein, Melmon, Henney, Weinstein, & Shearer, 1974; Hellstrand & Hermodsson, 1989; Katz et $al.$, 1982; Plaut, 1987), acute emotional stressors can lead to an increase in the number of NK cells in the peripheral circulation (Benschop, Oostveen, Heijnen, & Ballieux, 1993). In the resting state, NK cells are found along the endothelial cell layer of the arterial wall, which constitutes the vessel's marginal zone (Atherton & Born, 1972). In this state, the adhesive forces between the NK cell and the endothelium are greater than the shear forces that would allow the NK cells to move into the circulating pool of lymphocytes. During acute stress, increased flow in the systemic arterial system changes the equilibrium between adhesive and shear forces, permitting NK cells to leave the marginal zone. In addition, $β_2$-adrenoceptor stimulation of lymphocytes further disrupts interaction of NK cells with the endothelium, further decreasing adhesion (Benschop et $al.$, 1993). This would increase the number of NK cells recruited from the marginating pool to the circulating pool of lymphocytes. Whereas sympathetic nervous system activation may reduce NKCC activity (Bourne et $al.$, 1974; Hellstrand & Hermodsson, 1989), stress-induced recruitment of NK cells into the circulating pool would make an increased number of cells available for lysis once the peak concentration of catecholamines is diminished (Benschop, 1994, p. 119).

25.2.2. Anticipation and Responses to HIV Antibody Testing

When we assessed changes in psychological and immunological status during the 5 weeks both preceding and following HIV antibody testing in gay men, we observed significantly lower mitogen responses and NKCC in comparison with age- and sex-matched laboratory control subjects (Klimas et $al.$, 1991). Both the mitogen responses and NKCC, however, returned to normal values within 5 weeks after entry into the study and thereafter remained normal (Ironson, LaPerriere, Antoni, O'Hearn, Schneiderman, Klimas, & Fletcher, 1990). This suggested that for the high-risk men who ultimately turned out to be HIV seronegative, the decision to enroll in a study in which they would soon find out their serostatus was a potent stressor that significantly affected immunological function.

Examination of relationships among distress, cortisol level, and the mitogen response to PHA in the seronegative men indicated that at study entry both distress scores and plasma cortisol responses were elevated (Antoni, August, LaPerriere, Baggett, Klimas, Ironson, Schneiderman, & Fletcher, 1990a). In contrast, the mitogen response to PHA was depressed. We observed that during the 10-week study period, decreases in distress and cortisol level in the seronegative men were each reliably correlated with increases in the proliferative re-

PSYCHONEUROIMMUNOLOGY AND HIV/AIDS

sponse to PHA. Thus, there would seem to be systematic relationships among psychological distress, elevated cortisol levels, and diminished PHA responses in HIV seronegative men.

At entry into our study we had noted that the HIV seropositive men had a median CD4 count of $721/mm^3$ and diminished NKCC and proliferative responses to PWM and PHA compared with the men in the laboratory control condition (Klimas et al., 1991). Because the CD4 level was not extraordinarily low and because the seronegative men also had low NKCC and mitogen responses at the study outset, we expected that the seropositive men would show decreased blastogenic responses to their HIV diagnosis. We observed, however, that the seropositive men showed no change in their $CD4^+$ cell number or in lymphocyte responses to PHA and PWM following news of a positive HIV antibody test, despite significant increases in state anxiety and intrusive thoughts (Ironson et al., 1990). Instead, the seropositive men appeared to have a dampened immune response for mitogen stimulation. It is of considerable interest, however, that NKCC did decrease in response to the HIV seropositive diagnosis, suggesting important differences between CD4 and NK cell status. The NKCC decrease was correlated with increases in anxiety, suggesting that some important stressor–immune relationships still exist in HIV seropositive individuals.

The seropositive men also revealed a different pattern of endocrine–immune relationship than the seronegative men (Antoni, Schneiderman, Klimas, LaPerriere, Ironson, & Fletcher, 1991c). Whereas the HIV seronegative gay men revealed decreased distress after serostatus notification, accompanied by a decrease in cortisol and an increase in mitogen response to PHA, the seropositive men showed the expected increase in distress to serostatus notification, but also a paradoxical decrease in plasma cortisol. This decrease in plasma cortisol lasted for more than a week. The mitogen response to PHA was low both before and after serostatus notification.

In terms of stressor–endocrine–immune interactions, the seronegative men showed an expected positive correlation between distress and plasma cortisol, and an expected negative correlation between plasma cortisol and the proliferative lymphocyte response to PHA, when confronted with the stressor of entering a study in which they would find out their HIV diagnosis (Antoni et al., 1990a). The HIV seropositive men showed psychological distress responses to their HIV diagnosis, but in contrast to the seronegative men, they also showed an unexpected negative correlation between distress and cortisol and an unexpected positive relationship between cortisol and the mitogen response to PHA. This unanticipated pattern among HIV seropositive men was not attributable to differences in perceived risk prior to serostatus notification, extraneous environmental stressors, or differences in $CD4^+$ cell counts within the seropositive group.

Although our findings indicate that functional immune measures (i.e., NKCC, mitogen responses to PHA and PWM) are lower in HIV seropositive than in seronegative gay men before and after HIV diagnosis and that no significant changes in these measures occur as a function of diagnosis (Klimas et al., 1991), our data also indicate that the immune system of the seropositive men is not totally unresponsive. Namely, we observed that the seropositive men showed a decrease in NKCC following HIV serostatus notification that was related to anxiety responses at the time of notification (Ironson et al., 1990).

We also found a significant positive correlation between plasma cortisol and PHA responses ($r = 0.65$) in the seropositive men at the outset of the study (Antoni et al., 1991c). This contrasted with the reliable negative correlation ($r = -0.36$) displayed by the seronegative men. After HIV serostatus notification, the correlation between cortisol and the mitogen response to PHA in the seropositive men was $r = 0.56$. Also, in the seropositive group cortisol and PWM were not associated at baseline, but were significantly positively correlated ($r = 0.56$) after diagnosis. The picture that emerges among seronegative individuals is one of a strong relationship between increased distress scores and plasma cortisol on the one hand and diminished functional immune responses to a stressor (i.e., entry into the study) on the other. In contrast, among HIV seropositive gay men the picture is one of a "dampened" functional immune system and a paradoxical response to the stress of diagnosis. Thus, the seropositive men displayed increased distress that was associated with a decreased level of cortisol. Paradoxically, positive relationships were observed between cortisol and functional immune measures.

A conceivable explanation for our paradoxical findings may be understood in terms of disruption of the short-loop negative feedback system described by Axelrod and Reisine (1984). It can be seen in Fig. 1 that the release of ACTH by the anterior pituitary results in the release of cortisol, which produces negative feedback on the pituitary, terminating the further release of ACTH. Concomitantly, the release of ACTH by the anterior pituitary, when it results in the release of cortisol, facilitates the synthesis and release of plasma catecholamines (primarily epinephrine) from the adrenal medulla. This release of catecholamines from the adrenal medulla, in turn, provides positive feedback to the system.

To the extent that the distress of receiving an HIV seropositive diagnosis causes the CNS to activate the HPAC and SAM systems, the release of CRH from the hypothalamus should lead to increased release of ACTH from the pituitary and cortisol from the adrenal cortex. If, however, in the presence of increased central drive activating both the HPAC and SAM systems, the outflow from the SAM system is impaired, the decreased positive feedback from the SAM system could lead to a "safe signal," which initiates a decrease in cortisol release. This would lead to a decrease in cortisol level in response to psychological distress.

The positive relationship that we observed between decreases in cortisol level and diminished lymphocyte proliferation to mitogen stimulation would still require explanation. One possibility could be couched in terms of the adrenal medulla being more impaired than the neuronal innervation of the vasculature. Thus, while the impairment of the adrenal medulla would be expected to lead to a fall in cortisol in response to CNS-induced arousal, lesser impairment of the sympathetic innervation of the lymphoid organs could lead to a

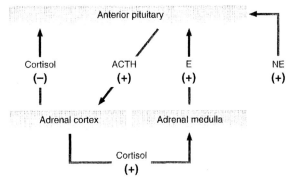

Figure 1.   Schematic diagram depicting how release of adrenocorticotropic hormone (ACTH) by the anterior pituitary results in the release of cortisol from the adrenal cortex, which produces negative feedback on the pituitary. The diagram also shows how the release of ACTH by the pituitary, when it results in the release of cortisol, facilitates the synthesis and release of catecholamines, primarily epinephrine (E) from the adrenal medulla. The release of E from the adrenal medulla and norepinephrine (NE) from nerve endings serves a positive feedback function. Adapted from Axelrod and Reisine (1984).

positive relationship between cortisol level and lymphocyte proliferation to plant mitogens. One would expect, however, that the mitogen responses would still be attenuated because of overall impairment of the SAM system.

Although the above line of reasoning is admittedly speculative, we have obtained some laboratory data that are consistent with the hypothesis. Thus, in our comparisons of autonomic function in HIV seronegative versus asymptomatic seropositive gay men, we have observed that the seropositive men had significantly less cardiac contractility and a lower stroke volume at rest (Starr, Antoni, Hurwitz, Rodriguez, Ironson, Fletcher, Kumar, Patarca, Lutgendorf, Quillian, Klimas, & Schneiderman, 1996) as well as decreased release of norepinephrine in response to the cold pressor stress test, a task in which the subjects placed a limb in ice water (Kumar, Morgan, Szapocznik, & Eisdorfer, 1991). In any event, the paradoxical findings that we have observed involving psychological distress, cortisol decreases, and immune function in HIV-infected individuals deserve further study.

25.2.3. Responses to Laboratory Stressors

In order to study the relationship between the HPAC and SAM systems on the one hand and immune system changes to stress on the other, we initiated a controlled laboratory study (Starr et al., 1996). Asymptomatic HIV seropositive versus seronegative gay men were subjected to an evaluative speech stressor task and to a mirror star tracing task. In the evaluative speech stress or task, the subjects were asked to prepare and present a story while being videotaped, using a scenario in which they were wrongfully accused of stealing a garment from a department store. The mirror tracing task required the subjects to trace the outline of a star-shaped pattern by using its mirror reflection, which horizontally inverts the image. Subjects were instructed to work as quickly and accurately as possible without lifting the pen from the pattern. Previous research in our laboratory, using impedance cardiography in conjunction with these tasks, determined that evaluative speech preparation raised blood pressure by increasing cardiac output via increased heart rate and contractility, whereas the mirror tracing task raised blood pressure by increasing systemic vascular resistance (Hurwitz, Nelesen, Saab, Nagel, Spitzer, Gellman, McCabe, Phillips, & Schneiderman, 1993).

Because our previous work on HIV-infected individuals had shown abnormalities in adrenocortical response to the natural stressor of HIV diagnosis (Antoni et al., 1991c), we examined ACTH and cortisol responses to our two laboratory behavioral challenges. Both ACTH and cortisol rose significantly to the speech stressor in the HIV seropositive and seronegative groups. These stressor-induced neurohumoral increases appeared to peak some 10 and 20 min following stressor onset, respectively, for ACTH and cortisol. This was similar to previously documented adrenocortical response latencies (Kuhn, 1989). Thus, it appears to be unlikely that the paradoxical decrease in cortisol that we previously found to the stress of serostatus notification in HIV-infected subjects (Antoni et al., 1991c) was the result of a specific defect in the HPAC axis.

In our laboratory behavior study, both ACTH and cortisol rose significantly to the speech stressor, but did not increase to the mirror tracing task (Starr et al., 1996). Both the preparation for speech and the mirror tracing task elicited comparable increases in blood pressure. The increase in blood pressure to the speech preparation was supported by an increase in cardiac output associated with increased myocardial drive (i.e., increased contractility). In contrast the increase in blood pressure to the mirror tracing task was associated with increased peripheral vasoconstriction (i.e., an increase in total peripheral resistance). Immune responses to the stressors occurred rapidly and closely in time with the cardiovascular responses, suggesting that stressor-induced sympathetic activation may be a critical mediator of the immune responses. These changes in immune response occurred too rapidly to be related to the ACTH or cortisol responses. Although in this study the immune response to stressful challenge appears to have been unaffected by the adrenocortical outflow, the findings do not address the possibility that important relationships may still exist under more chronically stressful conditions such as were present in our HIV serostatus notification study (Antoni *et al.*, 1991c).

Although no significant differences were found between HIV seronegative and seropositive men in task-induced ACTH and cortisol responses, there was a trend for a diminished ACTH response to the speech stressor in the HIV seropositive group. Previously, Mahendra Kumar in our group had shown that symptomatic HIV seropositive men produce a blunted ACTH response to the cold pressor test (Kumar, Kumar, Morgan, Szapocznik, & Eisdorfer, 1993). Thus, it appears likely that as HIV disease progresses, an abnormality in the HPAC axis may occur that results in diminished ACTH responsiveness.

The behavioral challenges employed in the laboratory behavior study induced distinctly different patterns of change in the enumerations of subpopulations of lymphocytes and in cellular immune function (Starr *et al.*, 1996). Specifically, during the speech preparation period both seropositive and seronegative groups exhibited increases in total number of lymphocytes, $CD4^+$, $CD8^+$, NK, and B cells as well as an increase in NKCC and a decrease in proliferation response to PHA. In contrast, the mirror tracing challenge elicited in both HIV seropositive and seronegative groups an increase in NK and B cells. The HIV seronegative men showed a significant decrease in $CD4^+$ cells whereas the seropositive men showed a significant increase in $CD4^+$ cells to the mirror tracing task.

The differences in the immune response patterns to the speech versus mirror tracing stressors cannot be accounted for by stressor duration as each lasted 4 min. Moreover, it is unlikely that the disparity in immune response patterns can be accounted for by differential stressor-induced response magnitudes, as the evoked blood pressure and heart rate elevations to these challenges did not differ. Thus, the present findings suggest that when a stressor induces a pattern of sympathetic nervous system activation that primarily challenges the heart, a concomitant immune response occurs that includes widespread cellular immune mobilization and alterations in cellular immune function; subsequent ACTH and cortisol responses follow. In contrast, when a stressor induces a pattern of increased sympathetic vascular drive, more limited immune alterations may be observed, which in this study included increased migration of NK and B cells and changes in $CD4^+$ cell migration. No subsequent pituitary–adrenocortical response occurs.

The finding that healthy HIV seropositive as well as seronegative individuals display immune responses to behavioral stressors has important implications. In our previous research, for example, we found that asymptomatic HIV seropositive men showed pronounced psychological stress responses to their seropositive diagnosis, but appeared to have a dampened immune response to mitogen stimulation. In contrast, the laboratory stress study seems to indicate that a considerable amount of immune responsiveness is still available to these men.

In a more recent study, however, conducted on 99 HIV seronegative and 132 seropositive asymptomatic and symptomatic men and women, we did provide evidence for a selective disruption of immunological responses to the speech stressor as a function of HIV serostatus (Brownley, Hurwitz, Fletcher, Kumar, Klimas, Milanovich, Motivala, Le Blanc, & Schneiderman, 1998). Compared with HIV seropositive subjects, seronegative subjects exhibited significantly greater increases in blood pressure caused in part by greater increases in cardiac output. As in the Starr *et al.* (1996) study, the immune response to the speech stressor included a widespread increase in lymphocyte expression into the circulation. The

HIV seronegative subjects, however, displayed significantly greater increases in NK cell count and NKCC and lesser increases in $CD3^+$, $CD8^+$, and HLA/DR (activated lymphoctye) cell counts than did the seropositive groups. Norepinephrine, epinephrine, cortisol, and ACTH responses to stressors did not differ significantly across groups.

In follow-up analyses, we examined whether the significant differences in immune response as a function of HIV serostatus could be accounted for by catecholamine, cortisol, or hemodynamic mechanisms. Controlling for cardiac output response and prevailing cortisol level, a series of differential relationships emerged. In the HIV seronegative group, stressor-induced increases in *both* norepinephrine and epinephrine were significantly related to NK, CD8 and HLA/DR, whereas in the HIV seropositive group, only norepinephrine was related to NK, CD8, and HLA/DR responses. Interestingly, in HIV seropositive, but not in seronegative subjects, epinephrine was related to NKCC responses. Our findings thus suggest a selective disruption in HIV seropositive individuals during acute stress, involving epinephrine and its relationship with immune trafficking and function, presumably implicating β-adrenergic receptors on lymphocytes, to which epinephrine has a strong affinity.

25.3. Psychosocial Predictors of HIV Disease Progression

In order to assess whether psychosocial variables predict immune changes and/or disease progression in HIV-infected gay men, we turned to our cohort of gay men who did not know their HIV serostatus on entering our study (Ironson, Friedman, Klimas, Antoni, Fletcher, LaPerriere, Simoneau, & Schneiderman, 1994). We monitored these men during the 5 weeks preceding and following their HIV diagnosis as well as 1 and 2 years thereafter.

A major potential confound in our analysis was that CD4 count at study entry was highly and significantly correlated with disease progression with symptoms ($r = 0.59$) and with disease progression to AIDS ($r = 0.59$). It is well known, of course, that decrease in CD4 enumeration is often used as a marker of disease progression itself. We therefore handled the potential confounds by making CD4 counts at study entry a covariate and then calculating partial correlations between our psychosocial variables and disease progression while controlling for CD4 number at entry to the study.

Magnitude of distress in response to the HIV seropositive diagnosis, magnitude of the increase in distress pre- to postdiagnosis, and denial increase pre- to postnotification were each positively and significantly correlated with disease progression to symptoms. The relationship between denial increase and symptoms was maintained even after controlling for $CD4^+$ cell number at entry into the study. An increase in denial pre- to postnotification was a significant predictor of disease progression to AIDS even after controlling for $CD4^+$ cell number at entry into the study. We also examined and found that magnitude of increase in denial pre- to postdiagnosis was significantly correlated with 1-year immune status. Thus, increase in denial correlated $r = -0.69$ with $CD4^+$ cell enumeration and $r = -0.68$ with lymphocyte response to PHA stimulation. We also found that 1-year immune status, in turn, was reliably correlated with 2-year disease progression. The correlation between CD4 with symptoms was $r = -0.58$ and with AIDS, $r = -0.73$; the correlation between PHA with symptoms was $r = -0.62$ and with AIDS, $r = -0.49$.

The study by Ironson *et al.* (1994) was the first published study to report significant relationships between psychological variables on the one hand and both changes in immune measures and HIV disease progression on the other. Previous studies that had attempted to relate psychosocial variables to immune status and/or disease progression in HIV-infected subjects produced inconsistent results. One study, for example, reported no relationship between depressive disorders, psychiatric distress, or psychosocial stressors on the one hand and changes in CD4 or CD8 on the other across a 6-month period (Rabkin, Williams, Remien, Goetz, Kertzner, & Gorman, 1991). The investigators did, however, observe a "suggestive pattern of association" between depression and an increased number of HIV-related symptoms. Conversely, another study reported that depressed HIV seropositive gay men revealed a faster decrease in CD4$^+$ cell counts over a 6-year period than a depressed comparison group, but depression did not predict an increase in disease progression (Burack, Barrett, Stall, Chesney, Ekstrand, & Coates, 1993). Complicating matters further, still another study reported that depression as measured by the Center for Epidemiological Studies—Depression Inventory did not predict faster CD4 decline, disease progression, or death in HIV-infected patients over an 8-year period (Lyketsos, Hoover, Guccione, Senterfitt, Dew, & Wesch, 1993).

There are numerous reasons why various studies may not have found a relationship between psychosocial factors on the one hand and both immune changes and disease progression on the other. These include (1) heterogeneity in the disease stage of subjects, (2) too limited a selection of immune markers, and/or (3) failure to assess individual differences in transactional variables such as coping strategies used to deal with specific stressors.

Because we have been able to demonstrate significant relationships between psychosocial variables on the one hand and changes in immune measures and HIV disease progression on the other (Ironson *et al.*, 1994), it would seem useful to consider some of the factors that might have contributed to differences in outcome. In the Ironson *et al.* (1994) study, we selected subjects who were asymptomatic and relatively homogeneous with regard to age and source of infection, thereby minimizing these potential sources of variance. We also looked at the lymphocyte proliferative response to PHA, which is a marker of CD4 states that had previously been associated with HIV disease progression (Page, Lai, Chitwood, Smith, Klimas, & Fletcher, 1990) and had been shown to differentiate HIV seropositive progressors from nonprogressors (Schellenkens, Roos, De Wolf, Lange, & Miedema, 1990). Also, by examining transactional factors such as denial and active coping strategies, which previously had been shown to impact on AIDS adjustment (Namir, Wolcott, Fawzy, & Alumbaugh, 1987) and immune status in HIV-infected gay men (Antoni, Goodkin, Goldstein, LaPerriere, Ironson, & Fletcher, 1991b), we extended the range of psychosocial factors studied in addition to the affective dimension. Of considerable importance also, we examined important psychological (i.e., distress, denial) and immune (e.g., CD4, PHA) responses to a potent stressor (i.e., serostatus notification). By selecting a sample of men who were asymptomatic and knew neither their HIV status nor their CD4 counts, we were able to disentangle the potential problem posed in prospective studies of not knowing which is causal in the constellation of having symptoms, immune system changes, and depression.

It is of interest that the above issues seem to be germane not only to the issue of disease progression, but also to the more general issue of relationships, as they exist, among psychosocial variables, immune status, and HIV spectrum disease. For example, one study found that an active coping style is associated with higher levels of NKCC in HIV seropositive gay men (Goodkin, Blaney, Feaster, Fletcher, Baum, Mantero-Atienza, Klimas, Millon, Szapocznik, & Eisdorfer, 1992) and another reported that in nonbereaved HIV seropositive gay men, higher levels of depressed mood were associated with fewer CD4$^+$ lymphocytes, more activated CD8$^+$ cells, and a lower proliferative response to PHA (Kemeny, Weiner, Taylor, Schneider, Visscher, & Fahey, 1994). In contrast, Perry, Fishman, Jacobsberg, and Frances (1992) found no relationship between 22 psychosocial variables—

except hopelessness—and CD4$^+$ cell count. In terms of this last study, it is worth noting that the subjects were quite heterogeneous in terms of stage of disease, age, and source of disease and that only CD4 enumeration, unrelated to stage of disease, was assessed.

25.4. Behavioral Interventions in HIV

There is presently no cure for AIDS although the recent advent of highly active antiretroviral combination therapy has brought about important improvements in the health care of some HIV-infected patients. Nevertheless, the therapy requires a complex combination of drugs that is expensive and poses major problems for adherence, so that only a tiny percentage of HIV-infected patients worldwide are likely to benefit. Thus, because only a small fraction of the more than 10 million HIV-infected people throughout the world have AIDS (Mann, 1992), as there is a relatively prolonged period between the onset of the infection and the development of AIDS, and because current pharmacotherapy regimens pose difficult problems for adherence, there is a compelling need to develop behavioral treatments to manage HIV.

The management of HIV spectrum disease involves psychosocial as well as biomedical considerations. Although the early asymptomatic stage after HIV diagnosis is undoubtedly stressful, studies conducted in Miami and Amsterdam, respectively, have reported that the levels of psychological distress in gay men were somewhat lower than might be expected (Blaney, Millon, Morgan, Eisdorfer, & Szapocznik, 1990; Mulder, Emmelkamp, Antoni, Mulder, Sandfort, & deVries, 1994). In both cases the asymptomatic HIV-infected gay men as a whole had higher levels of psychological distress compared with healthy noninfected norm groups, but considerably less than is commonly found in psychiatric patients. Similarly, low levels of depressive symptoms or syndromal psychiatric disorder in asymptomatic HIV-infected gay men in New York have been found (Rabkin, Williams, Neugebauer, Remien, & Goetz, 1990).

The anticipation and the impact of HIV antibody test notification among individuals at high risk for AIDS, however, can be highly stressful (Ironson *et al.*, 1990). Once a person is told that he or she is infected with HIV, the implications are profound. Feelings of anger and doom are usual as is the need to make major lifestyle changes (Christ & Weiner, 1985; Viney, Henry, Walker, & Crooks, 1989). There has also been reported to be an increase in the incidence of DSM-III Axis I affective and adjustment disorders (Jacobsen, Perry, & Hirsch, 1990). Rate of suicide in HIV seropositive individuals has been reported to be up to 36 times that of age-matched uninfected men (Marzuk, Tierney, Tardiff, Gross, Morgan, Hsu, & Mann, 1988). It would thus appear that given the large and increasing number of people who are infected with HIV, there is a substantial need to develop behavioral interventions to help infected individuals cope with the psychosocial aspects of their situation.

In our research we have conceptualized HIV infection as a long-term disease whose clinical course may be influenced by many factors. As is the case in other long-term diseases such as diabetes mellitus, it is useful to consider HIV spectrum disease within the framework of disregulation in one or more bodily systems. Progressive deterioration may occur across time as a function of the disorders occurring in these bodily systems. In fact, people suffering from HIV infection or diabetes die of complications of the disease rather than the disease itself. In HIV spectrum disease, AIDS occurs when the immune system has been sufficiently compromised for rapidly progressing cancer and opportunistic infections to occur. Because viral load and the extent to which the immune system becomes disordered appear to be the major predictors of the progression to symptoms and mortality, it would appear to be useful to develop pharmacological and behavioral interventions that can decrease the rate of viral replication and/or slow the decline of immune system status.

Because HIV infection is associated with increased distress (Antoni, Schneiderman, Fletcher, Goldstein, Ironson, & LaPerriere, 1990b; Kaisch & Anton-Culver, 1989), which in turn has adverse effects on the immune system (Ironson *et al.*, 1990; Irwin, Daniels, & Weiner, 1987d; Kiecolt-Glaser, Glaser, Dyer, Shuttleworth, Ogrocki, & Speicher, 1987b), behavioral interventions that decrease distress may beneficially impact immune status and thereby slow the course of disease. Furthermore, because HIV appears to be a multifactorial, multiphasic disease in which viral activity and immune status interact in a complex fashion (Pantaleo *et al.*, 1993), comprehensive therapy for HIV-infected people requires combination treatments that might well include a behavioral component (Schneiderman, Antoni, Ironson, Klimas, LaPerriere, Kumar, Esterling, & Fletcher, 1994).

One behavioral intervention approach that we have used with HIV-infected individuals has been group-based cognitive–behavioral stress management (CBSM). Our CBSM treatment includes the use of relaxation skills training, cognitive restructuring, instruction in self-monitoring of environmental stressors, and social skills training (Lichstein, 1988). In terms of psychosocial adjustments, our CBSM intervention is designed to help stressed individuals become aware of and reduce anxiety and distress, learn coping skills, and appropriately negotiate interpersonal conflict situations.

In our initial CBSM study, the goal was to assess the stress-buffering effects of our CBSM intervention on anxiety, depression, and immune system status in asymptomatic gay men learning of their HIV antibody test results (Antoni, Baggett, Ironson, August, LaPerriere, Klimas, Schneiderman, & Fletcher, 1991a). The 47 subjects enrolled in the study were assigned at entry into either a CBSM group or an assessment-only control condition. Subjects in the CBSM condition met twice weekly for 10 weeks in groups of 4–6 men led by two cotherapists. In one 90-min weekly session, subjects received training in cognitive restructuring, assertiveness skills, and behavior change strategies along with basic information on the psychological, social, and physiological aspects of stress responses, the nature of HIV transmission and associated risk behaviors, and a description of various safer sex behaviors. During these sessions, subjects were encouraged to generate examples of stressors recently experienced and to demonstrate the use of CBSM strategies as responses to these stressors by means of behavioral role play with other group members. In an additional 45 min session held each week, subjects received training in progressive muscle relaxation with the addition of an imagery component and a take-home imagery tape in Weeks 8–10. Because a key part of the CBSM intervention involved home practice of relaxation, subjects were required to self-monitor their daily practice frequency on cards that were distributed in packets of seven at each weekly relaxation session.

Psychological distress and immune measures were assessed at study entry, and at 4, 6, 8, and 10 weeks into the study. HIV antibody testing and notification of serostatus occurred between Weeks 5 and 6, and was given by a licensed clinical social worker, who had received extensive training in pre- and posttest counseling.

We observed that men in the control group who learned that they were HIV seropositive showed significant increases in anxiety and depression between pre- and postserostatus notification (see Fig. 2). In contrast, among the men who participated in the CBSM group, neither anxiety nor depression scores on the Profile of Mood States (POMS) questionnaire changed reliably during the pre- to post-HIV notification period. In fact, the depression scores following HIV serostatus notification for the men in the CBSM condition remained within the norms previously established for college students.

The immunological results paralleled the psychological findings. Thus, the men in the

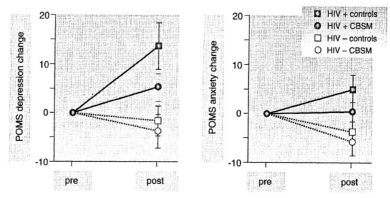

Figure 2. Changes in Profile of Mood States (POMS) anxiety and depression scores pre- to postnotification for seropositive (HIV+) and seronegative (HIV−) cognitive–behavioral stress management (CBSM) and control groups. Reprinted with permission from Antoni *et al.* (1991a).

control group who were told that they were HIV seropositive showed significant differences in CD4$^+$ and NK (CD56$^+$) cell counts as well as in lymphocyte proliferative responses to PHA in terms of pre- and postnotification change scores (see Fig. 3).

In a parallel study conducted on asymptomatic homosexual men, we examined the impact of aerobic exercise training as a buffer of the affective distress and immune changes that accompany HIV serostatus notification (LaPerriere, Antoni, Schneiderman, Ironson, Klimas, Caralis, & Fletcher, 1990). Fifty asymptomatic gay men with a fitness level of average or below (determined by predicted VO_{2max}) were randomly assigned either to an aerobic exercise training program or to a no-contact control condition. After 5 weeks of training, at a point 72 h before serostatus notification, we measured psychosocial, fitness, and immune measures. The psychosocial and immune measures were collected again 1-week after HIV serostatus notification.

We observed that HIV seropositive control subjects displayed significant increases in POMS anxiety and depression scores. They also revealed significant decrements in NK (CD56$^+$) cell number following HIV serostatus notification (see Fig. 4). In contrast, the HIV seropositive exercisers showed no similar changes, and in fact resembled both seronegative groups. These findings indicate that concurrent changes in some affective and immunological measures in response to an HIV serostatus notification stressor can be attenuated by an experimentally manipulated exercise training intervention.

The data that we have presented suggest that both CBSM and exercise can attenuate the distress associated with an HIV seropositive diagnosis and buffer some of the immune changes associated with the distress response (Antoni *et al.*, 1991a; LaPerriere *et al.*, 1990). We have also investigated the possibility of using CBSM and aerobic exercise to exert improved cellular control over latent herpesvirus activation under these same conditions in HIV-infected people (Esterling, Antoni, Schneiderman, Carver, LaPerriere, Klimas, Ironson, & Fletcher, 1992). The rationale for this study was based on previous findings that (1) both psychosocial stressors (Glaser, Pearson, Jones, Hillhouse, Kennedy, Mao, & Kiecolt-Glaser, 1991) and response styles (Esterling, Antoni, Kumar, & Schneiderman, 1990) have been associated with reactivation of latent herpesviruses such as Epstein–Barr virus (EBV) and herpes simplex virus (HSV) (Kemeny, Cohen, Zegans, & Conant, 1989); (2) herpes-

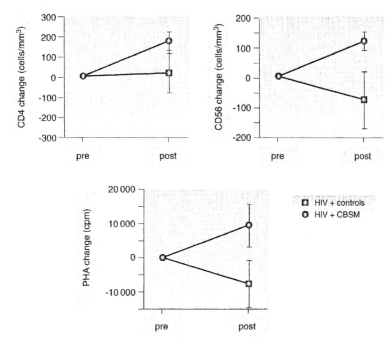

Figure 3. Helper-inducer (CD4$^+$) and natural killer (CD56$^+$) cell counts and lymphocyte proliferative responses to phytohemagglutinin (PHA) expressed as counts per minute (cpm) at pre- and postnotification time points for seropositive cognitive–behavioral stress management (HIV + CBSM) and HIV + control groups. Reprinted with permission from Antoni *et al.* (1991a).

viruses have been thought of as possible cofactors in the pathogenesis of AIDS because of their association with increased morbidity and mortality in patients with AIDS (Fletcher, 1992; Quinnan, Masur, & Rook, 1984) and their ability to specify transcriptional transactivators of HIV (Albrecht, DeLuca, Byrn, Schaffer, & Hammer, 1989); (3) even at early asymptomatic stages, HIV seropositive men have significantly higher EBV antibody titers

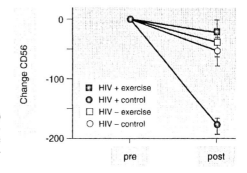

Figure 4. Changes in natural killer (CD56$^+$) cell counts pre- to postnotification for seropositive (HIV +) and seronegative (HIV −) aerobic exercise and control groups. Reprinted with permission from LaPerriere *et al.* (1990).

than matched HIV seronegative men (Sumaya, Boxwell, & Ench, 1986); (4) EBV reactivation may act as a cofactor for the progression of HIV to AIDS (Carbonari, Fiorilli, Mezzaroma, Cherchi, & Aiuti, 1989; Rosenberg & Fauci, 1991); and (5) relaxation training has been shown to decrease antibody titers of the herpesvirus, HSV (Kiecolt-Glaser, Glaser, Williger, Stout, Messick, Sheppard, Ricker, Romisher, Briner, Bonnell, & Donnerberg, 1985) (see Chapter 27).

A remarkable feature of both the herpesviruses and HIV is that when infection occurs, seroconversion and latency can result, with or without the presence of clinical disease. Thus, the herpesviruses persist in the host indefinitely (Rinaldo, 1990). Varicella and HSV establish latent infections in neurons, whereas CMV, human herpesvirus type 6 (HHV-6), and EBV persist in lymphocytes. Under certain conditions, the latent virus will lytically replicate, but the exact mechanisms causing reactivation are unknown. However, the latent state can be established in the presence of high levels of circulating antibody, suggesting that the cellular immune response is primarily responsible for controlling the latent infection.

When reactivation does occur, it is usually accompanied by a pronounced increase in specific antibody to the reactivating herpesvirus. Under sustained states of cellular immunodeficiency as occurs in people infected with HIV or in those undergoing immunosuppressive therapy for renal transplantation, reactivation of one or more of the latent herpesviruses results in severe morbidity and mortality (Ernberg, 1986; McVoy & Adler, 1989; Rinaldo, 1990). Moreover, herpesviruses are themselves known to suppress a wide number of immunological functions including cytokine production and lymphocyte proliferation to mitogens (Rinaldo, 1990). These phenomena suggest that HIV and herpesviruses may interact to perpetuate disruptions in the biological control of each respective infection and possibly in contributing to the severity and chronicity of related clinical signs and symptoms.

As CD4[+] cells become depleted with the advancement of HIV infection, innate immune functions such as NKCC may figure importantly in conferring protection against herpesviruses in the context of HIV-induced defects in major histocompatibility complex-restricted cytotoxic function (Biron, Byron, & Sullivan, 1989; Bukowski, Warner, Dennert, & Welsh, 1985). In similar fashion, stress-induced impairments in NKCC may facilitate herpesvirus reactivation in HIV-infected individuals (Glaser & Kiecolt-Glaser, 1987) with subsequent effects on HIV replication and progression to AIDS (Rosenberg & Fauci, 1991). Because several herpesviruses prevalent in HIV-infected people are known to have independent immunosuppressive effects, reactivation of these viruses could have implications for clinical disease progression (Griffiths & Grundy, 1987).

The notion that herpesviruses may be related to HIV progression is interesting and provocative. Because HHV-6 may reactivate during EBV infections and because both EBV and HHV-6 may reactivate following immunosuppression by HIV, these three viruses may potentiate one another in the progression of clinical disease. There is also some indication that once HHV-6 infects a lymphocyte, there are changes in the cell membrane fluidity and in receptor expression (i.e., CD2), allowing superinfection of HHV-6-infected cells by EBV as well as by HIV (Krueger, Schonnebeck, & Braun, 1990). Direct immunosuppression or transactivation of HIV by other herpesviruses such as CMV or HSV also provides a rationale for considering these concomitant infections as potential "cofactors" in the development of AIDS. If these viral interactions turn out to be valid and clinically meaningful, then it is highly possible the interventions designed to retard herpesvirus reactivation may be fruitful in the clinical management of HIV infection.

Within the context of our CBSM and aerobic exercise intervention (Esterling *et al.*, 1992), we have investigated EBV, which is the causative agent of infectious mononucleosis

(Henle, Henle, & Diehl, 1968) and a proposed etiological factor in Burkitt's lymphoma and nasopharyngeal carcinoma (de-The, 1980; Miller, 1980). Under usual conditions, EBV is under adequate intracellular and immunological control, which restricts viral replication and proliferation of EBV-positive B lymphocytes (Miller, 1980; Rickinson, Moss, Wallace, Rowe, Misko, Epstein, & Pope, 1981). Because HIV seems to bind to CD4$^+$ cells through the Leu3a/OKT4a epitope (Klatzmann, Champagne, Chouret, Grunest, Guetart, Hercend, Gluckman, & Montaganier, 1985), observation that EBV-transformed B cells also express the presumed attachment site for HIV (i.e., Leu3a epitope) (Alsip, Ench, Sumaya, & Boswell, 1988) is provocative. Evidence has also been presented that HIV can infect and multiply in EBV infected B lymphocytes (Montaganier, Gruest, Chamaret, Douguet, Axler, Guetard, Nugeyre, Barre-Sinoussi, Chermann, Klatzmann, & Gluckman, 1984; Pagano, Kenney, Markovitz, & Kamine, 1988). Thus, EBV-infected B cells may represent a potentially important reservoir for HIV, especially considering the proliferative capacity of these host cells (Rosenberg & Fauci, 1991).

Given the potential role of EBV and HHV-6 in the pathophysiology of HIV infection, we examined the effects of stress management (i.e., aerobic exercise or CBSM) on the HIV-infected host's ability to control the reactivation of these latent herpesviruses when the stress of HIV serostatus notification was involved. Sixty-five asymptomatic gay men were randomized into 10 weeks of aerobic exercise, CBSM, or a contact control condition. The subjects were tested for HIV serostatus, 5 weeks after entering the study, and received their diagnosis 72 h later. Potential immunomodulatory confounds such as nutritional status, alcohol use, drug use, and sleep were assessed at baseline and at the conclusion of the 10-week study period.

The HIV seropositive men had higher EBV viral capsid antigen (VCA) antibody titers than those diagnosed as seronegative at the beginning and end of the study as well as during all intermediate time points. In contrast, no significant differences were found between the HIV seropositive and seronegative men. Both the HIV seropositive and seronegative subjects in each of the interventions (i.e., aerobic exercise or CBSM) revealed significant decreases in both EBV VCA and HHV-6 antibody titers over the course of the intervention compared with the assessment-only seropositive and seronegative control subjects. These significant differences between intervention and control subjects were independent of total IgG levels and degree of polyclonal B-cell activation. It would thus appear that both the CBSM and aerobic exercise interventions can help to normalize immunological control over certain herpesviruses in both healthy individuals and those infected with HIV.

Having found that stress management interventions can help attenuate psychological distress and immune responses that occur in *asymptomatic* HIV seropositive gay men, we turned our attention to a CBSM intervention in *symptomatic* HIV seropositive homosexual men. We reasoned that if an HIV seropositive diagnosis represents a first major crisis in HIV-infected individuals, the emergence of symptoms in HIV may often present a second major crisis of adjustment (Bury, 1982; McDaniels, Hepworth, & Doherty, 1992). Several studies do indicate that distress and depression increase as HIV-infected men begin to experience the onset of symptoms (e.g., Belkin, Fleishman, Stein, Piette, & Mor, 1992; Burack et al., 1993; Hays, Turner, & Coates, 1992; Lyketsos et al., 1993; Ostrow, Monjan, Joseph, Van Raden, Fox, Kingsley, Dudley, & Phair, 1989; Rabkin et al., 1991; Singer, Zorilla, Fahy-Chandon, Chi, Syndulko, & Tourtellotte, 1993). Moreover, symptomatic HIV seropositive men may experience even more depression and anxiety than those with defined AIDS, because of the uncertainty they experience with regard to their eventual disease course (Chuang, Devins, Hunsley, & Gill, 1989; Dilley, Ochitill, Perl, & Volberding, 1985; Tross & Hirsch, 1988).

In our previous research we also found that CBSM can attenuate the latent reactivation of some herpesviruses in HIV seronegative people (Esterling *et al.*, 1992). Because earlier research also correlated increased distress and depression levels to HSV recurrences (Cappell, Gregoire, Thiry, & Sprecher, 1978; Goldmeier & Johnson, 1983; Katcher, Brightman, Luborsky, & Ship, 1973; Kemeny *et al.*, 1989) and increased HSV antibody titers (Kiecolt-Glaser *et al.*, 1985; Kiecolt-Glaser, Kennedy, Malkoff, Fisher, Speicher, & Glaser, 1988), we felt that this herpesvirus was a good candidate for stress management research. In addition, sexually active homosexual men are at high risk for HSV-1 and HSV-2 with the seroprevalence rate for HSV-2 in healthy gay men being over 90% (Nahmias & Josey, 1982).

Whereas HSV-1 tends to cause lesions in the oropharynx and latent infection in the trigeminal ganglia, HSV-2 causes lesions in the genital skin and mucosa and latent infection in the sacral ganglia. Both HSV-1 and HSV-2 can coinfect $CD4^+$ cells already infected with HIV (Albrecht *et al.*, 1989; Kucera, Leake, Iyer, Raben, & Myrvik, 1990). The viruses may also serve as antigens, which can result in increased HIV replication and increased susceptibility of noninfected cells to become infected (Rosenberg & Fauci, 1991). Moreover, HSV-1 and HSV-2 can upregulate the expression of HIV (Panataleo *et al.*, 1993) and the replication of HIV (Albrecht *et al.*, 1989; Kucera *et al.*, 1990). HSV ulcers are also thought to act as portals of entry for HIV (Stamm, Handsfield, Rompalo, Ashley, Roberts, & Corey, 1988). Finally, HSV infection has been associated with impaired lymphocyte and cytokine production (Wainberg, Portney, Clecner, Hubschman, Lagace-Simard, Rabinovitch, Remer, & Mendelson, 1985).

Based on the preceding considerations, we tested the effects of a 10-week, group-based, CBSM intervention using Beck Depression Inventory (BDI) scores, POMS scores, CD4 enumeration, HSV-1 and HSV-2 antibodies, and total IgG levels as outcome measures (Lutgendorf, Antoni, Ironson, Klimas, Kumar, Starr, McCabe, Cleven, Fletcher, & Schneiderman, 1997). The 22 intervention and 18 control subjects were *symptomatic* HIV seropositive gay men who did not have AIDS. The men's symptoms included lymphadenopathy, diarrhea, oral hairy leukoplakia, and chronic recurrent skin rash, among others. Subjects with AIDS-defining opportunistic infections such as toxoplasmosis of the brain, Pneumocystis carinii pneumonia, progressive multifocal leukoencephalopathy, or cancers such as Kaposi's sarcoma or high-grade B-cell lymphoma, were excluded from the study. Subjects taking azidothymidine (AZT) or another antiretroviral agent or combination had to have maintained a constant dosage for at least 2 months without change prior to the study. Twenty-one subjects (CBSM, $n = 12$; controls, $n = 9$) took antiretroviral drugs during the study. Six subjects took acyclovir or Zovirax during the study.

Control measures assessed included sleep, aerobic exercise, alcohol, tobacco, caffeine, other drugs, and nutrition. Subjects were also screened to eliminate those suffering from current major depression or other major psychiatric disorder.

Levels of depression and anxiety in the symptomatic men in this study exceeded those of the asymptomatic HIV seropositive men studied in our previous CBSM intervention study (Antoni *et al.*, 1991a). The mean BDI depression score of the symptomatic men at baseline (Lutgendorf *et al.*, 1997) corresponds to mild depression (Kendall, Hollon, Beck, Hammen, & Ingram, 1987). We found that the CBSM intervention significantly decreased depression as measured by the BDI and anxiety as measured by the POMS. Moreover, the intervention subjects who practiced relaxation the most consistently throughout the intervention period had the greatest drops in depression scores. Significant decreases in HSV-2 antibody levels across the 10-week study period were noted in the CBSM intervention group but not in the control group. This effect was independent of polyclonal IgG activation as assessed by total antibody to IgG. In contrast, the intervention did not produce significant changes in CD4$^+$ cell counts or in HSV-1 antibody levels. Decreased BDI depression scores were associated with decreased HSV-2 antibody levels, $r = 0.47$, but there were no other significant associations between mood and immunological variables. These findings extend our previous intervention work in *asymptomatic* HIV-infected men by showing that even in *symptomatic* seropositive men, a group-based CBSM intervention can reduce depression and help to normalize immunological control over a herpesvirus.

The present finding that CBSM can decrease HSV-2 antibody levels in symptomatic HIV seropositive individuals is particularly significant in light of the literature supporting a role for HSV-2 infection as a possible cofactor in HIV infection (Kucera *et al.*, 1990; Stamm *et al.*, 1988; Wainberg *et al.*, 1985) as well as its role in increased morbidity and mortality in HIV-infected people (Fletcher, 1992; Quinnan, Masur, & Rook, 1984). To the extent that viral cofactors may be implicated in the transformation from a predominantly Th1 to Th2 type of immune response (Ezzell, 1993; Salk, Bretcher, Salk, Clerici, & Shearer, 1993), which may be a key transition in the course of the infection (Shearer & Clerici, 1992), it is conceivable that HSV may be one such factor that serves in this capacity, thus weakening the body's resistance to HIV and possibly contributing to downward progression of the immune system and increased morbidity and susceptibility to opportunistic pathogens. Cytokine and genetic programs can be defined in terms of (1) T-helper 1 cells (Th1), which stimulate cytotoxic cell activity by production of interleukin 2 and γ-interferon and (2) T-helper 2 cells (Th2), which secrete interleukins 4, 5, 6, and 10. The predominant CD4$^+$ cell in asymptomatic HIV-infected humans is Th1, whereas the predominant CD4$^+$ cell in the later stages of HIV may be Th2. It remains to be investigated whether decreasing antibody titers to herpesviruses by CBSM or other treatments can delay or reverse Th1-to-Th2 changes.

25.5. Summary and Conclusions

The clinical course of HIV spectrum disease appears to be influenced by a wide variety of biological and psychosocial factors (Schneiderman, Antoni, Ironson, LaPerriere, & Fletcher, 1992). In addition to the direct burdens of HIV infection and its sequelae, individuals with HIV face a variety of social stigmas. They also must endure numerous psychosocial stressors including their own perceptions of progressive physical and neurological deterioration (Redfield & Burke, 1988), overwhelming medical costs (Bloom & Carliner, 1988), and multiple bereavements (Martin, 1988). In response to a host of psychosocial and environmental stressors as well as the fears of impending disability and death, infected people may also experience prolonged periods of distress and depression and may increase sexual risk behaviors and substance use. To the extent that these variables adversely impact on psychosocial functioning, immune status, and disease progression, there is an urgent need for further scientific investigation. Such investigation needs to focus on behavior–immune–disease interactions and the role that behavioral interventions may play in the management of HIV-infected patients.

One of the hallmarks of HIV spectrum disease is the progressive decline of CD4$^+$ cells in peripheral blood. It is the decline of CD4$^+$ cells destroyed by the virus that leaves the infected individual susceptible to opportunistic infections characteristic of AIDS. In our own research, we have observed that in addition to a reduced number of CD4$^+$ cells, even asymptomatic HIV-infected individuals show a decrease in CD4$^+$CD29$^+$ number, and decreased lymphocyte proliferative responses to both PHA and PWM (Klimas *et al.*, 1991).

Evidence has begun to accumulate that psychosocial factors may impinge on the immune status of HIV-infected individuals. Thus, for example, higher levels of depressed mood in nonbereaved HIV seropositive gay men have been associated with a diminished number of CD4$^+$ cells, an increased number of CD8$^+$ lymphocytes, and a lower proliferative response to PHA (Kemeny et al., 1994). We have also observed, at least in asymptomatic HIV seropositive individuals, that an increase in distress is associated with a decreased proliferative response to PHA (Antoni et al., 1991c). We also observed that in response to the stressor of HIV serostatus notification, seropositive gay men revealed a decrease in NKCC (Ironson et al., 1990). Conversely, an active coping style has been associated with higher levels of NKCC in HIV seropositive gay men (Goodkin et al., 1992). Although our study of the impact of serostatus notification has suggested that HIV seropositive individuals have a somewhat dampened immune response to mitogen stimulation (Ironson et al., 1990), our laboratory stress study indicates that asymptomatic HIV-infected individuals still have a considerable amount of immune responsiveness available to them (Starr et al., 1996).

The study in which we examined the impact of HIV serostatus notification on the course of HIV spectrum disease, indicated that significant relationships exist between psychosocial variables (e.g., denial) on the one hand and changes in both immune measures (i.e., decreases in CD4 count and lymphocyte proliferation to PHA stimulation) and HIV disease progression on the other (Ironson et al., 1994). Although this last finding must be interpreted with caution because of the small number of subjects studied, its success in comparison with investigations reporting negative results (Lyketsos et al., 1993; Rabkin et al., 1991) may be related to our use of (1) asymptomatic, relatively homogeneous subjects, (2) a reasonable selection of immune markers, (3) examination of transactional factors such as denial and coping as well as affect measures, and (4) a procedure to time-lock the psychological (distress, denial) and immune (CD4, PHA) responses to a potent stressor (serostatus notification).

There is presently no cure for HIV spectrum disease. Although pharmacological treatments have made progress in reducing viral load to very low levels, ameliorating symptoms, and dealing with intercurrent illnesses, the management of HIV spectrum disease involves psychosocial as well as biomedical considerations. In our initial psychosocial intervention study, we set out to assess the stress-buffering effects of a group-based CBSM intervention on distress and immune system status in asymptomatic gay men learning of their antibody results (Antoni et al., 1991a). We found that relative to HIV infected men in the control group, HIV seropositive subjects in the CBSM group revealed lower distress scores and less impairment in proliferative responses to PHA; significant differences between groups in CD4$^+$ and NK (CD56$^+$) cell enumeration were also observed.

In a parallel study using aerobic exercise as the intervention, we observed that relative to asymptomatic HIV seropositive control subjects, asymptomatic HIV-infected men in the exercise conditions showed less anxiety and depression and a significantly smaller decrement in NK cell number following HIV serostatus notification (LaPerriere et al., 1990). When we examined the effects of CBSM and aerobic exercise in terms of their ability to normalize immunological control over the herpesviruses EBV and HHV-6 in the face of the HIV notification stressor, we found that both interventions in asymptomatic HIV-infected subjects led to decreases in antibody titers over the course of the intervention when comparisons were made with asymptomatic HIV seropositive control subjects (Esterling et al., 1992). More recently, in a study conducted on symptomatic HIV-infected men, we found that relative to those in a control condition, only those men in the CBSM group revealed decreases in Beck depression and POMS anxiety scores as well as decreases in HSV-2 antibody titers (Lutgendorf *et al.*, 1997). The studies that have examined the impact of CBSM on psychosocial variables and antibody titers to herpesviruses in HIV-infected men strongly suggest that a group based stress management program that decreases distress may have an important impact on immune surveillance and control over herpesviruses.

The recent advent of combination therapies including nucleoside analogues (e.g., zidovudine, AZT), nonnucleoside reverse transcriptase inhibitors (e.g., nevirapine), and protease inhibitors (e.g., saquinavir) have brought about important improvements in the health care of many HIV-infected patients (Lewin, 1996). This has meant that now more than ever HIV/AIDS needs to be looked at as a chronic disease in which patient management is important, for once patients go on highly active antiretroviral therapy, missing surprisingly few doses can rapidly lead to viral escape and mutations that result in drug resistance and ineffective therapeutic response (Van Howe, Shapiro, Winters, Merigan, & Blaschke, 1996). At present little data are available concerning adherence to combination therapy, but experience derived from monotherapy indicates that many patients reveal poor compliance.

One study, for example, found that 25% of HIV seropositive men who have sex with men did not adhere to secondary prevention guidelines including those associated with antiviral treatment. Attitudes toward health care in general, and toward antiviral treatments in particular, differentiated adherent from nonadherent men (Stall, Hoff, Coates, Paul, Phillips, Ekstrand, Kegeles, Catania, Daigle, & Daz, 1996). Another group found that better adherence to AZT was associated with greater social stability, greater perceived social support, increased perceived benefits of AZT use, and reduced perceptions of barriers to use (Morse, Simon, Coburn, Hyslop, Greenspan, & Balson 1991). Still another group found that adherence to HIV antiretroviral therapy was associated with lower Beck depression scores, lower total mood disturbance on the POMS, higher social support, and more adaptive coping skills (Singh, Squier, Sivek, Wagener, Hong Nguyen, & Yu, 1996). Because our studies using CBSM show improvements on these measures (Antoni *et al.*, 1991a; Esterling *et al.*, 1992; Lutgendorf *et al.*, 1997), the addition of modules on adherence to highly active antiretroviral therapy, is likely to prove useful in improving medication adherence in HIV-infected patients. To the extent that the same interventions including relaxation training reduce antibody titers to herpesviruses and may reduce activated subsets of "at risk" CD4$^+$ cells, CBSM interventions may impact both directly and indirectly on important biological endpoints.

In summary, the exact relationships among behavior, psychosocial variables, immunological status, viral burden, and HIV spectrum disease are not yet understood, but there is now reason to believe that such relationships exist and are important. The present chapter examined some of these relationships in the context of psychosocial variables, responses to psychosocial stressors, disease progression, and behavioral interventions as secondary prevention. The study of psychoneuroimmunology in relation to HIV/AIDS appears to be an important, fertile ground for future research.

ACKNOWLEDGMENTS. This work was supported by NIMH program project MH49548.

26 Psychoneuroimmunology and Autoimmune Diseases

Annemieke Kavelaars and Cobi J. Heijnen

Annemieke Kavelaars and Cobi J. Heijnen • Department of Immunology, University Hospital for Children and Youth "Het Wilhelmina Kinderziekenhuis", 3501 CA Utrecht, The Netherlands.

26.1. Introduction

26.1.1. Autoimmunity

Autoimmune disease is the result of an immune reaction directed against components of the body, leading to tissue damage and organ dysfunction. Autoimmune diseases comprise a wide spectrum of diseases with, as a common factor, a chronic response against autoantigens. The etiopathogenesis of autoimmune diseases is not completely understood. Genetic predisposition associated with HLA haplotypes, inducing factors like inflammatory processes including molecular mimicry and dysregulation of the immune response, all contribute. Sex-related differences in the prevalence of autoimmune diseases have been recognized for a long time. More recent evidence derived predominantly from animal models suggests that the reactivity of the hypothalamus–pituitary–adrenal (HPA) axis system may also be a major factor in determining susceptibility to the development of an autoimmune disease. In addition, it has become evident that, during the course of an autoimmune disease, alterations in the interaction between the neuroendocrine and immune systems take place on the level of expression of receptors for neuroendocrine mediators.

In this chapter the contribution that psychoneuroimmunological research can give to the understanding of the etiopathogenesis of autoimmune diseases will be summarized.

26.1.2. Autoimmunity in Humans

In a healthy individual, the immune response is controlled in a way that results in tolerance for self structures. A number of different mechanisms are involved in the maintenance of immunological tolerance for self antigens: clonal deletion, which involves elimination of autoreactive cells; clonal energy, which involves inactivation of autoreactive cells; and active regulation resulting in the active suppression of autoimmune processes.

If mechanisms responsible for ongoing immunological tolerance to self structures fail, an autoimmune response can develop leading to tissue destruction and autoimmune disease. In humans, a large number of autoimmune diseases have been described. Organ-specific autoimmune diseases like insulin-dependent diabetes mellitus (IDDM) and multiple sclerosis involve activation of T cells and infiltration of these cells into the tissue and resultant tissue damage. In the case of more systemic diseases like systemic lupus erythematosus (SLE) or Graves' disease, the formation of immune complexes as a result of excessive formation of autoantibodies is mainly responsible for tissue damage.

Most autoimmune diseases in humans are characterized by periods of relative remission followed by exacerbations, suggesting that active regulation of control mechanisms is involved.

26.1.3. Psychoneuroimmunology and Autoimmunity in Humans

Clinical observations often note stressful life events before the onset of autoimmune disease. However, the validity of such observations is difficult to judge, as it often involves very subjective summaries and there are few well-controlled studies available. In rheumatoid arthritis, there are some indications that stressful events and/or failure of defense and adaptation mechanisms to compensate for it are related to disease onset (Koehler, 1985). With respect to the relationship between aggravation of the disease and psychological disturbances, contradictory results have been reported (Hendrie, Paraskevas, Baragar, & Adamson, 1971; Koehler, 1985; Lewis-Faning, 1950).

Sex steroids have been implicated in the pathogenesis of a number of autoimmune disorders because of the higher prevalence of certain autoimmune diseases in females than in males. Especially in various thyroid autoimmune diseases and SLE, the number of females affected is much higher than the number of males (Wilder, 1995). Rheumatoid arthritis, multiple sclerosis, and myasthenia gravis also occur more often in females than in males (Homo-Delarche, Fitzpatrick, Christeff, Nunez, Bach, & Dardenne, 1991). In contrast, in IDDM there is no evidence for a female predominance (Homo-Delarche *et al.*, 1991). In rheumatoid arthritis, SLE, and multiple sclerosis, it is a common phenomenon that the severity of autoimmune disease changes with pregnancy or during the use of contraceptive medication (Da Silva, 1995; Masi, Feigenbaum, & Chatterton, 1995; Wilder, 1995).

26.2. Animal Models for Autoimmune Disease

To enable the study of mechanisms involved in the pathogenesis of autoimmune diseases, a number of animal models have been developed. In this section, models used to investigate the contribution of the neuroendocrine system to the onset or course of experimental autoimmune diseases will be briefly described. It should be noted that most models of induced autoimmune disease have at least one major difference with the disease in humans: After a period of disease symptoms, the animals spontaneously recover. This feature makes it very difficult, if not impossible, to obtain insight into the pathogenic factors involved in the chronicity of the autoimmune response that is characteristic of human autoimmune disease.

26.2.1. Systemic Lupus Erythematosus

The most widely studied murine models for SLE are two murine models with a genetic basis: the (NZB/NZW) F1-NZB/W and MRL-lpr/lpr. These animals spontaneously develop an SLE-like syndrome during their life span. Animals spontaneously produce autoantibodies to DNA, poly(A), and a number of other antigens in association with an early onset and high incidence of renal disease similar to human lupus nephritis (Andrews, Eisenberg, Theofilopolous, Izui, Wilson, McConahey, Murphy, Roths, & Dixon, 1978).

26.2.2. Arthritis

The role of the neuroendocrine system in arthritis has also been studied in mice, but the major body of research focuses on rat models. Arthritis models involve induction of the disease by administration of antigen (mycobacterial antigens, collagen, avridine) emulsified in adjuvant in susceptible strains of rats, or by immunization with streptococcal cell walls. The induction of arthritis is T cell dependent and can be transferred from one animal to another by transferring antigen-specific T cells. Occurrence of arthritis is assessed using a clinical score based on degree of swelling, erythema, and deformation of the joints. Induction of arthritis in susceptible Lewis rats by administration of adjuvant will result in the first signs of disease after 8 to 10 days. The disease score then gradually increases to a maximum score at 2–4 weeks followed by spontaneous recovery and complete disappearance of clinical symptoms about 6 weeks after immunization (Pearson & Wood, 1959).

26.2.3. Experimental Autoimmune Encephalomyelitis (EAE)

EAE is used as a model for multiple sclerosis in both murine and rat models. The disease can be induced by administration of central nervous tissue homogenates emulsified in Freund's complete adjuvant into the footpad. The major encephalitogenic antigen studied is myelin basic protein. The T-cell response evoked is specific for the inducing antigen and antigen-specific T cells can transfer the disease to nonimmunized animals (Paterson, 1960). During the disease process, activated mononuclear leukocytes can be found in the CNS. This will lead to neurological deficits that represent the clinical symptoms of EAE, which in Lewis rats start about 9–10 days after immunization (Raine & Stone, 1977). The disease starts with a loss of tonicity in the tail or tail paresis. During the course of EAE, hind limb paresis or paralysis, and in some cases paralysis up to the diaphragm, will occur. Animals can die of EAE, but in most cases 5 to 7 days after appearance of the first clinical symptoms animals will recover completely. Moreover, as is the case for most animal models of arthritis, animals become resistant to reinduction of the disease.

26.2.4. Autoimmune Diabetes

The NOD mouse and the BB rat spontaneously develop immunologically mediated IDDM. During the life span of these animals, mononuclear cells, predominantly T cells and macrophages, infiltrate the islets of Langerhans. The animals will develop insulitis leading to the progressive destruction of beta cells and decreased insulin levels (Leiter, Prochazka, & Coleman, 1987).

26.3. Psychoneuroimmunology and Animal Models of Autoimmunity

26.3.1. Sex Steroids in Autoimmunity in Animal Models

The role of sex steroids in animal models of autoimmunity has been studied predominantly in animals that spontaneously develop SLE or IDDM. With respect to the SLE-like syndrome in NZB/NZW F1 mice, it is clear that the disease progresses more rapidly in females than in males. Disease progression in males is enhanced by prepubertal orchidectomy, whereas castration or administration of androgens can delay disease onset and decrease mortality in females (Roubinian, Talal, Greenspan, Goodman, & Siiteri, 1978). The earlier disease onset in females can also be observed in other murine SLE models like the MLR-lpr/lpr mouse. However, the effects of sex hormones in this model are less clear. Estrogen administration increases the B-cell-dependent features of the disease, such as formation of autoantibodies and immune complexes. In the same animals, however, T-cell-dependent lesions such as periarticular inflammation and renal vasculitis are decreased (Carlsten, Nilsson, & Jonsson, 1992).

In the NOD mouse, the expression of autoimmune steroids occurs earlier and more frequently in females (incidence 30–70%) than in males (less than 20%). Ovariectomy tends to have a protective effect, whereas castration of males strongly enhances the development of autoimmune diabetes. However, the exact role of putative immunomodulatory effects of sex steroids under these circumstances remains unclear (Lampeter, Signore, Gale, & Pozzilli, 1989).

26.3.2. How Can Sex Steroids Play a Role in Autoimmunity?

In general, sex steroids can modulate the activity of the immune system. Androgens suppress T- and B-cell proliferation *in vitro*, and increase the number of CD8$^+$ cells *in vivo.* In contrast, estrogens decrease the number of CD8$^+$ cells and increase antibody production (Ansar Ahmed, Dauphinee, & Talal, 1985; Lehman, Siebold, & Simmons, 1988). However, in light of the low abundance of specific receptors for sex steroids in mononuclear cells, the possibility exists that sex steroids exert their effects on autoimmune processes indirectly, e.g., via modulation of the reactivity of the HPA axis (Homo-Delarche *et al.*, 1991). Moreover, in the case of autoimmune diseases involving endocrine organs, e.g., diabetes or thyroiditis, sex steroids may have a profound effect on the activity of these organs thereby modulating clinical expression of the disease.

26.3.3. Hypothalamus–Pituitary–Adrenal Axis

From clinical practice it is well known that corticosteroids are potent anti-inflammatory drugs that can reduce the clinical symptoms of autoimmunity in most patients. Over the last 10 years it has become apparent from animal studies that the endogenous corticosteroids also play an important role in the development and course of autoimmune disease.

The major body of evidence for a role of the HPA axis in the onset of autoimmune disease comes from the comparison of strains or lines of rats with differences in susceptibility for disease. Initial studies by Sternberg and colleagues (Sternberg, Hill, Chrousos, Kamilaris, Listwak, Gold, & Wilder, 1989a) have shown that susceptibility for streptococcal cell wall arthritis is associated with a blunted reactivity of the HPA axis. Lewis rats develop arthritis in response to injection with streptococcal cell wall (SCW). In contrast, histocompatible Fischer rats develop only minimal alterations in the joints that rapidly disappear.

As described in Chapter 11, induction of an inflammatory response is associated with activation of the HPA-axis and results in increases in plasma corticosteroid levels. Studying the corticosteroid response to SCW injection in Fischer and Lewis rats, it was shown that the corticosterone response in Lewis rats was significantly lower. Further research along these lines has demonstrated that the reactivity of the HPA axis is reduced in levels of corticotropin-releasing hormone (CRH), and adrenocorticotropin hormone (ACTH) as well. Moreover, the reduced HPA-axis reactivity in Lewis rats relative to Fischer rats can also be observed after exposure to stress or a variety of inflammatory stimuli, and after injection of CRH (Sternberg, Young, Bernardini, Calogero, Chrousos, Gold, & Wilder, 1989b).

Interestingly, the development of SCW arthritis in Lewis rats can be prevented by administration of physiological doses of steroids. Conversely, administration of the glucocorticoid receptor antagonist RU486 to Fischer rats results in exacerbation of inflammation and the development of arthritis in this resistant strain (Sternberg *et al.*, 1989a).

When different strains of animals were compared, it became evident that the relation between HPA-axis reactivity and susceptibility to arthritis or EAE is not restricted to Lewis and Fischer rats. PVG rats, which are resistant to EAE, have higher basal corticosterone levels and higher stress-induced increases in corticosterone than susceptible rats (Mason, Macphee, & Antoni, 1990). Wistar rats, which have intermediate corticosterone responses to a number of stimuli, also display intermediary reactivity of the HPA axis.

Corticosterone is not only important in preventing the onset of T-cell-mediated autoimmunity in animal models, but also plays a role in the spontaneous recovery. In the EAE model it has been shown that endogenous corticosterone is indispensable for recovery. Removal of the adrenals as a source of corticosterone results in aggravation of the disease leading to death in all animals (Mason *et al.*, 1990).

A role for endogenous corticosteroids in autoimmunity has also been suggested in spontaneous models of autoimmunity. Studies in obese strain chickens that spontaneously develop thyroiditis revealed that these animals also have a blunted reactivity of the HPA axis (Wick, Hu, & Schwarz, 1993). After immunization or after injection of IL-1 obese strain chickens do not respond with detectable changes in corticosterone.

26.3.4. Possible Mechanisms Underlying the Relationship between HPA-Axis Reactivity and Susceptibility to Autoimmunity

In pharmacological doses, corticosteroids clearly have immunosuppressive effects. Both *in vitro* and *in vivo* corticosteroids inhibit T-cell proliferation, IL-2 production, and production of cytokines by monocytes/macrophages (Arya, Wong-Staal, & Gallo, 1984; Gillis, Wong-Staal, Crabtree, & Smith, 1994; Lew, Oppenheim, & Matsushima, 1988). However, there are indications suggesting that corticosteroid regulation of the balance between pro- and anti-inflammatory cytokines in helper T cells may be involved in determining susceptibility to autoimmune disease (Mason, 1991).

T-helper-1 (Th1)-type cytokines like IL-2 and IFN-γ support cell-mediated immune responses and promote the inflammatory response in EAE and arthritis. In contrast, Th2-type cytokines like IL-4 and IL-5 enhance the humoral immune response. Moreover, these Th2-type cytokines can inhibit the production of Th1-type cytokines (Fig. 1). *In vitro*, corticosteroids can promote the production of Th2-type cytokines and inhibit Th1 cytokine formation (Daynes, Araneo, Dowell, Huang, & Dudley, 1990). Our recent data (Kavelaars, Heijnen, Ellenbroek, van Loveren, & Cools, 1997) demonstrate that differences in HPA-axis reactivity, susceptibility to EAE, and the capacity of splenocytes to produce Th1 and Th2 cytokines are linked. We used two lines of Wistar rats (APO-sus and APO-unsus rats) that had been selected on the basis of the sensitivity to apomorphine. The neuroendocrine system of APO-sus and APO-unsus rats differs with respect to stress-induced activity of the HPA axis: APO-unsus rats have a low corticosterone and ACTH response to a number of stressful stimuli. As expected, APO-unsus animals are susceptible to EAE and arthritis, whereas APO-sus animals are resistant. In contrast, APO-sus animals generate a vigorous IgE response after infection with the nematode *Trichinella spiralis*. The APO-unsus animals

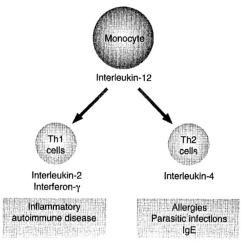

Figure 1.   Differentiation of T-helper cells into Th1 and Th2 cells, according to the cytokines they produce.

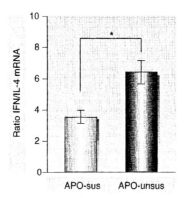

Figure 2. Relative production of IFN and IL-4 in APO-sus and APO-unsus rats.

do not show an IgE response. Investigation of the cytokine response of splenocytes revealed that APO-sus animals produce relatively more IL-4, a Th2 cytokine, than APO-unsus animals (Fig. 2).

In the mouse model, a similar association between differences in Th1 or Th2 dominance and HPA-axis reactivity has been described. Balb/c mice are Th2-type animals and C57bl/6 mice predominantly mount Th1-type responses (Heinzel, Sadick, Mutha, & Locksley, 1991; Scott, Pearce, Cheever, Coffman, & Sher, 1989). Of the two strains, the stress-induced increases in corticosterone are larger in the Balb/c mice (Shanks, Griffiths, & Anisman, 1994). Together, these results strongly suggest that a high reactivity of the HPA axis is associated with a high capacity to generate Th2-type responses, whereas low HPA-axis reactivity goes with a higher capacity to generate Th1-type responses.

26.3.5. Role of the Peripheral Nervous System in Autoimmunity

Much of the recent research on psychoneuroimmunology and autoimmunity in animal models has focused on the HPA axis with corticosteroids as the most important mediator. The contribution of the mediators of the peripheral nervous system should not be forgotten, however. The idea that the peripheral nervous system plays a role in autoimmune disease is mainly based on research in various models of arthritis and originates from clinical observations in patients with rheumatoid arthritis. In rheumatoid arthritis, the distribution of joints involved in the disease process is symmetrical, and distal joints that have a higher density of innervation are more involved in the disease process than proximal joints. Moreover, in hemiplegic patients, the joints in paralyzed limbs are generally not affected (Glick, 1967).

Studies on the role of the peripheral nervous system in autoimmunity focus on catecholamines as mediators of the sympathetic nervous system as well as on neuropeptides released from peptidergic neurons.

26.3.6. Catecholamines in Autoimmunity

In the arthritis model in rats as well as in dogs, it has been shown that sympathectomy or treatment with β_2-adrenergic receptor antagonists decreases the severity of joint injury (Levine, Codere, Helms, & Basbaum, 1988). When animals are treated before and during induction of disease, the onset of disease is delayed and the inflammatory symptoms reduced. In line with these findings, it has been shown that spontaneously hypertensive rats, which have an increased sympathetic tone, also have increased joint injury after induction of arthritis.

The mechanisms underlying these effects are largely unknown. β_2-Adrenergic receptor triggering is known to result in decreased production of the proinflammatory monocyte/macrophage-derived cytokine TNFα. Moreover, it has also been shown that β_2-agonists can favor Th2-type responses by inhibiting the production of IL-12. IL-12 is an important cytokine in directing T-cell responses toward a Th1-type response (Panina-Bordignon, Mazzeo, Di Luca, D'Ambrosio, Lang, & Fabbri, 1997). It is not known, however, how β_2-adrenergic agonist could have an immunoenhancing effect that could explain the effects on arthritis.

One complicating factor in the interpretation of the in vivo data could be that during the disease process the sensitivity of the immune system for catecholamines can be altered, although conflicting results have been presented. In some studies it was shown that the density of β_2-adrenergic receptors on peripheral blood mononuclear cells of patients with rheumatoid arthritis or SLE is lower than in healthy controls, whereas others presented evidence for increased levels of these receptors in arthritis patients (Baerwald, Grafee, Von Wichert, & Krause, 1992; Zoukos, Leonard, Thomaides, Thompson, & Cuzner, 1992).

By investigating children with juvenile rheumatoid arthritis (JRA), we have shown that the reactivity of the sympathetic nervous system in these patients is higher than in healthy controls. JRA patients have a higher turnover of central noradrenaline, heart rate at rest is increased, and the response to a head up tilt test is reduced (Kuis, de Jong-de Vos van Steenwijk, Sinnema, Kavelaars, Prakken, Helders, & Heijnen, 1996). The latter is presumably the result of a higher vasoconstrictive tone in the patients. Further investigation of the interaction between the sympathetic nervous system and immune system revealed that children with a more severe form of the disease also express a subtype of adrenergic receptor on peripheral blood mononuclear cells that is usually not found on these cells. Peripheral blood mononuclear cells of JRA patients express functional α_1-adrenergic receptors on the cell surface, the triggering of which leads to cytokine production (Heijnen, Rouppe van der Voort, Wulffraat, van der Net, Kuis, & Kavelaars, 1996). In this way, catecholamines could aggravate the disease process.

26.3.7. Neuropeptides in Autoimmunity

In arthritis models, it has been demonstrated that a number of neuropeptides are present at the site of inflammation. One of the first neuropeptides to be implicated in arthritis was substance P. This neuropeptide can be found in significant amounts in the synovial fluid of arthritis patients where it is released by primary sensory nerve fibers. Injection of substance P into the joint of arthritic rats results in increased inflammation and joint destruction, whereas substance P antagonist could decrease symptoms (Chapman & Tsao, 1980; Levine, Clark, Devor, Helms, Moskowitz, & Basbaum, 1984). As described in Chapters 8 and 9, substance P has a number of proinflammatory properties that are thought to be involved in the enhanced autoimmune reaction in the presence of substance P.

A number of other neuropeptides are present in synovial fluid or in the cells infiltrating the joint in arthritis like CRH, endorphin, and enkephalins. These neuropeptides have been implicated in both local regulation of the immune response as well as signaling to the CNS. Locally produced opioids are particularly thought to have an effect on opiate receptors expressed on local nerve terminals and can thereby modulate pain sensitivity.

27 The Effects of Stress on the Immune System

Implications for Reactivation of Latent Herpesviruses

Katherine L. Applegate, John Hay,
John T. Cacioppo, Janice K. Kiecolt-Glaser,
and Ronald Glaser

27.1. Introduction

There are presently seven known and characterized human herpesviruses: herpes simplex virus type 1 (HSV-1), HSV-2, cytomegalovirus (CMV), varicella zoster virus (VZV), Epstein–Barr virus (EBV), human herpesvirus-6 (HHV-6), and HHV-7. Recently, another human herpesvirus has been linked to Kaposi's sarcoma and designated HHV-8; studies are still under way to characterize this virus.

Five of these human herpesviruses have been well studied and characterized; e.g., HSV-1 has been associated with cold sores, keratitis, and encephalitis, HSV-2 with herpes

Katherine L. Applegate and Janice K. Kiecolt-Glaser • Department of Psychiatry, Ohio State University College of Medicine, Columbus, Ohio 43210.   John Hay • Department of Microbiology, State University of New York at Buffalo, Buffalo, New York 14260.   John T. Cacioppo • Department of Psychology, Ohio State University, Columbus, Ohio 43210-1222.   Ronald Glaser • Department of Medical Microbiology and Immunology, Ohio State University College of Medicine, Columbus, Ohio 43210.

genitalis, VZV is the etiological agent for chicken pox and shingles. EBV is a human tumor virus that has been associated with African Burkitt's lymphoma, nasopharyngeal carcinoma, and the most common form of infectious mononucleosis (IM), and CMV can cause a type of IM and pneumonia. Little is known of the pathophysiology and clinical manifestations of HHV-7. Somewhat more is known of HHV-6, which has been linked to an IM-like illness as well as exanthem subitum (roseola infantum), which is commonly observed in young children or infants.

After primary infection (which may or may not manifest clinical symptoms), each of these viruses latently infects one or more specific cells in the body where, for the remainder of the person's life, it is thought to persist in a latent state. Depending on the herpesvirus, a very high percentage (in some cases 100%) of individuals become infected by the time they reach middle age in industrialized countries. However, in poorly developed countries, individuals become infected with these viruses at a much earlier age. These differences have clinical implications, e.g., individuals in poorly developed countries who get infected with EBV generally do not develop IM, while individuals in North America get infected in their middle teens or early twenties and will often manifest clinical IM.

Stress has been implicated as a risk factor in the development, duration, and recurrence of herpesvirus infections. This chapter will briefly review research that addresses relationships between stress and herpesvirus infections.

Among the herpesviruses, HSV-1 and HSV-2 have been studied most intensively in association with psychosocial stressors; in addition, a number of studies have demonstrated psychosocial modulation of the steady-state expression of latent EBV. The impact of psychosocial stressors on VZV has not been as well studied as either HSV or EBV; however, stress-related alterations in the competency of the cellular immune response have implications for VZV as well.

The cellular immune response plays a central role in control of primary herpesvirus infections as well as in the subsequent control of viral latency (Glaser & Jones, 1994). Thus, this chapter will briefly review the broader evidence linking stress to immunological changes. Psychological and behavioral characteristics associated with increased risk for primary and recurrent herpesvirus infections will also be discussed, as well as data from psychosocial interventions.

27.2. Stress and the Herpesviruses: Clinical Manifestations

Psychosocial stressors have been linked to more frequent recurrences among individuals latently infected with HSV-1 and HSV-2. Both major, negative events and a high level of daily hassles may constitute psychosocial stress; moreover, these stressors may also be interpersonal, such as relationship difficulties, or individual problems relating to life satisfaction, one's ability to cope with challenges, or a sense of being supported by others. Concomitant increases in psychological distress or negative affective states may contribute to increased rates of infectious illness (Cohen & Williamson, 1991). For example, greater unhappiness was associated with more frequent cold sores in nurses during their first year of training (Luborsky, Mintz, Brightman, & Katcher, 1976). Consistent with the data from student nurses, 18 individuals with recurrent HSV-1 lesions reported increases in stressful life events, anxiety, and daily hassles in the week prior to the appearance of a cold sore (Schmidt, Zyzanski, Ellner, Kumar, & Arno, 1985).

Genital herpes recurrences also appear to be more prevalent among individuals who report feeling more distressed. Goldmeier and Johnson (1983) investigated the relationship between a cluster of dysphoric symptoms (including anxiety, depression, and hypo-chondriasis) and recurrence rates in 58 patients who were followed for up to 30 weeks after their first clinically confirmed HSV-2 lesion. More distressed patients experienced significantly more recurrences. Similarly, Stout and Bloom (1986) studied 37 individuals who reported recurrent genital herpes lesions. Participants who were above the median on recurrence rates had significantly higher scores on nine of the ten clinical scales of the Minnesota Multiphasic Personality Inventory, compared with subjects with fewer recurrences, suggesting greater depression, anger, anxiety, worry, and interpersonal troubles in the former.

Personality traits and characteristic response styles to stressful events may predispose individuals to psychological symptoms that place them at higher risk for recurrence of herpesvirus infections. Life event stress was associated with duration of reactivation episodes in a sample of male and female genital herpes patients (Silver, Auerbach, Vishniavsky, & Kaplowitz, 1986). Patients with an external locus of control, a proneness toward emotion-focused wishful thinking, and avoidance of cognitive strategies to cope with the stress of HSV-2 infection showed higher recurrence rates and symptom discomfort. In other work, psychosocial variables were better predictors of pain and itching symptoms in patients with recurrent HSV-2 than were somatic symptoms (Levenson, Hamer, Myers, Hart, & Kaplowitz, 1987). Somatization was the single best predictor of pain; interpersonal sensitivity and somatization were the most reliable predictors of itching.

Recurrence rates in patients with HSV-2 infection were examined prospectively to investigate psychological and immunological correlates (Kemeny, Cohen, Zegans, & Conant, 1989). This study explored the mechanisms linking potential stressors, physiological responses, and health outcomes by examining the interactions among stress, mood, immune function, and disease course simultaneously. Although previous research has examined ties between stress and immune function or between stress and some diseases, few studies have incorporated all three components at once. Kemeny *et al.* (1989) interviewed participants monthly for 6 months to assess current and past stressors, negative mood, health behaviors, and number of HSV-2 recurrences. Stressful life events were associated with lower proportions of $CD4^+$ and $CD8^+$ cells, and depressed mood was associated with a lower proportion of $CD8^+$ cells. Depressive mood was also linked to a higher HSV-2 recurrence rate.

Taken together, these studies suggest that individuals who report more unhappiness, anxiety, depression, and frustration are more likely to experience recurrent HSV than those who report less distress. Dysphoria and negative life events appear to have a detrimental effect on the course of HSV-1 and HSV-2 infections. The next section describes evidence linking stress with the downregulation of cellular immune function, the likely pathway for the observed clinical link between stress and more frequent recurrences.

27.3. Stress-Related Alterations in Cellular Immune Function

Stressful events can alter a wide range of immunological activities. For example, even commonplace aversive events like academic exams are associated with transient immunological changes (see Chapter 16). Comparisons of immunological data collected from medical students during a 3-day exam block in contrast to a "baseline" or lower-stress blood sample collected a month previously showed significant declines in natural killer (NK) cell activity; NK cells are thought to have important antiviral and antitumor functions (Glaser, Rice, Speicher, Stout, & Kiecolt-Glaser, 1986; Glaser, Rice, Sheridan, Fertel, Stout,

Speicher, Pinsky, Kotur, Post, Beck, & Kiecolt-Glaser, 1987; Kiecolt-Glaser, Speicher, Holliday, & Glaser, 1984b). IFNγ, a lymphokine, serves as a major regulator of NK cells by stimulating their growth and differentiation; it also enhances their ability to destroy target cells (Herberman, Ortaldo, Riccardi, Timonen, Schmidt, Maluish, & Djeu, 1982). Two separate studies also showed dramatic decreases in IFNγ production by lymphocytes *in vitro* during exams (Glaser *et al.*, 1986, 1987).

The proliferative response of lymphocytes cultured with a mitogen, a substance that stimulates cell replication, is thought to provide a laboratory model of the immune system's ability to respond to infectious agents such as bacteria or viruses. Medical students showed a poorer proliferative response to mitogens during exams compared with baseline (Glaser, Kiecolt-Glaser, Stout, Tarr, Speicher, & Holliday, 1985b). IL-2 is a lymphokine important for T-lymphocyte proliferation, and the IL-2 receptor, to which IL-2 binds, is an important component of this response. The percentage of peripheral blood T lymphocytes expressing the IL-2 receptor was lower during exams compared with lower-stress baseline periods in three independent medical student studies (Glaser, Pearson, Jones, Hillhouse, Kennedy, Mao, & Kiecolt-Glaser, 1991). Moreover, the level of mRNA to the IL-2 receptor in peripheral blood leukocytes (PBLs) decreased during exams in a subset of these students (Glaser *et al.*, 1991).

Thus, these studies demonstrate that even something as transient, predictable, and relatively benign as exam stress modulates a wide range of immunological activities. Other studies have addressed the question of whether longer-term adaptation occurs when a stressor is more chronic, such as living near a damaged nuclear reactor (McKinnon, Weisse, Reynolds, Bowles, & Baum, 1989), or caregiving for a family member with a progressive dementia (Kiecolt-Glaser, Glaser, Dyer, Shuttleworth, Ogrocki, & Speicher, 1987b). The weight of the evidence to date suggests that chronic stressors are associated with continued downregulation of immune function, rather than adaptation.

27.4. Stress and Herpesvirus Reactivation: Immunological Evidence

Although both the humoral and cellular arms of the immune system appear to be important in controlling herpesvirus infections, cell-mediated immunity may play a more critical role in controlling the reactivation of latent herpesviruses (Bonneau, 1994). For example, recovery from herpes zoster is normal in hypogammaglobulinemic patients who have little or no detectable VZV antibody (Ruckdeschel, Schimpff, & Smyth, 1977).

Herpesvirus-specific cytotoxic T lymphocytes (CTL) combat the spread of the herpesvirus infections by destroying virus-infected cells before progeny viruses are released. Memory CTL and helper T lymphocytes promote a more rapid response during subsequent reactivation of herpesvirus after the initial infection. When derivatives of these memory cells are later exposed to a herpesvirus, they lyse the infected cells, stimulate antibody, and secrete lymphokines to enhance the efficacy of the immune response. The ability of the immune system to fulfill these functions is mediated in part by signals from the nervous and endocrine systems. For further discussion of cellular and molecular mechanisms underlying reactivation of latent herpesviruses, see Bonneau (1994), Glaser and Jones (1994), and Jenkins and Baum (1995).

The large and reliable increases in antibody titers to latent herpesviruses during academic exams, particularly EBV and HSV-1, appear to reflect alterations in the competence of the cellular immune response. The characteristic elevations in EBV antibody titers are thought to occur in response to the increased synthesis of the virus or virus proteins (Glaser *et al.*, 1991); although counterintuitive, *elevated* antibody titers to a latent herpesvirus reflect *poorer* cellular immune system control over virus latency (Henle & Henle, 1981). Consistent with the elevations in herpesvirus antibody titers, specific T-cell killing of EBV-infected target cells decreased during exams, and a herpesvirus-relevant lymphokine was also altered (Glaser *et al.*, 1987). Moreover, medical students showed a lower memory T-cell proliferation response to five of six EBV polypeptides after a 3-day exam block compared with 3 weeks prior to exams (Glaser, Lafuse, Bonneau, Atkinson, & Kiecolt-Glaser, 1993a).

One study suggested that stress may also be associated with risk for EBV infection, as well as the severity of the primary infection. Kasl, Evans, and Neiderman (1979) followed cadets at West Point over 4 years who were seronegative for EBV on entry into the Naval Academy. They found that higher motivation for a future military career, poorer academic performance, and having a father who was an "overachiever" were associated with a greater chance for seroconversion, longer hospitalization with IM in the infirmary after seroconversion (i.e., more severe illness episodes), and higher antibody titers among those students who seroconverted but did not develop clinical symptoms. Stressful life situations can downregulate the cellular immune response, adversely affecting EBV and other herpesviruses.

Other studies have assessed alterations in herpesvirus latency associated with chronic or long-term stressors. Caregiving for a spouse with a progressive dementia such as Alzheimer's disease is clearly a chronic stressor; the progressive and degenerative disease course as well as the unpredictable and uncontrollable nature of symptom presentation represent significant challenges to caregivers over a period of years. Caregivers had higher antibody titers to EBV virus capsid antigen (VCA) compared with matched controls, as well as a poorer blastogenic response of PBLs to mitogenic stimulation (Kiecolt-Glaser, Dura, Speicher, Trask, & Glaser, 1991). In another study, caregivers had higher antibody titers to an HSV-1 total viral antigen and a lower HSV-1 specific T-cell response than controls (Glaser & Kiecolt-Glaser, 1997). Other researchers have also shown that a longer-term environmental stressor is associated with elevated HSV-1 antibody titers (McKinnon *et al.*, 1989).

27.5. Psychological Interventions

The research reviewed above provides good evidence that psychological distress and stressful life events can downregulate the immune response and increase the risk for recurrences among individuals latently infected with a herpesvirus. Based on these data, researchers have assessed the potential immunological benefits of psychological interventions directed toward stress reduction. For example, older adults were trained in progressive muscle relaxation and guided imagery over the course of 1 month (Kiecolt-Glaser, Glaser, Williger, Stout, Messick, Sheppard, Ricker, Romisher, Briner, Bonnell, & Donnerberg, 1985). Blood samples were taken at baseline, at the end of intervention, and at a 1-month follow-up. Following the intervention, participants in the intervention condition showed significant increases in NK cell activity and decreases in antibody titers to HSV-1 and self-reported distress; in contrast, subjects who did not receive the stress-reduction intervention did not show these significant immunological changes.

Several studies have assessed the tie between self-disclosure and immune function. A widespread notion within the psychotherapy literature suggests that individuals may best deal with trauma by trying to understand the event and assimilating this information (Pennebaker, Kiecolt-Glaser, & Glaser, 1988b). To accomplish this, individuals can describe the trauma in detail, either orally or in writing, which would allow for a reexamination of the event, incorporation of the information in new ways, and a general release of any thoughts and feelings they may have been holding back. In the first study 50 undergraduates were randomly assigned to one of two groups; half the subjects wrote about traumatic or troubling experiences for 20 min on 4 consecutive days, while the remainder wrote about trivial events and experiences (Pennebaker *et al.*, 1988b). The individuals who wrote about traumatic or upsetting events demonstrated a higher mitogen response following baseline compared with control subjects. "Trauma" subjects' average number of monthly health center visits dropped following the study, while control subjects' visits increased, replicating health data from similar studies. Most importantly, individuals who wrote about experiences they had not shared previously with other people had a better lymphocyte proliferative response than those subjects who had discussed the experiences previously.

Lutgendorf, Antoni, Kumar, and Schneiderman (1994) used a very similar paradigm to assess the effect of self-disclosure on control of latent EBV; they also measured antibody titers to EBV VCA. Although the students in their disclosure condition did not show significant decreases in antibody titers relative to controls, the degree to which students involved themselves in the disclosure process and abandoned avoidance of the stressful topic for the 3-week intervention period predicted decreases in antibody titers to EBV VCA.

Esterling, Antoni, Fletcher, Margulies, and Schneiderman (1994) demonstrated that students who either wrote or spoke about upsetting events showed decrements in EBV VCA antibody titers compared with students who were randomly assigned to write about trivial events. Together, these studies suggest that interventions like psychotherapy that enhance personal relationships, decrease distress, and/or enhance perceived self-efficacy could also have positive effects on at least certain components of the immune response. The two studies described below have assessed the effects of psychological interventions on an important health outcome, frequency and duration of HSV-2 recurrent lesions.

Behavioral intervention studies with HSV-2 patients with recurrent lesions have produced promising results. An intervention consisting of disease information, relaxation training, stress management techniques, and an imagery sequence produced significant improvements in recurrent genital herpes patients at follow-up (Longo, Clum, & Yaeger, 1988). The frequency of lesions, episode duration, and self-reported severity ratings for HSV-2 infections decreased in the intervention group compared with controls. Psychosocial measures showed lower levels of emotional distress and loneliness and elevated levels of internality for patients in the intervention group.

To test whether one particular element of the multicomponent intervention approach was effective, eight recurrent genital herpes patients were trained in progressive muscle relaxation (Burnette, Koehn, Kenyon-Jump, Hutton, & Carman, 1991). Five of these patients reported significant decreases in recurrence frequency, while a sixth showed a 30% decrease in episode frequency. Six of the eight patients also reported decreases in the duration of the episodes. Thus, collectively, these studies offer support for improving control of latent herpesvirus or decreasing the episode frequency via psychological interventions that reduce psychological distress.

27.6. Aging and Immunological Control of Latent Herpesviruses: Implications for Varicella Zoster

Immune function declines with age, particularly functional aspects of the cellular immune response (Wayne, Rhyne, Garry, & Goodwin, 1990). As one consequence, primary herpesvirus infections may become increasingly severe with age (Glaser & Jones, 1994). Not surprisingly, immunological data demonstrate age-related decrements in the control of latent herpesviruses. For example, Glaser *et al.* (1985c) found significantly higher antibody titers to latent EBV in the elderly than in a younger group of subjects.

Age-related immunological declines also appear to be linked to the greatly increased risk for VZV reactivation in older adults (Grose, 1994). Moreover, zoster is a much more common event in the immunosuppressed patient (Grose, 1994). Importantly, older adults show greater immunological impairments related to depression or stress than younger adults (Schleifer, Keller, Bond, Cohen, & Stein, 1989). Thus, it is not surprising that stress has been anecdotally reported as a risk factor for zoster, particularly among older adults (Schmader, Studenski, MacMillan, Grufferman, & Cohen, 1990).

To investigate the role of stressful life events in the reactivation of VZV, Schmader *et al.* (1990) compared individuals with acute herpes zoster with controls matched on age, sex, and race. Participants were asked about major life events that had occurred during the year prior to onset; the subjective meaning of these events was evaluated by asking "Did this have a negative, neutral, or positive effect on you?" The authors defined an event stressful only if the participant perceived it as negative. The number of life events defined as negative by the subject was significantly higher in zoster patients at 2, 3, and 6 months before onset compared with controls. Based on these findings, the authors suggest that negative life events may constitute a risk factor for VZV reactivation.

A prospective study by Dworkin, Hartstein, Rosner, Walther, Sweeney, and Brand (1992) examined the relationship between psychological variables and chronic pain before pain onset. Herpes zoster patients, assessed shortly after diagnosis of acute herpes zoster, were followed for 12 months to determine if psychological antecedents could predict development of chronic pain symptoms. Among the 19 patients in their sample, those who developed chronic pain over the following 12 months had higher pain intensity ratings at the initial assessment. Chronic pain sufferers also had higher levels of anxiety, greater depression, and lower life satisfaction at initial assessment than zoster patients who did not develop chronic pain. These findings suggest that greater distress or dysphoria during the early stages of an acute recurrence may be associated with more chronic herpes zoster pain.

In a recent study we measured antibody titers to VZV in 48 women who were providing care for a husband with a progressive dementia and 48 women from the community who were matched on age and education who had no caregiving responsibilities; the average age of both groups was 68, and caregivers reported significantly more stress and depression than controls. Antibody titers to a total viral antigen preparation were measured by enzyme-linked immunosorbent assay (ELISA); the average titers across four tests showed a trend for caregivers to have higher titers than controls. In a previous study (Glaser *et al.*, 1991) we found evidence for partial reactivation of latent EBV in medical students using the academic stress model; this conclusion was based on data obtained from experiments with four purified viral polypeptides that were used singularly to measure specific antibody levels to the respective polypeptide by ELISA. Antibody to only one of the four polypeptides (P52/50) showed changes in multiple serum samples tested across the academic year. The particular antibody that showed stress-related change is associated with viral DNA polymerase activity and is classified as an early nonstructural protein. As a follow-up to these observations, two purified VZV polypeptides (ORF 62 and ORF 63) were used with the caregivers and controls. Once again, using each viral polypeptide to probe for a specific antibody, ELISA tests were performed. The two groups did not differ on their antibody levels against the major structural viral polypeptide but caregivers had significantly higher titers to the VZV ORF 63 protein. ORF 63 appears to represent a major latency-associated transcript protein.

27.7. Conclusions

Collectively, the data obtained with EBV and VZV suggest that partial expression of the latent viral genome can occur for these two viruses (and perhaps other latent herpesviruses), and this can be measured by using purified viral polypeptides to probe for specific antibody titers by ELISA. Importantly, these data demonstrate that stress can modulate the steady-state expression of latent VZV as well as EBV.

In summary, there is ample evidence that stress can modulate reactivation of EBV, HSV-1, and HSV-2; VZV has not been studied as intensively. However, if psychological stress is associated clinically with both reactivation of VZV (Schmader *et al.*, 1990) as well as more chronic herpes zoster pain (Dworkin *et al.*, 1992), both psychological and pharmacological interventions that reduce distress may prove to be helpful.

ACKNOWLEDGMENTS. Ohio State University Comprehensive Cancer Center work on this chapter was supported by grants MH42096, MH50538, and AG11585 from the National Institutes of Health.

References

Aarons, R. D., Nies, A. S., Gerber, J. G., & Molinoff, P. B. (1983). Decreased β adrenergic receptor density on human lymphocytes after chronic treatment with agonists. *Journal of Pharmacology and Experimental Therapeutics, 224,* 1–6.

Abbas, A. K., Murphy, K. M., & Sher, A. (1996). Functional diversity of helper T lymphocytes. *Nature, 383,* 787–793.

Abbas, A. K., Lichtman, A. H., & Pober, J. S. (1997). *Cellular and molecular immunology.* Philadelphia: Saunders.

Abernethy, N. J., & Hay, J. B. (1992). The recirculation of lymphocytes from blood to lymph: Physiological considerations and molecular mechanisms. *Lymphology, 25,* 1–30.

Acheson, A., & Lindsay, R. M. (1994). Diverse roles of neurotrophic factors in health and disease. *Seminars in the Neurosciences, 6,* 333–341.

Ackerman, S. H., Keller, S. E., Schleifer, S. J., Shindledecker, R. D., Camerino, M., Hofer, M. A., Weiner, H., & Stein, M. (1988). Premature maternal separation and lymphocyte function. *Brain, Behavior, and Immunity, 2,* 161–165.

Ader, R. (1983). Developmental psychoneuroimmunology. *Developmental Psychobiology, 16,* 251–267.

Ader, R., & Cohen, N. (1975). Behaviorally conditioned immunosuppression. *Psychosomatic Medicine, 37,* 333–340.

Ader, R., & Cohen, N. (1981). Conditioned immunopharmacologic response. In R. Ader, & N. Cohen (Eds.), *Psychoneuroimmunology.* New York: Academic Press, pp. 611–646.

Ader, R., & Cohen, N. (1982). Behaviorally conditioned immunosuppression and murine systemic lupus erythematosus. *Science, 215,* 1534–1536.

Ader, R., & Cohen, N. (1993). Psychoneuroimmunology: Conditioning and stress. *Annual Review of Psychology, 44,* 53–85.

Ader, R., Cohen, N., & Grota, L. J. (1979). Adrenal involvement in conditioned immunosuppression. *International Journal of Immunopharmacology, 1,* 141–145.

Ader, R., Felten, D. L., & Cohen, N. (Eds.). (1991). *Psychoneuroimmunology* (2nd Ed.). San Diego: Academic Press.

Ader, R., Kelley, K., Moynihan, J. A., Grota, L., & Cohen, N. (1993). Conditioned enhancement of antibody production using antigen as the unconditioned stimulus. *Brain, Behavior, and Immunity, 7,* 334–343.

Afan, A. M., Broome, C. S., Nicholls, S. E., Whetton, A. D., & Miyan, J. A. (1997). Bone marrow innervation regulates cellular retention in the murine haemopoietic system. *British Journal of Haematology, 98,* 569–577.

Ahlborg, B., & Ahlborg, G. (1970). Exercise leukocytosis with and without β-adrenergic blockade. *Acta Medica Scandinavica, 187,* 241–246.

Ahmed, R., & Gray, D. (1996). Immunological memory and protective immunity: Understanding their relation. *Science, 272,* 54–60.

AIDS Feedback. (1991). 22 Ave Riant-Parc, 1209 Geneva, Switzerland, AF 09 91b/d.

Aird, F., Clevenger, C. V., Prystowski, M. B., & Redei, E. (1993). Corticotropin-releasing factor mRNA in rat thymus and spleen. *Proceedings of the National Academy of Sciences, USA, 90,* 7104–7108.

Akaike, A., Weihe, E., Schäfer, M. K.-H., Fu, Z. F., Zheng, Y. M., Vogel, W., Schmidt, H., Koprowski, H., & Dietzschold, B. (1995). Effect of neurotropic virus infection on neuronal and inducible nitric oxide synthase activity in rat brain. *Neurovirology, 1,* 118–125.

Alberts, S. C., Sapolsky, R. M., & Altmann, J. (1992). Behavioral, endocrine, and immunological correlates of immigration by an aggressive male into a natural primate group. *Hormones and Behavior, 26,* 167–178.

Albrecht, M. A., DeLuca, N. A., Byrn, R. A., Schaffer, P. A., & Hammer, S. M. (1989). The herpes simplex virus immediate-early protein, ICP4, is required to potentiate replication of human immunodeficiency virus in CD4 lymphocytes. *Journal of Virology, 63,* 1861–1868.

Aldenhoff, J. B., Dumais-Huber, C., Fritzsche, M., Sulger, J., & Vollmayr, B. (1997). Altered Ca(2^+)-homeostasis in single T-lymphocytes of depressed patients. *Journal of Psychiatry Research, 31,* 315–322.

Alexander, F. (1939). Psychological aspects of medicine. *Psychosomatic Medicine, 1,* 7–18.

525

Alsip, G. R., Ench, Y., Sumaya, C. V., & Boswell, R. N. (1988). Increased Epstein–Barr virus asymptomatic human immunodeficiency virus infections. *Journal of Infectious Diseases, 157,* 1072–1076.

Altmann, J. (1974). Observational study of behavior: Sampling methods. *Behaviour, 49,* 227–267.

Andersen, B. L. (1992). Psychological interventions for cancer patients to enhance the quality of life. *Journal of Consulting and Clinical Psychology, 60,* 552–568.

Andersen, B. L., Kiecolt-Glaser, J. K., & Glaser, R. (1994). A biobehavioral model of cancer stress and disease course. *American Psychologist, 49,* 389–404.

Andreoli, A., Keller, S. E., Rabaeus, M., Zaugg, L., Garrone, G., & Taban, C. (1992). Immunity, major depression, and panic disorder comorbidity. *Biological Psychiatry, 31,* 896–908.

Andrews, B. S., Eisenberg, R. A., Theofilopolous, A. N., Izui, S., Wilson, C. B., McConahey, P. J., Murphy, E. D., Roths, J. B., & Dixon, F. J. (1978). Spontaneous murine lupus-like syndromes, clinical and immunopathological manifestations in several strains. *Journal of Experimental Medicine, 148,* 1198–1207.

Anisman, H., Baines, M. G., Berczi, I., Bernstein, C. N., Blennerhassett, M. G., Gorczynski, R. M., Greenberg, A. H., Kisil, F. T., Mathison, R. D., Nagy, E., Nance, D. M., Perdue, M. H., Pomerantz, D. K., Sabbadini, E. R., Stanisz, A., & Warrington, R. J. (1996a). Neuroimmune mechanisms in health and disease: 1. Health. *CMAJ, 155,* 867–874.

Anisman, H., Baines, M. G., Berczi, I., Bernstein, C. N., Blennerhassett, M. G., Gorczynski, R. M., Greenberg, A. H., Kisil, F. T., Mathison, R. D., Nagy, E., Nance, D. M., Perdue, M. H., Pomerantz, D. K., Sabbadini, E. R., Stanisz, A., & Warrington, R. J. (1996b). Neuroimmune mechanisms in health and disease: 2. Disease. *CMAJ, 155,* 1075–1082.

Anonymous. (1993). Commentaries. *Psychological Inquiry, 4,* 65–69.

Ansar Ahmed, A., Dauphinee, M. J., & Talal, N. (1985). Effect of short term administration of sex hormones on normal and autoimmune mice. *Journal of Immunology, 134,* 204–210.

Antoni, M. (1987). Neuroendocrine influences in psychoimmunology and neoplasia: A review. *Psychology and Health, 1,* 3–24.

Antoni, M. H., August, S., LaPerriere, A., Baggett, H. L., Klimas, N., Ironson, G., Schneiderman, N., & Fletcher, M. A. (1990a). Psychological and neuroendocrine measures related to functional immune changes in anticipation of HIV-1 serostatus notification. *Psychosomatic Medicine, 52,* 495–510.

Antoni, M. H., Schneiderman, N., Fletcher, M. A., Goldstein, D. A., Ironson, G., & LaPerriere, A. (1990b). Psychoneuroimmunology and HIV-1. *Journal of Consulting and Clinical Psychology, 58,* 38–49.

Antoni, M. H., Baggett, L., Ironson, G., August, S., LaPerriere, A., Klimas, N., Schneiderman, N., & Fletcher, M. A. (1991a). Cognitive behavioral stress management intervention buffers distress responses and immunologic changes following notification of HIV-1 seropositivity. *Journal of Consulting and Clinical Psychology, 59,* 906–915.

Antoni, M. H., Goodkin, K., Goldstein, D. A., LaPerriere, A., Ironson, G., & Fletcher, M. A. (1991b). Coping responses to HIV-1 serostatus notification predict short-term affective distress and one-year immunologic status in HIV- 1 seronegative and seropositive gay men. *Psychosomatic Medicine, 53,* 227 (Abstract).

Antoni, M. H., Schneiderman, N., Klimas, N., LaPerriere, A., Ironson, G., & Fletcher, M. A. (1991c). Disparities in psychological, neuroendocrine and immunologic patterns in asymptomatic HIV-1 seropositive and seronegative gay men. *Biological Psychiatry, 29,* 1023–1041.

Antonica, A., Magni, F., Mearini, L., & Paolocci, N. (1994). Vagal control of lymphocyte release from rat thymus. *Journal of the Autonomic Nervous System, 48,* 187–197.

Antonica, A., Ayroldi, E., Magni, F., & Paolocci, N. (1996). Lymphocyte traffic changes induced by monolateral vagal denervation in mouse thymus and peripheral lymphoid organs. *Journal of Neuroimmunology, 64,* 115–122.

Appleton, I., Tomlinson, A., & Willoughby, D. A. (1994). Inducible cyclooxygenase (COX-2): A safer therapeutic target? [editorial]. *British Journal of Rheumatology, 33,* 410–412.

Ardawi, M. S., & Newsholme, E. A. (1984). Intracellular localization and properties of phosphate-dependent glutaminase in rat mesenteric lymph nodes. *Biochemical Journal, 217,* 289–296.

Arya, S. K., Wong-Staal, F., & Gallo, R. C. (1984). Dexamethasone mediated inhibition of human T-cell growth factor and gamma-interferon messenger RNA. *Journal of Immunology, 133,* 273.

Atherton, A., & Born, G. V. R. (1972). Quantitative investigations of the adhesiveness of circulating polymorphonuclear leucocytes to blood vessel walls. *Journal of Physiology (London), 222,* 447–474.

Athreya, B. H., Pletcher, J., Zulian, F., Weiner, D. B., & Williams, W. V. (1993). Subset-specific effects of sex hormones and pituitary gonadotropins on human lymphocyte proliferation *in vitro. Clinical Immunology and Immunopathology, 66,* 201–211.

Aubert, A., Goodall, G., & Dantzer, R. (1995a). Compared effects of cold ambient temperature and cytokines on macronutrient intake in rats. *Physiology and Behavior, 57,* 869–873.

Aubert, A., Vega, C., Dantzer, R., & Goodall, G. (1995b). Pyrogens specifically disrupt the acquisition of a task involving cognitive processing in the rat. *Brain, Behavior, and Immunity, 9,* 129–148.

Aubert, A., Goodall, G., Dantzer, R., & Gheusi, G. (1997a). Differential effects of lipopolysaccharide on pup retrieving and nest building in lactating mice. *Brain, Behavior, and Immunity, 11,* 107–118.

Aubert, A., Kelley, K. W., & Dantzer, R. (1997b). Differential effects of lipopolysaccharide on food hoarding behavior and food consumption in rats. *Brain, Behavior, and Immunity, 11,* 229–238.

Austyn, J. M. (1992). Antigen uptake and presentation by dendritic leukocytes. *Seminars in Immunology, 4,* 227–236.

Axelrod, J., & Reisine, T. (1984). Stress hormones: Their interaction and regulation. *Science, 224,* 452–459.

Babbini, M., Galardi, M., & Bartoletti, M. (1972). Changes in operant behavior as an index of withdrawal state from morphine in rats. *Psychonomic Sciences, 29,* 142–144.

Bachen, E. A., Manuck, S. B., Marsland, A. L., Cohen, S., Malkoff, S. B., Muldoon, M. F., & Rabin, B. S. (1992). Lymphocyte subset and cellular immune responses to a brief experimental stressor. *Psychosomatic Medicine, 54,* 673–679.

Baerwald, C., Grafee, C., Von Wichert, P., & Krause, A. (1992). Decreased density of β-adrenergic receptors on peripheral blood mononuclear cells in patients with rheumatoid arthritis. *Journal of Rheumatology, 19,* 204–210.

Bagby, G. J., Crouch, L. D., & Shepherd, R. E. (1996). Exercise and cytokines: Spontaneous and elicited response. In L. Hoffman-Goetz (Ed.), *Exercise and immune function* (pp. 55–78). New York: CRC Press.

Bandura, A. (1977). Self-efficacy: Toward a unifying theory of psychological change. *Psychological Review, 84,* 191–215.

Barnes, C. A., Forster, M. J., Fleshner, M., Ahanotu, E. N., Laudenslager, M. L., Mazzeo, R. S., Maier, S. F., & Lal, H. (1991). Exercise does not modify spatial memory, brain autoimmunity, or antibody response in aged F-344 rats. *Neurobiology of Aging, 12,* 47–53.

Bartholomew, S. A., & Hoffman, S. A. (1993). Effects of peripheral cytokine injections on multiple unit activity in the anterior hypothalamic area of the mouse. *Brain, Behavior, and Immunity, 7,* 301–316.

Bartrop, R. W., Luckhurst, E., Lazarus, L., Kiloh, L. G., & Penny, R. (1977). Depressed lymphocyte function after bereavement. *Lancet, 1,* 834–836.

Bauer, J., Hohagen, F., Ebert, T., Timmer, J., Ganter, U., Krieger, S., Lis, S., Postler, E., Voderholzer, U., & Berger, M. (1994). Interleukin-6 serum levels in healthy persons correspond to the sleep–wake cycle. *Clinical Investigation, 72,* 315.

Bauer, J., Hohagen, F., Bruns, F., Krieger, S., Lis, S., Riemann, D., & Berger, M. (1995). Induction of cytokine synthesis and fever suppresses REM sleep and improves mood in patients with major depression. *Journal of Sleep Research, 3,* 17.

Baum, M. K., Shor-Posner, G., Lu, Y., Rosner, B., Sauberlich, H. E., Fletcher, M. A., Szapocznik, J., Eisdorfer, C., Buring, J. E., & Hennekens, C. H. (1995). Micronutrients and HIV-1 disease progression. *AIDS, 9,* 1051–1056.

Baumann, H., & Gauldie, J. (1994). The acute phase response. *Immunology Today, 15,* 74–80.

Baxter, R. C. (1980). Simplified approach to confidence limits in radioimmunoassay. *Clinical Chemistry, 26,* 763–765.

Beahrs, J. O., Harris, D. R., & Hilgard, E. R. (1970). Failure to alter skin inflammation by hypnotic suggestion in five subjects with normal skin reactivity. *Psychosomatic Medicine, 32,* 627–631.

Beauchamp, G. K., Yamazaki, K., & Boyse, E. A. (1985). The chemosensory recognition of genetic individuality. *Scientific American, 253,* 86–92.

Beck, A. T. (1967). *Depression: Clinical, experimental, and theoretical aspects.* New York: Harper & Row.

Beck, A. T., Ward, C. H., Medelson, M., Mock, F., & Erbaugh, F. (1961). An inventory for measuring depression. *Archives of General Psychiatry, 4,* 561–571.

Behar, O., Ovadia, H., Polakiewicz, R. D., Abramsky, O., & Rosen, H. (1991). Regulation of proenkephalin messenger ribonucleic acid levels in normal B lymphocytes: Specific inhibition by glucocorticoid hormones and superinduction by cycloheximide. *Endocrinology, 129,* 649–655.

Behar, O., Ovadia, H., Polakiewicz, R. D., & Rosen, H. (1994). Lipopolysaccharide induces proenkephalin gene expression in rat lymph nodes and adrenal glands. *Endocrinology, 134,* 475–481.

Belkin, G. S., Fleishman, J. A., Stein, M. D., Piette, J., & Mor, V. (1992). Physical symptoms and depressive symptoms among individuals with HIV infection. *Psychosomatics, 33,* 416–427.

Bell, E. B., Sparshott, S. M., & Bunce, C. (1998). CD4[+] T-cell memory, CD45R subsets and the persistence of antigen—a unifying concept. *Immunology Today, 19,* 60–64.

Bellinger, D. L., Lorton, D., Brouxhon, S., Felten, S., & Felten, D. (1996). The significance of vasoactive intestinal polypeptide (VIP) in immunomodulation. *Advances in Neuroimmunology, 6,* 5–27.

Benca, R. M., Kushida, C. A., Everson, C. A., Kalski, R., Bergmann, B., & Rechtschaffen, A. (1989). Sleep deprivation in the rat: VII. Immune function. *Sleep, 12,* 47–52.

Benca, R., Obermeyer, W. H., Thisted, R. A., & Gillin, J. C. (1992). Sleep and psychiatric disorders. *Archives of General Psychiatry, 49,* 651–668.

Benedetto, N., Folgore, A., & Galdiero, M. (1993). Impairment of natural resistance to Toxoplasma gondii infection in rats treated with β-adrenergic, β-blockers, corticosteroids or total body irradiation. *Pathologie Biologie, 41,* 404–409.

Ben-Eliyahu, S., Yirmiya, R., Liebeskind, J. C., Taylor, A. N., & Gale, R. P. (1991). Stress increases metastatic spread of a mammary tumor in rats: Evidence for mediation by the immune system. *Brain, Behavior, and Immunity, 5,* 193–205.

Benestad, H. B., Strom, G. I., Ole, I. P., Haug, E., & Nja, A. (1998). No neuronal regulation of murine bone marrow function. *Blood, 91,* 1280–1287.

Benschop, R. J. (1994). *Characterisation of the effects of acute stress on the immune system.* Utrecht: University of Utrecht.

Benschop, R. J., Oostveen, F. G., Heijnen, C. J., & Ballieux, R. E. (1993). β₂-Adrenergic stimulation causes detachment of natural killer cells from cultured endothelium. *European Journal of Immunology, 23,* 3242–3247.

Benschop, R. J., Brosschot, J. F., Godaert, G. L. R., de Smet, M. B. M., Geenen, R., Olff, M., Heijnen, C. J., & Ballieux, R. E. (1994a). Chronic stress affects immunologic but not cardiovascular responsiveness to acute psychological stress in man. *American Journal of Physiology, 266,* R75–R80.

Benschop, R. J., Nieuwenhuis, E. E. S., Tromp, E. A. M., Godaert, G. L. R., Ballieux, R. E., & van Doornen, L. J. P. (1994b). Effects of β-adrenergic blockade on immunologic and cardiovascular changes induced by mental stress. *Circulation, 89,* 762–769.

Benschop, R. J., Godaert, G. L. R., Geenen, R., Brosschot, J. F., de Smet, M. B. M., Olff, M., Heijnen, C. J., & Ballieux, R. E. (1995). Relationships between cardiovascular and immunologic changes in an experimental stress model. *Psychological Medicine, 25,* 323–327.

Benschop, R. J., Jacobs, R., Sommer, B., Schuermeyer, T. H., Raab, H.-R., Schmidt, R. E., & Schedlowski, M. (1996a). Modulation of the immunologic response to acute stress in humans by β-blockade or benzodiazepines. *FASEB Journal, 10,* 517–524.

Benschop, R. J., Rodriguez-Feuerhahn, M., & Schedlowski, M. (1996b). Catecholamine-induced leukocytosis: Early observations, current research, and future directions. *Brain, Behavior, and Immunity, 10,* 77–91.

Benschop, R. J., Schedlowski, M., Wienecke, H., Jacobs, R., & Schmidt, R. E. (1997). Adrenergic control of natural killer cell circulation and adhesion. *Brain, Behavior, and Immunity, 11,* 321–332.

Benson, H. (1993). The relaxation response. In D. Goleman & J. Gurin (Eds.), *Mind body medicine.* New York: Consumer Reports Books, pp. 233–257.

Bentivoglio, M., Grassi-Zucconi, G., Olsson, T., & Kristensson, K. (1994). Trypanosoma brucei and the nervous system. *Trends in Neural Science, 17,* 325–329.

Bentley, P. J. (1980). *Endocrine pharmacology.* Cambridge: Cambridge University Press.

Berczi, I., & Nagy, E. (1991). Effects of hypophysectomy on immune function. In R. Ader, D. L. Felten, & N. Cohen (Eds.), *Psychoneuroimmunology* (pp. 339–376). San Diego: Academic Press.

Berkenbosch, F., Van Oers, J., del Rey, A., Tilders F., & Besedovsky, H. O. (1987). Corticotropin releasing factor producing neurons in the rat are activated by interleukin-1. *Science, 238,* 524–526.

Berkenbosch, F., de Goeij, D. E. C., del Rey, A., & Besedovsky, H. O. (1989). Neuroendocrine, sympathetic and metabolic responses induced by interleukin-1. *Neuroendocrinology, 50,* 570–576.

Bernard, C. (1879). *Lecons sur les phénomenes de la vie commune aux animaux et aux végétaux* (Vol. 2). Paris: Bailliere.

Bernton, E. W., Meltzer, M. S., & Holaday, J. W. (1988). Suppression of macrophage activation and T-lymphocyte function in hypoprolactinemic mice. *Science, 239,* 401–404.

Bernton, E. W., Bryant, H. U., & Holaday, J. W. (1991). Prolactin and immune function. In R. Ader, D. L. Felten, & N. Cohen (Eds.), *Psychoneuroimmunology* (pp. 403–428). San Diego: Academic Press.

Besedovsky, H. O., & del Rey, A. (1992). Immune-neuroendocrine circuits: Integrative role of cytokines. In W. F. Ganong & L. Martini (Eds.), *Frontiers in neuroendocrinology* (pp. 61–94). New York: Raven Press.

Besedovsky, H. O., & del Rey, A. (1996). Immuno-neuro-endocrine interactions: Facts and hypotheses. *Endocrine Reviews, 17,* 64–102.

Besedovsky, H. O., & Sorkin, E. (1977). Immune neuroendocrine network. *Clinical and Experimental Immunology, 27,* 1–12.

Besedovsky, H. O., Sorkin, E., & Mueller, J. (1975). Hormonal changes during the immune response. *Proceedings of the Society for Experimental Biology and Medicine, 150,* 466–479.

Besedovsky, H. O., Sorkin, E., Felix, D., & Haas, H. (1977). Hypothalamic changes during the immune response. *European Journal of Immunology, 7,* 323–325.

Besedovsky, H. O., Sorkin, E., & Keller, M. (1978). Changes in the concentration of corticosterone in the blood during skin-graft rejection in the rat. *Journal of Endocrinology, 76,* 175–176.

Besedovsky, H. O., del Rey, A., & Sorkin, E. (1979). Antigenic competition between horse and sheep red blood cells as a hormone-dependent phenomenon. *Clinical and Experimental Immunology, 37,* 106–113.

Besedovsky, H. O., del Rey, A., & Sorkin, E. (1981). Lymphokine containing supernatants from Con A-stimulated cells increase corticosterone blood levels. *Journal of Immunology, 126*, 385–387.

Besedovsky, H. O., del Rey, A., Sorkin, E., Da Prada, M., Burri, R., & Honegger, C. G. (1983). The immune response evokes changes in brain noradrenergic neurons. *Science, 221*, 564–566.

Besedovsky, H. O., del Rey, A., Sorkin, E., & Dinarello, C. A. (1986). Immunoregulatory feedback between interleukin-1 and glucocorticoid hormones. *Science, 233*, 652–654.

Besedovsky, H. O., del Rey, A., Klusman, I., Furukawa, H., Monge-Arditi, G., & Kabiersch, A. (1991). Cytokines as modulators of the hypothalamus–pituitary–adrenal axis. *Journal of Steroid Biochemistry and Molecular Biology, 40*, 613–618.

Besredka, A. (1926). *Die lokale Immunisierung.* Leipzig: Barth Verlag.

Bette, M., Schäfer, M. K.-H., van Rooijen, N., Weihe, E., & Fleischer, B. (1993). Distribution and kinetics of superantigen-induced cytokine gene expression in mouse spleen. *Journal of Experimental Medicine, 178*, 1531–1540.

Biron, G., Byron, K., & Sullivan, J. (1989). Severe herpes virus infections in an adolescent without natural killer cells. *New England Journal of Medicine, 320*, 1731–1735.

Black, S. (1963a). Inhibition of immediate-type hypersensitivity response under direct suggestion by hypnosis. *British Medical Journal, 1*, 925–929.

Black, S. (1963b). Shift in dose-response curve of Prausnitz-Kustner reaction under direct suggestion by hypnosis. *British Medical Journal, 1*, 989–992.

Black, S., & Friedman, M. (1965). Adrenal function and the inhibition of allergic response under hypnosis. *British Medical Journal, 1*, 562–567.

Black, S., Humphrey, J. H., & Niven, J. S. (1963). Inhibition of mantoux reaction by direct suggestion by hypnosis. *British Medical Journal, 1*, 1649–1652.

Blalock, J. E. (1994). The syntax of immune-neuroendocrine communication. *Immunology Today, 15*, 504–511.

Blalock, J. E., & Smith, E. M. (1985). A complete regulatory loop between the immune and neuroendocrine systems. *Federation Proceedings, 44*, 108–112.

Blaney, N. T., Millon, C., Morgan, R., Eisdorfer, C., & Szapocznik, J. (1990). Emotional distress, stress-related disruption and coping among healthy HIV-positive gay males. *Psychology and Health, 4*, 259–273.

Bleiker, E. M., van der Ploeg, H. M., Hendriks, J. H., & Ader, H. J. (1996). Personality factors and breast cancer development: A prospective longitudinal study. *Journal of the National Cancer Institute, 88*, 1478–1482.

Blom, J. M., Tamarkin, L., Shiber, J. R., & Nelson, R. J. (1995). Learned immunosuppression is associated with an increased risk of chemically-induced tumors. *Neuroimmunomodulation, 2*, 92–99.

Bloom, P., & Carliner, G. (1988). The economic impact of AIDS in the United States. *Science, 239*, 604–609.

Bluthé, R. M., Dantzer, R., & Kelley, K. W. (1989). Corticotropin releasing hormone is not involved in the behavioral effects of peripherally injected interleukin-1 in the rat. *Neurosciences Research Communications, 5*, 149–154.

Bluthé, R. M., Crestani, F., Kelley, K. W., & Dantzer, R. (1992a). Mechanisms of the behavioral effects of interleukin-1: Role of prostaglandins and CRF. *Annals of the New York Academy of Sciences, 650*, 268–275.

Bluthé, R. M., Dantzer, R., & Kelley, K.W. (1992b). Effects of interleukin-1 receptor antagonist on the behavioral effects of lipopolysaccharide in rat. *Brain Research, 573*, 318–320.

Bluthé, R. M., Sparber, S., & Dantzer, R. (1992c). Modulation of the behavioural effects of interleukin-1 in mice by nitric oxide. *Neuroreport, 3*, 207–209.

Bluthé, R. M., Pawlowski, M., Suarez, S., Parnet, P., Pittman, Q., Kelley, K. W., & Dantzer, R. (1994a). Synergy between tumor necrosis factor α and interleukin-1 in the induction of sickness behaviour in mice. *Psychoneuroendocrinology, 19*, 197–207.

Bluthé, R. M., Walter, V., Parnet, P., Layé, S., Lestage, J., Verrier, D., Poole, S., Stenning, B. E., Kelley, K. W., & Dantzer, R. (1994b). Lipopolysaccharide induces sickness behavior in rats by a vagal mediated mechanism. *Comptes Rendus de l'Académie des Sciences, Paris, Sciences de la Vie, 317*, 499–503.

Bluthé, R. M., Michaud, B., Kelley, K. W., & Dantzer, R. (1996a). Vagotomy attenuates behavioural effects of interleukin-1 injected peripherally but not centrally. *Neuroreport, 7*, 1485–1488.

Bluthé, R. M., Michaud, B., Kelley, K. W., & Dantzer, R. (1996b). Vagotomy blocks behavioural effects of interleukin-1 injected via the intraperitoneal route but not via other systemic routes. *Neuroreport, 7*, 2823–2827.

Bluthé, R. M., Michaud, B., Kelley, K. W., & Dantzer, R. (1997). Cholecystokinin receptors do not mediate behavioral effects of lipopolysaccharide in mice. *Physiology and Behavior, 62*, 385–389.

Boccia, M. L., Laudenslager, M. L., & Reite, M. L. (1988). Spacial distribution of food and dominance related behavior in bonnet macaques. *American Journal of Primatology, 16*, 123–130.

Boccia, M. L., Reite, M. L., & Laudenslager, M. L. (1989). On the physiology of grooming in a pigtail macaque. *Physiology and Behavior, 45*, 667–670.

Boccia, M. L., Laudenslager, M. L., Broussard, C. L., & Hijazi, A. S. (1992). Immune responses following competitive dominance water tests in two species of macaques. *Brain, Behavior, and Immunity, 6*, 201–213.

Boccia, M. L., Laudenslager, M. L., & Reite, M. L. (1994). Intrinsic and extrinsic factors affect infant responses to separation. *Psychiatry, 57*, 43–50.

Boccia, M. L., Scanlan, J. M., Laudenslager, M. L., Berger, C. L., & Reite, M. L (1997). Juvenile friends, behavior and immune responses to maternal separation in bonnet macaque infants. *Physiology and Behavior, 61*, 191–198.

Bode, L., Zimmermann, W., Ferszt, R., Steinbach, F., & Ludwig, H. (1995). Borna disease virus genome transcribed and expressed in psychiatric patients. *Nature Medicine, 1*, 232–236.

Bode, U., Wonigeit, K., Pabst, R., & Westermann, J. (1997). The fate of activated T cells migrating through the body: Rescue from apoptosis in the tissue of origin. *European Journal of Immunology, 27*, 2087–2093.

Bognar, I. T., Albrecht, S. A., Farasaty, M., Schmitt, E., Seidel, G., & Fuder, H. (1995). Inhibition by interleukin-1β of noradrenaline release in rat spleen: Involvement of lymphocytes, NO and opioid receptors. *Naunyn-Schmiedeberg's Archives of Pharmacology, 351*, 433–438.

Bongartz, W., Lyncker, L., & Kossman, K. T. (1987). The influence of hypnosis on white blood cell count and urinary levels of catecholamines and vanillyl mandelic acid. *Hypnosis, 14*, 52–61.

Bonneau, R. H. (1994). Experimental approaches to identify mechanisms of stress-induced modulation of immunity to herpes simplex virus infection. In R. Glaser & J. K. Kiecolt-Glaser (Eds.), *Handbook of human stress and immunity* (pp. 125–153). San Diego, Academic Press.

Booth, R. J., & Ashbridge, K. R. (1993). Four steps in the evolution of the defense model of the immune system. *Advances, 9*, 6–23.

Booth, R. J., Petrie, K. J., & Brook, R. J. (1995). Conditioning allergic skin responses in humans: A controlled trial. *Psychosomatic Medicine, 57*, 492–495.

Booth, R. J., Petrie, K. J., & Pennebaker, J. W. (1997). Changes in circulating lymphocyte numbers following emotional disclosure: Evidence of buffering? *Stress Medicine, 13*, 23–29.

Born, J., & Fehm, H. L. (1998). Hypothalamus–pituitary–adrenal activity during human sleep: A coordinating role for the limbic hippocampal system. *Experimental and Clinical Endocrinology and Diabetes, 106*, 153–163.

Born, J., & Hansen, K. (1997). Dependence of human cytokine production and mononuclear cell subset counts on circadian rhythm and sleep. In A. E. Henneberg & W. P. Kaschka (Eds.), *Immunological alterations in psychiatric diseases* (pp. 18–31). Basel: Karger.

Born, J., Späth-Schwalbe, E., Pietrowsky, R., Porzsolt, F., & Fehm, H. L. (1989). Neurophysiological effects of recombinant interferon-gamma and -alpha in man. *Clinical Physiology and Biochemistry, 7*, 119–127.

Born, J., Lange, T., Hansen, K., Mölle, M., & Fehm, H. L. (1997). Effects of sleep and circadian rhythm on human circulating immune cells. *Journal of Immunology, 158*, 4454–4464.

Bost, K. L., & Pascual, D. W. (1992). Substance P: A late-acting B lymphocyte differentiation cofactor. *American Journal of Physiology, 262*, C537–C545.

Bourne, H. R., Lichtenstein, L. M., Melmon, K. L., Henney, C. S., Weinstein, Y., & Shearer, G. M. (1974). Modulation of inflammation and immunity by cyclic AMP. *Science, 184*, 19–28.

Bovbjerg, D. H. (1994). Psychoimmunology: A critical analysis of the implications for clinical oncology in the 21st century. In C. E. Lewis, C. O'Sullivan, & J. Barraclough (Eds.), *The psychoimmunology of human cancer: Mind and body in the fight for survival* (pp. 417–426). New York: Oxford University Press.

Bovbjerg, D. H., & Redd, W. H. (1992). Anticipatory nausea and immune suppression in cancer patients receiving cycles of chemotherapy: Conditioned responses? In H. J. Schmoll, U. Tewes, & N. P. Plotnikoff (Eds.), *Psychoneuroimmunology: Interactions between brain, nervous system, behavior, endocrine and immune system* (pp. 237–250). Lewiston, NY: Hogrefe & Huber.

Bovbjerg, D. H., & Valdimarsdottir, H. (1993). Familial cancer, emotional distress, and low natural cytotoxic activity in healthy women. *Annals of Oncology, 4*, 745–752.

Bovbjerg, D. H., Cohen, N., & Ader, R. (1982a). The central nervous system and learning: A strategy for immune regulation. *Immunology Today, 3*, 287–291.

Bovbjerg, D., Ader, R., & Cohen, N. (1982b). Behaviorally conditioned suppression of a graft-vs-host response. *Proceedings of the National Academy of Sciences, USA, 79*, 583–585.

Bovbjerg, D., Ader, R., & Cohen, N. (1984). Acquisition and extinction of conditioned suppression of a graft-vs-host response in the rat. *Journal of Immunology, 132*, 111–113.

Bovbjerg, D., Cohen, N., & Ader, R. (1987). Behaviorally conditioned enhancement of delayed-type hypersensitivity in the mouse. *Brain, Behavior, and Immunity, 1*, 64–71.

Bovbjerg, D. H., Redd, W. H., Maier, L. A., Holland, J. C., Lesko, L. M., Niedzwiecki, D., Rubin, S. C., & Hakes, T. B. (1990). Anticipatory immune suppression and nausea in women receiving cyclic chemotherapy for ovarian cancer. *Journal of Consulting and Clinical Psychology, 58*, 153–157.

Bower, G. H. (1981). Mood and memory. *American Psychologist, 36*, 129–148.

Bower, J., Kemeny, M. E., Taylor, S. E., & Fahey, J. L. (1998). Cognitive processing, discovery of meaning, CD4 decline, and AIDS-related mortality among bereaved HIV seropositive men. *Journal of Consulting and Clinical Psychology, 66,* 979–986.

Bowlby, J. (1973). *Attachment and loss. Vol. II. Separation: Anxiety and anger.* New York: Basic Books.

Boyce, W. T., Jensen, E. W., Cassel, J. C., Collier, A. M., Smith, A. N., & Ramey, C. T. (1977). Influence of life events and family routines on childhood respiratory tract illness. *Pediatrics, 60,* 609–615.

Boyse, E. A., Beauchamp, G. K., Yamazaki, K., & Bard, J. (1990). Genetic components of kin recognition in mammals. In P. G. Hepper (Ed.), *Kin recognition* (pp. 148–161). Cambridge: Cambridge University Press.

Boyse, E. A., Beauchamp, G. K., Bard, J., & Yamazaki, K. (1991). Behavior and the major histocompatibility complex of the mouse. In R. Ader, D. L. Felten, & N. Cohen (Eds.), *Psychoneuroimmunology* (2nd ed., pp. 831–846). San Diego: Academic Press.

Boyce, W. T., Chesterman, E. A., Martin, N., Folkman, S., Cohen, F., & Wara, D. (1993). Immunologic changes at kindergarten entry predict respiratory illnesses following the Loma Prieta earthquake. *Journal of Developmental and Behavioral Pediatrics, 14,* 296–303.

Bozzola, M., Cisternino, M., Larizza, D., Maghnie, M., Moretta, A., Valtorta, A., Schimpff, R. M., & Severi, F. (1989). Letter to the editor. *Metabolism, 38,* 193.

Brahmi, Z., Thomas, J. E., Park, M., & Dowdeswell, I. R. (1985). The effect of acute exercise on natural killer-cell activity of trained and sedentary human subjects. *Journal of Clinical Immunology, 5,* 321–328.

Brandtzaeg, P., & Halstensen, T. S. (1992). Immunology and immunopathology of tonsils. *Advances in Otorhinolaryngology, 47,* 64–75.

Breder, C. D., Tsujimoto, M., Terano, Y., Scott, D. W., & Saper, C. B. (1993). Distribution and characterization of tumor necrosis factor-α like immunoreactivity in the murine central nervous system. *The Journal of Comparative Neurology, 337,* 543–567.

Breder, C. D., Hazuka, C., Ghayur, T., Klug, C., Huginin, M., Yasuda, K., Teng, M., & Saper, C. B. (1994). Regional induction of tumor necrosis factor α expression in the mouse brain after systemic lipopolysaccharide administration. *Proceedings of the National Academy of Sciences, USA, 91,* 11393–11397.

Bremond, A., Kune, G. A., & Bahnson, C. B. (1986). Psychosomatic factors in breast cancer patients: Results of a case control study. *Journal of Psychosomatic Obstetrics and Gynecology, 5,* 127–136.

Bret-Dibat, J. L., Kent, S., Couraud, J. Y., Creminon, C., & Dantzer, R. (1994). A behaviorally active dose of lipopolysaccharide increases sensory neuropeptide levels in mouse spinal cord. *Neuroscience Letters, 173,* 205–209.

Bret-Dibat, J. L., Bluthé, R. M., Kent, S., Kelley, K. W., & Dantzer, R. (1995). Lipopolysaccharide and interleukin-1 depress food-motivated behavior in mice by a vagal-mediated mechanism. *Brain, Behavior, and Immunity, 9,* 242–246.

Bret-Dibat, J. L., Creminon, C., Couraud, J. Y., Kelley, K. W., & Dantzer, R. (1997). Systemic capsaicin pretreatment fails to block the decrease in food-motivated behavior induced by lipopolysaccharide and interleukin-1β. *Brain Research Bulletin, 42,* 443–449.

Brines, R., Hoffman-Goetz, L., & Pedersen, B. K. (1996). Can you exercise to make your immune system fitter? *Immunology Today, 17,* 252–254.

Brittenden, J., Heys, S. D., Ross, J., & Eremin, O. (1996). Natural killer cells and cancer. *Cancer, 77,* 1226–1243.

Brosschot, J. F., Benschop, R. J., Godaert, G. L. R., de Smet, M. B. M., Olff, M., Heijnen, C. J., & Ballieux, R. E. (1992). Effects of experimental psychological stress on distribution and function of peripheral blood cells. *Psychosomatic Medicine, 54,* 394–406.

Brown, D. H., & Zwilling, B. S. (1994). Activation of the hypothalamic–pituitary–adrenal axis differentially affects the anti-mycobacterial activity of macrophages from BCG-resistant and susceptible mice. *Journal of Neuroimmunology, 53,* 181–187.

Brown, J. L., & Eklund, A. (1994). Kin recognition and the major histocompatibility complex: An integrative review. *American Naturalist, 143,* 435–461.

Brown, R. E. (1994). *An introduction to neuroendocrinology.* Cambridge: Cambridge University Press.

Brown, R. E., Singh, P. B., & Roser, B. (1987). The major histocompatibility complex and the chemosensory recognition of individuality in rats. *Physiology and Behavior, 40,* 65–73.

Brown, R., Pang, G., Husband, A. J., & King, M. G. (1989). Suppression of immunity to influenza virus infection in the respiratory tract following sleep disturbance. *Regional Immunology, 2,* 321–325.

Brown, R., Price, R. J., King, M. G., & Husband, A. J. (1990). Are antibiotic effects on sleep behavior in the rat due to modulation of gut bacteria? *Physiology and Behavior, 48,* 561–565.

Brown, R., Husband, A. J., King, M. G., & Bull, D. F. (1992). Sleep deprivation and the immune response to pathogenic and non-pathogenic antigens. In A. J. Husband (Ed.), *Behavior and immunity* (pp. 127–133). Boca Raton: CRC Press.

Brown, S. L., Smith, L. R., & Blalock, J. E. (1987). Interleukin-1 and interleukin-2 enhance pro-opiomelanocortin gene expression in pituitary cells. *Journal of Immunology, 139,* 3181–3184.

Brownley, K. A., Hurwitz, B. E., Fletcher, M. A., Kumar, M., Klimas, N., Milanovich, J. R., Motivala, S. J., Le Blanc, W. G., & Schneiderman, N. (1998). *Selective sympathetic disruption of the immunological response to speech stress in HIV spectrum disease.* Paper presented at the 19th Annual Meeting of the Society of Behavioral Medicine, New Orleans.

Bruunsgaard, H., Galbo, H., Halkjaer-Kristensen, J., Johansen, T. L., MacLean, D. A., & Pedersen, B. K. (1997). Exercise-induced increase in interleukin-6 is related to muscle damage. *Journal of Physiology (London), 499,* 833–841.

Bruunsgaard, H., Hartkopp, A., Halkjaer-Kristensen, J., & Pedersen, B. K. (1996). Decreased in vivo cell-mediated immunity, but normal vaccination response following intense, long-term exercise. *Medicine and Science in Sports and Exercise, 29,* 1176–1181.

Bryant, J., Day, R., Whiteside, T. L., & Herberman, R. B. (1992). Calculation of lytic units for the expression of cell-mediated cytotoxicity. *Journal of Immunological Methods, 146,* 91–103.

Büchler, M., Weihe, E., Friess, H., Malfertheiner, P., Bockman, E., Müller, S., Nohr, D., & Beger, H. G. (1992). Changes in peptidergic innervation in chronic pancreatitis. *Pancreas, 7,* 183–192.

Buck, R. W. (1979). Individual differences in non-verbal sending accuracy and electrodermal responding: The externalizing-internalizing dimension. In R. Rosenthal (Ed.), *Skill in non-verbal communication* (pp. 40–170). Cambridge, MA: Oelgeschlager, Gunn & Hain.

Buguet, A., Bert, J., Tapie, P., Tabaraud, F., Doua, F., Lonsdorfer, J., Bogui, P., & Dumas, M. (1993). Sleep–wake cycle in human African trypanosomiasis. *Journal of Clinical Neurophysiology, 10,* 190–196.

Bukowski, J., Warner, J., Dennert, G., & Welsh, R. (1985). Adaptive transfer studies demonstrating the antiviral effect of natural killer cells in vivo. *Journal of Experimental Medicine, 131,* 1531–1538.

Burack, J. H., Barrett, D. C., Stall, R. D., Chesney, M. A., Ekstrand, M. L., & Coates, T. J. (1993). Depressive symptoms and CD4 lymphocyte decline among HIV-infected men. *Journal of the American Medical Association, 270,* 2568–2573.

Burbach, J. P .H., Loeber, J. G., Verhoef, J., Wiegant, V. M., De Kloet, E. R., & De Wied, D. (1980). Selective conversion of β-endorphin into peptides related to α- and γ-endorphin. *Nature, 283,* 96–97.

Burgess, W., Gheusi, G., Yao, J., Johnson, R. W., Dantzer, R., & Kelley, K. W. (1998). Interleukin-1β converting enzyme deficient mice resist central but not systemic endotoxin-induced aphagia. *The American Journal of Physiology, 274,* R1829–R1833.

Burnette, M. M., Koehn, K. A., Kenyon-Jump, R., Hutton, K., & Carman, S. (1991). Control of genital herpes recurrences using progressive muscle relaxation. *Behavior Therapy, 22,* 237–247.

Bury, M. (1982). Chronic illness as a biographic disruption. *Sociology of Health and Illness, 4,* 167–182.

Buske-Kirschbaum, A., Kirschbaum, C., Stierle, H., Lehnert, H., & Hellhammer, D. (1992). Conditioned increase of natural killer cell activity in humans. *Psychosomatic Medicine, 54,* 123–132.

Busse, W. W., Anderson, C. L., Hanson, P. G., & Folts, J. D. (1980). The effect of exercise on the granulocyte response to isoproterenol in the trained athlete and unconditioned individual. *Journal of Allergy and Clinical Immunology, 65,* 358–364.

Butler, J., Kelly, J. G., O'Malley, K., & Pidgeon, F. (1983). Beta-adrenoceptor adaptation to acute exercise. *Journal of Physiology (London), 344,* 113–117.

Buttini, M., & Boddeke, H. (1995). Peripheral lipopolysaccharide stimulation induces interleukin-1β messenger RNA in rat brain microglial cells. *Neuroscience, 65,* 523–530.

Buzzetti, R., McLoughlin, L., Lavender, P. M., Clark, A. J. L., & Rees, L. H. (1989). Expression of proopiomelano-cortin gene and quantification of ACTH-like immunoreactivity in human normal peripheral mononuclear cells and lymphoid and myeloid malignancies. *Journal of Clinical Investigation, 83,* 733–738.

Cacioppo, J. T. (1994). Social neuroscience—Autonomic, neuroendocrine, and immune responses to stress. *Psychophysiology, 31,* 113–128.

Caine, N., & Reite, M. (1981). The effect of peer contact upon physiological response to maternal separation. *American Journal of Primatology, 1,* 271–276.

Caldwell, C. L., Irwin, M., & Lohr, J. (1991). Reduced natural killer cell cytotoxicity in depression but not in schizophrenia. *Biological Psychiatry, 30,* 1131–1138.

Camus, G., Pincemail, J., Ledent, M., Juchmes Ferir, A., Lamy, M., Deby Dupont, G., & Deby, C. (1992). Plasma levels of poly-morphonuclear elastase and myeloperoxidase after uphill walking and downhill running at similar energy cost. *International Journal of Sports Medicine, 13,* 443–446.

Cannon, J. G., & Kluger, M. J. (1983). Endogenous pyrogen activity in human plasma after exercise. *Science, 220,* 617–619.

Cannon, J. G., Evans, W. J., Hughes, V. A., Meredith, C. N., & Dinarello, C. A. (1986). Physiological mechanisms contributing to increased interleukin-1 secretion. *Journal of Applied Physiology, 61,* 1869–1874.

Cannon, W. B. (1929). *Bodily changes in pain, hunger, fear, and rage* (2nd ed.). New York: Appleton & Co.

Cannon, W. B. (1932). *The wisdom of the body*. New York: Norton.

Capitanio, J. P., & Lerche, N. W. (1991). Psychosocial factors and disease progression in simian AIDS: A preliminary report. *AIDS, 5*, 1103–1106.

Capitanio, J. P., & Lerche, N. W. (1998). Social separation, housing relocation, and survival in simian AIDS: A retrospective analysis. *Psychosomatic Medicine, 60*, 235–244.

Capitanio, J. P., & Reite, M. L. (1984). The roles of early separation experience and prior familiarity in the social relations of pigtail macaques: A descriptive multivariate study. *Primates, 25*, 475–484.

Capitanio, J. P., Rasmussen, K. L. R., Snyder, D. S., Laudenslager, M. L., & Reite, M. L. (1988). Long-term follow-up of previously separated pigtail macaques: Group and individual differences in response to novel situations. *Journal of Child Psychology and Psychiatry, 27*, 531–538.

Cappell, R., Gregoire, F., Thiry, L., & Sprecher, S. (1978). Antibody and cell-mediated immunity to herpes simplex virus in psychotic depression. *Journal of Clinical Psychiatry, 39*, 266–268.

Carbonari, M., Fiorilli, M., Mezzaroma, I., Cherchi, M., & Aiuti, F. (1989). CD4 as the receptor for retroviruses of the HTLV family: Immunopathogenetic implications. *Advances in Experimental Medicine and Biology, 257*, 3–7.

Carlson, S. L., Albers, K. M., Beiting, D. J., Parish, M., Conner, J. M., & Davis, B. M. (1995). NGF modulates sympathetic innervation of lymphoid tissues. *Journal of Neuroscience, 15*, 5892–5899.

Carlson, S. L., Johnson, S., Parish, M. E., & Cass, W. A. (1998). Development of immune hyperinnervation in NGF-transgenic mice. *Experimental Neurology, 149*, 209–220.

Carlsten, H., Nilsson, N., & Jonsson, R. (1992). Estrogen accelerates immune complex glomerulonephritis but ameliorates T cell-mediated vasculitis and sialadenitis in autoimmune MLR lpr/lpr mice. *Cellular Immunology, 144*, 190–202.

Carr, D. J. J., & Klimpel, G. R. (1986). Enhancement of the generation of cytotoxic T cells by endogenous opiates. *Journal of Neuroimmunology, 12*, 75–87.

Carr, D. J., DeCosta, B. R., Jacobson, A. E., Rice, K. C., & Blalock, J. E. (1990). Corticotropin-releasing hormone augments natural killer cell activity through a naloxone-sensitive pathway. *Journal of Neuroimmunology, 28*, 53–61.

Centers for Disease Control. (1981). Pneumocystis pneumonia—Los Angeles. *Morbidity and Mortality Weekly Report, 30*, 250–252.

Centers for Disease Control. (1992, October). HIV/AIDS surveillance report. *HIV/AIDS Surveillance*, 5–13.

Chandra, R. K., & Newberne, P. M. (1977). *Nutrition, immunity, and infection*. New York: Plenum Press.

Chao, C. C., Janoff, E. N., & Hu, S. (1991). Altered cytokine release in peripheral blood mononuclear cell cultures from patients with the chronic fatigue syndrome. *Cytokine, 3*, 292–295.

Chapman, J. P., & Tsao, M. U. (1980). Possible presence of substance P in the synovial fluid from the knee joint of an arthritic patient. *Federation Proceedings, 39*, 1789–1793.

Chiappelli, F., Gormley, G. F., Gwirstman, H. E., Lowy, M. T., Nguyen, L. D., Nguyen, L., Esmail, I., Strober, M., & Weiner, H. (1992). Effects of intravenous and oral dexamethasone on selected lymphocyte subpopulations in normal subjects. *Psychoneuroendocrinology, 17*, 145–152.

Chin, J., & Mann, J. M. (1990). HIV infections and AIDS in the 1990's. *Annual Review of Public Health, 11*, 127–142.

Chopra, D. (1989). *Quantum healing: Exploring the frontiers of mind/body medicine*. New York: Bantam Books.

Christ, G., & Weiner, L. (1985). Psychosocial issues for AIDS. In V. Devita, Jr., S. Hellman, & S. Rosenberg (Eds.), *AIDS* (pp. 275–297). Philadelphia: Lippincott.

Christensen, A. J., Edwards, D. L., Wiebe, J. S., Benotsch, E. G., McKelvey, L., Andrews, M., & Lubaroff, D. M. (1996). Effect of verbal self-disclosure on natural killer cell activity: Moderating influence of cynical hostility. *Psychosomatic Medicine, 58*, 150–155.

Chuang, H. T., Devins, G. M., Hunsley, J., & Gill, M. J. (1989). Psychosocial distress and well-being among gay and bisexual men with human immunodeficiency virus infection. *American Journal of Psychiatry, 146*, 876–880.

Chun, T. W., Curruth, L., Finzi, D., & Shen, X. (1997). *Quantitative analysis of latent integrated and unintegrated HIV-1 provirus in lymph nodes and peripheral blood: Implications for virus eradication*. Paper presented at the Fourth Conference on Retroviruses and Opportunistic Infections, Washington, DC.

Clark, J. I., & Weiner, L. M. (1995). Biologic treatment of human cancer. *Current Problems in Cancer, 19*, 187–261.

Clarke, A. S., Hedeker, D. R., Ebert, M. H., Schmidt, D. E., McKinney, W. T., & Kraemer, G. W. (1996). Rearing experience and biogenic amine activity in infant rhesus monkeys. *Biological Psychiatry, 40*, 338–352.

Clarkson, A. K. (1937). The nervous factor in juvenile asthma. *British Medical Journal, 2*, 845–860.

Clerici, M., Berzofsky, J. A., Shearer, G. M., & Tacket, C. O. (1991). Exposure to human immunodeficiency virus type 1-specific T helper cell responses before detection of infection by polymerase chain reaction and serum antibodies. *Journal of Infectious Diseases, 164*, 178–184.

Clifton, D. K., & Steiner, R. A. (1981). A new generalized method for analysing episodic hormone secretions. *Endocrinology, 108A,* 317–324.

Coates, T. J., McKusick, L., Kuno, R., & Stites, D. P. (1989). Stress reduction training changed number of sexual partners but not immune function in men with HIV. *American Journal of Public Health, 79,* 885–887.

Cochet, M., Chang, A. C. Y., & Cohen, S. N. (1982). Characterization of the structural gene and putative 5'-regulatory sequences for human proopiomelanocortin. *Nature, 297,* 335–337.

Coe, C. L. (1993). Psychosocial factors and immunity in nonhuman primates: A review. *Psychosomatic Medicine, 55,* 298–308.

Coe, C. L., Erschler, W. B., Champoux, M., & Olson, J. (1992a). Psychosocial factors and immune senescence in the aged primate. *Annals of the New York Academy of Sciences, 650,* 276–282.

Coe, C. L., Lubach, G. R., Schneider, M. L., Dierschke, D. J., & Erschler, W. B. (1992b). Early rearing conditions alter immune responses in the developing infant primate. *Pediatrics, 90,* 505–509.

Coe, C. L., Hou, F., & Clarke, A. S. (1996). Fluoxetine treatment alter leukocyte trafficking in the intrathecal compartment of the young primate. *Biological Psychiatry, 40,* 361–367.

Coe, C. L., Carlson, M., Gunnar, M. R., Lubach, G. R., Dragomir, C., Macovie, O., & Scripcaru, V. (1997). Salivary antibody levels in orphanage-raised children. *Psychosomatic Medicine, 59,* 98 (Abstract).

Cohen, J. (1960). A coefficient of agreement for nominal scales. *Education and Psychological Measurement, 20,* 37–46.

Cohen, J., Cullen, J., & Martin, L. (1982). *Psychosocial aspects of cancer.* New York: Raven Press.

Cohen, N., Ader, R., Green, N., & Bovbjerg, D. (1979). Conditioned suppression of a thymus-independent antibody response. *Psychosomatic Medicine, 41,* 487–491.

Cohen, S. (1988). Psychosocial models of the role of social support in the etiology of physical disease. *Health Psychology, 7,* 269–297.

Cohen, S., & Syme, S. L. (1984). *Social support and health.* San Diego: Academic Press.

Cohen, S., & Williamson, G. M. (1991). Stress and infectious disease in humans. *Psychological Bulletin, 109,* 5–24.

Cohen, S., & Wills, T. A. (1985). Stress, social support, and the buffering hypothesis. *Psychological Bulletin, 98,* 310–357.

Cohen, S., Tyrrell, D. A. J., & Smith, A. P. (1991). Psychological stress and susceptibility to the common cold. *New England Journal of Medicine, 325,* 606–612.

Cohen, S., Kaplan, J. R., Cunnick, J. E., Manuck, S. B., & Rabin, B. S. (1992). Chronic social stress, affiliation and cellular immune response in nonhuman primates. *Psychological Sciences, 3,* 301–304.

Cohen, S., Kessler, R. C., & Gordon, L. U. (1995). *Measuring stress. A guide for health and social scientists.* Oxford: Oxford University Press.

Cohen, S., Doyle, W. J., Skoner, D. P., Rabin, B. S., & Gwaltney, J. M., Jr. (1997). Psychological stress and susceptibility to the common cold. *Journal of the American Medical Association, 277,* 1940–1944.

Cole, S. W. (1999). Negative physical health effects of psychological inhibition in disclosure study control groups.

Cole, S. W., & Kemeny, M. E. (1997). The psychobiology of AIDS. *Critical Reviews in Neurobiology, 11,* 289–321.

Cooper, C. L., Cooper, R. F., & Faragher, E. B. (1989). Incidence and perception of psychosocial stress: The relationship with breast cancer. *Psychological Medicine, 19,* 415–422.

Coussons, M. E., Dykstra, L. A., & Lysle, D. T. (1992). Pavlovian conditioning of morphine-induced alterations of immune status. *Journal of Neuroimmunology, 39,* 219–230.

Cousins, N. (1976). Anatomy of an illness (as perceived by the patient). *New England Journal of Medicine, 295,* 1458–1463.

Cover, H., & Irwin, M. (1994). Immunity and depression: Insomnia, retardation and reduction of natural killer cell activity. *Journal of Behavioral Medicine, 17,* 217–223.

Crary, B., Borysenko, M., Sutherland, D. C., Kutz, I., Borysenko, J. Z., & Benson, H. (1983a). Decrease in mitogen responsiveness of mononuclear cells from peripheral blood after epinephrine administration in humans. *Journal of Immunology, 130,* 694–697.

Crary, B., Hauser, S. L., Borysenko, M., Kutz, I., Hoban, C., Ault, K. A., Weiner, H. L., & Benson, H. (1983b). Epinephrine-induced changes in the distribution of lymphocyte subsets in peripheral blood of humans. *Journal of Immunology, 131,* 1178–1181.

Crestani, F., Seguy, F., & Dantzer, R. (1991). Behavioral effects of peripherally injected interleukin-1: Role of prostaglandins. *Brain Research, 542,* 330–334.

Crist, D. M., Mackinnon, L. T., Thompson, R. F., Atterbom, H. A., & Egan, P. A. (1989). Physical exercise increases natural cellular-mediated tumor cytotoxicity in elderly women. *Gerontology, 35,* 66–71.

Crnic, L. S., & Segall, M. A. (1992). Prostaglandins do not mediate interleukin-1α effects on mouse behavior. *Physiology and Behavior, 51,* 349–352.

Crofford, L. J., Sano, H., Karalis, K., Webster, E. L., Goldmutz, E. A., Chrousos, G. P., & Wilder, R. L. (1992).

Local secretion of corticotropin-releasing hormone in the joints of Lewis rats with inflammatory arthritis. Journal of *Clinical Investigation, 90,* 2555–2564.

Cunnick, J. E., Lysle, D. T., Kucinski, B. J., & Rabin, B. S. (1990). Evidence that shock-induced immune suppression is mediated by adrenal hormones and peripheral β-adrenergic receptors. *Pharmacology, Biochemistry, and Behavior, 36,* 645–651.

Cunnick, J. E., Cohen, S., Rabin, B. S., Carpenter, A. B., Manuck, S. B., & Kaplan, J. R. (1991a). Alterations in specific antibody production due to rank and social instability. *Brain, Behavior, and Immunity, 5,* 357–369.

Cunnick, J. E., Lysle, D. T., Armfield, A., & Rabin, B. S. (1991b). Stressor-induced changes in mitogenic activity are not associated with decreased interleukin 2 production or changes in lymphocyte subsets. *Clinical Immunology and Immunopathology, 60,* 419–429.

Cunningham, E., Jr., Wada, E., Carter, D. B., Tracey, D. E., Battey, J. F., & De Souza, E. B. (1992). In situ histochemical localization of type I interleukin-1 receptor messenger RNA in the central nervous system, pituitary, and adrenal gland of the mouse. *Journal of Neuroscience, 12,* 1101–1114.

Cupps, T., & Fauci, A. (1982). Corticosteroid-mediated immunoregulation in man. *Immunology Review, 65,* 133–155.

Daar, E. S., Moudgil, T., Meyer, R. D., & Ho, D. D. (1991). Transient high levels of viremia in patients with primary human immunodeficiency virus type 1 infection. *New England Journal of Medicine, 324,* 961–964.

Dantzer, R. (1994). How do cytokines say hello to the brain? Neural versus humoral mediation. *European Cytokine Network, 5,* 271–273.

Dantzer, R., Bluthé, R. M., & Kelley, K. W. (1991). Androgen-dependent vasopressinergic neurotransmission attenuates interleukin-1 induced sickness behavior. *Brain Research, 557,* 115–120.

Dantzer, R., Bluthé, R. M., Kent, S., & Goodall, G. (1993). Behavioral effects of cytokines: An insight into mechanisms of sickness behavior. In E. De Souza (Ed.), *Methods in neuroscience* (pp. 130–149). Orlando: Academic Press.

Dantzer, R., Bluthé, R. M., Gheusi, G., Cremona, S., Layé, S., Parnet, P., & Kelley K. W. (1998). Molecular basis of sickness behavior. *Annals of the New York Academy of Sciences, 856,* 132–138.

Da Silva, J. A. P. (1995). Sex hormones, glucocorticoids and autoimmunity. *Annals of the Rheumatic Diseases, 54,* 6–16.

Dattore, P. J., Shontz, F. C., & Coyne, L. (1980). Premorbid personality differentiation of cancer and noncancer groups: A test of the hypothesis of cancer proneness. *Journal of Consulting and Clinical Psychology, 48,* 388–394.

Daun, J. M., & McCarthy, D. O. (1993). The role of cholecystokinin in interleukin-1 induced anorexia. *Physiology and Behavior, 54,* 237–241.

Daynes, R. A., Araneo, B. A., Dowell, T. A., Huang, K., & Dudley, D. (1990). Regulation of murine lymphokine production *in vivo.* III. The lymphoid tissue microenvironment exerts regulatory influences over T helper cell function. *Journal of Experimental Medicine, 171,* 979–996.

Deeks, S. G., Smith, M., Holodniy, M., & Kahn, J. O. (1997). HIV-1 protease inhibitors. A review for clinicians. *Journal of the American Medical Association, 277,* 145–153.

Delahanty, D. L., Dougall, A. L., Craig, K. J., Jenkins, F. J., & Baum, A. (1997). Chronic stress and natural killer cell activity after exposure to traumatic death. *Psychosomatic Medicine, 59,* 467–476.

Del Cerro, S., & Borrell, J. (1990). Interleukin-1 affects the behavioral despair response in rats by an indirect mechanism which requires endogenous CRF. *Brain Research, 528,* 162–164.

De Longis, A., Coyne, J., Dakof, G., Folkman, S., & Lazarus, R. (1982). Relationship of daily hassles, uplifts and major life events to health status. *Health Psychology, 1,* 119–136.

del Rey, A., & Besedovsky, H. O. (1989). Antidiabetic effects of interleukin-1. *Proceedings of the National Academy of Sciences, USA, 86,* 5943–5947.

del Rey, A., Besedovsky, H. O., Sorkin, E., Da Prada, M., & Bondiolotti P. (1982). Sympathetic immunoregulation: Difference between high- and low-responder animals. *American Journal of Physiology, 242,* R30–R33.

del Rey, A., Monge-Arditi, G., Besedovsky, H. O. (1998). Central and peripheral mechanisms contribute to the hypoglycemia induced by interleukin-1. *Annals of the New York Academy of Sciences, 840,* 153–161.

de Rijk, R., van Rooijen, N., Besedovsky, H., del Rey, A., & Berkenbosch, F. (1991). Selective depletion of macrophages prevents pituitary–adrenal activation in response to subpyrogenic, but not to pyrogenic, doses of bacterial endotoxin in rats. *Endocrinology, 129,* 330–338.

Demetrikopoulos, M. K., & Zhang, Z. (1998). *Effects of site specific electrical stimulation of the periaqueductal gray on immunity.* Paper presented at the Winter Conference on Brain Research, Snowbird, UT.

Demetrikopoulos, M. K., Siegel, A., Schleifer, S. J., Obedi, J., & Keller, S. E. (1994). Electrical stimulation of the dorsal midbrain periaqueductal gray suppresses peripheral blood natural killer cell activity. *Brain, Behavior, and Immunity, 8,* 218–228.

Demetrikopoulos, M. K., Weiss, J. M., & Goldfarb, R. H. (1998) . Environmental factors and disease: Stress and cancer. In B. S. McEwen (Ed.), *Handbook of physiology: Coping with the environment*. Oxford: Oxford University Press.

Dennis, M., & Philippus, M. J. (1965). Hypnotic and non-hypnotic suggestion and skin response in atopic patients. *American Journal of Clinical Hypnosis*, *7*, 342–345.

de-The, G. (1980). Role of Epstein–Barr virus in human diseases: Infectious mononucleosis, Burkitt's lymphoma, and nasopharyngeal carcinoma. In G. Klein (Ed.), *Viral oncology* (pp. 769–797). New York: Raven Press.

Devoino, L., Alperina, E., Galkina, O., & Ilyutchenok, R. (1997). Involvement of brain dopaminergic structures in neuroimmunomodulation. *International Journal of Neuroscience*, *91*, 213-228.

Devoino, L., Morozova, N., & Cheido, M. (1988). Participation of serotoninergic system in neuroimmunomodulation: Intraimmune mechanisms and the pathways providing an inhibitory effect. *International Journal of Neuroscience*, *40*, 111–128.

Dhabhar, F. S., Miller, A. H., McEwen, B. S., & Spencer, R. L. (1996). Stress-induced changes in blood leukocyte distribution: Role of adrenal steroid hormones. *The Journal of Immunology*, *157*, 1638–1644.

Diclemente, R. J., & Temoshok, L. (1985). Psychological adjustment to having cutaneous malignant melanoma as a predictor of follow-up clinical status. *Psychosomatic Medicine*, *47*, 81.

Dietzschold, B., Schwäble, W., Schäfer, M. K.-H., Hooper, C., Zheng, Y. M., Petry, F., Sheng, H., Fink, T., Loos, M., Koprowski, H., & Weihe, E. (1995). Expression of C1q, a subcomponent of the rat complement system, is dramatically enhanced in brains of rats with Borna disease or experimental allergic encephalomyelitis. *Journal of Neurological Science*, *130*, 11–16.

Dilley, J. W., Ochitill, H. N., Perl, M., & Volberding, P. A. (1985). Findings in psychiatric consultations with patients with acquired immune deficiency syndrome. *American Journal of Psychiatry*, *142*, 82–86.

Dillon, K. M., Minchoff, B., & Baker, K. H. (1985). Positive emotional states and enhancement of the immune system. *International Journal of Psychiatry in Medicine*, *15*, 13–17.

Dinarello, C. A. (1996). Biologic basis for interleukin-1 in disease. *Blood*, *87*, 2095–2147.

Dinges, D. F., Douglas, S .D., Zaugg, L., Campbell, D. E., McMann, J. M., Whitehouse, W. G., Orne, E.C., Kapoor, S. C., Icaza, E., & Orne, M. T. (1994). Leukocytosis and natural killer cell function parallel neurobehavioral fatigue induced by 64 hours of sleep deprivation. *Journal of Clinical Investigation*, *93*, 1930–1939.

Dinges, D. F., Douglas, S. D., Hamarman, S., Zaugg, L., & Kapoor, S. (1995). Sleep deprivation and human immune function. *Advances in Neuroimmunology*, *5*, 97–110.

Di Sebastiano P., Fink, T., Weihe, E., Friess, H., Innocenti, P., Beger, H. G., & Büchler, M. (1997). Immune cell infiltration and growth-associated protein 43 expression correlate with pain in chronic pancreatitis. *Gastroenterology*, *112*, 1648–1655.

Di Sebastiano, P., Fink, T., Weihe, E., Friess, H., Beger, H. G., & Büchler, M. (1995). Changes of protein gene product 9.5 (PGP 9.5) immunoreactive nerves in inflamed appendix. *Digestive Disease Science*, *40*, 366–372.

Di Sebastiano, P., Fink, T., di Mola, F. F., Weihe, E., Friess, H., Innocenti, P., Büchler, M. (1999). Neuroimmune appendicitis: A new pathological entity? *Lancet*, in press.

Dodd, J., & Role, L. W. (1991). The autonomic nervous system. In E. R. Kandel, J. H. Schwartz, & T. M. Jessell (Eds.), *Principles of neural science*. Amsterdam: Elsevier, pp. 663–679..

Dodt, C., Breckling, U., Derad, I., Fehm, H. L., & Born, J. (1997). Plasma epinephrine and norepinephrine concentrations of healthy humans associated with nighttime sleep and morning arousal. *Hypertension*, *30*, 71–76.

Dohrenwend, B. P., & Shrout, E. P. (1985). "Hassles" in the conceptualization of life stress variables. *American Psychologist*, *40*, 780–785.

Dopp, J. M., Mackenzie-Graham, A., Otero, G. C., & Merrill, J. E. (1997). Differential expression, cytokine modulation, and specific functions of type-1 and type-2 tumor necrosis factor receptors in rat glia. *Journal of Neuroimmunology*, *75*, 104–112.

Dufaux, B., & Order, U. (1989a). Plasma elastase-alpha 1-antitrypsin, neopterin, tumor necrosis factor, and soluble interleukin-2 receptor after prolonged exercise. *International Journal of Sports Medicine*, *10*, 434–438.

Dufaux, B., & Order, U. (1989b). Complement activation after prolonged exercise. *Clinica Chimica Acta*, *179*, 45–49.

Dufaux, B., Order, U., Geyer, H., & Hollmann, W. (1984). C-reactive protein serum concentrations in well-trained athletes. *International Journal of Sports Medicine*, *5*, 102–106.

Dunn, A. J. (1989). Neurochemistry of stress. In G. Adelman (Ed.), *Encyclopedia of neuroscience*. Boston: Birkhäuser, pp. 1146–1150.

Dunn, A. J., Powell, M. L., Meitin, C., & Small, P.A. (1989). Virus infection as a stressor: Influenza virus elevates plasma concentrations of corticosterone, and brain concentrations of MHPG and tryptophan. *Physiology and Behavior*, *45*, 591–594.

Dunn, A., Antoon, M., & Chapman, Y. (1991). Reduction of exploratory behavior by intraperitoneal injection of interleukin-1 involves brain corticotropin-releasing factor. *Brain Research Bulletin*, *26*, 539–542.

Dworkin, R. H., Hartstein, G., Rosner, H. L., Walther, R. R., Sweeney, E. W., & Brand, L. (1992). A high-risk method for studying psychosocial antecedents of chronic pain: The prospective investigation of herpes zoster. *Journal of Abnormal Psychology, 101*, 200–205.

Eggert, F., Luszyk, D., Westphal, E., Müller-Ruchholtz, W., & Ferstl, R. (1990). Vom Gen zum Geruch zum Verhalten: Über immunogenetische Grundlagen der chemosensorischen Identität und ihre psychobiologischen Effekte. *TW Neurologie Psychiatrie, 4*, 889–892.

Eggert, F., Wobst, B., Höller, C., Uharek, L., Luszyk, D., Zavazava, N., Westphal, E., Müller-Ruchholtz, W., & Ferstl, R. (1994). Determinanten der chemosensorischen Identität: Ein Modell für die olfaktorische Expression von MHC-Genen. *TW Neurologie Psychiatrie, 8*, 90–95.

Ehrke, M. J., Mihich, E., Berd, D., & Mastrangelo, M. J. (1989). Effects of anticancer drugs on the immune system in humans. *Seminars in Oncology, 16*, 230–253.

Eiden, L. E., Rausch, D. M., Da Cunha, A., Murray, E. A., Heyes, M., Sharer, L., Nohr, D., & Weihe, E. (1993). AIDS and the central nervous system. Examining pathobiology and testing therapeutic strategies in the SIV-infected rhesus monkey. *Annals of the New York Academy of Sciences, 693*, 229–244.

Ekman, P., Levenson, R. W., & Friesen, W. V. (1983). Autonomic nervous system activity distinguishes among emotions. *Science, 221*, 1208–1210.

Eller, L. S. (1995). Effects of two cognitive-behavioral interventions on immunity and symptoms in persons with HIV. *Annals of Behavioral Medicine, 17*, 339–348.

Ellis, A., & Grieger, R. (Eds.). (1977). *Handbook of rational-emotive therapy*. Berlin: Springer.

Erickson, J. D., Schäfer, M. K.-H., Bonner, T. I., Eiden, L. E., & Weihe, E. (1996). Distinct pharmacological properties and distribution in neurons and endocrine cells of two isoforms of the human vesicular monoamine transporter. *Proceedings of the National Academy of Sciences, USA, 93*, 5166–5171.

Ernberg, I. (1986). The role of Epstein–Barr virus in lymphomas of homosexual males. In E. Klein (Ed.), *Acquired immunodeficiency syndrome* (pp. 301–318). Basel: Karger.

Erschler, W. B., Coe, C. L., Gravenstein, S., Schultz, Kloop, R. G., Meyer, M., & Houser, D. (1998). Aging and immunity in nonhuman primates. I. Effects of age and gender on cellular immune function in rhesus monkeys (*Macaca mullatta*). *American Journal of Primatology, 15*, 181–188.7.

Essen, P., Wernerman, J., Sonnenfeld, T., Thunell, S., & Vinnars, E. (1992). Free amino acids in plasma and muscle during 24 hours post-operatively—A descriptive study. *Clinical Physiology, 12*, 163–177.

Esterling, B. A., Antoni, M. H., Kumar, M., & Schneiderman, N. (1990). Emotional repression, stress disclosure responses, and Epstein–Barr viral capsid antigen titers. *Psychosomatic Medicine, 52*, 397–410.

Esterling, B. A., Antoni, M. H., Schneiderman, N., Carver, C. S., LaPerriere, A., Klimas, N., Ironson, G., & Fletcher, M. A. (1992). Psychosocial modulation of antibody to Epstein-Barr viral capsid antigen and human herpesvirus type-6 in HIV-1 infected and at-risk gay men. *Psychosomatic Medicine, 54*, 354–371.

Esterling, B. A., Antoni, M. H., Fletcher, M. A., Marguilies, S., & Schneiderman, N. (1994). Emotional disclosure through writing or speaking modulates latent Epstein–Barr virus antibody titers. *Journal of Consulting and Clinical Psychology, 62*, 130–140.

Evans, D. L., Folds, J. D., Petitto, J. M., Golden, R. N., Pedersen, C. A., Corrigan, M., Gilmore, J. H., Silva, S. G., Quade, D., & Ozer, H. (1992). Circulating natural killer cell phenotypes in men and women with major depression. *Archives of General Psychiatry, 49*, 388–395.

Evans, P., Bristow, M., Hucklebridge, F., Clow, A., & Walters, N. (1993). The relationship between secretory immunity, mood and life-events. *British Journal of Clinical Psychology, 32*, 227–236.

Evans, W. J., Meredith, C. N., Cannon, J. G., Dinarello, C. A., Frontera, W. R., Hughes, V. A., Jones, B. H., and Knuttgen, H. G. (1986). Metabolic changes following eccentric exercise in trained and untrained men. *Journal of Applied Physiology, 61*, 1864–1868.

Everson, C. A. (1993). Sustained sleep deprivation impairs host defense. *American Journal of Physiology, 265*, R1148–R1154.

Exton, M. S., Bull, D. F., King, M. G., & Husband, A. J. (1995a). Modification of body temperature and sleep states using behavioral conditioning. *Physiology and Behavior, 57*, 723–729.

Exton, M. S., Bull, D. F., King, M. G., & Husband, A. J. (1995b). Paradoxical conditioning of the plasma copper and corticosterone responses to bacterial endotoxin. *Pharmacology, Biochemistry, and Behavior, 52*, 87–94.

Exton, M. S., von Hörsten, S., Schult, M., Vöge, J., Strubel, T., Donath, S., Steinmüller, C., Seeliger, H., Nagel, E., Westermann, J., & Schedlowski, M. (1998). Behaviorally conditioned immunosuppression using cyclosporin A: CNS reduces IL-2 production via splenic innervation. *Journal of Neuroimmunology, 88*, 182–191.

Eysenck, H. J. (1993). Prediction of cancer and coronary heart disease mortality by means of a personality inventory: Results of a 15-year follow-up study. *Psychological Reports, 72*, 499–516.

Eysenck, H. J. (1994). Synergistic interaction between psychosocial and physical factors in the causation of lung cancer. In C. E. Lewis, C. O'Sullivan, & J. Barraclough (Eds.), *The psychoimmunology of cancer: Mind and body in the fight for survival* (pp. 163–178). New York: Oxford University Press.

Ezzell, C. (1993). AIDS' unlucky strike: When T cells would rather switch than fight. *The Journal of NIH Research, 5,* 59–64.

Faith, R. E., Jurgo, A. J., Clinkscales, C. W., & Plotnikoff, N. P. (1987). Enhancement of host resistance to viral and tumor challenge by treatment with methionine-enkephalin. *Annals of the New York Academy of Sciences, 496,* 137–139.

Faragher, E. B., & Cooper, C. L. (1990). Type A stress prone behavior and breast cancer. *Psychological Medicine, 20,* 663–670.

Farris, E. J. (1938). Increase in lymphocytes in healthy persons under certain emotional states. *American Journal of Anatomy, 63,* 297–323.

Fawzy, F. I., Cousins, N., Fawzy, N. W., Kemeny, M. E., Elashoff, R., & Morton, D. (1990a). A structured psychiatric intervention for cancer patients. I. Changes over time in methods of coping and affective disturbance. *Archives of General Psychiatry, 47,* 720–725.

Fawzy, F. I., Kemeny, M. E., Fawzy, N. W., Elashoff, R., Morton, D., Cousins, N., & Fahey, J. L. (1990b). A structured psychiatric intervention for cancer patients. II. Changes over time in immunological measures. *Archives of General Psychiatry, 47,* 729–735.

Fawzy, F. I., Fawzy, N. W., Hyun, C. S., Elashoff, R., Guthrie, D., Fahey, J. L., & Morton, D. L. (1993). Malignant melanoma: Effects of an early structured psychiatric intervention, coping, and affective state on recurrence and survival 6 years later. *Archives of General Psychiatry, 50,* 681–689.

Fawzy, F. I., Fawzy, N. W., & Hyun, C. S. (1994). Short-term psychiatric intervention for patients with malignant melanoma: Effects on psychological state, coping, and the immune system. In C. E. Lewis, C. O'Sullivan, & J. Barraclough (Eds.), *The psychoimmunology of cancer: Mind and body in the fight for survival* (pp. 292–319). New York: Oxford University Press.

Fecho, K., Dykstra, L. A., & Lysle, D. T. (1993). Evidence for β-adrenergic receptor involvement in the immunomodulatory effects of morphine. *Journal of Pharmacology and Experimental Therapeutics, 265,* 1079–1087.

Felsner, P., Hofer, D., Rinner, I., Porta, S., Korsatko, W., & Schauenstein, K. (1995). Adrenergic suppression of peripheral blood T cell reactivity in the rat is due to activation of peripheral α_2-receptors. *Journal of Neuroimmunology, 57,* 27–34.

Felten, D., Felten, S., Carlson, S., Olschawka, J., & Livnat, S. (1985). Noradrenergic and peptidergic innervation of lymphoid tissue. *Journal of Immunology, 135* (Suppl. 2), 755s–765s.

Felten, D. L., Felten, S. Y., Bellinger, D. L., Carlson, S. L., Ackerman, K. D., Madden, K. S., Olschowski, J. A., & Livnat, S. (1987). Noradrenergic sympathetic neural interactions with the immune system: Structure and function. *Immunological Reviews, 100,* 225–260.

Felten, S. Y., & Felten, D. L. (1991). Innervation of lymphoid tissue. In R. Ader, D. L. Felten, & N. Cohen (Eds.), *Psychoneuroimmunology* (pp. 27–69). San Diego: Academic Press.

Felten, S. Y., & Felten, D. L. (1994). Neural–immune interactions. *Progress in Brain Research, 100,* 157–162.

Ferreira, S. H., Lorenzetti, B. B., Bristow, A. F., & Poole, S. (1988). Interleukin-1β is a potent hyperalgesic agent antagonized by a tripeptide analogue. *Nature, 334,* 698–700.

Ferstl, R., Eggert, F., Westphal, E., Zavazava, N., & Müller-Ruchholtz, W. (1992). MHC-related odors in humans. In R. L. Doty & D. Müller-Schwarze (Eds.), *Chemical signals in vertebrates VI* (pp. 205–211). New York: Plenum Press.

Fiatarone, M. A., Morley, J. E., Bloom, E. T., Benton, D., Makinodan, T., & Solomon, G. F. (1988). Endogenous opioids and the exercise-induced augmentation of natural killer cell activity. *Journal of Laboratory Clinical Medicine, 112,* 544–552.

Fiatarone, M. A., Morley, J. E., Bloom, E. T., Benton, D., Solomon, G. F., & Makinodan, T. (1989). The effect of exercise on natural killer cell activity in young and old subjects. *Journal of Gerontology, 44,* 37.

Field, C. J., Gougeon, R., & Marliss, E. B. (1991). Circulating mononuclear cell numbers and function during intense exercise and recovery. *Journal of Applied Physiology, 71,* 1089–1097.

Filteau, S. M., Menzies, R. A., Kaido, T. J., O'Grady, M. P., Gelderd, J. B., & Hall, N. R. S. (1992). Effects of exercise on immune functions of undernourished mice. *Life Sciences, 51,* 565–574.

Fink, T., & Weihe, E. (1988). Multiple neuropeptides in nerves supplying mammalian lymph nodes: Messenger candidates for sensory and autonomic neuroimmunomodulation? *Neuroscience Letters, 90,* 39–44.

Fink, T., Di Sebastiano, P. D., Büchler, M., Beger, H. G., & Weihe, E. (1994). Growth-associated protein-43 and protein gene-product 9.5 innervation in human pancreas: Changes in chronic pancreatitis. *Neuroscience, 63,* 249–266.

Fischler, B., Bocken, R., DeWaele, M., & Thielemans, K. (1990). Major depressive disorder, endogeneicity and natural killer cell numbers and activity. *International Journal of Neuroscience, 51,* 357–358.

Fitzgerald, L. (1988). Exercise and the immune system. *Immunology Today, 9,* 337–339.

Fitzgerald, L. (1991). Overtraining increases the susceptibility to infection. *International Journal of Sports Medicine, 12,* S5–S8.

Fleshner, M., Goehler, L. E., Hermann, J., Relton, J. K., Maier, S. F., & Watkins, L. R. (1995). Interleukin-1β induced corticosterone elevation and hypothalamic NE depletion is vagally mediated. *Brain Research Bulletin, 37,* 605–610.

Fletcher, C. V. (1992). Treatment of herpesvirus infections in HIV-infected individuals. *The Annals of Pharmacotherapy, 26,* 955–962.

Flores, L. R., Dretchen, K. L., & Bayer, B. M. (1996). Potential role of the autonomic nervous system in the immunosuppressive effects of acute morphine administration. *European Journal of Pharmacology, 318,* 437–446.

Ford, W. L. (1980). The lymphocyte—Its transformation from a frustrating enigma to a model of cellular function. In M. M. Wintrobe (Ed.), *Blood, pure and eloquent: A story of discovery of people and ideas* (pp. 457–508). New York: McGraw-Hill.

Forsen, A. (1991). Psychosocial stress as a risk for breast cancer. *Psychotherapy and Psychosomatics, 55,* 176–185.

Fox, B. H. (1989). Depressive symptoms and risk of cancer. *Journal of the American Medical Association, 262,* 1231.

Fox, B. H. (1995). The role of psychological factors in cancer incidence and prognosis. *Oncology, 9,* 245–255.

Fox, B. H., Ragland, D. R., Brand, R. J., & Rosenman, R. H. (1987). Type A behavior and cancer mortality: Theoretical considerations and preliminary data. *Annals of the New York Academy of Sciences, 496,* 620–627.

French, R. A., Zachary, J. F., Dantzer, R., Frawley, L. S., Chizzonite, R., Parnet, P., & Kelley, K. W. (1996). Dual expression of p80 type I and p68 type II interleukin-1 receptors on anterior pituitary cells synthesizing growth hormone. *Endocrinology, 137,* 4027–4036.

French, R. A., VanHoy, R. W., Chizzonite, R., Zachary, J. F., Dantzer, R., Parnet, P., Bluthé, R. M., & Kelley, K. W. (1999). Expression and localization of p80 and p68 interleukin-1 receptor proteins in the brain of adult mice. *Journal of Neuroimmunology, 93,* 194–202.

Frey, W. (1914). Der Einfluß des vegetativen Nervensystems auf das Blutbild. *Zeitschrift für die Gesamte Experimentelle Medizin, 2,* 38–49.

Friedman, H. S., & Booth-Kewley, S. (1987). The "disease prone personality": A meta-analytic view of the construct. *American Psychologist, 42,* 539–555.

Friedman, M., & Rosenman, R. H. (1974). *Type A behavior and your heart.* New York: Knopf.

Frieling, T., & Strohmeyer, G. (1995). Neuroimmune interactions in the gastrointestinal tract. *Zeitschrift für Gastroenterologie, 33,* 219–224.

Frimerman, A., Miller, H. I., Laniado, S., & Keren, G. (1997). Changes in hemostatic function at times of cyclic variation in occupational stress. *American Journal of Cardiology, 79,* 72–75.

Fry, L., Mason, A. A., & Pearson, R. S. B. (1964). Effect of hypnosis on allergic skin response in asthma and hay fever. *British Medical Journal, 1,* 1145–1148.

Fu, Z. F., Weihe, E., Zheng, Y. M., Schäfer, M. K.-H., Sheng, H., Corisdeo, S., Rauscher, F., Koprowski, H., & Dietzschold, B. (1993). Differential effects of rabies and borna disease viruses on immediate-, early-, and late-response gene expression in brain tissues. *Journal of Virology, 67,* 6674–6681.

Fukuoka, H., Kawatani, M., Hisamitsu, T., & Takeshige, C. (1994). Cutaneous hyperalgesia induced by peripheral injection of interleukin-1β in the rat. *Brain Research, 657,* 133–140.

Futterman, A. D., Kemeny, M. E., Shapiro, D., & Fahey, J. L. (1994). Immunological and physiological changes associated with induced positive and negative mood. *Psychosomatic Medicine, 56,* 499–511.

Galbo, H. (1983). *Hormonal and metabolic adaption to exercise.* Stuttgart: Thieme Verlag.

Galin, F. S., LeBoeuf, R. D., & Blalock, J. E. (1991). Corticotropin-releasing factor upregulates expression of two truncated pro-opiomelanocortin transcripts in murine lymphocytes. *Journal of Neuroimmunology, 31,* 51–58.

Ganea, D. (1996). Regulatory effects of vasoactive intestinal peptide on cytokine production in central and peripheral lymphoid organs. *Advances in Neuroimmunology, 6,* 61–74.

Garcia, J., & Hankins, W. G. (1977). On the origin of food aversion paradigms. In L. M. Barker, M. R. Best, & M. Domjan (Eds.), *Learning mechanisms in food selection* (pp. 3–19). New York: Baylor University Press.

Garcia, J., Hankins, W. G., & Rusiniak, K. W. (1974). Behavioral regulation of the milieu interne in man and rat. *Science, 185,* 824–831.

Garrido, E., Delgado, M., Martinez, C., Gomariz, R. P., & de la Fuente, M. (1996). Pituitary adenylate cyclase-activating polypeptide (PACAP38) modulates lymphocyte and macrophage functions: Stimulation of adherence and opposite effect on mobility. *Neuropeptides, 30,* 583–595.

Gatti, S., & Bartfai, T. (1993). Induction of tumor necrosis factor-α mRNA in the brain after peripheral endotoxin treatment: Comparison with interleukin-1 family and interleukin-6. *Brain Research, 624,* 291–294.

Gauci, M., Husband, A. J., Saxarra, H., & King, M. G. (1994). Pavlovian conditioning of nasal tryptase release in human subjects with allergic rhinitis. *Physiology and Behavior, 55,* 823–825.

Gaulton, G. N., Scobie, J. V., & Rosenzweig, M. (1997). HIV-1 and the thymus. *AIDS, 11,* 403–414.

Gebert, A. (1997). M cells in the rabbit palatine tonsil: The distribution, spatial arrangement and membrane subdomains as defined by confocal lectin histochemistry. *Anatomy and Embryology, 195,* 353–358.

Gebert, A., Rothkötter, H. J., & Pabst, R. (1996). M cells in Peyer's patches of the intestine. *International Review of Cytology, 167,* 91–159.

Geenen, V., Legros, J. J., Franchimont, P., Baudrihaye, M., Defresne, M. P., & Boniver, J. (1986). The neuroendocrine thymus: Coexistence of oxytocin and neurophysin in the human thymus. *Science, 232,* 508–510.

Geer, J. H., Davison, G. C., & Gatchel, R. I. (1970). Reduction of stress in humans through nonveridical perceived control of aversive stimulation. *Journal of Personal Psychology, 16,* 731–738.

Gennaro, S., Fehder, W., Nuamah, I. F., Campbell, D. E., & Douglas, S. D. (1997). Caregiving to very low birthweight infants: A model of stress and immune responses. *Brain, Behavior, and Immunity, 11,* 201–215.

Ghanta, V. K., Miura, T., Hiramoto, N. S., & Hiramoto, R. N. (1988). Augmentation of natural immunity and regulation of tumor growth by conditioning. *Annals of the New York Academy of Sciences, 521,* 29–42.

Ghanta, V. K., Rogers, C. F., Hsueh, C.-M., Demissie, S., Lorden, J. F., Hiramoto, N. S., & Hiramoto, R. N. (1994). Role of arcuate nucleus of the hypothalamus in the acquisition of association memory between the CS and US. *Journal of Neuroimmunology, 50,* 109–114.

Gibertini, M., Newton, C., Friedman, H., & Klein, T. W. (1995). Spatial learning impairment in mice infected with *Legionella pneumophila* or administered exogenous interleukin-1β. *Brain, Behavior, and Immunity, 9,* 113–128.

Gillardon, F., Moll, I., Michel, S., Benrath, J., Weihe, E., & Zimmermann, M. (1995). Calcitonin gene-related peptide and nitric oxide are involved in ultraviolet radiation-induced immuno-suppression. *European Journal of Pharmacology, 293,* 395–400.

Gillis, S., Wong-Staal, F., Crabtree, G. R., & Smith, K. A. (1994). Glucocorticoid-induced inhibition of T cell growth factor production. I. The effects on mitogen-induced lymphocyte proliferation. *Journal of Immunology, 123,* 1624.

Gilman, S. C., Schwarz, J. M., Milner, R. J., Bloom, F. E., & Feldman, J. D. (1982). β-Endorphin enhances lymphocyte proliferative responses. *Proceedings of the National Academy of Sciences, USA, 79,* 4226–4231.

Giorgi, J. V., & Detels, R. (1989). T-cell subset alterations in HIV infected homosexual men: NIAID Multicenter AIDS Cohort Study (MACS). *Clinical Immunology and Immunopathology, 52,* 10–18.

Glaser, R., & Jones, J. F. (Eds.). (1994). *Herpesvirus infections.* New York: Dekker.

Glaser, R., & Kiecolt-Glaser, J. (1987). Stress-associated depression in cellular immunity: Implications for acquired immune deficiency syndrome (AIDS). *Brain, Behavior, and Immunity, 1,* 107–112.

Glaser, R., Kiecolt-Glaser, J. K. (1997). Chronic stress modulates the virus-specific immune response to latent herpes simplex virus Type I. *Annals of Behavioral Medicine, 19,* 78–82.

Glaser, R., Kiecolt-Glaser, J. K., Speicher, C. E., & Holliday, J. E. (1985a). Stress, loneliness, and changes in herpes virus latency. *Journal of Behavioral Medicine, 8,* 249–260.

Glaser, R., Kiecolt-Glaser, J. K., Stout, J. C., Tarr, K. L., Speicher, C. E., & Holliday, J. E. (1985b). Stress-related impairments in cellular immunity. *Psychiatry Research, 16,* 233–239.

Glaser, R., Strain, E. C., Tarr, K., Holliday, J. E., Donnerberg, R. L., & Kiecolt-Glaser, J. K. (1985c). Changes in Epstein-Barr virus antibody titers associated with aging. *Proceedings of the Society of Experimental Biology and Medicine, 179,* 352–355.

Glaser, R., Rice, J., Speicher, C. E., Stout, J. C., & Kiecolt-Glaser, J. K. (1986). Stress depresses interferon production by leukocytes concomitant with a decrease in natural killer cell activity. *Behavioral Neuroscience, 100,* 675–678.

Glaser, R., Rice, J., Sheridan, J., Fertel, R., Stout, J., Speicher, C. E., Pinsky, D., Kotur, M., Post, A., Beck, M., & Kiecolt-Glaser, J. K. (1987). Stress-related immune suppression: Health implications. *Brain, Behavior, and Immunity, 1,* 7–20.

Glaser, R., Kennedy, S., Lafuse, W. P., Bonneau, R. H., Speicher, C., Hillhouse, J., & Kiecolt-Glaser, J. K. (1990). Psychological stress-induced modulation of interleukin 2 receptor gene expression and interleukin 2 production in peripheral blood leukocytes. *Archives of General Psychiatry, 47,* 707–712.

Glaser, R., Pearson, G. R., Jones, J. F., Hillhouse, J., Kennedy, S., Mao, H., & Kiecolt-Glaser, J. K. (1991). Stress-related activation of Epstein-Barr virus. *Brain, Behavior, and Immunity, 5,* 219–232.

Glaser, R., Kiecolt-Glaser, J. K., Bonneau, R. H., Malarkey, W., Kennedy, S., & Hughes, J. (1992). Stress-induced modulation of the immune response to recombinant hepatitis B vaccine. *Psychosomatic Medicine, 54,* 22–29.

Glaser, R., Lafuse, W. P., Bonneau, R. H., Atkinson, C., & Kiecolt-Glaser, J. K. (1993a). Stress-associated modulation of proto-oncogene expression in human peripheral blood leukocytes. *Behavioral Neuroscience, 107,* 525–529.

Glaser, R., Pearson, G. R., Bonneau, R. H., Esterling, B. A., Atkinson, C., & Kiecolt-Glaser, J. K. (1993b). Stress and the memory T-cell response to Epstein-Barr virus in healthy medical students. *Health Psychology, 12,* 435–442.

Glaser, R., Kiecolt-Glaser, J. K., Malarkey, W. B., & Sheridan, J. F. (1998). The influence of psychological stress on the immune response to vaccines. *Annals of the New York Academy of Sciences, 840,* 649–655.

Glick, E. N. (1967). Asymmetrical rheumatoid arthritis after poliomyelitis. *British Medical Journal, 3,* 26–31.

Goldmeier, D., & Johnson, A. (1983). Does psychiatric illness affect the recurrence rate of genital herpes? *British Journal of Venereal Disease, 58,* 40–43.

Goldstein, J. A. (1993). *Chronic fatigue syndrome: The limbic hypothesis.* Binghamton: Harworth Medical Press.

Goodkin, K., Antoni, M. H., & Blaney, P. H. (1986). Stress and hopelessness in the promotion of cervical intraepithelial neoplasia to invasive squamous cell carcinoma of the cervix. *Journal of Psychosomatic Research, 30,* 67–76.

Goodkin, K., Blaney, N. T., Feaster, D., Fletcher, M. A., Baum, M. K., Mantero-Atienza, E., Klimas, N. G., Millon, C., Szapocznik, J., & Eisdorfer, C. (1992). Active coping style is associated with natural killer cell cytotoxicity in asymptomatic HIV-1 seropositive men. *Journal of Psychosomatic Research, 36,* 635–650.

Gorczynski, R. M., Macrae, S., & Kennedy, M. (1982). Conditioned immune response associated with allogenic skin grafts in mice. *Journal of Immunology, 129,* 704–709.

Gorczynski, R. M., Macrae, S., & Kennedy, M. (1984). Factors involved in the classical conditioning of antibody responses in mice. In R. Ballieux, J. Fielding, & A. L'Abbatte (Eds.), *Breakdown in human adaption to stress: Towards a multidisciplinary approach.* The Hague: Nijhoff, pp. 704–712.

Gordon, T. P., & Gust, D. A. (1993). Return of juvenile rhesus monkeys (Macaca mulatta) to the natal social group following an 18 week separation. *Aggressive Behavior, 19,* 231–239.

Gordon, T. P., Gust, D. A., Wilson, M. E., Ahmed-Ansari, A., Brodie, A. R., & McClure, H. M. (1992). Social separation and reunion affects immune system in juvenile rhesus monkeys. *Physiology and Behavior, 51,* 467–472.

Goudie, P. (1987). Aversive stimulus properties of drugs. In A. J. Greenshaw & C. T. Dourish (Eds.), *Experimental psychopharmacology* (pp. 341–391). Clifton, NJ: Humana Press.

Goujon, E., Parnet, P., Aubert, A., Goodall, G., & Dantzer, R. (1995a). Corticosterone regulates behavioral effects of lipopolysaccharide and interleukin-1β in mice. *American Journal of Physiology, 269,* R154–R159.

Goujon, E., Parnet, P., Cremona, S., & Dantzer, R. (1995b). Endogenous glucocorticoids down regulate central effects of interleukin-1β on body temperature and behavior in mice. *Brain Research, 702,* 173–180.

Goujon, E., Parnet, P., Layé, S., Combe, C., & Dantzer, R. (1996). Adrenalectomy enhances proinflammatory cytokines gene expression in the spleen, pituitary and brain of mice in response to lipopolysaccharide. *Molecular Brain Research, 36,* 53–62.

Grant, I., Brown, G. W., Harris, T., McDonald, W. I., Patterson, T., & Trimble, M. (1989). Severely threatening events and marked life difficulties preceding onset or exacerbation of multiple sclerosis. *Journal of Neurology, Neurosurgery and Psychiatry, 53,* 8–13.

Green, M. L., Green, R. G., & Santoro, W. (1988). Daily relaxation modifies serum and salivary immunoglobulins and psychophysiologic symptom severity. *Journal of Behavioral Medicine, 13,* 187–199.

Green, R. G., & Green, M. L. (1987). Relaxation increases salivary immunoglobulin A. *Psychological Reports, 61,* 623–629.

Greenberg, M. A., Wortman, C. B., & Stone, A. A. (1996). Emotional expression and physical health: Revising traumatic memories or fostering self-regulation? *Journal of Personality and Social Psychology, 71,* 588–602.

Greenberg, P. D. (1991). Mechanisms of tumor immunology. In D. P. Stites & A. I. Terr (Eds.), *Basic and clinical immunology* (pp. 580–590). Norwalk, CT: Appleton & Lange.

Greenberg, P. D., & Riddell, S. R. (1992). Tumor-specific T-cell immunity: Ready for prime time? *Journal of the National Cancer Institute, 84,* 1059–1061.

Greene, W. A., & Miller, G. (1958). Psychological factors and reticuloendothelial disease. IV. Observations on a group of children and adolescents with leukemias: An interpretation of disease development in terms of mother–child unit. *Psychosomatic Medicine, 10,* 124–144.

Greer, S., & Morris, T. (1975). Psychological attributes of women who develop breast cancer: A controlled study. *Journal of Psychosomatic Research, 19,* 147–153.

Greer, S., Morris, T., & Pettingale, K. W. (1979). Psychological response to breast cancer: Effect on outcome. *Lancet, 2,* 785–787.

Greer, S., Pettingale, K. W., Morris, T., & Haybittle, J. (1985). Mental attitudes to cancer: An additional prognostic factor. *Lancet, 1,* 750.

Greer, S., Morris, T., Pettingale, K. W., & Haybittle, J. L. (1990). Psychological response to breast cancer and 15-year outcome. *Lancet, 335,* 49–50.

Griebel, P. J., & Hein, W. R. (1996). Expanding the role of Peyer's patches in B-cell ontogeny. *Immunology Today, 17,* 30–39.

Griffiths, P. D., & Grundy, J. E. (1987). Molecular biology and immunology of cytomegalovirus. *Journal of Biochemistry, 241,* 313–324.

Grose, C. (1994). Varicella zoster virus infections: Chickenpox, shingles, and varicella vaccine. In R. Glaser & J. F. Jones (Eds.), *Herpesvirus infections* (pp. 117–185). New York: Dekker.

Gross, J. J., & Levenson, R. W. (1993). Emotional suppression: Physiology, self-report, and expressive behavior. *Journal of Personality and Social Psychology, 64,* 970–986.

Gross, J. J., & Levenson, R. W. (1997). Hiding feelings: The acute effects of inhibiting negative and positive emotion. *Journal of Abnormal Psychology, 106,* 95–103.

Grossarth-Maticek, R. (1980). Psychosocial predictors of cancer and internal diseases. *Psychotherapy and Psychosomatics, 33,* 122–128.

Grossarth-Maticek, R., Kanazir, D. T., Schmidt, P., & Vetter, H. (1982a). Psychosomatic factors in the process of carcinogenesis: Theoretical models and empirical results. *Psychotherapy and Psychosomatics, 38,* 284–302.

Grossarth-Maticek, R., Siegrist, J., & Vetter, H. (1982b). Interpersonal repression as a predictor of cancer. *Social Sciences and Medicine, 16,* 493–498.

Gruber, B. L., Hall, N., Hersh, S. P., & Cubois, P. (1988). Immune system and psychologic changes in metastatic cancer patients while using ritualized relaxation and guided imagery: A pilot study. *Scandinavian Journal of Behavior Therapy, 17,* 25–46.

Gruber, B. L., Hersh, S. P., Hall, N., Waletzky, L. R., Kunz, J. F., Carpenter, J. K., Kverno, K. S., & Weiss, S. M. (1993). Immunological responses of breast cancer patients to behavioral interventions. *Biofeedback and Self-Regulation, 18,* 1–22.

Grunfeld, C., & Kotler, D. P. (1992). Pathophysiology of the AIDS wasting syndrome. *AIDS Clinical Reviews, 18,* 191–224.

Guilleminault, C., & Mondini, S. (1986). Mononucleosis and chronic daytime sleepiness: A longterm follow-up study. *Archives of Internal Medicine, 146,* 1333–1335.

Gust, D. A., Gordon, T. P., Wilson, M. E., Brodie, A. R., Ahmed-Ansari, A., & McClure, H. M. (1991). Formation of a new social group of unfamiliar female rhesus monkeys affects the immune and pituitary adrenocortical systems. *Brain, Behavior, and Immunity, 5,* 296–307.

Gust, D. A., Gordon, T. P., Wilson, M. E., Brodie, A. R., Ahmed-Ansari, A., & McClure, H. M. (1992). Removal from natal social group to peer housing affects cortisol levels and absolute numbers of T-cell subsets in juvenile rhesus monkeys. *Brain, Behavior, and Immunity, 6,* 189–199.

Gust, D. A., Gordon, T. P., & Hambright, M. K. (1993). Response to removal from and return to a social group in adult male rhesus monkeys. *Physiology and Behavior, 53,* 599–602.

Gust, D. A., Gordon, T. P., Brodie, A. R., & McClure, H. M. (1994). Effect of a preferred companion in modulating stress in adult female rhesus monkeys. *Physiology and Behavior, 55,* 681–684.

Gust, D. A., Gordon, T. P., Wilson, M. E., Brodie, A. R., Ahmed-Ansari, A., & McClure, H. M. (1996). Group formation of female pigtail macaques. *American Journal of Primatology, 39,* 263–273.

Gwaltney, J. M. (1990). The common cold. In G. L. Mandell, R. G. Douglas, & J. E. Bennett (Eds.), *Principles and practice of infectious diseases* (3rd ed.). New York: Churchill Livingstone, pp. 489–492.

Hack, V., Strobel, G., Rau, J. P., & Weicker, H. (1992). The effect of maximal exercise on the activity of neutrophil granulocytes in highly trained athletes in a moderate training period. *European Journal of Applied Physiology, 65,* 520–524.

Hack, V., Strobel, G., Weiss, M., & Weicker, H. (1994). PMN cell counts and phagocytic activity of highly trained athletes depend on training period. *Journal of Applied Physiology, 77,* 1731–1735.

Hahn, R. C., & Petitti, D. B. (1988). Minnesota Multiphasic Personality Inventory-rated depression and incidence of breast cancer. *Cancer, 61,* 845–848.

Hall, H. R., Minnes, L., Tosi, M., & Olness, K. (1992a). Voluntary modulation of neutrophil adhesiveness using a cyberphysiologic strategy. *International Journal of Neuroscience, 63,* 287–297.

Hall, H. R., Mumma, G., Longo, S., & Dixon, R. (1992b). Voluntary immunomodulation: A preliminary study. *International Journal of Neuroscience, 63,* 275–285.

Hall, H. R., Minnes, L., & Olness, K. (1993). The psychophysiology of voluntary immunomodulation. *International Journal of Neuroscience, 69,* 221–234.

Hamblin, A. S. (1993). *Cytokines and cytokine receptors.* Oxford: IRL Press at Oxford University Press.

Hanna, N., & Schneider, M. (1983). Enhancement of tumor metastasis and suppression of natural killer cell activity by β-estradiol treatment. *Journal of Immunology, 130,* 974–980.

Hansen, J. B., Wilsgard, L., & Osterud, B. (1991). Biphasic changes in leukocytes induced by strenuous exercise. *European Journal of Applied Physiology, 62,* 157–161.

Hanson, L. A., & Brandtzaeg, P. (1980). The mucosal defense system. In E. R. Stiehm & V. A. Fulginiti (Eds.), *Immunologic disorders in infants and children.* Philadelphia: Saunders, pp. 125–137.

Haour, F., Ban, E., Baran, D., Millon, G., & Fillion, G. (1990). Brain interleukin-1 receptors in the mouse: Characterization and modulation after lipopolysaccharide (LPS) injection. *Progress in Neuroimmunoendocrinology, 3,* 196–204.

Harbour, D. V., Smith, E. M., & Blalock, J. E. (1987). Novel processing pathway for proopiomelanocortin in lymphocytes: Endotoxin induction of a new prohormone-cleaving enzyme. *Journal of Neuroscience Research, 18,* 95–101.

Harbuz, M. S., Rees, R. G., Eckland, D., Jessop, D. S., Brewerton, D., & Lightman, S. L. (1992). Paradoxical responses of hypothalamic corticotropin-releasing factor (CRF) messenger ribonucleic acid (mRNA) and CRF-41 peptide and adenohypophysial proopiomelanocortin mRNA during chronic inflammatory stress. *Endocrinology, 130,* 1394–1400.

Hardy, C., Quay, J., Livnat, S., & Ader, R. (1990). Altered T-lymphocyte response following aggressive encounters in mice. *Physiology and Behavior, 47,* 1245–1251.

Hart, B. L. (1988). Biological basis of the behavior of sick animals. *Neuroscience and Biobehavioral Reviews, 12,* 123–137.

Hart, D. N. (1997). Dendritic cells: Unique leukocyte populations which control the primary immune response. *Blood, 90,* 3245–3287.

Hartmann, D. P., Holaday, J. W., & Bernton, E. W. (1989). Inhibition of lymphocyte proliferation by antibodies to prolactin. *FASEB Journal, 3,* 2194–2202.

Haus, E. (1992). Chronobiology of circulation blood cells and platelets. In Y. Touitou & E. Haus (Eds.), *Biologic rhythms in clinical and laboratory medicine* (pp. 504–526). Berlin: Springer-Verlag.

Hays, R. B., Turner, H., & Coates, T. J. (1992). Social support, AIDS-related symptoms and depression among gay men. *Journal of Consulting and Clinical Psychology, 60,* 463–469.

Heijnen, C. J., Bevers, C., Kavelaars, A., & Ballieux, R. E. (1985). Effect of α-endorphin on the antigen-induced primary antibody responses of human blood B cells in vitro. *Journal of Immunology, 136,* 213–216.

Heijnen, C.J., Zijlstra, J., Kavelaars, A., Croiset, G., & Ballieux, R.E. (1987). Modulation of the immune response by POMC-derived peptides: I. Influence on proliferation of human lymphocytes. *Brain, Behavior, and Immunity, 1,* 284–289.

Heijnen, C. J., Kavelaars, A., & Ballieux, R. E. (1991). β-Endorphin: Cytokine and neuropeptide. *Immunological Reviews, 119,* 41–63.

Heijnen, C. J., Rouppe van der Voort, C., Wulffraat, N., van der Net, J., Kuis, W., & Kavelaars, A. (1996). Functional α_1-adrenergic receptors on leukocytes of patients with polyarticular juvenile rheumatoid arthritis. *Journal of Neuroimmunology, 71,* 223–226.

Heilig, M., Irwin, M., Grewal, I., & Sercarz, E. (1993). Sympathetic regulation of T-helper cell function. *Brain, Behavior, and Immunity, 7,* 154–163.

Heinzel, F. P., Sadick, M. D., Mutha, S. S., & Locksley, R. M. (1991). Production on IFN-γ, IL-2, IL-4 and IL-10 by CD4$^+$ lymphocytes *in vivo* during healing in progressive murine leishmaniasis. *Proceedings of the National Academy of Sciences, USA, 88,* 7011–7015.

Heisel, J. S. (1972). Life changes as etiologic factors in juvenile rheumatoid arthritis. *Journal of Psychiatric Research, 16,* 411–420.

Hellstrand, K., & Hermodsson, S. (1989). An immunopharmacological analysis of adrenaline-induced suppression of human natural killer cell cytotoxicity. *International Archives of Allergy and Applied Immunology, 89,* 334–341.

Hellstrand, K., Hermodsson, S., & Strannegard, O. (1985). Evidence for a β-adrenoceptor-mediated regulation of human natural killer cells. *Journal of Immunology, 134,* 4095–4099.

Hendrie, H. C., Paraskevas, F., Baragar, F. D., & Adamson, J. D. (1971). Stress, immunoglobulin levels and early polyarthritis. *Journal of Psychosomatic Research, 15,* 337–342.

Henle, G., Henle, W., & Diehl, V. (1968). Relation of Burkitt's tumour-associated herpes-type virus to infectious mononucleosis. *Proceedings of the National Academy of Sciences, USA, 59,* 94–98.

Henle, W., & Henle, G. (1981). Epstein–Barr virus specific serology in immunologically compromised individuals. *Cancer Research, 41,* 4222–4225.

Hennig, J. (1994). *Die psychobiologische Bedeutung des sekretorischen Immunoglobulin A im Speichel.* Münster: Waxman.

Henry, J. P. (1986). Neuroendocrine patterns of emotional response. In R. Plutchick & H. Kellerman (Eds.), *Emotion: Theory, research and experiences.* San Diego: Academic Press, pp. 37–60.

Herberman, R., Ortaldo, J., Riccardi, C., Timonen, T., Schmidt, A., Maluish, A., & Djeu, J. (1982). Interferon and NK cells. In T. C. Merigan & R. M. Friedman (Eds.), *Interferons* (pp. 287–294). London: Academic Press.

Herbert, T. B., & Cohen, S. (1993a). Depression and immunity: A meta-analytic review. *Psychological Bulletin, 113,* 472–486.

Herbert, T. B., & Cohen, S. (1993b). Stress and immunity in humans: A meta-analytic review. *Psychosomatic Medicine, 55,* 364–379.

Hercend, T., & Schmidt, R. E. (1988). Characteristics and uses of natural killer cells. *Immunology Today, 9,* 291–293.

Hermann, G., Beck, F. M., Tovar, C. A., Malarkey, W. B., Allen, C., & Sheridan, J. F. (1994). Stress-induced changes attributable to the sympathetic nervous system during experimental influenza viral infection in DBA/2 inbred mouse strain. *Journal of Neuroimmunology, 53,* 173–180.

Hickie, I., Hickie, C., Lloyd, A., Silove, D., & Wakefield, D. (1993). Impaired in vivo immune responses in patients with melancholia. *British Journal of Medicine, 162,* 751–757.

Hickie, I., Hickie, C., Bennett, B., Wakefield, D., Silove, D., Mitchell, P., & Lloyd, A. (1995). Biochemical correlates of in vivo cell-mediated immune dysfunction in patients with depression: A preliminary report. *International Journal of Immunopharmacology, 17,* 685–690.

Hill, C. W., Greer, W. E., & Felsenfeld, O. (1967). Psychological stress, early response to foreign protein, and blood cortisol in vervets. *Psychosomatic Medicine, 29,* 279–283.

Hinze-Selch, D., Pollmächer, T., Wetter, T., Zhang, Y., Lu, H., & Holsboer, F. (1994). Cytokines in narcoleptic patients. *Journal of Sleep Research, 3,* 105.

Hiramoto, R. N., Hiramoto, N. S., Solvason, H. B., & Ghanta, V. K. (1987). Regulation of natural immunity (NK activity) by conditioning. *Annals of the New York Academy of Sciences, 496,* 545–552.

Hobson, A., & Stickgold, R. (1995). The conscious state paradigm: A neurocognitive approach to waking, sleeping and dreaming. In M. Gazzaniga (Ed.), *The cognitive neurosciences* (pp. 127–154). Cambridge, MA: MIT Press.

Hoefer, P. A., & Herzfeld, E. (1921). Zur Beeinflussung des Blutbildes durch Adrenalin. *Folia Haematologia, 27,* 77–95.

Hofer, M. A. (1994). Early relationships as regulators of infant physiology and behavior. *Acta Paediatrica Scandinavica, 397,* 9–18.

Hoffman, J., Benson, H., Arns, P., Stainbrook, G., Landsberg, L., Young, J., & Gill, A. (1982). Reduced sympathetic nervous system responsivity associated with the relaxation response. *Science, 54,* 190–192.

Hoffman-Goetz, L., & Pedersen, B.K. (1994). Exercise and the immune system: A model of the stress response? *Immunology Today, 15,* 382–387.

Hoffman-Goetz, L., Simpson, J. R., Cipp, N., Arumugam, Y., & Houston, M. E. (1990). Lymphocyte subset responses to repeated submaximal exercise in men. *Journal of Applied Physiology, 68,* 1069–1074.

Hoffmann, P., Terenius, L., & Thoren, P. (1996). Cerebrospinal fluid immunoreactive exercise in spontaneously hypertensive rat. *Regulatory Peptides, 28,* 233–239.

Holden, C. (1996). Small refugees suffer the effects of early neglect. *Science, 274,* 1076–1077.

Holland, J., Breitbart, W., Jacobsen, P., Lederberg, M., Loscaizo, M., Massie, M. J., & McCorkle, R. (1998). *Textbook of psycho-oncology.* New York: Oxford University Press.

Holmes, T. H., & Rahe, R. H. (1967). The social readjustment rating scale. *Journal of Psychosomatic Research, 11,* 213–218.

Homo-Delarche, F., Fitzpatrick, F., Christeff, N., Nunez, E. A., Bach, J. F., & Dardenne, M. (1991). Sex steroids, glucocorticoids, stress, and immunity. *Journal of Steroid Biochemistry and Molecular Biology, 40,* 619–637.

Honda, Y., Asaka, A., Tanaka, Y., & Juji, T. (1983). Discrimination of narcolepsy by using genetic markers and HLA. *Sleep Research, 12,* 254.

Hooper, C. D., Morimoto, K., Bette, M., Weihe, E., Koprowski, H., & Dietzschold, B. (1998). Collaboration of antibody and inflammation in the clearance of rabies virus from the central nervous system. *Journal of Virology, 72,* 3711–3719.

Hopkins, S. J., & Rothwell, N. J. (1995). Cytokines and the nervous system I: Expression and recognition. *Trends in Neurosciences, 18,* 83–87.

Hörsch, D., Kirsch, J. J., & Weihe, E. (1998). Elevated density of nerve fibres in anal fissures. *International Journal of Colorectoral Diseases, 13,* 134–140.

Hosoi, J., Murphy, G. F., Egan, C. L., Lerner, E. A., Grabbe, S., Asahina, A., & Granstein, R. D. (1993). Regulation of Langerhans cell function by nerves containing calcitonin gene-related peptide. *Nature, 363,* 159–163.

Housh, T. J., Johnson, G. O., Housh, D. J., Evans, S. L., & Tharp, G. D. (1991). The effect of exercise at various temperatures on salivary levels of immunoglobulin A. *International Journal of Sports Medicine, 12,* 498–500.

Howlett, T. H., & Rees, L. H. (1986). Endogenous opioid peptides and hypothalamo-pituitary function. *Annual Reviews of Physiology, 48,* 527–544.

Hrdina, P. D., von Kulmiz, P., & Stretch, R. (1979). Pharmacological modification of experimental depression in infant macaques. *Psychopharmacology, 64,* 89–93.

Hurwitz, B. E., Nelesen, R. A., Saab, P. G., Nagel, J. H., Spitzer, S. B., Gellman, M. D., McCabe, P. M., Phillips, D. J., & Schneiderman, N. (1993). Differential patterns of dynamic cardiovascular regulation as a function of task. *Biological Psychology, 36,* 75–95.

Husband, A. J., King, M. G., & Brown, R. (1987). Behaviorally conditioned modification of T cell subset ratios in rats. *Immunology Letters, 14,* 91–94.

Imai, S., Tokunaga, Y., Maeda, T., Kikkawa, M., & Hukuda, S. (1997). Calcitonin gene-related peptide, substance P, and tyrosine hydroxylase-immunoreactive innervation of rat bone marrows: An immunohistochemical and ultrastructural investigation on possible efferent and afferent mechanisms. *Journal of Orthopaedic Research, 15,* 133–140.

Ironson, G., LaPerriere, A., Antoni, M., OHearn, P., Schneiderman, N., Klimas, N., & Fletcher, M. A. (1990). Changes in immune and psychological measures as a function of anticipation and reaction to news of HIV-1 antibody status. *Psychosomatic Medicine, 52,* 247–270.

Ironson, G., Friedman, A., Klimas, N., Antoni, M., Fletcher, M. A., LaPerriere, A., Simoneau, J., & Schneiderman, N. (1994). Distress, denial, and low adherence to behavioral interventions predict faster disease progression in gay men infected with human immunodeficiency virus. *International Journal of Behavioral Medicine, 1,* 90–105.

Ironson, G., Antoni, M., & Lutgendorf, S. (1995). Can psychological interventions affect immunity and survival? Present findings and suggested targets with a focus on cancer and human immunodeficiency virus. *Mind/Body Medicine, 1,* 85–110.

Ironson, G., Wynings, C., Schneiderman, N., Baum, A., Rodriguez, M., Breenwood, D., Benight, C., Antoni, M., LaPerriere, A., Huang, H.-S., Klimas, N., & Fletcher, M. A. (1997). Post traumatic stress symptoms, intrusive thoughts, loss, and immune function after Hurricane Andrew. *Psychosomatic Medicine, 59,* 128–141.

Irwin, M. (1993). Brain corticotropin-releasing hormone- and interleukin-1β-induced suppression of specific antibody production. *Endocrinology, 133,* 1352–1360.

Irwin, M. R., & Hauger, R. L. (1988). Adaptation to chronic stress: Temporal pattern of immune and neuroendocrine correlates. *Neuropsychopharmacology, 1,* 239–242.

Irwin, M., Daniels, M., Bloom, E., Smith, T. L., & Weiner, H. (1987a). Life events, depressive symptoms and immune function. *American Journal of Psychiatry, 44,* 437–441.

Irwin, M., Smith, T. L., & Gillin, J. C. (1987b). Reduced natural killer cytotoxicity in depressed patients. *Life Sciences, 41,* 2127–2133.

Irwin, M. R., Vale, W., & Britton, K. T. (1987c). Central corticotropin-releasing factor suppresses natural killer cytotoxicity. *Brain, Behavior, and Immunity, 1,* 81–87.

Irwin, M., Daniels, M., & Weiner, H. (1987d). Immune and neuroendocrine changes during bereavement. *Psychiatric Clinics of North America, 10,* 449–465.

Irwin, M., Daniels, M., Smith, T. L., Bloom, E. T., & Weiner, H. (1987e). Impaired natural killer cell activity during bereavement. *Brain, Behavior, and Immunity, 1,* 98–104.

Irwin, M., Daniels, M., Risch, S. C., Bloom, E., & Weiner, H. (1988a). Plasma cortisol and natural killer cell activity during bereavement. *Biological Psychiatry, 24,* 173–178.

Irwin, M., Hauger, R. L., Brown, M. R., & Britton, K. T. (1988b). CRF activates autonomic nervous system and reduces natural killer cytotoxicity. *American Journal of Physiology, 255,* R744–R747.

Irwin, M. R., Segal, D. S., Hauger, R. L., & Smith, T. L. (1989). Individual behavioral and neuroendocrine differences in responsiveness to repeated audiogenic stress. *Pharmacology, Biochemistry, and Behavior, 32,* 913–917.

Irwin, M., Hauger, R. L., Jones, L., Provencio, M., & Britton, K. T. (1990a). Sympathetic nervous system mediates central corticotropin-releasing factor induced suppression of natural killer cytotoxicity. *Journal of Pharmacology and Experimental Therapeutics, 255,* 101–107.

Irwin, M., Caldwell, C., Smith, T. L., Brown, S., Schuckit, M. A., & Gillin, J. C. (1990b). Major depressive disorder, alcoholism, and reduced natural killer cell cytotoxicity: Role of severity of depressive symptoms and alcohol consumption. *Archives of General Psychiatry, 47,* 713–719.

Irwin, M., Patterson, T., Smith, T. L., Caldwell, C., Brown, S. A., Gillin, J. C., & Grant, I. (1990c). Reduction of immune function in life stress and depression. *Biological Psychiatry, 27,* 22–30.

Irwin, M. R., Vale, W., & Rivier, C. (1990d). Central corticotropin releasing factor mediates the suppressive effect of stress on natural killer cytotoxicity. *Endocrinology, 126,* 2837–2844.

Irwin, M., Brown, M., Patterson, T., Hauger, R., Mascovich, A., & Grant, I. (1991). Neuropeptide Y and natural killer cell activity: Findings in depression and Alzheimer caregiver stress. *FASEB Journal, 5,* 3100–3107.

Irwin, M., Hauger, R., & Brown, M. (1992a). Central corticotropin-releasing hormone activates the sympathetic nervous system and reduces immune function: Increased responsivity of the aged rat. *Endocrinology, 131,* 1047–1053.

Irwin, M., Lacher, U., & Caldwell, C. (1992b). Depression and reduced natural killer cytotoxicity: A longitudinal study of depressed patients and control subjects. *Psychological Medicine, 22,* 1045–1050.

Irwin, M., Mascovich, A., Gillin, C., Willoughby, R., Pike, J., & Smith, T. L. (1994). Partial sleep deprivation reduces natural killer cell activity in humans. *Psychosomatic Medicine, 56,* 493–498.

Irwin, M., McClintick, J., Costlow, C., Fortner, M., White, J., & Gillin, J. C. (1996). Partial night sleep deprivation reduces natural killer and cellular immune response in humans. *FASEB Journal, 10,* 643–653.

Irwin, M., Costlow, C., Williams, H., Artin, K. H., Chan, C. Y., Stinson, D., Levin, M., Hayward, A. R., & Oxman, M. (1998). Cellular immunity to varicella-zoster virus in patients with major depression. *Journal of Infectious Diseases, 178,* 104–108.

Ishimori, K. (1909). True causes of sleep—A hypnogenic substance as evidenced in the brain of sleep-deprived animals. *Tokyo Igaku Zasshi, 23,* 429–457.

Jacobs, R., Stoll, M., Stratmann, G., Leo, R., Link, H., & Schmidt, R. E. (1992). CD16⁻ CD56⁺ NK cells after bone marrow transplantation. *Blood, 79,* 3239–3244.

Jacobs, T. J., & Charles, E. (1980). Life events and the occurrence of cancer in children. *Psychosomatic Medicine, 42,* 11–24.

Jacobsen, P., Perry, S., & Hirsch, D. (1990). Behavioral and psychological responses to HIV antibody testing. *Journal of Consulting and Clinical Psychology, 58,* 31–37.

Jacobson, M. A., Kramer, F., Pavan, P. R., Owens, S., Pollard, R., & NIAID ACTG Protocol 266 Team. (1997). *Failure of highly active antiretroviral therapy (HAART) to prevent CMV retinitis despite marked CD4 count increase.* Paper presented at the Fourth National Conference on Retrovirus and Opportunistic Infection, Washington, DC.

Janeway, C. A. (1988). Frontiers of the immune system. *Nature, 333,* 804–806.

Janeway, C. A., & Travers, P. (1997). *Immunobiology: The immune system in health and disease.* New York: Churchill Livingstone.

Janz, L. J., Green-Johnson, J., Murray, L., Vriend, C. Y., Nance, D. M., Greenberg, A. H., & Dyck, D. (1996). Pavlovian conditioning of LPS-induced responses: Effects on corticosterone, splenic NE, and IL-2 production. *Physiology and Behavior, 59,* 1103–1109.

Jasnoski, M. L., & Kugler, J. (1987). Relaxation, imagery, and neuroimmunomodulation. *Annals of the New York Academy of Sciences, 496,* 772–780.

Jemmott, J. B., & Locke, S. E. (1984). Psychosocial factors, immunologic mediation, and human susceptibility to infectious disease: How much do we know? *Psychological Bulletin, 95,* 78–108.

Jemmott, J. B., & McClelland, D. C. (1989). Secretory IgA as a measure of resistance to infectious disease: Comments on Stone, Cox, Valdimarsdottir, and Neale. *Behavioral Medicine, 15,* 63–71.

Jemmott, J. B., Borysenko, J. Z., Borysenko, M., McClelland, D. C., Chapman, R., Meyer, D., & Benson, H. (1983). Academic stress, power motivation, and decrease in salivary immunoglobulin A secretion rate. *Lancet, 1,* 1400–1402.

Jenkins, F. J., & Baum, A. (1995). Stress and reactivation of latent herpes simplex virus: A fusion of behavioral medicine and molecular biology. *The Society of Behavioral Medicine, 17,* 116–123.

Jetschmann, J. U., Benschop, R. J., Jacobs, R., Kemper, A., Oberbeck, R., Schmidt, R. E., & Schedlowski, M. (1997). Expression and in vivo modulation of α- and β-adrenoceptors on human natural killer (CD16⁺) cells. *Journal of Neuroimmunology, 74,* 159–164.

Jiayi, D., Shikun, Y., & Renabo, X. (1989). The inhibitory effect of hydrocortisone on interferon production by rat spleen cells. *Journal of Steroid Biochemistry, 33,* 1139–1141.

Jonakait, G. M. (1993). Neural–immune interactions in the sympathetic ganglion. *Trends in Neuroscience, 16,* 419–423.

Jones, H. E. (1935). The galvanic skin reflex as related to overt emotional expression. *American Journal of Psychology, 47,* 241–251.

Jones, H. E. (1950). The study of patterns of emotional expression. In M. L. Reymer (Ed.), *Feelings and emotions: The Mooseheart Symposium* (pp. 161–168). New York: McGraw-Hill.

Jonsdottir, I. H., Asea, A., Hoffmann, P., Dahlgren, U. I., Andersson, B., Hellstrand, K., & Thoren, P. (1996a). Voluntary chronic exercise augments in vivo natural immunity in rats. *Journal of Applied Physiology, 80,* 1799–1803.

Jonsdottir, I. H., Johansson, C., Asea, A., Hellstrand, K., Thoren, P., & Hoffmann, P. (1996b). Chronic intra-cerebrovascular administration of β-endorphin augments natural killer cell cytotoxicity in rats. *Regulatory Peptides, 62,* 113–118.

Jung, W., & Irwin, M. (1998). Major depressive disorder, cigarette smoking, and reduced natural killer cell cytotoxicity. *Neuroimmunomodulation, 5,* 23.

Kabiersch, A., del Rey, A., Honegger, C. G., & Besedovsky, H. O. (1988). Interleukin-1 induces changes in noradrenaline metabolism in the rat brain. *Brain, Behavior, and Immunity, 2,* 267–274.

Kagan, J., Snidman, N., Julia-Sellers, J., & Johnson, M. O. (1991). Temperament and allergic symptoms. *Psychosomatic Medicine, 53,* 332–340.

Kaisch, K., & Anton-Culver, H. (1989). Psychological and social consequences of HIV exposure: Homosexuals in Southern California. *Psychology and Health, 3,* 63–75.

Kammer, G. M. (1988). The adenylate cyclase–cAMP–protein kinase A pathway and regulation of the immune response. *Immunology Today, 9,* 222–229.

Kamphuis, S., Kavelaars, A., Brooimans, R., Kuis, W., Zegers, B. J. M., & Heijnen, C.J. (1997). T helper 2 cytokines induce preproenkephalin mRNA expression and proenkephalin A in human peripheral blood mononuclear cells. *Journal of Neuroimmunology, 79,* 91–99.

Kamphuis, S., Eriksson, F., Kavelaars, A., Zijlstra, J., van de Pol, M., Kuis, W., & Heijnen, C. J. (1998). Role of endogenous pro-enkephalin A-derived peptides in human T cell proliferation and monocyte IL-6 production. *Journal of Neuroimmunology, 84,* 53–60.

Kang, D.-H., Coe, C. L., McCarthy, D. O., & Ershler, W. B. (1996). Academic exams significantly impact immune responses, but not lung function in healthy and well-managed asthmatic adolescents. *Brain, Behavior, and Immunity, 10,* 164–181.

Kang, D.-H., Coe, C. L., McCarthy, D. O., Jarjour, N. N., Kelly, E. A., Rodriguez, R. R., & Busse, W. B. (1997). Cytokine profiles of simulated blood lymphocytes in asthmatic and healthy adolescents across the school year. *Journal of Interferon and Cytokine Research, 17,* 481–487.

Kaplan, G. A., & Reynolds, P. (1988). Depression and cancer mortality and morbidity: Prospective evidence from the Alameda County Study. *Journal of Behavioral Medicine, 11,* 1–13.

Kaplan, J. R., Heise, E. R., Manuck, S. B., Shively, C. A., Cohen, S., Rabin, B. S., & Kasprowicz, A. L. (1991). The relationship of agonistic and affiliative behavior patterns to cellular immune function among cynomolgus monkeys (Macaca fascicularis) living in unstable social groups. *American Journal of Primatology, 25,* 157–173.

Kaplan, L. D., Wofsky, C. B., & Volberding, P. A. (1987). Treatment of patients with acquired immunodeficiency syndrome and associated manifestations. *Journal of the American Medical Association, 257,* 1367–1376.

Kappel, M., Tvede, N., Galbo, H., Haahr, P. M., Kjaer, M., Linstow, M., Klarlund, K., & Pedersen, B. K. (1991). Evidence that the effect of physical exercise on NK cell activity is mediated by epinephrine. *Journal of Applied Physiology, 70,* 2530–2534.

Kappel, M., Hansen, M. B., Diamant, M., Jorgensen, J. O. L., Gyhrs, A., & Pedersen, B. K. (1993). Effects of an acute bolus growth hormone infusion on the human immune system. *Hormone and Metabolic Research, 25,* 579–585.

Karacan, I., Wolff, S. M., Williams, R. L., Hursch, C. J., & Webb, W. B. (1968). The effects of fever on sleep and dream patterns. *Psychosomatics, 9,* 331–339.

Karalis, K., Sano, H., Redwine, J., Listwak, S., Wilder, R. L., & Chrousos, G. P. (1991). Autocrine or paracrine inflammatory actions of corticotropin releasing hormone in vivo. *Science, 254,* 421–423.

Karalis, K., Muglia, L. J., Bae, D., Hilderbrand, H., & Majzoub, J. A. (1997). CRH and the immune system. *Journal of Neuroimmunology, 72,* 131–136.

Kasl, S. V., Evans, A. S., & Niederman, J. C. (1979). Psychosocial risk factors in the development of infectious mononucleosis. *Psychosomatic Medicine, 41,* 445–466.

Katcher, A. H., Brightman, V., Luborsky, L., & Ship, I. (1973). Prediction of the incidence of recurrent herpes labialis and systemic illness from psychological measurements. *Journal of Dental Research, 5,* 49–58.

Katsuura, G., Arimura, A., Koves, K., & Gottschall, P. E. (1990 . Involvement of organum vasculosum of laminae terminalis and preoptic area in interleukin-1β induced ACTH release. *American Journal of Physiology, 258,* E163–E171.

Katz, P., Zaytoun, A. M., & Fauci, A. S. (1982). Mechanisms of human cell-mediated cytotoxicity. I. Modulation of natural killer cell activity by cyclic nucleotides. *Journal of Immunology, 129,* 287–296.

Kaufman, I. C., & Rosenblum, L. A. (1967). The reaction to separation in infant monkeys: Anaclitic depression and conservation-withdrawal. *Psychosomatic Medicine, 29,* 648–675.

Kavelaars, A., Ballieux, R. E., & Heijnen, C. (1988). Modulation of the immune response by proopiomelanocortin derived peptides. II. Influence of adrenocorticotropic hormone on the rise in intracellular free calcium concentration after T cell activation. *British Journal of Behavior and Immunology, 2,* 57–66.

Kavelaars, A., Ballieux, R. E., & Heijnen, C. J. (1989). The role of interleukin-1 in the CRF- and AVP-induced secretion of immunoreactive β-endorphin by human peripheral blood mononuclear cells. *Journal of Immunology, 142,* 2338–2342.

Kavelaars, A., Ballieux, R. E., & Heijnen, C. J. (1990a). β-Endorphin secretion by human peripheral blood mononuclear cells: Regulation by glucocorticoids. *Life Sciences, 46,* 1233–1240.

Kavelaars, A., Ballieux, R. E., & Heijnen, C.J . (1990b). In vitro β-adrenergic stimulation of lymphocytes induces the release of immunoreactive β-endorphin. *Endocrinology, 126,* 3028–3032.

Kavelaars, A., Berkenbosch, F., Croiset, G., Ballieux, R. E., & Heijnen, C. J. (1990c). Induction of β-endorphin secretion by lymphocytes after subcutaneous administration of CRF. *Endocrinology, 126,* 758–764.

Kavelaars, A., Eggen, B. J. L., De Graan, P. N. E., Gispen, W. H., & Heijnen, C. J. (1990d). The phosphorylation of the CD3 gamma chain of T lymphocytes is modulated by β-endorphin. *European Journal of Immunology, 20,* 943–945.

Kavelaars, A., Ballieux, R. E., & Heijnen, C. J. (1991). Two different signalling pathways for the induction of immunoreactive β-endorphin secretion by human peripheral mononuclear cells. *Endocrinology, 128,* 765–770.

Kavelaars, A., Jeurissen, F., & Heijnen, C. J. (1994). Substance P receptors and signal transduction in leukocytes. *Immunomethods, 5,* 41–48.

Kavelaars, A., Heijnen, C. J., Ellenbroek, B., van Loveren, H., & Cools, A. R. (1997). Apomorphine-susceptible and apomorphine-unsusceptible Wistar rats differ in their susceptibility for inflammatory and infectious diseases: A study on rats with group specific differences in structure reactivity of hypothalamic–pituitary–adrenal axis. *The Journal of Neuroscience, 17*, 2580–2584.

Kay, N., Allen, J., & Morley, J. E. (1982). Endorphins stimulate normal human peripheral blood lymphocyte natural killer activity. *Life Sciences, 35*, 53–59.

Keast, D., Arstein, D., Harper, W., Fry, R. W., & Morton, A. R. (1995). Depression of plasma glutamine concentration after exercise stress and its possible influence on the immune system. *Medical Journal of Australia, 162*, 1–8.

Keller, S. E., Stein, M., Camerino, M. S., Schleifer, S., & Sherman, J. (1980). Suppression of lymphocyte stimulation by anterior hypothalamic lesions in the guinea pig. *Cellular Immunology, 52*, 334–340.

Keller, S. E., Weiss, J. M., Schleifer, S. J., Miller, N. E., & Stein, M. (1983). Stress-induced suppression of immunity in adrenalectomized rats. *Science, 221*, 1301–1304.

Keller, S. E., Schleifer, S. J., Liotta, A. S., Bond, R. N., Farhoody, N., & Stein, M. (1988). Stress-induced alterations of immunity in hypophysectomized rats. *Proceedings of the National Academy of Sciences, USA, 85*, 9297–9301.

Keller, S. E., Schleifer, S. J., & Demetrikopoulos, M. K. (1991). Stress-induced changes in immune function in animals: Hypothalamo-pituitary-adrenal influences. In R. Ader, D. L. Felten, & N. Cohen (Eds.), *Psychoneuroimmunology* (2nd ed.). San Diego: Academic Press, pp. 771–784.

Kelley, K. W. (1991). Growth hormone in immunobiology. In R. Ader, D. L. Felten, & N. Cohen (Eds.), *Psychoneuroimmunology* (pp. 377–402). San Diego: Academic Press.

Kelley, K. W., & Dantzer, R. (1991). Growth hormone and prolactin as natural antagonists of glucocorticoids in immunoregulation. In N. Plotnikoff, A. Murgo, R. Faith, & J. Wybran (Eds.), *Stress and immunity* (pp. 433–452). Boca Raton: CRC Press.

Kelley, K.W., Arkins, S., & Lin, Y. M. (1992). Growth hormone, prolactin, and insulin-like growth factors: New jobs for old players. *Brain, Behavior, and Immunity, 6*, 317–326.

Kemeny, M. E. (1994). Psychoneuroimmunology of HIV infection. *Psychiatric Clinics of North America, 17*, 55–68.

Kemeny, M. E., & Gruenewald, T. L. (1999). Affect, cognition, the immune system, and health. In E. A. Mayer & C. Saper (Eds.), *The biological basis for mind body interactions. Progress in brain research series.* Amsterdam: Elsevier Science B.V., in press.

Kemeny, M. E., Cohen, F., Zegans, L. A., & Conant, M. A. (1989). Psychological and immunological predictors of genital herpes recurrence. *Psychosomatic Medicine, 51*, 195–208.

Kemeny, M. E., Weiner, H., Taylor, S. E., Schneider, S., Visscher, B., & Fahey, J. L. (1994). Repeated bereavement, depressed mood, and immune parameters in HIV seropositive and seronegative gay men. *Health Psychology, 13*, 14–24.

Kendall, M. D., & al-Shawaf, A. A. (1991). Innervation of the rat thymus gland. *Brain, Behavior, and Immunity, 5*, 9–28.

Kendall, P. C., Hollon, S. D., Beck, A. T., Hammen, C. L., & Ingram, R. E. (1987). Issues and recommendations regarding use of the Beck Depression Inventory. *Cognitive Therapy and Research, 11*, 289–299.

Kendall, R. A., & Targan, S. (1980). The dual effect of prostaglandin (PGE2) and ethanol on the natural killer cytolytic process: Effector activation and NK-cell–target cell conjugate lytic inhibition. *The Journal of Immunology, 125*, 2770–2777.

Kendell, R. E. (1991). Chronic fatigue, viruses and depression. *Lancet, 337*, 160–162.

Kent, S., Bluthé, R. M., Dantzer, R., Hardwick, A.J., Kelley, K. W., Rothwell, N. J., & Vannice, J. L. (1992a). Different receptor mechanisms mediate the pyrogenic and behavioral effects of interleukin-1. *Proceedings of the National Academy of Sciences, USA, 89*, 9117–9120.

Kent, S., Bluthé, R. M., Kelley, K. W., & Dantzer, R. (1992b). Sickness behavior as a new target for drug development. *Trends in Pharmacological Sciences, 13*, 24–28.

Kent, S., Bret-Dibat, J. L., Kelley, K. W., & Dantzer, R. (1995). Mechanisms of sickness-induced decreases in food-motivated behavior. *Neuroscience and Biobehavioral Reviews, 20*, 171–175.

Keppel, W. H., Regan, D. H., Heffeneider, S. H., McCoy, S., & Ramsey, F. (1993). Effects of behavioral stimuli on plasma interleukin-1 activity in humans at rest. *Journal of Clinical Psychology, 49*, 777–789.

Khan, M. M., Sansoni, P., Silverman, E. D., Engleman, E. G., & Melmon, K. L. (1986). β-Adrenergic receptors on human suppressor, helper, and cytolytic lymphocytes. *Biochemical Pharmacology, 35*, 1137–1142.

Kiecolt-Glaser, J. K., & Glaser, R. (1987). Psychosocial influences on herpesvirus latency. In E. Kurstak & Z. J. Lipowski (Eds.), *Viruses, immunity, and mental disorders* (pp. 403–411). New York: Plenum Press.

Kiecolt-Glaser, J. K., & Glaser, R. (1991). Stress and the immune function in hormone. In R. Ader, D. L. Felten, & N. Cohen (Eds.), *Psychoneuroimmunology* (2nd ed., pp. 849–857). San Diego: Academic Press.

Kiecolt-Glaser, J. K., Ricker, D., George, J., Messick, G., Speicher, C. E., Garner, W., & Glaser, R. (1984a). Urinary cortisol levels, cellular immunocompetency, and loneliness in psychiatric inpatients. *Psychosomatic Medicine, 46,* 15–24.

Kiecolt-Glaser, J. K., Speicher, C. E., Holliday, J. E., & Glaser, R. (1984b). Stress and the transformation of lymphocytes by Epstein–Barr virus. *Journal of Behavioral Medicine, 7,* 1–12.

Kiecolt-Glaser, J. K., Garner. W., Speicher, C., Penn, G. M., Holliday, J., & Glaser, R. (1984c). Psychosocial modifiers of immunocompetence in medical students. *Psychosomatic Medicine, 46,* 7–14.

Kiecolt-Glaser, J. K., Glaser, R., Williger, D., Stout, J., Messick, G., Sheppard, S., Ricker, D., Romisher, S. C., Briner, W., Bonnell, G., & Donnerberg, R. (1985). Psychosocial enhancement of immunocompetence in a geriatric population. *Health Psychology, 4,* 25–41.

Kiecolt-Glaser, J. K., Glaser, R., Strain, E., Stout, J. C., Tarr, K. L., Holliday, J. E., & Speicher, C. E. (1986). Modulation of cellular immunity in medical students. *Journal of Behavioral Medicine, 9,* 5–21.

Kiecolt-Glaser, J. K., Fisher, L., Ogrocki, P., Stout, J. C., Speicher, C. E., & Glaser, R. (1987a). Marital quality, marital disruption, and immune function. *Psychosomatic Medicine, 48,* 13–34.

Kiecolt-Glaser, J. K., Glaser, R., Dyer, C., Shuttleworth, E., Ogrocki, P., & Speicher, C. E. (1987b). Chronic stress and immunity in family caregivers of Alzheimer's disease victims. *Psychosomatic Medicine, 49,* 523–535.

Kiecolt-Glaser, J. K., Kennedy, S., Malkoff, S., Fisher, L. D., Speicher, C. E., & Glaser, R. (1988). Marital discord and immunity in males. *Psychosomatic Medicine, 50,* 213–229.

Kiecolt-Glaser, J. K., Dura, J. R., Speicher, C. E., Trask, O. J., & Glaser, R. (1991). Spousal caregivers of dementia victims: Longitudinal changes in immunity and health. *Psychosomatic Medicine, 53,* 345–362.

Kiecolt-Glaser, J. K., Malarkey, W. B., Chee, M., Newton, T., Cacioppo, J. T., Mao, H., & Glaser, R. (1993). Negative behavior during marital conflict is associated with immunological down-regulation. *Psychosomatic Medicine, 55,* 395–409.

Kiecolt-Glaser, J. K., Marucha, P. T., Malarkey, W. B., Mercado, A. M., & Glaser, R. (1995). Slowing of wound healing by psychological stress. *Lancet, 346,* 1194–1196.

Kiecolt-Glaser, J. K., Glaser, R., Gravenstein, S., Malarkey, W. B., & Sheridan, J. (1996). Chronic stress alters the immune response to influenza virus vaccine in the elderly. *Proceedings of the National Academy of Sciences, USA, 93,* 3043–3047.

Kimzey, S. L. (1975). The effects of extended spaceflight on hematologic and immunologic systems. *Journal of the American Medical Women's Association, 30,* 218–232.

King, M. G., Husband, A. J., & Kusnecov, A. W. (1987). Behaviourally conditioned immunosuppression using anti-lymphocyte serum: Duration of effect and role of corticosteroids. *Medical Science Research, 15,* 407–408.

Kinouchi, J., Brown, G., Pasternak, G., & Donner, D. B. (1991). Identification and characterization of receptors for tumor necrosis factor-α in the brain. *Biochemical and Biophysical Research Communications, 181,* 1532–1538.

Kirschbaum, C., Jabaaij, L., Buske-Kirschbaum, A., Hennig, J., Blom, M., Dorst, K., Bauch, J., DiPauli, R., Schmitz, G., Ballieux, R., & Hellhammer, D. (1992). Conditioning of drug-induced immunomodulation in human volunteers: A European collaborative study. *British Journal of Clinical Psychology, 31,* 459–472.

Kjaer, M. (1989). Epinephrine and some other hormonal responses to exercise in man: With special reference to physical training. *International Journal of Sports Medicine, 10,* 2–15.

Kjaer, M. (1992). Regulation of hormonal and metabolic responses during exercise in humans. *Exercise and Sports Science Reviews, 20,* 161–184.

Klatzmann, D. E., Champagne, E., Chouret, S., Grunest, J., Guetart, D., Hercend, T., Gluckman, J. C., & Montaganier, L. (1985). T lymphocyte T4 molecule behaves as the receptor for human retrovirus LAV. *Nature, 312,* 767–770.

Klein, J. (1991). *Immunologie.* Weinheim: VCH Verlagsgesellschaft.

Kleitman, N. (1963). *Sleep and wakefulness.* Chicago: University of Chicago Press.

Klerman, G. L., & Izen, J. E. (1977). The effects of bereavement and grief on physical health and general well-being. *Advances in Psychosomatic Medicine, 5,* 63–104.

Klimas, N., Caralis, P., LaPerriere, A., Antoni, M., Ironson, G., Simoneau, J., Ashman, N., Schneiderman, N., & Fletcher, M. A. (1991). Immunologic function in a cohort of HIV-1 seropositive and negative healthy homosexual men. *Journal of Clinical Microbiology, 29,* 1413–1421.

Klokker, M., Secher, N. H., Olesen, H. L., Madsen, P., Warberg, J., & Pedersen, B. K. (1997). Adrenergic β-1 and 1+2 receptor blockade suppresses the natural killer cell response during head up tilt. *Journal of Applied Physiology, 83,* 1492–1498.

Klosterhalfen, S., & Klosterhalfen, W. (1990). Conditioned cyclosporine effects but not conditioned taste aversion in immunized rats. *Behavioral Neuroscience, 104,* 716–724.

Kluger, M. J. (1979). *Fever: Its biology, evolution and function.* Princeton: Princeton University Press.

Kluger, M. J. (1991). Fever: Role of pyrogens and cryogens. *Physiological Reviews, 71,* 93–127.

Knapp, P. H., Levy, E. M., Giorgi, R. G., Black, P. H., Fox, B. H., & Heeren, T. C. (1992). Short-term immunological effects of induced emotion. *Psychosomatic Medicine, 54,* 133–148.

Kneier, A. W., & Temoshok, L. (1984). Repressive coping reactions in patients with malignant melanoma as compared to cardiovascular patients. *Journal of Psychosomatic Research, 28,* 145–155.

Kobasa, S. C. (1979). Stressful life events, personality, and health: An inquiry into hardiness. *Journal of Personality and Social Psychology, 31,* 1–11.

Koehler, T. (1985). Stress and rheumatoid arthritis: A survey of empirical evidence in human and animal studies. *Journal of Psychosomatic Research, 29,* 655–663.

Kokot, K., Schaefer, R. M., Teschner, M., Gilge, U., Plass, R., & Heidland, A. (1988). Activation of leukocytes during prolonged physical exercise. *Advances in Experimental Medicine and Biology, 240,* 57–63.

Kooijman, R., Hooghe-Peters, E. L., & Hooghe, R. (1996). Prolactin, growth hormone, and insulin-like growth factor-I in the immune system. *Advances in Immunology, 63,* 377–454.

Korth, C., Mullington, J., Schreiber, W., & Pollmächer, T. (1996). Influence of endotoxin on daytime sleep in humans. *Infection and Immunity, 64,* 1110–1115.

Kosco-Vilbois, M. H., Zentgraf, H., Gerdes, J., & Bonnefoy, J. Y. (1997). To "B" or not to "B" a germinal center? *Immunology Today, 18,* 225–230.

Kraemer, G. W., Ebert, M. H., Schmidt, D. E., & McKinney, W. T. (1991). Strangers in a strange land: A psychobiological study of infant monkeys before and after separation from real or inanimate mothers. *Child Development, 62,* 548–566.

Kranz A., Kendall, M. D., & von Gaudecker, B. (1997). Studies on rat and human thymus to demonstrate immunoreactivity of calcitonin gene-related peptide, tyrosine hydroxylase and neuropeptide Y. *Journal of Anatomy, 191,* 441–450.

Kreitler, S., Chaitchik, S., & Kreitler, H. (1993). Repressiveness: Cause or result of cancer. *Psycho-Oncology, 2,* 43–54.

Kreutzberg, G. W. (1995). Microglia, the first line of defence in brain pathologies. *Arzneimittelforschung, 45,* 357–360.

Kronfol, Z., Nair, M., Zhang, Q., Hill, E. E., & Brown, M. B. (1997). Circadian immune measures in healthy volunteers: Relationship to hypothalamic–pituitary–adrenal axis hormones and sympathetic neurotransmitters. *Psychosomatic Medicine, 59,* 42–50.

Krueger, G. R. F., Schonnebeck, M., & Braun, M. (1990). Changes in cell membrane fluidity and in receptor expression following infection with HHV-6 may influence superinfection with other viruses. *AIDS Research and Human Retroviruses, 6,* 148–149.

Krueger, J. M. (1990). Somnogenic activity of immune response modifiers. *Trends in Pharmacological Science, 11,* 122–126.

Krueger, J. M., & Karnovsky, M. L. (1987). Sleep and the immune response. *Annals of the New York Academy of Sciences, 496,* 510–516.

Krueger, J. M., & Karnovsky, M. L. (1995). Sleep as a neuroimmune phenomenon: A brief historical perspective. *Advances in Neuroimmunology, 5,* 5–12.

Krueger, J. M., & Majde, J. A. (1990). Sleep as a host defense: Its regulation by microbial products and cytokines. *Clinical Immunology and Immunopathology, 57,* 188–199.

Krueger, J. M., & Majde, J. A. (1994). Microbial products and cytokines in sleep and fever regulation. *Critical Reviews in Immunology, 14,* 355–379.

Krueger, J. M., Pappenheimer, J. R., & Karnovsky, M. L. (1982). The composition of sleep-promoting factor isolated from human urine. *J Biol Chem, 257,* 1664–1669.

Krueger, J. M., Takahashi, S., Kapas, L., Bredow, S., Roky, R., Guha-Thakurta, N., Novitsky, S., & Obál, F. (1995). Cytokines in sleep regulation. *Advances in Neuroimmunology, 5,* 171–188.

Krug, M., Tschernig, T., Holgate, S., & Pabst, R. (1998). How do lymphocytes get into the asthmatic airways? Lymphocyte traffic into and within the lung in asthma. *Clinical and Experimental Allergy, 28,* 10–18.

Kubicki, S., Henkes, H., Terstegge, K., & Ruf, B. (1988). AIDS related sleep disturbance—A preliminary report. In S. Kubicki, H. Henkes, U. Bienzle, & U. Pohle (Eds.), *HIV and the nervous system* (pp. 97–105). Jena: Fischer.

Kucera, L. S., Leake, E., Iyer, N., Raben, D., & Myrvik, Q. N. (1990). Human immunodeficiency virus type 1 (HIV-1) and herpes simplex virus type 2 (HSV-2) can coinfect and simultaneously replicate in the same human $CD4^+$ cell: Effect of coinfection on infectious HSV-2 and HIV-1 replication. *AIDS Research and Human Retroviruses, 6,* 641–647.

Kugler, J. (1995). Stress, salivary immunoglobulin A and susceptibility to upper respiratory tract infection: Evidence for adaptive immunomodulation. *Psychologische Beiträge, 36,* 175–182.

Kugler, J., Hess, M., & Haake, D. (1992). Secretion of salivary immunoglobulin A in relation to age, saliva flow, mood states, secretion of albumin, cortisol, and catecholamines in saliva. *Journal of Clinical Immunology, 12,* 45–49.

Kugler, J., Reintjes, F., Tewes, U., & Schedlowski, M. (1996). Competition stress in soccer coaches increases salivary immunoglobulin A and salivary cortisol concentration. *Journal of Sports Medicine and Physical Fitness, 36,* 117–120.

Kuhn, C. M. (1989). Adrenocortical and gonadal steroids in behavioral cardiovascular medicine. In N. Schneiderman, S. M. Weiss, & P. G. Kaufmann (Eds.), *Handbook of research methods in cardiovascular behavioral medicine* (pp. 185–204). New York: Plenum Press.

Kuis, W., Villiger, P. M., Laser, H. G., & Lotz, M. (1991). Differential processing of proenkephalin-A by human peripheral blood monocytes and T lymphocytes. *Journal of Clinical Investigation, 88,* 817–824.

Kuis, W., de Jong-de Vos van Steenwijk, C. C. E., Sinnema, G., Kavelaars, A., Prakken, B., Helders, P. J. M., & Heijnen, C. J. (1996). The autonomic nervous system and the immune system in juvenile chronic rheumatoid arthritis: An interdisciplinary study. *Brain, Behavior, and Immunity, 10,* 387–398.

Kumar, M., Morgan, R., Szapocznik, J., & Eisdorfer, C. (1991). Norepinephrine response in HIV+ subjects. *Journal of AIDS, 4,* 782–785.

Kumar, M., Kumar, A. M., Morgan, R., Szapocznik, J., & Eisdorfer, C. (1993). Abnormal pituitary–adrenocortical response in early HIV-1 infection. *Journal of Acquired Immune Deficiency Syndromes, 6,* 61–65.

Kusnecov, A. W., Sivyer, M., King, M. G., Husband, A. J., Cripps, A. W., & Clancy, R. L. (1983). Behaviorally conditioned suppression of the immune response by antilymphocyte serum. *Journal of Immunology, 130,* 2117–2120.

Kusnecov, A. W., Husband, A. J., King, M. G., & Smith, R. (1987). In vivo effects of β-endorphin on lymphocyte proliferation and interleukin 2 production. *Brain, Behavior, and Immunity, 1,* 88–97.

Kusnecov, A. W., Husband, A. J., & King, M. G. (1988). Behaviorally conditioned suppression of mitogen-induced proliferation and immunoglobulin production: Effect of time span between conditioning and reexposure to the conditioned stimulus. *Brain, Behavior, and Immunity, 2,* 198–211.

Kusnecov, A. W., King, M. G., & Husband, A. J. (1989). Immunomodulation by behavioural conditioning. *Biological Psychology, 28,* 25–39.

Labeur, M. S., Arzt, E., Wiegers, G. J., Holsboer, F., & Reul, J. M. H. M. (1995). Long-term intracerebroventricular corticotropin-releasing hormone administration induces distinct changes in rat splenocyte activation and cytokine expression. *Endocrinology, 136,* 2678–2688.

Laidlaw, T. M., Richardson, D. H., Booth, R. J., & Large, R. J. (1994). Immediate-type hypersensitivity reactions and hypnosis: Problems in methodology. *Journal of Psychosomatic Research, 38,* 569–580.

Laidlaw, T. M., Booth, R. J., & Large, R. J. (1996). Reduction in skin reactions to histamine after a hypnotic procedure. *Psychosomatic Medicine, 58,* 242–248.

Lampeter, E., Signore, A., Gale, E. A., & Pozzilli, P. (1989). Lessons from the NOD mouse for the pathogenesis and immunotherapy of human type 1 (insulin-dependent) diabetes melitus. *Diabetologica, 32,* 703–708.

Landmann, R. (1992). Beta-adrenergic receptors in human leukocyte subpopulations. *European Journal of Clinical Investigation, 22*(Suppl. 1), 30–36.

Landmann, R. M. A., Mueller, F. B., Perini, C. H., Wesp, M., Erne, P., & Buehler, F. R. (1984). Changes of immunoregulatory cells induced by psychological and physical stress: Relationship to plasma catecholamines. *Clinical and Experimental Immunology, 58,* 127–135.

LaPerriere, A., Antoni, M. H., Schneiderman, N., Ironson, G., Klimas, N., Caralis, P., & Fletcher, M. A. (1990). Exercise intervention attenuates emotional distress and natural killer cell decrements following notification of positive serologic status for HIV-1. *Biofeedback and Self-Regulation, 15,* 229–242.

LaPerriere, A., Fletcher, M. A., Antoni, M. H., Ironson, G., Klimas, N., & Schneiderman, N. (1991). Aerobic exercise training in an AIDS risk group. *International Journal of Sports Medicine, 12,* S53–S57.

Laudenslager, M. L., & Reite, M. L. (1984). Losses and separations: Immunological consequences and health implications. In P. Shaver (Ed.), *Reviews in personality and social psychology* (pp. 285–312). Beverly Hills: Sage.

Laudenslager, M. L., Reite, M., & Harbeck, R. J. (1982). Suppressed immune response in infant monkeys associated with maternal separation. *Behavioral and Neural Biology, 36,* 40–48.

Laudenslager, M. L., Capitanio, J. P., & Reite, M. L. (1985). Possible effects of early separation experiences on subsequent immune function in adult macaque monkeys. *American Journal of Psychiatry, 142,* 862–864.

Laudenslager, M. L., Held, P. E., Boccia, M. L., Reite, M. L., & Cohen, J. J. (1990). Behavioral and immunological consequences of brief maternal separation: A species comparison. *Developmental Psychobiology, 23,* 247–264.

Laudenslager, M. L., Rasmussen, K. L. R., Berman, C. M., Suomi, S. J., & Berger, C. L. (1993). Specific antibody levels in free-ranging rhesus monkeys: Relationships to plasma hormones, cardiac parameters, and early behavior. *Developmental Psychobiology, 26,* 407–420.

Laudenslager, M. L., Boccia, M. L., Berger, C. L., Gennaro-Ruggles, M. M., McFerran, B., & Reite, M. L. (1995). Total cortisol, free cortisol, and growth hormone associated with brief social separation experiences in young macaques. *Developmental Psychobiology, 28,* 199–212.

Laudenslager, M. L., Berger, C. L., Boccia, M. L., & Reite, M. L. (1996). Natural cytotoxicity toward K562 cells from infant through puberty: Effects of early social challenge. *Brain, Behavior, and Immunity, 10,* 275–287.

Laudenslager, M. L., Clarke, A. S., Kraemer, G. W., & Goldstein, M. A. (1997). Some long term physiological consequences of early separation experiences in rhesus monkeys. *American Journal of Primatology, 42,* 126 (Abstract).

Layé, S., Parnet, P., Goujon, E., & Dantzer, R. (1994). Peripheral administration of lipopolysaccharide induces the expression of cytokine transcripts in the brain and pituitary of mice. *Molecular Brain Research, 27,* 157–162.

Layé, S., Bluthé, R. M., Kent, S., Combe, C., Médina, C., Parnet, P., Kelley, K. W., & Dantzer, R. (1995). Subdiaphragmatic vagotomy blocks the induction of interleukin-1β mRNA in the brain of mice in response to peripherally administered lipopolysaccharide. *American Journal of Physiology, 268,* R1327–R1331.

Lazarus, R. S. (1966). *Psychological stress and the coping process.* New York: McGraw-Hill.

Lazarus, R. S. (1991). *Emotion and adaption.* Oxford: Oxford University Press.

Lazarus, R. S., & Folkman, S. (1989). *Manual for the study of daily hassles and uplifts scales.* Palo Alto, CA: Consulting Psychologists Press.

Legendre, R., & Pieron, H. (1913). Recherches sur le besoin de sommeil consecutif a une veille prologee. *Zeitschrift für Allgemeine Physiologie, 14,* 235–262.

Lehman, D., Siebold, K., & Emmons, L. R. (1988). Androgens inhibit proliferation of human peripheral blood lymphocytes *in vitro. Clinical Immunology and Immunopathology, 46,* 122–128.

Leiter, E. H., Prochazka, M., & Coleman, D. L. (1987). The nonobese diabetic (NOD) mouse. *American Journal of Pathology, 128,* 380–383.

Lekander, M., Furst, C., Rotstein, S., Hursti, T., & Fredrikson, M. (1997). Immune effects of relaxation during chemotherapy for ovarian cancer. *Psychotherapy and Psychosomatics, 66,* 185–191.

Lemp, G. F., Payne, S. F., Neal, D., Temelso, T., & Rutherford, G. W. (1990). Survival trends for patients with AIDS. *Journal of the American Medical Association, 263,* 402–406.

Leon, A., Buriani, A., Dal, T. R., Fabris, M., Romanello, S., Aloe, L., & Levi-Montalcini, R. (1994). Mast cells synthesize, store, and release nerve growth factor. *Proceedings of the National Academy of Sciences, USA, 91,* 3739–3743.

LeShan, L. (1966). An emotional life-history pattern associated with neoplastic disease. *Annals of the New York Academy of Medicine, 125,* 780–793.

Levenson, J. L., Hamer, R. M., Myers, T., Hart, R. P., & Kaplowitz, L. G. (1987). Psychological factors predict symptoms of severe recurrent genital herpes infection. *Journal of Psychosomatic Research, 31,* 153–159.

Levine, J. D., Clark, R., Devor, M., Helms, C., Moskowitz, M. A., & Basbaum, A. (1984). Intraneuronal substance P contributes to the severity of experimental arthritis. *Science, 226,* 547–549.

Levine, J. D., Coderre, T. J., Helms, C., & Basbaum, A. I. (1988). β₂-Adrenergic mechanisms in experimental arthritis. *Proceedings of the National Academy of Sciences, USA, 85,* 4553–4556.

Levine, M. I., Geer, J. H., & Kost, P. F. (1966). Hypnotic suggestion and the histamine wheal. *Journal of Allergy, 37,* 246–250.

Levine, S., & Wiener, S. G. (1988). Psychoendocrine aspects of mother–infant relationships in nonhuman primates. *Psychoneuroendocrinology, 13,* 143–154.

Levine, S., Wiener, S. G., Coe, C. L., Bayart, F. E., & Hayashi, K. T. (1987). Primate vocalization: A psychobiological approach. *Child Development, 58,* 1408–1419.

Levy, M. L., & Roberts, D. C. (1992). Clinical significance of psychoneuroimmunology: Prediction of cancer outcomes. In N. Schneiderman, P. McCabe, & A. Baum (Eds.), *Stress and disease processes* (pp. 165–174). New York: Erlbaum.

Levy, S., Herberman, R., Maluish, A., Schlien, B., & Lippman, M. (1985). Prognostic risk assessment in primary breast cancer by behavioral and immunological parameters. *Health Psychology, 4,* 99–113.

Levy, S., Herberman, R., Lippman, M., & d'Angelo, T. (1987). Correlation of stress factors with sustained depression of natural killer cell activity and predicted prognosis in patients with breast cancer. *Journal of Clinical Oncology, 5,* 348–353.

Lew, W., Oppenheim, J. J., & Matsushima, K. (1988). Analysis of the suppression of IL-1α and IL-1β production in human peripheral blood mononuclear adherent cells by a glucocorticoid hormone. *Journal of Immunology, 140,* 1895–1902.

Lewin, D. (1996). Protease inhibitors: HIV-1 summons a Darwinian defense. *Journal of NIH Research, 8,* 33–35.

Lewis, C. E., & McGee, J. O. (1992). Natural killer cells in tumour biology. In Anonymous, *The natural immune system: The natural killer cell* (pp. 175–203). Oxford: IRL Press.

Lewis, C. E., O'Sullivan, C., & Barraclough, J. (1994). *The psychoimmunology of cancer: Mind and body in the fight for survival?* New York: Oxford University Press.

Lewis, J. K., McKinney, W. T., Young, L. D., & Kraemer, G. W. (1976). Mother–infant separation in rhesus monkeys as a model of human depression. *Archives of General Psychiatry, 33,* 699–705.

Lewis, J. W., Cannon, J. T., & Liebeskind, J. C. (1980). Opioid and nonopioid mechanisms of stress analgesia. *Science, 208,* 623–625.

Lewis-Faning, E. (1950). Report on an enquiry into the etiology factors associated with rheumatoid arthritis. *Annals of Rheumatoid Diseases, 9,* 94–99.

Lichstein, K. (1988). *Clinical relaxation strategies.* New York: Wiley.

Licino, J., & Wong, M.-L. (1997). Pathways and mechanisms for cytokine signaling of the central nervous system. *Journal of Clinical Investigation, 100,* 2941–2947.

Licinio, J., Wong, M. L., & Gold, P. W. (1991). Localization of interleukin-1 receptor antagonist mRNA in rat brain. *Endocrinology, 129,* 562–564.

Liew, F. Y., Russell, S. M., Appleyard, G., Brand, C. M., & Beale, J. (1984). Cross-protection in mice infected with influenza A virus by the respiratory route is correlated with local IgA antibody rather than serum antibody or cytotoxic T cell reactivity. *European Journal of Immunology, 14,* 350–356.

Line, S. W., Kaplan, J. R., Heise, E. R., Hilliard, J. K., Cohen, S., Rabin, B. S., & Manuck, S. B. (1996). Effects of social reorganization on cellular immunity in male cynomolgus monkeys. *American Journal of Primatology, 39,* 235–249.

Linkins, R. W., & Comstock, G. W. (1990). Depressed mood and development of cancer. *American Journal of Epidemiology, 132,* 962–972.

Linner, K. M., Beyer, H. S., & Sharp, B. M. (1991). Induction of the messenger ribonucleotidic acid for proenkephalin A in cultured murine CD4-positive thymocytes. *Endocrinology, 128,* 717–724.

Linner, K. M., Quist, H. E., & Sharp, B. M. (1995). Met-enkephalin-containing peptides encoded by proenkephalin A mRNA expressed in activated murine thymocytes inhibit thymocyte proliferation. *Journal of Immunology, 154,* 5049–5060.

Liu, D., Diorio, J., Tannenbaum, B., Caldji, C., Francis, D., Freedman, A., Sharma, S., Pearson, D., Plotsky, P. M., & Meaney, M. J. (1997). Maternal care, hippocampal glucocorticoid receptors, and hypothalamic-pituitary-adrenal responses to stress. *Science, 277,* 1659–1662.

Locke, S., Ader, R., Besedovsky, H., Hall, N., Solomon, G., & Stom, G. (Eds.). (1985). *Foundations of psychoneuroimmunology.* New York: Aldine Publishing.

Locke, S. E., Ransil, B. J., Covino, N. A., Toczydolwski, J., Lohse, J. M., Dvorak, H. M., Arndt, K. A., & Franke, F. H. (1987). Failure of hypnotic suggestion to alter immune response to delayed-type hypersensitivity antigens. *Annals of the New York Academy of Sciences, 496,* 745–749.

Locke, S. E., Ransil, B. J., Zachariae, R., Molay, F., Tollins, K., Covino, N. A., & Danforth, D. (1994). Effect of hypnotic suggestion on the delayed-type hypersensitivity response. *Journal of the American Medical Association, 272,* 47–52.

Lockwood, L. L., Siebert, L. L., Laudenslager, M. L., Watkins, L. R., & Maier, S. F. (1993). Anesthesia-induced modulation of *in vivo* antibody levels. *Anesthesia and Analgesia, 77,* 769–774.

Loeber, J. G. (1984). Radioimmunoassay. In W. S. Hancock (Ed.), *CRC handbook of HPLC for the separation of amino acids, peptides, and proteins* (Vol. I) (pp. 91–104). Boca Raton: CRC Press.

Loeper, M., & Crouzon, O. (1904). L'action de L'adrenaline sur le sang. *Archives de Medicine Experimentale et d'Anatomie Path, 16,* 83–108.

Lolait, S. J., Clements, J. A., Markwick, A. J., Cheng, C., McNally, M., Smith, A. I., & Funder, J. W. (1986). Pro-opiomelanocortin mRNA and posttranslational processing of beta endorphin in spleen macrophages. *Journal of Clinical Investigation, 77,* 1776–1779.

Longo, D. J., Clum, G. A., & Yaeger, N. J. (1988). Psychosocial treatment for recurrent genital herpes. *Journal of Consulting and Clinical Psychology, 56,* 61–66.

Lorr, M., & McNair, D. M. (1982). *Profile of mood states.* San Diego: Educational and Industrial Testing Service.

Lovallo, W. R., Pincomb, G. A., Brackett, D. J., & Wilson, M. F. (1990). Heart rate as a predictor of neuroendocrine responses to aversive and appetitive challenges. *Psychosomatic Medicine, 52,* 17–26.

Lubach, G. R., Coe, C. L., & Erschler, W. B. (1995). Effects of early rearing environment on immune responses of infant rhesus monkeys. *Brain, Behavior, and Immunity, 9,* 31–46.

Luborsky, L., Mintz, J., Brightman, V. J., & Katcher, A. H. (1976). Herpes simplex virus and moods: A longitudinal study. *Journal of Psychosomatic Research, 20,* 543–548.

Lüdecke, D. K., Chrousos, G. P., & Tolis, G. (Eds.). (1990). *ACTH, Cushing's syndrome, and other hypercortisolemic states.* New York: Raven Press.

Lutgendorf, S. K., Antoni, M. H., Kumar, M., & Schneiderman, N. (1994). Changes in cognitive coping strategies predict EBV-antibody titre change following a stressor disclosure induction. *Journal of Psychosomatic Research, 38,* 63–78.

Lutgendorf, S., Antoni, M., Ironson, G., Klimas, N. G., Kumar, M., Starr, K. R., McCabe, P. M., Cleven, K., Fletcher, M. A., & Schneiderman, N. (1997). Cognitive behavioral stress management decreases dysphoric mood and herpes simplex virus-type 2 antibody titers in symptomatic HIV-seropositive gay men. *Journal of Consulting and Clinical Psychology, 65*, 31–43.

Lyketsos, C. G., Hoover, D. R., Guccione, M., Senterfitt, W., Dew, M. A., & Wesch, J. (1993). Depressive symptoms as predictors of medical outcomes in HIV infection. *Journal of the American Medical Association, 270*, 2563–2567.

Lyons, P. D., & Blalock, J. E. (1997). Pro-opiomelanocortin gene expression and protein processing in rat mononuclear leukocytes. *Journal of Neuroimmunology, 78*, 47–62.

Lysle, D. T., Cunnick, J. E., & Maslonek, K. A. (1991). Pharmacological manipulation of immune alterations induced by an aversive conditioned stimulus: Evidence for a β-adrenergic receptor-mediated Pavlovian conditioning process. *Behavioral Neuroscience, 105*, 443–449.

Lysle, D. T., Hoffman, K. E., & Dykstra, L. A. (1996). Evidence for the involvement of the caudal region of the periaqueductal gray in a subset of morphine-induced alterations of immune status. *Journal of Pharmacology and Experimental Therapeutics, 277*, 1533–1540.

Mackinnon, L.T., & Hooper, S. (1994). Mucosal (secretory) immune system responses to exercise of varying intensity and during overtraining. *International Journal of Sports Medicine, 15*, S179–S183.

Mackinnon, L.T., Chick, T. W., van As, A., & Tomasi, T. B. (1987). The effect of exercise on secretory and natural immunity. *Advances in Experimental Medicine and Biology, 216A*, 869–876.

MacNeil, B., & Hoffman-Goetz, L. (1993). Chronic exercise enhances in vivo and in vitro cytotoxic mechanisms of natural immunity in mice. *Journal of Applied Physiology, 74*, 388–395.

Macphee, I. A. M., Antoni, F. A., & Mason, D. W. (1989). Spontaneous recovery of rats from experimental allergic encephalomyelitis is dependent on regulation of the immune system by endogenous adrenal corticosteroids. *Journal of Experimental Medicine, 169*, 431–445.

Madden, K. S., & Felten, D. L. (1995). Experimental basis for neural–immune interactions. *Physiological Reviews, 75*, 77–106.

Maddison, D., & Viola, A. (1968). The health of widows in the year following bereavement. *Journal of Psychosomatic Research, 12*, 297–306.

Maes, M. (1995). Evidence for an immune response in major depression: A review and hypothesis. *Progress in Neuropsychopharmacology and Biological Psychiatry, 19*, 11–38.

Maes, M., Bosmans, E., Suy, E., Minner, B., & Raus, J. (1989). Impaired lymphocyte stimulation by mitogens in severely depressed patients a complex interface with HPA axis hyperfunction, noradrenergic activity and the ageing process. *British Journal of Psychiatry, 155*, 793–798.

Maes, M., Bosmans, E., Suy, E., Minner, B., & Raus, J. (1991a). A further exploration of the relationships between immune parameters and the HPA-axis activity in depressed patients. *Psychological Medicine, 21*, 313–320.

Maes, M., Bosmans, E., Suy, E., Vandervorst, C., DeJonckheere, C., & Raus, J. (1991b). Depression-related disturbances in mitogen-induced lymphocyte responses and interleukin-1 and soluble interleukin-2 receptor production. *Acta Psychiatrica Scandinavica, 84*, 379–386.

Maes, M., Lambrechts, J., Bosmans, E., Jacobs, J., Suy, E., Vandervorst, C., DeJonckheere, C., Minner, B., & Raus, J. (1992). Evidence for a systemic immune activation during depression: Results of leukocyte enumeration by flow cytometry in conjunction with monoclonal antibody staining. *Psychological Medicine, 22*, 45–53.

Maes, M., Scharpe, S., Meltzer, H. Y., Bosmans, E., Suy, E., Calabrese, J., & Cosyns, P. (1993). Relationships between interleukin-6 activity, acute phase proteins, and function of the hypothalamic–pituitary–adrenal axis in severe depression. *Psychiatry Research, 49*, 11–27.

Maes, M., Meltzer, H. Y., Bosmans, E., Bergmans, R., Vandoolaeghe, E., Ranjan, R., & Desnyder, R. (1995a). Increased plasma concentrations of interleukin-6, soluble interleukin-6, soluble interleukin-2 and transferrin receptor in major depression. *Journal of Affective Disorders, 34*, 301–309.

Maes, M., Smith, R., & Scharpe, S. (1995b). The monocyte–T-lymphocyte hypothesis of major depression. *Psychoneuroendocrinology, 20*, 111–116.

Maes, M., Vandoolaeghe, E., Ranjan, R., Bosmans, E., Bergmans, R., & Desnyder, R. (1995c). Increased serum interleukin-1-receptor-antagonist concentrations in major depression. *Journal of Affective Disorders, 36*, 29–36.

Maes, M., Bosmans, E., De Jongh, R., Kenis, G., Vandoolaeghe, E., & Neels, H. (1997). Increased serum IL-6 and IL-1 receptor antagonist concentrations in major depression and treatment resistant depression. *Cytokine, 9*, 853–858.

Maeurer, M. J., & Lotze, M. T. (1997). Tumor recognition by the cellular immune system: New aspects of tumor immunology. *International Review of Immunology, 14*, 97–132.

Maier, S. F., & Laudenslager, M. L. (1998). Inescapable shock, shock controllability, and mitogen stimulated lymphocyte proliferation. *Brain, Behavior, and Immunity, 2*, 87–91.

Maier, S. F., & Watkins, L. R. (1995). Intracerebroventricular interleukin-1 receptor antagonist blocks the enhancement of fear conditioning and interference with escape produced by inescapable shock. *Brain Research, 695*, 279–282.

Maier, S. F., & Watkins, L. (1998). Cytokines for psychologists: Implications of bi-directional immune-to-brain communication for understanding behavior, mood and cognition. *Psychological Reviews*, *105*, 83–107.

Maisel, A. S., Harris, C., Rearden, C. A., & Michel, M. C. (1990a). β-Adrenergic receptors in lymphocyte subsets after exercise. Alterations in normal individuals and patients with congestive heart failure. *Circulation*, *82*, 2003–2010.

Maisel, A. S., Knowlton, K. U., Fowler, P., Rearden, A., Ziegler, M. G., Motulsky, H. J., Insel, P. A., & Michel, M. C. (1990b). Adrenergic control of circulating lymphocyte subpopulations. Effects of congestive heart failure, dynamic exercise, and terbutaline treatment. *Journal of Clinical Investigation*, *85*, 462–467.

Mancia, G., & Zanchetti, A. (1980). Cardiovascular regulation during sleep. In J. Orem & C. D. Barnes (Eds.), *Physiology in sleep. Research topics in physiology* (pp. 1–55). New York: Academic Press.

Mann, J. M. (1990). Global AIDS into the 1990's. *Journal of Acquired Immune Deficiency Syndrome*, *3*, 438–442.

Mann, J. M. (1992). AIDS—The second decade: A global perspective. *Journal of Infectious Diseases*, *165*, 245–250.

Mann, J., Tarantola, D. J. M., & Netter, T. W. (1992). *AIDS in the world*. Cambridge, MA: Harvard University Press.

Martin, J. (1988). Psychological consequences of AIDS-related bereavement among gay men. *Journal of Consulting and Clinical Psychology*, *56*, 856–862.

Martin, J., Prystowski, M. B., & Angeletti, R. H. (1987). Preproenkephalin mRNA in T-cells, macrophages and mast cells. *Journal of Neuroscience Research*, *18*, 82–86.

Marzuk, P. M., Tierney, H., Tardiff, K., Gross, E. M., Morgan, E. B., Hsu, M. A., & Mann, J. J. (1988). Increased risk of suicide in persons with AIDS. *Journal of the American Medical Association*, *259*, 1333–1337.

Masi, A. T., Feigenbaum, S. L., & Chatterton, R. T. (1995). Hormonal and pregnancy relationships to rheumatoid arthritis convergent effects with immunological and microvascular systems. *Seminars in Arthritis and Rheumatism*, *25*, 1–27.

Mason, D. W. (1991). Genetic variation in the stress response: Susceptibility to experimental allergic encephalomyelitis and implications for human inflammatory disease. *Immunology Today*, *12*, 57–59.

Mason, D. W., Macphee, I. A. M., & Antoni, F. A. (1990). The role of the neuroendocrine system in determining genetic susceptibility to experimental allergic encephalomyelitis. *Immunology*, *70*, 1–9.

Matera, L. (1996). Endocrine, paracrine and autocrine actions of prolactin on immune cells. *Life Sciences*, *59*, 599–614.

Mayne, T. J., Vittinghoff, E., Chesney, M. A., Barrett, D. C., & Coates, T. J. (1996). Depressive affect and survival among gay and bisexual men infected with HIV. *Archives of Internal Medicine*, *156*, 2233–2238.

Mazzeo, R. S. (1996). Exercise, immunity, and aging. In L. Hoffman-Goetz (Ed.), *Exercise and immune function* (pp. 199–214). Boca Raton: CRC Press.

McCabe, P. M., & Schneiderman, N. (1985). Psychophysiologic responses to stress. In N. Schneiderman & J. T. Tapp (Eds.), *Behavioral medicine: The biopsychosocial approach*. Hillsdale, NJ: Erlbaum, pp. 99–131.

McCarthy, D. A., & Dale, M. M. (1988). The leucocytosis of exercise. A review and model. *Sports Medicine*, *6*, 333–363.

McClelland, D. C., Alexander, C., & Marks, E. (1982). The need for power, stress, immune functions, and illness among male prisoners. *Journal of Abnormal Psychology*, *91*, 61–70.

McClintick, J., Costlow, C., Fortner, M., White, J., Gillin, C., & Irwin, M. (1994). Partial sleep deprivation reduces natural killer cell activity, interleukin-2 production and lymphokine activated killer cell activity. *International Society of Psychoneuroendocrinology* (Abstract).

McDaniels, S., Hepworth, J., & Doherty, W. J. (1992). *Medical family therapy: A biopsychosocial approach to families with health problems*. New York: Basic Books.

McDowell, S. L., Chaloa, K., Housh, T. J., Tharp, G. D., & Johnson, G. O. (1991). The effect of exercise intensity and duration on salivary immunoglobulin A. *European Journal of Applied Physiology*, *63*, 108–111.

McDowell, S. L., Hughes, R. A., Hughes, R. J., Housh, D. J., Housh, T. J., & Johnson, G. O. (1992). The effect of exhaustive exercise on salivary immunoglobulin A. *Journal of Sports Medicine and Physical Fitness*, *32*, 412–415.

McEwen, B. S., & Stellar, E. (1993). Stress and the individual. Mechanisms leading to disease. *Archives of Internal Medicine*, *153*, 2093–2101.

McEwen, B. S., Biron, C. A., Brunson, K. W., Bulloch, K., Chambers, W. H., Dhabhar, F. S., Goldfarb, R. H., Kitson, R. P., Miller, A. H., Spencer, R. L., & Weiss, J. M. (1997). The role of adrenocorticoids as modulators of immune function in health and disease: Neural, endocrine and immune interactions. *Brain Research Reviews*, *23*, 79–133.

McGeer, P. L., Rogers, J., & McGeer, E. G. (1994). Neuroimmune mechanisms in Alzheimer disease pathogenesis. *Alzheimer Disease and Associated Disorders*, *8*, 149–158.

McGillis, J. P., Mitsuhashi, M., & Payan, D. G. (1991). Immunologic properties of substance P. In R. Ader, D. L. Felten, & N. Cohen (Eds.), *Psychoneuroimmunology* (pp. 209–233). San Diego: Academic Press.

McGrady, A., Woerner, M., Bernal, G. A. A., & Higgins, J. T. (1987). Effect of biofeedback assisted relaxation on blood pressure and cortisol levels in normotensives and hypertensives. *Journal of Behavioral Medicine, 10,* 301–310.

McGrady, A., Conran, P., Dickey, D., Garman, D., Farris, E., & Schumann-Brzezinski, C. (1992). The effects of biofeedback-assisted relaxation on cell-mediated immunity, cortisol, and white blood cell count in healthy adult subjects. *Journal of Behavioral Medicine, 15,* 343–354.

McKinney, W. T., & Bunney, W. E. (1969). Animal models of depression: Review of evidence and implications for research. *Archives of General Psychiatry, 21,* 240–248.

McKinnon, W., Weisse, C. S., Reynolds, C. P., Bowles, C. A., & Baum, A. (1989). Chronic stress, leukocyte subpopulations, and humoral response to latent viruses. *Health Psychology, 8,* 389–402.

McVoy, M. A., & Adler, S. P. (1989). Immunologic evidence for frequent age-related cytomegalovirus reactivation in seropositive immunocompetent individuals. *Journal of Infectious Disease, 160,* 1–10.

Meichenbaum, D. (1977). *Cognitive–behavior modification: An integrative approach.* New York: Plenum Press.

Mellors, J. W., Rinaldo, C. R., Jr., Gupta, P., White, R. M., Todd, J. A., & Kingsley, L. A. (1996). Prognosis in HIV-1 infection predicted by the quantity of virus in plasma. *Science, 272,* 1167–1170.

Merriam, G. R., & Wachter, K. W. (1984). Measurement and analysis of episodic hormone secretion. In D. Rodbard & G. Forti (Eds.), *Computers in endocrinology* (pp. 325–346). New York: Raven Press.

Mestecky, J. (1993). Saliva as a manifestation of the common mucosal immune system. *Annals of the New York Academy of Sciences, 694,* 184–194.

Mestecky, J., Russell, M. W., Jackson, S., & Brown, T. A. (1986). The human IgA system: A reassessment. *Clinical Immunology and Immunopathology, 40,* 105–114.

Metal'nikov, S., & Chorine, V. (1926). Role des reflexes conditionnels dans l'immunite. *Annales de l'Institut Pasteur, 40,* 893–900.

Metchnikoff, I. I. (1910). In P. Chalmers Mitchell (Ed.), *The prolongation of life, optimistic studies.* New York: Putnam.

Meyer, R. J., & Haggerty, R. J. (1962). Streptococcal infections in families. Factors altering individual susceptibility. *Pediatrics, 29,* 539–549.

Miettinen, S., Fusco, F. R., Yrjäneikki, J. K. R., Hirvonen, T., Roivainen, R., Närhi, M., Höckfelt, T., & Koistinaho, J. (1997). Spreading depression and focal brain ischemia induce cyclooxygenase-2 in cortical neurons through N-methyl-D-aspartic acid-receptors and phospholipase A2. *Proceedings of the National Academy of Sciences, USA, 94,* 6500–6505.

Millar, D. B., Hough, C. J., Mazorow, D. L., & Gootenberg, J. E. (1990). β-Endorphin's modulation of lymphocyte proliferation is dose, donor, and time dependent. *Brain, Behavior, and Immunity, 4,* 232–242.

Miller, A. H., Asnis, G. M., Lackner, C., Halbreich, U., & Norin, A. J. (1991). Depression, natural killer cell activity, and cortisol secretion. *Biological Psychiatry, 29,* 878–886.

Miller, G. (1980). Biology of Epstein–Barr virus. In G. Klein (Ed.), *Viral oncology* (pp. 713–738). New York: Raven Press.

Miller, G., Kemeny, M. E., Taylor, S. E., & Visscher, B. (1997). Social relationship and immune processes in HIV seropositive gay men. *Annals of Behavioral Medicine, 19,* 139–151.

Miller, N. E. (1964). Some psychophysiological studies of motivation and of the behavioral effects of illness. *Bulletin of the British Psychological Society, 17,* 1–20.

Mills, P. J., & Dimsdale, J. E. (1993). The promise of adrenergic receptor studies in psychophysiologic research II: Applications, limitations, and progress. *Psychosomatic Medicine, 55,* 448–457.

Mineka, S., & Suomi, S. J. (1978). Social separation in monkeys. *Psychological Bulletin, 85,* 1376–1400.

Moffit, K. H., Singer, J. A., Nelligan, D. W., Carlson, M. A., & Vyse, S. A. (1994). Depression and memory narrative type. *Journal of Abnormal Psychology, 103,* 581–583.

Moldofsky, H. (1995). Sleep, neuroimmune and neuroendocrine functions in fibromyalgia and chronic fatigue syndrome. *Advances in Neuroimmunology, 5,* 39–56.

Moldofsky, H., Lue, F. A., Eisen, J., Keystone, E., & Gorczynski, R. M. (1986). The relationship of interleukin-1 and immune functions to sleep in humans. *Psychosomatic Medicine, 48,* 309–318.

Moldofsky, H., Lue, F., Davidson, J., Jephthalh-Ochola, U., Carayanniotis, K., & Gorczynski, R. (1989a). The effect of 64 hours of wakefulness on immune functions and plasma cortisol in humans. In J. Horne (Ed.), *Sleep '88* (pp. 185–187). Stuttgart: Fischer Verlag.

Moldofsky, H., Lue, F., Davidson, J., & Gorczynski, R. (1989b). Effects of sleep deprivation on human immune functions. *FASEB Journal, 3,* 1972–1977.

Monjan, A. A., & Collector, M. I. (1977). Stress-induced modulation of the immune response. *Science, 196,* 307–308.

Montaganier, L., Gruest, S., Chamaret, C., Douguet, C., Axler, D., Guetard, D., Nugeyre, M. T., Barre-Sinoussi, F., Chermann, J. C., Klatzmann, D., & Gluckman, J. C. (1984). Adaptation of lymphoadenopathy associated virus (LAV) to replication in EBV-transformed B-lymphoblastoid cell lines. *Science, 226,* 63–66.

Moore, T. C., Spruck, C. H., & Said, S. I. (1988). Depression of lymphocyte traffic in sheep by vasoactive intestinal peptide (VIP). *Immunology, 64,* 475–478.

Moore-Ede, M. C., Sulzman, F. M., & Fuller, C. A. (1982). *The clocks that time us. Physiology of the circadian timing system.* Cambridge, MA: Harvard University Press.

Moos, R. H., & Solomon, G. F. (1965). Psychological comparisons between women with rheumatoid arthritis and their non-arthritic sisters. I. Personality tests and interview rating data. *Psychosomatic Medicine, 27,* 135–149.

Mora, J. M., Amtman, L. E., & Hoffman, S. J. (1926). Effect of mental and emotional states on the leukocyte count. *Journal of the American Medical Association, 86,* 945–946.

Morimoto, K., Hooper, D. C., Bornhorst, A., Corrisdeo, S., Bette, M., Fu, Z. F., Schäfer, M. K.-H., Koprowski, H., Weihe, E., & Dietzschold, B. (1996). Intrinsic responses to Borna disease virus infection of the central nervous system. *Proceedings of the National Academy of Sciences, USA, 93,* 13345–13350.

Morrow-Tesch, J. L, Mcglone, J. J., & Norman, R. L. (1993). Consequences of restraint stress on natural killer cell activity, behavior, and hormone levels in rhesus macaques (Macaca mulatta). *Psychoneuroendocrinology, 18,* 383–395.

Morse, E., Simon, P., Coburn, M., Hyslop, N., Greenspan, D., & Balson, P. N. (1991). Determinants of subject compliance within an experimental anti-HIV drug protocol. *Social Science and Medicine, 32,* 1161–1167.

Mosmann, T. R., & Sad, S. (1996). The expanding universe of T cell subset: Th 1, Th 2, and more. *Immunology Today, 17,* 138–146.

Mrazek, D. A., Klinnert, M. D., Mrazek, P., & Macey, T. (1991). Early asthma onset: Consideration of parenting issues. *Journal of the American Academy of Child and Adolescent Psychiatry, 30,* 277–282.

Mulder, C. L., Emmelkamp, P. M. G., Antoni, M. H., Mulder, J. W., Sandfort, T. G. M., & deVries, M. J. (1994). Cognitive–behavioral and experiential group psychotherapy for HIV infected homosexual men: A comparative study. *Psychosomatic Medicine, 56,* 423–431.

Mulder, C. L., Antoni, M. H., Emmelkamp, P. M. G., & Veugelers, P. J. (1995). Psychosocial group intervention and the rate of decline of immunological parameters in asymptomatic HIV-infected homosexual men. *Psychotherapy and Psychosomatics, 63,* 185–192.

Müller, S., & Weihe, E. (1991). Interrelation of peptidergic innervation with mast cells and ED1-positive cells in rat thymus. *Brain, Behavior, and Immunity, 5,* 55–72.

Mullington, J., Hermann, D., Holsboer, F., & Pollmächer, T. (1996a). Slow wave sleep and delta power are increased in healthy men following a sub-pyrogenic dose of endotoxin. *Journal of Sleep Research, 5*(Suppl. 1), 151.

Mullington, J., Hermann, D., Holsboer, F., & Pollmächer, T. (1996b). Human host response is not affected by sleep deprivation following experimental endotoxin challenge. *Journal of Sleep Research, 5*(Suppl. 1), 150.

Munck, A., & Guyre, P. M. (1991). Glucocorticoids and immune function. In R. Ader, D. L. Felten, & N. Cohen (Eds.), *Psychoneuroimmunology* (pp. 447–474). San Diego: Academic Press.

Munck, A., Guyre, P. M., & Holbrook, N. J. (1984). Physiological functions of glucocorticoids in stress and their relation to pharmacological actions. *Endocrine Reviews, 5,* 25–44.

Munoz, A., Wang, M. C., Good, R., Detels, H., Ginsberg, L., Kingsley, J., Phair, J., & Polk, B. F. (1988). *Estimation of the AIDS-free times after HIV-1 seroconversion.* Paper presented at the Fourth Annual Meeting of the International Conference on AIDS, Stockholm.

Muns, G., Liesen, H., Riedel, H., & Bergman, K. C. (1989). Influence of long-distance running on IgA in nasal secretion and saliva. *Deutsche Zeitschrift für Sportmedizin, 40,* 94–99.

Murphy, W. J., Durum, S. K., & Longo, D. L. (1993). Differential effects of growth hormone and prolactin on murine T cell development and function. *Journal of Experimental Medicine, 178,* 231–236.

Murray, D. R., Polizzi, S. M., Harris, T., Wilson, N., Michel, M. C., & Maisel, A. S. (1993). Prolonged isoproterenol treatment alters immunoregulatory cell traffic and function in the rat. *Brain, Behavior, and Immunity, 7,* 47–62.

Nahmias, A. J., & Josey, W. E. (1982). Herpes simplex viruses 1 and 2. In A. Evans (Ed.), *Viral infections of humans: Epidemiology and control* (pp. 351–372). New York: Plenum Press.

Naliboff, B. D., Benton, D., Solomon, G. F., Morley, J. E., Fahey, J. L., Bloom, E. T., Makinodan, T., & Gilmore, S. L. (1991). Immunological changes in young and old adults during brief laboratory stress. *Psychosomatic Medicine, 53,* 121–132.

Namir, S., Wolcott, D. L., Fawzy, F. I., & Alumbaugh, M. J. (1987). Coping with AIDS: Psychological and health implications. Special issue: Acquired immune deficiency syndrome. *Journal of Applied Social Psychology, 17,* 309–328.

Nash, H. L. (1987). Can exercise make us immune to disease? *Physiological Sportsmedicine, 1,* 250–253.

Nestel, P. J. (1969). Blood-pressure and catecholamine excretion after mental stress in labile hypertension. *Lancet, 1,* 692–694.

Neumann, H., Schmidt, H., Wilharm, E., Behrens, L., & Wekerle, H. (1997). Interferon γ gene expression in sensory neurons: Evidence for autocrine gene regulation. *Journal of Experimental Medicine, 12,* 2023–2031.

Newsholme, E. A. (1990). Psychoimmunology and cellular nutrition: An alternative hypothesis [editorial]. *Biological Psychiatry, 27*, 1–3.

Newsholme, E. A. (1994). Biochemical mechanisms to explain immunosuppression in well-trained and overtrained athletes. *International Journal of Sports Medicine, 15*, S142–S147.

Nielsen, H. B., Secher, N. H., Kappel, M., Hanel, B., & Pedersen, B. K. (1996a). Lymphocyte, NK, and LAK cell responses to maximal exercise. *International Journal of Sports Medicine, 17*, 60–65.

Nielsen, H. B., Secher, N., & Pedersen, B. K. (1996b). Lymphocytes and NK cell activity during repeated bouts of maximal exercise. *American Journal of Physiology, 271*, R222–R227.

Nieman, D. C. (1994a). Exercise, infection, and immunity. *International Journal of Sports Medicine, 15*, S131–S141.

Nieman, D. C. (1994b). Exercise, upper respiratory tract infection, and the immune system. *Medicine and Science in Sports and Exercise, 26*, 128–139.

Nieman, D. C. (1996). Prolonged aerobic exercise, immune response, and risk of infection. In L. Hoffman-Goetz (Ed.), *Exercise and immune function* (pp. 143–162). Boca Raton: CRC Press.

Nieman, D. C., & Henson, D. A. (1994). Role of endurance exercise in immune senescence. *Medicine and Science in Sports and Exercise, 26*, 172–181.

Nieman, D. C., Johanssen, L. M., & Lee, J. W. (1989). Infectious episodes in runners before and after a roadrace. *Journal of Sports Medicine and Physical Fitness, 29*, 289–296.

Nieman, D. C., Johanssen, L. M., Lee, J. W., & Arabatzis, K. (1990a). Infectious episodes in runners before and after the Los Angeles Marathon. *Journal of Sports Medicine and Physical Fitness, 30*, 316–328.

Nieman, D. C., Nehlsen Cannarella, S. L., Markoff, P. A., Balk Lamberton, A. J., Yang, H., Chritton, D. B., Lee, J. W., & Arabatzis, K. (1990b). The effects of moderate exercise training on natural killer cells and acute upper respiratory tract infections. *International Journal of Sports Medicine, 11*, 467–473.

Nieman, D. C., Henson, D. A., Gusewitch, G., Warren, B. J., Dotson, R. C., Butterworth, D. E., & Nehlsen Cannarella, S. L. (1993a). Physical activity and immune function in elderly women. *Medicine and Science in Sports and Exercise, 25*, 823–831.

Nieman, D. C., Miller, A. R., Henson, D. A., Warren, B. J., Gusewitch, G., Johnson, R. L., Davis, J. M., Butterworth, D. E., & Nehlsen Cannarella, S. L. (1993b). Effects of high- vs moderate-intensity exercise on natural killer cell activity. *Medicine and Science in Sports and Exercise, 25*, 1126–1134.

Nieman, D. C., Buckley, K. S., Henson, D. A., Warren, B. J., Suttles, J., Ahle, J. C., Simandle, S., Fagoaga, O. R., & Nehlsen Cannarella, S. L. (1995). Immune function in marathon runners versus sedentary controls. *Medicine and Science in Sports and Exercise, 27*, 986–992.

Nieschlag, E., & Wickings, F. J. (1977). A review of radioimmunoassays for steroids. *Zeitschrift für Klinische Chemie und Biochemie, 13*, 261–271.

Nistico, G., Caroleo, M. C., Arbitrio, M., & Pulvirenti, L. (1994). Evidence for an involvement of dopamine D1 receptors in the limbic system control of immune mechanisms. *Neuroimmunomodulation, 1*, 174–180.

Nohr, D., & Weihe, E. (1991). The neuroimmune link in the bronchus-associated lymphoid tissue (BALT) of cat and rat: Peptides and neural markers. *Brain, Behavior, and Immunity, 5*, 84–101.

Nohr, D., Michel, S., Fink, T., & Weihe, E. (1995). Pro-enkephalin opioid peptides are abundant in porcine and bovine splenic nerves, but absent from nerves of rat, mouse, hamster, and guinea-pig spleen. *Cell Tissue Research, 281*, 143–152.

Nohr, D., Buob, A., Gärtner, K., & Weihe, E. (1996). Changes in pulmonary calcitonin gene-related peptide and protein gene-product 9.5 innervation in mycoplasma pulmonis-infected rats. *Cell Tissue Research, 283*, 215–219.

Noldus, L. P. J. J. (1991). The observer: A software system for collection and analysis of observational data. *Behavioral Research Methods, 23*, 415–429.

Norman, S., Besedovsky, H. O., Schardt, M., & del Rey, A. (1988). Interactions between endogenous glucocorticoids and inflammatory responses in normal and tumor bearing mice: Role of T cells. *Journal of Leukocyte Biology, 44*, 551–558.

Norman, S. E., Chediak, H. D., Kiel, M., & Cohn, M. A. (1990). Sleep disturbances in HIV-infected homosexual men. *AIDS, 4*, 775–781.

Northoff, H., & Berg, A. (1991). Immunologic mediators as parameters of the reaction to strenuous exercise. *International Journal of Sports Medicine, 12*, S9–S15.

Nottet, H. S., & Gendelman, H. E. (1995). Unraveling the neuroimmune mechanisms for the HIV-1-associated cognitive/motor complex. *Immunology Today, 16*, 441–448.

O'Brien, W. A., Hartigan, P. M., Martin, D., & Esinhart, J. (1996). Changes in plasma HIV-1 RNA and $CD4^+$ lymphocyte count relative to treatment and progression to AIDS. *New England Journal of Medicine, 334*, 426–431.

Oates, E. L., Allaway, G. P., Armstrong, G. R., Boyajian, R. A., Kehr, H., & Prabhakar, B. S. (1988). Human lymphocytes produce pro-opiomelanocortin gene-related transcripts. *Journal of Biological Chemistry, 263*, 10041–10048.

Oerter, K. E., Guardabasso, V., & Rodbard, D. (1986). Detection and characterization of peaks and estimation of instantaneous secretory rate for episodic pulsatile hormone secretion. *Computers and Biomedical Research*, *19*, 170–191.

Oitzl, M. S., van Oers, H., Schöbitz, B., & De Kloet, E. R. (1993). Interleukin-1β, but not interleukin-6, impairs spatial navigation learning. *Brain Research*, *613*, 160–163.

Oka, T., Oka, K., Hosoi, M., Aou, S., & Hori, T. (1995). The opposing effects of interleukin-1β microinjected into the preoptic hypothalamus and the ventromedial hypothalamus on nociceptive behavior in the rat. *Brain Research*, *700*, 271–278.

O'Leary, A. (1990). Stress, emotion and human immune function. *Psychological Bulletin*, *108*, 363–382.

O'Leary, A., Shoor, S., Lorig, K., & Holman, H. R. (1988). A cognitive behavioral treatment for rheumatoid arthritis. *Health Psychology*, *7*, 524–544.

Olness, K. (1993). Hypnosis: The power of attention. In D. Goleman & J. Gurin (Eds.), Mind body medicine. New York: Consumer Reports Books, pp. 277–290.

Olness, K., Culbert, T., & Uden, D. (1989). Self-regulation of salivary immunoglobulin A by children. *Pediatrics*, *83*, 66–71.

Opp, M. R., Obál, F., & Krueger, J. M. (1991). Interleukin-1 alters rat sleep: Temporal and dose-related effects. *American Journal of Physiology*, *260*, R52–R58.

Opp, M. R., Hughes, T. K., & Smith, E. M. (1994). Interleukin-10 reduces sleep in rats. *Journal of Sleep Research*, *3*, 188.

Ormerod, M. G. (Ed.). (1990). *Flow cytometry. A practical approach.* Oxford: Oxford University Press.

Ortaldo, J. R., Gerard, J. P., Hendersson, L. E., Neubau, R. H., & Rabin, H. (1983). Responsiveness of purified natural killer cells to pure interleukin 2 (IL-2). In J. J. Oppenheim & H. Rabin (Eds.), *Interleukines, lymphokines and cytokines* (pp. 63–68). New York: Academic Press.

Ortega, E., Collazos, M. E., Maynar, M., Barriga, C., & De la Fuente, M. (1993). Stimulation of the phagocytic function of neutrophils in sedentary men after acute moderate exercise. *European Journal of Applied Physiology*, *66*, 60–64.

Oshida, Y., Yamanouchi, K., Hayamizu, S., & Sato, Y. (1988). Effect of acute physical exercise on lymphocyte subpopulations in trained and untrained subjects. *International Journal of Sports Medicine*, *9*, 137–140.

Osmond, D. G., Rolink, A., & Melchers, F. (1998). Murine B lymphopoieses: Towards a unified model. *Immunology Today*, *19*, 65–68.

Ostrow, D. G., Monjan, A., Joseph, J., Van Raden, M., Fox, R., Kingsley, L., Dudley, J., & Phair, J. (1989). HIV-related symptoms and psychosocial functioning in a cohort of homosexual men. *American Journal of Psychiatry*, *146*, 737–742.

Ottaway, C. A. (1991). Vasoactive intestinal peptide and immune function. In R. Ader, D. L. Felten, & N. Cohen (Eds.), *Psychoneuroimmunology* (pp. 225–262). San Diego: Academic Press.

Ottaway, C. A., & Husband, A. J. (1994). The influence of neuroendocrine pathways on lymphocyte migration. *Immunology Today*, *15*, 511–517.

Otterness, I. G., Golden, H. W., Seymour, P. A., Eksra, J. D., & Daumy, G. O. (1991). Role of prostaglandins in the behavioral changes induced by murine interleukin-1α in the rat. *Cytokine*, *3*, 333–338.

Pabst, R. (1994). Die Systeme und Organe der Abwehr. In D. Drenckhahn & W. Zenker (Eds.), *Benninghoff Anatomie* (pp. 733–756). Munich: Urban & Schwarzenberg.

Pabst, R., & Binns, R. M. (1989). Heterogeneity of lymphocyte homing physiology. Several mechanisms operate in the control of migration to lymphoid and non-lymphoid organs in vivo. *Immunological Reviews*, *108*, 83–109.

Pabst, R., & Tschernig, T. (1997). Lymphocyte dynamics: Caution in interpreting BAL numbers. *Thorax*, *52*, 1078–1080.

Pabst, R., & Westermann, J. (1994). Which steps of lymphocyte recirculation are regulated by interferon-gamma? *Research in Immunology*, *145*, 289–294.

Pabst, R., Westermann, J., & Rothkötter, H. J. (1991). Immunoarchitecture of regenerated splenic and lymph node transplants. *International Review of Cytology*, *128*, 215–260.

Pagano, J. S., Kenney, S., Markovitz, K., & Kamine, J. (1988). Epstein–Barr virus and interactions with human retroviruses. *Journal of Virological Methods*, *21*, 229–239.

Page, G. G., Ben-Eliyahu, S., Yirmiya, R., & Liebeskind, J.C. (1993). Morphine attenuated surgery-induced enhancement of metastatic colonization in rats. *Pain*, *54*, 21–28.

Page, J. B., Lai, S. H., Chitwood, D., Smith, P. J., Klimas, N., & Fletcher, M. A. (1990). HTLV-I/II seropositivity and mortality due to AIDS in HIV-1 seropositive IV drug users. *Lancet*, *335*, 1439–1441.

Palmo, J., Asp, S., Daugaard, J. R., Richter, E. A., Klokker, M., & Pedersen, B. K. (1995). Effect of eccentric exercise on natural killer cell activity. *Journal of Applied Physiology*, *78*, 1442–1446.

Palmblad, J., Cantell, K., Strander, H., Froberg, J., Karlsson, C., Levi, L., Granstrom, M., & Unger, P. (1976).

Stressor exposure and immunological responses in man: Interferon producing capacity and phagocytosis. *Psychosomatic Research, 20,* 193–199.

Palmblad, J., Petrini, B., Wasserman, J., & Akerstedt, T. (1979). Lymphocyte and granulocyte reactions during sleep deprivation. *Psychosomatic Medicine, 41,* 273–278.

Panerai, A. E., & Sacerdote, P. (1997). β-Endorphin in the immune system: A role at last? *Immunology Today, 18,* 317–319.

Panerai, A. E., Manfredi, B., Granucci, F., & Sacerdote, P. (1995). The β-endorphin inhibition of mitogen-induced splenocytes proliferation is mediated by central and peripheral paracrine/autocrine effects of the opioid. *Journal of Neuroimmunology, 58,* 71–76.

Panina-Bordignon, P., Mazzeo, D., Di Luca, P., D'Ambrosio, D., Lang, R., & Fabbri, L. (1997). β2-agonists prevent Th 1 development by selective inhibition of interleukin-12. *Journal of Clinical Investigation, 100,* 1513–1519.

Pantaleo, G., Graziosi, C., & Fauci, A. S. (1993). The immunopathogenesis of human immunodeficiency virus infection. *New England Journal of Medicine, 328,* 327–335.

Parnet, P., Brunke, D. L., Goujon, E., Demotes-Mainard, J., Biragyn, A., Arkins, S., Dantzer, R., & Kelley, K. W. (1993). Molecular identification of two types of interleukin-1 receptors in the murine pituitary gland. *Journal of Neuroendocrinology, 5,* 213–219.

Parnet, P., Amindari, S., Wu, C., Brunke-Reese, D. L., Goujon, E., Weyhenmeyer, J. A., Dantzer, R., & Kelley, K. W. (1994). Expression of type I and type II interleukin-1 receptors in the mouse brain. *Molecular Brain Research, 27,* 63–70.

Pascual, D. W., & Bost, K. L. (1990). Substance P production by P388D1 macrophages: A possible autocrine function for neuropeptide. *Immunology, 71,* 52–56.

Paterson, P. Y. (1960). Transfer of experimental allergic encephalomyelitis in rats by means of lymph node cells. *Journal of Experimental Medicine, 111,* 119–128.

Pavlidis, N., & Chirigos, M. (1980). Stress-induced impairment of macrophage tumoricidal function. *Psychosomatic Medicine, 4247–4254.*

Pavlov, I. P. (1927). *Conditioned reflexes.* Oxford: Oxford University Press.

Pawlikowski, M., Zelazowski, P., Dohler, K., & Stepien, H. (1988). Effects of two neuropeptides, somatoliberin (GRF) and corticoliberin (CRF), on human lymphocyte natural killer activity. *British Journal of Behavior and Immunology, 2,* 50–56.

Pearse, A. G. E. (1975). Neurocristopathy, neuroendocrine pathology and the APUD concept. *Zeitschrift für Krebsforschung, 84,* 1–18.

Pearson, C. M., & Wood, F. D. (1959). Studies of polyarthritis and other lesions induced in rats by injections of mycobacterial adjuvant: 1. General clinical and pathological characteristics and some modifying factors. *Arthritis and Rheumatism, 2,* 440–451.

Peavey, B. S., Lawlis, G. F., & Goven, A. (1985). Biofeedback-assisted relaxation: Effects on phagocytic capacity. *Biofeedback and Self-Regulation, 10,* 33–47.

Pedersen, B. K., & Clemmensen, I. H. (1997). Exercise and cancer. In B. K. Pedersen (Ed.), *Exercise immunology* (pp. 171–201). Texas: R.G. Landes.

Pedersen, B. K., & Nielsen, H. B. (1997). Acute exercise and the immune system. In B. K. Pedersen (Ed.), *Exercise immunology.* Austin, Texas: R.G. Landes, pp. 5–38.

Pedersen, B. K., & Ullum, H. (1994). NK cell response to physical activity: Possible mechanisms of action. *Medicine and Science in Sports and Exercise, 26,* 140–146.

Pedersen, B. K., Tvede, N., Christensen, L. D., Klarlund, K., Kragbak, S., & Halkjaer Kristensen, J. (1989). Natural killer cell activity in peripheral blood of highly trained and untrained persons. *International Journal of Sports Medicine, 10,* 129–131.

Pedersen, B. K., Kappel, M., Klokker, M., Nielsen, H. B., & Secher, N. H. (1994). The immune system during exposure to extreme physiologic conditions. *International Journal of Sports Medicine, 15,* S116–S121.

Peeke, H. V. S., Ellman, G., Dark, K., Salfi, M., & Reus, V. I. (1987). Cortisol and behaviorally conditioned histamine release. *Annals of the New York Academy of Sciences, 496,* 583–587.

Pennebaker, J. (1997). Writing about emotional experiences as a therapeutic process. *Psychological Science, 8,* 162–166.

Pennebaker, J. W., Kiecolt-Glaser, J. K., & Glaser, R. (1988a). Confronting traumatic experience and immunocompetence: A reply to Neale, Cox, Valdimarsdottir, and Stone. *Journal of Consulting and Clinical Psychology, 56,* 638–639.

Pennebaker, J. W., Kiecolt-Glaser, J. K., & Glaser, R. (1988b). Disclosure of traumas and immune function: Health implications for psychotherapy. *Journal of Consulting and Clinical Psychology, 56,* 239–245.

Pennebaker, J., & Beall, S. K. (1986). Confronting a traumatic event: Towards an understanding of inhibition and disease. *Journal of Abnormal Psychology, 95,* 274–281.

Péron, F. G., & Caldwell, B. V. (1972). *Immunologic methods in steroid determination*. New York: Appleton–Century–Crofts.

Perry, S., Fishman, B., Jacobsberg, L., & Frances, A. (1992). Relationships over 1 year between lymphocyte subsets and psychosocial variables among adults with infection by human immunodeficiency virus. *Archives of General Psychiatry, 49*, 396–401.

Persson, S., Jonsdottir, I., Thoren, P., Post, C., Nyberg, F., & Hoffmann, P. (1993). Cerebrospinal fluid dynorphin-converting enzyme activity is increased by voluntary exercise in the spontaneously hypertensive rat. *Life Sciences, 53*, 643–652.

Persson, S., Schäfer, M. K.-H., Nohr, D., Ekström, G., Post, C., Nyberg, F., & Weihe, E. (1994). Spinal prodynorphin gene expression in collagen-induced arthritis: Influence of the glucocorticosteroid budesonide. *Neuroscience, 63*, 313–326.

Peters, E. M., & Bateman, E. D. (1983). Ultramarathon running and upper respiratory tract infections. An epidemiological survey. *South African Medical Journal, 64*, 582–584.

Petrie, K. W., Booth, R. J., Pennebaker, J. W., Davison, K. P., & Thomas, M. G. (1995). Disclosure of trauma and immune response to a hepatitis B vaccination program. *Journal of Consulting and Clinical Psychology, 63*, 787–792.

Pettingale, K. W., Morris, T., Greer, H. S., & Haybittle, J. L. (1985). Mental attitudes to cancer: An additional prognostic factor. *Lancet, 1*, 750.

Pinel, J. P. J. (1993). *Biopsychology*. Boston: Allyn & Bacon.

Pirke, K. M. (1973). A comparison of three methods of measuring testosterone in plasma. Competitive protein binding, radioimmunoassay without chromatography and radioimmunoassay including thin layer chromatography. *Acta Endocrinologica, 74*, 168–176.

Pittius, C. W., Kley, N., Loeffler, J. P., & Hollt, V. (1985). Quantitation of proenkephalin A mRNA in bovine brain, pituitary and adrenal medulla: Correlation between mRNA and peptide levels. *EMBO Journal, 4*, 1257–1262.

Plata-Salaman, C. R. (1996a). Anorexia during acute and chronic disease. *Nutrition, 12*, 69–78.

Plata-Salaman, C. R. (1996b). Leptin (ob protein), neuropeptide Y, and interleukin-1 interactions as interface mechanisms for the regulation of feeding in health and disease. *Nutrition, 12*, 718–719.

Plaut, M. (1987). Lymphocyte hormone receptors. *Annual Review of Immunology, 5*, 621–669.

Plotnikoff, N., Murgo, A., Faith, R., & Wybran, J. (Eds.). (1991). *Stress and immunity*. Boca Raton: CRC Press.

Pollmächer, T., Schreiber, W., Gudewill, S., Trachsel, L., Galanos, C., & Holsboer, F. (1993). Influence of endotoxin on nocturnal sleep in humans. *American Journal of Physiology, 264*, R1077–R1083.

Pollmächer, T., Mullington, J., Korth, C., & Hinze-Selch, D. (1995). Influence of host defense activation on sleep in humans. *Advances in Neuroimmunology, 5*, 155–169.

Post, C., Salvati, P., Schäfer, M. K.-H., Schwäble, W., Cini, M., Calabresi, M., Vaghi, F., Wong, E. H. F., & Weihe, E. (1996). Early upregulation of complement factors and astrocyte dysfunction in animal models of global ischemia and stroke. *Society for Neuroscience Abstracts, 838.4*, 214.

Potts, W. K., & Wakeland, E. K. (1990). Evolution of diversity at the major histocompatibility complex. *Trends in Ecology and Evolution, 5*, 181–187.

Potts, W. K., & Wakeland, E. K. (1993). Evolution of MHC genetic diversity: A tale of incest, pestilence and sexual preference. *Trends in Genetics, 9*, 181–187.

Priestman, T. J., Priestman, S. G., & Bradshaw, C. (1985). Stress and breast cancer. *British Journal of Cancer, 51*, 493–498.

Prindull, G., & Ahmad, M. (1993). The ontogeny of the gut mucosal immune system and the susceptibility to infections in infants of developing countries. *European Journal of Pediatrics, 152*, 786–792.

Pross, H. F., Baines, M. G., Rubin, P., Shragge, P., & Patterson, M. S. (1981). Spontaneous human lymphocyte-mediated cytotoxicity against tumor target cells. IX. The quantitation of natural killer cell activity. *Journal of Clinical Immunology, 1*, 51–63.

Purcell, W. M., & Atterwill, C. K. (1995). Mast cells in neuroimmune function: Neurotoxicological and neuropharmacological perspectives. *Neurochemical Research, 20*, 521–532.

Pyne, D. B. (1994). Regulation of neutrophil function during exercise. *Sports Medicine, 17*, 245–258.

Quinnan, G. V., Masur, H., & Rook, A. H. (1984). Herpesvirus infections in the acquired immune deficiency syndrome. *Journal of the American Medical Association, 252*, 72–77.

Rabin, B. S., Moyna, M. N., Kusnecov, A., Zhou, D., & Shurin, M. R. (1996). Neuroendocrine effects of immunity. In L. Hoffman-Goetz (Ed.), *Exercise and immune function* (pp. 21–38). Boca Raton: CRC Press.

Rabkin, J. G., Williams, J. B. W., Neugebauer, R., Remien, R. H., & Goetz, R. (1990). Maintenance of hope in HIV-spectrum homosexual men. *American Journal of Psychiatry, 10*, 1322–1326.

Rabkin, J. G., Williams, J. B. W., Remien, R. H., Goetz, R., Kertzner, R., & Gorman, J. M. (1991). Depression, lymphocyte subsets, and human immunodeficiency virus symptoms on two occasions in HIV-positive homosexual men. *Archives of General Psychiatry, 48*, 111–119.

Ragland, D. R., Brand, R. J., & Fox, B. H. (1992). Type A/B behavior and cancer mortality: The confounding/ mediating effect of covariates. *Psycho-Oncology, 1,* 25–33.

Raine, C. S., & Stone, S. H. (1977). Animal model for multiple sclerosis–chronic experimental allergic encephalomyelitis in inbred guinea pigs. *New York State Journal of Medicine, 77,* 1693–1699.

Ramirez, A. J., Graig, T. K. J., Watson, J. P., Fentiman, I. S., North, W. R. S., & Rubens, R. D. (1989). Stress and relapse of breast cancer. *British Medical Journal, 298,* 291–293.

Ranft, U., Prank, K., & Brabant, G. (1988). Detection of episodic secretion by discrete deconvolution (DESADE)—A new method for the analysis of episodic hormone secretion. *Acta Endocrinologica, 117*(Suppl. 287), 79–80.

Rausch, D. M., Heyes, M. P., Murray, E. A., Lendvay, J., Sharer, L. R., Ward, J. M., Rehm, S., Nohr, D., Weihe, E., & Eiden, L. E. (1994). Cytopathologic and neurochemical correlates of progression to motor/cognitive impairment in SIV-infected rhesus monkeys. *Journal of Neuropathology and Experimental Neurology, 53,* 165–175.

Ravindran, A. V., Griffeths, J., Merali, Z., & Anisman, H. (1995). Lymphocyte subsets associated with major depression and dysthymia: Modification by antidepressant treatment. *Psychosomatic Medicine, 57,* 555–563.

Ravindran, A. V., Griffeths, J., Merali, Z., & Anisman, H. (1996). Variations in lymphocyte subsets associated with stress in depressive populations. *Psychoneuroendocrinology, 21,* 659–671.

Raynaert, C., Janne, P., Bosly, A., Staquet, P., Zdanowicz, N., Vause, M., Chatelain, B., & Lejeune, D. (1995). From health locus of control to immune control: Internal locus of control has a buffering effect on natural killer cell activity decrease in major depression. *Acta Psychiatrica Scandinavica, 92,* 294–300.

Rechtschaffen, A., & Kales, A. (1968). *A manual of standardized terminology, techniques and scoring system for sleep of human subjects* (NIH Publication 204). Washington, DC: U.S. Government Printing Office.

Rechtschaffen, A., Bergmann, B., Everson, C. A., Kushida, C. A., & Gilliland, M. A. (1989). Sleep deprivation in the rat: X. Integration and discussion of the findings. *Sleep, 12,* 68–87.

Redei, E. (1992). Immuno-reactive and bioactive corticotropin-releasing factor in rat thymus. *Neuroendocrinology, 55,* 115–118.

Redfield, R., & Burke, D. (1988). HIV infection: The clinical picture. *Scientific American, 259,* 90–98.

Reed, G. M., Kemeny, M. E., Taylor, S. E., Wang, H.-Y., & Visscher, B. (1994). Realistic acceptance as a predictor of decreased survival time in gay men with AIDS. *Health Psychology, 13,* 299–307.

Reite, M., Harbeck, R., & Hoffman, A. (1981a). Altered cellular immune response following peer separation. *Life Sciences, 29,* 1133–1136.

Reite, M., Short, R., Seiler, C., & Pauley, J. D. (1981b). Attachment, loss and depression. *Journal of Child Psychology and Psychiatry, 22,* 141–169.

Rickinson, A. B., Moss, D. J., Wallace, L. E., Rowe, M., Misko, I. S., Epstein, M. A., & Pope, J. H. (1981). Long-term T-cell mediated immunity to Epstein–Barr virus. *Cancer Research, 41,* 4216–4221.

Rider, M. S., & Achtenberg, J. (1989). Effect of music-assisted imagery on neutrophils and lymphocytes. *Biofeedback and Self-Regulation, 14,* 247–257.

Rider, M. S., Achtenberg, J., Lawlis, G. F., Goven, A., Toledo, R., & Butler, J. R. (1990). Effect of immune system imagery on secretory IgA. *Biofeedback and Self-Regulation, 15,* 317–333.

Rietschel, E. T., Kirikae, T., Schade, F. U., Ulmer, A .J., Zähringer, U., Schreier, M., & Brade, H. (1994). Bacterial endotoxin: Molecular relationships of structure to activity and function. *FASEB Journal, 8,* 217–225.

Rinaldo, C. R. (1990). Immune suppression by herpes viruses. *Annual Review of Medicine, 41,* 331–338.

Ritter, M. A., & Boyd, R. L. (1993). Development in the thymus: It takes two to tango. *Immunology Today, 14,* 462–469.

Rocha, B. (1985). The effects of stress in normal and adrenalectomized mice. *European Journal of Immunology, 15,* 1131–1135.

Rodin, J. (1986). Aging and health: Effects of the sense of control. *Science, 233,* 1271–1276.

Rogausch, H., del Rey, A., Kabiersch, A., & Besedovsky, H. O. (1995). Interleukin-1 increases splenic blood flow by affecting the sympathetic vasoconstrictor tonus. *American Journal of Physiology, 268,* R902–R908.

Rogausch, H., del Rey, A., Kabiersch, A., Reschke, W., Örtel, J., & Besedovsky, H. O. (1997). Endotoxin impedes vasoconstriction in the spleen: Role of endogenous interleukin-1 and sympathetic innervation. *American Journal of Physiology, 272,* R2048–R2054.

Rohde, T., MacLean, D. A., Hartkopp, A., & Pedersen, B. K. (1996). The relationship between plasma glutamine level and cellular immune responses in relation to triathlon race. *European Journal of Applied Physiology, 74,* 428–434.

Röhrenbeck, A., Bette, M., Dietzschold, B., & Weihe, E. (1997). *Up-regulation of CGRP and COX-2 expression in BDV-induced encephalitis.* Paper presented at the 7th European Neuropeptide Club Annual Meeting, Marburg, Germany.

Roitt, I. (1994). *Essential immunology* (8th ed.). Oxford: Blackwell Scientific.

Roitt, I. M., & Lehner, T. (1980). *Immunology of oral diseases.* Oxford: Blackwell Scientific.

Romeo, H. E., Fink, T., Yanaihara, N., & Weihe, E. (1994). Distribution and relative proportions of neuropeptide Y- and proenkephalin-containing noradrenergic neurones in rat superior cervical ganglion: Separate projections to submaxillary lymph nodes. *Peptides, 15,* 1479–1487.

Rose, S. (1993). Cognitive-behavioral group psychotherapy. In H. I. Kaplan & B. J. Sadock (Eds.), *Comprehensive group psychotherapy.* Baltimore: Williams and Wilkins, pp. 205–214.

Rosen, H., Behar, O., Abramsky, O., & Ovadia, H. (1989). Regulated expression of proenkephalin A in normal lymphocytes. *Journal of Immunology, 143,* 3703–3708.

Rosenberg, Z. F., & Fauci, A. S. (1991). Activation of latent HIV infection. *Journal of the National Institute of Health Research, 2,* 41–45.

Rosenman, R. H. (1986). Current and past history of type A behavior pattern. In T. H. Schmidt, T. M. Debroski, & G. Blümchen (Eds.), *Biological and psychological factors in cardiovascular disease.* Berlin: Springer, pp. 15–40.

Roser, B., Brown, R. E., & Singh, P. B. (1991). Excretion of transplantation antigens as signals of genetic individuality. In C. J. Wysocki & M. R. Kare (Eds.), *Chemical senses.* Vol. 3. *Genetics of perception and communications* (pp. 187–209). New York: Dekker.

Roth, J., LeRoith, D., Collier, E. S., Weaver, N. R., Watkinson, A., Cleland, C. F., & Glick, S. M. (1985). Evolutionary origins of neuropeptides, hormones and receptors: Possible applications to immunology. *Journal of Immunology, 135,* 816s–817s.

Rothwell, N. J., & Hopkins, S. J. (1995). Cytokines and the nervous system II. Actions and mechanisms of action. *Trends in Neuroscience, 18,* 130–136.

Rotter, J. B. (1975). Some problems and misconceptions related to the construct of internal versus external control of reinforcement. *Journal of Consulting and Clinical Psychology, 43,* 56–67.

Roubinian, J. R., Talal, N., Greenspan, J. S., Goodman, J. R., & Siiteri, P. K. (1978). Effect of castration and sex hormone treatment on survival, anti-nucleic acid antibodies and glomerulonephritis in NZB/NZW F$_1$ mice. *Journal of Experimental Medicine, 147,* 1568–1583.

Roudebush, R. E., & Bryant, H. U. (1991). Conditioned immunosuppression of a murine delayed type hypersensitivity response: Dissociation from corticosterone elevation. *Brain, Behavior, and Immunity, 5,* 308–317.

Rowland, R. R. R., & Tokuda, S. (1989). Dual immunomodulation by met-enkephalin. *Brain, Behavior, and Immunity, 3,* 171–178.

Ruckdeschel, J. C., Schimpff, S. C., & Smyth, P. C. (1977). Herpes zoster and impaired cell-mediated immunity to the varicella-zoster virus in patients with Hodgkins disease. *American Journal of Medicine, 62,* 77–82.

Russell, D., Peplau, D., & Cutrona, C. (1980). The revised UCLA loneliness scale: Concurrent and discriminant validity evidence. *Journal of Personality and Social Psychology, 39,* 472–480.

Russell, M., Dark, K. A., Cummins, R. W., Ellman, G., Callaway, E., & Peeke, H. V. S. (1984). Learned histamine release. *Science, 225,* 733–734.

Ruzyla-Smith, P., Barabasz, A., Barabasz, M., & Warner, D. (1995). Effects of hypnosis on the immune response: B-cells, T-cells, helper and suppressor cells. *American Journal of Clinical Hypnosis, 38,* 71–79.

Saah, A. J., Hoover, D. R., He, Y., Kingsley, L. A., & Phair, J. P. (1994). Factors influencing survival after AIDS: Report from the Multicenter AIDS Cohort Study (MACS). *Journal of Acquired Immune Deficiency Syndromes, 7,* 287–295.

Sacerdote, P., Bianchi, M., Ricciardi-Castagnoli P., & Panerai, A. E. (1992). Tumor necrosis factor alpha and interleukin-1 alpha increase pain thresholds in the rat. *Annals of the New York Academy of Sciences, 650,* 197–201.

Sachar, E. J. (1980). Hormonal changes in stress and mental illness. In D.T. Krieger & J. C. Hughes (Eds.), *Neuroendocrinology.* Sunderland, MA: Sinauer, pp. 129–135.

Sackett, G. (Ed.). (1978). *Observing behavior:* Vol. II. *Data collection and analysis methods.* Baltimore: University Park Press.

Salk, J., Bretscher, P. A., Salk, P. L., Clerici, M., & Shearer, G. M. (1993). A strategy for prophylactic vaccination against HIV. *Science, 260,* 1270–1272.

Samuels, A.J. (1951). Primary and secondary leucocyte changes following the intramuscular injection of epinephrine hydrochloride. *Journal of Clinical Investigation, 30,* 941–947.

Sandala, L., Lurie, P., Sunkutu, M. R., Chani, E. M., Hudes, E. S., & Hearst, N. (1995). "Dry sex" and HIV infection among women attending a sexually transmitted diseases clinic in Lusaka, Zambia. *AIDS, 9*(Suppl. 1), S61–S68.

Santen, R. J., & Barden, C. W. (1973). Episodic luteinizing hormone secretion in man: Pulse analysis, clinical interpretation, physiologic mechanism. *Journal of Clinical Investigation, 52,* 2617–2622.

Sapolsky, R. M. (1991). Testicular function, social rank and personality among wild baboons. *Psychoneuroendocrinology, 16,* 281–293.

Saraf, P., Frederich, R. C., Turner, E. M., Ma, G., Jaskowiak, N. T., River, D. J., III, Flies, J. S., Lowell, B. B., Fraker, D. L., & Alexander, H. R. (1997). Multiple cytokines and acute inflammation raise mouse leptin levels: Potential role in inflammatory anorexia. *Journal of Experimental Medicine, 185,* 171–175.

Sarason, I. G., Sarason, B. R., Potter, E. H., & Antoni, M. H. (1985). Life events, social support, and illness. *Psychosomatic Medicine, 47,* 156–163.

Saravia, F., Ase, A., Aloyz, R., Kleid, M., Ines, M., Vida, R., Nahmod, V. E., & Vindrola, O. (1993). Differential posttranslational processing of proenkephalin cleavage. *Endocrinology, 132,* 1431–1438.

Schäfer, M. K.-H., Nohr, D., Krause, J. E., & Weihe, E. (1993). Inflammation-induced upregulation of NK1 receptor mRNA in dorsal horn neurons. *Neuroreport, 4,* 1007–1010.

Schäfer, M. K.-H., Weihe, E., Erickson, J. D., & Eiden, L. E. (1995). Human and monkey cholinergic neurons visualized by immunoreactivity for VAChT, the vesicular acetylcholine transporter. *Journal of Molecular Neuroscience, 6,* 225–235.

Schäfer, M. K.-H., Weihe, E., Varoqui, H., Eiden, L. E., & Erickson, J. D. (1994). Distribution of the vesicular acetylcholine transporter (VAChT) in the central and peripheral nervous systems of the rat. *Journal of Molecular Neuroscience, 5,* 1–26.

Schäfer, M. K.-H., Schütz, B., Weihe, E., & Eiden, L. E. (1997). Target-independent cholinergic differentiation in the rat sympathetic nervous system. *Proceedings of the National Academy of Sciences, USA, 94,* 4149–4154.

Schäfer, M. K.-H., Eiden, L. E., & Weihe, E. (1998a). Cholinergic neurons and terminal fields of the central and peripheral nervous system revealed by immunohistochemistry for the vesicular acetylcholine transporter. I. Central nervous system. *Neuroscience, 84,* 331–359.

Schäfer, M. K.-H., Eiden, L. E., & Weihe, E. (1998b). Cholinergic neurons and terminal fields of the central and peripheral nervous system revealed by immunohistochemistry for the vesicular acetylcholine transporter. II. Peripheral nervous system. *Neuroscience, 84,* 361–376.

Schedlowski, M., Falk, A., Rohne, A., Wagner, T. O. F., Jacobs, R., Tewes, U., & Schmidt, R. E. (1993a). Catecholamines induce alterations of distribution and activity of human natural killer (NK) cells. *Journal of Clinical Immunology, 13,* 344–351.

Schedlowski, M., Jacobs, R., Stratmann, G., Richter, S., Hädicke, A., Tewes, U., Wagner, T. O. F., & Schmidt, R. E. (1993b). Changes of natural killer cells during acute psychological stress. *Journal of Clinical Immunology, 13,* 118–126.

Schedlowski, M., Jungk, C., Schimanski, G., Tewes, U., & Schmoll, H.-J. (1994). Effects of behavioral intervention on cortisol plasma levels and lymphocyte numbers in breast cancer patients: An exploratory study. *Psycho-Oncology, 3,* 181–187.

Schedlowski, M., Hosch, W., Oberbeck, R., Benschop, R. J., Jacobs, R., Raab, H.-R., & Schmidt, R. E. (1996). Catecholamines modulate human natural killer (NK) cell circulation and function via spleen-independent β_2-adrenergic mechanisms. *Journal of Immunology, 156,* 93–99.

Scheier, M. F., & Carver, C. S. (1985). Optimism, coping and health: Assessment and implications of generalized outcome expectancies. *Health Psychology, 4,* 219–247.

Schellenkens, P., Roos, M., De Wolf, F., Lange, J., & Miedema, F. (1990). Low T-cell responsiveness to activation via CD4/TCR is a prognostic marker for AIDS in HIV-1 infected men. *Journal of Clinical Immunology, 10,* 121–127.

Schleifer, S., Keller, S. E., Camerino, M., Thornton, J. C., & Stein, M. (1983). Suppression of lymphocyte stimulation following bereavement. *Journal of the American Medical Association, 250,* 374–377.

Schleifer, S. J., Keller, S. E., Meyerson, A. T., Raskin, M. J., Davis, K. L., & Stein, M. (1984). Lymphocyte function in major depressive disorder. *Archives of General Psychiatry, 41,* 484–486.

Schleifer, S. J., Keller, S. E., Siris, S. G., Davis, K. L., & Stein, M. (1985). Depression and immunity: Lymphocyte function in ambulatory depressed patients, hospitalized schizophrenic patients, and patients hospitalized for herniorraphy. *Archives of General Psychiatry, 42,* 129–133.

Schleifer, S. J., Keller, S. E., Bond, R. N., Cohen, J., & Stein, M. (1989). Major depressive disorder and immunity: Role of age, sex, severity, and hospitalization. *Archives of General Psychiatry, 61,* 81–87.

Schmader, K., Studenski, S., MacMillan, J., Grufferman, S., & Cohen, J. (1990). Are stressful life events risk factors for herpes zoster? *American Geriatrics Society, 38,* 1188–1194.

Schmale, A. H. (1972). Giving up as a final common pathway to changes in health. *Advances in Psychosomatic Medicine, 8,* 20–40.

Schmid-Ott, G., Jacobs, R., Jäger, B., Klages, S., Wolf, J., Werfel, T., Kapp, A., Schürmeyer, T. H., Lamprecht, F., Schmidt, R. E., & Schedlowski, M. (1998). Stress-induced endocrine and immunological changes in psoriasis patients and healthy controls. *Psychotherapy and Psychosomatics, 67,* 37–42.

Schmidt, D. D., Zyzanski, S., Ellner, J., Kumar, M. L., & Arno, J. (1985). Stress as a precipitating factor in subjects with recurrent herpes labialis. *The Journal of Family Practice, 20,* 359–366.

Schmidt, R. E., MacDermott, R. P., Bartley, G. T., Bertrovich, M., Amato, D. A., Austen, K. F., Schlossmann, S. F., Stevens, R. L., & Ritz, J. (1985). Specific release of proteoglycans from human natural killer cells during target lysis. *Nature, 318,* 289–291.

Schmidt, R. E., Michon, J. M., Woronicz, J., Schlossman, S.F., Reinherz, E. L., & Ritz, J. (1987). Enhancement of natural killer function through activation of the T11 E rosette receptor. *Journal of Clinical Investigation, 79,* 305–308.

Schmitt, D. A., Peres, C., Sonnenfeld, G., Tkackzuk, J., Arquier, M., Mauco, G., & Ohayon, E. (1995). Immune responses in humans after 60 days of confinement. *Brain, Behavior, and Immunity, 9,* 70–77.

Schneiderman, N., Antoni, M., Ironson, G., LaPerriere, A., & Fletcher, M. A. (1992). Applied psychosocial science and HIV-1 spectrum disease. *Journal of Applied and Preventive Psychology, 1,* 67–82.

Schneiderman, N., Antoni, M., Ironson, G., Klimas, N., LaPerriere, A., Kumar, M., Esterling, B., & Fletcher, M. A. (1994). HIV-1, immunity, and behavior. *Handbook of human stress and immunity.* San Diego: Academic Press.

Schöbitz, B., Voorhuis, D. A. M., & De Kloet, E. R. (1992). Localization of interleukin-6 mRNA and interleukin-6 receptor mRNA in rat brain. *Neuroscience Letters, 136,* 189–192.

Schreiber, H. (1993). Tumor immunology. In W. E. Paul (Ed.), *Fundamental immunology* (pp. 1143–1170). New York: Raven Press, Ltd.

Schürmeyer, T. H. (1989). Stress. In R. D. Hesch (Ed.), *Medizin der Gegenwart. Endokrinologie* (pp. 1221–1231). Munich: Urban & Schwarzenberg.

Schürmeyer, T. H. (1992). *Regulation der hypothalamisch–hypophysär–adrenalen Achse.* Stuttgart: Thieme Verlag.

Schwarz, M. S., & Schwartz, N. M. (1993). Biofeedback: Using the body's signals. In D. Goleman & J. Gurin (Eds.), *Mind body medicine.* New York: Consumer Reports Books, pp. 301–313.

Schwartz, R., & Geyer, S. (1984). Social and psychological differences between cancer and noncancer patients: Cause or consequence of the disease? *Psychotherapy and Psychosomatics, 41,* 195–199.

Schweizer, A., Feige, U., Fontana, A., Muller, K., & Dinarello, C.A. (1988). Interleukin-1 enhances pain reflexes. Mediation through increased prostaglandin E2 levels. *Agents and Actions, 25,* 246–251.

Scicchitano, R., Bienenstock, J., & Stanisz, A. M. (1988). In vivo immunomodulation by the neuropeptide substance P. *Immunology, 63,* 733–735.

Scott, A. M., & Cebon, J. (1997). Clinical promise of tumour immunology. *Lancet, 349,* 19–22.

Scott, P., Pearce, E., Cheever, A. W., Coffman, R. L., & Sher, A. (1989). Role of cytokines CD4⁺ T-cell subsets in the regulation of parasite immunity and disease. *Immunological Reviews, 112,* 161–182.

Seeman, T. E., Berkman, L. F., Blazer, D., & Rowe, J. W. (1994). Social ties and support and neuroendocrine function: The Macarthur studies of successful aging. *Annals of Behavioral Medicine, 16,* 95–105.

Seeman, T. E., Singer, B. H., Rowe, J. W., Horwitz, R. I., & McEwen, B. S. (1997). Price of adaptation (allostatic load) and its health consequences. *Archives of Internal Medicine, 157,* 2259–2268.

Segall, M. A., & Crnic, L. S. (1990). A test of conditioned taste aversion with mouse interferon-α. *Brain, Behavior, and Immunity, 4,* 223–231.

Seidel, A., Arolt, V., Hunstiger, M., Rink, L., Behnisch, A., & Kirchner, H. (1996a). Major depressive disorder is associated with elevated monocyte counts. *Acta Psychiatrica Scandinavica, 94,* 198–204.

Seidel, A., Arolt, V., Hunstiger, M., Rink, L., Behnisch, A., & Kirchner, H. (1996b). Increased CD56+ natural killer cells and related cytokines in major depression. *Clinical Immunology and Immunopathology, 78,* 83–85.

Seligman, M. E. P. (1975). *Helplessness.* San Francisco: Freeman.

Selye, H. (1936). A syndrome produced by diverse nocuous agents. *Nature, 138,* 32.

Selye, H. (1956). *The stress of life.* New York: McGraw-Hill.

Shanks, N., Griffiths, J., & Anisman, H. (1994). Central catecholamine alterations induced by stressor exposure: Analysis in recombinant inbred strains of mice. *Behavioral Brain Research, 63,* 25–33.

Sharp, B., & Linner, K. (1993). Editorial: What do we know about the expression of proopiomelanocortin transcripts and related peptides in lymphoid tissue? *Endocrinology, 133,* 1921A–1921B.

Shavit, Y. (1991). Stress-induced immune modulation in animals: Opiates and endogenous opioid peptides. In R. Ader, D. L. Felten, & N. Cohen (Eds.), *Psychoneuroimmunology* (2nd ed.). San Diego: Academic Press, pp. 789–806.

Shavit, Y., Lewis, J. W., Terman, G. W., Gale, R. P., & Liebeskind, J. C. (1983). Endogenous opioids may mediate the effects of stress on tumor growth and immune function. *Proceedings of the West Pharmacological Society, 26,* 53–56.

Shavit, Y., Depaulis, A., Martin, F. C., Terman, G. W., Pechnick, R. N., Zane, C. J., Gale, R. P., & Liebeskind, J. C. (1986). Involvement of brain opiate receptors in the immune-suppressive effect of morphine. *Proceedings of the National Academy of Science, USA, 83,* 7114–7117.

Shavit, Y., Martin, F. C., Yirmiya, R., Ben-Eliyahu, S., Terman, G. W., Weiner, H., Gale, R. P., & Liebeskind, J. C. (1987). Effects of a single administration of morphine or footshock stress on natural killer cell toxicity. *Brain, Behavior, and Immunity, 1,* 318–328.

Shearer, G. M., & Clerici, M. (1992). T helper cell immune dysfunction in asymptomatic, HIV-1 seropositive individuals: The role of TH1–TH2 cross-regulation. *Chemical Immunology, 54,* 21–43.

Shekelle, R. B., Raynor, W. J. J., Ostfeld, A. M., Garron, D. C., Bieliauskas, L. A., Liu, S. C., Maliza, C., & Paul, O. (1981). Psychological depression and seventeen-year risk of death from cancer. *Psychosomatic Medicine, 43,* 117–125.

Shibata, H., Fujiwara, R., Iwamoto, M., Matsuoka, H., & Yokoyama, M. (1991). Immunological and behavioral effects of fragrance in mice. *International Journal of Neuroscience, 57,* 151–159.

Shimizu, H., Uekara, Y., Shimomura, Y., & Kobayashi, I. (1991). Central administration of ibuprofen failed to block the anorexia induced by interleukin-1. *European Journal of Pharmacology, 195,* 281–284.

Shimizu, N., Kaizuka, Y., Hori, T., & Nakane, H. (1996). Immobilization increases norepinephrine release and reduces NK cytotoxicity in spleen of conscious rat. *American Journal of Physiology, 271,* R537–R544.

Shipp, M. A., Stefano, G. B., D'Adamio, L., Switzer, S. N., Howard, F. D., Sinisterra, J., Scharrer, B., & Reinherz, E. L. (1990). Downregulation of enkephalin-mediated inflammatory responses by CD10/neutral endopeptidase 24.11. *Nature, 347,* 394–396.

Shortman, K., & Scollay, R. (1994). Death in the thymus. *Nature, 372,* 44–45.

Shu, S., Plantz, G. E., Krauss, J. C., & Chang, A. E. (1997). Tumor immunology. *Journal of the American Medical Association, 278,* 1972–1981.

Sieber, W. J., Rodin, J., Larson, L., Ortega, S., Cummings, N., Levy, S., Whiteside, T., & Herberman, R. (1992). Modulation of human natural killer cell activity by exposure to uncontrollable stress. *Brain, Behavior, and Immunity, 6,* 141–156.

Siegel, R. E. (1968). *Galen's system of physiology and medicine.* New York: Karger Press.

Silver, P. S., Auerbach, S. M., Vishniavsky, N., & Kaplowitz, L. G. (1986). Psychological factors in recurrent genital herpes infection: Stress, coping style, social support, emotional dysfunction, and symptom recurrence. *Journal of Psychosomatic Research, 30,* 163–171.

Simon, R. H., Lovett, E. J., Tomaszek, D., & Lundy, J. (1980). Electrical stimulation of the midbrain mediates metastatic tumor growth. *Science, 209,* 1132–1133.

Simonton, C., Mathews-Simonton, S., & Creighton, J. (1978). *Getting well again.* Los Angeles: Tarcher.

Singer, E. J., Zorilla, C., Fahy-Chandon, B., Chi, S., Syndulko, K., & Tourtellotte, W. W. (1993). Painful symptoms reported by ambulatory HIV-infected men in a longitudinal study. *Pain, 54,* 15–19.

Singh, N., Squier, C., Sivek, C., Wagener, M., Hong Nguyen, M., & Yu, V. (1996). Determinants of compliance with antiretroviral therapy in patients with human immunodeficiency virus: Prospective assessment with implications for enhancing compliance. *AIDS Care, 8,* 261–269.

Singh, V. K. (1989). Stimulatory effects of corticotropin releasing neurohormone on human lymphocyte proliferation and interleukin-2 receptor expression. *Journal of Neuroimmunology, 23,* 257–262.

Singh, V. K., & Leu, S. J. C. (1990). Enhancing effect of corticotropin releasing neurohormone on the production of interleukin-1 and interleukin-2. *Neuroscience Letters, 120,* 151–154.

Smith, A. (1992). Sleep, colds, and performance. In R. J. Broughton & R. D. Ogilvie (Eds.), *Sleep, arousal, and performance* (pp. 233–242). Boston: Birkhäuser.

Smith, A., Tyrrell, D., Coyle, K., & Higgins, P. (1988). Effects of interferon alpha on performance in man: A preliminary report. *Psychopharmacology, 96,* 414–416.

Smith, C. K., Harrison, S. D., Ashworth, C., Montano, D., Davis, A., & Fefer, A. (1984). Life change and onset of cancer in identical twins. *Journal of Psychosomatic Research, 28,* 525–538.

Smith, E. M., & Blalock, J. E. (1981). Human leukocyte production of corticotropin and endorphin-like substances: Association with leukocyte interferon. *Proceedings of the National Academy of Sciences, USA, 77,* 5972–5976.

Smith, E. M., Galin, F. S., LeBoeuf, R. D., Coppenhaver, D. H., Harbour, D. V., & Blalock, J. E. (1990). Nucleotide and amino acid sequence of lymphocyte-derived corticotropin: Endotoxin induction of a truncated peptide. *Proceedings of the National Academy of Sciences, USA, 87,* 1057–1060.

Smith, E. M., Harbour, D. V., Hughes, T. K., Kent, T., Ebaugh, M. J., Jazayeri, A., & Meyer W. J, III. (1991). Neurohormones, serotonin, and their receptors in the immune system. In N. Plotnikoff, A. Murgo, R. Faith & J. Wybran (Eds.), *Stress and immunity* (pp. 453–479). Boca Raton: CRC Press.

Smith, G. R. (1989). Intentional psychological modulation of the immune system. In J. V. Basmajian (Ed.), *Biofeedback: Principle and practice for clinicians.* Baltimore: Williams & Wilkins, pp. 49–56.

Smith, G. R., & McDaniel, S. M. (1983). Psychologically mediated effect on the delayed-type hypersensitivity reaction to tuberculin in humans. *Psychosomatic Medicine, 45,* 65–70.

Smith, G. R., McKenzie, J. M., Marmer, D. J., & Steele, R. W. (1985). Psychological modulation of the immune system to varicella zoster. *Archives of Internal Medicine, 145,* 2210–2212.

Smith, G. R., Conger, C., O'Rourke, D. F., Steele, R. W., Charlton, R. K., & Smith, S. S. (1992). Psychological modulation of the delayed type hypersensitivity test. *Psychosomatics, 33,* 444–451.

Smith, J. A. (1994). Neutrophils, host defense, and inflammation: A double-edged sword. *Journal of Leukocyte Biology, 56,* 672–686.

Smith, J. A., Telford, R. D., Mason, I. B., & Weidemann, M. J. (1990). Exercise, training and neutrophil microbicidal activity. *International Journal of Sports Medicine, 11,* 179–187.

Smith, J. A., McKenzie, S. J., Telford, R. D., and Weidemann, M. J. (1992). Why does moderate exercise enhance, but intense training depress, immunity? In A. J. Husband (Ed.), *Behavior and immunity* (pp. 155–168). Boca Raton: CRC Press.

Smotherman, W. P., Brown, C. P., & Levine, S. (1977). Maternal responsiveness following differential pup treatment and mother–pup interactions. *Hormones and Behavior, 8,* 242–253.

Solomon, G. F., Levine, S., & Kraft, J. K. (1968). Early experience and immunity. *Nature, 220,* 821–822.

Solomon, G. F., Segerstrom, S. C., Grohr, P., Kemeny, M., & Fahey, J. (1997). Shaking up immunity: Psychological and immunologic changes after a natural disaster. *Psychosomatic Medicine, 59,* 114–127.

Solvason, H. B., Ghanta, V. K., & Hiramoto, R. N. (1993). The identity of the unconditioned stimulus to the central nervous system is interferon. *Journal of Neuroimmunology, 45,* 75–82.

Späth-Schwalbe, E., Porzolt, F., Digel, W., Born, J., Kloss, B., & Fehm, H. L. (1989). Elevated plasma cortisol levels during interferon-gamma treatment. *Immunopharmacology, 17,* 141–145.

Späth-Schwalbe, E., Hansen, K., Schmidt, F., Schrezenmeier, H., Marshall, L., Burger, K., Fehm, H. L., & Born, J. (1998). Acute effects of recombinant human interleukin-6 on endocrine and central nervous sleep functions in healthy men. *Journal of Clinical Endocrinology and Metabolism, 83,* 1573–1579.

Späth-Schwalbe, E., Perras, B., Fehm, H. L., & Born, J. (1999). Acute effects of recombinant interferon-alpha on nocturnal sleep in healthy men. Submitted for publication.

Spiegel, D. (1992). Effects of psychosocial support on women with metastatic breast cancer. *Journal of Psychosocial Oncology, 10,* 113–120.

Spiegel, D. (1997). Psychosocial aspects of breast cancer treatment. *Seminars in Oncology, 1,* S1–S36.

Spiegel, D., Bloom, J. R., Kraemer, H. C., & Gottheil, E. (1989). Effect of psychosocial treatment on survival of patients with metastatic breast cancer. *Lancet, 2,* 888–891.

Spiegel, H., & Spiegel, D. (1987). *Trance and treatment: Clinical uses of hypnosis.* Washington, DC: American Psychiatric Press.

Spielberger, C. D., Vagg, P. R., Barker, L. R., Donham, G. W., & Westberry, L. G. (1980). The factor structure of the State-Trait Anxiety Inventory. In I. G. Sarason & C. D. Spielberger (Eds.), *Stress and anxiety* (Vol. 7). Washington, DC: Hemisphere, pp. 95–109.

Sprenger, H., Jacobs, C., Nain, M., Gressner, A. M., Prinz, H., Wesemann, W., & Gemsa, D. (1992). Enhanced release of cytokines, interleukin-2 receptors, and neopterin after long-distance running. *Clinical Immunology and Immunopathology, 63,* 188–195.

Sprent, J., & Tough, D. F. (1994). Lymphocyte life-span and memory. *Science, 265,* 1395–1400.

Springer, T. A. (1990). Adhesion receptors of the immune system. *Nature, 346,* 425–434.

Srivastava, R., Ram, B. P., & Tyle, P. (1991). *Immunogenetics of the major histocompatibility complex.* New York: VCH.

Stall, R., Hoff, C., Coates, T., Paul, J., Phillips, K., Ekstrand, M., Kegeles, S., Catania, J., Daigle, D., & Daz, R. (1996). Decisions to get HIV tested and to accept antiretroviral therapies among gay/bisexual men: Implications for secondary prevention efforts. *Journal of Acquired Immune Deficiency Syndrome, 11,* 151–160.

Stamm, W. E., Handsfield, H. H., Rompalo, A. M., Ashley, R. L., Roberts, P. L., & Corey, L. (1988). The association between genital ulcer disease and acquisition of HIV infection in homosexual men. *Journal of the American Medical Association, 260,* 1429–1433.

Starr, K. R., Antoni, M. H., Hurwitz, B. E., Rodriguez, M. R., Ironson, G., Fletcher, M. A., Kumar, M., Patarca, R., Lutgendorf, S. K., Quillian, R. E., Klimas, N. G., & Schneiderman, N. (1997). Patterns of immune, neuroendocrine, and cardiovascular stress responses in asymptomatic human immunodeficiency virus seropositive and seronegative men. *Psychosomatic Medicine,* in press.

Stead, R., Tomioka, M., Pezzati, P., Marshall, J., Croitoru, K., Perdue, M., Stanisz, A., & Bienenstock, J. (1991). Interaction of the mucosal immune and peripheral nervous system. In R. Ader, D. L. Felten, & N. Cohen (Eds.), *Psychoneuroimmunology* (pp. 177–207). San Diego: Academic Press.

Stefano, G. B., Cadet, P., & Scharrer, B. (1989). Stimulatory effects of opioid neuropeptides on locomotor activity and conformational changes in invertebrate and human immunocytes: Evidence for a subtype of receptors. *Proceedings of the National Academy of Sciences, USA, 86,* 6307–6311.

Stein, C., Gramsch, C., & Herz, A. (1990a). Intrinsic mechanisms of antinociception in inflammation: Local opioid receptors and β-endorphin. *Journal of Neuroscience, 10,* 1292–1298.

Stein, C., Hassan, A. H. S., Prezwlocki, R., Gramsch, C., Peter, K., & Herz, A. (1990b). Opioids from immunocytes interact with receptors on sensory nerves to inhibit nociception in inflammation. *Proceedings of the National Academy of Sciences, USA, 87,* 5935–5939.

Stein, M., Miller, A. H., & Trestman, R. L. (1991). Depression, the immune system, and health and illness. *Archives of General Psychiatry, 48,* 171–177.

Steiniger, B., Barth, P., Herbst, B., Hartnell, A., & Crocker, P. R. (1997). The species-specific structure of microanatomical compartments in the human spleen: Strongly sialoadhesin-positive macrophages occur in the perifollicular zone, but not in the marginal zone. *Immunology, 92,* 307–316.

Stephanou, A., Jessop, D. S., Knight, R. A., & Lightman, S. L. (1990). Corticotropin-releasing factor-like immunoreactivity and mRNA in human leukocytes. *Brain, Behavior, and Immunity, 4,* 67–73.

Stephanou, A., Fitzharris, P., Knight, R. A., & Lightman, S. L. (1991). Characteristics and kinetics of proopio-melanocortin mRNA expression by human leukocytes. *Brain, Behavior, and Immunity, 5,* 319–327.

Sternberg, E. M. (1997). Neural–immune interactions in health and disease. *Journal of Clinical Investigation, 100,* 2641–2647.

Sternberg, E. M., Hill, J. M., Chrousos, G. P., Kamilaris, T., Listwak, S. J., Gold, P. W., & Wilder, R. L. (1989a). Inflammatory mediator-induced hypothalamic–pituitary–adrenal axis activation is defective in streptococcal cell wall arthritis-susceptible Lewis rat. *Proceedings of the National Academy of Sciences, USA, 86,* 2374–2378.

Sternberg, E. M., Young, W. S., Bernardini, R., Calogero, A. E., Chrousos, G. P., Gold, P. W., & Wilder, R. L. (1989b). A central nervous system defect in biosynthesis of corticotropin releasing hormone is associated with susceptibility to streptococcal cell wall-induced arthritis in Lewis rats. *Proceedings of the National Academy of Sciences, USA, 86,* 4771–4775.

Stitt, J. T. (1985). Evidence for the involvement of organum vasculosum laminae terminalis in the febrile response of rabbits and rats. *Journal of Physiology (London), 368,* 501–511.

Stone, A., Cox, D. S., Valdimarsdottir, H., & Neale, J. M. (1987). Secretory IgA as a measure of immunocompetence. *Journal of Human Stress, 13,* 136–140.

Stout, C. W., & Bloom, L. J. (1986). Genital herpes and personality. *Journal of Human Stress, 12,* 119–124.

Straub, R. H., Herrmann, M., Berkmiller, G., Frauenholz, T., Lang, B., Scholmerich, J., & Falk, W. (1997). Neuronal regulation of interleukin 6 secretion in murine spleen: Adrenergic and opioidergic control. *Journal of Neurochemistry, 68,* 1633–1639.

Straub, R. H., Westermann, J., Schölmerich, J., & Falk, W. (1998). Dialogue between the CNS and the immune system in lymphoid organs. *Immunology Today, 19,* 409–413.

Strauman, T. J., Lemieux, A. M., & Coe, C. L. (1993). Self-discrepancies and natural killer cell activity: The influence of negative psychological situations on stress physiology. *Journal of Personality and Social Psychology, 6,* 1042–1052.

Stroebe, M. S., Stroebe, W., & Hansson, R. O. (1993). *Handbook of bereavement: Theory, research, and intervention.* New York: Cambridge University Press.

Stumm, R., Schäfer, M. K.-H., & Weihe, E. (1997). *First analysis of NK1, NK2 and NK3 receptor mRNA expression in guinea-pig nervous system, gut and respiratory tract.* Paper presented at the 7th Annual Meeting of the European Neuropeptide Club, Marburg, Germany.

Sumaya, C. V., Boswell, R. N., & Ench, Y. (1986). Enhanced serological and virological findings of Epstein–Barr virus in patients with AIDS and AIDS-related complex. *Journal of Infectious Disease, 154,* 864–870.

Suomi, S. J., Seaman, S. F., Lewis, J. K., DeLizio, R. D., & McKinney, W. T. (1978). Effects of imipramine treatment on separation-induced social disorders in rhesus monkeys. *Archives of General Psychiatry, 35,* 321–325.

Tagasuki, M., Mickey, M. R., & Terasaki, P. I. (1973). Reactivity of lymphocytes from normal persons on cultured tumor cells. *Cancer Research, 33,* 2898–2902.

Takahashi, H., Hakamata, Y., Watanabe, Y., Kikuno, R., Miyata, T., & Numa, S. (1983). Complete nucleotide sequence of the human corticotropin-β-lipotropin precursor gene. *Nucleic Acids Research, 11,* 6847–6851.

Takao, T., Tracey, D. E., Mitchell, W. M., & De Souza, E. B. (1990). Interleukin-1 receptors in mouse brain: Characterization and neuronal localization. *Endocrinology, 127,* 3070–3078.

Tanneberger, S., & Hrelia, P. (1996). Interferons in precancer and cancer prevention: Where are we? *Journal of Interferon and Cytokine Research, 16,* 339–346.

Taylor, D. N. (1995). Effects of a behavioral stress-management program on anxiety, mood, self-esteem, and T-cell in HIV-positive men. *Psychological Reports, 76,* 451–457.

Taylor, G. R. (1993). Immune changes during short-duration missions. *Journal of Leukocyte Biology, 54,* 202–208.

Tazi, A., Dantzer, R., Crestani, F., & Le Moal, M. (1988). Interleukin-1 induces conditioned taste aversion in rats: A possible explanation for its pituitary–adrenal stimulating activity. *Brain Research, 473,* 369–371.

Tazi, A., Crestani, F., & Dantzer, R. (1990). Aversive effects of centrally injected interleukin-1 are independent of its pyrogenic activity. *Neurosciences Research Communications, 7*, 159–165.

Temoshok, L., & Dreher, H. (1992). *The type C connection*. New York: Random House.

Temoshok, L., & Fox, B. H. (1984). Coping styles and other psychological factors related to medical status and to prognosis in patients with cutaneous malignant melanoma. In B. H. Fox & B. H. Newberry (Eds.), *Impact of psychoendocrine systems in cancer and immunity* (pp. 258–287). New York: C. J. Hogrefe.

Tharp, G. D., & Barnes, M. W. (1990). Reduction of saliva immunoglobulin levels by swim training. *European Journal of Applied Physiology, 60*, 61–64.

The Endocrine Society. (1997). *Introduction to molecular and cellular research syllabus*. Baltimore: The Endocrine Society Press.

Thomas, C. B., Duszynski, K. R., & Shaffer, J. W. (1979). Family attitudes reported in youth as potential predictors of cancer. *Psychosomatic Medicine, 41*, 287–302.

Thomas, L. (1975). Symbiosis as an immunologic problem. In E. Neter & F. Milgrom (Eds.), *The immune system and infectious diseases* (pp. 2–11). Basel: Karger.

Thoren, P., Floras, J. S., Hoffmann, P., & Seals, D. R. (1990). Endorphins and exercise: Physiological mechanisms and clinical implications. *Medicine and Science in Sports and Exercise, 22*, 417–428.

Tindall, B., & Cooper, D. A. (1991). Primary HIV infection: Host responses and intervention strategies. *AIDS, 5*, 1–14.

Toellner, K. M., Gulbranson-Jugde, A., Taylor, D. R., Man-Yuen Sze, D., & MacLennan, I. C. M. (1996). Immunoglobulin switch transcript production in vivo related to the site and time of antigen-specific B cell activation. *Journal of Experimental Medicine, 183*, 2303–2312.

Tomasi, T. B., Trudeau, F. B., Czerwinski, D., & Erredge, S. (1982). Immune parameters in athletes before and after strenuous exercise. *Journal of Clinical Immunology, 2*, 173–178.

Tønnesen, E., Christensen, N. J., & Brinkløv, M. M. (1987). Natural killer cell activity during cortisol and adrenaline infusion in healthy volunteers. *European Journal of Clinical Investigation, 17*, 497–503.

Toth, L. A. (1995). Sleep, sleep deprivation and infectious disease: Studies in animals. *Advances in Neuroimmunology, 5*, 79–92.

Toth, L. A., Tolley, E. A., & Krueger, J. M. (1993). Sleep as a prognostic indicator during infectious disease in rabbits. *Proceedings of the Society for Experimental Biology and Medicine, 203*, 179–192.

Travis, J. (1994). Glia: The brain's other cells. *Science, 266*, 970–972.

Tross, S., & Hirsch, D. A. (1988). Psychological distress and neuropsychological complications of HIV infection and AIDS. *American Psychologist, 43*, 929–934.

Tvede, N., Heilmann, C., Halkjaer-Kristensen, J., & Pedersen, B. K. (1989). Mechanisms of B-lymphocyte suppression induced by acute physical exercise. *Journal of Clinical and Laboratory Immunology, 30*, 169–173.

Tvede, N., Steensberg, J., Baslund, B., Halkjaer-Kristensen, J., & Pedersen, B. K. (1991). Cellular immunity in highly trained elite racing cyclists during periods of training with high and low intensity. *Scandinavian Journal of Medicine and Science in Sports, 1*, 163–166.

Tvede, N., Kappel, M., Klarlund, K., Duhn, S., Halkjaer-Kristensen, J., Kjaer, M., Galbo, H., & Pedersen, B. K. (1994). Evidence that the effect of bicycle exercise on blood mononuclear cell proliferative responses and subsets is mediated by epinephrine. *International Journal of Sports Medicine, 15*, 100–104.

Uciechowski, P., Gessner, J. E., Schindler, R., & Schmidt, R. E. (1992). Fc gamma RIII activation is different in CD16⁺ cytotoxic T lymphocytes and natural killer cells. *European Journal of Immunology, 22*, 1635–1638.

Ueda, N., & Shah, S.V. (1994). Apoptosis. *Journal of Laboratory and Clinical Medicine, 124*, 169–177.

Uehara, A., Sekiya, C., Takasugi, Y., Namiki, M., & Arimura, A. (1989). Anorexia induced by interleukin-1: Involvement of corticotropin releasing factor. *American Journal of Physiology, 257*, R613–R617.

Ullum, H., Haahr, P. M., Diamant, M., Palmo, J., Halkjaer-Kristensen, J., & Pedersen, B. K. (1994). Bicycle exercise enhances plasma IL-6 but does not change IL-1alpha, IL-1beta, IL-6, or TNF-alpha pre-mRNA in BMNC. *Journal of Applied Physiology, 77*, 93–97.

Ursin, A., Baude, E., & Levine, S. (1978). *Psychobiology of stress: A study of coping men*. New York: Academic Press.

Ursin, H., & Olff, M. (1993). The stress response. In S. C. Stanford & P. Salmon (Eds.), *Stress: From synapse to syndrome*. San Diego: Academic Press.

Uthgenannt, D., Schoolmann, D., Pietrowsky, R., Fehm, H. L., & Born, J. (1995). Effects of sleep on the release of cytokines in humans. *Psychosomatic Medicine, 57*, 97–104.

Valdimarsdottir, H. B., & Bovbjerg, D. H. (1994). Psychological distress in women with a familial risk of breast cancer. *Psycho-Oncology, 4*, 133–141.

Van Dam, A. M., Brouns, M., Louisse, S., & Berkenbosch, F. (1992). Appearance of interleukin-1 in macrophages and ramified microglia in the brain of endotoxin-treated rats: A pathway for the induction of non specific symptoms of sickness. *Brain Research, 588*, 291–296.

van den Bergh, P., Dobber, R., Ramlal, S., Rozing, J., & Nagelkerken, L. (1994). Role of opioid peptides in the regulation of cytokine production by murine CD4+ T cells. *Cellular Immunology, 154*, 109–122.

van der Poll, T., Coyle, S. M., Barbosa, K., Braxton, C. C., & Lowry, S. F. (1996). Epinephrine inhibits tumor necrosis factor-α and potentiates interleukin 10 production during endotoxemia. *Journal of Clinical Investigation, 97*, 713–719.

Vanhelder, W. P., Radomski, M. W., & Goode, R. C. (1984). Growth hormone responses during intermittent weight lifting exercise in men. *European Journal of Applied Physiology, 53*, 31–34.

Van Hove, G. F., Shapiro, J. M., Winters, M. A., Merigan, T. C., & Blaschke, T. F. (1996). Patient compliance and drug failure in protease inhibitor monotherapy. *Journal of the American Medical Association, 276*, 1955–1956.

Van Rood, Y. R., Bogaards, M., Goulmy, E., & Houwelingen, H. C. (1993). The effects of stress and relaxation on the *in vitro* immune response in man: A meta-analytic study. *Journal of Behavioral Medicine, 16*, 163–181.

van Rooijen, N. (1990). Antigen processing and presentation in vivo: The microenvironment as a crucial factor. *Immunology Today, 11*, 436–439.

van Tits, L. J., & Graafsma, S. J. (1991). Stress influences CD4+ lymphocyte counts [letter]. *Immunology Letters, 30*, 141–142.

Van Woudenberg, A. D., Metzelaar, M. J., Van der Kleij, A. A. M., De Wied, D., Burbach, J. P. H., & Wiegant, V. M. (1993). Analysis of proopiomelanocortin (POMC) messenger ribonucleic acid and POMC-derived peptides in human peripheral blood mononuclear cells: No evidence for a lymphocyte-derived POMC system. *Endocrinology, 133*, 1922–1933.

Veldhuis, J. D., Evans, W. S., Rogol, A. D., Drake, C. R., Thorner, M. O., Merriam, G. R., & Johnson, M. L. (1984). Performance of LH pulse-detection algorithms at rapid rates of venous sampling. *American Journal of Physiology, 247*, E554–E559.

Viney, L., Henry, R., Walker, B., & Crooks, L. (1989). The emotional reactions of HIV antibody positive men. *British Journal of Medical Psychology, 62*, 153–161.

Vintners, H. V., & Anders, K. H. (1990). *Neuropathology of AIDS*. Boca Raton: CRC Press.

Vitkovic, L. (1997). Neuropathogenesis of HIV-1 infection: Interactions between interleukin-1 and transforming growth factor-β1. *Molecular Psychiatry, 2*, 111–112.

Vlajkovic, S., Milavovic, S., Cvijanovic, V., & Jankovic, B. D. (1994). Behavioral and immunological events induced by electrical stimulation of the rat midbrain periaqueductal gray region. *International Journal of Neuroscience, 77*, 287–302.

von Economo, C. (1930). Sleep as a problem of localization. *Journal of Nervous and Mental Disorders, 71*, 249–259.

Wainberg, M. A., Portney, J. D., Clecner, B., Hubschman, S., Lagace-Simard, J., Rabinovitch, N., Remer, Z., & Mendelson, J. (1985). Viral inhibition of lymphocyte proliferative responsiveness in patients suffering from recurrent lesions caused by herpes simplex virus. *Journal of Infectious Diseases, 152*, 441–448.

Walterhöfer, G. (1933). Die Veränderungen des weißen Blutbildes nach Adrenalinjektionen. *Deutsches Archiv für Klinische Medizin, 135*, 208–223.

Waltman, T. J., Irwin, M., Harris, T. J., & Maisel, A. S. (1992). Cell mediated immunity in rats with congestive heart failure. *Proceedings of the American Heart Association, 65*, 19203.

Wan, W., Vriend, C. Y., Wetmore, L., Gartner, J. G., Greenberg, A. H., & Nance, D. M. (1993). The effects of stress on splenic immune function are mediated by the splenic nerve. *Brain Research Bulletin, 30*, 101–105.

Wan, W., Wetmore, L., Sorensen, C. M., Greenberg, A. H., & Nance, D. M. (1994). Neural and biochemical mediators of endotoxin and stress-induced c-fos expression in the rat brain. *Brain Research Bulletin, 34*, 7–14.

Wang, J., Whetsell, M., & Klein, J. R. (1997). Local hormone networks and intestinal T cell homeostasis. *Science, 275*, 1937–1939.

Watkins, L. R., Wiertelak, E. P., Goehler, L.E., Mooney-Heiberger, K., Martinez, J., Furness, L., Smith, K. P., & Maier, S. F. (1994a). Neurocircuitry of illness-induced hyperalgesia. *Brain Research, 639*, 283–299.

Watkins, L. R., Wiertelak, E. P., Goehler, L. E., Smith, K. P., Martin, D., & Maier, S. F. (1994b). Characterization of cytokine-induced hyperalgesia. *Brain Research, 654*, 15–26.

Watkins, L. R., Goehler, L. E., Relton, J. K., Tartaglia, N., Silbert, L., Martin, D., & Maier, S. F. (1995a). Blockade of interleukin-1 induced hyperthermia by subdiaphragmatic vagotomy: Evidence for vagal mediation of immune–brain communication. *Neuroscience Letters, 183*, 27–31.

Watkins, L. R., Maier, S. F., & Goehler, L. E. (1995b). Immune activation: The role of proinflammatory cytokines in inflammation, illness responses and pathological pain states. *Pain, 63*, 289–302.

Watson, M., Greer, S., Blake, S., & Shrapnell, K. (1984). Reaction to a diagnosis of breast cancer: Relationship between denial delay and rate of psychological morbidity. *Cancer, 53*, 2008–2012.

Wayne, S. J., Rhyne, R. L., Garry, P. J., & Goodwin, J. S. (1990). Cell mediated immunity as a predictor of morbidity and mortality in subjects over sixty. *Journal of Gerontology, Medical Sciences, 45*, M45–M48.

Wayner, E. A., Flannery, G. R., & Singer, G. (1978). Effects of taste aversion conditioning on the primary antibody response to sheep red blood cells and *Brucella abortus* in the albino rat. *Physiology and Behavior, 21,* 995–1000.

Weber, R. J., & Pert, A. (1989). The periaqueductal gray matter mediates opiate-induced immunosuppression. *Science, 245,* 188–190.

Weber, R. J., & Pert, A. (1990). Suppression of natural killer cell activity following electrical stimulation of the rat mesencephalon. *Society for Neuroscience Abstracts, 16,* 403.6.

Weber, R. J., Suo, J. L., Gomez-Flores, R., & Hall D. M. (1997). *Immunosuppression following PAG microinjection of morphine is correlated with elevated splenic catecholamines, plasma ACTH and corticosterone, and is not blocked by miferpristone (RU486).* Paper presented at Research Prospectives in PsychoNeuroImmunology, Boulder, CO.

Webster, E. L., Tracey, D. E., Jutila, M. A., Wolfe, S. A., & de Souza, E. B. (1990). Corticotropin-releasing factor receptors in mouse spleen: Identification of receptor-bearing cells as resident macrophages. *Endocrinology, 127,* 440–452.

Weidemann, M. J., Smith, J. A., Gray, A. B., McKenzie, S. J., Pyne, D. B., Kolbuch Braddon, M. E., & Telford, R. D. (1992). Exercise and the immune system. *Today's Life Sciences, 7,* 24–33.

Weihe, E. (1998). Neuropeptides for neuroimmune, endocrine, developmental and pain control: New ligands, new receptors and the "knock out connection." *Experimental and Clinical Endocrinology and Diabetes, 106,* 89–91.

Weihe, E., & Hartschuh, W. (1988). Multiple peptides in cutaneous nerves: Regulators under physiological conditions and a pathogenetic role in skin disease? *Seminars in Dermatology, 7,* 284–300.

Weihe, E., & Krekel, J. (1991). The neuroimmune connection in human tonsils. *Brain, Behavior, and Immunity, 5,* 41–54.

Weihe, E., & Nohr, D. (1992). A neuroimmune sensory–sympathetic link in the pathophysiology and chronic pain cycle of "reflex sympathetic dystrophy" (RSD). In W. Jänig & R. F. Schmidt (Eds.), *Reflex sympathetic dystrophy* (pp. 281–301). New York: VCH.

Weihe, E., & Schäfer, M. K.-H. (1994). Regulation of cellular phenotype in the nociceptive pathway. *NATO ASI Series II, 79,* 337–360.

Weihe, E., Müller, S., Fink, T., & Zentel, H. J. (1989a). Tachykinins, calcitonin gene-related peptide and neuropeptide Y in nerves of the mammalian thymus: Interactions with mast cells in autonomic and sensory neuroimmunomodulation? *Neuroscience Letters, 100,* 77–82.

Weihe, E., Millan, M. J., Höllt, V., Nohr, D., & Herz, A. (1989b). Induction of the gene encoding pro-dynorphin by experimentally induced arthritis enhances staining for dynorphin in the spinal cord of the rats. *Neuroscience, 31,* 77–95.

Weihe, E., Nohr, D., Michel, S., Müller, S., Zentel, H. J., Fink, T., & Krekel, J. (1991a). Molecular anatomy of the neuro-immune connection. *International Journal of Neuroscience, 59,* 1–23.

Weihe, E., Nohr, D., Müller, S., Büchler, M., Friess, H., & Zentel, H. J. (1991b). The tachykinin neuroimmune connection in inflammatory pain. *Annals of the New York Academy of Sciences, 632,* 283–295.

Weihe, E., Schäfer, M. K.-H., Nohr, D., & Persson, S. (1994a). Expression of neuropeptides, neuropeptide receptors and neuropeptide processing enzymes in spinal neurons and peripheral non-neural cells and plasticity in models of inflammatory pain. In T. Hökfelt, R. F. Schmidt, & H. G. Schaible (Eds.), *Neuropeptides, nociception and pain* (pp. 43–69). London: Chapman & Hall.

Weihe, E., Schäfer, M. K.-H., Erickson, J. D., & Eiden, L. E. (1994b). Localization of vesicular monoamine transporter isoforms (VMAT1 and VMAT2) to endocrine cells and neurons in rat. *Journal of Molecular Neuroscience, 5,* 149–164.

Weihe, E., Nohr, D., Schäfer, M. K.-H., Persson, S., Ekström, G., Källström, J., Nyberg, F., & Post, C. (1995). CGRP gene expression in collagen-induced arthritis. *Canadian Journal of Physiological Pharmacology, 73,* 1015–1019.

Weihe, E., Tao-Cheng, J.-H., Schäfer, M. K.-H., Erickson, J., & Eiden, L. E. (1996). Visualization of the vesicular acetylcholine transporter in cholinergic nerve terminals and its targeting to a specific population of small synaptic vesicles. *Proceedings of the National Academy of Sciences, USA, 93,* 3547–3552.

Weihe, E., Hartschuh, W., Schäfer, M. K., Romeo, H., & Eiden, L. E. (1998). Cutaneous Merkel cells of the rat contain both dynorphin and vesicular monoamine transporter type 1 (VMAT1) immunoreactivity. *Canadian Journal of Physiological Pharmacology, 76,* 334–339.

Weiner, H. (1992). *Perturbing the organism: The biology of stressful experience.* Chicago: University of Chicago Press.

Weiner, H., Thaler, M., Reiser, M. F., & Mirsky, I. A. (1957). Etiology of duodenal ulcer. *Psychosomatic Medicine, 19,* 1–10.

Weinstock, C., Konig, D., Harnischmacher, R., Keul, J., Berg, A., & Northoff, H. (1997). Effect of exhaustive exercise stress on the cytokine response. *Medicine and Science in Sports and Exercise, 29,* 345–354.

Weiss, J. M., & Sundar, S. (1992). Effects of stress on cellular immune responses in animals. In A. Tasman & M. Riba (Eds.), *Review of psychiatry* (Vol. 11, pp. 145–168). San Diego: Academic Press.

Weisse, C. S. (1992). Depression and immunocompetence: A review of the literature. *Psychological Bulletin, 111,* 475–489.

Wekerle, H. (1993). T-cell autoimmunity in the central nervous system. *Intervirology, 35,* 95–100.

Westermann, J., & Pabst, R. (1990). Lymphocyte subsets in the blood: A diagnostic window on the lymphoid system? *Immunology Today, 11,* 406–410.

Westermann, J., & Pabst, R. (1992). Distribution of lymphocyte subsets and natural killer cells in the human body. *Clinical Investigator, 70,* 539–544.

Westermann, J., & Pabst, R. (1996). How organ-specific is the migration of "naive" and "memory" T lymphocytes? *Immunology Today, 17,* 278–282.

Westermann, J., Smith, T., Peters, U., Tschernig, T., Pabst, R., Steinhoff, G., Sparshott, S. M., & Bell, E. B. (1996). Both activated and non-activated leukocytes from the periphery continuously enter the thymic medulla of adult rats: Phenotypes, sources and magnitude of traffic. *European Journal of Immunology, 26,* 1866–1874.

Westermann, J., Geismar, U., Sponholz, A., Bode, U., Sparshott, S. M., & Bell, E. B. (1997). CD4$^+$ T cells of both naive and memory phenotype enter rat lymph nodes and Peyer's patches via high endothelial venules: Within the tissue their migratory behavior differs. *European Journal of Immunology, 27,* 3174–3181.

Westermann, J., Michel, S., Lopez-Kostka, S., Bode, U., Rothkötter, H. J., Bette, M., Weihe, E., Straub, R. H., & Pabst, R. (1998). Regeneration of implanted splenic tissue in the rat: Re-innervation is host age-dependent and necessary for tissue development. *Journal of Neuroimmunology, 88,* 67–76.

Westly, H. J., Kleiss, A. L., Kelley, K. W., Wong, P. K. Y., & Yuen, P. H. (1986). Newcastle disease virus-infected splenocytes express the proopiomelanocortin gene. *Journal of Experimental Medicine, 163,* 1589–1594.

Whitehouse, W. G., Dinges, D. F., Orne, E., Keller, S. E., Bates, B. L., Bauer, N. K., Morahan, P., Haupt, B. A., Carlin, M. M., Bloom, P. B., Zaugg, L., & Orne, M. T. (1996). Psychosocial and immune effects of self-hypnosis training for stress management throughout the first semester of medical school. *Psychosomatic Medicine, 58,* 249–263.

Wick, G., Hu, Y., & Schwarz, S. (1993). Immunoendocrine communication via the hypothalamo-pituitary-adrenal axis in autoimmune diseases. *Endocrine Reviews, 14,* 539–563.

Wiedenfeld, S. A., O'Leary, A., Bandura, A., Brown, S., Levine, S., & Raska, K. (1990). Impact of perceived self-efficacy in coping with stressors on components of the immune system. *Journal of Personality and Social Psychology, 59,* 1082–1094.

Wiedermann, C. J., Reinisch, N., Kähler, C., Geisen, F., Zilian, U., Herold, M., & Braunsteiner, H. (1992). In vivo activation of circulating monocytes by exogenous growth hormone in man. *Brain, Behavior, and Immunity, 6,* 387–393.

Wiegand, M., Möller, A. A., Schreiber, W., Krieg, J. C., Fuchs, D., Wachter, H., & Holsboer, F. (1991). Nocturnal sleep EEG in patients with HIV infection. *European Archives of Psychiatric and Clinical Neuroscience, 240,* 153–158.

Wiegers, G. J., Croiset, G., Reul, J. M. H. M., Holsboer, F., & De Kloet, E. R. (1993). Differential effects of corticosteroids on rat peripheral blood T-lymphocyte mitogenesis in vivo and in vitro. *American Journal of Physiology, 265,* E825–E830.

Wiegers, G. J., Labeur, M. S., Stec, I. E., Klinkert, W. E., Holsboer, F., & Reul, J. M. (1995). Glucocorticoids accelerate anti-T cell receptor-induced T cell growth. *Journal of Immunology, 155,* 1893–1902.

Wiertelak, E. P., Smith, K. P., Furness, L., Mooney-Heiberger, K., Mayr, T., Maier, S. F., & Watkins, L. R. (1994). Acute and conditioned hyperalgesic response to illness. *Pain, 56,* 227–234.

Wilder, R. L. (1995). Neuroendocrine–immune systems interactions and autoimmunity. *Annual Review of Immunology, 13,* 307–338.

Williams, P.L. (1995). *Gray's anatomy: The anatomical basis of medicine and surgery.* New York: Churchill Livingstone.

Williams, R. H. (Ed.). (1996). *Textbook of endocrinology.* Philadelphia: Saunders.

Williams, R. M., Bienenstock, J., & Stead, R. H. (1995). Mast cells: The neuroimmune connection. *Chemical Immunology, 61,* 208–235.

Williamson, S. A., Knight, R. A., Lightman, S. A., & Hobbs, J. R. (1987). Differential effects of β-endorphin fragments on human natural killing. *Brain, Behavior, and Immunity, 1,* 329–335.

Wiltschke, C., Krainer, M., Budinsky, A. C., Berger, A., Muller, C., Zeillinger, R., Speiser, P., Kubista, E., Eibl, M., & Zielinski, C. C. (1995). Reduced mitogenic stimulation of peripheral blood mononuclear cells as a prognostic parameter for the course of breast cancer: A prospective longitudinal study. *British Journal of Cancer, 71,* 1292–1296.

Wolff, C. T., Friedman, S. B., Hofer, M. A., & Mason, J. W. (1964). Relationship between psychological defenses and mean urinary 17-hydroxycorticosteroid excretion rates. *Psychosomatic Medicine, 26,* 576–609.

World Health Organization (ed.). (1994). *World Health Organization Statistics Annual*. Geneva: WHO.

Worlein, J. M., Eaton, G. G., Johnson, D. F., & Glick, B. B. (1988). Mating season effects on mother–infant conflict in Japanese macaques (*Macaca fuscata*). *Animal Behavior, 36*, 1472–1481.

Worlein, J. M., Berger, C. L., & Laudenslager, M. L. (1995). Effect of ketamine anesthesia on immune parameters in pigtail macaques. *American Journal of Primatology, 35*, 166 (Abstract).

Wybran, J., Appelboom, T., Famaey, J. P., & Govaerts, A. (1979). Suggestive evidence for receptors for morphine and methionine-enkephalin on normal human blood T lymphocytes. *Journal of Immunology, 123*, 1068–1071.

Yada, T., Sakurada, M., Ishihara, H., Nakata, M., Shioda, S., Yaekura, K., Hamakawa, N., Yanagida, K., Kikuchi, M., & Oka, Y. (1997). Pituitary adenylate cyclase-activating polypeptide (PACAP) is an islet substance serving as an intra-islet amplifier of glucose-induced insulin secretion in rats. *Journal of Physiology (London), 505*, 319–328.

Yamazaki, K., Boyse, E. A., Mike, V., Thaler, H. T., Mathieson, B. J., Abbott, J., Boyse, J., Zayas, Z. A., & Thomas, L. (1976). Control of mating preferences in mice by genes in the major histocompatibility complex. *Journal of Experimental Medicine, 144*, 1324–1335.

Yamazaki, K., Beauchamp, G. K., Bard, J., Boyse, E. A., & Thomas, L. (1991). Chemosensory identity and immune function in mice. In C. J. Wysocki & M. R. Kare (Eds.), *Chemical senses*: Vol. 3. *Genetics of perception and communications* (pp. 211–225). New York: Dekker.

Yirmiya, R. (1996). Endotoxin produces a depressive-like episode in rats. *Brain Research, 711*, 163–174.

Yirmiya, R., Rosen, H., Donchin, O., & Ovadia, H. (1994). Behavioral effects of lipopolysaccharide in rats: Involvement of endogenous opioids. *Brain Research, 648*, 80–86.

Zachariae, R. (1996). *Mind and immunity: Psychological modulation of immunological and inflammatory parameters*. Copenhagen: Munksgaard & Rosinante.

Zachariae, R., & Bjerring, P. (1990). The effect of hypnotically induced analgesia on the flare reaction of the cutaneous histamine prick test. *Archives of Dermatological Research, 282*, 539–543.

Zachariae, R., & Bjerring, P. (1993). Increase and decrease of cutaneous reactions obtained by hypnotic suggestions during sensitization studies on dinitrochlorobenzene and diphenylcyclopropenone. *Allergy, 48*, 6–11.

Zachariae, R., Bjerring, P., & Arendt-Nielsen, A. (1989). Modulation of type I immediate and type IV delayed immunoreactivity using direct suggestion and guided imagery during hypnosis. *Allergy, 44*, 537–542.

Zachariae, R., Kristensen, J. S., Hokland, P., Ellegaard, J., Metze, E., & Hokland, M. (1990). Effect of psychological intervention in the form of relaxation and guided imagery on cellular immune function in normal healthy subjects. *Psychotherapy and Psychosomatics, 54*, 32–39.

Zachariae, R., Bjerring, P., Zachariae, C., Arendt-Nielsen, L., Nielsen, T., Eldrup, E., Larsen, C. S., & Gotliebsen, K. (1991). Monocyte chemotactic activity in sera after hypnotically induced emotional states. *Scandinavian Journal of Immunology, 34*, 71–79.

Zachariae, R., Hansen, J. B., Andersen, M., & Jinquan, T. (1994). Changes in cellular immune function after immune specific guided imagery and relaxation in high and low hypnotizable healthy subjects. *Psychotherapy and Psychosomatics, 61*, 74–92.

Zacharko, R. M., Zalcman, S., Macneil, G., Andrews, M., Mendalla, P. D., & Anisman, H. (1997). Differential effects of immunologic challenge on self-stimulation from the nucleus accumbens and the substantia nigra. *Pharmacology, Biochemistry and Behavior, 58*, 881–886.

Zalcman, S., Shanks, N., & Anisman, H. (1991). Time-dependent variations of central norepinephrine and dopamine following antigen administration. *Brain Research, 557*, 69–76.

Zeier, H., Brauchli, P., & Joller-Jemelka, H. I. (1996). Stress-induced enhancement of salivary IgA in air traffic controllers. *Biological Psychology, 42*, 413–423.

Zeller, M. (1944). The influence of hypnosis on passive transfer and skin tests. *Annals of Allergy, 2*, 515–517.

Zentel, H. J., & Weihe, E. (1991). The neuro-B cell link of peptidergic innervation in the bursa Fabricii. *Brain, Behavior, and Immunity, 5*, 132–147.

Zentel, H. J., Nohr, D., Albrecht, R., Jeurissen, S. H., Vainio, O., & Weihe, E. (1991). Peptidergic innervation of the bursa Fabricii: Interrelation with T-lymphocyte subsets. *International Journal of Neuroscience, 59*, 177–188.

Ziegler, K. (1924). Über die Verteilung der Blutzellen in der Blutbahn. *Klinische Wochenschrift, 3*, 1481.

Zonderman, A. B., Costa, P. T., & McCrae, R. R. (1989). Depression as a risk for cancer morbidity and mortality in a nationally representative sample. *Journal of the American Medical Association, 262*, 1191–1195.

Zoukos, Y., Leonard, T., Thomaides, T. N., Thompson, A. J., & Cuzner, M. L. (1992). β-Adrenergic receptor density and function of peripheral blood mononuclear cells are decreased in multiple sclerosis: A regulatory role for cortisol and interleukin-1. *Annals of Neurology, 31*, 657–662.

Zukin, R. S., & Zukin, S. R. (1981). Multiple opiate receptors: Emerging concepts. *Life Sciences, 29*, 2681–2685.

Index

Printed in the United Kingdom
by Lightning Source UK Ltd.
116471UKS00003B/3